Buildings' Thermal Behaviour and Energy Efficiency for a Sustainable Construction

Buildings' Thermal Behaviour and Energy Efficiency for a Sustainable Construction

Editors

Paulo Santos
Mark Bomberg

MDPI

Basel • Beijing • Wuhan • Barcelona • Belgrade • Novi Sad • Cluj • Manchester

Editors
Paulo Santos
University of Coimbra
Coimbra
Portugal

Mark Bomberg
Clarkson University
Potsdam, NY
USA

Editorial Office
MDPI
St. Alban-Anlage 66
4052 Basel, Switzerland

This is a reprint of articles from the Topical Collection published online in the open access journal *Buildings* (ISSN 2075-5309) (available at: https://www.mdpi.com/journal/buildings/topical_collections/Build_Therm_Behav_Energ_Eff_Sus_Const).

For citation purposes, cite each article independently as indicated on the article page online and as indicated below:

Lastname, A.A.; Lastname, B.B. Article Title. *Journal Name* **Year**, *Volume Number*, Page Range.

ISBN 978-3-7258-0719-2 (Hbk)
ISBN 978-3-7258-0720-8 (PDF)
doi.org/10.3390/books978-3-7258-0720-8

Contents

About the Editors

Paulo Santos

Paulo Santos currently is Associate Professor at the Department of Civil Engineering (DEC) of the Faculty of Sciences and Technology, University of Coimbra (FCTUC). Research member of the Institute for Sustainability and Innovation in Structural Engineering (ISISE), Functional Performance (FP) research group.

The main actual research topics are Thermal Behaviour, Energy Efficiency in Buildings and Sustainable Construction, with main focus on the Lightweight Steel-Framed (LSF) construction system and sustainable construction materials. As a result of his research work he is author of around 190 publications, among 4 books, 2 dissertations, 2 book editions, 15 journal special editions, 3 book chapters, 64 articles published in peer-reviewed scientific journals, and 107 papers published in proceedings of scientific conferences, being 81 publications indexed and listed in the Web of Science Core Collection.

He has around 2800 citations and 77,500 reads on Research Gate (Berlin), with a h-index of 26. He was supervisor of 54 doctoral and master theses already completed within his research topics.

He participated in around 16 funded European and national research projects within his research areas, being the Principal Investigator of Tyre4BuildIns research project. He was member of the scientific committee and / or organizer of several national and international scientific conferences. He is editorial board member of 20 international scientific journals, for which he already has 288 verified Web of Science editor records. He reviewed, by request of the editors, around 125 articles submitted to international journals. He was also member of the scientific evaluation committee of several national and international research projects proposed for funding.

Mark Bomberg

Mark Bomberg, Technology D. (Lund U., Sweden 1974), D. Science in Engineering (Warsaw), technical coordinator of the virtual network Environmental Quality Management (EQM), is a Research Prof. at Mechanical Eng., Clarkson U, Potsdam NY, RD Manager of DFIE, Inc., Honorary Member of Building Enclosure Technology and Environment (BETEC) Committee of the National Institute of Building Science (NIBS) in Washington, DC. He worked at National Research Council of Canada (1975–2000) and was an Editor-in-Chief of J. Building Physics (1984–2018). He lives in Canada but works in the US and Europe. He was teaching in, the US, Canada, Mexico, Germany, Poland and China. He is a Guest Editor in Journals: Energies and Buildings (MPDI, Switzerland). Frontiers (Switzerland). He received the highest awards in building physics in the US and Canada. His research background includes heat, air and moisture, material science and durability of construction materials. He wrote more than 220 peer reviewed papers and 7 books, has 88,000 reads, on the Research Gate.

Preface

In the pursuit of sustainable development, the built environment is undergoing a profound transformation. With a pressing need to address environmental concerns and energy efficiency, the discourse surrounding buildings' thermal behavior has become increasingly significant. This topical collection, titled *"Buildings' Thermal Behaviour and Energy Efficiency for a Sustainable Construction"*, stands as a testament to the growing importance of sustainable practices in the construction industry.

Within these published papers, readers will find a comprehensive exploration of the intricate relationship between buildings, their thermal dynamics, and energy efficiency. Authored by leading experts and scholars in the field, each chapter delves into various facets of this critical subject matter, offering invaluable insights, innovative approaches, and practical solutions.

From the fundamentals of heat transfer and building envelope design to advanced technologies such as passive heating and cooling systems, this collection encapsulates the latest advancements shaping the future of sustainable construction. Moreover, it examines the multifaceted impact of thermal behavior on occupant comfort, indoor air quality, and overall building performance.

At its core, this collection serves as a beacon for architects, engineers, policymakers, and stakeholders alike, guiding them towards informed decision-making and responsible practices in building design and construction. By fostering a deeper understanding of buildings' thermal behavior and energy efficiency, we aim to catalyze positive change and propel the industry towards a more sustainable and resilient future.

We extend our heartfelt gratitude to the contributors whose expertise and dedication have enriched this compilation. Their unwavering commitment to advancing knowledge and driving innovation is a testament to the collective effort required to tackle the challenges of the 21st century.

As editors, it is our sincere hope that this collection inspires dialogue, sparks creativity, and serves as a catalyst for transformative action in the pursuit of sustainable construction practices.

Paulo Santos and Mark Bomberg
Editors

buildings

MDPI

Article

Techno-Economic Analysis of the Energy Resilience Performance of Energy-Efficient Buildings in a Cold Climate and Participation in the Flexibility Market

Ashish Saini *, Ala Hasan * and Jari Shemeikka

VTT Technical Research Centre of Finland, Otaniemi, 02044 Espoo, Finland; jari.shemeikka@vtt.fi
* Correspondence: ashish.saini@vtt.fi (A.S.); ala_hasan@hotmail.com or ala.hasan@vtt.fi (A.H.)

Abstract: Unexpected power outages and extreme weather encouraged research on energy-resilient buildings throughout the world. Resilient building research mainly focuses on hot weather rather than cold extremes. This study defines resilience terminologies based on the available literature and discusses the impact of energy efficiency on energy resilience performance in energy-efficient buildings due to abrupt power outages in an extremely cold climate. The assessment involves the case simulation of a multistory apartment located in southern Finland at design outdoor conditions (−26 °C) in IDA-ICE 4.8, a dynamic building simulation software, and its techno-economic assessment to ensure building resilience for up to 7 days of power outages. The assessment shows the efficient building envelope can enhance the time taken by the building to drop the indoor temperature to the threshold by approximately 15%. Additionally, the efficient heating system along with the building envelope can reduce the instantaneous power demand by up to 5.3 times, peak power demand by up to 3.5 times, and on average power consumption by 3.9 times. Similarly, the study finds that the total energy requirement during a blackout can be reduced by 4.1 times. The study concludes that enhanced building resilience is associated with energy-efficient parameters such as an efficient energy system and an efficient building envelope that has low thermal losses and high thermal inertia retention. The batteries contribute the maximum proportion to the overall retrofitting cost, and the proportion can go up to 70% in baseline configurations and 77% in efficient configurations of buildings. The analysis concludes that the required investment varies largely with the technologies involved and the combination of components of these energy systems. The assessment finds that the high investment costs associated with batteries and battery recharging costs are the main bottlenecks to feasible flexibility in market participation.

Keywords: building resilience; energy-efficient building; energy flexibility; energy resilience; habitability; survivability

Citation: Saini, A.; Hasan, A.; Shemeikka, J. Techno-Economic Analysis of the Energy Resilience Performance of Energy-Efficient Buildings in a Cold Climate and Participation in the Flexibility Market. *Buildings* **2023**, *13*, 2936. https://doi.org/10.3390/buildings13122936

Academic Editor: Paulo Santos

Received: 25 October 2023
Revised: 17 November 2023
Accepted: 21 November 2023
Published: 24 November 2023

1. Introduction

A prominent approach to decarbonization is the increasing penetration of renewables in energy generation. An Intergovernmental Panel on Climate Change (IPCC) report warned about the serious health risks for more than 350 million people due to the aftereffects of climate change [1]. Recent worldwide events show the susceptibility of energy infrastructure to failure during extreme situations, such as extreme weather, natural disasters, and international conflicts [2]. Lack of planning for resource utilization in development strategies can lead to tragic consequences during extreme weather.

Residential, commercial, and public services encompass approximately 30% of Europe's total energy consumption. Among this major portion are buildings and common areas [3]. Abrupt weather can lead to abnormal temperatures within buildings and common areas. Thus, the buildings require retrofitting with insulators and energy systems that can satisfy the energy demand to maintain indoor conditions during the disruption.

This integration of energy systems, or retrofitting, involves extra investment and operational and maintenance costs. The cost involved varies with the technology used and is constrained by numerous factors, including technological maturity, geographical locations, local availability, and policies.

2. State of the Art

Although resilience research has been ongoing for a while, it primarily focuses on the grid system rather than the building system. Among building resilience research, overheating scenarios studies fill the major research proportion, and very limited research is available on extremely cold weather conditions. Lisa and Graham examined passive survivability assessment protocols and metrics and proposed an evaluation methodology for building resilience. The research includes power outage simulation and analyzing the results of building design in a multifamily building [4]. Hamdy et al. introduced the cost-effective flexibility index (CEFI) and active survivability index (ASI) as comparison indicators of building design to analyze survivability from an economic viewpoint in fully electrified buildings in the cold climate of Norway [5]. Homaei and Hamdy experimented with the quantification of thermal resilience in buildings for prolonged power outages and formulated a standard framework for the cold climate. The study introduces weighted unmet thermal performance as an indicator to benchmark the building resilience class [6]. Zhivov intensified the role of thermal mass in building resilience. It scrutinizes the system's resilience through a quantitative approach derived from analytical and experimental studies of the extremely cold climate [7].

Ozkan and Good established the positive impact of the building envelope on maintaining thermal resilience in both temperature extremes. They performed a comparative evaluation of different envelope conditions within the same building with defined performance indicators such as energy use intensity, thermal energy demand intensity, and thermal autonomy [8]. Kesik et al. used thermal autonomy, passive habitability metrics, and other critical building parameters to establish a standard framework for building resilience benchmarking using a common set of conventions and protocols [9]. Nik et al. discussed the different resilience definitions, emphasizing the energy systems and the associated framework of the resilient energy systems. The paper highlights that the lack of a standardized framework and ambiguity in resilience definitions exhibit challenges in designing climate-resilient energy systems [10]. Attia et al. examined resilience through varied terminologies such as vulnerability, resistance, robustness, and recovery from a disruptive event. The difference in these terminologies thwarts reaching a consensus over resilience calculation methodology and framework for buildings [11]. Referring to the literature considered, it is clear that an opportunity exists to analyze building resilience in an extremely cold climate, which will enrich the existing research and help in achieving more resilient buildings.

As discussed above, research on resilience has been going on for some time. Along with cold climate resilience research, there have been research studies about resilience targeting overheating in buildings. Wang et al. developed a forecast model for heating and cooling and analyzed the overheating severity and duration in high-thermal-performance buildings across varied climate zones in China [12]. Roostaie and Nawari built a comprehensive building assessment framework, integrating the sustainability framework and resilience indicators. The paper assessed the impact of resilience indicators on building sustainability through a decision-making trial and evaluation laboratory (DEMATEL) approach [13]. Zhang et al. investigated and highlighted the importance of resilience criterion inclusion in the design phase in achieving resilient cooling. The technological performance was analyzed for different cooling technologies for heatwave and power outage situations [14]. Tariq et al. scrutinized the efficacy of natural ventilation to achieve resilience during a heatwave for locations with future warm climate scenarios by maintaining thermal comfort and acceptable indoor air quality [15]. Shady et al. analyzed critical resilience cooling parameters within the building and suggested performance-driven thermal environmental quality indicators for resilience against power outages and heatwaves [11].

3. Objective and Novelty

As seen in the state of the art, very little research focuses on resilience in cold climates. Even though the population residing in cold climates is small compared to hot or tropical climates, extreme weather in cold regions is equally fatal and needs research and assessment. The logistic disruption in extreme cold exacerbates the situation, apart from infrastructure challenges. This research aims to define resilience and establish survivability conditions for building parameters during power cut-offs in extremely cold climates. The study paper performs a case simulation in an energy-efficient multiapartment building in Southern Finland to present the impact of energy-efficient systems on building resilience. Furthermore, the assessment shows the type and extent of the system required for different resilience durations and the associated investment for retrofitting or installation. The study explores the possibilities of the system's participation in the flexibility market to achieve a return on investment and increase the system's operational time.

4. Methodology

This study first establishes its definition of resilience, habitability, and survivability based on the technical literature in the scientific community. Similarly, this study establishes the minimum threshold boundary condition of relevant parameters based on the academic literature and discussions with experts.

After the parameter setup, the system proceeds to the case study simulation. The case study progresses by developing both configurations of the buildings within IDA-ICE 4.8, a dynamic building simulation software. The building model resembles the real-life multiapartment positive energy building located in Southern Finland [16]. IDA-ICE provides the localization of Finland, which encompasses the standard climate file, wind profiles, and relative humidity. Since the study aims for blackout analysis in extremely cold climates, there was no radiation during the simulation duration, and the outdoor temperature was modified to −26 °C, the design temperature condition of Southern Finland according to the Finnish decree [17]. After establishing the model and the parameters within the software, the study proceeds to simulate the building operations for two weeks and record the temperature within the typical apartments. During the first week, the building maintains the normal temperature, and grid power is available. During the start of the second week, a blackout occurs. The study includes the simulation of two weeks to smooth the simulation process.

The study moves ahead with the processing of simulation results in Microsoft Excel (version 1808) to plot them as graphs. To obtain the energy and power matrix, the system is simulated again to turn on the power supply as soon as the setpoint touches the survivability threshold. The new simulation results are plotted as graphs to provide the energy and power matrix sizing points for different resilience durations. Further on, the assessment explores different energy system components that fulfill the technical requirements to ensure building resilience during the blackout while following the pertinent constraints. The relevant technical specifications of selected components are assumed based on real-life components available on the market. After the selection of technologies and the sizing of the energy system, the study performs an economic assessment of the participating energy systems. The costs of components are comparable to those of real-life components with similar technical specifications, and specific caution is considered for the application of value-added tax.

Toward the end of the study, it explores the participation of the proposed energy system in the flexibility market under two scenarios to check the return on investment of the integrated system. The formulae used and assumptions are discussed in the respective sections. The study performs only theoretical analysis and simulation due to the unavailability of blackout data for the Finnish region. The study finishes with the findings of the assessment and the possible future work.

5. Material and Simulation Setup

5.1. Definitions of Resilience in Building

The state-of-the-art section depicts the variations in building resilience terminologies. A consensus over definitions, parameters, and terminologies can expedite further research on the same. This section presents the different definitions of building resilience and represents the author's definitions.

UN (United Nations) General Assembly resolution 71/276 describes resilience as "the ability of a system, community or society to resist, absorb, accommodate, adapt to, transform and recover from the effects of a hazard in a timely and efficient manner, including through the preservation and restoration of its essential basic structures and functions through risk management" [11]. The resilience definition and the associated characteristics vary with the domain of study performing the investigation.

From an engineering perspective, a resilient building is a building that can withstand a power outage while maintaining safe indoor environmental conditions such as operational temperature and ventilation rate, along with the supply of minimum energy required for a definite time or before being recovered. Table 1 shows the different definitions of building resilience among different engineering disciplines.

Table 1. Different definitions of resilient buildings.

S. No.	Definition	Resilience Characteristics
1	Building resilience is the ability of a building to cope with severe weather disruptions and recover in a timely and efficient manner [18]	Withstand, mitigation, recoverability, rapidity
2	A resilient building is a building that is not only robust but can also fulfill its functional requirements (withstand) during a major disruption. Its performance might even be disrupted, but it must recover to an acceptable level promptly in order to avoid disaster impacts [19]	Withstand, absorption, recoverability, rapidity
3	A resilient built environment is one that is designed, located, built, operated, and maintained in a way that maximizes the ability of built assets, associated support systems, and the people that reside or work within the built asset to withstand, recover from, and mitigate the impacts of threats [14]	Withstand, recoverability, mitigation
4	Resilience is the intrinsic ability of the system to proactively react to the disruption (external or internally generated), adapt, and recover to reach a new state of the system to serve the normal functionalities [20]	Vulnerability, adaptation, recoverability
5	Resilient urban energy systems need to be capable of planning and preparing for, absorbing, recovering from, and adapting to any adverse events that may happen in the future. The complex, dynamic, and adaptive system (for example, cities) would not necessarily return to an equilibrium state [21]	Preparation, absorption, recoverability, adaptation

From a thermally resilient building standpoint, resilience is the ability of a building to adapt thermally and maintain safe thermal conditions indoors during power disruptions. Thermal conditions can be classified as habitable conditions or survivable conditions while assessing the building's resilience. Different terminologies, such as habitability or survivability, define the building's resilience based on the conditions maintained within the building. The next paragraph presents the authors' definitions of habitability and survivability.

"Habitability refers to the time duration for which the building remains habitable in case of energy supply disruption to the building due to seen or unseen circumstances". Habitability can further be classified as active or passive based on the components used to maintain the indoor conditions. Passive habitability indicates the situation in which the thermal inertia retained via the building envelope maintains the thermal conditions within the building. In the case of active habitability, an on-site energy system or an external energy source assists in maintaining indoor conditions.

Moreover, "Survivability indicates the ability of a building to maintain thermal conditions along with minimum operational capabilities during a power outage". Habitability ascribes to thermal resilience, while survivability employs thermal resilience along with minimum operational capabilities such as lighting, appliances, sewage, and domestic hot water. Apart from minimum operational capabilities, survivability encompasses a broader temperature range compared to the comfortable range.

5.2. Boundary Conditions

This section explores the parameters used in building resilience and establishes a recommended range to maintain survivability conditions within the building based on the different literature.

According to the American Society of Heating, Refrigerating, and Air-Conditioning Engineers (ASHRAE), the acceptable temperature range for naturally ventilated buildings ranges from 10 to 34 °C. The ASHRAE-55 model confers a prevalent mean outdoor temperature range of 10 °C to 33 °C for passive survivability. Additionally, the ASHRAE Thermal Environmental Conditions for Human Occupancy standard 55-2004 [22] defines a comfort temperature range of 19–26 °C for winters at a maximum humidity ratio of 0.12 Kg_{water}/kg_{dryair} [4]. The guide for resilient thermal energy systems design in cold and arctic climates prescribes a minimum temperature of 16 °C indoors during emergency situations such as blackouts for human comfort and maintaining dexterity in critical operations [23]. Homaei and Hamdy define the lower threshold of the habitability range in the Norwegian environment as 15 °C in their work on the quantification of energy flexibility and survivability with batteries. The habitability defined in their work is analogous to survivability in this study [5]. Similarly, Kesik et al. define the lower temperature threshold for passive habitability as 15 °C in mechanically cooled buildings. Their habitability temperature corresponds to the survivability temperature threshold of this study [9]. Compiling all the work, the advocated lower temperature threshold is 15 °C. This study only defines the lower temperature threshold because the study focuses on resilience in a cold climate.

High relative humidity also impacts human health with extended exposure, but this study considers a short-duration exposure in exceptional circumstances (one-week blackout). Thus, this study employs standard relative humidity values, which are shown in Table 2. The Federation for European Heating, Ventilation, and Air Conditioning Associations prescribes 1000–2000 ppm CO_2 concentration in an indoor environment and suggests intervention if concentration goes out of this range [24]. Contrary to this, the Finnish decree advocates an upper threshold limit in the range of 1500–1600 ppm in an indoor environment [25]. Incorporating recommendations, the study considers the upper threshold limit of CO_2 as 1500 ppm and the lower ventilation setpoint as 1.26 $m^3/h/m^2$. Air change within the building maintains the CO_2 concentration and is depicted by the ventilation setpoint.

The expected operational household equipment during the blackout includes a fridge, freezer, microwave oven, communication devices (laptop, Wi-Fi, and mobile charger), and minimum lighting. There was no usage of dishwashers, saunas, laundry, or showers during the blackout. Figure 1 shows the typical occupants' household behavior in the Finnish region in terms of the proportion of average peak power used per hour of the day. The profile incorporates the usage of the above-mentioned necessities only and is modified based on online tools and the available literature [26,27]. Survivability conditions for a short duration require a minimum of 5 to 7 L of hot water per day, according to the Safe Drinking Water Foundation [28]. The study assumes an even distribution of hot water over 24 h to simplify the simulation. Table 2 represents the building's indoor parameters for survivability and normal conditions.

Table 2. Building indoor parameters (setpoints) for survivability and normal conditions.

Parameters	Survivability Condition	Normal Condition	Units
Lower temperature threshold	15	21	°C
CO_2 upper threshold concentration	1500	800	ppm
Ventilation lower threshold	1.26	1.8	m^3/h
Outside temperature	−26	−26	°C
Domestic hot water consumption	7	48	liters/person/day
Domestic hot water idling load	1.2	1.2	W/m^2 [29]
Relative humidity	NA, restricted by microbial growth	20–80	%
Occupancy presence	100	60	%/24 h

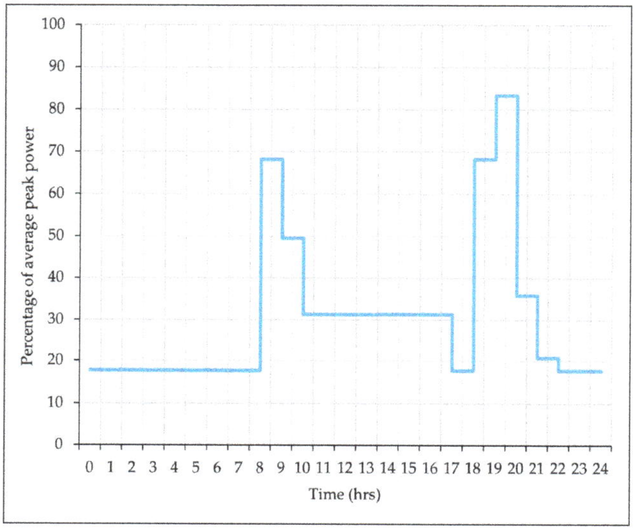

Figure 1. Assumed power consumption profile for residents during the blackout.

Based on assumed occupants' behavior and power consumption during a blackout, a typical apartment consumes an average power of 113 W and 341 W of peak power. The typical energy consumption per day, excluding space heating and domestic hot water energy, is 2.7 kWh. Table 3 shows the space heating and domestic hot water temperatures for the blackout duration, which follow the Finnish standards for net-zero energy buildings.

Table 3. Temperature maintained for heating and hot water within the building [29].

S. No.	System	Temperature Range
1	Space heating	20–35 °C
2	Domestic hot water	58 °C

This study scrutinizes an urban building block located in the Kalasatama urban area, Helsinki, in Southern Finland, containing residential apartments and commercial spaces. It comprises two towers with 5 floors and 13 floors, respectively, with an aggregated area of 7391.5 m^2 and an effective apartment area of 5863.84 m^2. Figure 2 shows the 3D (3-dimensional) diagram of the reference building in IDA-ICE simulation software, Version: 4.8.

Figure 2. A 3D model of the building in IDA-ICE software [29].

The study performs a resilience simulation for two sets of energy efficiency parameters for buildings. The two sets are named baseline configuration and efficient configuration. The baseline configuration building derives its U-values and glazing values for the building envelope from the Finnish building code. Similarly, the efficient configuration building acquires these values from the Flexible user-centric energy-positive houses (EXCESS) project under EU Horizon 2020. The district heating system ratifies the heating and domestic hot water requirements for the baseline building configuration. However, a heat pump suffices all the heating energy requirements in an efficient configuration building. Table 4 illustrates the parameter values for both configuration buildings.

Table 4. Building parameters for simulation scenarios [29,30].

Building Parameters	Baseline Configuration	Efficient Configuration
External walls (W/m^{2}k)	0.17	0.15
Internal walls (W/m^{2}k)	4.02	4.02
Internal floors (W/m^{2}k)	2.37	2.37
Roof (W/m^{2}k)	0.09	0.09
Slab toward ground (W/m^{2}k)	0.18	0.16
Glazing (W/m^{2}k)	0.70	0.60
Heating type	District	Electric
Heating Coefficient of Performance (COP)	0.97	4.60
Domestic hot water type	District	Electric
Domestic hot water COP	0.97	2.50
Ventilation control	Constant air volume	Constant air volume
VHR efficiency [%]	60	75

The default configuration of IDA-ICE employs ideal heaters and coolers for heating and cooling and operates on a proportional–integral (PI) controller. These controllers cannot record the data for the individual zones. Thus, both buildings employ a custom macro-controller that can document individual zone temperatures. The macro-controller operates on PI logic and maintains the normal condition and survivability condition setpoints accordingly. After setting up the building parameters, survivability parameters, and macro-controller, the system simulates for two weeks in both building configurations and records the temperature profiles for all typical apartments. Both buildings withstand the blackout in the same conditions, except for their building envelope and heating system.

5.3. Energy and Power Required

The indoor temperature can drop below the threshold value during the blackout if no external power is supplied. Thus, to calculate the required power to maintain survivability temperature thresholds, the system is modified to turn on the power supply within the building as soon as the indoor temperature touches the survivability threshold.

The hourly time-series data provide the power required at each hour to maintain the survivability conditions. The product of the power supplied and the duration for which the power is supplied calculates the energy required for the specific duration. In this way, the assessment provides the power and energy matrix, which specifies the power and energy required for each resilience duration.

5.4. Energy System for Building Resilience

The scenario under consideration includes a blackout; hence, no power is available from the grid. Thus, external energy systems need to support the energy supply post thermal inertia depletion of the building. Multiple energy systems possess the potential to fulfill the energy requirements but are thwarted by practical constraints such as availability at the site, space availability for installation, technological maturity, and local policies. Therefore, diesel generators, oil boilers, and lithium-ion batteries turn out to be the most feasible solutions considering different constraints, such as technical, economical, geographical, and local constraints.

Diesel generators produce one-phase or three-phase electrical energy, depending on the type of generator. The reference building employs a three-phase diesel generator and lithium-ion batteries coupled with the inverter to supply electrical energy while the oil boiler and heat pump deliver the heating, depending on the building configuration. The technical specifications of the considered components reflect the technical specifications of real-life components that can be used in the building.

As shown in Table 4, both configuration buildings require electricity as an input for necessities other than space heating and domestic hot water. The baseline configuration building fulfills the heat energy requirement through district heating, while the efficient configuration building generates the heat energy through the heat pump. Since the baseline configuration building connects to the district heating system, it needs heat as an energy input for space heating and domestic hot water. The efficient configuration building requires electricity as an energy input, even for space heating and domestic hot water, due to the connection with the heat pump.

Since the baseline configuration building requires heating and electrical energy separately as input, the sizing requires two types of duration curves. On the contrary, only electrical energy duration curves show the energy required within the efficient configuration building. The delivered power simulation results plot the duration curve after sorting. The baseline configuration plots the electrical and heating energy duration curves separately, while the efficient configuration building only plots the electrical duration curve.

5.5. Energy Systems Cost

The energy system tries to simulate real-life work. Thus, the cost of system components also matches real-life values. The cost includes all product costs, shipment costs, and installation costs, with special consideration of appropriate value-added tax (VAT) applicability. Relevant assumptions are made for installation costs in cases of data unavailability as per reference product installation instructions. Appendix A shows the reference costs of the products. The prices of diesel fuel and heating oils are dynamic; thus, the final price in the study is the average market price from 18 July to 24 October 2022. Table 5 depicts the additional costs and fuel costs associated with the energy system integration.

Table 5. Additional costs and fuel costs associated with the integration of energy systems.

Parameter	Value	Unit
VAT	24	%
Average monthly salary—skilled labor	3311	EUR/month [31]
Diesel cost	2.14	EUR/L [32]
Heating oil cost	0.096528	EUR/L [33]
Heating installation cost	10% of equipment	[34]
Occupancy (area of component/total area)	70	%
Other electronics cost	0.05	EUR/W
Battery system operation and maintenance cost	5	EUR/W_p/a

5.6. Energy Flexibility

The integration of the energy system is a precautionary measure to withstand unplanned power outages. Since the integrated system includes different components, with strategic implementation, some components can operate during normal duration without impacting the building's resilience potential. The expanded operation opens the possibility of increased operational time for the system and helps in achieving a return on investment.

The increased renewable energy penetration in electricity and energy-intensive equipment usage has increased power quality disruptions in the form of harmonics, voltage, and frequency disruptions [35]. A new market emerged due to these disruptions, ensuring the grid's operation according to the National Code. Energy reserves with rapid power delivery capabilities can participate in this market and earn revenue for their service. These energy markets are called ancillary services markets or flexibility markets. The different types of ancillary markets include frequency containment reserve (FCR), frequency restoration reserve (FRR), and fast frequency reserve (FFR), depending upon the activation time from the start of disruption. These energy markets operate in the form of a yearly market or hourly market, based on which participation requirements change. Each market requires a minimum size of bid to participate [36]. The systems with less power reserve availability can participate through an aggregator who charges a fee in exchange for its service. The aggregator combines the smaller power reserves and regulates them according to the TSO's requirements.

Due to the smaller size of building power reserves, the study considers participation in the hourly market. This study performs a perfunctory calculation to assess the impact of the participation of energy systems in the flexibility market on achieving a return on investment. Appendix B describes the assumptions for participation in the flexibility market in both scenarios. It describes the system characteristics participating in the flexibility market for both building configurations. The first scenario considers daily participation with full reserve activation, while the second scenario considers daily participation with no reserve activation. Economic calculations assess the system's feasibility to participate in the flexibility market and achieve a return on investment using the system's characteristics. The revenue calculation employs the capacity fee calculation formula prescribed by FINGRID for participation in the respective flexibility markets, which are shown in Equations (1) and (2) [37]. The return-on-investment time calculation and net yearly revenue calculations employ Equations (3) and (4).

$$\text{Capacity fee } (€) = \text{maintained reserve capacity } (\text{MW,h}) \times \text{hourly market price } (€/\text{MW,h}) - \text{sanctions } (€) \quad (1)$$

$$\text{Sanctions } (€) = \text{reserve capacity not delivered } (\text{MW,h}) \times 3 \times \text{hourly market price for specific hour } (€/\text{MW,h}) \quad (2)$$

$$\text{Return on investment (years)} = \text{system cost}/\text{net yearly revenue} \quad (3)$$

$$\text{Net yearly revenue } (€) = \text{yearly revenue } (€) - \text{yearly recharging cost } (€) \quad (4)$$

6. Simulation Results and Discussion

The simulation duration is two weeks, during which the building operates at normal conditions for the first week. The blackout occurs at the start of the second week, and the building tries to maintain the survivability conditions indoors. This section discusses the results of the building simulations performed in IDA-ICE with the above-mentioned boundary conditions.

6.1. Temperature Decay Profile

The custom macro-controller records the temperature profile in all typical apartments for both building configurations. The temperature decay profiles are represented in Figures 3 and 4 for efficient and baseline building configurations, respectively. The building operates at normal conditions during the first week, and then the blackout occurs at the 168th hour. The indoor temperature within both buildings starts to drop and follows the curve as shown in the figures. The multiple lines imply the variation in temperature drop rate in different apartments due to the number of parameters such as the size of the apartment, location within the building, and so on. The apartments' names follow the building drawing convention for both configuration buildings [38].

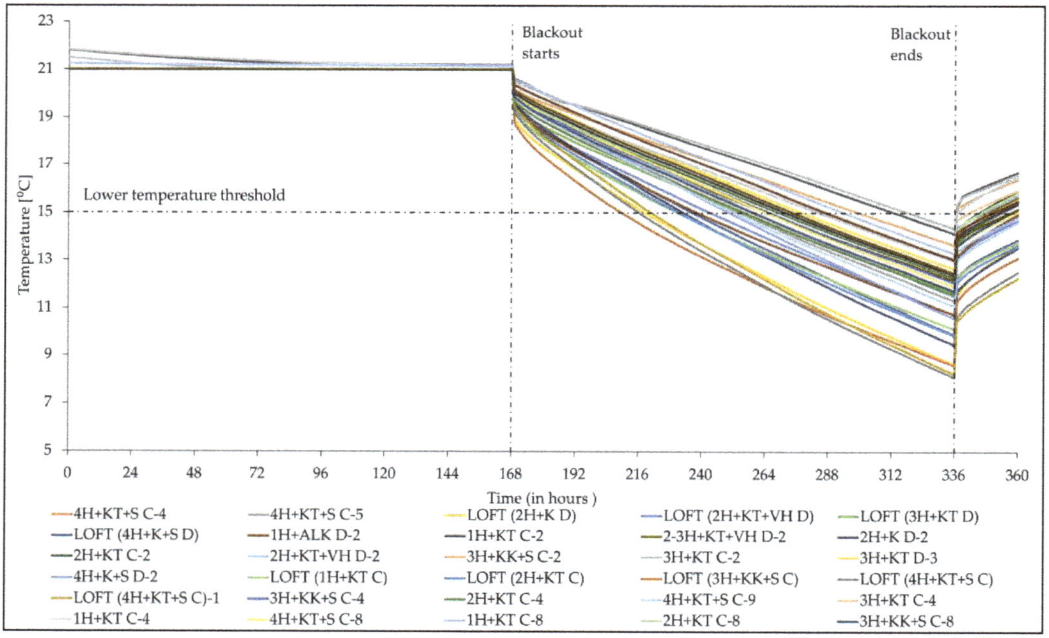

Figure 3. Temperature decay profiles of all typical zones in the efficient configuration building.

Table 6 presents the maximum, minimum, and average time taken by the apartments to drop the indoor temperature to the survivability condition threshold. From Figures 3 and 4, and Table 6, it is clear that the typical apartments with an efficient configuration building take more time to drop the indoor temperature to the lower temperature threshold value of the survivability parameters. The analysis shows that on an average basis, a typical apartment with the efficient configuration takes 15% more time than the baseline configuration building apartment to drop the indoor temperature to the lower temperature threshold.

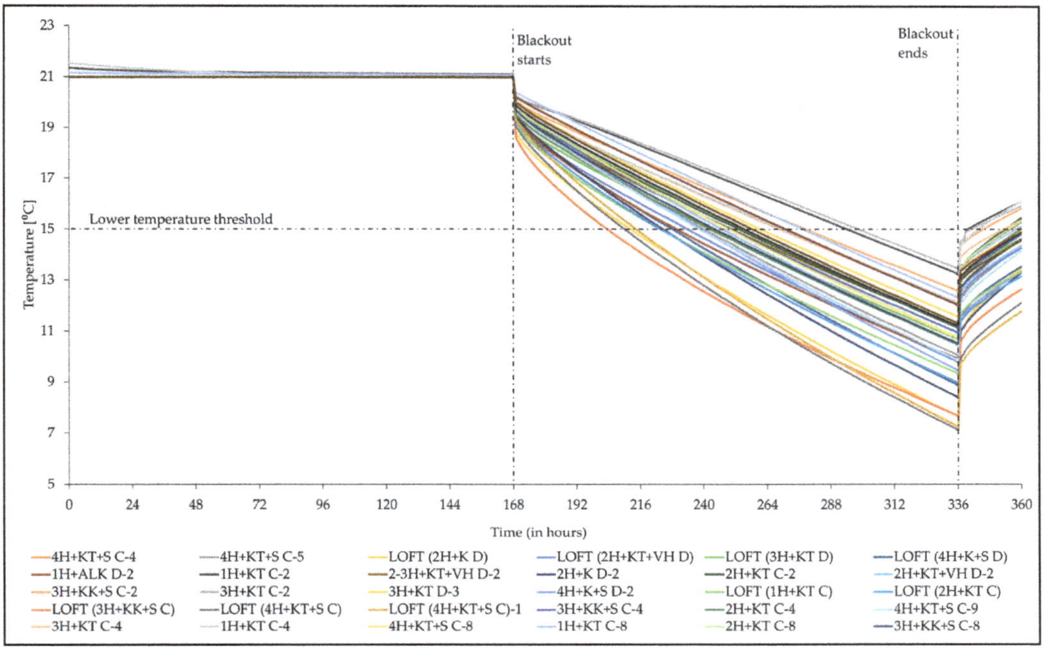

Figure 4. Temperature decay profiles of all typical zones in the baseline configuration building.

Table 6. Maximum, minimum, and average time taken by the typical zones to drop the zone temperature to 15 °C.

Time Taken in Hours to Drop the Temperature to 15 °C	Baseline Configuration Building (in Hours)	Efficient Configuration Building (in Hours)
Minimum time	36	42
Maximum time	131	151
Average time	79	93

Figure 5 represents the temperature profile of apartments, which takes the maximum, minimum, and closest to the average time to drop the indoor temperature to 15 °C. The efficient configuration building apartments take more decay time. The blackout occurs at the same temperature in both configuration buildings while maintaining the same survivability conditions. The only difference among buildings lies in the building envelope parameters and energy system for heating. Since no heat supply occurs during the blackout, it is clear that the improved building envelope parameters, higher thermal inertia retention, and lower thermal loss rate enhance the decay time in efficient configuration building apartments.

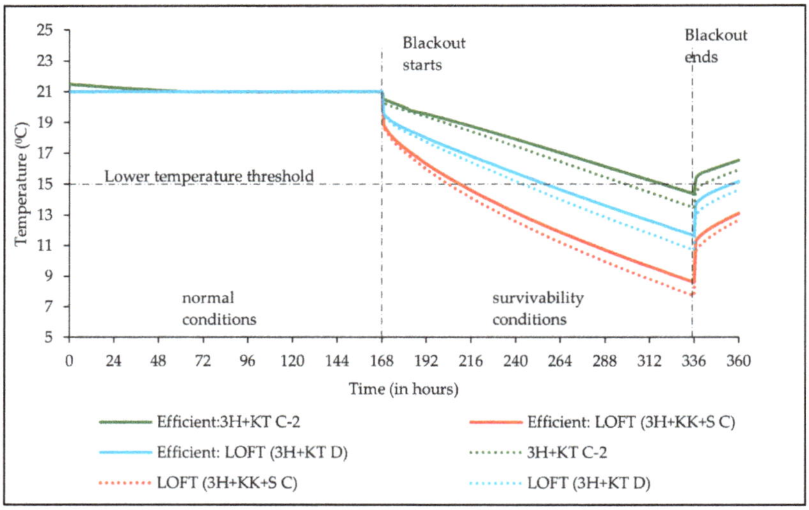

Figure 5. Temperature decay profile for apartments with maximum, minimum, and closest to the average time in both configuration buildings (dotted lines represent baseline configuration building zones).

6.2. Delivered Power

As evident from Figure 5, the temperature within the building keeps dropping below the survivability threshold if no power is supplied during the blackout. Hence, the system is simulated with the macro-controller to turn on the power supply as soon as the setpoints touch the survivability threshold. The simulated results provide the temperature and power required to maintain the survivability parameters within the building. Additionally, IDA-ICE can classify the total power consumption among different components.

As discussed previously, both buildings differ in the building envelope and the heating system used. Thus, space heating–electric and domestic hot water–electric represent typical electrical energy consumption in maintaining thermal setpoints of space heating and domestic hot water production in an efficient configuration building. Similarly, space heating–thermal and domestic hot water–thermal illustrate heat energy input for the baseline configuration building. HVAC implies the ventilation system power that maintains the CO_2 concentration and moisture level within the building. Equipment resident and lighting resident refer to the electrical energy consumption for household amenities and light bulbs, respectively. Figure 6 shows the power consumption among different components during the simulation period in both configuration buildings.

Both configuration buildings have almost equal power requirements for equipment resident and lighting resident because of the same occupant behavior and the same household appliance usage. The slight difference observed in equipment energy consumption is due to the variation in internal gains due to the different thermal envelopes and thermal loss rates. HVAC electricity consumption in an efficient configuration building is less due to the lower ventilation setpoints. The efficient configuration building's peak power consumption in space heating and domestic hot water is much less compared to the baseline configuration building. This is due to the highly efficient heat pump, better thermal envelope, and low thermal losses. The heat pump operates at a COP ranging from 3 to 5 against the standard district heating COP of 0.98 as per the Finnish localization of IDA-ICE.

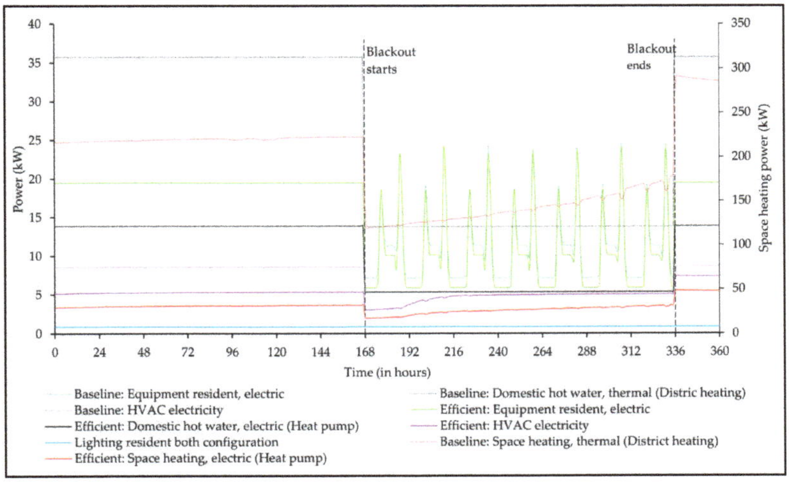

Figure 6. Power delivered during the simulation period in both configuration buildings (dotted lines represent the baseline configuration building).

The combined effect of the improved building envelope and highly efficient heat pump leads to 3 to 5.3 times less peak power requirement in the efficient configuration building compared to the baseline configuration building, which is shown in Figure 7. On an average consumption basis, the baseline configuration building consumes 3.9 times more power. This indicates that more energy-efficient heating systems, along with improved building parameters, help achieve building resilience.

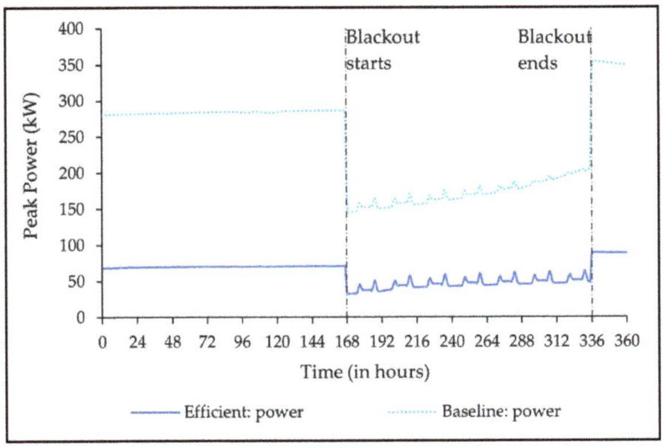

Figure 7. Power required to maintain survivability parameters within both configuration buildings (the dotted line represents the baseline configuration building).

6.3. Power and Energy Matrix for Building Resilience

The section explores the peak power and total energy required to maintain the survivability conditions for different resilience durations within both configuration buildings based on the simulation results from the delivered power section. Figure 8 depicts the peak power and peak power per m^2 needed in both configuration buildings to maintain the survivability conditions for resilience duration varying from 1 to 7 days.

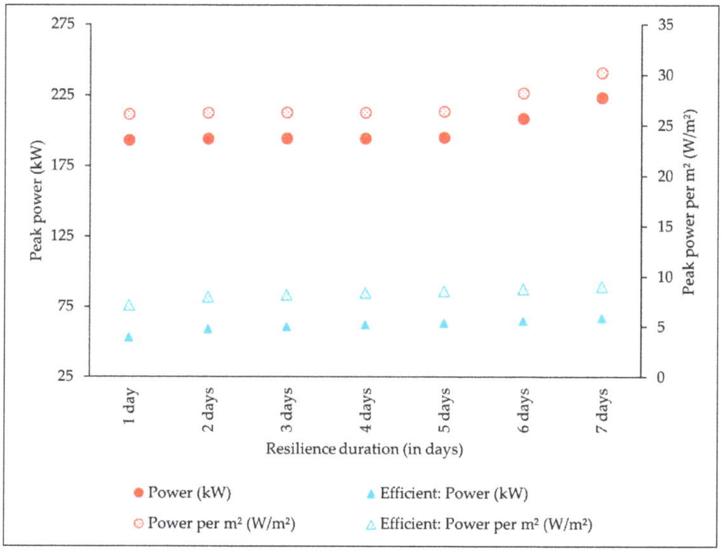

Figure 8. Peak power and power per m^2 needed in both configuration buildings for different resilience durations (circle represents the baseline configuration building).

The peak power required in both configuration buildings increases with the resilience duration. The rate of increment is higher in baseline configuration building because of the less efficient heating systems, reduced thermal inertia retention, and higher thermal losses from the building envelope. From Figure 8, it is clear that the baseline configuration building dissipates 3.1 to 3.6 times more peak power than the efficient configuration building. The peak power per m^2 follows the same trend as peak power, as both buildings occupy the same area. The use of a less efficient heating system and a less thermal-resistant building envelope causes increased power consumption within the baseline configuration building.

Therefore, an efficient heating system and an improved building envelope, which help in higher thermal inertia retention and fewer thermal losses, enhance the building's resilience prospects.

Table 7 depicts the peak power requirement among different components for both configuration buildings. Analyzing Table 7 shows that within the baseline configuration building, space heating consumes the maximum power amount followed by equipment resident and domestic hot water for all resilience durations. Less efficient heating systems, higher thermal losses, and lower thermal inertia retention cause the maximum power requirement in space heating, hence this power consumption trend.

Considering the efficient configuration building's power consumption trend, the order for the first two days is equipment resident, followed by space heating and domestic hot water. From the third day, the power dissipation order among components is the same as in the baseline configuration building. In an efficient configuration building, the power consumption trend among components during the initial days varies due to higher thermal inertia retention for the increased duration due to the efficient building envelope. Table 8 shows the peak power per m^2 among different components in both configuration buildings, and the consumption trend is the same as observed in peak power for the respective buildings.

Table 7. Resilience matrix for peak power required by different components in the apartment to maintain survivability conditions in both building configurations.

Resilience Duration	Peak Power in the Efficient Configuration Building					Peak Power in the Baseline Configuration Building				
	Efficient: Equipment Resident, Electric	Efficient: Domestic Hot Water, Electric (Heat Pump)	Efficient: HVAC Electricity	Efficient: Space Heating, Electric (Heat Pump)	Efficient: Lighting Resident	Baseline: Equipment Resident	Baseline: Domestic Hot Water, Thermal (District Heating)	Baseline: HVAC Electricity	Baseline: Space Heating, Thermal (District Heating)	Baseline: Lighting Resident
	kW	kW	kW	kW	kW	kW	kW	kW	kW	kW
1 day	23.21	5.35	3.62	19.93	0.86	23.07	13.78	5.36	149.98	0.86
2 days	24.01	5.35	4.76	24.26	0.86	24.00	13.78	5.38	149.98	0.86
3 days	24.01	5.35	4.93	25.68	0.86	24.14	13.78	5.39	149.98	0.86
4 days	24.01	5.35	4.95	27.04	0.86	24.14	13.78	5.40	149.98	0.86
5 days	24.01	5.35	4.99	28.43	0.86	24.14	13.78	5.41	150.81	0.86
6 days	24.01	5.35	5.00	29.77	0.86	24.46	13.78	5.41	164.12	0.86
7 days	24.01	5.35	5.02	31.64	0.86	24.46	13.78	5.42	178.82	0.86

Table 8. Resilience matrix for peak power per m^2 required by different components in the apartment to maintain survivability conditions in both configuration buildings.

Resilience Duration	Peak Power per m^2 in Efficient Configuration Building					Peak Power per m^2 in Baseline Configuration Building				
	Efficient: Equipment Resident, Electric	Efficient: Domestic Hot Water, Electric (Heat Pump)	Efficient: HVAC Electricity	Efficient: Space Heating, Electric (Heat Pump)	Efficient: Lighting Resident	Baseline: Equipment Resident	Baseline: Domestic Hot Water, Thermal (District Heating)	Baseline: HVAC Electricity	Baseline: Space Heating, Thermal (District Heating)	Baseline: Lighting Resident
	W/m^2	W/m^2	W/m^2	W/m^2	W/m^2	W/m^2	W/m^2	W/m^2	W/m^2	W/m^2
1 day	3.14	0.72	0.49	2.70	0.12	3.12	1.86	0.73	20.29	0.12
2 days	3.25	0.72	0.64	3.28	0.12	3.25	1.86	0.73	20.29	0.12
3 days	3.25	0.72	0.67	3.47	0.12	3.27	1.86	0.73	20.29	0.12
4 days	3.25	0.72	0.67	3.66	0.12	3.27	1.86	0.73	20.29	0.12
5 days	3.25	0.72	0.68	3.85	0.12	3.27	1.86	0.73	20.40	0.12
6 days	3.25	0.72	0.68	4.03	0.12	3.31	1.86	0.73	22.20	0.12
7 days	3.25	0.72	0.68	4.28	0.12	3.31	1.86	0.73	24.19	0.12

Figure 9 shows the total energy and energy per m^2 required in both configuration buildings for different resilience durations. Similar to the peak power, total energy and energy per m^2 needed in both configurations increase with enlarged resilience duration. The rate of increment in the baseline configuration is much higher than in the efficient configuration building. Through analysis of Figure 9, it is evident that the baseline configuration building consumes 3.7 to 4.1 times more energy than the baseline configuration building. The energy consumption ratio and rate of energy consumption increase with resilience duration more in the baseline configuration building because of the less efficient heating system and higher thermal losses. The less efficient heating system and high thermal losses mandate the steady usage of higher average power for a longer duration, which causes this energy requirement trend. Tables 9 and 10 depict the classification of total energy and energy per m^2 consumed by different components in both configuration buildings, respectively.

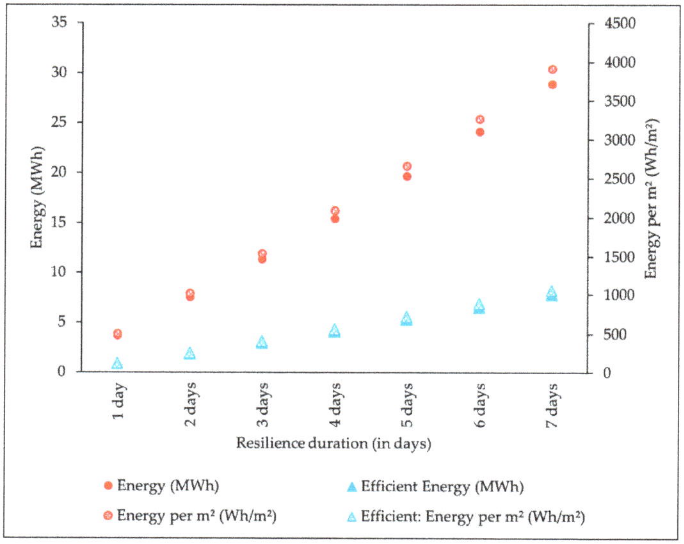

Figure 9. Total energy and energy per m^2 needed in both building configurations for different resilience durations (circle represents the baseline building configuration).

On analyzing Tables 7–10 together, it is clear that the same components occupy the maximum proportion of total energy consumption as in peak power consumption for the baseline configuration building. The order of energy consumption in an efficient configuration building is space heating, followed by equipment resident and domestic hot water. It is clear from Figure 6 that the equipment resident consumes more peak power than space heating for the first two days, but the duration of peak power is relatively small during the entire day. The total energy is the average power consumed throughout the resilience duration. Thus, on an average power basis, space heating consumes more energy throughout the day, which leads to the above-discussed trend of energy consumption within an efficient configuration building.

Table 9. Resilience matrix for total energy required by different components in the apartment to maintain survivability conditions in both configuration buildings.

Resilience Duration	Energy in the Efficient Configuration Building					Energy in the Baseline Configuration Building				
	Efficient: Equipment Resident, Electric	Efficient: Domestic Hot Water, Electric (Heat Pump)	Efficient: HVAC Electricity	Efficient: Space Heating, Electric (Heat Pump)	Efficient: Lighting Resident	Baseline: Equipment Resident	Baseline: Domestic Hot Water, Thermal (District Heating)	Baseline: HVAC Electricity	Baseline: Space Heating, Thermal (District Heating)	Baseline: Lighting Resident
	MWh	MWh	MWh	MWh	MWh	MWh	MWh	MWh	MWh	MWh
1 day	0.24	0.13	0.08	0.43	0.02	0.27	0.33	0.13	2.93	0.02
2 days	0.48	0.26	0.18	0.97	0.04	0.53	0.66	0.26	5.96	0.04
3 days	0.72	0.39	0.30	1.57	0.06	0.80	0.99	0.39	9.11	0.06
4 days	0.95	0.51	0.42	2.20	0.08	1.07	1.32	0.52	12.43	0.08
5 days	1.19	0.64	0.54	2.87	0.10	1.33	1.65	0.65	15.92	0.10
6 days	1.43	0.77	0.66	3.56	0.12	1.60	1.98	0.78	19.64	0.12
7 days	1.67	0.90	0.78	4.28	0.14	1.87	2.32	0.91	23.67	0.14

Table 10. Resilience matrix for total energy per m^2 needed by different components in the apartment to maintain survivability conditions in both building configurations.

Resilience Duration	Energy per m^2 in the Efficient Configuration Building					Energy per m^2 in the Baseline Configuration Building				
	Efficient: Equipment Resident, Electric	Efficient: Domestic Hot Water, Electric (Heat Pump)	Efficient: HVAC Electricity	Efficient: Space Heating, Electric (Heat Pump)	Efficient: Lighting Resident	Baseline: Equipment Resident	Baseline: Domestic Hot Water, Thermal (District Heating)	Baseline: HVAC Electricity	Baseline: Space Heating, Thermal (District Heating)	Baseline: Lighting Resident
	Wh/m^2	Wh/m^2	Wh/m^2	Wh/m^2	Wh/m^2	Wh/m^2	Wh/m^2	Wh/m^2	Wh/m^2	Wh/m^2
1 day	32.27	17.36	10.31	58.54	2.78	36.13	44.75	17.48	396.48	2.78
2 days	64.67	34.73	24.39	131.80	5.55	72.30	89.50	34.92	806.26	5.55
3 days	96.85	52.09	40.17	212.59	8.33	108.36	134.25	52.39	1232.61	8.33
4 days	129.01	69.45	56.22	297.98	11.10	144.34	179.00	69.91	1681.51	11.10
5 days	161.37	86.81	72.40	387.63	13.88	180.49	223.75	87.46	2153.45	13.88
6 days	193.82	104.18	88.62	480.98	16.66	216.77	268.50	105.03	2656.55	16.66
7 days	226.24	121.54	104.89	579.21	19.43	252.93	313.25	122.61	3202.45	19.43

6.4. Energy System Sizing for Building Resilience

The Energy System for Building Resilience section discussed the different energy systems that possess the capability to fulfill the energy requirement following the different constraints. This section deals with the sizing of those components to ensure building resilience for different resilience durations.

As mentioned before, efficient configuration building plots the electrical energy duration curve, which is shown in Figure 10. The duration curve in Figure 10 shows that 75% of the total time requires less than 40 kW of peak power for 1-day resilience. Similarly, for 2- and 3-day resilience, 84% and 70% of total time require less than 45 kW of peak power; 80% of total time requires less than 50 kW of peak power for a resilience duration of 4 to 7 days.

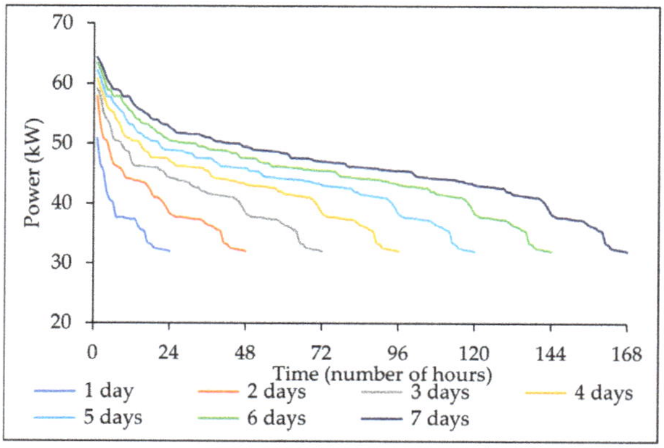

Figure 10. Electrical power duration curves for an efficient building configuration.

Figure 11 shows the electrical power consumption within the efficient configuration building. It also describes the average power consumption for each resilience duration. Along with these, it also depicts the duration for which the power requirement is fulfilled if different generators (with technical specifications similar to real life) provide power supply for the mentioned resilience duration.

Peak power consumption occurs for a very short duration, and average power is more than instantaneous power for the majority of the time, as shown in Figure 11. Therefore, the generator fulfills the energy supply until the above-mentioned power consumption point to avoid excess oversizing of the generator. The battery assists in fulfilling the remaining energy peak.

Similarly, Figure 12 shows the electrical and thermal power duration curves for each resilience duration for the baseline configuration building. Figure 13 represents the instantaneous heating and electrical power curve for the baseline configuration building. The electrical power duration curves have steep peak power for a short duration, as shown in Figure 13. Thus, the electrical power is divided among batteries and generators in such a way that the generator compensates for more than the average power and the battery assists with the remaining peak power. The heating power is separately provided using the boilers.

The combined analysis of Figures 10–13 gives the sizing points for power distribution among electrical (generator and battery) and heating. Table 11 represents the discussed total power distribution among heating and electrical components for both configuration buildings.

The technical specifications, such as the rated power of selected generators, are closest to the average required power and resemble the power rating of real-life equipment. The battery and power conversion device's technical specifications also resemble those of a real-life product with the potential to fulfill energy requirements. Major technical specifications for generators and batteries are shown in Appendix A.

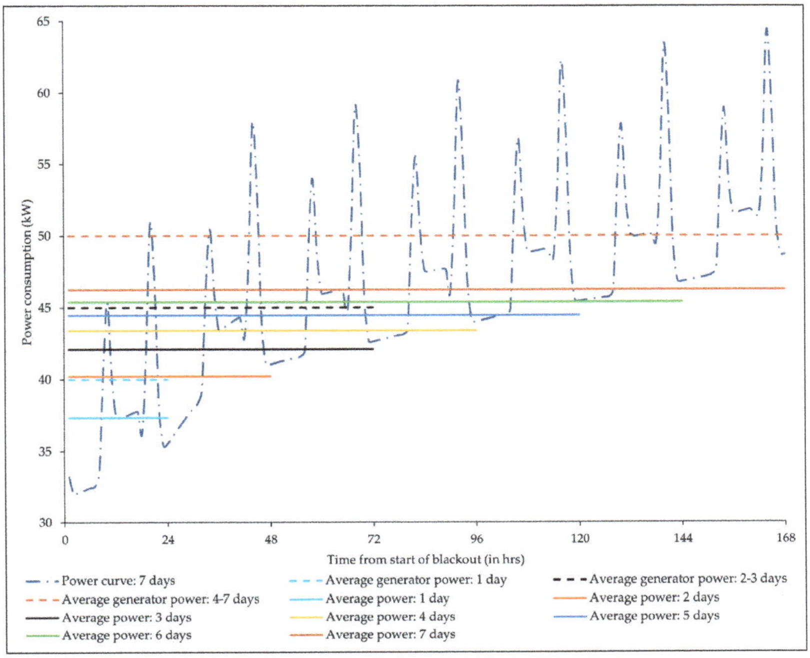

Figure 11. Peak power, average power needed, and average generator power for different resilience durations of an efficient configuration building.

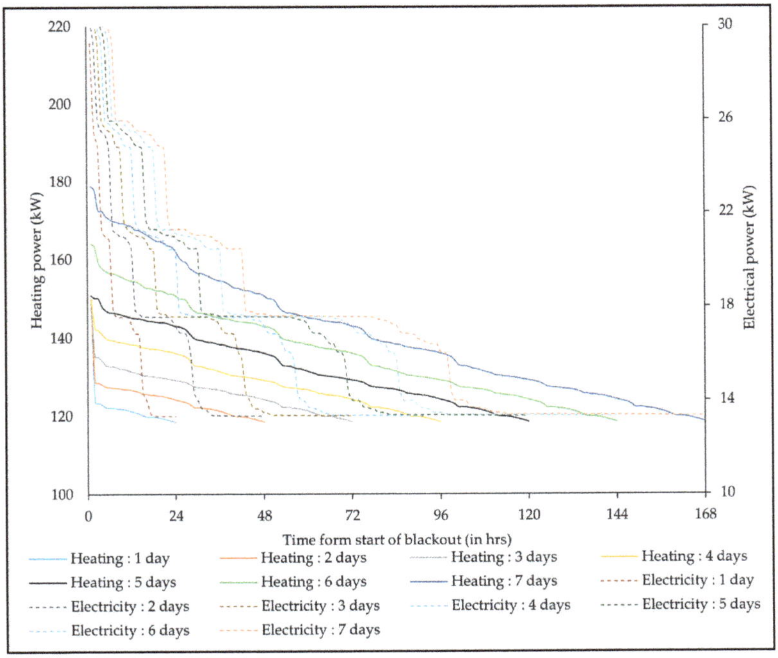

Figure 12. Electrical and heating power duration curves for the baseline configuration building (electrical power is represented by dotted lines).

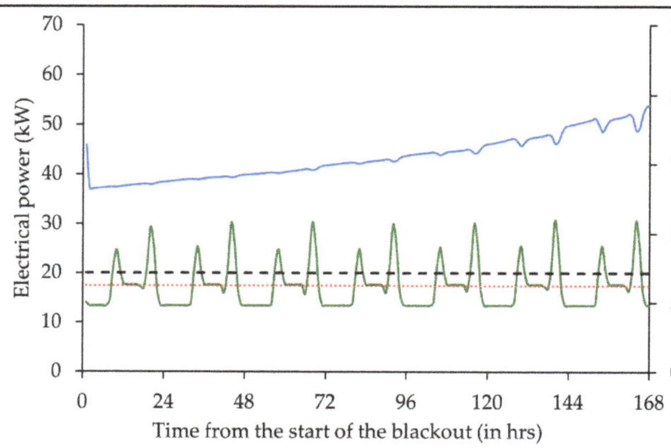

Figure 13. Peak power, average power needed, and average generator power for different resilience durations in the baseline configuration building.

Table 11. Peak power distribution among heating and electrical components for both building configurations.

Resilience Duration	Efficient Configuration Building				Baseline Configuration Building			
	Total Power	Generator-Rated Power	Generator Power	Battery Power	Total Power	Heating Power Required	Electrical Power Required Generator	Battery
	kW	kW	kW	kW	kW	kW	kW	kW
1 day	53	64	40	13	167	137	20	10
2 days	59	64	45	14	173	143	20	10
3 days	60	64	45	16	180	149	20	11
4 days	62	80	50	12	187	156	20	11
5 days	63	80	50	14	195	165	20	10
6 days	65	80	50	15	209	178	20	11
7 days	67	80	50	17	223	193	20	10

The system assesses the combination of energy system components that can ensure building resilience for different durations for both configuration buildings. The assessment of energy systems assesses the different scenarios of system components for both configuration buildings. The baseline configuration includes a system containing all boilers, generators, and batteries, a system containing only boilers and generators, and a system containing only boilers and batteries. Similarly, the efficient configuration building assessment includes scenarios combining both generators and batteries, only generators, and only batteries. The scenarios in both configuration buildings give the range of components needed for any resilience duration in each configuration building, for any resilience duration of up to 7 days. The assessment ensures that the combination of components in each system fulfills the power and energy requirements for each duration while following the electrical technical constraint.

Table 12 shows the different combinations of components of energy systems in an efficient configuration and baseline configuration building, respectively, that fulfill the energy requirements during the power outage.

Table 12. Number of components needed in both building configurations for different combinations of the energy system.

Resilience Duration	Efficient Configuration Building								Baseline Configuration Building											
	System Contains Both Generators and Batteries				If Only Generator Is Used for Resilience, No Battery		If Only Battery Is Used for Resilience, No Generator		System Contains All Boilers, Generators, and Batteries					If Only Boilers and Generators Are Used, No Battery			If Only Boilers and Batteries Are Used, No Generator		
	Number of Generators	Generator-Prime-Rated Power (kW)	Number of Battery Modules	Number of Inverters	Number of Generators	Generator-Prime-Rated Power (kW)	Number of Battery Modules	Number of Inverters	Number of Generators	Generator-Prime-Rated Power (kW)	Number of Boilers	Number of Batteries	Number of Inverters	Number of Generators	Generator-Prime-Rated Power (kW)	Number of Boilers	Number of Battery Modules	Number of Inverters	Number of Boilers
1 day	2	64	3	2	2	80	32	8	2	24	3	3	2	2	48	3	15	8	3
2 days	2	64	5	3	2	80	69	18	2	24	3	6	3	2	48	3	30	15	3
3 days	2	64	13	7	2	80	108	27	2	24	3	8	4	2	48	3	45	23	3
4 days	2	80	7	4	2	80	148	37	2	24	3	11	6	2	48	3	60	30	3
5 days	2	80	10	5	2	80	190	48	2	24	3	13	7	2	48	3	74	37	3
6 days	2	80	16	8	2	80	232	58	2	24	3	16	8	2	48	3	89	45	3
7 days	2	80	27	14	2	80	276	69	2	24	3	19	10	2	48	3	104	52	3

Analyzing Table 12, within the efficient configuration building, the number of battery modules used observes a dip after 3 days due to a change in the prime-rated power of generators used. The baseline configuration building shows an increasing trend in the number of batteries used with the resilience duration due to the constant rated power of the generator for all resilience durations. Two sets of generators in both buildings ensure that the system can withstand the worst scenario, which is a blackout when the batteries are discharged.

6.5. Energy System Cost Matrix for Resilience Duration

This section discusses the retrofitting cost of energy systems within the buildings calculated using the product and the additional costs mentioned in Appendix A and Table 5. Table 13 represents the overall cost needed and cost per m^2 needed to maintain resilience through different combinations of components within both configuration buildings. Compared to the system with no batteries, the retrofitting cost is low in the efficient configuration building. The same component ensures the energy supply for both energy types (heat and electricity). Hence, the observed trend is due to the single energy input in the efficient configuration building versus two separate energy inputs in the baseline configuration building. The two separate energy requirements imply the usage of two types of components, which means more initial investment. Both configuration buildings consume similar amounts of energy, except for thermal temperature maintenance. Thus, the cost increment rate with prolonged resilience duration is low in the efficient configuration building because of fewer thermal losses and a more efficient heating system, which together lead to less fuel consumption to maintain survivability conditions.

Analyzing the no-generator systems, the batteries supply electrical energy in this case. The baseline configuration building directly obtains heat from the boiler and has less associated investment and operational costs. In an efficient configuration building, electricity is required as an input, even for heating. The heat pump consumes electricity to produce heat energy, thus requiring additional energy conversion. More batteries are required to fulfill the energy demand, thus resulting in more investment costs. Hence, the single type of energy input leads to a higher retrofitting cost in an efficient configuration building. The trend in the proposed system (containing both batteries and a generator) for retrofitting costs varies with the rated power of the generator included in the system to supply electricity.

Figure 14 depicts the distribution of overall cost among different components of the system containing both generators and batteries for both building configurations. Both graphs show that the batteries occupy the maximum proportion of the overall cost for all resilience durations, and the proportion varies from 40% to 77% of the overall cost. In an efficient configuration building, for a small resilience duration, the generator cost is higher than the safety and storage cost at constant generator power, while for a higher resilience duration, the safety and storage cost is higher. This trend appears due to the increasing area required by the system and the additional safety compliance arrangements associated with that area. For short-period resilience, the system requires a smaller number of batteries and generators, resulting in less area and a low proportion of overall cost. With increased resilience duration, the number of batteries required increases even with a constant number of generators. Therefore, with increased resilience duration, safety and storage cost constitute the second-largest proportion of the overall cost. A similar trend is observed in the baseline configuration building: for smaller resilience durations, the generator and boiler occupy a higher proportion of the overall cost against the safety and storage cost. With the increased resilience duration, the area requirement increases, thus safety and storage cost become the largest contributor to the overall cost next to the battery in retrofitted energy systems.

Table 13. Cost matrix showing overall cost and cost per m^2 needed in both configuration buildings for different combinations of energy systems.

Resilience Duration	Efficient Configuration Building						Baseline Configuration Building					
	System Contains Both Generators and Batteries		Only Generator System, No Battery		Only Battery System, No Generator		System Contains Both Generators and Batteries		Only Boiler and Generator System, No Battery		Only Boiler and Battery System, No Generator	
	Overall Cost (1000 × EUR)	Cost per m^2 (EUR/m^2)	Overall Cost (1000 × EUR)	Cost per m^2 (EUR/m^2)	Overall Cost (1000 × EUR)	Cost per m^2 (EUR/m^2)	Overall Cost (1000 × EUR)	Cost per m^2 (EUR/m^2)	Overall Cost (1000 × EUR)	Cost per m^2 (EUR/m^2)	Overall Cost (1000 × EUR)	Cost per m^2 (EUR/m^2)
1 day	152	21	75	10.09	793	107	171	23	118	15.97	433	59
2 days	206	28	76	10.24	1709	231	250	34	124	16.78	815	110
3 days	413	56	77	10.36	2669	361	304	41	130	17.60	1202	163
4 days	270	37	78	10.49	3655	494	386	52	136	18.41	1584	214
5 days	349	47	78	10.61	4693	635	441	60	142	19.22	1945	263
6 days	505	68	79	10.74	5727	775	519	70	148	20.03	2331	315
7 days	790	107	80	10.86	6811	922	601	81	154	20.84	2716	367

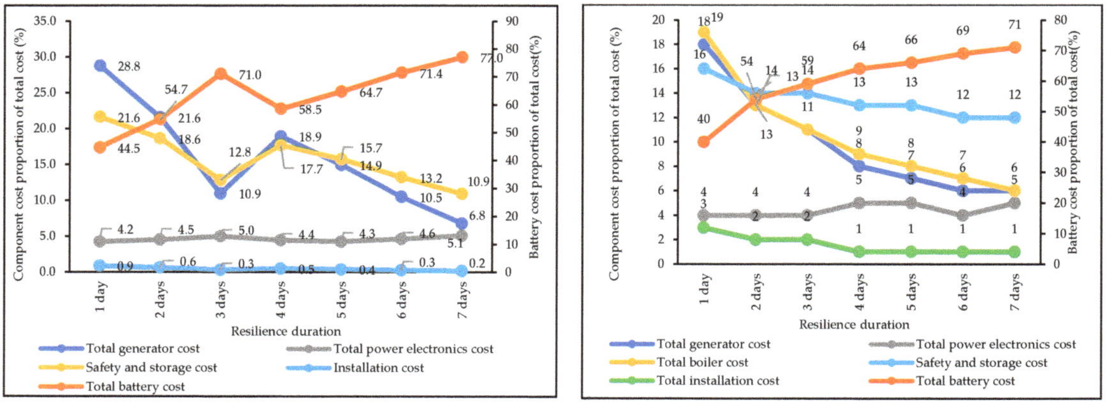

(**a**) (**b**)

Figure 14. Component distribution in the overall cost of the proposed system ((**a**): efficient configuration building and (**b**): baseline configuration building).

Figure 14 shows that batteries constitute the maximum proportion of the overall cost, and the associated initial investment cost is also higher. Thus, the efficient configuration building has a higher resilience cost because the heating energy required is supplied through a heat pump, which requires electrical input. However, the boiler supplies the heat directly in the baseline configuration building, thus resulting in less retrofitting costs.

6.6. Flexibility Market Participation

This section assesses the participation of both configuration building energy systems in the flexibility market under two scenarios. The first scenario involves daily participation with full reserve activation all the time, and the second scenario considers daily participation with no reserve activation. Appendix B describes the assumptions for both scenarios' participation in the flexibility market. Additionally, it discusses the characteristics of integrated energy systems for both configuration buildings.

Table 14 shows the yearly recharging cost and the revenue generated through participation in different flexibility markets at the full activation scenario at March 2022 electricity prices for both configuration buildings. The analysis of Table 14 shows that the yearly recharging cost of participation in the flexibility market is more than the revenue earned through participation in the respective flexibility market for efficient configuration building. While in the baseline configuration building, the revenue is higher than the recharging cost in the case of the aFRR market with the most favorable bidding price. Even in this case of higher revenue, the return-on-investment time is higher than 10 years, which is more than the lithium-ion battery shelf life. Hence, the system becomes non-feasible in a full activation scenario at the March 2022 electricity prices. Since the electricity price in October 2022 is almost 3–4 times higher than March 2022 electricity prices, there is therefore no return on investment at higher electricity prices in both configuration buildings' energy systems under the full reserve activation scenario.

The above-discussed energy market compensation is capacity-based. Thus, the scenario with an equal number of hours of participation per day with no reserve activation will not have any recharging costs. Hence, the system will have a return on investment even within the feasible time limit. Table 15 shows the revenue earned through participation in the energy market in the case of no activation of the reserve.

Table 14. Yearly recharging cost and potential revenue from participation in various flexibility markets by the energy systems of both configuration buildings at March 2022 electricity prices in full activation scenario.

Resilience Duration	Efficient Configuration Building								Baseline Configuration Building							
	Yearly Recharging Cost	Yearly Recharging Cost for FCR-N Market	Potential Revenue						Yearly Recharging Cost	Yearly Recharging Cost for FCR-N Market	Potential Revenue					
			Participation in FCR-N		Participation in FCR-D		Participation in aFRR				Participation in FCR-N		Participation in FCR-D		Participation in aFRR	
			60% of Average Price	200% of Average Price	60% of Average Price	200% of Average Price	60% of Average Price	200% of Average Price			60% of Average Price	200% of Average Price	60% of Average Price	200% of Average Price	60% of Average Price	200% of Average Price
	1000×EUR	1000×EUR	1000×EUR	1000×EUR	1000×EUR	1000×EUR	1000×EUR	1000×EUR	1000×EUR	1000×EUR	1000×EUR	1000×EUR	1000×EUR	1000×EUR	1000×EUR	1000×EUR
1 day	29.0		2.3	7.7	0.2	0.8	4.6	15.3		26.6	2.3	7.7	0.5	1.7	10.5	35.0
2 days	48.4		3.8	12.8	0.4	1.3	7.7	25.5		53.2	4.6	15.3	1.0	3.5	28.0	70.0
3 days	125.7	115.2	10.0	33.2	1.0	3.3	19.9	66.4	77.4	70.9	6.1	20.4	1.4	4.6	37.3	93.3
4 days	67.7	62.1	5.4	17.9	0.5	1.8	10.7	35.7	106.4	97.5	8.4	28.1	1.9	6.4	51.3	128.3
5 days	96.7	88.6	7.7	25.5	0.8	2.5	15.3	51.0	125.7	115.2	10.0	33.2	2.3	7.5	60.7	151.7
6 days	154.7	141.8	12.3	40.8	1.2	4.1	24.5	81.7	164.4	141.8	13.0	43.4	3.0	9.8	79.3	198.3
7 days	261.1	239.3	20.7	68.9	2.1	6.8	41.3	137.8	183.7	168.4	14.6	48.5	3.3	11.0	88.7	221.7

Table 15. Yearly recharging cost and potential revenue from participation in various flexibility markets by the energy systems of both configuration buildings at March 2022 electricity prices in no activation scenario.

Resilience Duration	Efficient Configuration Building							Baseline Configuration Building						
	System Cost	Potential Revenue						System Cost	Potential Revenue					
		Participation in FCR-N		Participation in FCR-D		Participation in aFRR			Participation in FCR-N		Participation in FCR-D		Participation in aFRR	
		60% of Average Price	200% of Average Price	60% of Average Price	200% of Average Price	60% of Average Price	200% of Average Price		60% of Average Price	200% of Average Price	60% of Average Price	200% of Average Price	60% of Average Price	200% of Average Price
	1000×EUR	1000×EUR	1000×EUR	1000×EUR	1000×EUR	1000×EUR	1000×EUR	1000×EUR	1000×EUR	1000×EUR	1000×EUR	1000×EUR	1000×EUR	1000×EUR
1 day	152.2	2.6	8.6	0.2	0.8	6.9	23.0	171.0	2.6	8.6	0.2	0.8	6.9	23.0
2 days	206.3	4.3	14.3	0.4	1.4	11.5	38.3	249.6	5.1	17.1	0.5	1.7	13.8	45.9
3 days	413.1	11.1	37.1	1.1	3.6	29.9	99.5	304.4	6.9	22.9	0.7	2.2	18.4	61.2
4 days	270.0	6.0	20.0	0.6	1.9	16.1	53.6	386.1	9.4	31.4	0.9	3.0	25.3	84.2
5 days	348.5	8.6	28.6	0.8	2.8	23.0	76.5	440.9	11.1	37.1	1.1	3.6	29.9	99.5
6 days	504.9	13.7	45.7	1.3	4.4	36.7	122.5	546.5	14.6	48.6	1.4	4.7	39.0	130.1
7 days	790.1	23.1	77.2	2.2	7.4	62.0	206.7	601.2	16.3	54.3	1.6	5.2	43.6	145.4

From Tables 14 and 15, it is clear that the possible reasons for no return on investment in energy systems include the high electricity cost of battery recharge and the high initial investment of battery modules. However, an increasing number of battery energy storage system plants as reserves in the flexibility market indicates the probable future feasibility of the technology. The reduction in battery investment costs may help make this concept more feasible. Additionally, an onsite renewable energy system also supports the flexibility and market participation of energy reserves.

7. Conclusions

The recent trends indicate the challenge of climate change to human health and dwellings. The study shows the lack of research on resilience in extremely cold climates. The existing research does not assume consensus over terminologies, characteristics, or thresholds for relevant parameters. This study establishes the terminology and threshold parameters by compiling the available literature. This will further support resilience analysis in extremely cold climates.

Simulation results from IDA-ICE show that the time taken by the building to drop the indoor temperature to the threshold value from the normal temperature increases by 15%. The efficient building envelope, having higher thermal inertia retention and low thermal losses, enhances the building's resilience. The integration of an efficient heating system along with the building envelope reduces instantaneous power consumption by 3 to 5.3 times and, on an average basis, by 3.9 times. While peak power reduces by a range of 3.1 to 3.6 times, depending on the resilience duration, the total energy required similarly reduces by 3.7 to 4.1 times due to the efficient building envelope and efficient heating system.

Withstanding all the technical requirements and applicable constraints for energy systems—diesel generators, lithium-ion batteries, and oil boilers—are found apt and sized for usage in both buildings. Real-life products' analogous technical specifications and prices simulate the system closer to reality. The economic assessment shows that battery proportions in overall cost can go up to 70% and 77% in the baseline and efficient building configuration, respectively. The range of investment needed highly depends upon the number of batteries involved in the system. The range of investment varies from 10.09 EUR/m^2 to 10.86 EUR/m^2 for the efficient configuration building and from 15.97 EUR/m^2 to 20.84 EUR/m^2 for the baseline configuration building in the case of no battery system. For the system with no generator, the efficient configuration building's resilience cost varies from 107 EUR/m^2 to 922 EUR/m^2. The baseline configuration building's resilience cost varies from 59 EUR/m^2 to 367 EUR/m^2 depending upon the duration of resilience. The system containing no batteries has a very low retrofitting cost because of the lower initial investment and the lower associated fuel cost.

Ancillary services markets emerged as the outcome of grid fluctuations. The reserves can participate within them to ensure the grid's operation according to the National Code and earn revenue for their service. This study assessed the participation of building reserves in different hourly ancillary markets for two scenarios, including full activation of reserves and no activation of reserves. In the full activation scenario, the recharging cost of the battery is higher than the potential revenue in both configuration buildings. Thus, no return is feasible in the situation of recharging from grid electricity. The scenario with no activation of reserves implies a feasible return on investment because no battery recharging cost is incurred during participation. The increasing number of renewable-integrated battery reserve pilot plants participating in the flexibility market also corroborates the same. Thus, more of these systems can participate in flexibility markets in the future if battery costs are reduced, or an onsite renewable energy generation system is installed.

8. Future Work

This work can be further extended to assess and establish the resilience characteristics of buildings in extreme cold climates. The assessment included the case simulation in the multiapartment building, which primarily comprises the newly built buildings. Thus, the study can be expanded to the different building stocks available. The energy system retrofitting varies with the type of building and location of the site, thus also extending the techno-economic assessment to the different building stocks. This study assesses the participation in the flexibility market for a range of revenue with appropriate assumptions to simplify the calculations. Thus, participation in the flexibility market can be assessed for the more realistic operational scenario.

Author Contributions: Conceptualization, J.S. and A.H.; methodology, A.S., A.H. and J.S.; original draft preparation, A.S.; writing—review and editing, A.H. and J.S.; supervision, A.H. and J.S.; funding, A.H. and J.S. All authors have read and agreed to the published version of the manuscript.

Funding: The authors acknowledge the financial support from the Academy of Finland's project "Integration of building flexibility into future energy systems (FlexiB)" grant No. 333364 and the VTT Technical Research Centre of Finland's self-funded project. The first author also acknowledges EIT InnoEnergy, the KTH Royal Institute of Technology in Sweden, and Aalto University in Finland for the Master studies during which the main research work of this study was performed.

Data Availability Statement: All data is contained within the article.

Conflicts of Interest: The authors declare no conflict of interest.

Appendix A

- Battery cell specifications

Parameter	Value	Unit
Rated power	7	kW
Peak power	11	kW
Battery efficiency	90	%
Area	0.15	m^2
Product cost	8987	EUR
Shipment cost	95	EUR

- Inverter

Parameter	Value	Unit
Maximum power input	45	kW
Maximum current input	43.5	A_{dc}
Input voltage	680–1000	V_{dc}
Maximum power output	29.9	kW
Maximum current output	43.5	A_{ac}
Output voltage (line to neutral)	230	V_{ac}
Output voltage (line to line)	400	V_{ac}
Inverter efficiency	98	%
Area	0.12	m^2
Product cost	2190	EUR
Shipment cost	70	EUR
Maximum power input	45	kW

- Generators

Parameter	Type 1	Type 2	Type 3	Type 4	Type 5	Unit
Rated power	24	32	48	64	80	kW
Fuel consumption at full	6.7	9.4	19.2	19.2	23.8	L/h
Fuel consumption at (3/4)th	5.9	7.1	14.6	14.6	17.9	L/h
Generator area	1.26	1.97	2.79	2.90	2.90	m^2
Product cost	12,350	13,400	16,700	17,150	18,950	EUR
Shipment cost	155	155	155	155	155	EUR
Rated power	24	32	48	64	80	kW

- Boilers

Parameter	Value	Unit
Thermal power	70	kW
Effective thermal power	65.43	kW
Electrical power	6	kW
Efficiency	93.47	%
Area	0.49	m^2

- Boiler burner

Parameter	Value	Unit
Power rating (up to)	120	kW
Area	0.18	m^2

Appendix B

The following are the assumptions made in the calculation of potential revenue through participation in the flexibility market:

1. During analysis, the battery reserve participates in only one type of flexibility market for the entire year;
2. The time duration of January 2022 to October 2022 has been analyzed to calculate the potential number of bids;
3. The bid is procured daily for the entire year, and procurement of the bid is conducted in the same proportion as it was conducted from January 2022 to October 2022;
4. In the case of full activation, the reserve is activated all the times for which the bid is accepted;
5. In case of no activation, the bid is accepted, and the reserve will provide the flexibility capacity, but the reserve will not be activated;
6. The price for which the bid is accepted is considered constant, which is the average of the actual bidding price in the respective market for the mentioned time duration;
7. Aggregator facilitates the small reserves' participation in the flexibility market and charges 20% of revenue for the service;
8. The electricity price in the year 2022 varies a lot. Thus, the calculations are performed with the average electricity price for March 2022. Sensitivity analysis was performed with an electricity price of October 2022;
9. No replacement of batteries is considered. Hence, no replacement cost is included in the calculations;
10. The participation of batteries is considered in the complete number of hours;
11. Batteries' charging and discharging times are considered equal;
12. In the case of participation in the FCR-N market, the batteries are to be maintained at a 50% state of charge at all times due to their symmetric nature. Therefore, only half of the energy reserve can be used.

Table A1. Efficient building configuration energy system's characteristics for participation in the flexibility market.

Resilience Duration	Number of Generator	Number of Battery Modules	Number of Inverters	Max Power Rating Possible for Participation in Energy Markets	Total Energy Possible	Number of Hours Possible	Charging Time with Fast Charger	Daily Available Hours	Number of Daily Participations in Flexibility Market in Hours	Daily Energy Used	Cost of Daily Recharging	Total Energy Possible for FCR-N	Number of Hours Possible for FCR-N	Battery Recharging Time (Fast Charging)	Number of Daily Participations in FCR-N	Daily Hours of Participation in FCR-N	Daily Energy Used	Cost of Daily Recharging
				kW	kWh	hours	hours	hours		MWh	EUR	MWh	hours	hours		hours	MWh	EUR
1 day	2	3	2	42	96	2.3	2.3	12	12	0.5	79	48	1.1	1.1	10	11	0.5	73
2 days	2	5	3	70	160	2.3	2.3	12	12	0.8	132	80	1.1	1.1	10	11	0.8	121
3 days	2	13	7	182	416	2.3	2.3	12	12	2.2	344	208	1.1	1.1	10	11	2	316
4 days	2	7	4	98	224	2.3	2.3	12	12	1.2	185	112	1.1	1.1	10	11	1.1	170
5 days	2	10	5	140	320	2.3	2.3	12	12	1.7	265	160	1.1	1.1	10	11	1.5	243
6 days	2	16	8	224	512	2.3	2.3	12	12	2.7	424	256	1.1	1.1	10	11	2.5	389
7 days	2	27	14	378	864	2.3	2.3	12	12	4.5	715	432	1.1	1.1	10	11	4.2	656

Table A2. Baseline building configuration energy system's characteristics for participation in flexibility market.

Resilience Duration	Number of Generator	Number of Boiler	Number of Battery	Number of Inverters	Max Power Rating Possible for Participation in Energy Markets	Total Energy Possible	Number of Hours Possible	Charging Time with Fast Charger	Daily Available Hours for Flexibility	Number of Daily Participations in Flexibility Market	Cost of Daily Recharging	Yearly Recharging Cost	Total Energy Possible for FCR-N	Number of Hours Possible for FCR-N	Battery Recharging Time (Fast Charging)	Number of Daily Praticipations in FCR-N	Daily Hours of Participation in FCR-N	Daily Energy Used	Cost of Daily Recharging	Yearly Recharging Cost
					kW	kWh	hours	hours	hours		EUR	1000 × EUR	kWh	hours	hours		hours	MWh	EUR	1000 × EUR
1 day	2	3	3	2	42	96	2.3	2.3	12	12	79.5	29.0	48	1.1	1.1	10	11	0.5	73	26.6
2 days	2	3	6	3	84	192	2.3	2.3	12	12	159.0	58.0	96	1.1	1.1	10	11	0.9	146	53.3
3 days	2	3	8	4	112	256	2.3	2.3	12	12	211.9	77.4	128	1.1	1.1	10	11	1.2	194	70.8
4 days	2	3	11	6	154	352	2.3	2.3	12	12	291.4	106.4	176	1.1	1.1	10	11	1.7	267	97.5
5 days	2	3	13	7	182	416	2.3	2.3	12	12	344.4	125.7	208	1.1	1.1	10	11	2.0	316	115.3
6 days	2	3	17	9	238	544	2.3	2.3	12	12	450.4	164.4	256	1.1	1.1	10	11	2.5	389	142.0
7 days	2	3	19	10	266	608	2.3	2.3	12	12	503.4	183.7	304	1.1	1.1	10	11	2.9	461	168.3

References

1. Intergovernmental Panel on Climate Change (IPCC). Impacts of 1.5 °C Global Warming on Natural and Human Systems. In *Global Warming of 1.5 °C*; Cambridge University Press: Cambridge, UK, 2022; pp. 175–312. [CrossRef]
2. King, C.W.; Zarnikau, J.; Funkhouser, E. Committee Chairs Committee Members Other Contributors the Timeline and Events of the February 2021 Texas Electric Grid Blackouts. 2021. Available online: https://energy.utexas.edu/sites/default/files/UTAustin%20%282021%29%20EventsFebruary2021TexasBlackout%2020210714.pdf (accessed on 14 November 2023).
3. IEA. IEA—Energy Consumption. Available online: https://www.iea.org/data-and-statistics/data-tools/energy-statistics-data-browser?country=WORLD&fuel=Energy%20consumption&indicator=TFCShareBySector (accessed on 20 October 2022).
4. White, L.M.; Wright, G.S. Assessing Resiliency and Passive Survivability in Multifamily Buildings. 2020. Available online: www.ibpsa.us (accessed on 8 November 2022).
5. Homaei, S.; Hamdy, M. Quantification of energy flexibility and survivability of all-electric buildings with cost-effective battery size: Methodology and indexes. *Energies* **2021**, *14*, 2787. [CrossRef]
6. Homaei, S.; Hamdy, M. Thermal resilient buildings: How to be quantified? A novel benchmarking framework and labelling metric. *Build. Environ.* **2021**, *201*, 108022. [CrossRef]
7. Zhivov, A.M. Parameters for Thermal Energy Systems Resilience. *E3S Web Conf.* **2021**, *246*, 08001. [CrossRef]
8. Ozkan, A.; Good, J. Evaluation of the Thermal Resilience of a Community Hub. 2022. Available online: www.ashrae.org (accessed on 27 March 2023).
9. Kesik, T.; O'brien, W.; Ozkan, A. Toward a standardized framework for thermal resilience modelling and its practical application to futureproofing. *Sci. Technol. Built Environ.* **2022**, *28*, 742–756. [CrossRef]
10. Nik, V.M.; Perera, A.T.D.; Chen, D. Towards climate resilient urban energy systems: A review. *Natl. Sci. Rev.* **2021**, *8*, nwaa134. [CrossRef] [PubMed]
11. Attia, S.; Levinson, R.; Ndongo, E.; Holzer, P.; Kazanci, O.B.; Homaei, S.; Zhang, C.; Olesen, B.W.; Qi, D.; Hamdy, M.; et al. Resilient cooling of buildings to protect against heat waves and power outages: Key concepts and definition. *Energy Build.* **2021**, *239*, 110869. [CrossRef]
12. Wang, R.; Lu, S.; Zhai, X.; Feng, W. The energy performance and passive survivability of high thermal insulation buildings in future climate scenarios. *Build. Simul.* **2022**, *15*, 1209–1225. [CrossRef]
13. Roostaie, S.; Nawari, N. The DEMATEL approach for integrating resilience indicators into building sustainability assessment frameworks. *Build. Environ.* **2022**, *207*, 108113. [CrossRef]
14. Zhang, C.; Kazanci, O.B.; Levinson, R.; Heiselberg, P.; Olesen, B.W.; Chiesa, G.; Sodagar, B.; Ai, Z.; Selkowitz, S.; Zinzi, M.; et al. Resilient cooling strategies—A critical review and qualitative assessment. *Energy Build.* **2021**, *251*, 111312. [CrossRef]
15. Ahmed, T.; Kumar, P.; Mottet, L. Natural ventilation in warm climates: The challenges of thermal comfort, heatwave resilience and indoor air quality. *Renew. Sustain. Energy Rev.* **2021**, *138*, 110669. [CrossRef]
16. EXCESS Project. Demo Case—Finland. Available online: https://positive-energy-buildings.eu/demo-cases/finland (accessed on 15 November 2023).
17. Ministry of the Environment. Unofficial Translation. Legally Binding only in Finnish and Swedish Decree of the Ministry of the Environment on the Energy Performance of New Buildings. 2018. Available online: https://ym.fi/documents/1410903/35099218/1010+2017+YMA_uuden_rakennuksen_energiatehokkuus+EN.pdf/87c893a5-08cf-cc7b-b2a1-90032ebb9368/1010+2017+YMA_uuden_rakennuksen_energiatehokkuus+EN.pdf?t=1680607785707 (accessed on 14 November 2023).
18. Mallawarachchi, H.; Rameezdeen, R.; De Silva, L.; Rameezedeen, R. Green Buildings, Resilience Ability and the Challenge of Disaster Risk Collaboration of Formal and Informal Sector in MSWM View Project Green Buildings, Resilience Ability and the Challenge of Disaster Risk. 2013. Available online: https://www.researchgate.net/publication/277892802 (accessed on 8 November 2022).
19. Moazami, A.; Carlucci, S.; Geving, S. Robust and resilient buildings: A framework for defining the protection against climate uncertainty. *IOP Conf. Ser. Mater. Sci. Eng.* **2019**, *609*, 072068. [CrossRef]
20. Cer, G.; Rezgui, Y.; Zhao, W. Critical review of existing built environment resilience frameworks: Directions for future research. *Int. J. Disaster Risk Reduct.* **2017**, *25*, 173–189. [CrossRef]
21. Sharifi, A.; Yamagata, Y. Principles and criteria for assessing urban energy resilience: A literature review. *Renew. Sustain. Energy Rev.* **2016**, *60*, 1654–1677. [CrossRef]
22. *ANSI/ASHRAE Standard 55-2004;* Thermal Environmental Conditions for Human Occupancy. ASHRAE Standard: Atlanta, GA, USA, 2004.
23. Alexander, Z. *Guide for Resilient Thermal Energy Systems Design in Cold and Arctic Climates;* ASHRAE: Atlanta, GA, USA, 2021; pp. 8–20, ISBN 978-1-947192-84-3.
24. Küçükhüseyin, Ö. CO_2 Monitoring and Indoor Air Quality. *REHVA Eur. HVAC J.* **2021**, *58*, 54–59.
25. Finlex. Decree of the Ministry of Social Affairs and Health the Health Conditions of the Dwelling and Other Living Area, as well as the Qualifications of External Experts. Available online: https://www.finlex.fi/fi/laki/alkup/2015/20150545#Pidm45237817260688 (accessed on 8 November 2022).
26. Paatero, J.V.; Lund, P.D. A model for generating household electricity load profiles. *Int. J. Energy Res.* **2006**, *30*, 273–290. [CrossRef]

27. Reda, F.; Fatima, Z. Northern European nearly zero energy building concepts for apartment buildings using integrated solar technologies and dynamic occupancy profile: Focus on Finland and other Northern European countries. *Appl. Energy* **2019**, *237*, 598–617. [CrossRef]
28. Safe Water Drinking Foundation. Water Consumption. Available online: https://www.safewater.org/fact-sheets-1/2017/1/23/water-consumption (accessed on 8 November 2022).
29. Wallin, A. Trilateral. Master's Degree Programme in Energy Technology Simulation Study of a Semi-Deep Ground Source Heat Pump System for a New Residential Building. Master's Thesis, LUT University, Lappeenranta, Finland, 2020.
30. ur Rehman, H.; Hasan, A.; Reda, F. Challenges in reaching positive energy building level in apartment buildings in the Nordic climate: A techno-economic analysis. *Energy Build.* **2022**, *262*, 111991. [CrossRef]
31. Statistics Finland. Work, Wages and Livelihood. Available online: https://www.stat.fi/tup/suoluk/suoluk_palkat_en.html#Index%20of%20wage%20and%20salary%20earnings,%20average%20annual%20changes (accessed on 10 November 2022).
32. Helsinki Diesel Prices. Available online: https://www.globalpetrolprices.com/Finland/Helsinki/diesel_prices/#:~:text=Helsinki%20Diesel%20prices%2C%20liter.%20Diesel%20prices%3A%20We%20show,the%20world%20for%20this%20period%20is%200.95%20Euro (accessed on 24 October 2022).
33. Finland Heating Oil Prices. Available online: https://www.globalpetrolprices.com/Finland/heating_oil_prices/ (accessed on 26 October 2022).
34. Lindberg, R.; Kivimäki, C.; Heinolainen, P. *Rakennusosien Kustannuksia 2022*; Rakennustieto: Helsinki, Finland, 2022. Available online: https://www.adlibris.com/fi/kirja/rakennusosien-kustannuksia-2022-9789522674005 (accessed on 8 November 2022).
35. Liang, X. Emerging power quality challenges due to integration of renewable energy sources. In Proceedings of the IEEE Industry Application Society, 52nd Annual Meeting: IAS 2016, Portland, OR, USA, 2–6 October 2016; Institute of Electrical and Electronics Engineers Inc.: Piscataway, NJ, USA, 2016. [CrossRef]
36. Modig, N.; Eriksson, R.; Ruokolainen, P.; Ødegård, J.N.; Weizenegger, S.; Fechtenburg, T.D. Overview of Frequency Control in the Nordic Power System. Available online: https://www.epressi.com/media/userfiles/107305/1648196866/overview-of-frequency-control-in-the-nordic-power-system-1.pdf (accessed on 8 November 2022).
37. Oyj, F. Appendix 1 to the Market Agreement of Frequency Containment Reserves Unofficial Translation Terms and Conditions for Providers of Frequency Containment Reserves (FCR). Available online: https://www.fingrid.fi/globalassets/dokumentit/fi/sahkomarkkinat/reservit/terms-and-conditions-for-providers-of-frequency-containment-reserves-fcr-as-of-22-may-2023.pdf (accessed on 12 August 2023).
38. Rakentaja.fi. Building Drawings. Available online: https://www.rakentaja.fi/artikkelit/11940/rakennuspiirustukset.htm (accessed on 27 March 2023).

buildings

MDPI

Article

Experimental Study of Indoor Air Quality in Educational Buildings: A Spanish Case Study

Alberto Leal Matilla [1], Jorge Pablo Diaz Velilla [2], Alicia Zaragoza-Benzal [1], Daniel Ferrández [1] and Paulo Santos [3,*]

[1] Departamento de Tecnología de la Edificación, Universidad Politécnica de Madrid, Avda. Juan de Herrera, 6, 28040 Madrid, Spain; alberto.leal.matilla@upm.es (A.L.M.); alicia.zaragoza@upm.es (A.Z.-B.); daniel.fvega@upm.es (D.F.)

[2] Departamento de Ingeniería de Organización, Universidad Politécnica de Madrid, Administración de Em-Presas y Estadística, C. de los Ciruelos, 28660 Madrid, Spain; jorge.diaz.velilla@upm.es

[3] Department of Civil Engineering, University of Coimbra, ISISE, ARISE, Rua Luís Reis Santos—Pólo II, 3030-788 Coimbra, Portugal

* Correspondence: pfsantos@dec.uc.pt; Tel.: +351-239797199

Abstract: Ensuring good indoor air quality in the spaces within educational centres is essential for the health and academic performance of students. In this sense, studying the evolution of health pollutants and their relationship with the environmental parameters of indoor humidity and temperature presents a challenge for the design of more efficient and comfortable buildings with a lower risk of virus infection. In this work, the relationship between pollution levels and SARS-CoV-2 virus infections in the academic year 2021/22 is shown, comparing the pollution values measured on-site with the value of the official measuring stations of the Community of Madrid. In addition, the impact of ventilation measures implemented during this period is assessed, aiming to establish guidelines for ensuring a safer and healthier school environment. It was found that during winter months, when there is less outdoor ventilation, pollution levels exceeded the recommended limits, according to reference regulations. This highlights the need for interior conditioning strategies in educational spaces. Thus, this multidimensional approach, considering both airborne pollutants and weather conditions, provides a comprehensive perspective on indoor air quality in school buildings in the central area of a metropolitan city, such as the Community of Madrid.

Keywords: indoor air quality (IAQ); educational buildings; classrooms; contagion risk; pollutants

Citation: Matilla, A.L.; Velilla, J.P.D.; Zaragoza-Benzal, A.; Ferrández, D.; Santos, P. Experimental Study of Indoor Air Quality in Educational Buildings: A Spanish Case Study. *Buildings* **2023**, *13*, 2780. https://doi.org/10.3390/buildings13112780

Academic Editor: Christopher Yu-Hang Chao

Received: 1 October 2023
Revised: 20 October 2023
Accepted: 2 November 2023
Published: 5 November 2023

1. Introduction

Sustainability is nowadays a major concern for the building and civil engineering industries, which have been embracing three major performance assessment criteria: (a) environmental; (b) economic; and (c) social [1]. Regarding buildings' life-cycle social assessment [2], there are several performance evaluation categories, namely: (i) Accessibility; (ii) Adaptability; (iii) Health and Comfort; (iv) Impacts on the Neighbourhood; (v) Maintenance and Maintainability; and (vi) Safety and Security. Concerning the (iii) Health and Comfort category, Santos et al. [3] assessed the indoor air quality (IAQ) and thermal characteristics of three Portuguese schools for higher education, making use of numerical simulation models, instead of in situ measurements.

In another research work about the same educational buildings, Santos et al. [4] performed a more complete social performance evaluation, where, besides the abovementioned sub-criteria (thermal characteristics and IAQ), three additional buildings' characteristics were evaluated, specifically: acoustic, visual comfort and spatial. It was concluded that one of the evaluated school buildings had a much better IAQ, i.e., smaller indoor CO_2 concentration above the external environment, due to increased natural ventilation (larger windows' area) and a consequent higher air renewal rate.

There are several studies showing how poor IAQ in classrooms has a negative impact on student learning [5]. More specifically, it has been corroborated that low IAQ levels decrease school performance [6] and can have harmful effects on the physical and mental health of students, because of prolonged periods of time spent inside school buildings [7,8]. In Spain, many educational centres are outdated and do not have support systems and controlled indoor air renewal, which makes it difficult to establish efficient ventilation and appropriate thermal comfort conditions [9]. For this reason, the development of studies aimed at guaranteeing the well-being of students and improving their academic performance is of vital importance today [10]. In this regard, following the experiences of the SARS-CoV-2 (Severe Acute Respiratory Syndrome Coronavirus 2) pandemic that affected the world in 2020, multiple research projects have been launched that have tried to address this issue from different application perspectives [11].

Firstly, from a practical point of view, Harvard University established a guide that compiled measures to improve ventilation in classrooms and prevent contagion [12]. In the Spanish case, this guide was considered for its application together with the guidelines established by the Ministry of Health to minimise infections in the classroom [13]. In these guidelines, among other things, natural or forced ventilation of rooms and periodic monitoring of the concentration of carbon dioxide in the environment were recommended. In general terms, as shown in simulation studies by Na et al. [8] and Chen et al. [14], the establishment of effective cross-ventilation flows allows the indoor air in classrooms to be renewed and the level of pathogens in the environment to be reduced. In this regard, multiple studies have shown that without effective ventilation, CO_2 concentrations in excess of 1500 ppm can be reached [15,16], which can lead to breathing difficulties for students, as well as headaches, fatigue and reduced ability to concentrate and learn [17].

On the other hand, although CO_2 is not responsible for SARS-CoV-2 infection, it is an effective indicator of indoor air renewal rate [18]. Thus, by setting maximum CO_2 concentrations as a function of occupancy rate, a rough idea of IAQ and the associated risk to students can be established [19]. Other studies have established a direct relationship between the value of fine particulate matter (PM) in the air and the risk of viral infection indoors [12]. In line with this idea, Ramalho et al. observed that airborne dust particles from Saharan winds contributed directly to the spread of respiratory diseases [20]. Thus, the higher the exposure to PM 2.5 and PM 10, the higher the long-term mortality risk. The European Union (EU) and the World Health Organisation (WHO) have taken this into account by setting daily and annual maximum values for exposure to PM 2.5 and PM 10, as shown in Table 1.

Table 1. Limit values for PM 2.5 and PM 10 [21,22].

	WHO		European Environment Agency	
	PM 2.5	PM 10	PM 2.5	PM 10
Annual (μg/m^3)	10	20	25	40
24 h (μg/m^3)	15	40	—	50

Thus, as stated by Tikul et al., microparticles are strong pollutants that contribute to the toxicity of cities, and they recorded a strong correlation between health problems and the concentration of ambient PM 2.5 in the city of Chiang Mai (Thailand), where students had great difficulty maintaining attention in the classroom [10]. Other studies have corroborated how in large cities, despite the incorporation of natural ventilation systems in schools, having a highly polluted outdoor environment can increase the concentration of fine particles in the classroom and can be harmful, making it necessary to use filtration systems [23]. In this sense, architectural design that considers effective ventilation based on quantified and reliable data, together with the use of simulation software, is crucial to obtaining effective and economical strategies to improve student health in schools [24].

Formaldehyde (HCHO) is another of the most abundant pollutants in the indoor classroom environment and has been classified as carcinogenic by the International Agency for Research on Cancer (IARC) [25]. Concern about this indoor pollutant has increased as a result of the increased use of hydroalcoholic gels since the COVID-19 pandemic [26]. Yu et al. demonstrated that short-term exposure to HCHO causes respiratory tract irritation and eye reddening, while prolonged exposure can lead to asthma and even cancer [27]. In general terms, HCHO concentration is directly related to temperature and relative humidity and is inversely proportional to the age of the building and outdoor air exchange [28,29]. In this regard, it is worth noting the study carried out in France by Hu et al. where a decrease of more than 25% in the concentration of formaldehyde in classrooms was observed if windows were opened regularly and the abusive use of cleaning products was controlled [30].

The main goal of this research is to study the evolution of CO_2, PM 2.5, PM 10 and HCHO parameters in a typical school in the Community of Madrid during a complete academic year. In this way, the aim is to establish a relationship between the values determined for these pollutants with the environmental conditions of ambient temperature and relative humidity, as well as with the pollution values determined by the reference measurement stations of the Madrid City Council. The intention is to show in detail, and with the support of various statistical analyses, the correlation between the contagions produced during the course and these pollutants. Thus, we analyse the possible causes, as well as preventive measures that can be extrapolated from this case study to other schools located in large urban centres.

2. Methodology

This section describes both the building under study and the process of taking measurements for data collection and subsequent analysis.

2.1. Contextualisation and Description of the Educational Establishment

Currently, the Community of Madrid has a population exceeding 6.75 million inhabitants in an area of 604.45 km^2 [31] and has become one of the European metropolitan areas with the highest mortality rate attributed to pollution. The education centre used as a reference for this research is located in the Carabanchel district, in the San Isidro neighbourhood (Madrid, Spain) with a population density of 183.55 inhabitants/km^2. Regarding the origin of PM 2.5 suspended particles, these would be attributed to road traffic (55% of total emissions), industrial activities (35.9%) and airport traffic (0.9%), with the remaining sources of the emission classified as "unknown" [32]. On the other hand, the city also has a high source of CO_2 emissions into the atmosphere and PM 10 particles of natural origin, which include wind-transported silt-sized sediments. Finally, the ageing of the Carabanchel district and the lack of renovation of the building stock significantly contribute to increasing the concentration of formaldehyde in the environment.

According to current regulations [33], the metropolitan area of the Community of Madrid is characterised by a temperate type C climate. All the classrooms in the building have the same orientation of windows to the east and access on the west side. In order to carry out the monitoring and periodic follow-up of the evolution of pollutants in its interior, periodic data have been collected in two classrooms located on the third floor of the building. The distribution and measurement points can be seen in Figure 1a, where the data were collected at a height of 1.2 m. In Figure 1a, it can be seen that Classroom 1 is a large space that provides access to an outdoor terrace and is ventilated at four distinct points: the entrance door to the classroom, the access door to the terrace and the classroom windows on both sides of the building, whose orientation is east–west, thus ensuring a continuous flow of fresh air. On the other hand, Classroom 2, adjacent to the previous one and with independent access through the hall, has four windows on the east side and, on the west side, only the access door to the classroom. Figure 1b shows an image of the building studied. In addition, Table 2 shows some of the most relevant data for this research regarding the configuration of the classrooms studied.

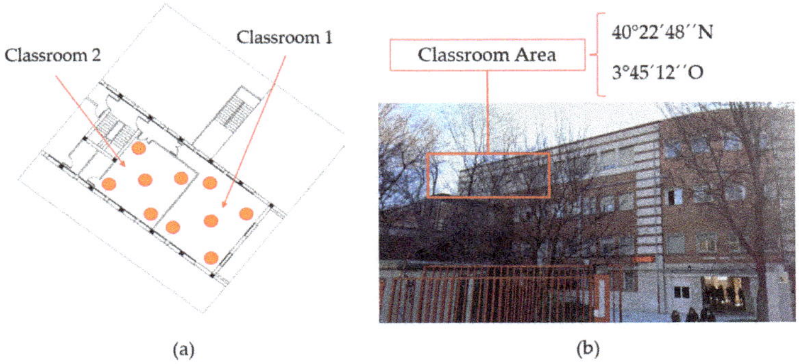

Figure 1. (**a**) Classrooms under study and locations of monitoring equipment (orange circles), (**b**) exterior image of the building.

Table 2. Characteristics of the studied classrooms.

Classroom	Students	Age (Year)	S (m^2)	V (m^3)	m^2/Student	m^3/Student
1	10	20 ± 1	80.32	200.80	8.03	20.08
2	15	22 ± 2	57.17	173.16	3.84	11.54

Neither of the two classrooms included in this study had forced ventilation systems, making it necessary to open doors and windows to establish indoor air renewal flows. The Spanish Regulation of Thermal Installations in Buildings (RITE) [34] establishes that classrooms must have an indoor air quality type IDA-2, which implies a renewal flow of 12.5 dm^3/s and maximum CO_2 concentrations of 500 ppm above the concentration in the outside air. However, the necessary hygrothermal conditions to meet these requirements were not met throughout the course due to the climatic and pandemic requirements.

2.2. Collection and Analysis of Experimental Data

Figure 2a shows a detail of the measuring equipment used for periodic data collection (Temtop M2000 2nd model), with records taken between 8:00 a.m. and 02:00 p.m. at a frequency of one measurement per minute. This measuring equipment uses laser scattering sensors to irradiate airborne particles, and then collects the scattered light to obtain the light change curve over time. The microprocessor calculates the equivalent particle diameter and the number of particles with different diameters per unit volume. This Temtop M2000 2nd equipment has allowed real-time data capture of PM 2.5, PM 10, CO_2 and HCHO particles, as well as being equipped with a thermo-hygrometer to monitor indoor temperature and humidity conditions. During the measurement process, the recommendations of the European Commission were followed, as set out in the *EMEP/EEA air pollutant emission inventory guidebook—'the Guidebook'*, which establishes the methodology for data collection and time series consistency [35].

On the other hand, in the case of suspended particles PM 2.5 and PM 10, the values of the two meteorological stations closest to the educational centre under study were compiled daily. Thus, in the case of the Community of Madrid, these stations are Plaza Elíptica (Figure 2b) and Farolillo (Figure 2c), with the latter measuring only PM 10 particles. These stations were equipped with a microbalance system to analyse the amount of pollutants in the air and report their data periodically with a frequency of one measurement per hour. In this way, by collecting these values, it was possible to establish differences between the measurements provided by official sources and the actual measurements taken in situ.

(a) (b) (c)

Figure 2. (**a**) Temtop M2000 2nd measuring equipment used for indoor data collection; (**b**) Farolillo measuring station; (**c**) Plaza Elíptica measuring station.

It should be noted that the values of CO_2, HCHO, temperature and humidity were taken as univariable, so that their comparison was carried out in relation to the values established by the different applicable regulations. The measuring equipment located inside the classroom was distributed in such a way that it did not interrupt the teaching activity, avoiding areas of direct radiation and with the clear objective of determining the state of the air at the height of the working plane. Once the data from the stations in the city of Madrid had been collected, they were exported to an Excel spreadsheet to be organised together with the values collected in the classroom; this process was carried out during the eight months of the school year in which the measurements were collected. The data were then exported to SPSS version 29.0 software for statistical analysis of the results.

3. Results and Discussion

In this section, the results obtained for the different parameters evaluated in this research are presented in an organised manner, including a critical discussion of the values presented in each analysis.

3.1. Indoor Hygrothermal Conditions

Firstly, Figures 3 and 4 show the descriptive values for interior humidity and room temperature in the classrooms studied. These graphs have been made with the help of Box-Plot diagrams that allow us to visualise the monthly average evolution of these parameters throughout the academic year.

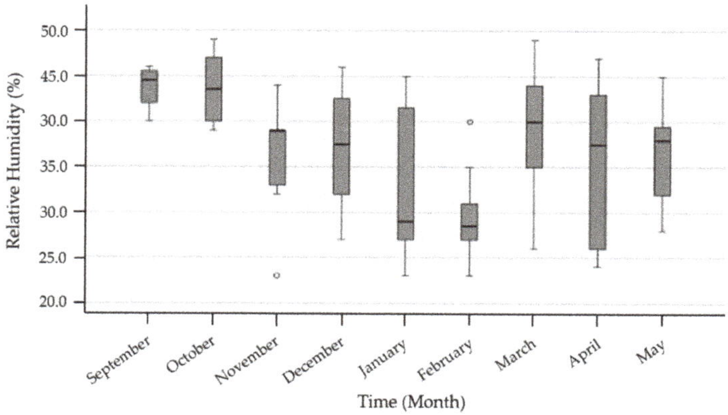

Figure 3. Evolution of indoor relative humidity throughout the academic year.

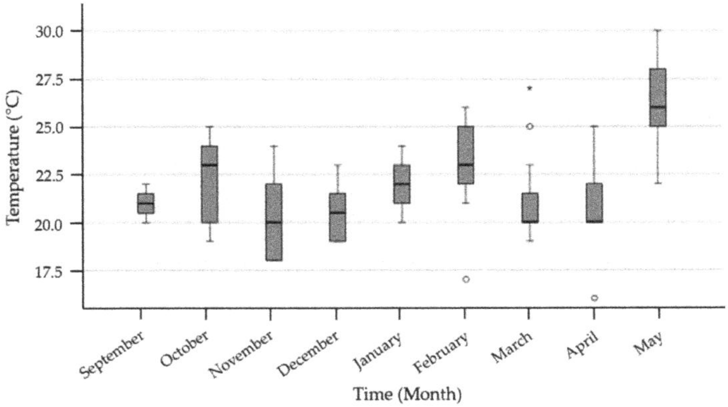

Figure 4. Evolution of the indoor temperature throughout the academic year.

The relative humidity values established in the RITE as recommended for occupied rooms are set between 40.0 and 50.0% in winter and 45.0 and 60.0% in summer [34]. Thus, if we look at Figure 3, except for the months of September and October, the average relative humidity values collected throughout the academic year were below the recommended minimum.

After analysing Figure 4, it can be seen that in the winter months of November (mean temperature 20.0 °C) and December (mean temperature 20.5 °C), the indoor temperature in the classrooms studied did not meet the minimum values recommended by the current standard of 21–23 °C in winter conditions [34]. Similarly, in the month of May (mean temperature 26.3 °C), the temperature exceeded the maximum recommended by the standard of 25 °C. This shows how the need for ventilation caused by COVID-19 reduced the thermal comfort in the classrooms, even though each classroom was equipped with a 10 kW heat pump air-conditioning unit.

In view of the results, it is possible to reflect on the need to incorporate forced ventilation systems inside these educational buildings to improve student comfort. Likewise, as Varela et al. have shown, a commitment to energy rehabilitation of the façade by means of external thermal insulation systems (ETICS) would be very beneficial and would have a positive impact on improving the indoor thermal comfort and quality of life of the inhabitants of the Carabanchel district (Madrid) [36]. The average annual values of 21.9 °C temperature and 36.5% relative humidity are below the minimum recommended for the interior of educational centres.

3.2. CO_2 Concentration Measurements

In this section we first show the average evolution of CO_2 in the interior of the classrooms analysed during the academic year. These values are shown with the help of a Box-Plot diagram in Figure 5. When interpreting these results, it is necessary to point out that the usual average values for outdoor atmospheric CO_2 range between 420 ppm and 450 ppm, depending on whether the environment is rural or urban [34]. For the specific case of this study, it is considered appropriate to use the upper limit of 450 ppm, since the school is located in the central area of the Community of Madrid, being an urban centre with a high population density and vehicle traffic. Likewise, at indoor CO_2 concentrations 300 ppm higher than outdoor atmospheric CO_2, it is understood that indoor air quality may be compromised and affect student performance [37].

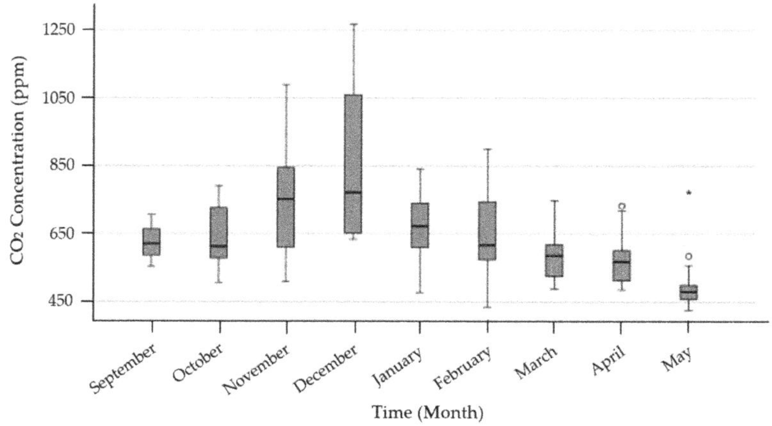

Figure 5. Indoor CO_2 concentration evolution during the academic year.

First, Figure 5 shows how the average CO_2 value inside the classrooms was higher than 450 ppm in all months. These elevated average CO_2 levels, despite having natural ventilation flows to the outside, are due to the high atmospheric pollution in the city, with many buildings adjacent to the educational centre with fuel oil or gas boilers and high permanent vehicular traffic. Likewise, there is an upward trend in CO_2 levels in the period September–December, and then decreases in the period January–May. This trend has a clear explanation in the closing of windows in winter during class time and ventilation between sessions, which contributed to increasing the indoor temperature and thus CO_2 levels, since as Zhang et al. have established, the human CO_2 emission rate increases by approximately 3.5% when the temperature increases by 1 °C [37]. On the other hand, particularly alarming are the maximum values above 1000 ppm reached in the months of November and December, which far exceed the recommended maximum value of 750 ppm indoor concentration.

Table 3 presents a more generalised analysis with a grouping of the daily measurements by quarterly assessments, corresponding to the usual academic year of Vocational Training in Spain. In this way, a univariate test is presented, where the mean difference between the CO_2 concentrations of each assessment compared to the mean outdoor CO_2 value set at 450 ppm is established.

Table 3. Test for difference of means at 99.5% confidence for CO_2 concentration.

Evaluation Period	Time Interval (Days)	Difference of Means (ppm)
First (September–December)	46	287
Second (January–March)	55	179
Third (April–May)	26	86

Thus, in the analysis by academic evaluations, it can be seen that the first evaluation period, corresponding to the winter months, showed a higher average indoor CO_2 concentration. In fact, the value obtained for the mean difference (287 ppm) is close to the maximum permitted value of 300 ppm above the outdoor CO_2 concentration. Table 3 shows how indoor air quality decreases in the colder months when ventilation was less pronounced. However, it is necessary to highlight that, as a limitation of this study, the evolution of CO_2 in the outdoor environment could not be monitored. Knowing these outdoor values would be of great utility for interpreting the results in order to implement efficient natural ventilation systems. Otherwise, with high levels of atmospheric CO_2, ventilation measures without prior purification of the air blown in could be harmful and counterproductive.

3.3. Concentration Measurements of Suspended Particulate Matter PM 2.5 and PM 10

The analysis of suspended particles is fundamental to understand viral propagation through the air. Thus, first of all, the monthly daily evolution over the course of the year for PM 2.5 and PM 10 is shown in Figures 6 and 7.

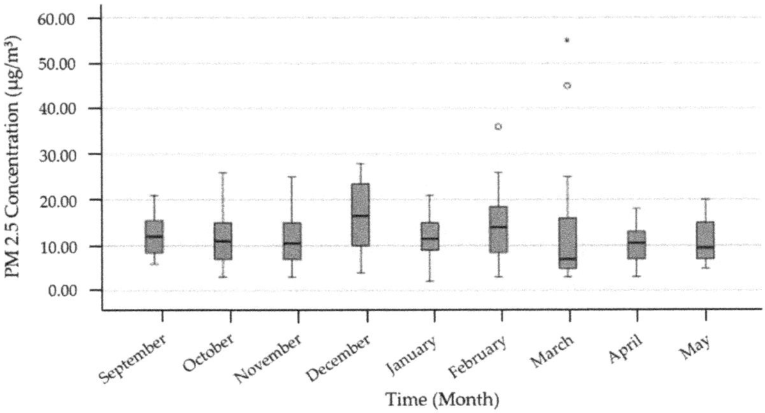

Figure 6. PM 2.5 concentration trend over the academic year.

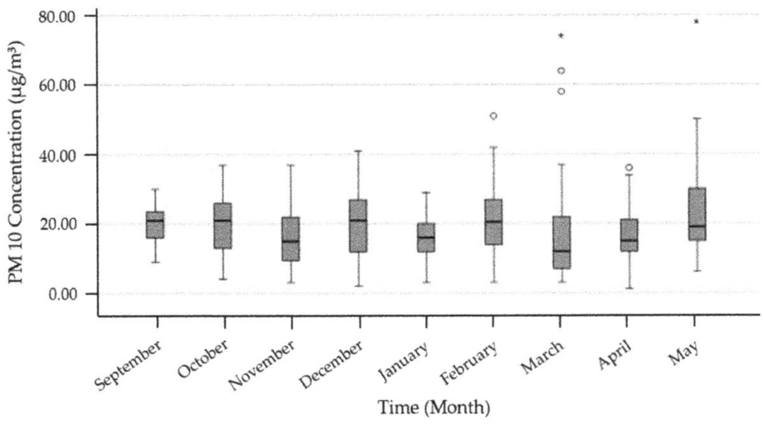

Figure 7. PM 10 concentration trend over the academic year.

It can be seen that both the mean concentration values obtained for PM 2.5 in Figure 6 and for PM 10 in Figure 7 maintained a constant tendency throughout the course of the year of around 10 $\mu g/m^3$ and 20 $\mu g/m^3$, respectively. However, when analysing the annual evolution, the most critical months correspond to December, February and March. The increase in concentration during the month of December is associated with the Christmas holiday period and the increase in vehicle traffic. In addition, a continuous trend was observed throughout the school year, with concentration peaks occurring in the first hour of the school day, corresponding to the arrival of students and workers, and during the last hour of the school day at dismissal time when there was a greater traffic flow. Similarly, the increase in suspended particulate concentration during February and March is due to the wave of haze from Morocco that affected the Community of Madrid at that time, leading to severe asthma episodes and an increase in viral contamination. This wind of haze took place in the first week of February 2022 and in mid-March of the same year.

Thus, although the WHO sets the maximum permitted 24 h air pollutant concentration at 15 μg/m^3 for PM 2.5 and 45 μg/m^3 for PM 10, the students involved in this study were frequently subjected to episodes wherein the average concentration exceeded these recommended health limits. Concentrations of PM 2.5 were higher than 30 μg/m^3 in February and 40 μg/m^3 in March, which are high risk levels for people with respiratory diseases. Likewise, PM 10 exceeded the concentration of 60 μg/m^3 on more than one occasion, which affected the normal course of classes.

Next, the results measured in situ in the classroom were cross-checked with the values collected from the official measuring stations of the Community of Madrid (Farolillo and Plaza Elíptica). Firstly, as shown in Table 4, the normality test was performed for the measurements of suspended particulate matter taken in each of the three school evaluation periods studied. In Table 4, the Kolmogorov–Smirnov–Lilliefors test (for sample size $n > 30$) and the Shapiro–Wilk test (for sample size $n < 30$) were performed.

Table 4. Normality test for suspended particulate matter concentration by school evaluation periods.

Particle	Evaluation Period	Measuring Station	Number of Days	Kolmogorov–Smirnov–Lilliefors		Shapiro–Wilk	
				Statistic	*p*-Value	Statistic	*p*-Value
PM 2.5	First	Educational Centre	46	0.127	0.055	0.931	0.008
		Plaza Elíptica		0.095	0.200	0.967	0.211
	Second	Educational Centre	55	0.132	0.016	0.848	0.000
		Plaza Elíptica		0.121	0.040	0.824	0.000
	Third	Educational Centre	26	0.138	0.199	0.955	0.287
		Plaza Elíptica		0.213	0.003	0.914	0.029
PM 10	First	Educational Centre	46	0.115	0.146	0.932	0.009
		Plaza Elíptica		0.121	0.082	0.974	0.375
		Farolillo		0.113	0.175	0.960	0.105
	Second	Educational Centre	55	0.151	0.003	0.809	0.000
		Plaza Elíptica		0.265	<0.001	0.469	0.000
		Farolillo		0.312	<0.001	0.380	0.287
	Third	Educational Centre	26	0.106	0.200	0.972	0.654
		Plaza Elíptica		0.140	0.188	0.954	0.266
		Farolillo		0.195	0.010	0.848	0.001

From the analysis of Table 4, it can be seen that only the period comprising the first assessment complies with the assumptions of normality for PM 10 and PM 2.5, i.e., they had a significance level $\alpha > 0.05$. Similarly, the second and third assessments did not meet the assumption of normality for at least one of the analysed measurement stations (i.e., they had a significance level of at least $\alpha < 0.05$). This initial analysis justifies the application of the different analysis tests (parametric or non-parametric) used in this section. For ease of interpretation, PM 2.5, for which only two measurement stations are available, will be studied first, followed by PM 10, with three measurement stations.

Thus, in Tables 5 and 6, the comparative study between the values collected by the Plaza Elíptica station and the values measured in situ for PM 2.5 particles is carried out. In the case of the first evaluation, which complies with the assumptions of normality, a parametric comparison of means is performed using Student's *t*-test (Table 5). On the other hand, the second and third assessments did not meet the normality test, so a non-parametric study was performed using the Mann–Whitney U test, comparing medians instead of mean values (Table 6).

Table 5. Independent samples test for PM 2.5 in the first assessment.

Equality of Variances	Levene's Test for Equality of Variances		*t*-Test for Equality of Means	
	F	*p*-Value	*t*	N. Combinations
Assumed	32.700	<0.001	−3.750	92.000
Not-Assumed	—	—	−3.750	69.702

Table 6. Mann–Whitney test and descriptive statistics for PM 25 in the second and third assessments.

Evaluation Period	Mann–Whitney U	*p*-Value	Educational Centre (μg/m^3)		Plaza Elíptica (μg/m^3)	
			Mean	Interquartile Range Rank	Mean	Interquartile Range Rank
Second	1274.500	0.087	12.5	11.0	10.0	8.0
Third	296.500	0.238	11.0	8.0	9.0	7.0

In the analysis presented in Table 5, the homogeneity of variances is rejected, as well as the hypothesis that the means behave in the same way (Welch = −3.75; p-Value < 0.001). In this way, we proceeded to a descriptive analysis, verifying that the mean value determined for the concentration of PM 2.5 inside the school was significantly higher than that collected at the nearest measuring station (Plaza Elíptica). These mean values for the first evaluation were 15.25 μg/m^3 for the school, compared to 10.34 μg/m^3 at the official reference station.

On the other hand, Table 6 does not show statistically significant differences between medians in the second evaluation (U = 1274.5; p-Value = 0.087), nor in the third evaluation (U = 269.5; p-Value = 0.238). For this reason, Table 6 also includes a descriptive study of the medians determined for the two measurement points analysed, the educational centre and Plaza Elíptica. In this sense, the median concentration of PM 2.5 in these school periods is significantly higher at the school than at the reference station, being 20% and 18% higher for the second and third evaluations, respectively.

Hence, the results show that in the school period analysed, in the first evaluation, the maximum permitted values for exposure to PM 2.5 set by the WHO were exceeded by 15 μg/m^3, with the average values of the other evaluations being very close to this limit. This alarming situation reflects the reality of the educational centres in the area, with an ageing building stock that is inefficiently ventilated, keeping the air polluted for long periods of time.

In the following, we proceed with the analysis of PM 10 particulate matter, including in Table 7 the analysis for the first evaluation and in Table 8 the study corresponding to the second and third evaluations. In this way, the analysis in Table 7 is carried out using an ANOVA table, since it was verified in Table 4 that the assumptions of normality were fulfilled for this school period. On the other hand, in Table 8 the non-parametric study is carried out using the Kruskal–Wallis test for comparison with more than one assumption, since for these larger particles, the measurement data from two stations close to the school (Farolillo and Plaza Elíptica) were available.

Table 7. PM 10 ANOVA for the first evaluation period and post hoc test for multiple comparisons.

ANOVA between Groups		Multiple Comparison (Plaza Elíptica—Educational Centre)		
F (Statistic)	*p*-Value	Mean Difference	Standard Deviation	*p*-Value
3.403	0.036	−4.809	1.904	0.036

In Table 7, it can be seen that there are statistically significant differences between the daily measurement of PM 10 in the centre and at least one of the reference stations in the Community of Madrid (F = 3.403; p-Value = 0.036). For this reason, a post hoc test of multiple comparisons between in situ measurements and those obtained from the Plaza Elíptica station (which turned out to be the only statistically significant comparison) has been presented in Table 7. Thus, it was found that during the first evaluation, the daily monitoring of the classroom showed an average concentration of 21.4 μg/m^3, compared to the 16.5 μg/m^3 measured by the station.

Table 8. Kruskal–Wallis test for PM 10 in the second and third evaluation periods and pairwise comparisons of stations.

Evaluation Period	Kruskal–Wallis H	*p*-Value	Plaza Elíptica—Farolillo Statistic	*p*-Value	Farolillo—Educational Centre Statistic	*p*-Value	Plaza Elíptica—Educational Centre Statistic	*p*-Value
Second	0.341	0.843	0.000	1.000	0.143	0.705	0.143	0.705
Third	6.212	0.045	0.074	0.785	3.711	0.054	4.800	0.028

Table 8 shows that there are statistically significant differences in medians when comparing at least one pair of measurement stations during the third evaluation (H = 6.212; *p*-Value = 0.045). For this reason, a post hoc test was performed to check this effect from an analytical point of view. Table 8 shows that during the period corresponding to the third evaluation, the stations of the Community of Madrid presented statistically significant differences between medians with the values measured in situ (Plaza Elíptica, *p*-Value = 0.054; and Farolillo, *p*-Value = 0.028). These differences amounted to 19 μg/m^3 for the concentration at the school, compared to 15 μg/m^3 for the median of the reference measurement centres.

As for PM 2.5, the values collected for PM 10 were on average higher than those collected by the reference measurement centres. For this reason, it was decided to construct 95% confidence intervals for the school period analysed in each quarter and considering the value of suspended particulate matter measured in situ. These results are presented in Table 9.

Table 9. Confidence intervals for PM 2.5 and PM 10 during the course.

Particle Type	Evaluation Period	Mean	Error	95% Interval Confidence Lower Bound	Upper Bound
PM 2.5	First	12.764	1.203	10.328	15.147
	Second	11.317	1.734	7.884	14.750
	Third	15.695	1.366	12.990	18.400
PM 10	First	17.131	1.942	13.286	20.975
	Second	15.366	2.595	10.226	20.505
	Third	21.974	2.044	17.927	26.022

It can be seen in Table 9 that, although the average values sometimes did not exceed the maximum permitted by the WHO at 24 h, if the upper limits of the intervals are observed, in the case of PM 2.5, the value of 15 μg/m^3 was exceeded in two evaluations. Furthermore, this type of airborne particulate matter is more harmful to health and causes more respiratory problems than PM 10 particles [38], which, although they did not exceed the WHO limit, also had concentrations above 15 μg/m^3 on average over the whole school year. Figure 8 shows a graph with the daily evolution of PM 2.5 and PM 10, together with the number of SARS-CoV-2 virus infections during the same period in the Community of Madrid.

Figure 8 shows that, in general terms, the trend in the evolution of suspended particles was similar to that of the number of COVID-19 infections. However, it is true that during the Christmas period, due to the festive season, the number of infections shot up. The effect of the haze can be seen in the graph with point values for PM 10 measured at 87 μg/m^3 inside the classroom, which doubles the maximum value permitted by the WHO and highlights the vulnerability of the current building stock to this type of situation, which is repeated annually.

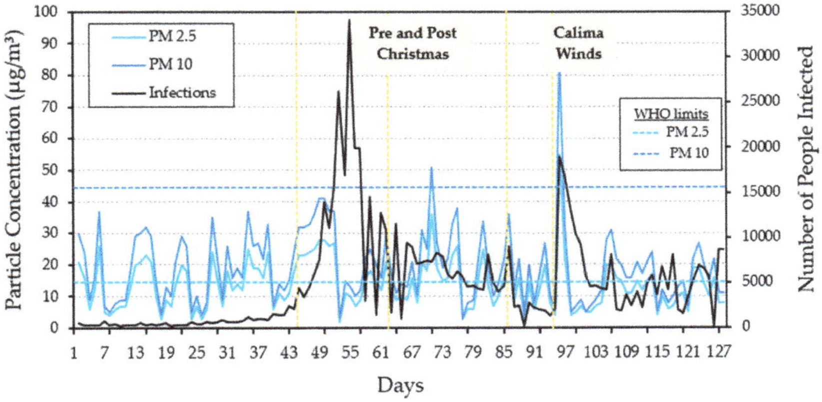

Figure 8. Daily evolution of the concentrations of PM 2.5 and PM 10 during the school period Sep–May and number of infected people in Madrid due to SARS-CoV-2 virus.

Analysis of the Effect of Hygrothermal Conditions on PM 2.5 and PM 10

The aim of this section is to check whether there is an effect of ambient temperature and relative humidity on the evolution of suspended particulate matter. Some studies in the literature show the importance of controlling these hygrothermal parameters to obtain good indoor air quality, with the concentration of PM 2.5 and PM 10 increasing as the values of these comfort-related variables increase [39,40]. Thus, first, the normality, homoscedasticity and independence of the values collected during the course were tested, and the results obtained for the normality test are presented in Table 10.

Table 10. Normality test for the effect of hygrothermal conditions on PM 2.5 and PM 10.

Particle	Evaluation Period	N.	Kolmogorov–Smirnov (p-Value)	Shapiro–Wilk (p-Value)
PM 2.5	First	46	0.016	<0.001
	Second	55	0.199	0.287
	Third	26	0.055	0.008
PM 10	First	46	0.003	<0.001
	Second	55	0.200	0.654
	Third	26	0.146	0.009

Table 11 determines the strength of this correlation in the daily measurements of PM 2.5 and PM 10, together with the effect of relative humidity and indoor temperature analysed individually and combined. This study has been carried out monthly rather than by evaluations, as the correlation obtained in this case was higher and better accuracy is obtained.

Table 11. Analysis of interactions between particle concentration and hygrothermal effects.

Particle	Cross-Variable *	Sum of Squares	Mean Square	F	p-Value
PM 2.5	Relative humidity	589.354	65.484	1.079	0.384
	Indoor temperature	396.239	44.027	0.725	0.685
	Relative humidity and indoor temperature	542.341	60.260	0.993	0.451
PM 10	Relative humidity	1394.511	154.946	1.131	0.347
	Indoor temperature	821.166	91.241	0.666	0.738
	Relative humidity and indoor temperature	1274.024	141.558	1.034	0.418

* The average daily values measured during the nine months of the school year (df = 9) were taken into consideration.

After analysing Table 11, it can be seen that all the factors included were statistically significant and showed a correlation with the particulate matter measurements. These results make physical sense, as the values measured in situ in the classroom correspond to the same conditions of ventilation and student concentration. On the other hand, the temperature factor showed the highest correlation for both PM 2.5 (p-Value = 0.685) and PM 10 (p-Value = 0.738), values much higher than those obtained when analysing the effect of relative humidity. This effect has already been corroborated by other researchers, who have demonstrated a positive correlation between temperature and the accumulation of atmospheric pollutants, especially in buildings located in urban areas with high traffic flow [10,41].

3.4. Measurements of HCHO Concentration

The last parameter analysed in this study was the evolution of the concentration of formaldehyde (HCHO). The monitoring of this atmospheric agent has experienced a growing interest among the scientific community as a result of the widespread use of hydroalcoholic gels in classrooms. This sanitary measure, in principle beneficial for sanitising and reducing viral contagion, leads to an increase in HCHO concentration in inadequately ventilated spaces [42]. Figure 9 shows the evolution of this parameter over the academic year under study with the help of a Box-Plot diagram.

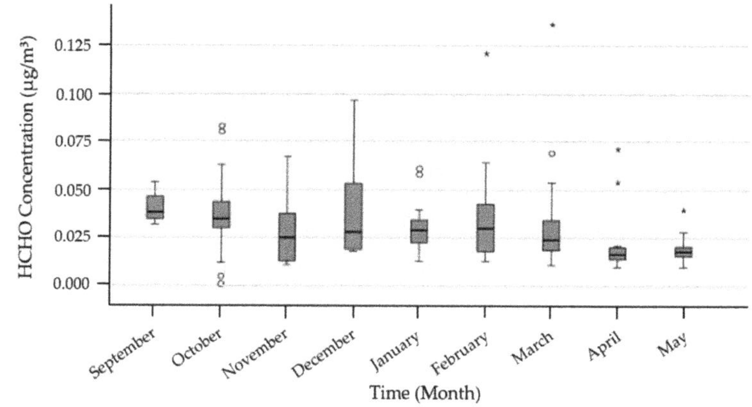

Figure 9. Evolution of formaldehyde concentration over the academic year.

To discuss Figure 9, it is useful to consider the maximum permissible exposure values for HCHO set by the Scientific Committee for Occupational Exposure Limits (SCOEL), which were published by the European Commission in 2016. These values are set at a maximum exposure of 0.3 ppm in 8 h and 0.6 ppm in short exposure periods, which is equivalent to 0.37 $\mu g/m^3$ and 0.74 $\mu g/m^3$, respectively [43]. Thus, when analysing Figure 9, it can be seen that in general terms the average values remained below the maximum permitted, although it is true that in all months there were specific situations that exceeded the optimal exposure range for health. In the specific case of the school studied, these values were the consequence of the repeated use of hydroalcoholic gel in interiors and other cleaning products. Therefore, in line with other studies, hand washing with toilet soap and the use of non-alcohol-based products is considered more favourable, as these solutions, which are increasingly used after the COVID-19 pandemic, may be carcinogenic in the long term [44].

Table 12 shows the results obtained for the daily average for each of the three evaluations.

Table 12. Analysis of the average evolution of HCHO in each school evaluation period.

Evaluation Period	Sample	HCHO Mean Concentration (μg/m^3)
First	46	0.3295 < 0.37 *
Second	55	0.2115 < 0.37 *
Third	26	0.3417 < 0.37 *

* Maximum short exposure value.

The overall analysis by evaluations in Table 12 shows that, although the maximum permitted values were not exceeded, concentrations very close to 0.37 μg/m^3 were obtained in both the first and third evaluations, thus requiring a reduction in the use of hydroalcoholic solutions to minimise health risks to students. These results are in line with those obtained by other researchers [45].

4. Critical Discussion and Implications of the Obtained Results

In recent decades, the construction industry has been striving to develop new solutions and eco-friendly designs to advance towards the European sustainability targets for buildings included in the 2030 Agenda [46]. Some of these initiatives are based on the design of climatically functional buildings using passive bioclimatic systems [47]; other studies focus on the development of new, more sustainable building materials produced under circular economy criteria [48], while another large number of studies seek to optimise the design of the urban environment by showing the possibilities of construction systems such as green façades [49], green roofs [50] or water-reserved roofs [51], among many others.

Hence, the current climate emergency and the growing concern about the deterioration of the urban and rural environment are shaping the course of building research in the attempt to mitigate the ecological impact of buildings and considering their economic and social context [52]. In this sense, large European cities are suffering a progressive increase in their average temperature (heat island effect) [53], while increasing their pollution levels and decreasing air quality [54] and experiencing increasingly frequent violent weather events such as acid rain, floods, droughts or heat waves [55].

This research is a clear example of the need to rehabilitate and refurbish the ageing building stock, highlighting the damage caused to students' health by deficiencies in the current design of some urban educational centres. In the specific case of Spain, 60% of existing housing was built before the 1980s and is still in use without having undergone any refurbishment [56]. This implies that more than half of today's buildings were built without taking into consideration any standards that include relevant issues such as the design of air renewal systems or effective thermal insulation. For this reason, in order to improve the IAQ of existing schools located in the urban centre of the city of Madrid, it is necessary to promote a retrofitting initiative taking into consideration some relevant aspects such as those included in Figure 10.

However, until actions such as those shown in Figure 10 are implemented, teachers and school management teams are the key actors in improving IAQ in classrooms. In view of the results obtained in this research, it is clear that in-situ measurements are necessary to carry out a critical analysis of the air situation in occupied classrooms, using the values obtained by the reference centres as a guide and initial information. In addition, the need to promote more effective and regular ventilation practices is highlighted, in order to reduce a double risk linked to a decrease in academic performance and, worse still, the appearance of respiratory diseases in students.

Finally, it is necessary to appreciate the increases in danger generated by the increment in suspended dust. For this reason, concentrations of harmful gases for health, formation of stale air, increased risk of diseases and other key points linked to IAQ can be effectively mitigated by establishing an efficient ventilation system equipped with soundproofing systems and professional filters, including IR filters, located at the entrance and exit of the classrooms.

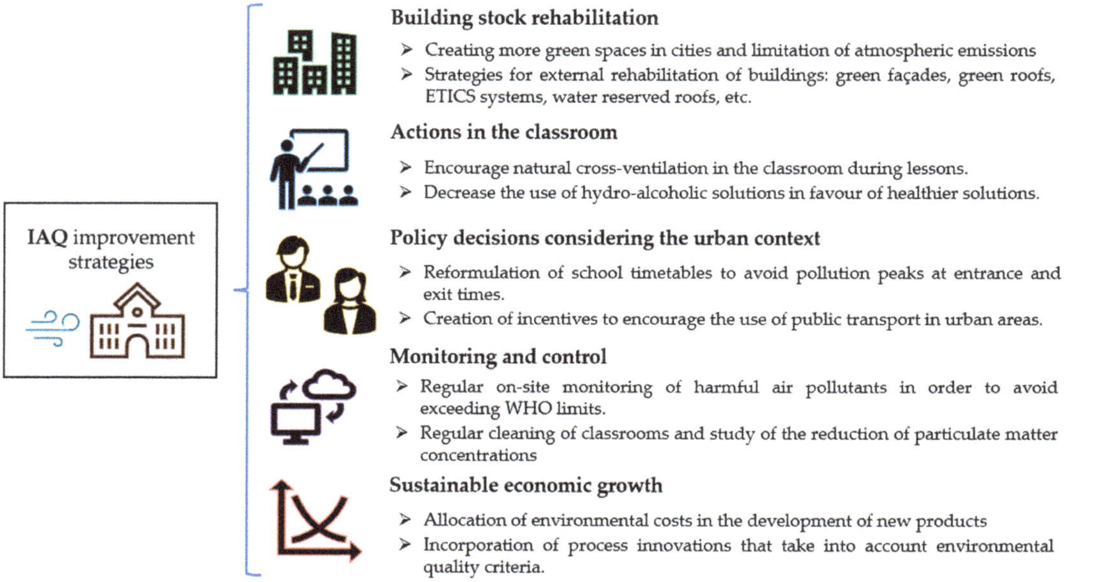

Figure 10. Schematic compilation of possible alternatives for improving indoor air quality in schools.

5. Conclusions

In this work, a monitoring of the IAQ of two classrooms belonging to an educational centre located in the urban centre of the city of Madrid was carried out during a complete academic year. Measurements of the daily evolution of airborne particles harmful to health such as CO_2, PM 10, PM 2.5 and HCHO, as well as a control of the hygrothermal comfort parameters of relative humidity and ambient temperature, were collected during the monitoring. It was also possible to visualise the trend in the number of SARS-CoV-2 virus infections and the concentration of suspended particles detected in the classroom, yielding alarming results that call for an urgent renovation of the current building stock. The most relevant conclusions that can be drawn from this work are as follows:

- The lack of refurbishment of the school centre since its foundation in the 1970s has been reflected in a decrease in hygrothermal comfort, with relative humidity values below those recommended by the RITE regulations and an indoor temperature lower than recommended in winter and higher in summer.
- Regarding CO_2 concentration, during the school period of the first evaluation, values close to the maximum permissible level of 700 ppm were recorded, sometimes exceeding 1000 ppm inside the classroom.
- In the analysis of PM 2.5 particles, values close to the maximum recommended by the WHO of 15 $\mu g/m^3$ were observed during the months studied, although it was only exceeded in specific situations. PM 10 particles showed a behaviour close to 20 $\mu g/m^3$ on average during the academic year. For both types of suspended particles, statistically significant differences were observed between the values measured in situ and those obtained from the two reference stations available in the Community of Madrid, highlighting the need for regular monitoring in the classroom.
- In line with other studies, a strong correlation has been observed between the evolution of hygrothermal conditions in the classroom and the concentration of PM 2.5 and PM 10 suspended particulate matter, which underlines the importance of energy retrofitting of the building.
- Finally, high formaldehyde concentration values were found in the classroom. On average, values of 0.34 $\mu g/m^3$ were obtained during the period corresponding to the

third evaluation, very close to the risk concentration set at 0.37 μg/m^3 for this type of space.

The main limitation of this work has been the impossibility of monitoring a larger number of classrooms in the building in order to make a comparison between different orientations, room volumes and student concentrations, which would have enriched the analysis of the results. On the other hand, a limitation of the measurements is the absence of data corresponding to NO_X concentrations, CO concentrations (whose high volumes of accumulation in cities can promote diseases such as Parkinson's) and the lack of measurements outside school hours. As future lines of work, we propose that a CFD simulation model be carried out based on the data obtained and use different constructive solutions to establish strategies to improve the IAQ in the monitored classrooms. At the same time, we intend that a numerical analysis be carried out to establish statistical models that describe the behaviour of the air throughout the normal academic year under different assumptions about particle concentrations and environmental conditions.

Author Contributions: Conceptualization, A.L.M. and J.P.D.V.; methodology, A.L.M. and J.P.D.V.; software, A.L.M.; validation, A.L.M., J.P.D.V. and D.F.; formal analysis, A.L.M.; investigation, A.L.M. and J.P.D.V.; resources, A.L.M. and D.F.; data curation, A.L.M.; writing—original draft preparation, A.L.M., D.F. and A.Z.-B.; writing—review and editing, D.F., P.S. and A.Z.-B.; visualization, A.Z.-B.; supervision, J.P.D.V. and D.F. All authors have read and agreed to the published version of the manuscript.

Funding: This research received no external funding.

Data Availability Statement: Not applicable.

Acknowledgments: The authors would like to thank the students and teachers at the school involved in the study for their willingness and for the facilities offered to carry out this research work.

Conflicts of Interest: The authors declare no conflict of interest.

References

1. *EN 15643*; Sustainability of Construction Works—Framework for Assessment of Buildings and Civil Engineering Works. CEN—European Committee for Standardisation: Brussels, Belgium, 2021.
2. *EN 16309 + A1*; Sustainability of Construction Works—Assessment of Social Performance of Buildings—Calculation Methodology. CEN—European Committee for Standardisation: Brussels, Belgium, 2014.
3. Santos, P.; Gervasio, H.; Pereira, A.; Simoes da Silva, L.; Bettencourt, A. Comparative life cycle social assessment of buildings: Health and comfort criterion. *Matériaux Tech.* **2017**, *104*, 601. [CrossRef]
4. Santos, P.; Carvalho Pereira, A.; Gervásio, H.; Bettencourt, A.; Mateus, D. Assessment of health and comfort criteria in a life cycle social context: Application to buildings for higher education. *Build. Environ.* **2017**, *123*, 625–648. [CrossRef]
5. Stafford, T.M. Indoor air quality and academic performance. *J. Environ. Econ. Manag.* **2015**, *70*, 34–50. [CrossRef]
6. Kabirikopaei, A.; Lau, J.; Nord, J.; Bovaird, J. Identifying the K-12 classrooms' indoor air quality factors that affect student academic performance. *Sci. the Total Environ.* **2021**, *786*, 147498. [CrossRef]
7. Kosonen, R.; Tan, F. The effect of perceived indoor air quality on productivity loss. *Energy Build.* **2004**, *36*, 981–986. [CrossRef]
8. Na, H.; Choi, H.; Kim, H.; Park, D. Optimizing indoor air quality and noise levels in old school classrooms with air purifiers and HRV: A CONTAM simulation study. *J. Build. Eng.* **2023**, *73*, 106645. [CrossRef]
9. Rodríguez, D.; Urbieta, I.R.; Velasco, A.; Campano-Laborda, M.A.; Jímenez, E. Assessment of indoor air quality and risk of COVID-19 infection in Spanish secondary school and university classrooms. *Build. Environ.* **2022**, *226*, 109717. [CrossRef] [PubMed]
10. Tikul, N.; Hokpunna, A.; Chawana, P. Improving indoor air quality in primary school buildings through optimized apertures and classroom furniture layouts Author links open overlay panel. *J. Build. Eng.* **2022**, *62*, 105324. [CrossRef]
11. Greenhalgh, T.; Jimenez, J.L.; Prather, K.A.; Tufekci, Z.; Fisman, D.; Schooley, R. Ten scientific reasons in support of airborne transmission of SARS-CoV-2. *Lancet* **2021**, *397*, 1603–1605. [CrossRef] [PubMed]
12. Harvard, C. Guía en 5 Pasos Para Medir la Tasa de Renovación en Las Aulas. 2020. Available online: https://alara.es/guia-para-medir-la-tasa-de-renovacion-de-aire-en-aulas/ (accessed on 15 February 2023).
13. Evaluación del Riesgo de la Transmisión de SARS-CoV- Mediante Aerosoles. Medidas de Prevención y Recomendación. 2020. Available online: https://www.mscbs.gob.es/profesionales/saludPublica/ccayes/alertasActual/nCov/documentos/COVID19_Aerosoles.pdf (accessed on 15 November 2022).

14. Chen, Y.H.; Tu, Y.P.; Sung, S.Y.; Weng, W.C. A comprehensive analysis of the intervention of a fresh air ventilation system on indoor air quality in classrooms. *Atmos. Pollut. Res.* **2022**, *13*, 101373. [CrossRef]
15. Stabile, L.; Frattolillo, A.; Dell'Isola, M.; Massimo, A.; Russi, A. Air Permeability of Naturally Ventilated Italian Classrooms. *Energy Procedia* **2015**, *78*, 3150–3155. [CrossRef]
16. Schibuola, L.; Scarpa, M.; Tambani, C. Natural Ventilation Level Assessment in a School Building by CO2 Concentration Measures. *Energy Procedia* **2016**, *1101*, 257–264. [CrossRef]
17. Bogdanovica, S.; Zemitis, J.; Bogdanovics, R. The Effect of CO2 Concentration on Children's Well-Being during the Process of Learning. *Energies* **2020**, *13*, 6099. [CrossRef]
18. Azuma, K.; Kagi, N.; Yanagi, U.; Osawa, H. Effects of low-level inhalation exposure to carbon dioxide in indoor environments: A short review on human health and psychomotor performance. *Environ. Int.* **2018**, *121*, 51–56. [CrossRef] [PubMed]
19. Poirier, B.; Guyot, G.; Woloszyn, M.; Geoggroy, H.; Ondarts, M.; Gonze, E. Development of an assessment methodology for IAQ ventilation performance in residential buildings: An investigation of relevant performance indicators. *J. Build. Eng.* **2021**, *43*, 103140. [CrossRef]
20. Ramalho, O.; Wyart, G.; Mandin, C.; Blondeau, P.; Cabanes, P.A.; Leclerc, N.; Mullot, J.U.; Boulanger, G.; Redaelli, M. Association of carbon dioxide with indoor air pollutants and exceedance of health guideline values. *Build. Environ.* **2015**, *93*, 115–124. [CrossRef]
21. World Health Organization. Calidad del Aire Ambiente (Exterior) y Salud. Available online: https://www.who.int/es/news-room/fact-sheets/detail/ambient-(outdoor)-air-quality-and-health (accessed on 15 March 2023).
22. Air Quality in Europe—2020 Report. European Environment Agency. Available online: https://www.eea.europa.eu/publications/air-quality-in-europe-2020-report (accessed on 15 February 2023).
23. Rivas, I.; Viana, M.; Moreno, T.; Bouso, L.; Pandlfi, M.; Alvarez-Pedrerol, M.; Forns, J.; Alastuey, A.; Sunyer, J.; Querol, X. Outdoor infiltration and indoor contribution of UFP and BC, OC, secondary inorganic ions and metals in PM2.5 in schools. *Atmos. Environ.* **2015**, *196*, 129–138. [CrossRef]
24. Mukhtar, A.; Yusoff, M.Z.; Ng, K.C. The potential influence of building optimization and passive design strategies on natural ventilation systems in underground buildings: The state of the art. *Tunn. Undergr. Space Technol.* **2019**, *92*, 103065. [CrossRef]
25. IARC, W. Formaldehyde, 2-Butoxyethanol and 1-tert-Butoxypropan-2-ol. *IARC Monogr. Eval. Carcinog. Risks Hum.* **2006**, *88*, 39–325.
26. Brdarić, D.; Kovač-Andrić, E.; Šapina, M.; Kramarić, K.; Lutz, N.; Perković, T.; Egorov, A. Indoor air pollution with benzene, formaldehyde, and nitrogen dioxide in schools in Osijek, Croatia. *Air Qual. Atmos. Health* **2019**, *12*, 963–968. [CrossRef]
27. Yu, L.; Wang, B.; Cheng, M.; Yang, M.; Gan, S.; Fan, L.; Wang, D.; Chen, W. Association between indoor formaldehyde exposure and asthma: A systematic review and meta-analysis of observational studies. *Indoor Air* **2020**, *30*, 682–690. [CrossRef] [PubMed]
28. Parthasarathy, S.; Maddalena, R.L.; Russel, M.L.; Apte, M.G. Effect of Temperature and Humidity on Formaldehyde Emissions in Temporary Housing Units. *J. Air Waste Manag. Assoc.* **2011**, *61*, 689–695. [CrossRef] [PubMed]
29. Qin, D.; Guo, B.; Zhou, J.; Cheng, H.; Chen, X. Indoor air formaldehyde (HCHO) pollution of urban coach cabins. *Sci. Rep.* **2020**, *10*, 332. [CrossRef]
30. Hu, D.; Tobon, Y.; Agostini, A.; Grosselin, B.; Chen, Y.; Robin, C.; Yahyaoui, A.; Colin, P.; Mellouki, A.; Daële, V. Diurnal variation and potential sources of indoor formaldehyde at elementary school, high school and university in the Centre Val de Loire region of France. *Sci. Total Environ.* **2022**, *811*, 152271. [CrossRef]
31. Madrid: Población Por Municipios y Sexo. 2022. Available online: https://www.ine.es/jaxiT3/Datos.htm?t=2881 (accessed on 15 November 2022).
32. Khomenko, S.; Cirach, M.; Pereira-Barboza, E.; Mueller, N.; Barrera-Gómez, J.; Rojas-Rueda, D.; Hoogh, K.; Hoek, G.; Nieuwenhuijsen, M. Premature mortality due to air pollution in European cities: A health impact assessment. *Lancet Planet. Health* **2021**, *5*, e121–e134. [CrossRef]
33. *Real Decreto 314/2006*; Ministerio de Vivienda: Código Técnico de la Edificación (in Spanish). Madrid, Spain. 2016. Available online: https://www.codigotecnico.org/pdf/Documentos/HE/DccHE.pdf (accessed on 1 September 2023).
34. *Real Decreto 238/2013*; Ministerio de Industria, Energía y Turismo: Reglamento de Instalaciones Térmicas en los Edificios (RITE) (in Spanish). Madrid, Spain. 2013. Available online: https://www.boe.es/buscar/doc.php?id=BOE-A-2013-3905 (accessed on 1 September 2023).
35. European Environment Agency. *EMEP/EEA Air Pollutant Emission Inventory Guidebook*; European Environment Agency (EEA): Luxembourg, 2019; ISSN 1977-8449.
36. Varela Luján, S.; Viñas, C.; Rodríguez, A.; Aguilera, P.; González, M. Experimental comparative study of the thermal performance of the façade of a building refurbished using ETICS, and quantification of improvements. *Sustain. Cities Society* **2019**, *51*, 101713. [CrossRef]
37. Zhang, X.; Zhao, C.; Zhang, T.; Xie, J.; Liu, J.; Zhang, N. Association of indoor temperature and air quality in classrooms based on field and intervention measurements. *Build. Environ.* **2023**, *229*, 109925. [CrossRef]
38. Das, K.; Chatterjee, N.D.; Jana, D.; Bhattacharya, R.K. Application of land-use regression model with regularization algorithm to assess PM2.5 and PM10 concentration and health risk in Kolkata Metropolitan. *Urban Clim.* **2023**, *49*, 101473. [CrossRef]
39. Lewis, S.L.; Russel, L.M.; McKinsey, J.A.; Harris, W.J. Small contributions of dust to PM2.5 and PM10 concentrations measured downwind of Oceano Dunes. *Atmos. Environ.* **2023**, *294*, 119515. [CrossRef]

40. Lin, Z.; Lawrence, W.R.; Gong, W.; Lin, L.; Hu, J.; Zhu, S.; Meng, R.; He, G.; Xu, X.; Liu, T.; et al. The impact of mortality underreporting on the association of ambient temperature and PM10 with mortality risk in time series study. *Heliyon* **2023**, *9*, e14648. [CrossRef]
41. Lu, C.; Xu, H.; Meng, W.; Hou, W.; Zhang, W.; Shen, G.; Cheng, H.; Wang, X.; Wang, X.; Tao, S. A novel model for regional indoor PM2.5 quantification with both external and internal contributions included. *Environ. Int.* **2020**, *145*, 106124. [CrossRef]
42. Malayeri, M.; Bahri, M.; Haghighat, F.; Shah, A. Impact of air distribution on indoor formaldehyde abatement with/without passive removal material: A CFD modeling. *Build. Environ.* **2022**, *212*, 108792. [CrossRef]
43. INSHT. *NTP 590: Prevención de la Exposición a Formaldehído*; Ministerio de Trabajo y Asuntos Sociales: Madrid, Spain, 2001.
44. Figueroa, D.; Nishio, S.; Yamazaki, R.; Ohta, E.; Hamaguchi, S.; Utsumi, M. Recognition of hand disinfection by an alcohol-containing gel using two-dimensional imaging in a clinical setting. *J. Hosp. Infect.* **2023**, *135*, 157–162. [CrossRef] [PubMed]
45. Aodah, A.H.; Bakr, A.A.; Booq, R.Y.; Rahman, M.J.; Alzahrani, D.A.; Alsulami, K.A.; Alshaya, H.A.; Alsuabeyl, M.; Alyamani, E.; Twafik, E.A. Preparation and evaluation of benzalkonium chloride hand sanitizer as a potential alternative for alcohol-based hand gels. *Saudi Pharm. J.* **2021**, *29*, 807–814. [CrossRef]
46. Mikulić, D.; Slijepčević, S.; Buturac, G. Energy renovation of multi apartment buildings: Contributions to economy and climate changes. *Energy Build.* **2020**, *224*, 110247. [CrossRef]
47. Elaouzy, Y.; El Fadar, A. Sustainability of building-integrated bioclimatic design strategies depending on energy affordability. *Renew. Sustain. Energy Rev.* **2023**, *179*, 113295. [CrossRef]
48. Zaragoza-Benzal, A.; Ferrández, D.; Atanes-Sánchez, E.; Saíz, P. Study of the hygroscopic properties of environmentally friendly lightened composites through waste recovery. *Constr. Build. Mater.* **2023**, *404*, 133219. [CrossRef]
49. Bakhshoodeh, R.; Ocampo, C.; Oldham, C. Impact of ambient air temperature, orientation, and plant status on the thermal performance of green façades. *Energy Build.* **2023**, *296*, 113389. [CrossRef]
50. Zambrano-Prado, P.; Pons-Gumí, D.; Tobos-Chavero, S.; Parada, F.; Josa, A.; Gabarrell, X.; Rieradevall, J. Perceptions on barriers and opportunities for integrating urban agri-green roofs: A European Mediterranean compact city case. *Cities* **2021**, *114*, 103196. [CrossRef]
51. Yang, W.; Wang, Z.; Zhao, X. Experimental investigation of the thermal isolation and evaporative cooling effects of an exposed shallow-water-reserved roof under the sub-tropical climatic condition. *Sustain. Cities Soc.* **2015**, *14*, 293–304. [CrossRef]
52. Borrás, J.G.; Lerma, C.; Mas, A.; Vercher, J.; Gil, E. Contribution of green roofs to energy savings in building renovations. *Energy Sustain. Dev.* **2022**, *71*, 212–221. [CrossRef]
53. Susca, T.; Zanghirella, F.; del Fatto, V. Building integrated vegetation effect on micro-climate conditions for urban heat island adaptation. Lesson learned from Turin and Rome case studies. *Energy Build.* **2023**, *295*, 113233. [CrossRef]
54. Belias, E.; Licina, D. Influence of outdoor air pollution on European residential ventilative cooling potential. *Energy Build.* **2023**, *289*, 113044. [CrossRef]
55. Laino, E.; Iglesias, G. Extreme climate change hazards and impacts on European coastal cities: A review. *Renew. Sustain. Energy Rev.* **2023**, *184*, 113587. [CrossRef]
56. Monzón-Chavarrías, M.; López-Mesa, B.; Resende, J.; Corvacho, H. The nZEB concept and its requirements for residential buildings renovation in Southern Europe: The case of multi-family buildings from 1961 to 1980 in Portugal and Spain. *J. Build. Eng.* **2021**, *34*, 101918. [CrossRef]

buildings

MDPI

Article

Chopped Straw as an Insulation Material: The Influence of Different Blow-In Technologies and Flame Retardants on Hygrothermal Properties

Tarja Salonen *, Henriette Fischer and Azra Korjenic

Research Unit of Ecological Building Technologies, Institute of Material Technology, Building Physics and Building Ecology, Faculty of Civil and Environmental Engineering, Vienna University of Technology, A-1040 Vienna, Austria; henriette.fischer@tuwien.ac.at (H.F.); azra.korjenic@tuwien.ac.at (A.K.)
* Correspondence: tarja.salonen@tuwien.ac.at

Abstract: The demand for renewable resources in building construction is increasing, and wheat straw is an excellent option due to its superior environmental performance compared to traditional insulation materials. However, the hygrothermal properties of chopped wheat straw insulation have remained largely unexplored. At the moment, blown-in straw is only blown in vertically, although horizontal blowing would be more efficient depending on the situation. This study investigates the effect of different blowing techniques on the thermal properties of chopped wheat straw insulation, focusing on the difference between vertical and horizontal blowing techniques. In-situ-measured thermal conductivities were compared with design values used in energy balances. In addition, the long-term hygrothermal behavior of chopped wheat straw insulation treated with flame retardants was investigated. The methodology included heat flow plate measurements, needle probe measurements and laboratory measurements using the hot plate method. The results show that there is no significant difference in thermal performance between the blowing techniques. The measured thermal conductivities were lower than expected, challenging the current general normative moisture surcharge on the thermal conductivity of natural fiber insulation. The addition of the flame retardant had no noticeable effect on the hygrothermal properties of the chopped straw. Chopped wheat straw can be regarded as a highly ecological insulation material with great potential for the future.

Keywords: bio-based building materials; blow-in insulation; chopped straw; ecological insulation; flame retardants; hygrothermal behavior

Citation: Salonen, T.; Fischer, H.; Korjenic, A. Chopped Straw as an Insulation Material: The Influence of Different Blow-In Technologies and Flame Retardants on Hygrothermal Properties. *Buildings* **2023**, *13*, 2555. https://doi.org/10.3390/buildings13102555

Academic Editors: Paulo Santos and Florence Collet

Received: 16 August 2023
Revised: 1 October 2023
Accepted: 4 October 2023
Published: 10 October 2023

1. Introduction

The use of environmentally sustainable building elements is becoming increasingly important in the context of the current climate crisis. The urgency of addressing climate change and its impacts is demonstrated by the United Nations Sustainable Development Goal No. 13, which calls for immediate action to combat these issues, along with the European Green Deal. Within this broader context, there is a growing need to implement sustainability in the construction industry and to use building materials with low primary energy consumption in their manufacture [1]. Sustainable construction aims to reduce environmental impact throughout the whole life cycle. Achieving this goal involves various strategies, including the selection of environmentally friendly building materials, the utilization of renewable energy sources, and a reduction in overall energy consumption. It is crucial that our "built environment" meets both current requirements and those of the future generations.

The use of biomass waste as a substitute for traditional materials appears to be a viable solution to address pollution issues, reduce CO_2 emissions and promote the development of energy-efficient, cost-effective and durable building materials [2,3]. One of the agricultural wastes that is produced in large quantities and is available in most parts of Europe is wheat straw. Straw, a renewable by-product of grain production, offers a low primary energy content and minimal environmental impact. It is biodegradable and can be seamlessly returned to the natural cycle.

Many studies have focused on wheat straw as a by-product. Liu et al. investigated wheat straw as an aggregate and geopolymer as a binder and showed that such a novel kind of bio-insulation material exhibits satisfactory thermal and mechanical performances, allowing it to be applied as a thermal insulation material in building, especially focusing on prefabricated buildings [4]. Researchers also focus on the usage of earth mixed with bio-based materials [5]. Agave and wheat straw fibers in hybrid boards with different compositions and densities are an environmentally friendly material that can contribute to energy-efficient building insulation [6]. Date palm tree leaves and wheat straw fibers demonstrate low thermal conductivity, excellent sound absorption, and potential suitability for building insulation and soundproofing applications [7].

When pure, untreated wheat straw is used, there are currently two options: straw bales and chopped straw for blow-in insulation. Straw bales have been proven in numerous studies to be cost-effective and energy efficient as insulation materials that are easy to handle [8–10]. Typically installed between wooden frame structures, straw bales have demonstrated their capacity to serve as effective thermal and moisture buffer materials [2]. Moisture buffering materials are able to absorb and release moisture through interaction with the environment, which is strongly influenced by the hygrothermal properties of the material [11]. Active moisture management through hygroscopic building materials can reduce the energy demand and stabilize the indoor climate of buildings [12].

While straw bales may require thicker layers than conventional insulation materials such as extruded polystyrene (XPS) or expanded polystyrene (EPS) to achieve similar insulation performance, their low embodied emissions make them an environmentally friendly alternative building material [13]. The usage of blow-in insulation in timber frame construction, even though still not as studied as the usage of straw bales, is gaining traction, with materials such as wood fiber and chopped straw experiencing a surge in popularity [14]. The main advantage of this insulation type over the usage of straw bales is the ability to prefabricate. The focus of production shifts from the job site to the factory. All timber components can be manufactured here with great care and expertise, on time and at low cost. This allows the straw to be blown into prefabricated wooden frames, whereas straw bale construction usually involves long and tedious manual labor.

In order to ensure the continuous improvement of manufacturing processes, it is imperative to address concerns regarding the orientation of straw fibers during horizontal blowing. Previous studies have indicated that vertically oriented straw bales and similar elements exhibit higher thermal conductivities compared to those with a horizontal or random orientation [15,16]. The OIB (Vienna, Austrian Institute for Building Technology) provides specifications for the manufacturing process of blow-in insulation in Austria, which must be properly executed. This procedure only allows blow-in for vertical wall elements [17]. Although horizontal blowing is sometimes more efficient for the factory production of building components, research on the effectiveness of this approach is currently lacking.

According to ÖNORM B6015-2, a higher thermal conductivity is attributed to natural fibers by adding 20% due to their moisture characteristics and it is generally assumed that a condition of 80% humidity is reached in their steady state [18]. Current research suggests that a higher moisture content and, therefore, higher thermal conductivity could be due to latent heat effects during measurements with a plate-measuring device [19], which do not occur under real conditions. For this reason, measurements of the thermal conductivity under real conditions are of great importance.

Striking a balance between ecological building practices, the use of renewable and biodegradable materials, and fire safety requirements presents a formidable challenge, as these aspects often appear contradictory. Optimizing the fire properties of organic building materials frequently requires the incorporation of flame retardants, which can impact the moisture behavior of blow-in insulation. Flame retardants may elevate the concentration of salts, leading to hygroscopic effects that hinder the material's ability to absorb and release moisture [8]. The long-term consequences of such effects can include mold growth and the deterioration of insulation material. Therefore, the inclusion of flame retardants demands careful consideration, taking into account their potential ramifications [9].

To address these research gaps, this study investigated the following research questions:

- What is the difference in the thermal conductivity due to the blowing technology (horizontal and vertical)?
- What difference occurs between real-life measurements of the thermal conductivity and measurements in laboratory experiments?
- Does the addition of a flame retardant affect the hygrothermal properties of chopped straw?

2. Materials and Methods

The following investigations were carried out to answer the research questions:

- In situ measurement and calculation of the U-value and thermal conductivity of the different blow-in technologies.
- Needle probe measurement of the thermal conductivity on the in situ test bench.
- Sampling of the in situ measurement and investigation of the thermal conductivity in the laboratory.
- In situ hygrothermal long-term measurement of the straw insulation with flame retardants.

2.1. Investigation of Different Chopped Straw Blow-In Techniques

The test stand is located on the premises of the Technical University, Science Center, Vienna, and was erected in the year 2020. The outdoor test rig for in situ measurements is designed to perform measurements in real environments, allowing data to be collected under natural conditions and providing real-world insights in addition to laboratory measurements. It stands in a mostly unprotected area where it is exposed to the climatic conditions of south-eastern Vienna.

A timber-framed exterior wall was constructed for the tests and installed in this outdoor test facility (Figure 1). The outer wall measured 2.09 × 2.18 m, was made of a solid spruce timber frame with 0.20 × 0.06 m beams and was oriented to the east.

Figure 1. Outdoor test stand in Vienna (**left**), vertical blowing technology (**middle**), horizontal blowing technology (**right**).

The investigated building material was untreated wheat straw with a fiber length between approx. 5 and 30 mm. The straw was installed using a conventional blowing machine, without the need for additional blowers. This means that the prepared straw was blown and compacted into the respective component through a long hose using air from a nozzle. The processing speed varies from approximately 5.5 m^3 to 10 m^3 per hour, at a density of 90 to 105 kg/m^3. During compaction tests in accordance with EN 15101-1 [20], the volume of the insulation increased by up to 3%, but remained permanently stable. This property is retained by the straw used here throughout its entire service life due to its special fiber processing and structure.

The outer wall has three cavities, each filled using a different blowing method: one filled horizontally, one filled vertically, and one filled vertically with used straw. The straw already used was removed from another construction of a research project and blown in again. Different measurements are taken on all three samples. The vertical element represents the standard blowing technology, while the other two methods and the influence on their hygrothermal characteristics are unexplored.

As shown in Figure 2, the wall element was internally clad with an airtight glued oriented strand board (OSB) with a thickness of 0.019 m. Spruce squared lumber with dimensions of 0.20 × 0.06 m was placed between the individual compartments in the insulation layer. The outside of the wall was clad in a 0.02 m thick diffusion-open wood fiber board with a windproof membrane. Normally, a ventilated façade is used for such a structure. However, due to the air flow not significantly affecting the measurements, it was omitted in this case. A heat flow plate was installed on the inner surface and temperature sensors were placed on both surfaces.

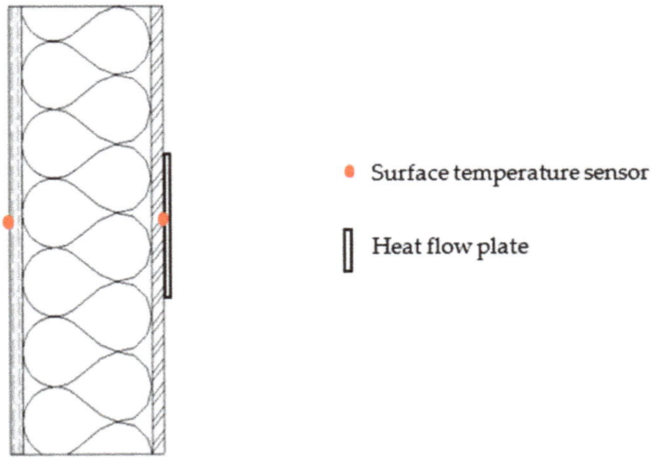

Figure 2. Examined wall structure, vertical section.

2.1.1. Measurement Technique

Two air conditioners, RAV-GM301ATP-E from Toshiba, were used for conditioning inside the outdoor test stand. The indoor air temperature was maintained between 21 °C and 23 °C throughout the experiments, while the relative humidity was maintained at approximately 60%. PT1000 sensors from RS Components were utilized to measure the internal and external surface temperatures. The heat-flow-measuring plates, type 8, with dimensions of 250 mm × 250 mm × 1.5 mm, were provided by Phymeas.

Data recording and storage were carried out using the Keysight 34980A (HP34980A) switch system/data logger. Measurements were taken and stored at 10 min intervals.

Furthermore, thermal conductivity measurements were conducted using the needle probe THERM 2227-2 from Ahlborn.

Figure 3 displays the positions of the measuring points, where measurements were performed with the needle probe at two points per compartment, labeled as P1 and P2, respectively. The lower edge of the heat flux plates was situated at a height of 0.4 m from the floor, with the surface temperature sensors positioned in the center of the heat flux plate.

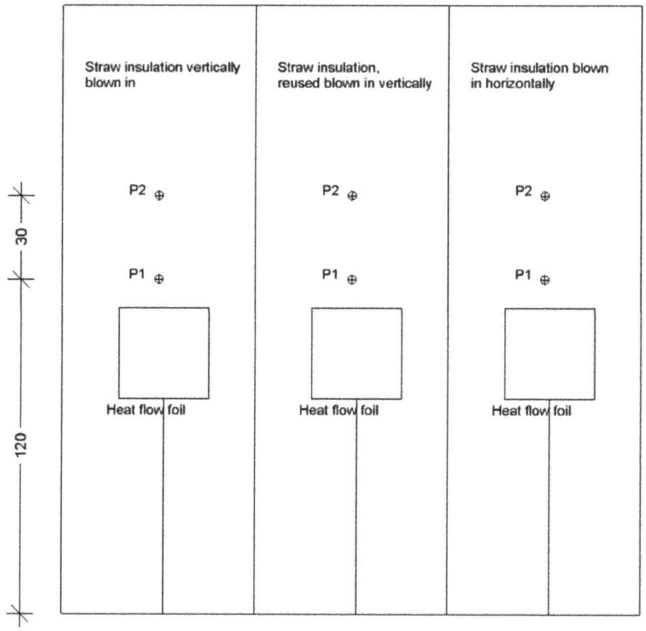

Figure 3. Position of the measuring points (dimensions displayed in cm).

2.1.2. Determination of the Thermal Insulation Properties by Means of Heat Flow Measurement

The in situ determination of U-values is the subject of ongoing research [21]. In addition to the three ISO 9869-1 criteria [22], a number of other criteria have been added in an attempt to determine a steady state as accurately as possible. The measured data were filtered in an Excel spreadsheet according to these criteria in order to obtain only the data that met these criteria. The filter method using temperatures is based on [23]. Normally, this method is used to measure U-values, but in this case, it is used to calculate thermal conductivity from measured values.

1. The temperature difference between indoor air and outdoor air must be greater than 15 K. A high temperature difference is recommended for appropriate results in lightweight constructions [24].
2. The surface temperature difference between the inside and outside of the façade must be greater than 15 K.
3. The surface temperature difference between the inside and outside of the façade must be greater than 15 K on average over the last 24 h.
4. In order to ensure that the thermal conditions were relatively constant [22], the temperature difference between the inside air and the outside air must not have fluctuated more than 2 K in the last 24 h.
5. To meet the requirement of [22], that solar radiation has no effect on the measurements, the surface temperatures of the interior and exterior façade surfaces must not have fluctuated by more than 2 K in the last 24 h.
6. The calculated R-value must not differ by more than ±5% from that of the previous 24 h (thermal resistance) [22].

The thermal conductivity of the straw insulation layer was determined from the data to enable a direct comparison of values without the influence of the other layers.

Equation (1) is taken from ISO 9869-1 and is part of the Average Method for the in situ measurement of thermal resistance and thermal transmittance [22]. First, the thermal conductance Λ, expressed in the unit W/(m²·K), of the entire wall structure was calculated:

$$\Lambda = \frac{\sum_{j=1}^{n} q_j}{\sum_{j=1}^{n} \left(T_{si,j} - T_{se,j}\right)} \tag{1}$$

The symbol q represents the heat flux (in W/m²). $T_{si,j}$ represents the interior surface temperature of the building element and $T_{se,j}$ shows the exterior surface temperature, both in degrees Celsius. Fourier's Law states that the rate of heat transfer (q q) through a material is directly proportional to the temperature difference across the material (Δ T) and inversely proportional to the thermal resistance (R) of the material [25].

Next, the thermal resistance (R-value) in m²·K/W of the entire wall structure R_{Totall} was calculated using:

$$R_{Total} = \frac{1}{\Lambda} \tag{2}$$

This value is derived from the reciprocal of the thermal conductance (Λ) calculated in Equation (1). It serves as a critical metric for assessing the overall insulating capacity of the wall assembly.

To calculate the overall thermal conductivity λ_{Total}, D_{Total} (thickness of the wall in meters) was divided by R_{Total}:

$$\lambda_{Total} = \frac{D_{Total}}{R_{Total}} \tag{3}$$

This equation illustrates how quickly heat is conducted through the wall assembly, considering both its thermal resistance (R-value) and thickness.

Now that the thermal conductivity of the entire structure is known, the thermal conductivities of the OSB board and DHF board were proportionally subtracted from λ-total to obtain $\lambda_{Insulation}$:

$$\lambda_{Insulation} = \frac{\lambda_{Total} * D_{Total} - \lambda_{OSB} * D_{OSB} - \lambda_{DHF} * D_{DHF}}{D_{Insulation}} \tag{4}$$

2.1.3. Needle Probe Measurement of Thermal Conductivity

A needle probe is a thermal measuring device used to determine the thermal conductivity of a material and is similar to the transient hot-wire method. It consists of a heat-resistant wire whose tip is inserted into the sample. A temperature difference is created between the needle probe and the sample by heating the wire. The temperature changes are continuously recorded throughout the process and are incorporated into the calculation of thermal conductivity. It is particularly suitable for loose-fill insulation materials.

In this investigation, small holes were drilled from the inside to measure the thermal conductivity within the insulation layer. Material inhomogeneity, incorrect calibration, and environmental factors such as pressure and humidity can introduce sources of error into the results. To minimize these errors, measurements were taken at two positions, respectively, and at different timings. The measurements were taken over a four-month period at approximately one-month intervals.

2.1.4. Sampling and Examination in the Laboratory

The following measurements were made in the laboratory to determine the differences between the measurements on real exterior walls and the laboratory measurements.

The samples were directly taken from the test wall using special sheet metal frames (Figure 4), accurately reflecting the actual insulation situation. The location of each sampling point is shown in Figure 5. The sampling procedure was specified by the OIB (Austrian Institute for Building Technology).

Figure 4. Sampling by means of sheet metal frame.

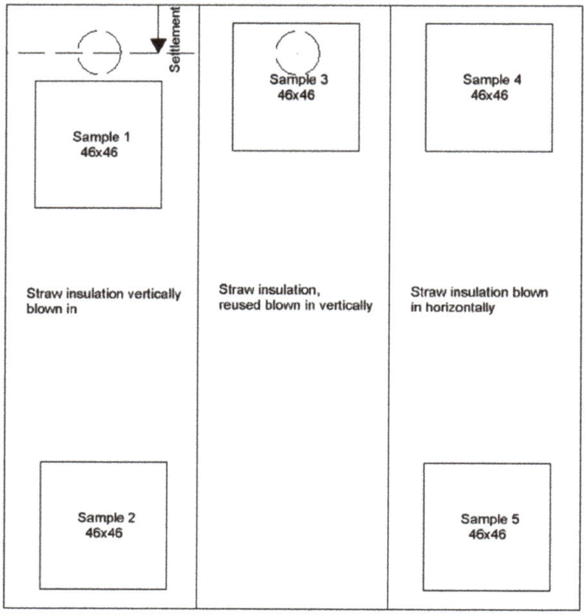

Figure 5. Position and dimension of the samples taken.

A total of five specimens with the dimensions of 460 × 460 × 100 mm were examined. As the primary focus of this study is on the difference between conventionally blown straw and straw blown into a horizontal wall element, only one sample was taken from the reused straw, while two samples were taken from each of the others. To obtain a comprehensive range of measurement results and capture the greatest variation in layer density, one sample was taken from the top and another from the bottom.

For preparation, the specimens were dried at 65 °C until reaching constant weight. The thermal conductivity was measured using a single-plate device measuring 500 × 500 mm, following the EN 1946-2 [26] standards (Guarded Hot-Plate).

2.2. *Investigation of the Hygrothermal Influence of a Flame Retardant*

The in situ measurements were conducted on the same wall element used in the initial investigation, this time filled with blow-in straw treated with the flame retardant MK46—an aqueous solution containing various acidic organic and inorganic compounds or salts. The flame retardant was added in an "Eirich Intensive Mixer" equipped with a battery-powered spray pump. The functionality of the mixture was tested as part of a research project called "Urban Straw" [27]. To investigate the variation in hygrothermal properties resulting from the treated straw, both horizontal and vertical wall elements were used for the blowing process. One compartment was filled with straw in a horizontal position, while the two remaining compartments were filled in a vertical position. Figure 6 shows the exterior wall elements before blowing in the flame-retardant-treated straw, in both horizontal and vertical positions.

Figure 6. Horizontal position before blowing in (**left**) and vertical position before blowing in (**right**).

Sensors were installed to determine the thermal conductivity and were compared to those used with untreated straw. Additionally, moisture behavior was observed by placing relative humidity sensors in the insulation layer.

To prevent potential structural damage caused by condensation, a simulation was conducted using the Archiphysik software. The simplified verification procedure following the ÖNORM B 8110-2:2020-01-01 was applied.

During the in situ measurements, two temperature and two humidity sensors were installed at the boundary layers inside each compartment of the component. For temperature and humidity measurements, Retronic HC2A-S3 sensors were used, which have a measurement accuracy of ±0.8% for the relative humidity and ±0.1 K for the temperature. The sensors are protected by a probe housing and measure the relative air humidity of their location. Additionally, two temperature sensors were placed on the inner and outer surface, and a heat flow foil was applied on the inner surface of the component. The positions of the sensors are illustrated in Figure 7. The same measurement technology as in the initial investigations was used, and the indoor conditions were set up in a similar manner. Furthermore, a temperature and humidity sensor from Retronic was integrated into the component.

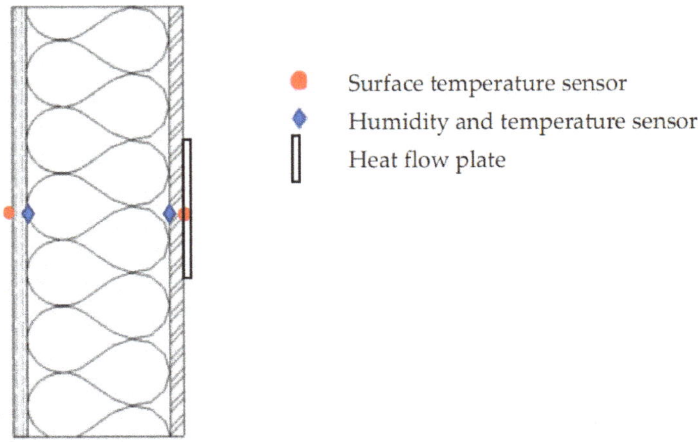

Figure 7. Arrangement of the sensors.

3. Results and Discussion

3.1. Impact of Different Blowing Techniques on the Thermal Conductivity of Untreated Chopped Straw

Measurements were taken over the course of a full year, from summer 2021 to summer 2022, covering an entire winter period. During the cold season, the thermal insulation properties are particularly relevant, as they provide the temperature difference necessary to calculate the insulation properties.

3.1.1. Results of the Heat Flow Measurement

Table 1 presents the mean value of the determined thermal conductivity by means of the heat flow measurements.

Table 1. Mean values of the thermal conductivity of chopped straw (heat flow measurement).

	Mean Value (W/m*K)
Vertical	0.0408
Horizontal	0.0401
Quotient horizontal/vertical	0.984 (1.6%)

The result of the mean value of the vertical blown-in specimen is 0.0408 and of the horizontal blown-in specimen is 0.0401 W/mK. The differences between the values are very small. The mean value differs by only 1.6%. The measurement inaccuracies of the measuring instruments provided by the manufacturer are partly dependent on the prevailing temperature and have been calculated, resulting in a combined uncertainty of 8.9%. This implies that the measured values can deviate from the actual values by up to 8.9%. Consequently, no significant difference in thermal conductivity, depending on the blowing technique, can be observed.

The results of the two different blowing techniques are so close that a preference for one method is not justified. If there are advantages in individual cases, the preferred method should be used.

3.1.2. Determination of the Thermal Insulation Properties Using Needle Probe Measurements

The measurement of the thermal conductivity by means of the needle probe yielded the following results, as shown in Table 2.

Table 2. Thermal conductivity measured with the needle probe at two points for each compartment.

		Thermal Conductivity (W/mK)		
Date		**Vertical**	**Vertical Reused**	**Horizontal**
18.02.2022	P1	0.044	0.047	0.042
18.03.2022	P1	0.045	0.046	0.047
	P2	0.043	0.045	0.046
02.05.2022	P1	0.045	0.047	0.046
	P2	0.044	0.047	0.047
05.05.2022	P1	0.046	0.048	0.048
	P2	0.045	0.047	0.048
12.05.2022	P1	0.044	0.048	0.046
	P2	0.045	0.048	0.047
01.06.2022	P1	0.046	0.047	0.046
	P2	0.045	0.048	0.047

Table 3 shows the statistical values for the thermal conductivity measured with the needle probe. The mean value, the median and the standard deviation as well as their quotient and difference were calculated.

Table 3. Statistical values of the thermal conductivity measured with the needle probe.

	Statistical Values		
	Mean Value (W/mK)	**Median (W/mK)**	**Standard Deviation (W/mK)**
Vertical	0.0447	0.045	0.0009
Vertical reused	0.0471	0.047	0.0009
Horizontal	0.0463	0.047	0.0016
	Quotient of the statistical values (-)		
Vertical/vertical Reused	0.949 (5%)	0.957 (4%)	0.958 (4%)
Vertical/horizontal	0.965 (3%)	0.957 (4%)	0.555 (44%)
Vertical reused/ Horizontal	1.017 (2%)	1.000 (0%)	0.579 (42%)
	Difference in the statistical values (W/mK)		
Vertical/standing reused	0.002	0.002	0.000
Vertical/horizontal	0.002	0.002	0.001
Vertical reused/horizontal	0.001	0.000	0.001

The mean value of the vertical blown-in specimen is 0.0447 W/mK, of the vertical re-used specimen is 0.0471 W/mK and the horizontal blown-in specimen 0.0463 W/mK.

The needle probe measurements in Table 2 also confirm that there is no significant difference between the different blowing techniques.

None of the three variants deviate from each other in mean or median by more than 5%. The measurement accuracy of the needle probe is ±0.001 W/(mK) per measurement, which means that a deviation of 0.002 W/(mK) can occur when comparing two measurement results due to the measuring device. However, the difference in statistical values among all three variants never exceeds 0.002 W/(mK).

As a result, the findings of both measurement methods are consistent with each other. The conclusion that the position of the wall element, whether vertical or horizontal, has no significant influence on the thermal performance of the insulation material is supported by the results of both measurements. Additionally, the reused straw showed similar results in the second measurement.

3.1.3. Results of the Laboratory Experiments

Table 4 contains the values measured in the laboratory for the samples taken from the outdoor test rig, such as thermal conductivity, installation thickness, raw density at 23 °C and 50% relative humidity, and moisture content after measurement. The moisture content was measured gravimetrically in the laboratory. The samples are from the wall element of the first test of the untreated straw with different blowing techniques. The exact positions are shown in Figure 5. The boundary conditions for the conversion to the dry state are from $\lambda_{10,g}$ to $\lambda_{10,dry}$ according to ETA-17/0559 from "ISO-Stroh" [28] with the moisture conversion coefficient f_u = 0.574 kg/kg and mass-specific moisture content $u_{23,50}$ = 0.075 kg/kg.

Table 5 shows that the thermal conductivity of the straw blown in the upright position is 11% higher than that of the straw blown in horizontally. However, it is important to note that the two results for the samples of horizontal straw already differ by 24%, which indicates that this difference cannot solely be attributed to the blowing technique.

Table 4. Thermal conductivity of the samples taken in the laboratory with the single-plate instrument.

Wall Element	Nr.	Installation Thickness d (mm)	Bulk Density (23/50) ρ (kg/m³)	Moisture Content after Measurement u_g (%)	Measured Value Thermal Conductivity $\lambda_{10,g}$ (W/mK)	Thermal Conductivity $\lambda_{10,dry}$ (W/mK)
Vertical	1	98.6	104	0.21	0.05137	0.0513
	2	99.4	103	0.25	0.05125	0.0512
Reused	3	99.6	115	0.19	0.04095	0.0409
Horizontal	4	99.2	118	0.19	0.05103	0.0510
	5	98.1	121	0.17	0.04128	0.0412
Measure variabilities of samples with the same blow-in techniques Thermal conductivity (W/mK)						
Quotient vertical 1/vertical 2						1.00 (0%)
Quotient horizontal 4/horizontal 5						1.24 (24%)

Table 5. Comparison of the average values of thermal conductivity for the different positions, measured with the single-plate instrument.

Mean Value (W/mK)	
Vertical	0.0513
Vertical reused	0.0409
Horizontal	0.0461
Quotient of the mean value (-)	
Vertical/vertical reused	1.25 (25%)
Vertical/horizontal	1.11 (11%)
Vertical reused/horizontal	0.89 (11%)
Difference in the mean value (W/mK)	
Vertical/vertical reused	0.0103
Vertical/horizontal	0.0051
Vertical reused/horizontal	0.0052

The assumption that the horizontal position of the outer wall element during the injection results in poorer thermal insulation properties is also not supported by this measurement.

It is known that air has lower thermal conductivity compared to straw, making air pockets within the insulation desirable. Increasing the straw density reduces the presence of air pockets between the straw particles, thereby reducing heat transfer through convection and conduction, which, in turn, enhances the insulating effect of the material. However, if the density is increased excessively, the thermal conductivity also increases as the compressed air in the voids becomes more conductive.

A previous study [16] demonstrated a relationship between density and thermal conductivity, depicted as a curve that initially decreases and then rises again. Up to a density of approximately 65 kg/m^3, the insulating properties improve, and beyond that point, the thermal conductivity increases with density. However, these laboratory measurements (Table 4) revealed a different relationship between density and thermal conductivity. In the results shown, samples with higher density, such as positions 3 and 5, exhibited the best insulating performance. It is worth noting that the previous study only measured up to a density of 85 kg/m^3, and further research is needed to understand the implications of higher densities, as these findings seem to differ.

3.2. Difference in the Thermal Conductivity Using Different Measurement Methods

The design value of the thermal conductivity is usually obtained by multiplying $\lambda_{10,dry}$ with the normative factor $F_{m(23.80)}$. This design value is usually used in energy balances to calculate the U-value. The results of the measured and calculated thermal conductivities of chopped straw are shown in Table 6. It is shown that a general addition of 20% to the $\lambda_{10,dry}$ does not agree with the values measured in situ.

Table 6. Comparison of thermal conductivity using different measurement methods.

Thermal Conductivity (W/mK)	Vertical	Horizontal
Heat flow measurement	0.0408	0.0401
Needle probe measurement	0.0447	0.0463
$\lambda_{10,dry}$—Single-plate measurement	0.0513	0.046
$\lambda_{10,dry}$ adding 20%	0.062	0.055

3.3. *Impact of a Flame Retardant on Hygrothermal Properties of Chopped Straw*

3.3.1. Comparison of Thermal Conductivity of Horizontal and Vertical Blown-In Straw

The following table shows the thermal conductivities of untreated straw and straw treated with a flame retardant (Table 7). As in the first test, this was determined by means of heat flow measurement (U-value).

Table 7. Mean values of the thermal conductivity of the untreated straw and treated straw (heat flow measurement).

	Untreated Straw	Treated Straw
	Mean Value (W/mK)	**Mean Value (W/mK)**
Vertical	0.0408	0.0412
Horizontal	0.0401	0.0425
Quotient horizontal/vertical	0.983 (1.7%)	1.03 (3.0%)

In Table 7, the quotients of lying to standing show that the differences in the mean value (1.7% and 3%) are relatively small.

The measurement inaccuracy of the measuring devices here is 9.6%; thus, no significant difference between lying and standing blown-in straw in terms of its thermal properties can be determined from the measurements.

Table 8 compares the differences in the thermal conductivity of the untreated and the treated straw. The thermal conductivity is higher for the treated straw in both the upright and lying positions. The mean value of the vertical element is higher by 1% and the mean value of the horizontal element is higher by 6%. It is important to note that the measurements were conducted during two different winter periods, leading to varying temperatures and humidity levels. Additionally, the sum of the measurement inaccuracies of both measurements is 18.5%. Therefore, the measurement results do not show a clear difference, and the small deviation can be attributed to the measurement inaccuracy and external influences. The investigation revealed no evidence that the flame retardant leads to a measurably higher thermal conductivity due to an increased salt concentration.

Table 8. Comparison of the differences between the mean values of the thermal conductivity of the treated and untreated chopped straw insulation (heat flow measurement).

	Mean Value (W/mK)
Quotient vertical with/without flame retardant	1.010 (1%)
Quotient horizontal with/without flame retardant	1.060 (6%)

3.3.2. Long-Term Hygrothermal Behavior of Straw Constructions with Flame Retardants

Considering the fact that flame retardants can influence the hygroscopic properties of the treated straw, the present study aimed to examine whether this leads to moisture-related issues.

Both sensors are located within the insulation layer, with one adjacent to the outer planking and the other adjacent to the inner planking.

The temperature and humidity curves of the two sensors for each compartment of the wall element are presented in the following diagrams. The inner sensor (Figure 8) never reaches a relative humidity value above 55% in all three compartments, while the outer sensor (Figure 9) consistently exceeds 70%.

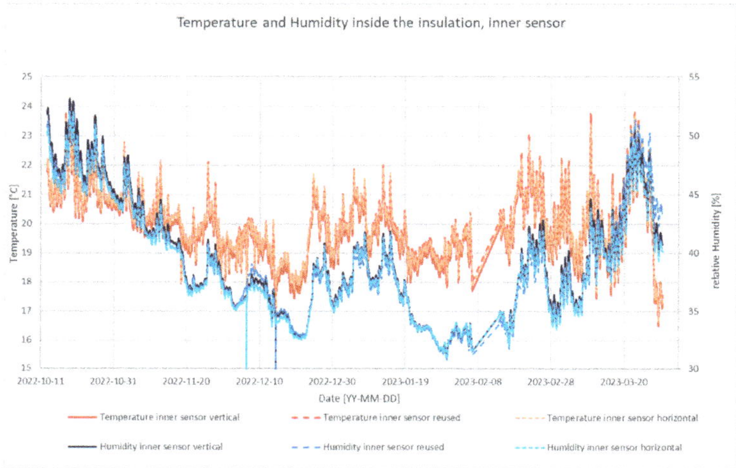

Figure 8. Temperature and relative humidity inside the insulation, inner sensor, one winter period.

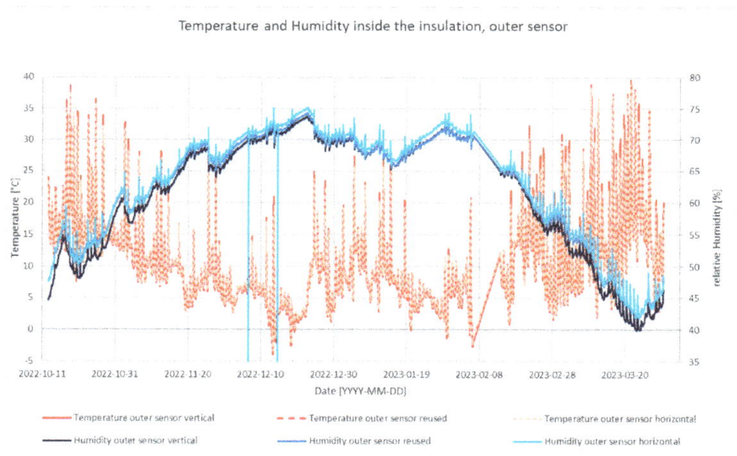

Figure 9. Temperature and relative humidity inside the insulation, outer sensor, one winter period.

A relative humidity of 70% over an extended period is considered a critical value for mold growth in building materials [29,30]. Natural insulation materials made from organic substances, such as lignocellulosic materials, carry a higher risk of pest infestation compared to inorganic or synthetic insulation materials. Nevertheless, using bio-based insulation materials from an ecological perspective is desirable. However, it is crucial to ensure that the chosen material does not pose an increased risk of mold infestation. In some cases, a fungicidal treatment of the material may be necessary to prevent this.

Despite the higher risk of infestation with organic insulation materials, they possess the advantage of being able to temporarily absorb and buffer high moisture levels [26].

According to research conducted by the Fraunhofer Institute for Building Physics, the three primary growth criteria for fungi are substrate, humidity, and temperature over a specific period. They present these criteria in relation to each other using an isopleth system. The critical condition arises when the temperature remains continuously above 20 °C with humidity exceeding 70% for an extended period (several days). However, in our case, the critical condition is only reached for a few hours, indicating that the structure is functional in terms of mold growth.

As 100% relative humidity is never reached, it is assumed that no condensation occurs in the entire component. To verify this, an additional simulation was conducted using the Archiphysik program. The structure was simulated in the program with the same component layers and material parameters as built in the test rig. These data were obtained from the manufacturers, and the measured thermal conductivity was used for the insulation layer.

First, the measured and simulated U-values were compared to check whether the simulation model corresponds to the real structure. Then, according to ÖNORM B 8110-2:2020-01-01, the simplified verification procedure was conducted to ensure the prevention of damage-causing amounts of condensate inside the building component. The calculations show that there is no condensate formation, meaning that the verification is fulfilled.

4. Conclusions

All of the tests carried out have consistently shown that there is no deterioration in the thermal insulation properties of the blow-in technology in horizontally positioned wall elements compared with conventional blow-in technology in vertically positioned wall elements. Horizontal blowing can lead to a simplification of the working process, and, thus, to greater efficiency on the construction site or in prefabrication.

This observation could be due to the hypothesis that the fibers exhibit fluid-like behavior when blown in the air and the general flow behavior is very turbulent. There was also no apparent difference in the orientation of the fibers when the module was opened.

Further investigations would be interesting here; for example, two wall elements could be built with a glass front and filmed during the blow-in and analyzed afterwards.

Although it was expected that re-blowing would cause the straw chaff to have a finer microstructure, resulting in higher thermal conductivity, i.e., poorer insulating properties, the needle probe measurement for the re-blown straw showed similar thermal conductivity to the new straw. In the case of the heat flow measurement, it was not possible to obtain comparable values due to the middle position. Here, further investigations with other measurement methods would be interesting to support the hypothesis that it has similar insulating properties as the unused straw.

Both blow-in technologies lead to better results than the data sheet indicates. In ETA-18/0305 [18], the design value of thermal conductivity for blow-in straw is given as $\lambda = 0.058$ (W/m*K), while the results of this study show better values by about 0.01 (W/m*K). This may be due to the fact that the general moisture allowance of 20% is not justified. Further research in this area is highly relevant for all bio-based insulation materials.

The correlation of density and thermal conductivity shows different results in our case in comparison to those in the compared study [16]. However, the measurements of the study only went up to 85 kg/m^3. Since there are not yet many studies on chopped straw, further in situ and laboratory tests would be interesting to analyze this relationship more precisely.

The recommendations for further research include more in situ and laboratory measurements of chopped straw in general, as well as detailed investigations of the different hygrothermal behavior of biobased insulation materials. This could lead to more widespread use of these energy-efficient, low-emission building materials.

Author Contributions: Conceptualization, T.S. and H.F.; methodology, T.S.; validation, T.S.; writing—original draft preparation, T.S.; writing—review and editing, H.F.; visualization, T.S.; supervision, A.K.; project administration, A.K. All authors have read and agreed to the published version of the manuscript.

Funding: This research was part of the natuREbuilt project and was funded by the Austrian Research Promotion Agency (FFG) under the COIN funding program. FFG grant number 881230. The APC was funded by Open Access Funding by TU Wien.

Data Availability Statement: Data will be made available on request.

Acknowledgments: We would like to express our gratitude to the DPM Group and the manufacturers ISO-Stroh for their partnership and contribution to this research project. The authors acknowledge TU Wien Bibliothek for financial support through its Open Access Funding Programme.

Conflicts of Interest: The authors declare no conflict of interest.

References

1. Linh, D.; Salomone, R.; Nguyen, Q.T. Circular bio-based building materials: A literature review of case studies and sustainability assessment methods. *Build. Environ.* **2023**, *244*, 110774.
2. Cintura, E.; Nunes, L.; Esteves, B.; Faria, P. Agro-industrial wastes as building insulation materials: A review and challenges for Euro-Mediterranean countries. *Ind. Crops Prod.* **2021**, *171*, 113833. [CrossRef]
3. Paiva, A.; Pereira, S.; Sá, A.; Cruz, D.; Varum, H.; Pinto, J. A contribution to the thermal insulation performance characterization of corn cob particleboards. *Energy Build.* **2012**, *45*, 274–279. [CrossRef]
4. Liu, L.; Zou, S.; Li, H.; Deng, L.; Bai, C.; Zhao, X.; Wang, S.; Li, N. Experimental physical properties of an eco-friendly bio-insulation material based on wheat straw for buildings. *Energy Build.* **2019**, *201*, 19–36. [CrossRef]
5. Labat, M.; Magniont, C.; Oudhof, N.; Aubert, J.E. From the experimental characterization of the hygrothermal properties of straw-clay mixtures to the numerical assessment of their buffering potential. *Build. Environ.* **2016**, *97*, 69–81. [CrossRef]
6. Ali, M.; Alabdulkarem, A.; Nuhait, A.; Al-Salem, K.; Almuzaiqer, R.; Bayaquob, O.; Salah, H.; Alsaggaf, A.; Algafri, Z. Thermal Analyses of Loose Agave, Wheat Straw Fibers and Agave/Wheat Straw as New Hybrid Thermal Insulating Materials for Buildings. *J. Nat. Fibers* **2021**, *18*, 2173–2188. [CrossRef]
7. Ali, M.; Alabdulkarem, A.; Nuhait, A.; Al-Salem, K.; Iannace, G.; Almuzaiqer, R. Characteristics of Agro Waste Fibers as New Thermal Insulation and Sound Absorbing Materials: Hybrid of Date Palm Tree Leaves and Wheat Straw Fibers. *J. Nat. Fibers* **2022**, *19*, 6576–6594. [CrossRef]
8. Cascone, S.; Rapisarda, R.; Cascone, D. Physical properties of straw bales as a construction material: A review. *Sustainability* **2019**, *11*, 3388. [CrossRef]
9. Sun, C.; Gu, J.; Dong, Q.; Qu, D.; Chang, W.; Yin, X. Developments in the Built Environment Are straw bales better insulation materials for constructions? A review. *Dev. Built Environ.* **2023**, *15*, 100209. [CrossRef]
10. Zhou, Y.; Trabelsi, A.; El Mankibi, M. A review on the properties of straw insulation for buildings. *Constr. Build. Mater.* **2022**, *330*, 127215. [CrossRef]
11. Brambilla, A.; Sangiorgio, A. Moisture buffering of building materials. *Moisture Build.* **2021**, 99–128. [CrossRef]
12. Kreiger, B.K.; Srubar, W.V. Moisture buffering in buildings: A review of experimental and numerical methods. *Energy and Buildings* **2019**, *202*, 109394. [CrossRef]
13. Koh, C.H.; Kraniotis, D. A review of material properties and performance of straw bale as building material. *Constr. Build. Mater.* **2020**, *259*, 120385. [CrossRef]
14. Marktanteil von Dämmstoffen aus Nachwachsenden Rohstoffen Steigt. Umfrage zum Einsatz Biobasierter Baustoffe. 2021. Available online: https://www.bba-online.de/news/marktanteil-von-daemmstoffen-aus-nachwachsenden-rohstoffen-waechst/ (accessed on 10 April 2023).
15. Sabapathy, K.A.; Gedupudi, S. Straw bale based constructions: Measurement of effective thermal transport properties. *Constr. Build. Mater.* **2019**, *198*, 182–194. [CrossRef]
16. Vėjelienė, J. Processed Straw as Effective Thermal Insulation for Building Envelope Constructions. *Eng. Struct. Technol.* **2012**, *4*, 96–103. [CrossRef]
17. Deutsches Institut für Bautechnik. *Europäische Technische Bewertung ETA-17/0559*; Deutsches Institut für Bautechnik: Berlin, Germany, 2011.
18. *ÖNORM B6015-2*; Bestimmung der Wärmeleitfähigkeit mit dem Plattengerät Teil 2: Ermittlung des Nennwertes und des Bemessungswertes der Wärmeleitfähigkeit für Homogene Baustoffe. Austrian Standards Institute: Wien, Austria, 2009.

19. Zirkelbach, D.; Tieben, J.; Tanaka, E.; Pfabigan, N.; Andresen, N.; Bachinger, J.; Nusser, B. *ThermNat—Building Components with Insulation from Sustainable Raw Materials: Focus (Hygro-)Thermal Conditions*; ThermNat: Holzkirchen, Germany; Wien, Austria; Dresden, Germany, 2023.
20. *ÖNORM EN 15101-1*; Thermal Insulation Products for Buildings—In-Situ Formed Loose Fill Cellulose (LFCI) Products—Part 1: Specification for the Products before Installation. Austrian Standards Institute: Wien, Austria, 2019.
21. Anders, S.; Iravani, A.; Hoffmann, C.; Geissler, A.; Voss, K. In-Situ U-Wert-Messung—Teil 1: Vergleich von Verfahren und Einflüsse in Heizkammerversuchen. *Bauphysik* **2023**, *45*, 1–10. [CrossRef]
22. *ISO 9869-1*; Thermal insulation—Building elements—In-Situ Measurement of Thermal Resistance and Thermal Transmittance—Part 1: Heat Flow Meter Method. International Organization for Standardization: Geneva, Switzerland, 2014; pp. 1–11.
23. Tudiwer, D. *Influences of Living Wall Systems on Thermal Protection, Summery Overheating and Hydrothermal Comfort*; Technical University Vienna: Vienna, Austria, 2019.
24. O'Hegarty, R.; Kinnane, O.; Lennon, D.; Colclough, S. In-Situ U-value monitoring of highly insulated building envelopes: Review and experimental investigation. *Energy Build.* **2021**, *252*, 111447. [CrossRef]
25. Incropera, F.P.; Dewitt, D.P. *Fundamentals of Heat and Mass Transfer*; John Wiley & Sons: New York, NY, USA, 1998.
26. *EN 1946-2*; Thermal Performance of Building Products and Components—Specific Criteria for the Assessment of Laboratories Measuring Heat Transfer Properties—Part 2: Measurements by the Guarded Hot Plate Method. European Committee for Standardization: Brussels, Belgium, 1999.
27. Schubert, P.; Korjenic, A.; Fischer, H.; Kirchengast, I. *URBAN STRAW—Brandschutztechnische Konditionierung von Einblas-Stroh zur Gebäudedämmung der Urbanen Gebäudeklassen 4 und 5*; STADT der Zukunft: Wien, Austria, 2023.
28. Teil, A.; Bewertungsstelle, T.; Bauprodukts, H. *Europäische Technische Bewertung ETA-17/0559*; Deutsches Institut für Bautechnik: Berlin, Germany, 2011; No. 305.
29. Grimm, R. Schimmel im Gebäude: Was Wächst da Eigentlich? 2018. Available online: https://www.baustoffwissen.de/baustoffe/baustoffknowhow/forschung_technik_trends/was-ist-schimmel-mikroorganismen-naehrstoffe-baustoffe-umweltbundesamt-pilze-bakterien-milben/ (accessed on 18 May 2023).
30. Jan Küver, R.M.; Peterschewski, J. *Untersuchungen zum Verhalten von Konventionellen und Ökologischen Dämmstoffen Gegenüber Mikrobiellem Befall unter Verschiedenen Klimatischen Bedingungen und Bewertung der Mikrobiellen Kontamination für die Wohnhygiene und Effizienz der Energieeinsparung*; Fraunhofer IRB: Stuttgart, Germany, 2004.

buildings

MDPI

Systematic Review

Insights and Evidence on Energy Retrofitting Practices in Rural Areas: Systematic Literature Review (2012–2023)

Ahmed Abouaiana * and Alessandra Battisti

Department of Planning, Design, and Technology of Architecture, Sapienza, University of Rome, Via Flaminia 72, 00196 Rome, Italy; alessandra.battisti@uniroma1.it
* Correspondence: abouaiana.ahmed@gmail.com; Tel.: +2-2011-1225-8628

Abstract: Rural commons face extraordinary challenges like fragility and sensitivity due to climate change. Retrofitting rural built environments affords benefits that could overcome these challenges and support sustainable development. However, notwithstanding the vast energy retrofitting interventions available, the associated aspects require investigation, particularly in distinct rural contexts with all their valuable, cultural, and historical inheritance. Hence, this study aimed to examine energy retrofitting practices in rural settlements worldwide over a decade to diagnose the goals that are being undertaken, stakeholder engagement, and finally, the bi-correlation between rural contexts and interventions, and retrofitting contributions to valorizing the place's identity. This study is a systematic literature review (SLR) considering the items of the PRISMA checklist (Preferred Reporting Items for Systematic Reviews and Meta-Analyses). An SLR of published peer-reviewed studies between January 2012 and March 2023 in 16 electronic databases in all available languages, using a combination of seven keywords within three domains, was conducted. The initial search resulted in 397; after applying the inclusion/exclusion criteria, there were 60 eligible articles. The academic progress and tendencies in the energy retrofitting domain of rural built environments are discussed and summarized into four major thematic classifications (energy efficiency strategies, energy efficiency planning, policy evaluation, and occupant behavior). Briefly, rural buildings lack energy-saving designs. Simulation tools are essential; however, they should be calibrated with on-site conditions, showing the reasons for selecting the applied retrofitting measures and correlation with the surrounding context. Successful implementation requires cross-disciplinary collaboration, engaging decision makers, and providing energy education for the local community. Regulations should include micro-context-specific environmental performance indicators. These insights could help map out future academic pursuits and help the stakeholders better understand their nature. Simultaneously, this study assists early-stage researchers in conducting systematic literature reviews utilizing different tools. However, the SLR protocol may have limited findings due to the specific search terms used, so the authors believe the more the literature search scope is broadened, the more discoveries could be made.

Keywords: qualitative approach; energy efficiency; rural commons; evidence-based practices; content analysis

Citation: Abouaiana, A.; Battisti, A. Insights and Evidence on Energy Retrofitting Practices in Rural Areas: Systematic Literature Review (2012–2023). *Buildings* **2023**, *13*, 1586. https://doi.org/10.3390/buildings13071586

Academic Editors: Paulo Santos and Mark Bomberg

Received: 29 May 2023
Revised: 16 June 2023
Accepted: 20 June 2023
Published: 22 June 2023

1. Introduction

Traditional rural settlements emerged vernacularly (without architects [1]) or with architects and planners, while respecting the rural identity [2]. They have a distinguished architectural typology and urban materials through transferred knowledge between generations, to achieve dwellers' satisfaction. They are characterized by a high population density, relying on economic activities using the integration of primitive small-scale and modern techniques. Dwellers can solve problems, such as mobility, energy efficiency, and efficient use of space in their ways [3]. Rural areas play a crucial role in our daily lives by supporting the production of food and raw materials, offering recreational opportunities, contributing to our overall well-being and environmental health, enhancing ecosystem services, and providing aesthetic value. They comprise the vast majority of the global territories and host half of the population in developing countries and one-third of the population in EU countries.

Nowadays, rural areas have undergone significant changes since the mid-20th century due to many reasons, like industrialization [4], philosophical shifts [5], and socioeconomic aspects [6], that produced modern built environment patterns. that produced modern built environment patterns, shifted the productive rural nature to consumerism which led to the consumption of natural resources and increases the demand for energy and water resources, all irreversible adverse impacts on the environment, making them confront distinct environmental challenges due to climate change as one of the most fragile areas [7–10].

Energy has played a crucial role in human civilization, but the energy industry is a major contributor to greenhouse gas emissions (e.g., nearly 80% in the EU). Scholars have highlighted the negative effects of this sector on the environment and emphasized its crucial role in achieving sustainable development goals (goal 7: supply affordable, sustainable, and reliable energy to all by 2030) [11]. In this context, the global community has intensified its efforts since the Paris Agreement in 2015 (an international treaty on climate change) to fulfill, at the national scale, the commitment to mitigate climate effects, particularly in rural contexts, where energy is a decisive factor [12]. For instance, the European Commission aims to support sustainable rural development through a number of initiatives in the framework of the European Green Deal [13]; rural specificity of a country is an essential characteristic influencing energy poverty [14]. The current policy in the United States of America (USA) is to finance Rural Clean Energy Initiative projects that focus on clean energy transitions [15]. This is regardless of the current global energy crisis due to shared global challenges, such as the Russian–Ukrainian conflict, which brings into mind the energy crisis that resulted in the aftermath of the Egyptian–Israeli war in the 1970s.

The building sector is a predominant aspect of the built environment, and is responsible for about 40% of global energy consumption. It has a substantial carbon footprint regarding the associated greenhouse gas emissions (GHG) from electricity consumption and anthropological activities. Energy retrofitting refers to adapting the latest technologies or features to obsolete systems [16]. Therefore, building energy retrofitting and utilizing existing clean energy solutions in rural areas can play a pivotal role in mitigating rural commons' challenges, such as meeting climate change goals [17], achieving low-carbon transitions [18], fulfilling sustainable development objectives [19], decreasing natural resource exploitation [20], and enhancing the quality of life [21].

Energy retrofitting practices are characterized by intricacy, unlike any domains, which was confirmed in the past two decades [22] and by a recent study [23]. This was through the efforts of the scholars, representing a significant increase in the scholarly publications in the field within this period [24]. Numerous studies defined this complexity in obstacles accompanying energy retrofitting domains related to different perspectives, like policies, user behavior and culture, techno-economic factors, and business engagement [25–27]. Moreover, the nature of the process requires cross-disciplinary collaboration among various stakeholders, as emphasized in Sibilla and Kurul's systematic review titled "Transdisciplinarity in energy retrofit" [28].

Briefly, cross-disciplinarity has different forms. Multi-disciplinarity: coherence of the conceptual frameworks associated with various disciplines, focusing on a particular topic from different angles (working together but without significant contributions). Intra-disciplinary: combines two or more diverse sub-fields under the major one, concentrating participants' efforts to derive concepts. Inter-disciplinarity: builds a mutual framework for the collaborating disciplines under the same field to create a synthesis, working on common questions to achieve shared results (the results are more than the sum of each discipline). Trans-disciplinarity: multi-level coordination incorporating academic researchers from diverse fields with non-academic participants and professional practices to flourish the epistemology and theory for knowledge generation [29–31]. Figure 1 visualizes the Latin origin of the concepts for a better understanding.

Original Latin Word	Uni *"mono"*	Multi	Intĕr	Intra	Trans
Meaning	One; Single; alone	Multitude; Many; Mass	Between; Among; During	Within; Inside; Under	Across; Over; Beyond

Figure 1. The Latin origin of the cross-disciplinarity forms.

Rural commons have specific socio-cultural values, which is higher than that in urban contexts [32], which can be represented by, the strong bond between residents and their home [33], regardless of the physical identity of the rural buildings (despite their steady transformations [34]). Consequently, many studies have emphasized the necessity of preserving the cultural and natural values of rural places while developing them [35–38], including energy retrofitting.

Therefore, this study aimed to identify the retrofitting practices for energy efficiency in rural settlements in the past decade to highlight the progress in academic research, synthesize the evidence into insights [39], and provide a meticulous, holistic summary through evidence-based practices, such as a systematic literature review (SLR) [40,41]. SLR is a holistic search for relevant scientific research on the investigated topic using structured methodologies [42]. We aim to provide an explicit model of the status of global research on rural built environments beyond the technical interventions and the numerical results (energy saving quantifying). We hypothesize that rural contexts differ from others, represented by the reciprocal correlation between interventions and the surrounding contexts. This can be addressed by answering three research questions (RQs):

- (RQ1) What are the thematic classifications of the activities undertaken within these practices, considering cross-disciplinarity and stakeholder engagement?
- (RQ2) What is the relationship between rural contexts and interventions?
- (RQ3) To what extent do the interventions contribute to valorizing the rural identity?

2. Materials and Methods

The aim was accomplished and the research questions were addressed by conducting an SLR. We used the PRISMA (Preferred Reporting Items for Systematic Reviews and Meta-Analyses) checklist [43], which emerged in 2009 [44]. The checklist was employed to identify the research question, search keywords that meet the objectives, and synthesize the qualitative results.

The proposed SLR steps were revised based on references [45,46]. The SLR was executed based on the modified protocol, using explicit inclusion and exclusion criteria (Section 2.1) and employing the relevant keywords (retrofitting rural built environments) (Section 2.2) which led to a comprehensive examination of the literature to identify retrofitting factors globally. Figure 2 depicts the SLR process and structure.

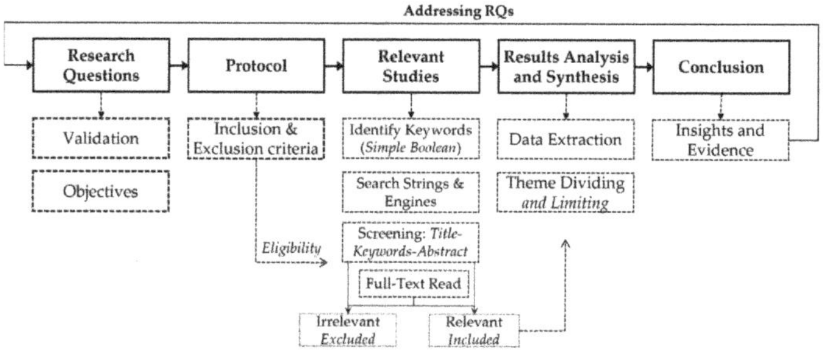

Figure 2. The SLR procedure and structure.

2.1. SLR Protocol: Inclusion and Exclusion Criteria

The search selected peer-reviewed articles in journals and conferences to ensure their quality. They should be primary sources and published within the last decade (between 2012 and March 2023) in all available languages and electronic databases. Books, grey literature (e.g., M.Sc. and Ph.D. theses), and secondary sources were excluded.

Meanwhile, the articles should address the specific problem and answer at least one research question. Namely, they had to report energy efficiency improvement strategies or practices in the buildings or surrounding environment or discuss retrofitting policies in rural contexts. Conversely, the articles that distinctly discuss non-rural contexts (i.e., urban and suburban areas) or undefined contexts or provide generic guidelines or evaluated policies were excluded. Table 1 summarizes the exclusion and inclusion criteria.

Table 1. The inclusion and exclusion criteria for the SLR.

Factor	Inclusion Criteria	Exclusion Criteria
Document Type	Peer-reviewed journal articles; Peer-reviewed conference articles; Primary research.	Grey literature (e.g., M.Sc. and Ph.D. theses); Books and book chapters; Secondary research.
Year Range	Between January 2012–March 2023	Before January 2012 and after February 2023
Ultimate context and intimate context	All kinds of rural settlements (e.g., historical and abandoned villages); Discusses retrofitting (policies/practices) in generic contexts, including rural ones.	Not rural or undefined context (e.g., cities, urban or suburban contexts)
Relevance to the objectives	The articles address "retrofitting" for energy efficiency and answer one or more research question(s).	The article discusses a specific topic not relevant to the research questions.
Language	All available languages	Not applicable (n/a)
Research topic	Retrofitting built environment; Case studies and best practices; Review policies or energy efficiency programs that specified rural contexts;	Review policies or energy efficiency programs with general guidelines that are not specific to rural contexts; Specific topic (e.g., material development); Does not include retrofitting or any synonyms.

2.2. Research Strategy

The research context's keywords were specified to construct the search string. Three strings of search keywords were used. The first term is "retrofitting" and its synonyms "refurbishment" and "renovation" as they are three commonly used expressions to describe physical changes executed to improve existing buildings or built environments [47]; the asterisk is used to give a different form of the keyword, for instance (renovat* may indicate renovation, renovated, renovated, or renovating). The second term is "energy efficiency" or "energy saving" which is the core of the research objective. The third term is "rural." Meanwhile, in order to broaden the obtained results, related built environment aspects like buildings and the context classifications (e.g., village, settlement, community) were excluded. However, these aspects were extracted during the review process. The search string implemented using simple Boolean (AND and OR) logic:

(retrofit* OR refurbishment OR renovat***)** **AND** **(**"energy efficiency" OR "energy saving" OR "energy-efficien*"**)** **AND** rural

This study used the Rome Digital Library System of Sapienza University—SBS (Discovery Sapienza) powered by EBSCOhost (https://web.uniroma1.it/sbs/en/discovery sapienza, accessed on 22 June 2023), as well as EBSCOhost which is one of the commonly employed databases by scholars in different disciplines [48]. It searches various electronic sources simultaneously using the maximum number of keywords; of note is that the platform can be utilized without registration, and only a few additional features are exclusive to users with an institutional email, like saving the research result in folders under the account.

2.3. Results and Study Selections

The initial search process on SBS was conducted between January 2012 and March 2023, applying the Peer Review filter; 353 studies were found, in 16 electronic databases, Figure 3, all available in English. The SBS engine auto-detected duplications; these were removed and the number decreased to 221 studies. These were exported in (RIS) format (available in Supplementary File S1) to a reference manager software Zotero V.6 to check for further duplications, omitting 53 more articles, leaving 168 studies which were exported for screening using the Rayyan online platform (https://www.rayyan.ai, accessed on 15 March 2023). At this stage, one duplicated article was eliminated (provided in two languages).

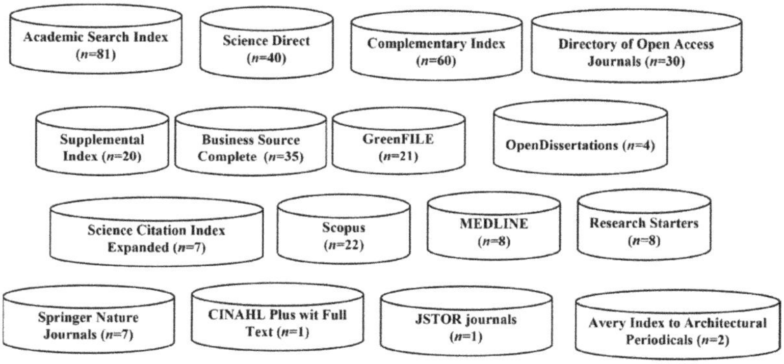

Figure 3. The number of publications from each database.

Rayyan is a common tool for SLR in many domains [49], it allows users to review each paper (displaying the title, abstract, keywords, publisher, and authors) and add notes and PDF files. The interface shows a summary, for instance, of screening time and the number of sessions. Each paper can be labeled and selected to be included, excluded, or labelled as maybe (to be decided later). It enables collaboration among the reviewer team which accelerates the study selection process and decreases the screening time by 40% compared to similar tools [50].

The evaluation process was implemented in two steps. Firstly, the initial screening process of the title, keywords, and abstract was implemented via the Rayyan platform. The screening process excluded 92 articles that could not be related to the research questions. The remaining articles (75) qualified for the second step. It should be noted that if one author or both authors had reservations about approving the study, the article was accepting for the next step (full-paper analysis).

In the second phase, the papers were downloaded and read one by one. Some manuscripts had limited access, but the authors could download them through institutional accounts. In this second evaluation process, 18 papers were excluded. Figure 4 illustrates the PRISMA flowchart of the selection process.

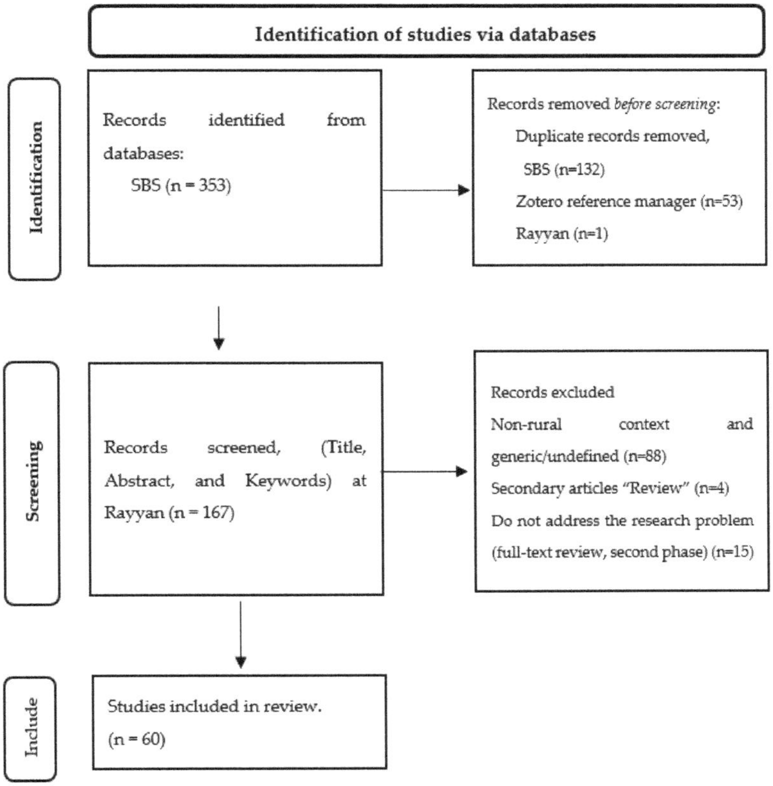

Figure 4. The PRISMA flow diagram of the selection of articles.

2.4. Data Extraction

After selecting the eligible articles, the authors developed a data extraction form in an Excel sheet (available in Supplementary File S2). The extracted data are divided as follows:

- The bibliographical data, authors' names, numbers, citation score, and keywords (Section 3.1);
- Authors' affiliations, affiliation countries, and cross-disciplinarity (Section 3.2);
- Characteristics of the geographical and micro-contexts (Section 3.3);
- Scope of analysis, focus, and theme classifications (Section 3.4);
- Result synthesis (Section 3.5).

In conclusion, the SLR resulted in 60 peer-reviewed papers (36% of the total articles). This occurred because of the rigid restriction of the eligibility criteria, which searched a combination of 7 keywords within three domains between January 2012 and March 2023. The included articles are shown in Table 2.

Table 2. The included SLR studies (in ascending order by year of publication).

ID	Ref	Title
ID01	[51]	State of the Irish housing stock—Modelling the heat losses of Ireland's existing detached rural housing stock & estimating the benefit of thermal retrofit measures on this stock
ID02	[52]	Evaluating fuel poverty policy in Northern Ireland using a geographic approach
ID03	[53]	Renovation alternatives to improve energy performance of historic rural houses in the Baltic Sea region
ID04	[54]	Overview of rural building energy efficiency in China

Table 2. *Cont.*

ID	Ref	Title
ID05	[55]	Azioni e strumenti per il recupero e la valorizzazione dell'architettura e del paesaggio rurale e montano
ID06	[56]	A comprehensive analysis of building energy efficiency policies in China: status quo and development perspective
ID07	[57]	Impact of Civil Envelope on Energy Consumption based on EnergyPlus
ID08	[58]	Analysis on building energy performance of Tibetan traditional dwelling in cold rural area of Gannan
ID09	[59]	Material flow accounting for an Irish rural community engaged in energy efficiency and renewable energy generation
ID10	[60]	Retrofitting domestic appliances for PV powered DC Nano-grid and its impact on net zero energy homes in rural India
ID11	[61]	An examination of energy efficiency retrofit depth in Ireland
ID12	[62]	Role of Self-Efficacy in Reducing Residential Energy Usage
ID13	[63]	Evaluation of refurbishment alternatives for an Italian vernacular building considering architectural heritage, energy efficiency and costs
ID14	[64]	Analysis and Optimization on Energy Performance of a Rural House in Northern China Using Passive Retrofitting
ID15	[65]	Effect of Building Roof Insulation Measures on Indoor Cooling and Energy Saving in Rural Areas in Chongqing
ID16	[66]	Energy retrofit and environmental sustainability improvement of a historical farmhouse in Southern Italy
ID17	[67]	Thermal comfort optimisation of vernacular rural buildings: passive solutions to retrofit a typical farmhouse in central Italy
ID18	[68]	Redesign of a Rural Building in a Heritage Site in Italy: Towards the Net Zero Energy Target
ID19	[69]	Analysis of Passive Energy-saving Retrofitting of Rural Residential Houses in Southern Anhui Province—A case in Hongcun
ID20	[70]	Sustainability evaluation of retrofitting solutions for rural buildings through life cycle approach and multi-criteria analysis
ID21	[71]	SWOT Analysis for the Promotion of Energy Efficiency in Rural Buildings: A Case Study of China
ID22	[72]	Geometric classification method of rural residences at regional scale
ID23	[73]	China's building stock estimation and energy intensity analysis
ID24	[74]	An exploration about the Solar Energy Utilization and the Enclosure System Renovation for Rural Residential Buildings in Cold Areas of Northern China—Taking the rural residential renovation design in Zhujialin Village, Linyi as an example
ID25	[75]	Residential energy transition and thermal efficiency in an arid environment of northeast Patagonia, Argentina
ID26	[76]	Towards a cleaner domestic heating sector in China: Current situations, implementation strategies, and supporting measures
ID27	[77]	Energy, carbon, and cost analysis of rural housing retrofit in different climates
ID28	[78]	Indoor Temperature Improvement and Energy-Saving Renovations in Rural Houses of China's Cold Region—A Case Study of Shandong Province
ID29	[79]	Facility Energy Management Application of HBIM for Historical Low-Carbon Communities: Design, Modelling and Operation Control of Geothermal Energy Retrofit in a Real Italian Case Study
ID30	[80]	An Integrated HBIM Simulation Approach for Energy Retrofit of Historical Buildings Implemented in a Case Study of a Medieval Fortress in Italy
ID31	[81]	Reduced biodiversity in modernized villages: A conflict between sustainable development goals
ID32	[82]	The Economic Effects of New Patterns of Energy Efficiency and Heat Sources in Rural Single-Family Houses in Poland
ID33	[83]	Preliminary Energy Evaluations for the Retrofit of Rural Protected Buildings in a Peripheral Context of Milan
ID34	[84]	Heat consumption scenarios in the rural residential sector: the potential of heat pump-based demand-side management for sustainable heating
ID35	[85]	Integrated assessment of the environmental and economic effects of "coal-to-gas conversion" project in rural areas of northern China.
ID36	[86]	Renewable Energy Utilization in Rural Residential Housing: Economic and Environmental Facets
ID37	[87]	The role of personal and environmental factors in rural homeowner decision to insulate; an example from Poland
ID38	[88]	Active–passive combined energy-efficient retrofit of rural residence with non-benchmarked construction: A case study in Shandong province, China
ID39	[89]	Life Cycle Carbon Emission Analyzing of Rural Residential Energy Efficiency Retrofit-A Case Study of Gansu province
ID40	[90]	Framework for design and optimization of a retrofitted light industrial space with a renewable energy-assisted hydroponics facility in a rural northern Canadian community

Table 2. *Cont.*

ID	Ref	Title
ID41	[91]	Life-Cycle Assessment of a Rural Terraced House: A Struggle with Sustainability of Building Renovations
ID42	[92]	Evaluation of Rural Dwellings' Energy-Saving Retrofit with Adaptive Thermal Comfort Theory
ID43	[93]	Retrofitting Rural Dwellings in Delta Region to Enhance Climate Change Mitigation in Egypt
ID44	[94]	Theoretical Study on the Relationship of Building Thermal Insulation with Indoor Thermal Comfort Based on APMV Index and Energy Consumption of Rural Residential Buildings
ID45	[95]	Passive Energy-Saving Optimal Design for Rural Residences of Hanzhong Region in Northwest China Based on Performance Simulation and Optimization Algorithm
ID46	[96]	Analysis of Energy Performance and Integrated Optimization of Tubular Houses in Southern China Using Computational Simulation.
ID47	[97]	Mitigating heat demand peaks in buildings in a highly renewable European energy system
ID48	[98]	Evaluation of energy-saving retrofit projects of existing rural residential envelope structures from the perspective of rural residents: the Chinese case
ID49	[99]	Net-zero energy retrofit of rural house in severe cold region based on passive insulation and BAPV technology
ID50	[100]	Environment improvement and energy saving in Chinese rural housing based on the field study of thermal adaptability
ID51	[101]	Evaluation of energy-saving retrofits for sunspace of rural residential buildings based on orthogonal experiment and entropy weight method
ID52	[102]	Opportunities stemming from retrofitting low-resource East African dwellings by introducing passive cooling and daylighting measures
ID53	[103]	Estimating the impact of energy efficiency on housing prices in Germany: Does regional disparity matter?
ID54	[104]	Evolutionary Game Analysis of Energy-Saving Renovations of Existing Rural Residential Buildings from the Perspective of Stakeholders
ID55	[105]	Exploring pathways of phasing out clean heating subsidies for rural residential buildings in China
ID56	[106]	Energy Saving and Thermal Comfort Performance of Passive Retrofitting Measures for Traditional Rammed Earth House in Lingnan, China
ID57	[107]	A study on influencing factors of optimum insulation thickness of exterior walls for rural traditional dwellings in northeast of Sichuan hills, China
ID58	[108]	Green retrofit of existing residential buildings in China: An investigation on residents' perceptions
ID59	[109]	Sustainable Energy Development and Climate Change Mitigation at the Local Level through the Lens of Renewable Energy: Evidence from Lithuanian Case Study
ID60	[110]	Improving Building Envelope Efficiency Lowers Costs and Emissions from Rural Residential Heating in China

3. Results and Analysis

3.1. Publications, Authors, Affiliations

The distribution of papers by year demonstrates a steady increase of about 9% each year in publication rate in the latest three years (Figure 5a). China contributed half of the studies on this subject. Studies in Italy came in second with 15%, followed by Poland, and the United Kingdom (UK) and Ireland with 7% and 5%, respectively. The remaining articles were implemented in 13 countries (6 in Europe, 2 in Asia, 2 in Africa, 1 in South America, and 2 in North America), Figure 5b.

A total number of 259 authors affiliated with 80 countries produced these articles. Chinese affiliations came in first place with 31%, followed by Italy at 11%, the United Kingdom (UK) and Ireland at 9%, the United States of America (USA) and Poland at 8% each, and 2% of each for Turkey, Sweden, the Netherlands, Germany, and Japan; the remaining countries had one affiliation each (Figure 5c).

Twelve authors have collaborated twice and produced eight articles: Grohmann, D., and Menconi, M.E. [67,70]; Cotana, F., Piselli, C., and Romanelli, J. [79,80]; Klepacka. A.M. [86,87]; and Arıcı, M., Jiang, W., Li, Q., Li, D., and Qi, H. [99,101]. As remarked from the timeline of publications, it seems these authors split/developed the same research line for two publications. The collaboration rate among authors was predominated by 2–5 authors, at a frequency of 35 times, followed by 6–10 authors that collaborated in 23% of the papers. Only one article was conducted by one author, showing the need for more multi- and inter-disciplinarity, supporting the findings of Sibilla and Kurul.

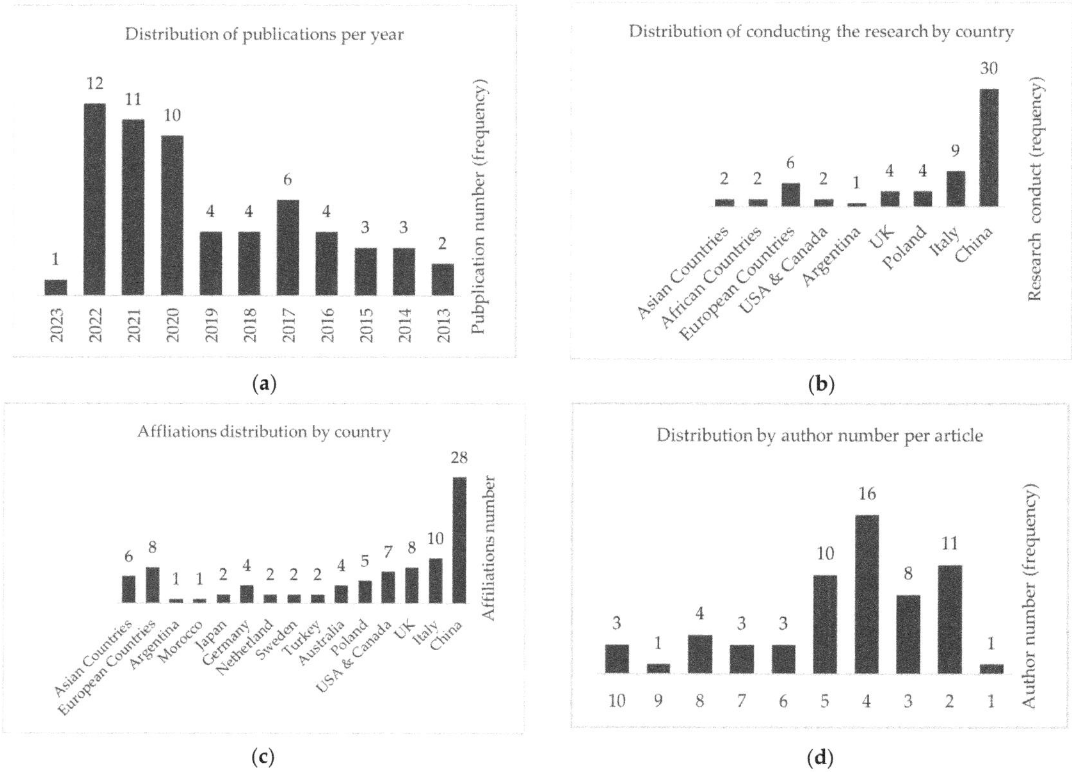

Figure 5. Quantitative analysis of included articles. (**a**) Distribution of articles by year; (**b**) distribution of research by country; (**c**) distribution by affiliations' country; (**d**) distribution by author count of paper.

The citations from three databases indicated a positive correlation between the year of publication and citation number, with more citations in later years (Figure 6a). The top five cited papers are [51,53,54,56,73], all published by Elsevier. In this logic, Huo et al.'s recent study (in 2019) [73] had the most citations, double that of the next highest cited paper in the same reference year (but no observed correlation between author count and citation numbers). Google Scholar (GS) had the highest citation number for all publications compared to the other two databases (Web of Science and Scopus); both almost completely overlapped with GS. The average overlap between Web of Science (WoS) and Scopus with GS was 73% and 63%, respectively, and WoS coincided with Scopus by two-thirds.

In terms of publisher (in Figure 6b) 53% of the articles were published in Journals by Elsevier (Figure 6c) between 2013 and 2022 and achieved a total citation of 954, 720, and 616 in GS, Scopus, and WoS, respectively. The average citation per paper, respectively, was 30, 23, and 19. MDPI's journals (Figure 6d) was second with 25% (published between 2017 and 2022) and achieved citation numbers of 184, 150, and 144 in GS, Scopus, and WoS, respectively, with an average of 12, 10, and 10 for each paper. This was followed by Springer with two articles. The remaining articles were distributed equally among ten publishers. In terms of journals, *Energies* was first with eight articles, mostly discussing energy efficiency planning and scenarios. In second, *Energy & Buildings* articles mainly discussed bottom-up retrofitting strategies. Next were *Energy for Sustainable Development*, *Energy Procedia*, and *Journal of Cleaner Production*, each with four articles, followed by *Buildings* and *Energy Policy* with three each.

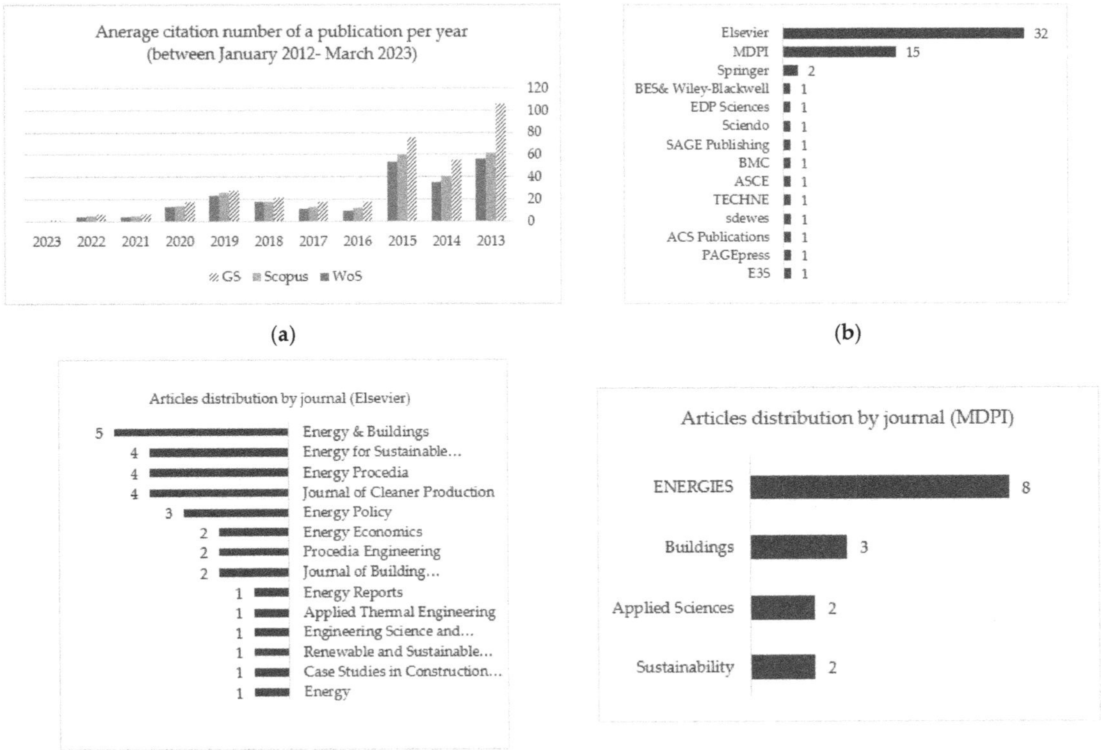

(a) (b) (c) (d)

Figure 6. (**a**) Annual average citation per article for each database; (**b**) distribution of publishers; (**c**) distribution of journals published by Elsevier; (**d**) distribution of journals published by MDPI.

3.2. Affiliations and Cross-Disciplinarity (Supplementary File S2)

In terms of the affiliations for the included articles, almost all the papers were implemented by academic bodies and research centers; only the authors of five papers (9%) were affiliated with governmental bodies [61,66,76,104,105]. Additionally, 9% of papers included authors from the private sector in the domains of technology, e2nergy, and energy economics [56,80,89,98,102]. It was apparent that there was an absence of local communities and social contributions to the publications, apart from a few papers, that engaged the owner/landlords in early decision making [70]. The institution that produced the most articles was the University of Perugia in Italy with four articles, followed by the Polish Academy for Science in Poland with three papers, 12 institutions which produced a few papers, and the remaining institutions published one article each. Figure 7 shows the top institutions by author affiliation.

To identify the primary domains that produced the included studies, we examined the affiliated departments and then collected, classified, and grouped them under 15 domains, as seen in Table 3. The top contributions were architecture and planning at 38% of the total, followed by applied and environmental sciences at about 22%, the energy domain at 15%, and building and construction management and economics at 12%.

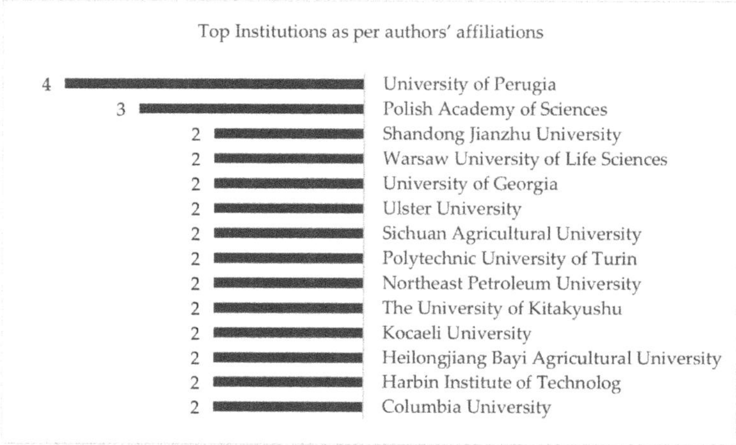

Figure 7. Most frequent institutions by author affiliation.

Table 3. Classification of the scientific domain of authors' departments extracted from the included papers (available in Supplementary Files S3 and S4).

Discipline	Frequency	The Description Extracted from the Department's Names
Architectural and urban/rural planning	23	Architecture, urban or rural planning technology for architecture, architecture and built environment, landscape architecture, and maritime architecture
Applied Sciences	14	Civil engineering, chemical engineering, and mechanical engineering
Environmental Sciences	12	Environmental engineering and hydropower engineering
Energy	9	Energy technology, energy efficiency economics, and energy
Economics	7	Finance
Building and Construction	7	Building physics, building and real state, and construction management
Agricultural science	5	Rural and agriculture development and agricultural and applied economics
Political science	3	Public and international affairs and environmental policies
Technology	3	Technology development
Biology	2	Biological and geoenvironmental technologies and zoology
Business and management	2	Management and economics, sustainable development, and smart decision making
Computer Science	2	
Art History	1	Art and history (conservation)
Psychology	1	
Mathematics	1	

The collaboration level in the included studies varied. Half of the articles were conducted by one discipline, predominated by architecture and planning. Meanwhile, 42% of the studies were produced by two domains, mainly the combination of architecture and urban/rural planning, followed by architecture and applied engineering (4 articles), architectural and environmental sciences (3 articles), and architecture and planning with building and construction (2 papers). Figure 8 maps out the cross-disciplinarity of the included papers.

The total number of keywords, as stated in the papers, was 297 (Figure 9), with an average of 5/article. As seen in the word cloud, the most frequently used keyword was energy efficiency, and in second was building energy efficiency, energy consumption, retrofit, and China. The third group included energy retrofit, rural residence, sustainability, and multi-objective optimization. The fourth group included cultural heritage, historical buildings, thermal insulation, rural dwellings, renewable energy, and economic analysis, which indicates an observed correlation between the SLR protocol and keywords.

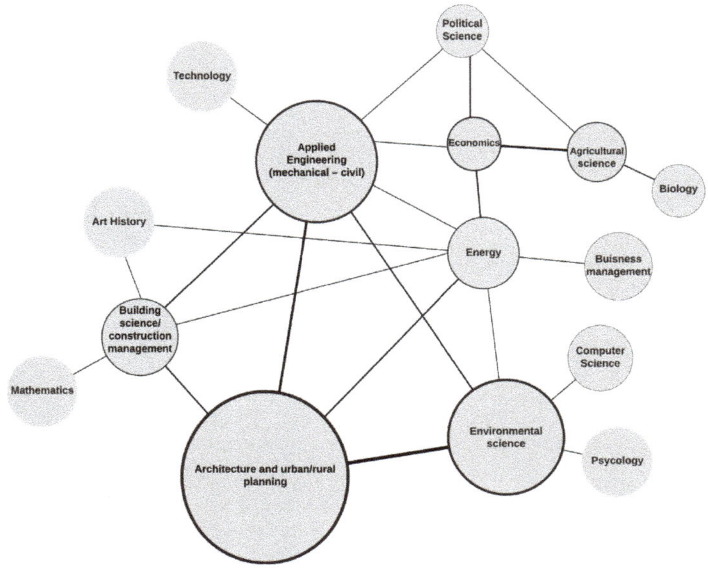

Figure 8. Visual map illustrating the connectivity and strength of collaboration among the different domains in the included studies (the frequencies of each discipline and the cross-disciplinary collaborations are available in Supplementary File S4).

Figure 9. A visual summary of the keywords provided by the included studies (available in Supplementary File S3), generated by https://www.jasondavies.com/wordcloud/, accessed on 15 May 2023.

3.3. Micro-Context Patterns

In terms of micro-context, all the articles have focused on only rural contexts, apart from a few papers that evaluated energy policies in generic contexts (urban and rural) that clearly indicated applications to rural contexts (Figure 10). Approximately 70% did not define a particular context or settlement pattern; 7% indicated agriculture-based settlements; 12% applied to historical and listed buildings in rural areas, except for protected rural buildings in a city [83] and a rural building in an archeological park in a city [68]; 4% were conducted in remote and abandoned villages in India [60], China [69], and Argentina [75];

4% mentioned the adaptive reuse of a touristic building in Italy and China [66,107]; 2% were implemented in urban areas, namely, the countryside in a city [71]; and one article was implemented in mountainous areas [69]. Figure 10 summarizes the pattern characteristics of the investigated micro-context.

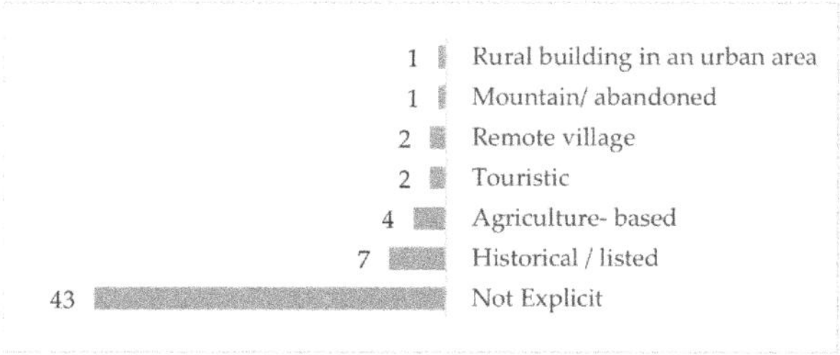

Figure 10. Characteristics and patterns of the examined context scope of the included articles.

3.4. *Analysis Scope and Theme Classification*

In terms of analysis scope, the articles have analyzed at least one or more aspects as the main aim of the paper (Table 4) which can be summarized as:

- One-factor analysis: Eighteen articles (30%) investigated one aspect, i.e., energy (energy efficiency and clean heating), and only one paper [67] examined retrofitting to attain thermal comfort in a traditional farmhouse building.
- Two-factor analysis: A total of 32 papers (53%) analyzed two aspects, namely, energy/cost (15 articles), energy/thermal comfort (6 articles), energy/carbon (4 articles), energy/valorization (2 articles), energy/human behavior (1 article), and carbon/cost (1 article).
- Three-factor analysis: Nine papers have provided multi-objective optimization, namely, energy–thermal comfort with valorization and air quality (three papers) and the others investigated energy, cost, and thermal comfort.
- Four-factor analysis: one paper discussed energy, cost, carbon, and air quality.

Energy analyses predominated in 97% of papers and overlapped with all factors apart from two articles that focused on retrofitting for achieving thermal comfort, vernacular dwellings for thermal comfort, and investigating carbon-associated energy saving and cost. Economic analyses (economic assessment, cost, and energy economics) came second with 40% and third was thermal comfort (17%). Only 7% clearly associated valorizing the traditional identity with the aims, and the same percentage considered indoor/outdoor air quality. Hydroponics farming was discussed as a retrofitting solution in one paper. Finally, the latest paper discussed an unusual approach led by biodiversity: investigating the impact of existing building retrofitting on birds. Table 4 summarizes the analyzed factors. Note that some papers examined different aspects as supportive elements to the main objective. For instance, reference [93] discussed users' perspectives on the implemented retrofitting, while the main goal was to quantify the potential energy saving.

The content review and analysis of the included studies led to their classification based on the central theme with each containing sub-themes. They can be classified into four categories: (1) providing energy efficiency strategies (57%), mainly quantifying potential energy savings of suggested retrofitting measures; (2) discussing energy efficiency planning (17%) to help guide the future for stakeholders via delivering energy efficiency and/or retrofitting guidelines for developing policies and supporting decision making [111]; (3) policy evaluation (18%) to assess the energy policies related to building retrofitting and reviewing national retrofitting schemes; and (4) discussing social and human behavioral aspects (8%), like the perception of the practices and willingness to retrofit.

Table 4. Analysis scope according to the aim of included studies.

Analyzed Factors	Frequent	Energy	Economics/Cost	Thermal Comfort	Carbon and/or GHG	Air Quality	Valorization	Human Behavior	Hydroponics Farming	Biodiversity
One	18	x								
	1			x						
Two	16	x	x							
	6	x		x						
	2	x					x			
	4	x			x					
	1	x							x	
	1	x								x
	1	x						x		
	1		x		x					
Three	2	x	x			x				
	2	x	x		x					
	1	x		x		x				
	1	x	x				x			
	1	x		x			x			
	1	x	x	x						
Four	1	x	x		x	x				
Total		58	24	10	8	4	4	1	1	1

3.5. Results Analysis and Synthesis

A quantitative analysis has been provided, showing the characteristics of the included articles; based on this and the four main classifications, a qualitative analysis was used to synthesize the results. In line with the study hypotheses, each pattern affects/is affected by the retrofitting strategy to add value to the area and diagnose the main implemented activities beyond the in-depth discussion of the technical solutions and numerical results.

3.5.1. Theme 1: Energy Efficiency Strategies

Generally, almost all research under this theme provided retrofitting solutions utilizing field and experimental methods (case studies) and simulation tools.

Historical Contexts

The majority of these studies were accomplished in Italy; Piselli, Guastaveglia et al., and Piselli, Romanelli et al. retrofitted the infrastructure of historical buildings in Italy (offshore and educational buildings) within a European Project (GEOFIT Horizon 2020), namely with geothermal systems for cooling and heating [79,80]. They promoted utilizing emerging technologies, namely building information modeling (BIM), as a human-centered operational management approach supporting energy and environmental performance. Meanwhile, technologies with low architectural impact can preserve the cultural heritage identity.

Cellura et al. [68] retrofitted a rural building in a park in an ancient village in Italy to optimize the original materials and thermal comfort and reach the net zero energy target (reducing cooling, heating, and lighting loads). For this last aspect, they integrated active and passive solutions. The management authority was engaged in the early stages of decision making. To preserve the original building configurations, they renovated building systems, altering windows, internal insulation, and the renewable energy source (RES); however, the type of RES can be excluded due to visual reasons (i.e., wind turbine). The retrofitting solutions were restricted to national guidelines and legalization.

In this light, Caputo et al. [83] stated that the guideline does not provide adequately efficient solutions. In other words, "heritage preservation and energy efficiency measures are often conceived as mutually exclusive purpose" [63], regardless of the associated challenges of refurbishing protected buildings regarding architectural constraints. Menconi et al. [67] retrofitted dwellings using internal insulation to preserve the original features of the building, that are considered an asset of the historical landscape [53]. Eventually, Alev et al. [52] provided an interdisciplinary study to analyze renovation alternatives to improve the energy performance of historic rural dwellings in three Baltic Sea countries. The study revealed that older rural buildings have poor insulation and airtightness, ⋯, which require considerable upgrades. Among the various structural components, external wall insulation had the most significant energy-saving potential due to its large area and high thermal transmittance. With this, it is feasible to meet energy performance standards while maintaining the building's original appearance.

Agricultural, Remote, Tourist, and Mountainous Contexts

Muthuvel et al. [60] relied on technological solutions to achieve zero-energy homes by retrofitting dwelling building systems (home appliances) related to the residents' agricultural activities in a remote village. It found that the control is as simple as the system. For the tourism contexts, Congedo et al. [66] studied an adaptively reused farmhouse in Italy, by employing local construction materials and traditional practices to improve energy/environmental performance. This approach boosted the local economy of the cultural, tourism, and construction sectors, all as a valorization strategy to preserve traditional identity. They demonstrated how the building has been affected by the surrounding agri-economic activities (olive groves and vineyards) that add value by supporting agriculture tourism. Simultaneously, they highlighted the importance of linking the interventions with the top-down processes supporting Cellura et al.'s outcomes.

Sun and Leng [58] provided multi-objective optimization of rural dwellings in ethnic minority areas; these areas had a low standard of living and, consequently, poor energy performance of buildings. However, there are possible environmental benefits like less air pollution, higher atmospheric transparency, and solar radiation. From the relationship between thermal comfort and energy, Yanru Li et al. provided guidelines for refurbishing rural buildings to attain optimum indoor environment enhancement for elderly people [100]. In the studies that focused on optimizing these factors, it is inevitable to adapt retrofitting measures regarding different thermal comfort factors for occupants, especially for improving infrastructure, like energy sources, as mentioned by Cardoso and González: "thermal comfort levels in residential buildings not suitable for the children to perform school".

For industrial buildings, Udovichenko et al. [90] developed a design framework for retrofitting a building with a hybrid renewable energy-assisted hydroponics farming system to produce fresh food in a harsh climate in rural Canada. For this building type, the lead constraint is environmental factors (e.g., humidity and water), requiring suitable environmental control; in other words, improving building systems is an essential retrofitting solution.

In mountainous areas in China, J. Han and Yang [69] quantified the energy savings due to a passive retrofitting solution without demonstrating the relationship to or influence of the context, similar to what was implemented by Yanru Li et al. [100], who indicated that the air conditioning load was the most significant proportion of the total construction load. Limited land affects the building footprint and retrofitting.

Cao et al. [92] proposed energy-saving retrofit assessment techniques for natural ventilation because they noted that the existing research on energy-saving retrofit is aimed at air-conditioned buildings and is not suitable for rural dwellings. The authors concur with this observation as many papers have made this same assumption in different contexts [84,93,94,106].

In the same context, the authors discovered that many studies have concentrated on assessing energy conservation through retrofitting techniques for building systems or envelopes in certain situations. They employed simulation tools to evaluate the effectiveness of these methods, which resulted in improved energy efficiency, among other benefits. However, the impact of implementing these interventions in rural contexts was not thoroughly explained, as referenced in references [57,58,74,78,84,88,89,94,96,99]. In other words, the authors argue that these practices can be implemented in any context, particularly the social aspects that have not been explored. Thus, different approaches should be presented. Some of these studies conducted on-site surveys to calibrate the simulation model to minimize the discrepancy between simulated energy use and observed data, such as in references [68,88].

Other studies observed the same behavior but explained the relationship with the surrounding environment or with the building morphology itself, like the role of ventilation as a passive cooling measure with great potential for sustainable renovation of rural buildings [65]. Rural buildings can benefit from less air pollution, more daylight, and integration with the natural environment, namely shading with trees and wall vegetation [102], especially in forest-based settlements. Selecting and combining the renewable energy source (i.e., biomass) with a fuelwood source available due to the surrounding context for heating and or cooking activities [75,85] or selecting specific heating sources like the ground-source brine-to-water heat pumps are only suitable for rural areas because of land requirements [97]. Other studies discussed the impact of building typologies on energy consumption, such as detached dwellings having the largest heat loss because of the high surface area to volume ratio or construction typology of cavity walls allowing for more cavity insulation retrofitting [57,63], decides emphasizing the negative correlation between building age and readiness to retrofit.

3.5.2. Theme 2: Energy Efficiency Planning

Tahsildoost and Zomorodian [77] provided multi-criteria decision-making and defined optimum retrofit strategies for rural buildings based on their economic viability for four buildings in different climate zones in Iran. They highlighted the importance of government-supported programs and incentive methods to educate and subsidize rural buildings, especially in low-income areas, to appeal to the building owners.

In a Chinese context, Qi et al. [72] provided a geometric classification method to facilitate energy performance evaluation and stated the necessity to analyze the potential for energy savings and the effects of retrofitting measures. Liu et al. [110] evaluated potential energy savings, costs, greenhouse gas emission reductions, and adoption strategies for improving building envelope efficiency in rural residential buildings. Replacing current fuel subsidies with retrofit subsidies is a win–win–win for rural households, local governments, and the environment, as it is a more efficient approach, supporting Siudek et al.'s finding, which stated that easy and moderate retrofits could effectively reduce the operating heating costs of clean heating.

Cardoso and González [75] evaluated the impact of altering the energy source in rural Argentina. The shift from using fuelwood alone to a combination of energy sources (liquid petroleum gas and fuelwood) positively impacted the quality of life. This transition was also a part of a social assistance program to support vulnerable sectors. However, many households continue to rely on wood stoves for their high caloric power and the sense of security they provide, as they are not dependent on external sources of energy. This traditional practice remains prevalent.

In a European context, Zeyen et al. [97] discussed cost-effective ways to decrease space heating demand peaks; they found that retrofitting solutions and energy efficiency measures should be applied to manage thermal peak demands rather than reducing energy demand. In Poland, Ksiezopolski et al. [82] investigated the correlations between the thermal modernization of a rural single-family house (SFH) by altering the energy source from coal to a clean energy source and examining its impact on household income. It was found that the process of changing supply and consumption patterns would enhance energy and environmental security. Siudek et al.'s [86] investigated the correlations between the thermal modernization of SFHs, the changes in energy sources to clean energy, and the disposable income of households in Polish rural areas. It was found that switching from coal to alternative energy sources for new house construction can significantly increase the cost of building, especially when installing renewable equipment for heating and water.

Byrne and O'Regan [59] provided an approach to delivering relevant and accessible information to stakeholders, showing how community efforts have progressed in renewable energy production. The rural residents who rely on energy sources that produce more pollution (e.g., wood and heating oil) have a higher carbon footprint than that of urban residents. In Germany, Taruttis and Weber [103] explored the correlation between energy efficiency and the market value of SFHs to uncover the potential financial gains associated with investing in energy retrofits. Surplus for enhanced energy efficiency can attain a price that is superior to that in urban contexts. Rosin et al. [81] promoted a novel approach to modernizing the building envelope to maintain energy efficiency and biodiversity of species highly dependent on building structures. They advocated that retrofitting measures should preserve ecological values and mitigate adverse impacts on biodiversity.

3.5.3. Theme 3: Policy Evaluation

In the UK and Ireland, Ahern et al. [51] evaluated the energy retrofitting national scheme and realized characterization of rural dwelling typologies, energy efficiency, and carbon emissions. They utilized a market-based approach that found individuals can enhance the energy efficiency of their house without any upfront financial burden or added stress of debt obligations. The cost of retrofitting external insulation varied based on building size, insulation depth, and market conditions. Government involvement may be needed for exterior wall insulation retrofitting schemes, as it poses a financial risk to utility companies. Walker et al. [52] examined the effectiveness of a national retrofitting scheme. They found that most retrofits are minor solutions that may not reduce fuel poverty, reflecting the importance of developing national schemes to tackle this. Collins and Curtis [61] evaluated the energy efficiency measurements of a national scheme, providing insights and recommendations for retrofitting strategies correlated with household characteristics.

The study of Žičkienė et al. [109] highlighted the significant political barriers in rural Lithuania as the local authorities are not directly involved in promoting RES usage, and there is insufficient support, exorbitant initial costs, and a lack of collaboration among stakeholders. In addition, they pointed out that the lack of adequate funding for research and innovation and sluggish academic research adversely affect the process. For these reasons, they engaged experts to provide policy implications, as they found that retrofitting public and residential buildings and energy infrastructure (e.g., energy-saving and controllable lighting systems) is more effective than switching energy generation from fossil fuels to RES as governments may turn a blind eye due to the high cost of RESs to achieve national energy security.

In Chinese contexts, references [54,56,71,76,105] have examined different approaches to assess building energy efficiency policies for retrofitting new and existing buildings and infrastructure to address challenges of clean energy reformation and energy efficiency. In general, in the study of He et al. in 2015, they stated that Chinese policies are usually more supportive of rural areas than urban ones [54]; in contrast, Huo et al. and T. Han et al. emphasized that the energy intensity of rural residential buildings has been doubled, because of national building energy efficiency projects.

Their findings indicated that rural policy development has largely been left behind at the expense of the welfare of significant household numbers. Meanwhile, the policies' implementation is too weak, and the applicable strategic plan remains unclear (despite the availability of provincial rural energy security guidelines [98]), especially with the difficulty of developing clean heating in more impoverished areas (with low-income residents). Thus, they emphasized the importance of adapting building energy efficiency policies and codes to consider the specific features of regions (especially for new buildings) like microclimate, house typology, ethnic characteristics, and ownership type. Moreover, there should be responsibility for financing and facilitating the market mechanisms managing long-term investments, and close coordination between various social departments to create awareness. Huo et al. [73] provided additional recommendations that the government should consider energy efficiency retrofitting plans according to floor space, and plan the construction period based on building stock (rural dwelling stock represents 40% of the total).

In Italy, Bosia and Savio [55] reviewed three regional cultural heritage guidelines; they showed that the guidelines provided an adequate diagnosis of the values and traditional features of the built environment to help the stakeholders better understand them. In addition, the policies focused on delivering local materials and traditional technologies, which is common in all Italian regions [66] (considering that Italian regulations exclude the listed buildings from the minimum energy requirements, even after retrofitting [68]); however, there is a lack of local materials and know-how. They showed the importance of having materials available and the building capacity, which require much effort due to the high population, which supports J. Li and Shui's findings [56].

3.5.4. Theme 4: Human Behavioral Aspects

Energy efficiency retrofit projects were investigated from the perspective of farmers and various stakeholders. In China, references [98,104,108] conducted three field studies in the households of several villages. The reasons that these rural Chinese residents were less enthusiastic and unwilling to retrofit can be summed up as the high initial cost, lack of appropriate fines and subsidies, sluggish promotion of the retrofitting program, and low local farmer acceptance; in addition, there may be a lack of knowledge about energy-saving technologies [71]. Meanwhile, the willingness to retrofit varied significantly due to various housing characteristics and building typologies. Meanwhile, engaging all stakeholders is essential; the intention to participate of the government, energy-saving service enterprises, and village residents directly affect the promotion of energy-saving renovations of existing buildings. Top-down support remains pivotal by affording appropriate penalties and aid to help rural residents mobilize their passion for contributing to energy-saving.

According to Lai et al. [108], the condition of the resident's housing may impact their willingness to retrofit. While safety concerns may encourage retrofitting, individuals living in well-maintained farmhouses or detached villas may hesitate to do so. These findings contradict previous studies [95,105] that suggest that low income limits individuals' investment in improving building performance, as viewed from both bottom-up and political perspectives. The authors would agree with the notion that there is a negative correlation between income and the decision to retrofit. Collins and Curtis stated that those living in a modern building are more willing to invest in retrofitting than those in older ones.

A study conducted by Barry et al. [62] examined the impact of self-efficacy on energy usage in 647 households in the USA. The study found that individual habits and perceptions of energy significantly affected consumption. Additionally, occupant behavior plays a crucial role in reducing energy usage. In a study conducted in Poland by Kaya et al. [87], a different approach was taken that linked education level with the willingness to invest in retrofitting. The study found that residents with higher education levels showed a minor interest in retrofitting, whereas farmers showed a significantly lower interest in retrofitting compared to other economic activities. Furthermore, the study observed a negative correlation between family size and willingness to retrofit. Lastly, residents who purchased energy-efficient home appliances had a significantly higher probability of retrofitting compared to those who opted for exterior insulation.

4. Discussion and Conclusions

Implementing energy retrofitting methods is crucial for attaining sustainable development objectives and reducing the impact of climate change. Mainly, when working in rural areas, it is essential to take into account their specific characteristics: cultural and social aspects, economic patterns, and infrastructure challenges. Conversely, the complexity of energy retrofitting practices (the associated internal and external factors) sets them apart from other domains. That can be grouped into two main aspects: technical issues and issues related to the stakeholders themselves. These require an intensive focus from various standpoints and cross-disciplinarity collaboration to generate the knowledge that can help to effectively tackle and mitigate these uncertainties.

In this view, this study attempted to investigate the academic progress and trends and diagnose the activities for retrofitting the built environment of rural settlements (RQ1) beyond the numerical and quantified results and technical details. In addition to explore the factors that can influence these practices and their impact on the rural area's traditional identity (RQ2) and (RQ3), which was addressed through a systematic literature review.

4.1. Summary of Main Findings

The SLR used the keywords retrofitting, rural, and built environment. Afterward, 16 scholarly databases were searched for peer-reviewed articles published between 2012 and March 2023. The study was conducted; consequently, a restricted protocol was developed (Table 1). Application of the inclusion criteria resulted in a total number of 60 eligible studies. The search and analysis were conducted between March and May 2023, utilizing various search, screening, and data mining tools. The data were extracted and content analysis was conducted in line with the research problem; these aspects included bibliographical data (e.g., author, citation, journal, and affiliation), cross-disciplinarity among the collaborators, the macro context of implementation, micro-context rural patterns, analysis scope of articles, and thematic classification and focus. Then, a synthesis of the results was provided to address three research questions (Section 1).

A framework for addressing areas of concern related to retrofitting rural built environments was developed by classifying papers in a publication pool. The articles were classified based on their publication year and journal. The analysis revealed that the years from 2020 to 2022 saw the highest number of published papers, with a total of 11 and 10, respectively (almost doubled in 2017). This might reflect the increased research interest in this domain or it might have been because of the lockdown.

The citation number positively correlates with the publishing time; the older the article, the higher the number of citations. Google Scholar had the highest citation number for each paper, one-third more than those of the Scopus and Web of Science databases. The top-cited paper was J. Li and Shui's study [56] in 2015, which was 63% more than the other papers in the same year. The most frequent number of authors was four (15 times), and the highest number of authors (ten) occurred in two papers [80,101]; the average number of authors was 5–6 authors per paper. The total number of keywords was 297; the most frequent keywords were energy efficiency building, energy efficiency, energy consumption, renewable energy, China, retrofit and multi-objective optimization, sustainability, and rural residences.

Among the journals, *Energies* published the most papers with eight articles. At the same time, Elsevier published the highest number of publications, with 32 papers. The highest number of contributor affiliations was from China, with 28 affiliations, followed by Italy with 10. The most common affiliation was the University of Perugia, Italy (four times), followed by the Polish Academy of Sciences, Poland. Twelve affiliations were listed in two papers, and the remaining affiliation were distributed equally. Consequently, the majority of implementations were conducted in China, with 30 articles, and Italy with 9 articles. In terms of settlement pattern, in Italy, the majority of cultural heritage and historical publications had been conducted by seven papers. These facts might provide insights into the high academic interest of Chinese scholars in regenerating rural China.

Many papers provided multi-objective optimization approaches. Techno-economic analyses predominated to reduce energy and assess the feasibility. This was followed by the technical interventions to optimize energy efficiency and attain thermal comfort or reduce the associated GHG emissions, demonstrating the importance of providing an economic analysis of any interventions and emphasizing the positive correlation with thermal comfort. It is evident that the architecture and planning domain predominated as the most frequent domain. At the same time, the highest connectivity was among architecture planning with both environmental and applied science, whereas the energy domain had the most interactions with other domains (Figure 8).

4.2. Addressing Research Questions

The research was classified into four main themes: (1) applying energy efficiency strategies by promoting retrofitting measures with the aim of energy conservation, using field methods (e.g., case studies, questionnaires, and on-site monitoring), and employing simulation tools; the majority of studies were conducted from a bottom-up approach. (2) Planning for energy efficiency including providing guidelines and scenario planning, utilizing methods such as multi-criteria decision-making, econometric, analytical, and field, in addition to different tools like assessment tools, remote sensing, and simulations. Bottom-up and mixed approaches characterized these studies. (3) Evaluating and examining the content of existing retrofitting schemes or clean energy policies. (4) The papers investigating the social and behavioral aspects that can affect the willingness to retrofit.

Almost all the studies confirmed the low thermal conditions of rural buildings, which can be interpreted as the farmers and residents typically constructing their own houses using rural traditions and habits, resulting in envelope structures that do not meet energy-saving building design standards (the contemporary buildings' typologies are not exempted). Simulation tools are one of the most effective assessment tools for exploring the energy conservation potential of different retrofitting scenarios for better decision-making and pre-renovation planning for existing buildings. However, most studies only focused on pre-retrofitting scenarios without accounting for calibrating the simulation results with in situ situations or providing a rational background behind applying specific retrofitting measures and the logical correlation with the intimate context; otherwise, it does not make sense to devote more retrofitting resources in this way.

Regarding this, a set of questions should be raised and investigated about the reliability of the obtained results; what if the same intervention is implemented in urban areas? In this case, it is enough to use simulation tools and alter weather conditions and locations. Therefore, the authors advocate for the essential need to provide more innovative/applicable solutions that cover techno-economic and socio-cultural aspects and develop these implementations into on-ground ones. In addition, there is a crucial need to monitor the impact of post-refurbishing. For example, when providing internal insulation to preserve the original features of an occupied dwelling, what is the acceptance rate of this solution, what is the time frame for implementation, and how will it affect the residents' daily life? Another question is, while providing policy implications like proposing retrofitting energy sources, to what extent will these recommendations be applied in line with the decision makers' priorities.

This can occur through collaborations between academic bodies, the private sector, experts, and practitioners, to obtain insightful findings and widen the perspective of the results. Conversely, how to engage the decision maker and the local community, which is a real challenge and ambitious goal, should be adequately investigated in order to align the various stakeholders on the same page and as active contributors. Hence, a vital question could be asked before any intervention: who are the right stakeholders, and what are the academic capabilities and qualifications of the research team to lead the retrofitting process? In particular, the high support by scientific and academic bodies and the essential role of energy efficiency scholars as knowledge brokers is needed to solve real-world problems. In the same context, the researchers should be open to inter- and trans-disciplinary collaboration.

In the same domain, it is evident that there is an absence of locals' engagement, apart from engaging the landlord in early decision-making. Hence, the residents should participate in retrofitting processes (at least be kept informed). They should be aware of the benefit of conserving energy and educated about energy efficiency through awareness campaigns via on-site visits, the media, or even social media, with particular reference to adjusting behavior for energy efficiency to reduce consumption patterns. In addition, the economic aspect makes them reluctant to invest in retrofit so we can alternatively show them how to apply simple and affordable techniques like altering home appliances to energy-efficient ones. Considering that building capacities are more complex in densely populated areas [55,56] and vice versa, these approaches are more effective in sparsely populated ones, like remote areas.

In general, policies vary by the different contexts, but the common finding was the vital role of decision makers and top-down development in fulfilling building energy efficiency. Energy code regulations for sustainable buildings should be updated with more environmental performance indicators concerning micro-context specificities (e.g., microclimate, economic activities, and building morphologies), providing long-term energy and carbon emission savings goals and an appropriate regulatory environment. The national guidelines should be considered during any bottom-up intervention and should to linked to the findings.

The decision making might not be energy-efficiency-oriented because of the national priorities that may focus on improving residents' quality of life by renovating the deteriorated rural dwellings and high poverty rates, as discussed in reference [112], or concentrating on the dwellings' visual appearance [78]. This does not preclude local authorities, for example, from employing cost-effective building retrofitting solutions rather than changing conventional energy sources, as mentioned in the Lithuanian case. Meanwhile, enabling the societal debate on domestic energy policies can support the whole scene. Promoting building energy efficiency in rural areas is a daunting challenge that requires financing and training in building industry skills [56].

The micro-context provides constraints to retrofitting implementations; the visual aspects and original materials are essential when dealing with culturally significant buildings or historical rural dwellings. It is crucial to prioritize preserving their heritage value and exploring the surrounding activities and operations like management lighting systems/natural ventilation [67,79]. Improving building systems can offer practical solutions to address preserving historical value or replace the less visible envelope elements. This requires more investigations using emerging technologies. The external landscape is redesigned to respect internal activities and vice versa in enhancing the local identity. The authors argue that effectively engaging the locals to retrofit their built environment and putting them in the decision-making position can enhance the sense of belonging, promoting another aspect of settlement valorization.

Additionally, micro-contexts correlated with energy consumption patterns, such as the living habits of rural residents [54,62]. Different building shapes and orientations affect the context [71,72]. The type of renewable energy can be determined, such as relying on passive solar techniques with high solar irradiance, or likewise using wooden biomass in forest settlements. Clearly, the location affects or should affect the retrofitting implementations, supporting the study's hypothesis; this is, despite the obscurity of this part in many of the included articles, such as what has been discussed in Section 3.5.1.

To conclude, this research contributes to a better understanding of retrofitting practices in specific rural contexts beyond the numerical results of energy savings. It provided insights to interested parties, from broader perspectives for scholars who work in energy efficiency and sustainable rural development, in addition to local authorities, especially those dedicated to energy savings or historical heritage, to promote integrated retrofitting within energy communities' key concept in rural commons [7,113,114]. Moreover, the study provided a method for conducting SLRs, employing different tools to streamline the process that may be beneficial for early-stage researchers on a given issue. This review also offers crucial systematic details such as the authors' country and affiliations, data collection methods, methods, tools, analysis scope, journals and publishers, and citation numbers (presented in the Supplementary Editable Excel Files S2–S4). This information is essential for future researchers who work in retrofitting rural areas and serves as a valuable reference that could be developed further; therefore, scholars are highly encouraged to investigate other elements and provide additional insights and trajectories to regenerate the rural heritage. Contemporaneously, further investigations into the national action plans, providing innovative retrofitting solutions and methods to involve the right stakeholder wills be beneficial.

However, despite the SLR being based on a quantitative and qualitative examination, the authors declare that the restricted SLR protocol constitutes a limitation of the finding. The search terms focused on "retrofitting rural built environment," and altering it may affect the choice of the articles to be included in the research; this indicates that some articles were not included as they do not have the term "rural" in the title, abstract, keywords, or text, like what has been implemented in references [34,115]. Consequently, analogous studies can be accomplished utilizing a broader search strategy and considering different analysis scopes, which might lead to the finding of new themes.

Supplementary Materials: The following supporting information can be downloaded at: https://www.mdpi.com/article/10.3390/buildings13071586/s1, File S1: The exported RIS file that includes the 168 papers before applying the screening criteria. S2, S3, and S4 are presented in the same Excel file, where S2 is the analysis of the included SLR studies in line with the RQs, and S3 presents the analysis of authors and affiliations. S4 presents the supported table for analysis.

Author Contributions: Conceptualization, A.A.; methodology, A.A. and A.B.; software, A.A.; investigation, A.A.; resources, A.A.; writing—original draft preparation, A.A.; writing—review and editing, A.A. and A.B.; visualization, A.A.; supervision, A.B. All authors have read and agreed to the published version of the manuscript.

Funding: This research received no external funding.

Data Availability Statement: The data presented in this study are openly available in the Supplementary Files.

Conflicts of Interest: The authors declare no conflict of interest.

References

1. May, J.; Reid, A. *Architettura Senza Architetti [Architecture without Architects. Guide to Spontaneous Constructions from around the World]*; Rizzoli: New York, NY, USA, 2010; ISBN 8817037850.
2. Margione, E. Italians New towns as an experimental territory for the modern movement in Italy. The case study of Oriolo Frezzotti and his architecture for public facilities in Littoria, Sabaudia and Pontinia. In Proceedings of the Regionalism, Nationalism & Modern Architecture, Porto, Portugal, 25–27 October 2018; Pimentel, J.C., Trevisan, A., Cardoso, A., Eds.; CEAA | Centro de Estudos Arnaldo Araújo and Escola Superior Artistica do Porto: Porto, Portogal, 2018; pp. 202–220.
3. Rapoport, A. Environmental quality, metropolitan areas and traditional settlements. *Habitat Int.* **1983**, *7*, 37–63. [CrossRef]
4. Torreggiani, D.; Tassinari, P. Landscape quality of farm buildings: The evolution of the design approach in Italy. *J. Cult. Herit.* **2012**, *13*, 59–68. [CrossRef]
5. Al-Din, S.S.M. The influence of Mediterranean modernist movement of architecture in Lefkosa: The first and early second half of 20th century. *J. Contemp. Urban Aff.* **2017**, *1*, 10–23. [CrossRef]
6. Zeghlache, H.; Alıkhodja, N. Retrofitting Assessment of Berber Dwelling: Case of Setif, Algeria. *Open House Int.* **2017**, *42*, 108–116. [CrossRef]
7. Abouaiana, A.; Battisti, A. Multifunction Land Use to Promote Energy Communities in Mediterranean Region: Cases of Egypt and Italy. *Land* **2022**, *11*, 673. [CrossRef]
8. Delgado-Artés, R.; Garófano-Gómez, V.; Oliver-Villanueva, J.-V.; Rojas-Briales, E. Land use/cover change analysis in the Mediterranean region: A regional case study of forest evolution in Castelló (Spain) over 50 years. *Land Use Policy* **2022**, *114*, 105967. [CrossRef]
9. Cherif, S.; Doblas-Miranda, E.; Lionello, P.; Borrego, C.; Giorgi, F.; Iglesias, A.; Jebari, S.; Mahmoudi, E.; Moriondo, M.; Zittis, G.; et al. Drivers of change. In *Climate and Environmental Change in the Mediterranean Basin—Current Situation and Risks for the Future*; First Mediterranean Assessment ReportFuture, First Mediterranean Assessment Report, Guiot, J., Cramer, W., Marini, K., Eds.; Union for the Mediterranean: Barcelona, Spain; Plan Bleu: Valbonne, France; UNEP/MAP: Marseille, France, 2020; pp. 59–180.
10. Eurostat Urban and Rural Living in the EU. Available online: https://ec.europa.eu/eurostat/web/products-eurostat-news/-/EDN-20200207-1 (accessed on 10 February 2022).
11. Łukasiewicz, K.; Pietrzak, P.; Kraciuk, J.; Kacperska, E.; Cieciora, M. Sustainable Energy Development—A Systematic Literature Review. *Energies* **2022**, *15*, 8284. [CrossRef]
12. Gkoltsiou, A.; Athanasiadou, E.; Paraskevopoulou, A.T. Agricultural Heritage Landscapes of Greece: Three Case Studies and Strategic Steps towards Their Acknowledgement, Conservation and Management. *Sustainability* **2021**, *13*, 5955. [CrossRef]
13. Hunkin, S.; Krell, K. *Approaches for a Rural Lowcarbon Economy*; Interrreg Europe: Lille, France, 2022.
14. Cyrek, M.; Cyrek, P. Rural Specificity as a Factor Influencing Energy Poverty in European Union Countries. *Energies* **2022**, *15*, 5463. [CrossRef]
15. IEA Rural Clean Energy Initiative & Support to Rural Energy Transition. Available online: https://www.iea.org/policies/16609-rural-clean-energy-initiative-support-to-rural-energy-transition?s=1 (accessed on 4 May 2023).
16. Asarpota, K.; Nadin, V. Energy strategies, the Urban dimension, and spatial planning. *Energies* **2020**, *13*, 3642. [CrossRef]
17. IEA. *The Critical Role of Buildings Perspectives for the Clean Energy Transition*; IEA: Paris, France, 2019.
18. Cucchiella, F.; Rotilio, M. Planning and prioritizing of energy retrofits for the cities of the future. *Cities* **2021**, *116*, 103272. [CrossRef]
19. Liang, X.; Shen, G.Q.; Guo, L. Improving management of green retrofits from a stakeholder perspective: A case study in China. *Int. J. Environ. Res. Public Health* **2015**, *12*, 13823–13842. [CrossRef] [PubMed]
20. Gajić, D.; Peulić, S.; Mavrič, T.; Sandak, A.; Tavzes, Č.; Malešević, M.; Slijepčević, M. Energy Retrofitting Opportunities Using Renewable Materials—Comparative Analysis of the Current Frameworks in Bosnia-Herzegovina and Slovenia. *Sustainability* **2021**, *13*, 603. [CrossRef]
21. Jia, L.; Qian, Q.K.; Meijer, F.; Visscher, H. Exploring key risks of energy retrofit of residential buildings in China with transaction cost considerations. *J. Clean. Prod.* **2021**, *293*, 126099. [CrossRef]
22. Fox, C. *The Engineering and Management of Retrofit Projects in the Process Industries*; European Construction Institute (ECI): Watford, UK; Loughborough University: Loughborough, Leicestershire, 2003; ISBN 1873844549.

23. Murto, P.; Jalas, M.; Juntunen, J.; Hyysalo, S. Devices and strategies: An analysis of managing complexity in energy retrofit projects. *Renew. Sustain. Energy Rev.* **2019**, *114*, 109294. [CrossRef]
24. de la Cruz-Lovera, C.; Perea-Moreno, A.-J.; de la Cruz-Fernández, J.L.; Montoya, F.G.; Alcayde, A.; Manzano-Agugliaro, F. Analysis of Research Topics and Scientific Collaborations in Energy Saving Using Bibliometric Techniques and Community Detection. *Energies* **2019**, *12*, 2030. [CrossRef]
25. Prabatha, T.; Hewage, K.; Karunathilake, H.; Sadiq, R. To retrofit or not? Making energy retrofit decisions through life cycle thinking for Canadian residences. *Energy Build.* **2020**, *226*, 110393.
26. Palumbo Fernández, M.; Bosch González, M.; Fernández Renna, A.I.; Simó Solsona, M. Why it's so hard? Exploring social barriers for the deployment of thermal energy storage in spanish buildings. *Energy Res. Soc. Sci.* **2021**, *76*, 102057.
27. Butt, B.; Jones, R.V.; Fuertes, A. Opportunities and barriers to business engagement in the UK domestic retrofit sector: An industry perspective. *Build. Serv. Eng. Res. Technol.* **2021**, *42*, 293–305. [CrossRef]
28. Sibilla, M.; Kurul, E. Transdisciplinarity in energy retrofit. A Conceptual Framework. *J. Clean. Prod.* **2020**, *250*, 119461. [CrossRef]
29. OECD. Addressing societal challenges using transdisciplinary research. *OECD Sci. Technol. Ind. Policy Pap.* **2020**. [CrossRef]
30. Schary, D.P.; Cardinal, B.J. Interdisciplinary and intradisciplinary research and teaching in kinesiology: Continuing the conversation. *Quest* **2015**, *67*, 173–184. [CrossRef]
31. Lawrence, R.J. Deciphering interdisciplinary and transdisciplinary contributions. *Transdiscipl. J. Eng. Sci.* **2010**, *1*, 111–116. [CrossRef]
32. Mormont, M. Rural nature and urban natures. *Sociol. Rural.* **1987**, *27*, 3–20. [CrossRef]
33. Belanche, D.; Casaló, L.V.; Rubio, M.Á. Local place identity: A comparison between residents of rural and urban communities. *J. Rural Stud.* **2021**, *82*, 242–252. [CrossRef]
34. Abouaiana, A.; Mendonça, P. Retrofitting Dwellings in Traditional Coastal Settlements in Egypt and Portugal Using Nature-Based Solutions and Conventional Thermal Insulation Materials: Technical and Economic Assessment. *J. Archit. Eng.* **2022**, *28*, 05022005. [CrossRef]
35. Conticelli, E.; De Luca, C.; Egusquiza, A.; Santangelo, A.; Tondelli, S. Inclusion of migrants for rural regeneration through cultural and natural heritage valorization. In *Planning Natureand Ecosystem Services*; FedOAPress: Naples, Italy, 2019.
36. Gebrati, F. Rural innovation and the valorization of local resources in the High Atlas of Marrakesh. In *Rural Transformations*; Routledge: Oxfordshire, UK, 2022; pp. 186–199, ISBN 1003110096.
37. Battisti, A. *Resilience of Inner Areas: Regeneration and Enhancement Strategies in Small Towns*; Technische Universität München Fakultät für Architektur: München, Germany, 2020; ISBN 978-3-948278-20-5.
38. *Naturopa European Rural Heritage*; Ballester, J.-M. (Ed.) No. 95; Council of Europe: Strasbourg, France, 2001. Available online: https://rm.coe.int/090000168093e8b0 (accessed on 28 May 2023).
39. Munn, Z.; Peters, M.D.J.; Stern, C.; Tufanaru, C.; McArthur, A.; Aromataris, E. Systematic review or scoping review? Guidance for authors when choosing between a systematic or scoping review approach. *BMC Med. Res. Methodol.* **2018**, *18*, 143.
40. Grant, M.J.; Booth, A. A typology of reviews: An analysis of 14 review types and associated methodologies. *Health Inf. Libr. J.* **2009**, *26*, 91–108. [CrossRef]
41. Jesson, J.; Matheson, L.; Lacey, F.M. *Doing Your Literature Review: Traditional and Systematic Techniques*; SAGE Publications Ltd.: Thousand Oaks, CA, USA, 2011.
42. Cook, D.J.; Mulrow, C.D.; Haynes, R.B. Systematic reviews: Synthesis of best evidence for clinical decisions. *Ann. Intern. Med.* **1997**, *126*, 376–380. [CrossRef]
43. Page, M.J.; McKenzie, J.E.; Bossuyt, P.M.; Boutron, I.; Hoffmann, T.C.; Mulrow, C.D.; Shamseer, L.; Tetzlaff, J.M.; Akl, E.A.; Brennan, S.E.; et al. The PRISMA 2020 statement: An updated guideline for reporting systematic reviews. *BMJ* **2021**, *372*, n71. [CrossRef]
44. Liberati, A.; Altman, D.G.; Tetzlaff, J.; Mulrow, C.; Gøtzsche, P.C.; Ioannidis, J.P.A.; Clarke, M.; Devereaux, P.J.; Kleijnen, J.; Moher, D. The PRISMA statement for reporting systematic reviews and meta-analyses of studies that evaluate health care interventions: Explanation and elaboration. *J. Clin. Epidemiol.* **2009**, *62*, e1–e34. [CrossRef]
45. Snyder, H. Literature review as a research methodology: An overview and guidelines. *J. Bus. Res.* **2019**, *104*, 333–339. [CrossRef]
46. Muka, T.; Glisic, M.; Milic, J.; Verhoog, S.; Bohlius, J.; Bramer, W.; Chowdhury, R.; Franco, O.H. A 24-step guide on how to design, conduct, and successfully publish a systematic review and meta-analysis in medical research. *Eur. J. Epidemiol.* **2020**, *35*, 49–60. [CrossRef] [PubMed]
47. Iralde, N.S.I.; Pascual, J.; Salom, J. Energy retrofit of residential building clusters. A literature review of crossover recommended measures, policies instruments and allocated funds in Spain. *Energy Build.* **2021**, *252*, 111409. [CrossRef]
48. Xiao, Y.; Watson, M. Guidance on Conducting a Systematic Literature Review. *J. Plan. Educ. Res.* **2019**, *39*, 93–112. [CrossRef]
49. Scott, A.M.; Forbes, C.; Clark, J.; Carter, M.; Glasziou, P.; Munn, Z. Systematic review automation tools improve efficiency but lack of knowledge impedes their adoption: A survey. *J. Clin. Epidemiol.* **2021**, *138*, 80–94. [CrossRef]
50. Ouzzani, M.; Hammady, H.; Fedorowicz, Z.; Elmagarmid, A. Rayyan—A web and mobile app for systematic reviews. *Syst. Rev.* **2016**, *5*, 1–10. [CrossRef]
51. Ahern, C.; Griffiths, P.; O'Flaherty, M. State of the Irish housing stock—Modelling the heat losses of Ireland's existing detached rural housing stock & estimating the benefit of thermal retrofit measures on this stock. *Energy Policy* **2013**, *55*, 139–151.

52. Walker, R.; Liddell, C.; McKenzie, P.; Morris, C. Evaluating fuel poverty policy in Northern Ireland using a geographic approach. *Energy Policy* **2013**, *63*, 765–774. [CrossRef]
53. Alev, Ü.; Eskola, L.; Arumägi, E.; Jokisalo, J.; Donarelli, A.; Siren, K.; Broström, T.; Kalamees, T. Renovation alternatives to improve energy performance of historic rural houses in the Baltic Sea region. *Energy Build.* **2014**, *77*, 58–66. [CrossRef]
54. He, B.; Yang, L.; Ye, M.; Mou, B.; Zhou, Y. Overview of rural building energy efficiency in China. *Energy Policy* **2014**, *69*, 385–396. [CrossRef]
55. Bosia, D.; Savio, L. Azioni e strumenti per il recupero e la valorizzazione dell'architettura e del paesaggio rurale e montano = Actions and tools for the conservation and valorisation of rural architecture and landscape. *Techne J. Technol. Archit. Environ.* **2014**, *7*, 87–93.
56. Li, J.; Shui, B. A comprehensive analysis of building energy efficiency policies in China: Status quo and development perspective. *J. Clean. Prod.* **2015**, *90*, 326–344. [CrossRef]
57. Yu, S.; Cui, Y.; Xu, X.; Feng, G. Impact of Civil Envelope on Energy Consumption based on EnergyPlus. *Procedia Eng.* **2015**, *121*, 1528–1534. [CrossRef]
58. Sun, H.; Leng, M. Analysis on building energy performance of Tibetan traditional dwelling in cold rural area of Gannan. *Energy Build.* **2015**, *96*, 251–260. [CrossRef]
59. Byrne, S.; O'Regan, B. Material flow accounting for an Irish rural community engaged in energy efficiency and renewable energy generation. *J. Clean. Prod.* **2016**, *127*, 363–373. [CrossRef]
60. Muthuvel, P.; Daniel, S.A.; Yazhini, D.G. Retrofitting domestic appliances for PV powered DC Nano-grid and its impact on net zero energy homes in rural India. *Eng. Sci. Technol. Int. J.* **2016**, *19*, 1836–1844. [CrossRef]
61. Collins, M.; Curtis, J. An examination of energy efficiency retrofit depth in Ireland. *Energy Build.* **2016**, *127*, 170–182. [CrossRef]
62. Barry, N.A.; Harper, C.M.; Berryman, C.; Farley, C. Role of Self-Efficacy in Reducing Residential Energy Usage. *J. Archit. Eng.* **2016**, *22*, 1–8. [CrossRef]
63. Cristina, B.; Paolo, C.S.; Spigliantini, G. Evaluation of refurbishment alternatives for an Italian vernacular building considering architectural heritage, energy efficiency and costs. *Energy Procedia* **2017**, *133*, 401–411. [CrossRef]
64. Liang, X.; Wang, Y.; Zhang, Y.; Jiang, J.; Chen, H.; Zhang, X.; Guo, H.; Roskilly, T. Analysis and Optimization on Energy Performance of a Rural House in Northern China Using Passive Retrofitting. *Energy Procedia* **2017**, *105*, 3023–3030. [CrossRef]
65. Ran, J.; Tang, M.; Jiang, L.; Zheng, X. Effect of Building Roof Insulation Measures on Indoor Cooling and Energy Saving in Rural Areas in Chongqing. *Procedia Eng.* **2017**, *180*, 669–675. [CrossRef]
66. Congedo, P.M.; Baglivo, C.; Zacà, I.; D'Agostino, D.; Quarta, F.; Cannoletta, A.; Marti, A.; Ostuni, V. Energy retrofit and environmental sustainability improvement of a historical farmhouse in Southern Italy. *Energy Procedia* **2017**, *133*, 367–381. [CrossRef]
67. Menconi, M.E.; Chiappini, M.; Hensen, J.L.M.; Grohmann, D. Thermal comfort optimisation of vernacular rural buildings: Passive solutions to retrofit a typical farmhouse in central Italy. *J. Agric. Eng.* **2017**, *48*, 127–136. [CrossRef]
68. Cellura, M.; Ciulla, G.; Guarino, F.; Longo, S. Redesign of a rural building in a heritage site in Italy: Towards the net zero energy target. *Buildings* **2017**, *7*, 68. [CrossRef]
69. Han, J.; Yang, X. Analysis of Passive Energy-saving Retrofitting of Rural Residential Houses in Southern Anhui Province—A case in Hongcun. *Energy Procedia* **2018**, *152*, 470–474. [CrossRef]
70. Rocchi, L.; Menconi, M.E.; Grohmann, D.; Paolotti, L.; Boggia, A.; Kadziński, M.; Miebs, G. Sustainability evaluation of retrofitting solutions for rural buildings through life cycle approach and multi-criteria analysis. *Energy Build.* **2018**, *173*, 281–290. [CrossRef]
71. Zhang, L.; Guo, S.; Wu, Z.; Alsaedi, A.; Hayat, T. SWOT Analysis for the Promotion of Energy Efficiency in Rural Buildings: A Case Study of China. *Energies* **2018**, *11*, 851. [CrossRef]
72. Qi, F.; Musonda, B.M.; Shen, H.; Wang, Y. Geometric classification method of rural residences at regional scale. *Energy Build.* **2018**, *172*, 170–180. [CrossRef]
73. Huo, T.; Cai, W.; Ren, H.; Feng, W.; Zhu, M.; Lang, N.; Gao, J. China's building stock estimation and energy intensity analysis. *J. Clean. Prod.* **2019**, *207*, 801–813. [CrossRef]
74. Liu, L.; Tong, H.; Li, H. An exploration aboutthe Solar Energy Utilization and the Enclosure SystemRenovation for Rural Residential Buildings in Cold Areas of Northern China—Taking the rural residential renovation design in Zhujialin Village, Linyi as an example. *E3S Web Conf.* **2019**, *136*, 2031. [CrossRef]
75. Cardoso, M.B.; González, A.D. Residential energy transition and thermal efficiency in an arid environment of northeast Patagonia, Argentina. *Energy Sustain. Dev.* **2019**, *50*, 82–90. [CrossRef]
76. Wang, J.; Zhou, Z.; Zhao, J.; Zheng, J.; Guan, Z. Towards a cleaner domestic heating sector in China: Current situations, implementation strategies, and supporting measures. *Appl. Therm. Eng.* **2019**, *152*, 515–531. [CrossRef]
77. Tahsildoost, M.; Zomorodian, Z. Energy, carbon, and cost analysis of rural housing retrofit in different climates. *J. Build. Eng.* **2020**, *30*, 101277. [CrossRef]
78. Cui, Y.; Sun, N.; Cai, H.; Li, S. Indoor temperature improvement and energy-saving renovations in rural houses of China's cold region—A case study of Shandong Province. *Energies* **2020**, *13*, 870. [CrossRef]
79. Piselli, C.; Guastaveglia, A.; Romanelli, J.; Cotana, F.; Pisello, A.L. Facility Energy Management Application of HBIM for Historical Low-Carbon Communities: Design, Modelling and Operation Control of Geothermal Energy Retrofit in a Real Italian Case Study. *Energies* **2020**, *13*, 6338. [CrossRef]

80. Piselli, C.; Romanelli, J.; Di Grazia, M.; Gavagni, A.; Moretti, E.; Nicolini, A.; Cotana, F.; Strangis, F.; Witte, H.J.L.; Pisello, A.L. An Integrated HBIM Simulation Approach for Energy Retrofit of Historical Buildings Implemented in a Case Study of a Medieval Fortress in Italy. *Energies* **2020**, *13*, 2601. [CrossRef]
81. Rosin, Z.M.; Hiron, M.; Żmihorski, M.; Szymański, P.; Tobolka, M.; Pärt, T.; McKenzie, A. Reduced biodiversity in modernized villages: A conflict between sustainable development goals. *J. Appl. Ecol.* **2020**, *57*, 467–475. [CrossRef]
82. Ksiezopolski, K.; Drygas, M.; Proninska, K.; Nurzynska, I. The Economic Effects of New Patterns of Energy Efficiency and Heat Sources in Rural Single-Family Houses in Poland. *Energies* **2020**, *13*, 6358. [CrossRef]
83. Caputo, P.; Ferrari, S.; Ferla, G.; Zagarella, F. Preliminary Energy Evaluations for the Retrofit of Rural Protected Buildings in a Peripheral Context of Milan. *J. Sustain. Dev. Energy Water Environ. Syst.* **2020**, *8*, 715–734. [CrossRef]
84. Campos, J.; Csontos, C.; Harmat, Á.; Csüllög, G.; Munkácsy, B. Heat consumption scenarios in the rural residential sector: The potential of heat pump-based demand-side management for sustainable heating. *Energy. Sustain. Soc.* **2020**, *10*, 40. [CrossRef]
85. Li, Y.; Yuan, X.; Tang, Y.; Wang, Q.; Ma, Q.; Mu, R.; Fu, J.; Hong, J.; Kellett, J.; Zuo, J. Integrated assessment of the environmental and economic effects of "coal-to-gas conversion" project in rural areas of northern China. *Environ. Sci. Pollut. Res.* **2020**, *27*, 14503–14514. [CrossRef] [PubMed]
86. Siudek, A.; Klepacka, A.M.; Florkowski, W.J.; Gradziuk, P. Renewable Energy Utilization in Rural Residential Housing: Economic and Environmental Facets. *Energies* **2020**, *13*, 6637. [CrossRef]
87. Kaya, O.; Klepacka, A.M.; Florkowski, W.J. The role of personal and environmental factors in rural homeowner decision to insulate; an example from Poland. *Renew. Sustain. Energy Rev.* **2021**, *150*, 111474. [CrossRef]
88. Hu, X.; Xiang, Y.; Zhang, H.; Lin, Q.; Wang, W.; Wang, H. Active–passive combined energy-efficient retrofit of rural residence with non-benchmarked construction: A case study in Shandong province, China. *Energy Rep.* **2021**, *7*, 1360–1373. [CrossRef]
89. Qiangnian, L.; Tongze, H.; Changlin, N.; Ping, L. Life Cycle Carbon Emission Analyzing of Rural Residential Energy Efficiency Retrofit-A Case Study of Gansu province. *E3S Web Conf.* **2021**, *329*, 1063. [CrossRef]
90. Udovichenko, A.; Fleck, B.A.; Weis, T.; Zhong, L. Framework for design and optimization of a retrofitted light industrial space with a renewable energy-assisted hydroponics facility in a rural northern canadian community. *J. Build. Eng.* **2021**, *37*, 102160. [CrossRef]
91. Struhala, K.; Ostry, M. Life-Cycle Assessment of a Rural Terraced House: A Struggle with Sustainability of Building Renovations. *Energies* **2021**, *14*, 2472. [CrossRef]
92. Cao, W.; Yang, L.; Zhang, Q.; Chen, L.; Wu, W. Evaluation of Rural Dwellings' Energy-Saving Retrofit with Adaptive Thermal Comfort Theory. *Sustainability* **2021**, *13*, 5350. [CrossRef]
93. Abouaiana, A. Retrofitting Rural Dwellings in Delta Region to Enhance Climate Change Mitigation in Egypt. *Environ. Clim. Technol.* **2021**, *25*, 136–150. [CrossRef]
94. Nie, J.; Pang, Y.; Wang, C.; Zhang, H.; Yin, K. Theoretical Study on the Relationship of Building Thermal Insulation with Indoor Thermal Comfort Based on APMV Index and Energy Consumption of Rural Residential Buildings. *Appl. Sci.* **2021**, *11*, 8565. [CrossRef]
95. Shao, T.; Zheng, W.; Cheng, Z. Passive Energy-Saving Optimal Design for Rural Residences of Hanzhong Region in Northwest China Based on Performance Simulation and Optimization Algorithm. *Buildings* **2021**, *11*, 421. [CrossRef]
96. Wang, J.; Gao, W.; Wang, Z.; Zhang, L. Analysis of Energy Performance and Integrated Optimization of Tubular Houses in Southern China Using Computational Simulation. *Appl. Sci.* **2021**, *11*, 9371. [CrossRef]
97. Zeyen, E.; Hagenmeyer, V.; Brown, T. Mitigating heat demand peaks in buildings in a highly renewable European energy system. *Energy* **2021**, *231*, 120784. [CrossRef]
98. Han, T.; Liu, P.; Niu, C.; Li, Q. Evaluation of energy-saving retrofit projects of existing rural residential envelope structures from the perspective of rural residents: The Chinese case. *Environ. Dev. Sustain. A Multidiscip. Approach Theory Pract. Sustain. Dev.* **2022**, *39*, 1–28. [CrossRef]
99. Jiang, W.; Ju, Z.; Tian, H.; Liu, Y.; Arıcı, M.; Tang, X.; Li, Q.; Li, D.; Qi, H. Net-zero energy retrofit of rural house in severe cold region based on passive insulation and BAPV technology. *J. Clean. Prod.* **2022**, *360*, 132198. [CrossRef]
100. Li, Y.; Zhou, T.; Wang, Z.; Li, W.; Zhou, L.; Cao, Y.; Shen, Q. Environment improvement and energy saving in Chinese rural housing based on the field study of thermal adaptability. *Energy Sustain. Dev.* **2022**, *71*, 315–329. [CrossRef]
101. Li, Q.; Hu, H.; Ma, L.; Wang, Z.; Arıcı, M.; Li, D.; Luo, D.; Jia, J.; Jiang, W.; Qi, H. Evaluation of energy-saving retrofits for sunspace of rural residential buildings based on orthogonal experiment and entropy weight method. *Energy Sustain. Dev.* **2022**, *70*, 569–580. [CrossRef]
102. Kebir, N.; Miranda, N.D.; Sedki, L.; Hirmer, S.; McCulloch, M. Opportunities stemming from retrofitting low-resource East African dwellings by introducing passive cooling and daylighting measures. *Energy Sustain. Dev.* **2022**, *69*, 179–191. [CrossRef]
103. Taruttis, L.; Weber, C. Estimating the impact of energy efficiency on housing prices in Germany: Does regional disparity matter? *Energy Econ.* **2022**, *105*, 105750. [CrossRef]
104. Huang, M.-Q.; Lin, R.-J. Evolutionary Game Analysis of Energy-Saving Renovations of Existing Rural Residential Buildings from the Perspective of Stakeholders. *Sustainability* **2022**, *14*, 5723. [CrossRef]
105. Chen, S.-Y.; Xue, M.-T.; Wang, Z.-H.; Tian, X.; Zhang, B. Exploring pathways of phasing out clean heating subsidies for rural residential buildings in China. *Energy Econ.* **2022**, *116*, 106411. [CrossRef]

106. Li, S.; Wang, M.; Shen, P.; Cui, X.; Bu, L.; Wei, R.; Zhang, L.; Wu, C. Energy Saving and Thermal Comfort Performance of Passive Retrofitting Measures for Traditional Rammed Earth House in Lingnan, China. *Buildings* **2022**, *12*, 1716. [CrossRef]
107. Hou, J.; Zhang, T.; Liu, Z.; Hou, C.; Fukuda, H. A study on influencing factors of optimum insulation thickness of exterior walls for rural traditional dwellings in northeast of Sichuan hills, China. *Case Stud. Constr. Mater.* **2022**, *16*, e01033. [CrossRef]
108. Lai, Y.; Li, Y.; Feng, X.; Ma, T. Green retrofit of existing residential buildings in China: An investigation on residents' perceptions. *Energy Environ.* **2022**, *33*, 332–353. [CrossRef]
109. Žičkienė, A.; Morkunas, M.; Volkov, A.; Balezentis, T.; Streimikiene, D.; Siksnelyte-Butkiene, I. Sustainable Energy Development and Climate Change Mitigation at the Local Level through the Lens of Renewable Energy: Evidence from Lithuanian Case Study. *Energies* **2022**, *15*, 980. [CrossRef]
110. Liu, S.; Liu, H.; Mauzerall, D.L. Improving Building Envelope Efficiency Lowers Costs and Emissions from Rural Residential Heating in China. *Environ. Sci. Technol.* **2023**, *57*, 595–605. [CrossRef]
111. Doe Doe Office of Indian Energy Energy Planning. Available online: https://www.energy.gov/sites/prod/files/2016/03/f30/leslie_01_EnergyPlanning.pdf (accessed on 11 May 2023).
112. Abouaiana, A. Rural Energy Communities as Pillar towards Low Carbon Future in Egypt: Beyond COP27. *Land* **2022**, *11*, 2237. [CrossRef]
113. Abouaiana, A.; Battisti, A. Agile-Transdisciplinary Conceptual Framework for Retrofitting Mediterranean Built Environments. [Forthcoming]. In *Mediterranean Architecture and the Green-Digital Transition Selected Papers from the World Renewable Energy Congress Med Green Forum 2022*; Sayigh, A., Ed.; Springer Nature: Basel, Switzerland, 2023; ISBN 978-3-031-33147-3.
114. Abouaiana, A. *Agile Methodology as a Transdisciplinary Retrofitting Approach for Built Environment in Traditional Settlements in Mediterranean Region*; Sapienza University of Rome: Rome, Italy, 2022.
115. Principi, P.; Roberto, F.; Carbonari, A.; Lemma, M. Evaluation of energy conservation opportunities through Energy Performance Contracting: A case study in Italy. *Energy Build.* **2016**, *128*, 886–899. [CrossRef]

buildings

MDPI

Article

Sustainability Assessment of Cementitious Ceramic Tile Adhesives

Jacek Michalak

Research and Development Center, Atlas sp. z o.o., 2 Kilińskiego St., 91-421 Lodz, Poland; jmichalak@atlas.com.pl

Abstract: This article presents the results of analyzing environmental impact indicators of thirteen ceramic tile adhesives (CTAs). The analyzed data came from ten third-party-verified Environmental Product Declarations (EPDs) created in 2016–2022. The paper examines seven environmental impact indicators for modules A1–A3 (cradle-to-gate). Significant differences were observed between the values of environmental indicators, which, in the case of Global Warming Potential (GWP), differed by almost 270% in the most extreme case. For the depletion of abiotic resources (elements) (ADP_e), the values of products differed by nearly fourteen thousand times. Results are discussed from the perspective of the CTAs' manufacturer assessing the product. The analysis focused on issues such as the historical dimension of data, which is the basis for Life Cycle Assessment (LCA), the need for their constant updating, and the subject of uncertainty—usually wholly omitted in the considerations on the environmental impact of construction products. The results of the analysis were also evaluated in terms of the planned introduction of the new 3+ assessment system in connection with the future amendment of the Construction Products Regulation (CPR). The results of the CTAs' analysis of environmental indicators showed that, despite the EPDs functioning for a decade, the obligatory assessment of construction products in terms of sustainability using the 3+ system did not create the conditions for its proper occurrence. This analysis showed that, without obtaining reliable data on the environmental impact of CTAs, correct AVCP is not possible, and the consumer is not able to make proper choices.

Keywords: sustainability; construction products; ceramic tile adhesive (CTA); assessment and verification of constancy of performance (AVCP); Environmental Product Declaration (EPD)

Citation: Michalak, J. Sustainability Assessment of Cementitious Ceramic Tile Adhesives. *Buildings* **2023**, *13*, 1326. https://doi.org/10.3390/buildings13051326

Academic Editors: Paulo Santos and Mark Bomberg

Received: 4 April 2023
Revised: 11 May 2023
Accepted: 14 May 2023
Published: 19 May 2023

1. Introduction

The construction sector, which consumes vast amounts of raw materials, produces large quantities of waste, and emits significant amounts of greenhouse gases, is one of the branches of the economy with a fundamental impact on the natural environment. Official statistics say that construction consumes the most considerable quantity of raw materials, emits the most greenhouse gases, and produces the most waste among all branches of the economy [1,2]. In 2020, buildings and all related operations in the 27 European Union (EU) countries were responsible for 511 Mt CO_2 eq., which accounted for 15% of global emissions. EU 2020 emissions were 24% lower than in 2005, but by 2030 they must decrease by 20% to meet the targets set by the European Commission (EC) [3,4]. Achieving the goals set by the EC is possible through further reduction in energy consumption of buildings, decarbonization of energy sources, optimization of the use of renewable energy sources, reduction in emissions resulting from the operation of buildings and facilities, and reduction in the negative environmental impact of construction products and construction works at the erection, renovation, and disassembly stages. In 2017, building materials were responsible for 11% of global CO_2 emissions [5].

1.1. Construction Products in the European Union

In EU countries, the principles of construction product assessment and verification of constancy of performance (AVCP) inform the Construction Products Regulation (CPR). According to the CPR, construction works must be designed, constructed, and dismantled in such a way that the use of natural resources is sustainable and ensures their re-use or recycling, the construction works are durable, and environmentally friendly raw materials and recycled materials are used in their construction [6]. Despite the importance of environmental assessment of construction products [7], positive political trends [8,9], and the need to protect the environment, environmental impact assessment is still not mandatory in the AVCP process [6]. It is common to expect that environmental assessment should be required, and such expectations are powerfully articulated by the scientific community [10]. The European Commission, taking into account scientists' and ecologists' expectations, and in line with its environmental policy, published a proposal on 29 March 2022 for a regulation laying down harmonized conditions for the marketing of construction products, amending Regulation (EU) 2019/1020 on market surveillance and compliance of products, and repealing CPR [11]. The CPR, and, thus, its novelization, is of fundamental importance for the construction products market in the EU. It assumes a significant change involving introducing a new assessment system to the AVCP—the 3+ system. In this system, the manufacturer shall assess construction product performance concerning essential characteristics or product requirements related to environmental sustainability and keep it updated. The notified body controls environmental sustainability assessment, namely, it givesn input values, assumptions, and compliance with applicable generic or product category-specific rules. The notified body also verifies the manufacturer's initial and updated assessment and validates the process to generate that assessment. On the one hand, changes are needed and expected, but revolutionary changes also raise questions: are they feasible for the manufacturer, and possible for notified bodies?

1.2. Sustainability and Construction Products

One needs to consider the environmental impact issue in the broader aspect of developing contemporary strategies that directly link core businesses and sustainability issues and clearly articulate their limitations [12]. Years of industry experience with certified management systems are primarily associated with bureaucracy, and often with the abandonment of innovative thinking [13]. All of the construction sector's partner actors and governments must understand and participate in the new sustainable business development [14]. The need to change the attitudes of scientists has also been postulated [15], especially when considering the scenarios in which economic growth is a lower priority than sustainability. This, and even more critical ideas, are currently far from being understood [16]. Various pathways for sustainability transformation are found in macro-level discussions on policy, science, and business [17]. One important issue is the mutual relationship between a circular versus a linear building element [18]. The European Commission, recognizing the issue's importance, already in 2004 gave CEN a mandate to develop standards for integrated environmental building performance [19]. One of the results of the work of CEN experts is the possibility of using Life Cycle Assessment (LCA) to assess the impact of a construction product on the environment using type III Environmental Product Declaration (EPD) [20]. In 2006, the ISO organization defined the procedures and requirements for preparing EPDs for all product types [21].

In 2011, the EU initiated work under the Single Market for Green Products Initiative, aimed, among other things, at using known methods for measuring and informing about environmental performance in the life cycle of products—Product Environmental Footprint (PEF) [22,23]. Both approaches (EPD and PEF) have since been amended [24–27].

EPDs should provide quantifiable, reliable, coherent, and comparable environmental information on construction products on a scientific basis. Ten years after the publication of EN 15804, which defined the guidelines for developing EPDs for construction products, in January 2023, over sixteen thousand verified EPDs for construction products were registered worldwide [28]. From this number, the most significant number of EPDs for construction products was issued by the French FDES—4585—followed by International EPD—2945—Norwegian EPD Norge—1977—and IBU (Germany)—1302. The Polish Research Building Institute (ITB) took thirteenth place in this ranking, with 263 issued EPDs [28]. Still, the distribution of EPDs is not homogeneous across EU countries and European regions, remaining limited in terms of the type of construction products [29]. Three Nordic countries, i.e., Denmark, Finland, and Sweden, are leaders in implementing environmental policies and regulations [30]. The analysis of construction product EPDs in Spain showed that 20% of construction products used in buildings could use data from EPDs to conduct a building LCA [31]. Additionally, the view, supported by the results of the analyses, that environmental data are considered valid only for comparative purposes since they differ from one region/country to another, is critical [32]. For this reason, the institutional role of industry associations is essential, which can partially explain cross-national or cross-regional variations [33]. It is also important that small and medium-sized enterprises may not meet the requirements for the environmental assessment of their products due to insufficient financial, technical, and human resources [34].

Even when operators of EPD programs work according to the same guidelines based on identical Product Category Rules (PCR), the published EPDs differ. Often these differences are so significant that they prevent proper product assessment and confuse users [35]. This is the case even though PCR determines the LCA calculation rules (allocation, collection, system boundaries, environmental indicators, EPD format). Europe is a leading continent in the field of knowledge on the environmental impact of construction products because of the activities of the ECO Platform EPD, thanks to which a greater consensus on communicating environment information has been developed than on other continents [36]. However, global scale harmonization is still the primary challenge faced by all who want to increase the significance and comparability of EPDs [37]. The first information on differences in the quality of LCA studies appeared shortly after the publication of the first EPDs [38]. Differences between EPDs for different groups of products are the subject of research by many researchers [32,39]. The defective part of the current EPD scheme is the interpretation of the results [40]. It is also worth mentioning that the multitude of EPDs impact categories and the units in which individual environmental indicators are expressed are challenging to communicate and, thus, difficult to understand, creating a complex decision-making process [41]. Comparison of data contained in EPDs is often tricky because environmental impact values are calculated for different functional units [42]. The current state of the comparability of data contained in the EPD is illustrated by the analysis of 436 EPDs verified by two renowned operators (International EPD System, Insitute Bauen und Umwelt e.v.) for the categories boards, thermal insulation, and floor covering. Of the analyzed EPDs, only 0.04% of documents were fully comparable, 2.75% could be compared with caution, as much as 89.15% should be treated as incomparable, and 8.06% EPDs could not be compared in any aspect [43]. It should also be remembered that the LCA analysis of construction products ignores the share of components whose content is below 0.5%, some impacts are avoided, and different inclusions of recycled materials are applied [44].

It is also worth noting that environmental indicator analysis primarily concerns the Global Warming Potential (GWP) indicator [45]. Importantly, EPDs often did not cover the mandatory scope of the scheme [46]. Other studies have indicated that as long as the databases are representative of the context, methodological choices may be a minor concern [47]. In connection with the problems described above, attempts have been made to develop a new environmental scoring methodology for construction products based on LCA [48]. Benchmarking methods are also used regarding their applicability to the EPD's interpretation [49,50].

PEF—the second approach to the environmental assessment of products—is less popular and, like EPDs, is not free from imperfections [51]. As if that were not enough, comparing the results obtained with the EPD and PEF methods is complex and sometimes even impossible due to the different system boundaries and source data used. Thus, the idea of alternative use of the two mentioned methods cannot be implemented in practice [52].

Summing up the decade of functioning type III EPDs, one can say that the use of these documents is still limited and is primarily used for B2B communication, and preparation of offers for tenders and voluntary building certification systems, such as BAMB, BREEAM, CASBEE, DGNB, Green Globes, HQE, LEED, ÖGNI, SBTool, and TQB [53–55]. Construction product weights in the previously mentioned certification systems are between 12.5% and 15.0% [56]. It is also important to note the importance of assessing the environmental impact using the certification systems mentioned above for buildings under renovation, and not only the construction of new ones [57]. EPDs should also be implemented in building information modeling (BIM), which is the foundation of digital transformation in the architectural, engineering, and construction industries, including identifying the main interaction problems between BIM and the LCA [58]. In this aspect, it is crucial to integrate databases with BIM to extract quantities and calculate the construction products' environmental impact at the design stage [59]. Published in 2022, ISO 22057 provides the principles and requirements to enable environmental data in EPDs for construction products to be used in BIM [60]. Additionally, ISO 22057 describes the weaknesses of existing digitized EPD approaches [61].

Most EU countries do not have central LCA repositories. This lack creates barriers to developing benchmarks [62].

Despite the shortcomings mentioned above in the environmental assessment of products using EPDs, it is clear that it is necessary to develop this path—more EPDs must be available [63]. The analysis of data contained in EPDs over the years allows the identifying of current trends or relationships between various environmental indicators [64]. The more EPDs become available, the more accurate the results will be and, thus, the easier it will be to implement artificial intelligence techniques to predict environmental impacts, including construction products [65]. Another view, presented by Wittmayer and Schäpke, is that it is necessary to move to more process-oriented approaches [15]. Without proper, i.e., high-quality, EPDs, it will not be possible to significantly improve the environmental performance of construction projects [66].

1.3. *Sustainability of Construction Products by Manufacturers*

Most studies on EPD, LCA, and the circular economy do not consider the industry, or do so only to a small extent. However, symbiosis in the industrial dimension is necessary between feedstock, technology, products, side streams, downstream valorization, and long-term circularity [67]. An analysis of literature on the subject clearly shows that the discussion on making EPDs more coherent and comparable does not involve industry, but academia [53]. Additionally, other earlier studies prove that construction sector practices are often removed from academic research [68]. In addition, when considering the perception of the industry in terms of broadly understood environmental protection by various stakeholders, including science, it is worth mentioning "greenwashing", i.e., a situation in which, taking into account the current trends in terms of being sustainable, the companies themselves, when reporting data, present them not necessarily as they are, and more as commonly expected [69]. EPDs as a communication tool are believed to give companies a competitive advantage [70].

A completely different issue than the issues raised above is environmental impact assessment as a part of the manufacturer's AVCP of construction products. It is an entirely different issue from the challenges faced by building LCA practitioners for whom recommendations can be found in the literature [71]. It is essential to add that, in the wealthy scientific literature on the environmental assessment of construction products, there is practically no work on this issue from the perspective of the manufacturer of the building material.

1.4. Research Hypothesis

This article analyzes third-party-verified EPDs for thirteen cementitious ceramic tile adhesives (CTAs). CTAs are essential construction products, and the reason for selecting them for the research presented in this article is, above all, the fact that they contain relatively few ingredients in their composition, and their performance properties are clearly defined in EN 12004:A1:2012 [72]. The primary ingredient is Portland cement, for which the environmental impact is the subject of intensive research by many centers and is well recognized [73]. Of course, there are differences in the environmental impact ratings of Portland types of cement depending on who conducted the analysis. Still, there is general agreement that the reasons for the differences are known [73]. Given the above, CTAs of different origins can be defined as comparable products.

The study results were analyzed from the perspective of the manufacturer performing AVCP of CTAs, with particular emphasis on the conditions that will arise in the event of introducing a new assessment system, i.e., the 3+ system.

The research hypothesis is that the model for assessing the environmental impact of construction products using EPDs does not provide consistent and comparable data that all participants in the construction market can use.

2. Materials and Methods

CTAs are commonly used to install ceramic tiles on walls and floors, indoors and outdoors [74]. Global production of CTAs in 2020 was approximately 65 million tons [75]. The basis of the AVCP of CTAs in the EU area is the requirements contained in EN 12004:2007 + A1:2012 [72]. Similar provisions have been adopted by the ISO organization [76]. Thus, these exact requirements apply to CTAs around the world. It is also worth mentioning that the requirements underlying the AVCP of the CTA process have been practically unchanged in the EU since 2001 [74].

The subject of the analysis presented in this article was data for CTAs coming only from third-party-verified EPDs. The following extracting criteria were taken into account when collecting data: EPD owner—company/association, EPD program, product identification—type according to EN 12004/ISO 13007-1, validation from/to date, geographical representativeness, including the origin of the data used, temporal coverage, compliance with EN 15804 and ISO 14025, PCR, declared functional unit, LCA database, LCA software, system boundaries, end of life scenarios, and environmental impacts.

In addition to the EPD databases search and literature research, this study was conducted based on the author's almost thirty years of professional experience with the assessment and verification of constancy of performance of construction products.

3. Results

The systematic search of the EPD programs resulted in 10 EPDs. In this article, data on thirteen CTAs were analyzed—in the case of one EPD, data for three CTAs were selected for analysis, and in the case of another EPD, data for two CTAs. From the remaining eight EPDs, data for one CTA from each declaration were used for analysis. All 13 analyzed CTAs contained gray Portland cement as a mineral binder, and CTAs containing white cement were not analyzed.

Table 1 summarizes the primary data characterizing the EPDs of the thirteen analyzed CTAs, which for this study have been marked with letters of the alphabet from A to M.

Table 1. Primary data on EPDs, of which data on 13 CTAs are the subject of the analysis presented in this article.

CTA	Valid (from–to)	Dataset	Geographical Coverage	EPD Operator	EPD Owner	Ref.
A	05.2016–05.2021	2011–2015 generic and manufacturer data + literary research	EU	IBU	assoc.	[77]
B	09.2016–10.2023 *	2004–2018 generic data + 2017–2018 data from 2 locations in Italy	Int.	EPD Int. AB	mfr.	[78]
C	08.2017–06.2022	2016 data from 2 locations in Italy + generic data	Global	EPD Int. AB	mfr.	[79]
D	09.2016–06.2024 *	2005–2017 generic data + 2015–2017 data from 2 locations in Italy	Int.	EPD Int. AB	mfr.	[80]
E	12.2019–12.2024	2013–2018 generic data + 2018 data from 6 locations in Turkey	TR	EPD Int. AB	mfr.	[81]
F, G, H	01.2017–01.2022	2015 data from 4 locations in Turkey + generic data	WW	EPD TR	mfr.	[82]
I	11.2020–11.2025	2019 data from 5 location in Poland + 2017 generic data	PL	ITB	mfr.	[83]
J	11.2020–11.2025	2019 data from 5 location in Poland + 2017 generic data	PL	ITB	mfr.	[84]
K	09.2016–12.2022	2014–2015 data from ten producers, except cement—2004 data + generic data	CA, MX, US	UL Env.	assoc.	[85]
L, M	03.2022–no date	2021 data from one location in Israel + 2011–2018 generic data	IL	IIS	mfr.	[86]

Abbreviations: EU (European), Int. (International), TR (Turkey), WW (Worldwide), PL (Poland), CA (Canada), MX (Mexico), US (United States), IL (Israel), IIS (The Israeli Institute of Standards), assoc. (association), mfr. (manufacturer), * EPDs revised in 2019.

Of the ten analyzed EPDs, only one followed ISO 14025, and the remaining nine were under ISO 14025 and EN 15804. Among the analyzed CTAs, in the case of the product marked as F, no classification was given following EN 12004/ISO 13007-1, but from other entries in the EPD it can be concluded that it is a class C1 CTA. In the case of the product marked as K, the CTA marking per ISO 13007-1 was not provided, although the EPD contains a record that the product complies with this standard's requirements, but without specifying which ones. For the CTAs marked A, B, D, E, F, G, H, I, J, and K, 1 kg is given as the functional unit. In the case of CTAs marked as C, L, and M, 1 m^2 was indicated as the functional unit while stating in the EPD that in the case of product C, it corresponds to 3.13 kg of dry product, which is equivalent to 2.5 mm thickness, and for the CTA marked as K equals 4.2 kg of dry product and 3 mm thickness. For the CTA marked M, 3.9 kg of dry product is required to cover 1 m^2 of the surface with a 3 mm layer. The information provided allows for the calculation of the environmental impact for a functional unit defined as 1 kg of CTAs.

Table 2 summarizes the environmental impact indicators—six mandatory categories that shall be included in an EPD according to clauses 6.5 and 7.2.3.1—Table 3 of EN 15804:2012+A2:2019, i.e., Global Warming Potential (GWP), Ozone Depletion Potential (ODP), Acidification Potential (AP), Eutrophication Potential (EP), Photochemical Ozone Creation Potential (POCP), depletion of abiotic resources (elements) (ADP_e), and depletion of abiotic resources (fossil) (ADP_{ff}). In the analysis presented in this article, only the production stage, covering cradle-to-gate—A1–A3 modules, i.e., A1—extraction, and processing of raw materials and the processing of secondary materials, A2—transport of the materials to the manufacturer, and A3—production processes, was analyzed.

Table 2. Summary of environmental indicators of thirteen analyzed CTAs.

CTA	GWP	ODP	AP	EP	POCP	ADP_e	ADP_{ff}
	[kg CO_2 eq.]	[kg CFC11 eq.]	[kg SO_2 eq.]	[kg $(PO_4)^{3-}$ eq.]	[kg ethene eq.]	[kg Sb eq.]	[MJ]
A	6.38×10^{-1}	5.14×10^{-9}	2.03×10^{-3}	1.82×10^{-4}	1.92×10^{-4}	1.06×10^{-6}	7.09
B	4.75×10^{-1}	1.76×10^{-8}	4.25×10^{-4}	1.58×10^{-4}	3.13×10^{-4}	1.46×10^{-7}	5.60
C	5.69×10^{-1}	3.45×10^{-8}	1.77×10^{-3}	4.09×10^{-4}	1.16×10^{-4}	6.74×10^{-7}	6.93
D	3.35×10^{-1}	1.50×10^{-8}	2.42×10^{-4}	1.24×10^{-4}	1.50×10^{-4}	8.88×10^{-8}	2.72
E	2.98×10^{-1}	1.58×10^{-8}	8.21×10^{-4}	3.36×10^{-4}	9.06×10^{-7}	1.15×10^{-7}	1.79
F	3.06×10^{-1}	1.47×10^{-8}	3.01×10^{-5}	7.00×10^{-4}	1.94×10^{-4}	2.18×10^{-7}	1.67
G	4.71×10^{-1}	2.87×10^{-8}	9.41×10^{-5}	1.28×10^{-3}	3.54×10^{-4}	9.68×10^{-7}	3.86
H	3.55×10^{-1}	1.93×10^{-8}	5.59×10^{-5}	8.85×10^{-4}	2.49×10^{-4}	6.28×10^{-7}	2.49
I	4.35×10^{-1}	1.90×10^{-8}	4.75×10^{-4}	3.64×10^{-4}	7.73×10^{-5}	1.23×10^{-3}	2.47
J	4.31×10^{-1}	2.00×10^{-8}	5.31×10^{-4}	2.92×10^{-4}	9.71×10^{-5}	7.30×10^{-4}	2.62
K	5.19×10^{-1}	1.08×10^{-9}	2.33×10^{-3}	1.87×10^{-4}	1.68×10^{-4}	7.11×10^{-7}	4.25
L	2.37×10^{-1}	1.72×10^{-8}	9.38×10^{-4}	1.06×10^{-4}	7.86×10^{-4}	2.19×10^{-6}	4.26
M	5.77×10^{-1}	3.28×10^{-8}	1.85×10^{-3}	3.43×10^{-4}	1.59×10^{-3}	3.05×10^{-6}	6.49

Abbreviations: CTA (Ceramic Tile Adhesive), GWP (Global Warming Potential), ODP (Ozone Depletion Potential), AP (Acidification Potential), EP (Eutrophication Potential), POCP (Photochemical Ozone Creation Potential), ADP_e (Abiotic Depletion Potential (elements)), ADP_{ff} (Abiotic Depletion Potential (fossil)).

4. Discussion

It should first be noted that 10 EPDs for CTAs is a small number. As stated in the introduction, in January 2023, there were about 16,000 third-party-verified EPDs for construction products [28], which is not too much, especially considering the global dimension of construction and the variety of materials used. These data correspond well with the number of scientific publications on EPDs, CTAs, and construction products. Therefore, the result of the search in the Scopus database of scientific journals for the query "EPD" + "construction products" is the identification of 2962 articles (Scopus database accessed on 10 March 2023). In the same study, only 13 articles were available for the query "EPD" + "ceramic tile adhesive". When the query concerned only "EPD", 36,818 articles were identified in the database. For the query "EPD" + "mortar", the search identified 305 articles. The reason for the search with the word "mortar" was that CTAs are often referred to as mortars, although they are a particular case of a large group of products called mortars. CTAs are characterized by strictly defined parameters and purposes in EN 12004/ISO 13007-1. In the case of product A, analyzed in this article, which is a CTA, its environmental impact data were calculated based on data for a broadly defined group referred to as mortars (repair mortars, adhesives, joint fillers, screeds, floor leveling compounds, grouts, and waterproofing slurries), in which modified mineral mortars meet criteria relevant for CTAs [77]. Additionally, for product K, the term "cement mortar for tile installation" is used instead of CTA in the EPD. Santos et al., in a literature review on the LCA of mortars published in 2021, also stated that there is a lack of quantification of mortar's environmental impacts. However, more scientific papers on LCA and mortars have since been published [87].

The analysis described in this article used data for CTAs from 10 EPDs, but only six of them were valid at the time of the investigation, being the subject of this article. Of the four EPDs that were no longer valid, three had expired in 2022, and one a year earlier, i.e., in 2021. It has not been identified that these documents have been amended, and their validity has been extended, which is worth noting. This result may be surprising at first, because the number of EPDs issued for construction products is growing [28]. On the other hand, there are claims that EPD owners are less interested in this document [49,88].

This lack of interest is due to various reasons, including the inability of the manufacturer to position its products among similar products to find out whether it is comparatively environmentally friendly [49,89]. This is important, because thinking about the environmental impact of manufacturers as reducing negative impact has been formatted over the years [90]. Additionally, as noted by Dijkstra-Silva et al., "being "less bad" still means harming the environment" [90]. For this reason, various authors point to the necessity and importance of recognizing the positive impact of manufacturers on the environment [90]. All this indicates the need to develop new sustainable business models [14].

In addition, one should remember that the sustainability assessment of construction products is still voluntary [6,9]. Additionally, one more important point to note. So far, researchers who have extended responsibility for their work do not create the basis for accurate decisions and policy formulation, but create for other researchers [91]. An analysis of 1246 articles related to environmental labeling topics from over 22 years showed the growth of theoretical considerations parallel to the development of environmental labeling in the world [92].

The small number of EPDs for CTAs identified in this analysis and the fact that as much as 40% of the EPDs analyzed in the article have expired, with no evidence of their amendment and renewal, proves the existence of a crisis in this area. The situation identified in the study regarding the decreasing number of valid EPDs for CTAs contradicts the expectation of a higher demand for relevant data on construction products. It may also prove that we are still not dealing with a significant change in the approach and transition from predominantly qualitative to predominantly quantitative assessment of construction products, which manufacturers recommend [91].

Figure 1 shows the analysis results of seven environmental impact indicators for 13 CTAs, which are listed in Table 2. Due to the different units in which individual environmental indicators are expressed, and the essence of presenting them all in one figure, Figure 1 shows every single indicator as the quotient of the value of this indicator to the lowest indicator value in a given group. Due to the significant differences between the analyzed values, a logarithmic scale was used to present the results, enabling data analysis from a wide range.

Significant differences exist between the values of all seven environmental impact indicators in the range of modules A1–A3 of the analyzed CTAs. Minor discrepancies exist for GWP, where the most significant value for CTA marked as A is 2.69 times greater than the smallest value for GWP for CTA marked as L. For ADP_{ff}, the value for CTA marked as A is 4.25 times greater than that for CTA marked as F. The most significant differences were observed for ADP_e and POCP indicators. For ADP_e, the highest value (CTA-I) is 13.851 times higher than the value for CTA marked as D. Such significant differences between the values of environmental impact indicators between the analyzed CTAs prove that it is impossible to make comparisons between products, which is a natural expectation from all participants of the construction product market.

When analyzing the differences between CTAs' environmental impact indicators and trying to understand the state of affairs, various issues need to be considered. One important aspect is the type of binder, i.e., Portland cement, that is used to produce CTAs. In the past, CEM I [93] was primarily or even exclusively used to make CTAs. Due to the significant environmental impact of cement production, a trend has been observed for several years to increase the use of types of cement with a lower clinker content, i.e., CEM II, CEM III, CEM IV, and CEM V instead of CEM I. The difference in environmental impact between CEM II, III, IV, V, and CEM I is significant. For example, the GWP for CEM I was 0.889 kg CO_2 eq., while for CEM II—0.704, CEM III—0.482, CEM IV—0.568, and CEM V—0.518 kg CO_2 eq. [94]. The above data were determined based on production data of all cement plants in Poland from 2017, and the average environmental impact of types of cement produced in Poland is slightly lower than the moderate environmental impact of European types of cement [73].

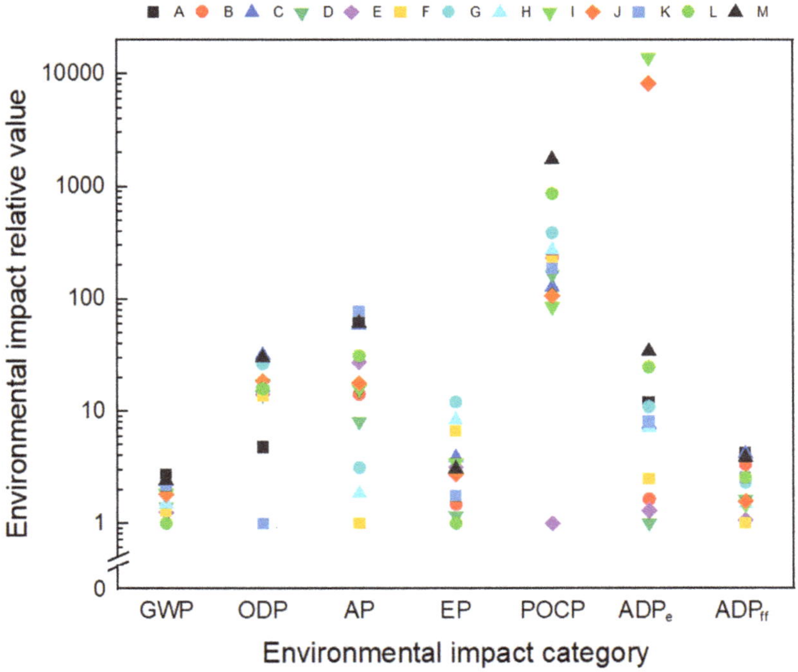

Figure 1. Relative values of environmental impact indicators GWP (Global Warming Potential), ODP (Ozone Depletion Potential), AP (Acidification Potential), EP (Eutrophication Potential), POCP (Photochemical Ozone Creation Potential), ADP_e (Abiotic Depletion Potential (elements)), ADP_{ff} (Abiotic Depletion Potential (fossil)) for modules A1–A3 for the 13 analyzed CTAs marked with letters from A to M.

In the EPDs analyzed in this article, apart from the CTAs marked as I and J, the type of cement used for production was not disclosed, and the data for what kind of cement was included in the calculations. In addition, it is worth noting that the data came from different years. Therefore, for the CTA marked as K, the data came from 2004, as indicated in the declaration, which was published twelve years later in September 2016. Considering that the cement content in CTAs is usually in the range of 30–40% by mass, this means a significant share in all environmental indicators. In addition, due to the substantial share of the cement industry in global energy consumption and considerable carbon dioxide emissions, the cement industry started actions to reduce its negative impact on the environment much earlier than other branches of the economy, which today results in a much more considerable amount of environmental data. However, comparing existing EPDs for cement shows differences as high as 300,000% between the highest and the lowest reported value for ADP_e [73]. In the case of the CTAs analyzed in the article, the differences in this indicator were also huge—CTAs marked as D and E differed by 30% in the ADP_e range for modules A1–A3, but the value for CTA marked as J was 8.220 times higher than product D, and as much as 13.851 times higher ADP_e value was recorded for product I compared to product D.

When analyzing the values of CTAs' environmental impact indicators, one should also pay attention to the various cut-off criteria used in the LCA analyses, which are included in the EPDs analyzed in this article. For example, for the CTAs marked F, G, H, and K, no cut-off criteria were applied, and all reported data were incorporated [83,86]. For products marked L and M, data for some raw materials were not found in the available databases. Due to this, a contribution from those materials (less than 1% of the product mass) was excluded from the calculation [86]. In the composition of CTAs, apart from cement and fillers, which comprise the majority of the product, there are ingredients of 0.3–0.4% of the product mass, such as methylcellulose ethers. Often, data on these ingredients are not included in the calculations. It should be remembered that, due to the high-energy process by which methylcellulose ethers are obtained, and in how few places in the world they are produced, which often means long transport to the location of their use for the production of CTA, not taking into account their impact additionally reduces the credibility of the obtained data. The use of different cut-off criteria is one of the factors that cause difficulties when comparing products.

It is also worth noting that the use of secondary raw materials was invisible in this study, which is vital for various reasons, including for the circular economy [92,95].

When evaluating each product, including a construction product, it is necessary to determine the criterion whose fulfillment determines whether the product complies with the requirements. In the case of construction products, the 3+ system proposed in the draft CPR, which will apply to construction products assessed in terms of sustainability, states that "the manufacturer shall carry out the assessment of the performance of the product concerning essential characteristics or product requirements related to environmental sustainability and keep it updated" [96]. In the proposed 3+ system, the notified body shall, given input values, assumptions made, and compliance with appropriate generic or product category-specific rules (i) verify the manufacturer's initial and updated assessment, and (ii) validate the process applied to generate that assessment [96]. Measurements or calculations that are the basis for assessing each product must be reliable, and the variability of the results resulting from uncertainty is an inherent part of the assessment process. It is also the case with CTAs' evaluation against the existing criteria [97]. The results of this analysis should also be considered in this aspect. Among the seven environmental impacts analyzed for the thirteen CTAs, the most negligible dispersion of results was observed in the case of GWP and the largest in the case of ADP_e. Thus, the mean value for the thirteen analyzed CTAs for the GWP indicator is 4.34×10^{-1} kg CO_2 eq., and the standard deviation from the analyzed sample is 1.23×10^{-1} kg CO_2 eq. Thus, one standard deviation, a classic measure of volatility, is 28% of the mean GWP's value. For ADP_e, the mean value of the thirteen analyzed results is 1.52×10^{-4} kg Sb eq., and the standard deviation is as much as 3.82×10^{-4} kg Sb eq., i.e., one standard deviation is 252% of the mean value. Significant differences between the analyzed CTAs in the ADP_e category are not particularly surprising, because the depletion of abiotic resources is the most controversial environmental indicator subject to LCA [98]. The results presented in this analysis and the literature on the EPDs also indicate that EPDs and the environmental indicators contained therein are not analyzed in terms of uncertainty [99,100].

Rasmussen et al., analyzing 81 third-party-verified EPDs of cross-laminated timber, glulam, laminated veneer lumber, and timber developed mainly in Europe but also in North America, Australia, and New Zealand, proposed that life cycle assessment practitioners can use median values from their study as the generic data [71]. In the case of CTAs, in light of the data analyzed in this paper and considering the future potential AVCP of these products, it should be stated that using mean values as generic data can lead to many misunderstandings.

As mentioned earlier, following the provisions of the CPR draft amendments, the notified body should assess the manufacturer's documentation in the field of sustainability. Questions arises about this verification—how to relate it to the data collected over a decade? Is it possible? Based on the data presented in this analysis, there is only one answer—unfortunately, it is impossible.

The draft amendment to the CPR assumes that the producer will systematically update environmental indicators. Currently, third-party-verified EPDs are primarily valid for five years. Additionally, for this reason, the situation will create a field for conflict between producers. Comparing data from this year with data from five years ago will pose problems. The data for CTAs collected in Table 2 does not clearly show the changing trend related to the passage of time. However, for other construction products, where the data associated with the production of the same product by the same manufacturer at the exact production locations in different years, differences were reported [101]. This critical observation about change over time is consistent with the fact that, in the current model, the assessment concerns the "static" manner, but the world is changing dynamically [102].

Another issue regarding the future assessment of construction products in terms of sustainability is the human resources of the notified body. Of course, this issue is known and discussed in the literature [103]. As mentioned many times, the inconsistency between published EPDs is known. The fact that years have passed and this issue remains unresolved is also a sign of the lack of availability of properly educated specialists.

The introduction of a mandatory 3+ system in the field of sustainability, however justified, also generates additional costs. These costs may be difficult for smaller entrepreneurs to cover. In the case of smaller producers, there will also be a problem with access to suitably qualified staff. Additionally, from this perspective, the question arises about the legitimacy and sense of the proposed formula for assessing construction products in the field of sustainability. In this connection, there is the already raised issue of the possible role of associations of producers [33,34].

A view can also be formulated that, due to the lack of comparability of EPDs, it is difficult to talk about public trust in the issue of proper assessment of construction products in terms of sustainability. However, public trust is fundamental in building matters [104]. Research on the conscious consumption of such construction products as interior wall paints and coatings showed a higher importance of making environmental assessment transparent to end-users [105]. It is also crucial in terms of the fact that consumers today do not have the willingness to pay the premium price for sustainable products [106]. Additionally, all this must be considered by the manufacturer when creating a strategy promoting environmentally friendly products. This is not easy, although, from the perspective of science, there are views that it is a business opportunity [107]. Of course, it can be, but it has to be based on solid foundations, including consistent and comparable data from EPDs.

As the study results described in this article show, data on the environmental impact of CTAs do not make it easier for construction market participants to make the right decisions when choosing a product. Additionally, considering that other research referring only to the knowledge of the role of EPDs in the environment of people professionally related to the construction industry showed that as many as 75.9% of the surveyed sellers of construction products and 74.0% of investors, as well as 64.6% of contractors, believe that EPDs are a mandatory document, but in fact they are voluntary documents [108]. Only in the professional group of architects did a minority of respondents (43.5%) consider the EPD a required by-law document [108]. Although these answers are incorrect, they indicate that sustainability issues are perceived as significant. In the same study, the share of people considering the environmental impact (by a declarative "yes" or "no") when selecting construction products varies from 39.1% in the group of architects to 59.8% in the group of investors [108]. However, to another question that only required a "yes" or "no" answer, i.e., the question of knowing the GWP value of 1 m^2 of External Thermal Insulation Composite System, responses of "yes" ranged from between 27.1% and 49.7% in the surveyed groups of professionals [109]. Other studies, in which 55 respondents gave answers regarding use of EPDs by architects, showed as many as 76% declare their use [54]. Above all, however, it is worth mentioning that the results of the examination of sustainability knowledge to understand sustainable behavior indicated that knowledge had a significant, albeit weak, bivariate correlation with behavior [110].

It is also important to mention that, although CTAs belong to the mortars group and are perceived and classified in this way in many studies, which is also visible in the EPDs analyzed in this article, they are a group of specialist construction products. For this reason, environmental impacts related to the production of CTAs cannot be considered at the general level of the relationship between mortars and sustainability, treating mortars as a mixture of three components: aggregates, binders, and water [111]. CTAs are produced in specialized dry-mix mortar plants. Thus, modules A1–A3 can be or are based on existing or historical data, so it is possible to estimate the impacts in principle. However, it should be remembered that the given value of environmental indicators in the declaration of performance of construction products will always be only an approximate/estimate/historical value.

The analysis presented in this article is a contribution to knowledge transfer from industry to science. The flow of knowledge from industry to science is often the missing link for properly developing a topic. Of course, issues such as those discussed in this article, such as the unclear and sometimes confusing values of environmental indicators, reliability of LCA, LCA input data availability, the necessity of harmonization, inadequate legislative framework, and necessity of coherence of EPDs, are the subject of scientific research and publications [112]. However, still, the producers' point of view is rarely taken into account. However, due to the plans to amend the CPR and introduce a new 3+ conformity assessment system, it is necessary to consider the manufacturers' position. Otherwise, further development of construction products will be disturbed, including the dissemination of information on the environmental impact of the product. Finally, it is reasonable to write that the conclusions drawn from the analysis described in this paper are consistent with the recently formulated recommendations [113].

However, to the author's knowledge, no analysis has been published on the consequences of introducing a mandatory CTA environmental impact assessment in the event of the CPR amendment. Results clearly show that notified bodies are not prepared to verify EPDs, along with a significant increase in their number due to the introduction of the obligatory sustainability assessment of construction products.

5. Conclusions

This study analyzed the values of seven environmental impact indicators of thirteen CTAs. The author discussed specific conditions related to the assessment of a construction product, which has not been a subject of particular interest so far, such as the issue of uncertainty. The work indicates many factors influencing this state of affairs:

- The analysis of the values of seven environmental impact indicators (GWP, ODP, AP, EP, POCP, ADP_e, ADP_{ff}) showed significant differences between them, the smallest being for GWP and the largest for ADP_e;
- The difference between the lowest and highest value of the GWP indicator amounted to almost 270%;
- The difference of nearly fourteen thousand times between the lowest and highest value of the ADP_e indicator make it impossible to compare CTAs;
- The analysis described in this article clearly shows that the planned introduction of the environmental assessment of construction products has not been properly prepared. The results obtained over a decade on the environmental impact of CTAs may be of little use for the future assessment of these products. In addition, it should be noted that due to permanently going on a benchmark that takes place on the market of construction products, many manufacturers may practice greenwashing to show that their products are not "worse" than others.

6. Future Directions and Limitations

The analysis showed that the environmental impacts of CTAs collected over a decade are inconsistent, and their use in AVCP does not help consumers choose products. It is necessary to specify the requirements for qualitative data used in LCA precisely, understood as narrowing the boundaries, e.g., to data from one year preceding the year of EPDs creation. In light of the obtained results of the analysis and considering other experiences gained over the decade, it seems reasonable to develop product category rules used in developing EPD by producer organizations at the local/national level. It appears that the participation of producer organizations should facilitate the creation of a transparent situation for all participants in the construction products market.

The analysis described in this paper concerned data for 13 CTAs presented in 10 EPDs, i.e., on a relatively limited sample. As mentioned earlier, sixteen thousand EPDs were published in January 2023 for construction products in general, which is also not a representative number for the construction products market. Undoubtedly, one of the reasons why CTA producers do not develop EPDs en masse is that the results contained therein are not easily presented to consumers. Above all, it should be remembered that EPDs are voluntary documents, which is also the reason for their limited number.

Funding: This research received no external funding.

Data Availability Statement: Not applicable.

Conflicts of Interest: The author declares no conflict of interest.

Abbreviations

Abbreviation	Meaning
ADP_e	Abiotic Depletion Potential (elements)
ADP_{ff}	Abiotic Depletion Potential (fossil)
AP	Acidification Potential
AVCP	Assessment and Verification of Constancy of Performance
CPR	Construction Products Regulation
CTA	Ceramic Tile Adhesive
EP	Eutrophication Potential
EPD	Environmental Product Declaration
GWP	Global Warming Potential
LCA	Life Cycle Analysis
ODP	Ozone Depletion Potential
PCR	Product Category Rule
PEF	Product Environmental Footprint
POCP	Photochemical Ozone Creation Potential

References

1. European Academies Science Advisory Council. *Decarbonisation of Buildings: For Climate, Health and Jobs*; European Academies Science Advisory Council: Halle, Germany, 2021; Available online: https://easac.eu/publications/details/decarbonisation-of-buildings-for-climate-health-and-jobs/ (accessed on 10 February 2023).
2. European Environment Agency. *Trends and Projections in Europe 2021*; European Environment Agency: Copenhagen, Denmark, 2021; Available online: https://www.eea.europa.eu/publications/trends-and-projections-in-europe-2021 (accessed on 10 February 2023).
3. European Commission. *The European Green Deal*; European Commission: Brussels, Belgium, 2019. Available online: https://eur-lex.europa.eu/legal-content/EN/TXT/?uri=CELEX:52019DC0640 (accessed on 10 February 2023).
4. European Commission. *Fit for 55': Delivering the EU's 2030 Climate Target on the Way to Climate Neutrality*; European Commission: Brussels, Belgium, 2021. Available online: https://eur-lex.europa.eu/legal-content/EN/TXT/?uri=CELEX%3A52021DC0550 (accessed on 10 February 2023).
5. Anderson, J.; Moncaster, A. Using an analysis of concrete and cement EPD: Verification, selection, assessment, benchmarking and target setting. *Acta Polytech. CTU Proc.* **2022**, *33*, 20–26. [CrossRef]
6. Regulation (EU) No 305/2011 of the European Parliament and of the Council. *Brussels, Belgium*. 2011. Available online: https://eur-lex.europa.eu/legal-content/EN/TXT/?uri=CELEX:32011R0305 (accessed on 10 February 2023).
7. Scherz, M.; Wieser, A.A.; Passer, A.; Kreiner, H. Implementation of Life Cycle Assessment (LCA) in the Procurement Process of Buildings: A Systematic Literature Review. *Sustainability* **2022**, *14*, 16967. [CrossRef]
8. Kylili, A.; Fokaides, P.A. Policy trends for the sustainability assessment of construction materials: A review. *Sustain. Cities Soc.* **2017**, *35*, 280–288. [CrossRef]
9. Wall, S. CE Marking of Construction Products—Evolution of the European Approach to Harmonisation of Construction Products in the Light of Environmental Sustainability Aspects. *Sustainability* **2021**, *13*, 6396. [CrossRef]
10. Moncaster, A.; Malmqvist, T.; Forman, T.; Pomponi, F.; Anderson, J. Embodied carbon of concrete in buildings, Part 2: Are the messages accurate? *Build. Cities* **2022**, *3*, 334–355. [CrossRef]
11. European Commission. *Proposal for a Regulation Laying Down Harmonised Conditions for the Marketing of Construction Products, Amending Regulation (EU) 2019/1020 and Repealing Regulation (EU) 305/2011*; European Commission: Brussels, Belgium, 2022. Available online: https://ec.europa.eu/docsroom/documents/49315?locale=en (accessed on 12 February 2023).

12. Martinuzzi, A.; Schönherr, N. Introduction: The Sustainable Development Goals and the Future of Corporate Sustainability. In *Business and the Sustainable Development Goals*; Schönherr, N., Martinuzzi, A., Eds.; Palgrave Pivot: Cham, Switzerland, 2019; pp. 1–17.
13. D'Amato, D.; Toppinen, A.; Kozak, R. *The Role of Business in Global Sustainability Transformations*; Routledge: London, UK, 2022.
14. Shasi; Centobelli, P.; Cerchione, R.; Ertz, M.; Oropallo, E. What we learn is what we earn from sustainable and circular construction. *J. Clean. Prod.* **2023**, *382*, 135183. [CrossRef]
15. Wittmayer, J.M.; Schäpke, N. Action, research and participation: Roles of researchers in sustainability transitions. *Sustain. Sci.* **2014**, *9*, 483–496. [CrossRef]
16. Svenfelt, Å.; Alfredsson, E.C.; Bradley, K.; Fauré, E.; Finnveden, G.; Fuehrer, P.; Gunnarsson-Östling, U.; Isaksson, K.; Malmaeus, M.; Malmqvist, T.; et al. Scenarios for sustainable futures beyond GDP growth 2050. *Futures* **2019**, *111*, 1–14. [CrossRef]
17. D'Amato, D.; Korhonen, J. Integrating the green economy, circular economy and bioeconomy in a strategic sustainability framework. *Ecol. Econ.* **2021**, *188*, 107143. [CrossRef]
18. Van Gulck, V.; Wastiels, L.; Steeman, M. How to evaluate circularity through an LCA study based on the standards EN 15804 and EN 15978. *Int. J. Life Cycle Assess.* **2022**, *27*, 1249–1266. [CrossRef]
19. European Commission. M/350 EN Standardisation Mandate to CEN. In *Development of Horizontal Standardized Methods for Assessment of the Integrated Environmental Performance of Buildings*; European Commission: Brussels, Belgium, 2004.
20. *EN 15804:2012*; Sustainability of Construction Works—Environmental Product Declarations—Core Rules for the Product Category of Construction Products. European Committee for Standardization (CEN): Brussels, Belgium, 2012.
21. *ISO 14025:2006*; Environmental Labels and Declarations—Type III Environmental Declarations—Principles and Procedures. International Organization for Standardization (ISO): Geneva, Switzerland, 2006.
22. European Commission. *Product Environmental Footprint (PEF) Guide. Consolidated Version*; European Commission: Ispra, Italy, 2012.
23. European Commission. *2013/179/EU: Commission Recommendation of 9 April 2013 on the Use of Common Methods to Measure and Communicate the Life Cycle Environmental Performance of Products and Organisations*; European Commission: Brussels, Belgium, 2013. Available online: http://data.europa.eu/eli/reco/2013/179/oj (accessed on 16 February 2023).
24. *EN 15804:2012+A2:2019*; Sustainability of Construction Works—Environmental Product Declarations—Core Rules for the Product Category of Construction Products. European Committee for Standardization (CEN): Brussels, Belgium, 2019.
25. *EN 15804:2012+A2:2019/AC:2021*; Sustainability of Construction Works—Environmental Product Declarations—Core Rules for the Product Category of Construction Products. European Committee for Standardization (CEN): Brussels, Belgium, 2021.
26. European Commission. *Commission Recommendation of 16.12.2021 on the Use of the Environmental Footprint Methods to Measure and Communicate the Life Cycle Environmental Performance of Products and Organisations*; European Commission: Brussels, Belgium, 2021. Available online: https://environment.ec.europa.eu/publications/recommendation-use-environmental-footprint-methods_en (accessed on 16 February 2023).
27. European Commission. *Understanding Product Environmental Footprint and Organisation Environmental Footprint Methods*; European Commission: Brussels, Belgium, 2021. Available online: https://ec.europa.eu/environment/eussd/smgp/pdf/EF%20simple%20guide_v7_clen.pdf (accessed on 11 March 2023).
28. Anderson, J. Over 130,000 Construction Product EPD Available Globally; ConstructionLCA; 2023. Available online: https://constructionlca.co.uk/2023/03/01/over-130000-construction-product-epd-available-globally/ (accessed on 12 March 2023).
29. Soust-Verdaguer, B.; Palumbo, E.; Llatas, C.; Acevedo, Á.V.; Hoxha, E.; Passer, A. Environmental Product Declarations (EPDs) of construction products in Spain: Current status and future challenges. *IOP Conf. Ser. Earth Environ. Sci.* **2022**, *1078*, 012128. [CrossRef]
30. Attia, S.; Santos, M.C.; Al-Obaidy, M.; Baskar, M. Leadership of EU member States in building carbon footprint regulations and their role in promoting circular building design. *IOP Conf. Ser. Earth Environ. Sci.* **2021**, *855*, 012023. [CrossRef]
31. Soust-Verdaguer, B.; Palumbo, E.; Llatas, C.; Acevedo, Á.V.; Fernández Galvéz, M.D.; Hoxha, E.; Passer, A. The Use of Environmental Product Declarations of Construction Products as a Data Source to Conduct a Building Life-Cycle Assessment in Spain. *Sustainability* **2023**, *15*, 1284. [CrossRef]
32. Dias, A.; Nezami, S.; Silvestre, J.; Kurda, R.; Silva, R.; Martins, I.; de Brito, J. Environmental and economic comparison of natural and recycled aggregates using LCA. *Recycling* **2022**, *7*, 43. [CrossRef]
33. Jordan, N.D. How coordinated sectoral responses to environmental policy increase the availability of product life cycle data. *Int. J. Life Cycle Assess.* **2021**, *26*, 692–706. [CrossRef]
34. Adibi, N.; Mousavi, M.; Escobar, M.M.; Glachant, M.; Adibi, A. Mainstream Use of EPDs in Buildings: Lessons Learned from Europe. In Proceedings of the ISBS 2019 4th International Sustainable Buildings Symposium, Dallas, TX, USA, 18–20 July 2019; Gültekin, A.B., Ed.; IntechOpen Limited: London, UK, 2016; pp. 137–146.
35. Waldman, B.; Huang, M.; Simonen, K. Embodied carbon in construction materials: A framework for quantifying data quality in EPDs. *Build. Cities* **2020**, *1*, 625–636. [CrossRef]
36. Fenga, H.; Hewageb, K.; Sadiqc, R. Comparative Analysis of Environmental Product Declarations on Building Materials—Softwood Lumbers under Different Product Category Rules. In Proceedings of the 1st International Conference on New Horizons in Green Civil Engineering (NHICE-01), Victoria, BC, Canada, 25–27 April 2018.
37. Minkov, N.; Schneider, L.; Lehmann, A.; Finkbeiner, M. Type III environmental declaration programmes and harmonization of product category rules: Status quo and practical challenges. *J. Clean. Prod.* **2015**, *94*, 235–246. [CrossRef]

38. Passer, A.; Lasvaux, S.; Allacker, K.; De Lathauwer, D.; Spirinckx, C.; Wittstock, B.; Kellenberger, D.; Gschösser, F.; Wall, J.; Wallbaum, H. Environmental product declarations entering the building sector: Critical reflections based on 5 to 10 years experience in different European countries. *Int. J. Life Cycle Assess.* **2015**, *20*, 1199–1212. [CrossRef]
39. Anderson, J.; Moncaster, A. Embodied carbon of concrete in buildings, Part 1: Analysis of published EPD. *Build. Cities* **2020**, *1*, 198–217. [CrossRef]
40. Božiček, D.; Kunič, R.; Košir, M. Interpreting environmental impacts in building design: Application of a comparative assertion method in the context of the EPD scheme for building products. *J. Clean. Prod.* **2021**, *279*, 123399. [CrossRef]
41. Stapel, E.; Tozan, B.; Sørensen, C.; Birgisdottir, H. Environmental Product Declarations–an extensive collection of availability, EN15804 revision and the ILCD+ EPD format. *IOP Conf. Ser. Earth Environ. Sci.* **2022**, *1078*, 012108. [CrossRef]
42. Kerr, J.; Rayburg, S.; Neave, M.; Rodwell, J. Comparative Analysis of the Global Warming Potential (GWP) of Structural Stone, Concrete and Steel Construction Materials. *Sustainability* **2022**, *14*, 9019. [CrossRef]
43. Moré, F.B.; Galindro, B.M.; Soares, S.R. Assessing the completeness and comparability of environmental product declarations. *J. Clean. Prod.* **2022**, *375*, 133999. [CrossRef]
44. Bayram, B.; Greiff, K. Life cycle assessment on construction and demolition waste recycling: A systematic review analyzing three important quality aspects. *Int. J. Life Cycle Assess.* **2023**, 1–23. [CrossRef]
45. Röck, M.; Baldereschi, E.; Verellen, E.; Passer, A.; Sala, S.; Allacker, K. Environmental modelling of building stocks–An integrated review of life cycle-based assessment models to support EU policy making. *Renew. Sustain. Energy Rev.* **2021**, *151*, 111550. [CrossRef]
46. Del Rosario, P.; Palumbo, E.; Traverso, M. Environmental product declarations as data source for the environmental assessment of buildings in the context of level(s) and DGNB: How feasible is their adoption? *Sustainability* **2021**, *13*, 6143. [CrossRef]
47. AzariJafari, H.; Guest, G.; Kirchain, R.; Gregory, J.; Amor, B. Towards comparable environmental product declarations of construction materials: Insights from a probabilistic comparative LCA approach. *Build. Environ.* **2021**, *190*, 107542. [CrossRef]
48. Bahrar, M.; Jusselme, T. Development of a new environmental scoring methodology for building products, a French case study. *IOP Conf. Ser. Earth Environ. Sci.* **2022**, *1078*, 012129. [CrossRef]
49. Carstens, A.; Brinkmann, T.; Rapp, B. Deriving Benchmarks for Construction Products Based on Environmental Product Declarations. In *Advances and New Trends in Environmental Informatics Progress: Digital Twins for Sustainability*; Kamilaris, A., Wohlgemuth, V., Karatzas, K., Athanasiadis, I.S., Eds.; Springer International Publishing: Cham, Switzerland, 2021; pp. 35–50.
50. Welling, S.; Ryding, S.O. Distribution of environmental performance in life cycle assessments—Implications for environmental benchmarking. *Int. J. Life Cycle Assess.* **2021**, *26*, 275–289. [CrossRef]
51. Röck, M.; Sørensen, A.; Steinmann, J.; Lynge, K.; Horup, L.H.; Tozan, B.; Le Den, X.; Birgisdottir, H. *Towards Embodied Carbon Benchmarks for Buildings in Europe: #1 Facing the Data Challenge*; Rambøl: Copenhagen, Denmark, 2022.
52. Pedersen, E.; Remmen, A. Challenges with product environmental footprint: A systematic review. *Int. J. Life Cycle Assess.* **2022**, *27*, 342–352. [CrossRef]
53. Durão, V.; Silvestre, J.D.; Mateus, R.; de Brito, J. Assessment and communication of the environmental performance of construction products in Europe: Comparison between PEF and EN 15804 compliant EPD schemes. *Resour. Conserv. Recycl.* **2020**, *156*, 104703. [CrossRef]
54. Galindro, B.M.; Welling, S.; Bey, N.; Olsen, S.I.; Soares, R.; Ryding, S.-O. Making use of life cycle assessment and environmental product declarations. A survey with practitioners. *J. Ind. Ecol.* **2020**, *24*, 965–975. [CrossRef]
55. Andersen, S.C.; Larsen, H.F.; Raffnsøe, L.; Melvang, C. Environmental product declarations (EPDs) as a competitive parameter within sustainable buildings and building materials. *IOP Conf. Ser. Earth Environ. Sci.* **2019**, *323*, 012145. [CrossRef]
56. Gelowitz, M.D.C.; McArthur, J.J. Investigating the effect of environmental product declaration adoption in LEED® on the construction industry: A case study. *Procedia Eng.* **2016**, *145*, 58–65. [CrossRef]
57. Park, J.; Yoon, J.; Kim, K.H. Critical review of the material criteria of building sustainability assessment tools. *Sustainability* **2017**, *9*, 186. [CrossRef]
58. Ferreira, A.; Pinheiro, M.D.; de Brito, J.; Mateus, R. A critical analysis of LEED, BREEAM and DGNB as sustainability assessment methods for retail buildings. *J. Build. Eng.* **2023**, *66*, 105825. [CrossRef]
59. Almeida, R.; Chaves, L.; Silva, M.; Carvalho, M.; Caldas, L. Integration between BIM and EPDs: Evaluation of the main difficulties and proposal of a framework based ON ISO 19650:2018. *J. Build. Eng.* **2023**, *68*, 106091. [CrossRef]
60. Olanrewaju, O.I.; Enegbuma, W.I.; Donn, M. Data quality assurance in Environmental Product Declaration Electronic Database: An integrated Clark-Wilson Model, machine learning and blockchain conceptual framework. In *Architectural Science and User Experience: How can Design Enhance the Quality of Life, In Proceedings of the 55th International Conference of the Architectural Science Association*; Izadpanahi, P., Perugia, F., Eds.; Curtin University: Perth, Australia, 2022; pp. 199–208.
61. *ISO 22057:2022*; Sustainability in Buildings and Civil Engineering Works—Data Templates for the Use of Environmental Product Declarations (EPDs) for Construction Products in Building Information Modelling (BIM). International Organization for Standardization (ISO): Geneva, Switzerland, 2022.
62. Anderson, J.; Rønning, A. Using standards to maximise the benefit of digitisation of construction product Environmental Product Declaration (EPD) to reduce Building Life Cycle Impacts. *E3S Web Conf.* **2022**, *349*, 10003–10008. [CrossRef]
63. Tozan, B.; Stapel, E.; Sørensen, C.; Birgisdottir, H. The influence of EPD data on LCA results. *IOP Conf. Ser. Earth Environ. Sci.* **2022**, *1078*, 012105. [CrossRef]

64. Anderson, J.; Moncaster, A. Embodied carbon, embodied energy and renewable energy: A review of environmental product declarations. *Proc. Inst. Civ. Eng. Struct. Build.* **2022**, 1–12. [CrossRef]
65. Koyamparambath, A.; Adibi, N.; Szablewski, C.; Adibi, S.A.; Sonnemann, G. Implementing artificial intelligence techniques to predict environmental impacts: Case of construction products. *Sustainability* **2022**, *14*, 3699. [CrossRef]
66. Crawford, R.H.; Stephan, A.; Prideaux, F. The EPiC database: Hybrid embodied environmental flow coefficients for construction materials. *Resour. Conserv. Recycl.* **2022**, *180*, 106058. [CrossRef]
67. Talwar, N.; Holden, N.M. The limitations of bioeconomy LCA studies for understanding the transition to sustainable bioeconomy. *Int. J. Life Cycle Assess.* **2022**, *27*, 680–703. [CrossRef] [PubMed]
68. Moncaster, A.M.; Hinds, D.; Cruickshank, H.; Guthrie, P.M.; Crishna, N.; Baker, K.; Beckmann, K.; Jowitt, P.W. A key issue: Knowledge exchange between academia and industry. *Proc. Inst. Civ. Eng. Eng. Sustain.* **2010**, *163*, 167–174.
69. Guerra, B.C.; Shahi, S.; Mollaei, A.; Skaf, N.; Weber, O.; Leite, F.; Haas, C. Circular economy applications in the construction industry: A global scan of trends and opportunities. *J. Clean. Prod.* **2021**, *324*, 129125. [CrossRef]
70. Bajramović, E.; Bajramović, B.; Hodžić, D. Environmental Product Declaration as proof of producers awareness of the product impact on the environment. In Proceedings of the 8th International Professional and Scientific Conference Occupational Safety and Health, Zadar, Croatia, 21–24 September 2022; pp. 539–545.
71. Rasmussen, F.N.; Andersen, C.E.; Wittchen, A.; Hansen, R.N.; Birgisdóttir, H. Environmental product declarations of structural wood: A review of impacts and potential pitfalls for practice. *Buildings* **2021**, *11*, 362. [CrossRef]
72. *EN 12004:2007+A1:2012*; Adhesives for Tiles—Requirements, Evaluation of Conformity, Classification and Designation. European Committee for Standardization (CEN): Brussels, Belgium, 2012.
73. Juarez, R.I.C.; Finnegan, S. The environmental impact of cement production in Europe: A holistic review of existing EPDs. *Clean. Environ. Syst.* **2021**, *3*, 100053. [CrossRef]
74. Michalak, J. Ceramic tile adhesives from the producer's perspective: A literature review. *Ceramics* **2021**, *4*, 378–390. [CrossRef]
75. Stancu, C.; Dębski, D.; Michalak, J. Construction products between testing laboratory and market surveillance: Case study of cementitious ceramic tile adhesives. *Materials* **2022**, *15*, 6167. [CrossRef]
76. *13007-1:2004*; Ceramic Tiles—Grouts and Adhesive—Part. 1: Terms, Definitions and Specifications for Adhesives. International Organization for Standardization (ISO): Geneva, Switzerland, 2004.
77. FEICA—Association of the European Adhesive and Sealant Industry. *Environmental Product Declaration as per ISO 14025 and EN 15804. Modified Mineral Mortars, Group 2*; FEICA: Brussels, Belgium, 2016.
78. Mapei SpA. *Environmental Product Declaration in Accordance with ISO 14025 for Keraflex Maxi S1 Zero, Keraflex Maxi Ultra White*; Mapei SpA: Milan, Italy, 2017.
79. Kerakoll Spa. *Environmental Product Declaration for H40 No Limits adhesive Gel with SAS Technology for Ceramic Tiles and Natural stone*; Kerakoll Spa: Sassuolo, Italy, 2017.
80. Mapei SpA. *Environmental Product Declaration in Accordance with ISO 14025 for Granirapid (Grey & White), Elastorapid Igrey & White), Kerabond (Grey & White), Isolastic*; Mapei SpA: Milan, Italy, 2016.
81. SG Weber Yapi. *Environmental Product Declaration in Accordance with EN 15804 and ISO 14025 for Cement Based XL Tiles*; SG Weber Yapi: Kemalpaşa, Turkey, 2019.
82. KYK Yapi Kimyasallari. *Environmental Product Declaration in Accordance with ISO 14025 and EN 15804 for Adhesive Mortar*; KYK Yapi Kimyasallari: Eskişehir, Turkey, 2017.
83. Atlas. *Environmental Product Declaration for Atlas Atut*; ITB: Warsaw, Poland, 2020.
84. Atlas. *Environmental Product Declaration for Atlas Zaprawa Klejąca Uelastyczniona*; ITB: Warsaw, Poland, 2020.
85. Tile Council for North America (TCNA). *Environmental Product Declaration. Cement Mortar for Tile Installation. Industry-Wide Report Products Manufactured in North America*; TCA: Anderson, SC, USA, 2016.
86. Termokir. *Environmental Product Declaration. Termokir Tile Adhesives*; AD Series; Termokir: Horshim, Israel, 2022.
87. Santos, T.; Almeida, J.; Silvestre, J.D.; Faria, P. Life cycle assessment of mortars: A review on technical potential and drawbacks. *Constr. Build. Mater.* **2021**, *288*, 123069. [CrossRef]
88. Kägi, T.; Dinkel, F.; Frischknecht, R.; Humbert, S.; Lindberg, J.; De Mester, S.; Ponsioen, T.; Sala, S.; Schenker, U.W. Session "Midpoint, endpoint or single score for decision-making?"—SETAC Europe 25th Annual Meeting. *Int. J. Life Cycle Assess.* **2016**, *21*, 129–132. [CrossRef]
89. Galindro, B.M.; Zanghelini, G.M.; Soares, S.R. Use of benchmarking techniques to improve communication in life cycle assessment: A general review. *J. Clean. Prod.* **2019**, *213*, 143–157. [CrossRef]
90. Dijkstra-Silva, S.; Schaltegger, S.; Beske-Janssen, P. Understanding positive contributions to sustainability. A systematic review. *J. Environ. Manag.* **2022**, *320*, 115802. [CrossRef] [PubMed]
91. Lützkendorf, T. Assessing the environmental performance of buildings: Trends, lessons and tensions. *Build. Res. Inf.* **2018**, *46*, 594–614. [CrossRef]
92. Dórea, R.J.D.S.; Lopes Silva, D.A.; de Almeida Neto, J.A.; Rodrigues, L.B. Environmental Labeling: An Analysis of the Past 22 Years of Research. *J. Int. Consum. Mark.* **2022**, *34*, 184–200. [CrossRef]
93. Lutz, H.; Bayer, R. *Dry Mortars. Ullmann's Encyclopedia of Industrial Chemistry*; Willey Online Library: Hoboken, NJ, USA, 2015.
94. Stowarzyszenie Producentów Cementu. *Deklaracja Środowiskowa III typu—EPD. Cementy CEM I, CEM II, CEM III, CEM IV, CEM V Produkowane w Polsce*; ITB: Warsaw, Poland, 2020.

95. Rahla, K.M.; Mateus, R.; Bragança, L. Selection criteria for building materials and components in line with the circular economy principles in the built environment—A review of current trends. *Infrastructures* **2021**, *6*, 49. [CrossRef]
96. European Commission, Annexes to the Proposal for a Regulation of the European Parliament and of the Council Laying down Harmonised Conditions for the Marketing of Construction Products, Amending Regulation (EU) 2019/1020 and Repealing Regulation (EU) 305/2011. Available online: https://ec.europa.eu/docsroom/documents/49315 (accessed on 10 March 2023).
97. Łukasik, M.; Michałowski, B.; Michalak, J. Assessment of the constancy of performance of cementitious adhesives for ceramic tiles: Analysis of the test results Commissioned by Polish Market Surveillance Authorities. *Appl. Sci.* **2020**, *10*, 6561. [CrossRef]
98. Van Oers, L.; Guinée, J. The abiotic depletion potential: Background, updates, and future. *Resources* **2016**, *5*, 16. [CrossRef]
99. Marsh, E.; Allen, S.; Hattam, L. Tackling uncertainty in life cycle assessments for the built environment: A review. *Build. Environ.* **2022**, *231*, 109941. [CrossRef]
100. Marsh, E.; Orr, J.; Ibell, T. Quantification of uncertainty in product stage embodied carbon calculations for buildings. *Energy Build.* **2021**, *251*, 111340. [CrossRef]
101. Michałowski, B.; Michalak, J. Sustainability-oriented assessment of external thermal insulation composite systems: A case study from Poland. *Cogent Eng.* **2021**, *8*, 1943152. [CrossRef]
102. Andersen, S.C.; Birkved, M. Reconsidering the assessment method of Environmental implications of Circular Economy in the Built Environment. *IOP Conf. Ser. Earth Environ. Sci.* **2022**, *1078*, 012007. [CrossRef]
103. Czarnecki, L.; Kaproń, M. Sustainable construction as a research area. *Int. J. Soc. Mater. Eng. Resour.* **2010**, *17*, 99–106. [CrossRef]
104. Czarnecki, L.; Gemert, D. Innovation in construction materials engineering versus sustainable development. *Bull. Pol. Acad. Sci. Tech. Sci.* **2017**, *65*, 765–771. [CrossRef]
105. Rochikashvili, M.; Bongaerts, J.C. How eco-labelling influences environmentally conscious consumption of construction products. *Sustainability* **2018**, *10*, 351. [CrossRef]
106. Shao, J.; Ünal, E. What do consumers value more in green purchasing? Assessing the sustainability practices from demand side of business. *J. Clean. Prod.* **2019**, *209*, 1473–1483. [CrossRef]
107. Calderon-Monge, E.; Pastor-Sanz, I.; Garcia, F.J.S. Analysis of sustainable consumer behavior as a business opportunity. *J. Bus. Res.* **2020**, *120*, 74–81. [CrossRef]
108. Michalak, J.; Michałowski, B. Understanding of Construction Product Assessment Issues and Sustainability among Investors, Architects, Contractors, and Sellers of Construction Products in Poland. *Energies* **2021**, *14*, 1941. [CrossRef]
109. Michalak, J.; Michałowski, B. Understanding Sustainability of Construction Products: Answers from Investors, Contractors, and Sellers of Building Materials. *Sustainability* **2022**, *14*, 3042. [CrossRef]
110. Heeren, A.J.; Singh, A.S.; Zwickle, A.; Koontz, T.M.; Slagle, K.M.; McCreery, A.C. Is sustainability knowledge half the battle? An examination of sustainability knowledge, attitudes, norms, and efficacy to understand sustainable behaviours. *Int. J. Sustain. High. Educ.* **2016**, *17*, 613–632. [CrossRef]
111. Farinha, C.B.; de Brito, J.; Veiga, R. Mortars and sustainability. In *Eco-Efficient Rendering Mortars. Used of Recycled Materials*; Woodhead Publishing Series in Civil and Structural Engineering; Farinha, C.B., de Brito, J., Veiga, R., Eds.; Woodhead Publishing: Duxford, UK, 2021; pp. 1–6.
112. Fořt, J.; Černý, R. Limited interdisciplinary knowledge transfer as a missing link for sustainable building retrofits in the residential sector. *J. Clean. Prod.* **2022**, *343*, 131079. [CrossRef]
113. Lützkendorf, T. LCA of building materials within the framework of the Construction Products Regulation (CPR) in Europe. *Ce/Papers* **2022**, *5*, 43–47. [CrossRef]

buildings

MDPI

Article

Application of Spectrometry for Determining the Solar Radiation of Deciduous Trees' Shade: A Passive Energy Conservation Approach for Mediterranean Climates

María Luisa del Campo-Hitschfeld [1,2,†], Nicolás Arenas [3,†], Marco Rivera [4,5,†] and Pablo Ballesteros-Pérez [6,*,†]

1 Kipus Technological Centre, Faculty of Engineering, Universidad de Talca, Campus Curicó, Curicó 3340000, Chile; mdelcampo@kipus.cl
2 Department of Construction Engineering, Faculty of Engineering, Universidad de Talca, Campus Curicó, Curicó 3340000, Chile
3 Alumni of Construction Engineering, Faculty of Engineering, Universidad de Talca, Campus Curicó, Curicó 3340000, Chile; narenas.calderon@gmail.com
4 Department of Electrical Engineering, Faculty of Engineering, Universidad de Talca, Campus Curicó, Curicó 3340000, Chile; marcoriv@utalca.cl or marco.rivera@nottingham.ac.uk
5 Power Electronics and Machine Centre, Faculty of Engineering, University of Nottingham, Nottingham NG7 2RD, UK
6 Project Management, Innovation and Sustainability Research Centre (PRINS), Universitat Politècnica de València, Camino de Vera s/n, 46022 Valencia, Spain
* Correspondence: pabbalpe@dpi.upv.es
† These authors contributed equally to this work.

Abstract: Deciduous trees are well known for controlling solar gains in buildings, contributing to energy savings in a sector that consumes 35% of global energy. However, there is still a lack of information about the real thermal impact that deciduous trees have. This work proposes a new method that is cheap and easy to implement to quantify the shading efficiency of different types of deciduous trees in hot seasons. The results can be applied in energy evaluations of buildings. The trees selected belong to the central valley of Chile, which is characterized by hot summers and cold winters. The trees selected can also be found in other parts of the world. A spectrometer is used for measuring the amount of solar radiation (irradiance) that is present in the shadow of trees, measuring wavelengths between 339 nm and 750 nm (mostly within the visible light range). The full referential irradiance spectrum of the site is obtained by calibrating the standard ASTM G-173-03. At the site, the spectrometer is used to obtain the visible light range, while the infrared radiation (IR) and ultraviolet (UV) radiation ranges are obtained from the literature. Our results indicate that the analyzed deciduous trees reduce an average of 82% of the solar radiation. This information will help project designers during the building energy efficiency design phase by representative modeling of the solar radiation gains allowed by deciduous trees.

Keywords: deciduous trees; energy conservation; spectrometer; passive technology; solar radiation control; spectrometry

Citation: del Campo-Hitschfeld, M.L.; Arenas, N.; Rivera, M.; Ballesteros-Pérez, P. Application of Spectrometry for Determining the Solar Radiation of Deciduous Trees' Shade: A Passive Energy Conservation Approach for Mediterranean Climates. *Buildings* **2023**, *13*, 1130. https://doi.org/10.3390/buildings13051130

Academic Editor: Kian Jon Chua

Received: 4 April 2023
Revised: 16 April 2023
Accepted: 21 April 2023
Published: 23 April 2023

1. Introduction

Climate change is a consequence of global warming, generated by excessive amounts of carbon dioxide (CO_2) emitted to the environment [1]. Some consequences of climate change are an increasing number of droughts, floods, and melting glaciers [2].

CO_2 emissions are released to the atmosphere mostly as a result of human activities, where fossil fuels combustion for power generation and energy consumption are the main emitters [3–9]. Acaroglu et al. [7] observed the direct relationship between climate change and energy consumption. They showed that if energy came from renewable sources, energy consumption would not increase the Earth's temperature while maintaining economic growth.

Other authors in [6] evidenced a positive direct relationship between the CO_2 increase in emissions, and global growth of the gross domestic product (GDP). They also showed how this emissions increment comes from a more intensive use of energy to keep higher living standards, but also causes the degradation of many environmental systems [8].

The built environment is one of the biggest energy consumers these days. According to Nejat et al. [10], global energy consumption is approximately distributed as follows: 35% from residential and commercial buildings, 29% from the industry, 33% from transportation, and 3% for other uses. Similar percentages can be found in Chile, where this study was developed: 23% from buildings, 40% from the manufacturing industry (including mining operations), 33% from transportation, and 4.2% from other uses [11].

To reduce energy consumption, increasing energy efficiency is one of the most effective approaches with potentially lower environmental impact [12]. Energy efficiency measures in buildings can be classified into active and passive strategies. Active strategies involve energy consumption to achieve some predefined level of indoor environmental comfort. This is the case with heating, ventilation, and air conditioning (HVAC) or artificial lighting. Active strategies could be very efficient in their energy use but still involve energy consumption for their operation.

Conversely, passive strategies do not consider energy consumption to achieve a reduction of the building energy demand [13,14]. Some examples of passive strategies are elements for solar control gains, thermal insulation, natural light, air infiltration mechanisms, and natural ventilation [15,16]. Most of these elements need to be considered at the architectural and planning stage of the building. They can also involve complementary elements such as landscaping or surrounding objects (sometimes other buildings). Most of these elements and strategies also entail lower investments when considered at the building design stage.

Some passive strategies related to landscaping are the incorporation of water elements (pools, wetlands), hardscape materials (bricks, stones), and vegetation (trees, bushes) [17]. These elements significantly contribute to the hygrothermal comfort of their surrounding spaces, but also protect human beings from solar ultraviolet (UV) radiation [18]. Particularly, trees decrease the temperature around them while providing visual comfort and mental health. They also protect biodiversity and reduce urban heat islands. Hence, trees are especially beneficial in locations with high temperatures [19–22].

Among them, deciduous trees have advantages in climates with hot summers and cold winters [17]. Deciduous trees lose their leaves in autumn, renewing them again in spring. This way, they do not impede solar radiation when more thermal energy is needed in the building (autumn, winter), yet they partially block the solar radiation when the temperature is high (spring, summer) [23]. However, although this contribution is recognized, neither the reduction of solar radiation they generate nor their variation for different types of deciduous trees have been fully addressed. That is why, in this paper, we propose a low-cost, fast, and simple method to measure how effective deciduous trees are when they are blocking solar radiation. Our results will help building designers when anticipating solar control strategies in building energy assessments.

1.1. Literature Review

As mentioned earlier, deciduous trees can be used for passive energy conservation in buildings. During the winter, they allow solar irradiance to trespass on the empty spaces between their branches. In summer, their leaves partially block solar radiation, providing a cooler thermal sensation, while they also contribute to reducing the carbon footprint. Trees also increase soil humidity around them as a consequence of the temperature reduction wherever their shade is projected.

However, what is the real impact of deciduous trees on solar radiation control during summer? It is difficult to quantify, mostly because of their organic nature [24–26]. Some variables are the tree's age, biochemical trails, canopy structure, phenological events (flowering), and other disturbances (such as droughts), which make it difficult to obtain consistent results [27–30]. This great variability suggests the need to study the average contribution of deciduous trees in solar control. However, this contribution is expected to suffer from high variability, depending on the tree. This is why we analyze different types of deciduous trees and use an easy-to-implement and low-cost equipment method to measure that contribution. This way, other researchers and building practitioners will be able to replicate our measurements in different contexts.

Studies developed by some researchers determined that the effect on building energy savings as a result of tree shades remains between 8.8% and 40% [23,24,31,32]. More specifically, authors in [23] determined that deciduous trees generate 15% cooling savings on the west and east façades of a building, 7% on the north one, and a maximum of 40% heating energy savings. Other authors have investigated the benefits of tree shades in building energy savings [33,34]. For example, some of them studied some shade characteristics (extension and quality) to predict the temperature of exterior walls [24]. Others studied the influence of long-wave and short-wave thermal radiation influence of trees at different locations around the west wall of buildings in summer [35]. Similarly, [31] evaluated the impact of trees on the indoor and outdoor thermal comfort and energy demand in courtyard buildings in Iran. All this research generated interesting advances in the effect generated by the shade of trees on the energy efficiency of buildings. However, when it comes to building energy assessments, those results can be hardly transferred to other climates, especially when the building envelope properties and the internal loads are different.

Namely, there is still a significant lack of information about the spectral irradiance blocked by a deciduous tree shade. By knowing the different wavelengths of the light spectrum trespassing the tree's branches and leaves, it should be possible to determine the amount of energy that is effectively controlled by the tree and arrives at its shade. This is why our study involves all light wavelengths to produce representative results from a varied set of deciduous trees. This will allow future studies to derive other variables, such as the temperature or spectrum range under the shade of different types of deciduous trees.

Additionally, when performing previous research studies, different data collection and processing methods have been used (field measurements, numerical simulation methods, thermal infrared remote sensing, etc.) [35–39].

1.1.1. Processing Methods

In general, there are two types of numerical simulation methods commonly used when modeling urban climates and microclimates: the energy balance model and computational fluid dynamics. Energy balance models are based on heat flow stabilization, whereas fluid dynamics models use fluid dynamics equations such as conservation of energy, mass, and momentum. Among the latter, a software widely used to simulate the impact of vegetation in cities is the ENVI-met simulation software [36,40–42]. ENVI-met includes a microclimate model for simulating the interactions between buildings and green areas in a urban environment. However, its accuracy ultimately depends on the input data quality. In the absence of solar control values to be applied as a function of the type of the tree, the margin of error of the results is uncertain.

1.1.2. Data Collection Methods

Authors in [43] used a mobile, up–down lifting tower, and an autoleveling instrument-mounting platform with UV radiation sensors to gather representative spectral light information projected on a building. The authors of [18] measured the UV light with electronic dosimeters between 11:00 to 14:00 to obtain a UV protection factor. However, these approaches failed to capture the effect of a tree shade across all light wavelengths; hence, they were unable to obtain the amount of irradiance present in the shade of the tree. That is why, in our study, we use a spectrometer to measure the solar irradiance trespassing on a tree shade, and with it, we are able to quantify the solar radiation controlled by trees when they are located around a building. This information allows building designers to feed existing software with more realistic data and to estimate the temperature more accurately beneath a tree shade. It will also help them to select the most suitable deciduous tree according to the climate conditions and energy efficiency strategies adopted in each building project.

2. Materials and Methods

2.1. Location

This research was conducted in the central valley of Chile. Chile is a country with different climatic zones due to its 3800 km length. In this country, some effects of climate change have become evident: droughts in most of its territory, melting glaciers, stronger wind gusts, and intense rains in short time periods [44,45].

The central valley of Chile has a Mediterranean climate, with hot temperatures during summer and cold winters. Summer lasts around 5 months, with an average maximum temperature of 30 °C, including an average day temperature of nearly 20 °C [46]. In this region, the insolation rate in January (summer in the southern hemisphere) is 7047 Wh/m^2 per day [47]. Nevertheless, climate change is currently extending the hot seasons' duration [45].

The season with low temperatures in the central valley of Chile takes place between May and August. In those months, the average temperature remains around 7 °C and the average minimum temperature around 4 °C. Additionally, in July (the central month in the winter), the insolation rate drops to 1825 Wh/m^2 per day [47]. Under these conditions, buildings require solar control in months with high temperatures and solar gains in colder months. Thus, deciduous trees represent a suitable alternative to achieve both aims.

There is a wide variety of trees in every climate. That is why this piece of research analyzes the shade of five deciduous trees that can commonly be found in Mediterranean climates (and also in the central valley of Chile). As described later, a spectrometer is used for performing field measurements. The measurements are taken from noon to afternoon, for a future application of these results in west façade building energy simulations. However, other factors such as tree size, nutrition, growth rate, age, and soil quality, which can affect the physical conditions of the tree, are not analyzed in this work. The reason is that these factors could considerably expand the scope of this research and so will be left for future research. Yet, an important contribution of this study is that our method approach can be easily reproduced in any other conditions and locations.

2.2. Instruments

A convenient device for measuring the amount of energy of the light spectral irradiance is the spectroradiometer, or spectrometer for short. Spectrometers measure the amount of electromagnetic radiation (EMR) passing through a substance. The EMR reduction measured by this device represents the reflection and interferences that occur in the body of the substance.

Spectrometers are widely used in chemistry, biochemistry, genetics, and molecular biology, as they allow quantifying the amount of light absorbed at a specific wavelength [48]. Similarly, they also allow indirectly quantifying the amount of chemical substances contained in a solution [48,49]. Additionally, most spectrometers allow the analysis of a wide range of energy frequencies of spectral irradiance, generally between gamma rays and radio waves [50].

The advantages of spectrometers are their ease of use, and that they allow quick and quantitative measurements [51,52]. For example, the authors in [27] used a spectrometer to measure the leaf reflectance of different trees, determining the leaf pigments variation among species. However, to the best of our knowledge, a spectrometer has never been used to determine the spectral irradiance of a tree shade.

A spectrometer is capable of breaking down the light it receives into different wavelengths. Each wavelength has a determined irradiance level depending on its frequency. This allows measuring the total amount of energy of the light spectrum. In the case of trees, their leaves mostly use energy from the visible spectrum to perform photosynthesis. These are known as photosynthetically active regions (PARs), which absorb (instead of reflecting, as do most inorganic materials) most of the energy during this process.

However, within the spectral irradiance, the PAR (corresponding to the visible spectrum between 400 nm and 650 nm) represents approximately 45% from the global solar radiation (GR) [53], whereas IR radiation (between 650 nm and 4000 nm) represents approximately 50% of solar radiation. Hence, more than 70% of light is reflected by and transmitted from leaves. These amounts are similar across different tree species [54].

The authors in [55] used a spectrometer to measure the spectral distribution in a thinned Sitka spruce (Picea sitchensis) forest at noon for three days. They proved the effectiveness of solar radiation to penetrate the tree canopy during cloudy, overcast, and clear sky days, allowing the trees to keep their photosynthesis activity.

More research attention has been given to the analysis of UV light (between 280 nm and 400 nm). There are a few researcherswho have performed studies focusing on the UV end of the light spectrum [56–60]. For example, the authors of [58] developed a UV radiation mathematical model under the shades of Korean trees, while in [59], they determined the percentage of the UV spectrum that tree leaves reflect and transmit. On the other hand, in [60], the authors resorted to polysulphone films adhered to the ground where the shade of a tree was projected to estimate a UV protection factor on a clear sky day. However, the study that most resembles our work is the impact analysis of the UVB (280 nm–320 nm) and UVA (320 nm–400 nm) spectrum presented in [57] regarding the shade of five typical Australian trees. The authors of that study took three measurements in the shade of the tree with a spectrometer on a sunny day in summer. They determined that the average ratio of the UV irradiance was 26% lower under the shade compared to the same spectrum in the sun. However, the objective of that study was to determine whether the UV present in the shade exceeds the occupational limit for UV exposure; thus, it lacked many operational and applicable details for buildings energy efficiency design.

2.3. *Research Method*

The method proposed below seeks to present an alternative that is easy to implement, cheap, and without the need for advanced knowledge to quantify the shading efficiency of different types of deciduous trees during the hot season. As shown in the introduction, the solar control effect of trees is recognized, hence its impact as a passive energy efficiency measured in buildings. The use of the spectrometer with the ASTM G-173-03 standard (created by the American society for testing and materials) allows the replicability of the proposed method, allowing its application in other trees and in other geographical locations. The instruments (spectrometer), tool (the ASTM G-173-03 standard), and the trees analyzed are described below. The workflow summary of the study is also included at the end.

2.3.1. Spectrometer

For performing this research, we used the "GL Optic Mini-Spectrometer" and the "GL Optic SpectroSoft" software. The selected spectrometer has a handy size, which makes it appropriate for field measurements (Figure 1). It has a spectral range of 339–750 nm, and a physical resolution of 1.7 nm–1.8 nm. This equipment also receives and analyzes the UV–visible light, converting it to an electric signal [50]. The GL Optic SpectroSoft software is a spectral analysis tool for color calculation conforming to CIE 1931 2° XYZ; xy; CIE 1964 10°, uv; u′v′. With this equipment, we analyzed the EMR present in the tree shades, which are made up of diffuse radiation, but mainly of the radiation trespassing the tree canopy. We also used this equipment to obtain the GR by measuring directly pointing at the sun. Hence, by comparing EMR measurements directly facing the sun with those that have the tree canopy in between (under the shade), we managed to estimate the amount of energy that a tree can control.

Figure 1. GL optic mini-spectrometer.

2.3.2. On-Site Spectral Solar Irradiance

We also used the ASTM G-173-03 standard as a reference for calculating the solar spectral irradiance on site. This standard emerged intending to normalize the information gathered from the solar light spectrum for photovoltaic systems, determining the relative optical performance of materials. This standard was drafted by the American Society for Testing and Materials (ASTM) and the US government. They developed two terrestrial solar spectral irradiance distribution standards. The two spectra encompass a direct normal spectral irradiance model and a standard total (global, hemispherical, within 2-pi steradian field of view of the tilted plane) spectral irradiance model. Both reference spectra incorporated in the ASTM G-173-03 standard (see Figure 2) can provide solar radiance data for our context and allow comparison with our measurements.

Namely, we used the global normal spectral irradiance model, as it involves the direct and indirect components contributing to the total global (hemispherical) light spectrum. The validation of its use in this paper is carried out by comparing the global radiation curve of the ASTM with the measurement of the spectrometer pointing at the sun, obtaining a difference of 5% regarding the GR measured with the spectrometer. This normalized spectrum allowed us to complete the IR and UV ends of the spectrum that could not be measured by our optical spectrometer on site (Figure 3).

2.3.3. Types of Studied Trees

The research was conducted on five types of deciduous trees: *Liriodendron tulipifera* (Tulip), *Liquidambar styraciflua* (Liquidambar), *Acer negundo* (Maple), *Melia azedarach* (Melia or White Cedar), and *Prunus cerasifera Erhr* (Plum). Images of these trees can be found in Table 1.

These species were chosen because they are common in regions with Mediterranean climates and also in the central valley region of Chile. However, these trees can also be found in different parts of the world, such as coastal, mountainous, and valley regions. These trees are common in areas with an abundance of solar radiation in the summer, whose soils vary from dry to high humidity [61–64]. These species are also frequently used for solar control in buildings. The sample trees we used in our measurements comprised adult trees with no elements nearby that could affect our measurements.

Figure 2. ASTM G-173-03 standard graphic on site, including the extraterrestrial irradiance (solar spectrum at top of atmosphere), global tilted irradiance, and direct (+ circumsolar) irradiance.

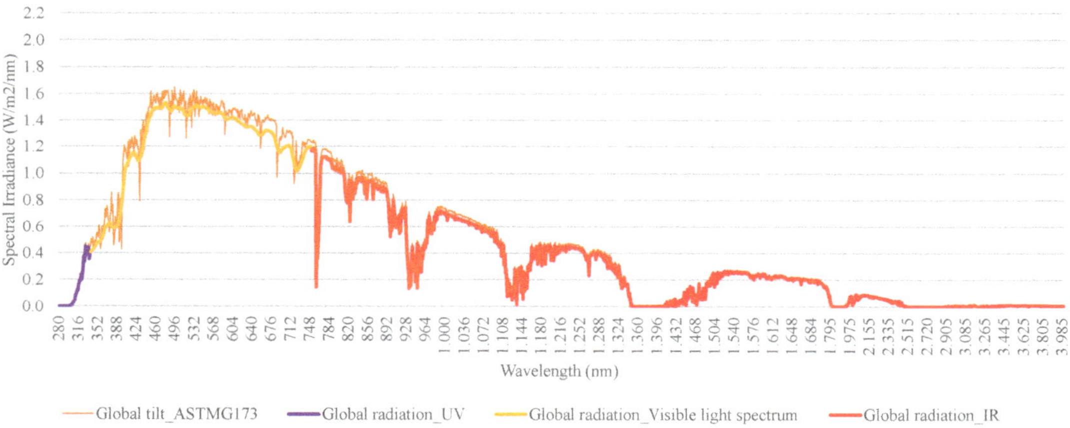

Figure 3. Graphic of the spectral irradiance built from the ASTM G-173-03 standard and the measurements made with the spectrometer.

Table 1. Images of the studied trees.

Liriodendron tulipifera	*Liquidambar styraciflua*	*Acer negundo*	*Melia azedarach*	*Prunus cerasifera*

2.3.4. Measurements of the Solar Control Carried Out by the Tree

To ensure consistency, the measurements in the shadow of each type of tree were always taken by orienting the spectrometer directly to the sun, always on a cloudless day, and between 12:00 to 16:00 h as reported in the supplementary data. The latitude and longitude of the place were approximately −34.81771° and −71.23644°, respectively. The measurements were taken between 17–20 February 2017, with a clear sky. The altitude and azimuth in which the spectrometer was oriented corresponded to 53.82° and 55.42°, respectively, at 12:00 h, until reaching 53.30° and 303.82° at 16:00 h. This time interval was defined as starting at 12:00 (noon) as this is the time (for future applicability of these results) when trees usually cover the west façade of a nearby building, hence enabling solar gain control. From 16:00, however, the sun position casts shadows from nearby elements overlapping with the trees' shadows on most buildings. From 16:00 onwards, the temperature also usually starts to decrease (though slowly).

Hence, 12:00–16:00 with hourly measurement intervals was deemed a representative sampling range. Hourly measurement intervals were also deemed appropriate as they captured significant time deviations in the EMR, but did not entail excessive fieldwork.

To consider the potential shade heterogeneity (because of the leafless areas of the canopy), several shots were taken under the shade of the tree while pointing at the sun holding the spectrometer on a tripod (see Figure 4) to ensure homogeneous measurement conditions (azimuth, altitude). To facilitate this task, the shadow of each tree was divided into quadrants with a 1 m^2 grid. Every hour, a solar radiation measurement with the spectrometer was taken for each quadrant. It should be noted that the grid was drawn on the ground where the shade of the tree was present, as shown in Figure 4. Thus, the measurements are representative of the reduction generated by the tree. Namely, measurements were taken covering the central part of the tree shadow, neglecting the borders (see Figure 5). An average of 14 quadrants were obtained for each tree that was measured. Then, an average value of solar radiation was calculated for every hour for each tree.

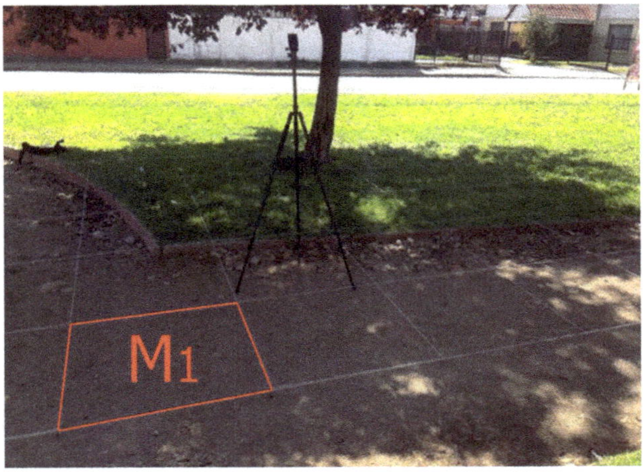

Figure 4. Location of the spectrometer in a quadrant to take the measurements.

As mentioned earlier, to determine the GR, one shot was made by pointing directly at the sun every hour from 12:00 to 16:00. This allowed us to obtain, by its difference with the average solar radiation values obtained under the tree shade, the EMR value blocked by that tree at that particular time of the day. To calculate this difference, the GL SpectroSoft software was used to extract and process the measurement data.

The GL SpectroSoft software displays the solar light radiation measured of each shot in a two-dimension graph on a Cartesian plane such as the one shown in Figure 6. In this figure, the X-axis represents the wavelengths, mainly of the visible light spectrum (from 339 nm to 750 nm), and the Y-axis represents the global irradiance magnitude in milliwatts (mW) per square meter (m^2), that is, mW/m^2.

Figure 5. Measurement grid with canopy edges removed.

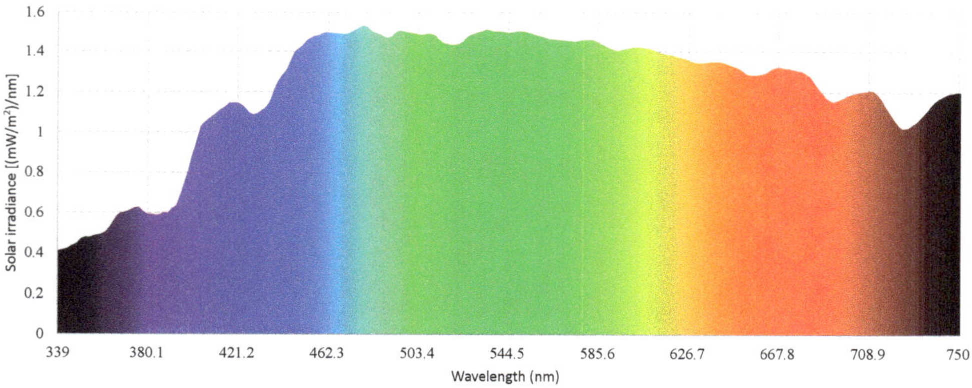

Figure 6. Example graph form SpectroSoft software generated from one shot with the spectrometer.

Hence, all measurements had to be stored numerically (one (X, Y) point per wavelength and global irradiance) through several spreadsheets resembling the one shown in Table 2, where:

- Tree: scientific name of the species.
- Hour: time of the day at which the measurements were performed.
- Measurement date: calendar date of the measurements.
- Direct measurement to the sun: represents the GR, capturing the direct and diffuse solar radiation.
- Measurements 1. . . n: measurements collected under the shade of the tree in quadrants 1. . . n.
- Wave: spectrum wavelength, measured in nanometers (nm).
- Value: global irradiance delivered by the spectrometer based on each wavelength, and measured in milliwatts/square meter (mW/m^2).

Table 2. Database for data processing for each tree.

Tree									
Hour									
Measurement Date									
Direct measurement to the sun		Measurement 1		Measurement 2		Measurement 3		Measurement 4	
Wave (nm)	Value (mW/m^2)	Wave (nm)	Value (mW/m^2)	Wave (nm)	Value (mW/m^2)	Wave (nm)	Value (mW/m^2)	Wave (nm)	Value (mW/m^2)

This way, the visible light spectrum (between 400 nm and 650 nm) was obtained with the spectrometer, along with some wavelengths of the UV and IR spectra (between 339 nm and 400 nm for the UV, and between 650 nm to 750 nm for the IR). The remaining UV and IR tails of the spectrum that were not measured were completed from the literature. Namely, the UV wavelengths were obtained by subtracting 74% from the onsite spectral irradiance as shown in [57], whereas, to determine the EMR of the IR wavelengths, we used the spectral irradiance onsite measurement and subtracted 70% of the irradiance values according to [54].

As a summary, the previous research steps are represented in sequential boxes and shown in Figure 7. In the next section, we will present the main research results.

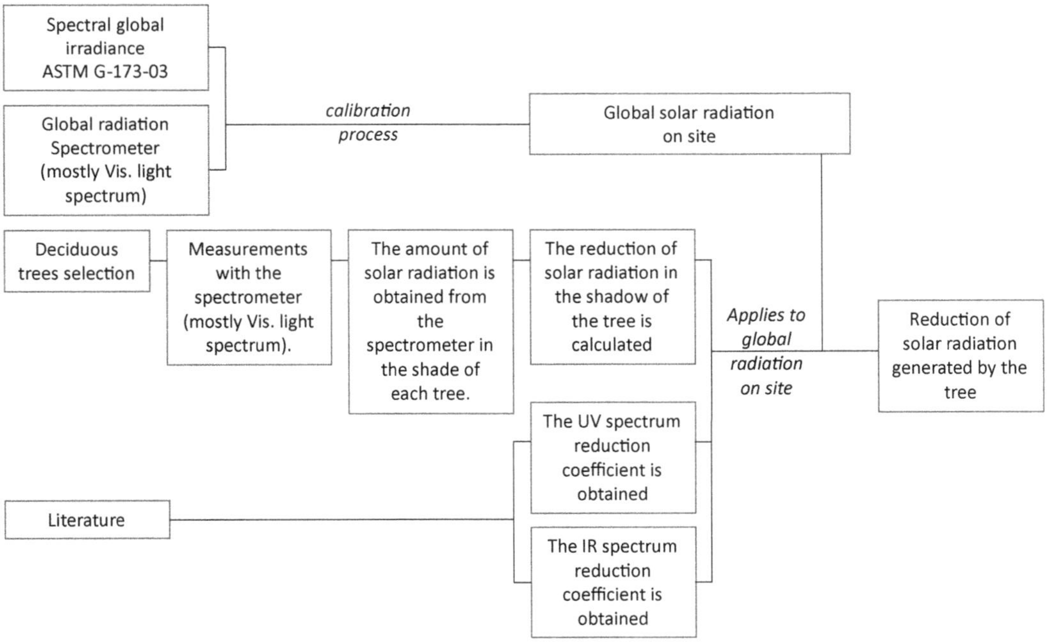

Figure 7. Workflow of this piece of research.

3. Analysis and Results

In this research, all five types of trees were evaluated under the conditions described in the previous section. The overall average solar radiation results measured with the spectrometer are presented later in Table 3. All measurement results can also be found as supplementary material. However, for clarity, the data processing steps followed are presented for one of the trees (Liriodendron Tulipifera). Basically, the GR was obtained first by pointing directly at the sun at hourly intervals (12:00, 13:00, 14:00, 15:00, or 16:00).

This is represented in Figure 8 which represents a measurement taken on 20 February for one of the sample trees. Then, the graph that represents the average measurements of different quadrants under the three (different shots) at that same time of the day for the same tree can also be represented (12:00, as superimposed in Figure 8 for a better comparison). To determine the amount of solar radiation controlled by the tree, the Y values of average spectral irradiance are subtracted from the values of the solar spectrum (Figure 8).

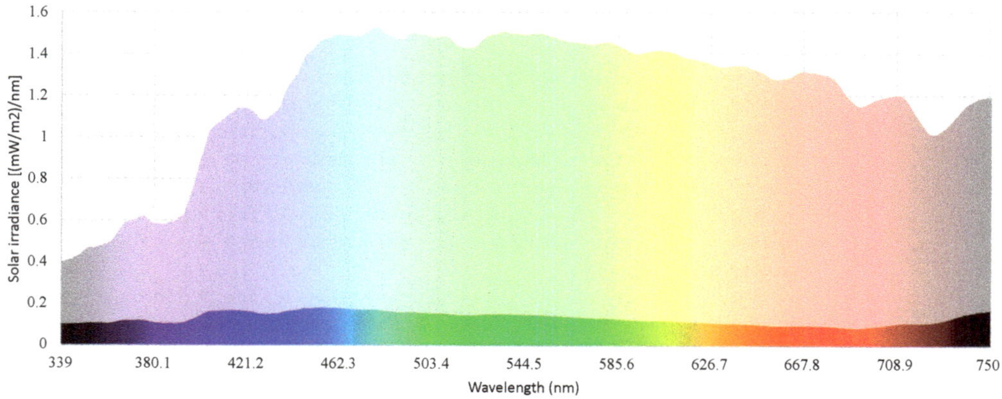

Figure 8. Solar spectrum of the GR measured at 12:00 vs. average spectral irradiance from all quadrants under the tree shade measured at 12:00.

Our final comparisons were calculated through the relative difference of intensities of the GR on site and the radiation value obtained under the shade of the tree (SR), as shown in Equation (1):

$$Relative\ difference = \frac{GR - SR}{GR} * 100\% \quad (1)$$

Equation (1) was evaluated for every nanometer of the spectrum (from 339 nm to 750 nm). Then, an average value was obtained for that tree and that time of the day, repeating these steps for the five hours of a day and the rest of the trees.

Table 3. Solar radiation reduction percentage for the five types of trees calculated with the spectrometer's measurements (from 339 to 750 nm).

	Liquidambar styraciflua L. (%)	*Acer negundo* L. (%)	*Liriodendron tulipifera* L. (%)	*Melia azedarach* L. (%)	*Prunus cerasifera Ehrh* (%)
12:00	90.42	98.29	88.50	89.74	91.25
13:00	92.84	92.96	89.61	79.02	93.60
14:00	86.48	94.50	90.05	87.03	94.67
15:00	90.61	92.54	86.18	88.82	94.58
16:00	88.95	94.44	90.96	87.06	96.06
Average	89.41	95.50	90.01	86.70	93.53

Table 3 and Figure 9 show that the Acer Negundo (Maple) controls the highest amount of solar radiation, with 95.5% on average. A total of 93.53% on average was obtained for the Prunus Cerasifera (Plum) tree. The Melia Azedarach (Melia) tree controls less radiation than the rest (86.70% on average).

It was also found that the solar radiation reduction measured with the spectrometer under the trees does not vary significantly over time (average values at each time of the day vary both and below the average value without following any visible pattern). This means that the radiation control capacity of a tree is not that affected by the solar azimuth and altitude.

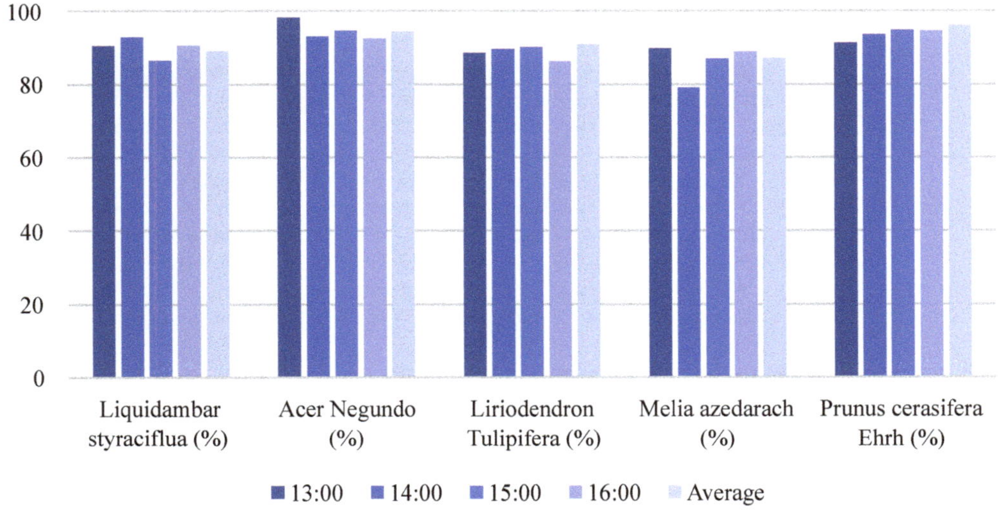

Figure 9. Solar radiation reduction percentage per hour for all trees calculated with the spectrometer's measurements (from 339 to 750 nm).

The greatest data dispersion can be found in the Melia Azendarach (Melia) tree, with a difference of 10.72% between 12:00 and 13:00 h. This tree also presents the second-largest difference of 8.01% between 13:00 and 14:00 h. This higher dispersion may be due to the greater separation between the leaves of this tree, thus allowing larger spaces through which the solar radiation can penetrate.

Furthermore, to obtain the solar radiation in the spectral ranges not measured by the spectrometer, a reduction of 74% was applied to the UV range, as defined in [56,57]; and a 70% reduction to the IR range, as defined in [54]. These reductions allowed obtaining the average results shown in Figure 10.

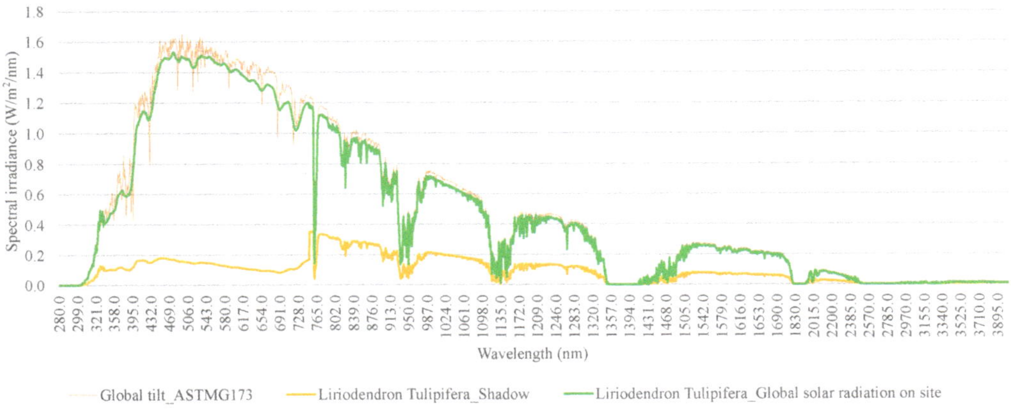

Figure 10. Difference between GR on site and the average solar radiation under the tree shade.

The total solar radiation reduction results, considering the UV, visible light, and IR spectra, are presented in Tables 4–7. The most noteworthy amount of solar radiation reduction was achieved in all trees within the visible light spectrum, with percentages exceeding 86% over the other spectrum ranges. The percentage of average reduction generated by the studied trees was 82%, ranging from 935.45 W/m^2, on average, of GR on site to 167 W/m^2, on average, under the shade of the trees. More specifically, the reduction of each tree is distributed as follows: 79% for the Melia Azedarach, 81% for the Liquidambar styraciflua and Liriodendron Tulipifera, 84% for the Prunus Cerasifera, and 85% for the Acer Negundo.

The largest amount of solar radiation present in the shade of the tree is provided by the IR spectrum (78% on average with respect to the total solar radiation in the shade), while the visible is 18%, and the UV just 4%. Thus, we can notice the great impact of the chlorophyll generation in the leaves, since the GR without the shade of the tree is distributed as follows: 43% of IR, 56% of visible, and 1% of UV, and always reducing, in greater proportion, the spectrum of visible light (between 86% and 95%).

Table 4. Range magnitude of the solar spectrum in shade (SSR)(W/m^2).

Spectral Range	Liriodendron Tulipifera	Prunus	Liquidambar	Melia	Acer
UV	9.03	6.17	8.98	9.57	6.92
VIS	35.85	20.03	35.21	48.30	19.09
IR	130.77	126.56	129.85	134.47	125.23
Total	175.65	152.76	174.04	192.35	151.24

Table 5. Range magnitude of the solar spectrum measured to the sun (GR) (W/m^2).

Spectral Range	Liriodendron Tulipifera	Prunus	Liquidambar	Melia	Acer
UV	43.46	44.36	44.64	42.32	49.97
VIS	347.29	361.12	362.40	355.63	395.21
IR	520.83	525.60	525.35	525.10	533.98
Total	911.59	931.08	932.394	923.06	979.15

Table 6. Reduction percentage of solar radiation (SSR/GR).

Spectral Range	Liriodendron Tulipifera	Prunus	Liquidambar	Melia	Acer
UV	79%	86%	80%	77%	86%
VIS	90%	94%	90%	86%	95%
IR	75%	76%	75%	74%	77%
Total	81%	84%	81%	79%	85%

Finally, it is also worth noting that, although measurements were made on clear (cloudless) days to reduce the noise produced by other uncontrolled weather variables, similar percentages of reduction would have been found under different weather conditions. Hence, in the presence of clouds, the indirect solar radiation controlled should be almost the same. However, in the case of rain, the direct solar radiation component would not be present and the diffuse component should be very low. Therefore, the GR impact of the tree shade on a rainy day would be irrelevant.

Table 7. Range percentage of the solar spectrum in shade (% SSR).

Spectral Range	Liriodendron Tulipifera	Prunus	Liquidambar	Melia	Acer
UV	5%	4%	5%	5%	5%
VIS	20%	13%	20%	25%	13%
IR	74%	83%	75%	70%	83%
Total	100%	100%	100%	100%	100%

4. Discussion

The use of the proposed method was proven to be valid for obtaining the reduction of solar radiation generated by a deciduous tree. The combination of instruments, such as ASTM G-173-03 standard and spectrometry, was demonstrated to be a valid option, easy to implement, with fast results, low calculation requirement, and without the requirement of specific knowledge in the use of evaluation software.

This means that the new method can be easily replicated for other geographical locations and other types of trees, testing the versatility of the proposed method. The level of detail obtained in the results (total irradiance, spectral range, wavelengths) allows for other future applications to be explored.

The use of a spectrometer for measuring solar radiation under a tree shade was shown to be a useful and handy tool for field measurements. Additionally, the software is easy to use and allows almost instant results processing. This is how solar irradiance reduction measurements for other tree shapes, sizes, or canopy densities in other locations or climate conditions, as indicated in [24], could also be calculated by means of spectrometry. The level of detail obtained in the results (total irradiance, spectral range, wavelengths) allows for other future applications to be explored.

On the other hand, the use of the ASTM G-173-03 standard as a basis for constructing the solar irradiance of the site is validated. The solar irradiance results obtained in this study with the spectrometer (between 339 nm and 750 nm) turned out to be almost identical to the ASTM G-173-03 standard for that range, being, on average, 5% lower than ASTM, allowing to adjust and build all the spectral irradiance with that value. This confirms what was observed in [53], obtaining differences below 10% in the waveband of 400 nm to 700 nm.

As suggested in [54], a high percentage of the solar radiation within the visible spectrum is absorbed by the tree, regardless of the tree type. Our work measured an average reduction within the visible spectrum of 91% for the five trees (from 95% of the Acer Negundo to 86% of the Melia Azendarach). This also confirms that the trees capture most of the visible light to perform photosynthesis.

The IR spectrum was measured in [55] with a reduction of 70%, on average, under the shade, a similar value to our study, where the average reduction reached 75% (from 77% of the Acer Negundo and Prunus Cerasifera to 74% of the Melia Azendarach).

The UV spectrum was observed with a reduction of 89% in [60], while authors in [57] obtained an average reduction of 74%. Our study determined that, on average, the UV spectrum was reduced by 82%, with Prunus Cerasifera and the Acer Negundo having 86% and 77% for Melia Azendarach.

From the results obtained, it can be concluded that the trees studied controlled between 79% and 85% of the solar radiation under their shade. These percentages mean an average irradiance reduction from 935.5 W/m^2 to 169.2 W/m^2 by the studied deciduous trees. These results are valid as highly effective elements for passive solar control for adjacent buildings. In that sense, it can be concluded that 169.2 W/m^2 on average would be reaching the external envelope (the shadowed surface) of the building. In other words, the shaded areas of the building envelope would receive only 18% of the solar irradiance of the place, while surfaces without shadow would have the entire irradiance, that is, 935.5 W/m^2.

This reduction of energy gain will generate energy savings, as well as an improvement in the indoor thermal comfort. However, to determine the magnitude of this improvement, the characteristics of the building envelope, thermal inertia, internal loads, occupants' behaviors, and shade surface percentage on the building, will also have significant impacts. This is why the shadow effect will be different for each building according to its characteristics. This is demonstrated in [65], where a statistical model linked with the hedonic characteristics of the building (structure characteristics, occupants' behaviors, and shade density) was developed and in which 3.8% to 20% electricity reduction was achieved. However, there is scarce research addressing these aspects, due to the variability that all these factors generate in the results. Arguably, that is why, unlike [65], other authors consider small samples that do not allow generating general conclusions that can be extended to other building characteristics [34,66].

Thanks to these measurements, it is possible to determine the total energy transferred to a nearby building when performing building energy assessments. The amount of solar energy effectively transmitted and reflected, though, will depend on other factors of the building (exterior surface emissivity, thermal transmittance, and wind speed, among others), which must also be incorporated as input in the energy efficiency analysis software.

5. Conclusions

Deciduous trees constitute a well-known passive energy measure for controlling solar gains in buildings. However, there is a lack of information about the real impact they have. In this work, we quantified the reduction of solar radiation captured by five types of deciduous trees that are easily found in Mediterranean climates and which are characterized by hot summers and cold winters.

To perform the measurements, we used a spectrometer. This is a validated tool that, in a practical and simple way, allows obtaining quantitative values of solar radiation in the field. However, besides direct measurements, we also used the ASTM G-173-03 standard to complete the data from the IR and UV tails of the light spectrum.

Our results indicate that the deciduous trees analyzed dissipate an average of 82% of the UV radiation, 91% of the visible light spectrum, and 75% of the IR radiation. It was also found that the average solar radiation absorption did not vary significantly either by type of tree or by the time of the day. This was unexpected as all selected trees had different branch structures, as well as different leaf shapes and densities. However, the presence of larger spaces between leaves and branches resulted in higher absorption variability over measurements.

This information will help project designers during the building energy efficiency design phase by more representatively modeling the solar radiation gains allowed by deciduous trees. Specifically, this work contributes to providing a first set of quantitative data on the effectiveness of the tree shade for absorbing solar radiation energy. These data can be used by any building energy simulation software to obtain more accurate and realistic outputs. Additionally, our results confirm again the important effect that deciduous trees have on nearby buildings; this is on top of other benefits of trees in urban climates, such as absorbing CO_2 and reducing the carbon footprint.

In future works, we expect to analyze the impact of solar radiation control that evergreen trees may have on different seasons, as photosynthesis activity may affect the solar radiation absorption of the tree leaves during the year. Similarly, we will study the optimum distance of trees to a nearby building depending on the canopy size and the desired level of heat reduction in the building. In this sense, it is also interesting as a future work to evaluate the impact of the shade of deciduous trees in high-rise urban buildings, regarding the energy demand reduction. Just as this study considered the assumption of measuring during the afternoon hours, thinking about the effect on the west façade, one could consider evaluating the solar reduction achieved from the morning to afternoon to evaluate the reduction achieved by a deciduous tree located on a green roof. Finally, as the spectrometer turned out to be a very versatile instrument, future work could be carried out to calculate the energy transmitted inside the building and the indoor thermal comfort achieved, considering the building envelope characteristics.

Supplementary Materials: The following supporting information can be downloaded at: https://www.mdpi.com/article/10.3390/buildings13051130/s1, Measurement data of the studied trees.

Author Contributions: Conceptualization, M.L.d.C.-H.; data curation, N.A.; formal analysis, M.L.d.C.-H., N.A., M.R., P.B.-P.; validation, M.L.d.C.-H., M.R., P.B.-P.; writing—original draft, M.L.d.C.-H., M.R.; writing—review and editing, M.R. and P.B.-P. All authors have read and agreed to the published version of the manuscript.

Funding: This research was funded by FIC-R Maule Project BIP N°40.047.270-0; ANID/ATE220023 Project; FONDECYT Regular Research Project 1220556; CLIMAT AMSUD 21001 and FONDAP SERC Chile 15110019.

Data Availability Statement: The data presented in this study are available as Supplementary Material.

Acknowledgments: The authors acknowledge the technical support given by Miguel Lagos and Claudio Tenreiro from the Faculty of Engineering and the Technological Center Kipus Universidad de Talca, Chile. The authors also thank the support of Alejandra Schueftan from the Universidad Austral de Chile, and Maureen Trebilcock from the Universidad del Bío-Bío for their contribution to the review of this work.

Conflicts of Interest: The authors declare no conflict of interest.

Abbreviations

The following abbreviations are used in this manuscript:

ASTM	American society for testing and materials
IR	Infrared radiation
UV	Ultraviolet
CO_2	Carbon dioxide
GDP	Global domestic product
HVAC	Heating, ventilation, and air conditioning
UVA	Ultraviolet A
UVB	Ultraviolet B
EMR	Electromagnetic radiation
PAR	Photosynthetically active region
US	United States
GR	Global solar radiation on site
SR	Shade of the tree
SSR	Solar spectrum in shadow
VIS	Visible light

References

1. Heun, M.K.; Brockway, P.E. Meeting 2030 primary energy and economic growth goals: Mission impossible? *Appl. Energy* **2019**, *251*, 112697. [CrossRef]
2. Dodman, D.; Hayward, B.; Pelling, M.; Broto, V.C.; Chow, W.; Chu, E.; Dawson, R.; Khirfan, L.; McPhearson, T.; Prakash, A.; et al. Cities, Settlements and Key Infrastructure. In *Climate Change 2022: Impacts, Adaptation, and Vulnerability. Contribution of Working Group II to the Sixth Assessment Report of the Intergovernmental Panel on Climate Change*; Cambridge University Press: Cambridge, UK; New York, NY, USA, 2022; pp. 907–1040. [CrossRef]
3. Ortiz, P.; Kubler, S.; Rondeau, É.; McConky, K.; Shukhobodskiy, A.A.; Colantuono, G.; Georges, J.P. Greenhouse gas emission reduction in residential buildings: A lightweight model to be deployed on edge devices. *J. Clean. Prod.* **2022**, *368*, 133092. [CrossRef]
4. Kongboon, R.; Gheewala, S.H.; Sampattagul, S. Greenhouse gas emissions inventory data acquisition and analytics for low carbon cities. *J. Clean. Prod.* **2022**, *343*, 130711. [CrossRef]
5. Islam, A.; Teo, S.H.; Ng, C.H.; Taufiq-Yap, Y.H.; Choong, S.Y.T.; Awual, M.R. Progress in recent sustainable materials for greenhouse gas (NOx and SOx) emission mitigation. *Prog. Mater. Sci.* **2023**, *132*, 101033. [CrossRef]
6. Karaaslan, A.; Çamkaya, S. The relationship between CO_2 emissions, economic growth, health expenditure, and renewable and non-renewable energy consumption: Empirical evidence from Turkey. *Renew. Energy* **2022**, *190*, 457–466. [CrossRef]

7. Acaroğlu, H.; Güllü, M. Climate change caused by renewable and non-renewable energy consumption and economic growth: A time series ARDL analysis for Turkey. *Renew. Energy* **2022**, *193*, 434–447. [CrossRef]
8. Kober, T.; Schiffer, H.W.; Densing, M.; Panos, E. Global energy perspectives to 2060–WEC's World Energy Scenarios 2019. *Energy Strategy Rev.* **2020**, *31*, 100523. [CrossRef]
9. Zhang, J.; Lu, G.; Skitmore, M.; Ballesteros-Pérez, P. A critical review of the current research mainstreams and the influencing factors of green total factor productivity. *Environ. Sci. Pollut. Res.* **2021**, *28*, 35392–35405. [CrossRef]
10. Nejat, P.; Jomehzadeh, F.; Taheri, M.M.; Gohari, M.; Majid, M.Z.A. A global review of energy consumption, CO_2 emissions and policy in the residential sector (with an overview of the top ten CO_2 emitting countries). *Renew. Sustain. Energy Rev.* **2015**, *43*, 843–862. [CrossRef]
11. Comisión Nacional de Energía, Chile. *Anuario Estadístico de Energía*; Comision Nacional de Energía: Santiago de Chile, Chile, 2021.
12. Himeur, Y.; Alsalemi, A.; Al-Kababji, A.; Bensaali, F.; Amira, A. Data fusion strategies for energy efficiency in buildings: Overview, challenges and novel orientations. *Inf. Fusion* **2020**, *64*, 99–120. [CrossRef]
13. Mujeebu, M.A.; Bano, F. Integration of passive energy conservation measures in a detached residential building design in warm humid climate. *Energy* **2022**, *255*, 124587. [CrossRef]
14. Yao, R.; Costanzo, V.; Li, X.; Zhang, Q.; Li, B. The effect of passive measures on thermal comfort and energy conservation. A case study of the hot summer and cold winter climate in the Yangtze River region. *J. Build. Eng.* **2018**, *15*, 298–310. [CrossRef]
15. Al-Marri, W.; Al-Habaibeh, A.; Abdo, H. Exploring the relationship between energy cost and people's consumption behaviour. *Energy Procedia* **2017**, *105*, 3464–3470. [CrossRef]
16. Delzendeh, E.; Wu, S.; Lee, A.; Zhou, Y. The impact of occupants' behaviours on building energy analysis: A research review. *Renew. Sustain. Energy Rev.* **2017**, *80*, 1061–1071. [CrossRef]
17. Rouhollahi, M.; Whaley, D.; Behrend, M.; Byrne, J.; Boland, J. The role of residential tree arrangement: A scoping review of energy efficiency in temperate to subtropical climate zones. *Renew. Sustain. Energy Rev.* **2022**, *158*, 112155. [CrossRef]
18. Sivarajah, S.; Thomas, S.C.; Smith, S.M. Evaluating the ultraviolet protection factors of urban broadleaf and conifer trees in public spaces. *Urban For. Urban Green.* **2020**, *51*, 126679. [CrossRef]
19. Shahidan, M.F.; Jones, P.J.; Gwilliam, J.; Salleh, E. An evaluation of outdoor and building environment cooling achieved through combination modification of trees with ground materials. *Build. Environ.* **2012**, *58*, 245–257. [CrossRef]
20. Salata, F.; Golasi, I.; de Lieto Vollaro, R.; de Lieto Vollaro, A. Urban microclimate and outdoor thermal comfort. A proper procedure to fit ENVI-met simulation outputs to experimental data. *Sustain. Cities Soc.* **2016**, *26*, 318–343. [CrossRef]
21. Kaloustian, N.; Diab, Y. Effects of urbanization on the urban heat island in Beirut. *Urban Clim.* **2015**, *14*, 154–165. . [CrossRef]
22. Duarte, D.H.; Shinzato, P.; dos Santos Gusson, C.; Alves, C.A. The impact of vegetation on urban microclimate to counterbalance built density in a subtropical changing climate. *Urban Clim.* **2015**, *14*, 224–239. . [CrossRef]
23. Rouhollahi, M.; Whaley, D.; Byrne, J.; Boland, J. Potential residential tree arrangement to optimise dwelling energy efficiency. *Energy Build.* **2022**, *261*, 111962. [CrossRef]
24. Berry, R.; Livesley, S.J.; Aye, L. Tree canopy shade impacts on solar irradiance received by building walls and their surface temperature. *Build. Environ.* **2013**, *69*, 91–100. [CrossRef]
25. Schlerf, M.; Atzberger, C.; Hill, J. Remote sensing of forest biophysical variables using HyMap imaging spectrometer data. *Remote Sens. Environ.* **2005**, *95*, 177–194. [CrossRef]
26. Alavipanah, S.K.; Karimi Firozjaei, M.; Sedighi, A.; Fathololoumi, S.; Zare Naghadehi, S.; Saleh, S.; Naghdizadegan, M.; Gomeh, Z.; Arsanjani, J.J.; Makki, M.; et al. The Shadow Effect on Surface Biophysical Variables Derived from Remote Sensing: A Review. *Land* **2022**, *11*, 2025. [CrossRef]
27. Wang, R.; Gamon, J.A.; Cavender-Bares, J. Seasonal patterns of spectral diversity at leaf and canopy scales in the Cedar Creek prairie biodiversity experiment. *Remote Sens. Environ.* **2022**, *280*, 113169. [CrossRef]
28. Xie, X.; Yang, Y.; Li, W.; Liao, N.; Pan, W.; Su, H. Estimation of Leaf Area Index in a Typical Northern Tropical Secondary Monsoon Rainforest by Different Indirect Methods. *Remote Sens.* **2023**, *15*, 1621. [CrossRef]
29. Kamarianakis, Z.; Panagiotakis, S. Design and Implementation of a Low-Cost Chlorophyll Content Meter. *Sensors* **2023**, *23*, 2699. [CrossRef]
30. Zou, X.; Jin, J.; Mõttus, M. Potential of Satellite Spectral Resolution Vegetation Indices for Estimation of Canopy Chlorophyll Content of Field Crops: Mitigating Effects of Leaf Angle Distribution. *Remote Sens.* **2023**, *15*, 1234. [CrossRef]
31. Darvish, A.; Eghbali, G.; Eghbali, S.R. Tree-configuration and species effects on the indoor and outdoor thermal condition and energy performance of courtyard buildings. *Urban Clim.* **2021**, *37*, 100861. [CrossRef]
32. Han, H.; Han, C.; Huang, L.; Lan, T.; Xue, X. Irradiance Restoration Based Shadow Compensation Approach for High Resolution Multispectral Satellite Remote Sensing Images. *Sensors* **2020**, *20*, 6053. [CrossRef]
33. Morakinyo, T.E.; Balogun, A.A.; Adegun, O.B. Comparing the effect of trees on thermal conditions of two typical urban buildings. *Urban Clim.* **2013**, *3*, 76–93. [CrossRef]
34. Akbari, H.; Kurn, D.M.; Bretz, S.E.; Hanford, J.W. Peak power and cooling energy savings of shade trees. *Energy Build.* **1997**, *25*, 139–148. [CrossRef]
35. Zhang, T.; Spence, C.; Qi, F.; Yang, W.; Song, T. Influence of tree location on thermal radiation disturbance of the west wall of summer buildings. *Energy Build.* **2022**, *273*, 112359. [CrossRef]

36. Ouyang, W.; Sinsel, T.; Simon, H.; Morakinyo, T.E.; Liu, H.; Ng, E. Evaluating the thermal-radiative performance of ENVI-met model for green infrastructure typologies: Experience from a subtropical climate. *Build. Environ.* **2022**, *207*, 108427. [CrossRef]
37. Neinavaz, E.; Schlerf, M.; Darvishzadeh, R.; Gerhards, M.; Skidmore, A.K. Thermal infrared remote sensing of vegetation: Current status and perspectives. *Int. J. Appl. Earth Obs. Geoinf.* **2021**, *102*, 102415. [CrossRef]
38. Zhang, L.; Sun, X.; Wu, T.; Zhang, H. An Analysis of Shadow Effects on Spectral Vegetation Indexes Using a Ground-Based Imaging Spectrometer. *IEEE Geosci. Remote Sens. Lett.* **2015**, *12*, 2188–2192. [CrossRef]
39. Kováč, D.; Ač, A.; Šigut, L.; Peñuelas, J.; Grace, J.; Urban, O. Combining NDVI, PRI and the quantum yield of solar-induced fluorescence improves estimations of carbon fluxes in deciduous and evergreen forests. *Sci. Total Environ.* **2022**, *829*, 154681. [CrossRef]
40. Sinsel, T.; Simon, H.; Ouyang, W.; dos Santos Gusson, C.; Shinzato, P.; Bruse, M. Implementation and evaluation of mean radiant temperature schemes in the microclimate model ENVI-met. *Urban Clim.* **2022**, *45*, 101279. [CrossRef]
41. Crank, P.J.; Sailor, D.J.; Ban-Weiss, G.; Taleghani, M. Evaluating the ENVI-met microscale model for suitability in analysis of targeted urban heat mitigation strategies. *Urban Clim.* **2018**, *26*, 188–197. [CrossRef]
42. Liu, Z.; Cheng, W.; Jim, C.; Morakinyo, T.E.; Shi, Y.; Ng, E. Heat mitigation benefits of urban green and blue infrastructures: A systematic review of modeling techniques, validation and scenario simulation in ENVI-met V4. *Build. Environ.* **2021**, *200*, 107939. [CrossRef]
43. Yang, X.; Miller, D.R.; Montgomery, M.E. Vertical distributions of canopy foliage and biologically active radiation in a defoliated/refoliated hardwood forest. *Agric. For. Meteorol.* **1993**, *67*, 129–146. [CrossRef]
44. Ricalde, I.; Vicuña, S.; Melo, O.; Tomlinson, J.E.; Harou, J.J.; Characklis, G. Assessing tradeoffs in the design of climate change adaptation strategies for water utilities in Chile. *J. Environ. Manag.* **2022**, *302*, 114035. [CrossRef] [PubMed]
45. Alaniz, A.J.; Smith-Ramírez, C.; Rendón-Funes, A.; Hidalgo-Corrotea, C.; Carvajal, M.A.; Vergara, P.M.; Fuentes, N. Multiscale spatial analysis of headwater vulnerability in South-Central Chile reveals a high threat due to deforestation and climate change. *Sci. Total Environ.* **2022**, *849*, 157930. [CrossRef] [PubMed]
46. Mora-Melià, D.; López-Aburto, C.S.; Ballesteros-Pérez, P.; Muñoz-Velasco, P. Viability of Green Roofs as a Flood Mitigation Element in the Central Region of Chile. *Sustainability* **2018**, *10*, 1130. [CrossRef]
47. Instituto Nacional de Normalización, Chile. *NCh 1079 Architecture and Construction—Climatic Zoning for Dwellings for Chile and Recommendations for Architectural Design. Norma Chilena 1079*; Instituto Nacional de Normalización: Santiago de Chile, Chile, 2019.
48. Parmar, A.; Sharma, S. Derivative UV-vis absorption spectra as an invigorated spectrophotometric method for spectral resolution and quantitative analysis: Theoretical aspects and analytical applications: A review. *TrAC Trends Anal. Chem.* **2016**, *77*, 44–53. [CrossRef]
49. Cavaco, A.M.; Utkin, A.B.; Marques da Silva, J.; Guerra, R. Making Sense of Light: The Use of Optical Spectroscopy Techniques in Plant Sciences and Agriculture. *Appl. Sci.* **2022**, *12*, 997. [CrossRef]
50. Passos, M.L.; Saraiva, M.L.M. Detection in UV-visible spectrophotometry: Detectors, detection systems, and detection strategies. *Measurement* **2019**, *135*, 896–904. [CrossRef]
51. Alexandre-Franco, M.F.; Fernández-González, C.; Reguero-Padilla, G.; Cuerda-Correa, E.M. Olive-tree polyphenols and urban mining. A greener alternative for the recovery of valuable metals from scrap printed circuit boards. *Environ. Res.* **2022**, *214*, 114112. [CrossRef] [PubMed]
52. Worsfold, P.J.; Zagatto, E.A. Spectrophotometry Overview. In *Encyclopedia of Analytical Science*, 3rd ed.; Worsfold, P., Poole, C., Townshend, A., Miró, M., Eds.; Academic Press: Oxford, UK, 2019; pp. 244–248. [CrossRef]
53. McCree, K. The measurement of photosynthetically active radiation. *Sol. Energy* **1973**, *15*, 83–87. [CrossRef]
54. Deng, J.; Pickles, B.J.; Kavakopoulos, A.; Blanusa, T.; Halios, C.H.; Smith, S.T.; Shao, L. Concept and methodology of characterising infrared radiative performance of urban trees using tree crown spectroscopy. *Build. Environ.* **2019**, *157*, 380–390. [CrossRef]
55. Dengel, S.; Grace, J.; MacArthur, A. Transmissivity of solar radiation within a Picea sitchensis stand under various sky conditions. *Biogeosciences* **2015**, *12*, 3825–3853. [CrossRef]
56. Parisi, A.; Kimlin, M.; Wong, J.; Wilson, M. Diffuse component of solar ultraviolet radiation in tree shade1Paper presented at the 2nd Online Conference for Photochemistry and Photobiology 1. *J. Photochem. Photobiol. Biol.* **2000**, *54*, 116–120. [CrossRef] [PubMed]
57. Parisi, A.V.; Kimlin, M.G. Comparison of the spectral biologically effective solar ultraviolet in adjacent tree shade and sun. *Phys. Med. Biol.* **1999**, *44*, 2071. [CrossRef] [PubMed]
58. Na, H.R.; Heisler, G.M.; Nowak, D.J.; Grant, R.H. Modeling of urban trees' effects on reducing human exposure to UV radiation in Seoul, Korea. *Urban For. Urban Green.* **2014**, *13*, 785–792. [CrossRef]
59. Yoshimura, H.; Zhu, H.; Wu, Y.; Ma, R. Spectral properties of plant leaves pertaining to urban landscape design of broad-spectrum solar ultraviolet ra-diation reduction. *Int. J. Biometeorol.* **2010**, *54*, 179–191. [CrossRef]
60. Gies, P.; Elix, R.; Lawry, D.; Gardner, J.; Hancock, T.; Cockerell, S.; Roy, C.; Javorniczky, J.; Henderson, S. Assessment of the UVR Protection Provided by Different Tree Species. *Photochem. Photobiol.* **2007**, *83*, 1465–1470. [CrossRef]
61. NC State University, N. A. S. U. North Carolina Extension Gardener Plant Toolbox. Available online: https://plants.ces.ncsu.edu/ (accessed on 10 June 2022).
62. Ministerio del Medio Ambiente. National Geographic Society, California Academy of Science. iNaturalist Chile. 2022. Available online: https://inaturalist.mma.gob.cl (accessed on 12 June 2022).

63. Seifert, J R.; Jacobs, D.F.; Selig, M.F. Influence of seasonal planting date on field performance of six temperate deciduous forest tree species. *For. Ecol. Manag.* **2006**, *223*, 371–378. [CrossRef]
64. GBIF. Global Biodiversity Information Facility. 2022. Available online: https://www.gbif.org (accessed on 11 June 2022).
65. Pandit, R.; Laband, D.N. A Hedonic Analysis of the Impact of Tree Shade on Summertime Residential Energy Consumption. *Arboric. Urban For. Online* **2010**, *36*, 73–80. . [CrossRef]
66. Chagolla, M.; Álvarez-García, G.; Simá, E.; Tovar, R.; Huelsz, G. Effect of Tree Shading on the Thermal Load of a House in a Warm Climate Zone in Mexico. In Proceedings of the ASME International Mechanical Engineering Congress and Exposition, Proceedings (IMECE), Houston, TX, USA, 9–15 November 2012. [CrossRef]

buildings

MDPI

Article

Evaluating BIPV Façades in a Building Envelope in Hot Districts for Enhancing Sustainable Ranking: A Saudi Arabian Perspective

Esam M. H. Ismaeil [1,2,*] and Abu Elnasr E. Sobaih [3,4,*]

1 Civil and Environment Department, College of Engineering, King Faisal University, Al-Ahsa 31982, Saudi Arabia
2 Architecture and Urban Planning Department, Faculty of Engineering, Port Said University, Port Said 42526, Egypt
3 Management Department, College of Business Administration, King Faisal University, Al-Ahsa 31982, Saudi Arabia
4 Hotel Management Department, Faculty of Tourism and Hotel Management, Helwan University, Cairo 12612, Egypt
* Correspondence: emohamed@kfu.edu.sa (E.M.H.I.); asobaih@kfu.edu.sa (A.E.E.S.)

Abstract: Enhancing contractual construction project documents with sustainability and green building requirements reflects growing concerns for the majority of organizations in hot zone districts. The aim is to provide a healthy, best functional performance, safe environment with occupant comfort, and an efficient building performance as an environmental-friendly building. This research study develops a holistic evaluation system for the façade composite of contractual documents. The aim of the current study was to enhance building energy performance under the sustainability rating system focusing on adapting active envelope energy applications. The research used technical evaluation with energy simulation based PVsyst V7.1.0 software and contractual status evaluation for an ongoing unique case study project in Saudi Arabia. Feasibility analysis was carried out for a sustainable active envelope using the adopted specifications of the Building Integrated Photovoltaics (BIPV) façade item instead of the contractual passive item in the Giftedness and Creativity Center project. The project was registered in the sustainability rating system called Leadership in Energy and Environmental Design (LEED). The results showed that using BIPV facades as an active renewable energy source enhances building energy performance over the project life cycle. Additionally, it generates 68% of energy demand as a nearly-zero energy project. Several other advantages include lower cost than tender cost without any contractual conflicts, energy savings per year, project upgrade to the platinum certificate, added value to the public investment, CO_2 emission reduction, and barrels of oil saved.

Keywords: sustainable rating system; building integrated photovoltaics (BIPV) facades; sustainable building; nearly-zero energy; energy performance

Citation: Ismaeil, E.M.H.; Sobaih, A.E.E. Evaluating BIPV Façades in a Building Envelope in Hot Districts for Enhancing Sustainable Ranking: A Saudi Arabian Perspective. *Buildings* **2023**, *13*, 1110. https://doi.org/10.3390/buildings13051110

Academic Editors: Paulo Santos and Mark Bomberg

Received: 7 March 2023
Revised: 11 April 2023
Accepted: 17 April 2023
Published: 22 April 2023

1. Introduction

Ensuring sustainable future is a key objective of the UN 2030 vision, which is highly acknowledged by the UN General Assembly under the Sustainable Development Goals (SDGs) which aim at achieving a comfortable and safe life, planet protection, development plans to prevent hunger and poverty, and reducing environmental degradation [1]. The SDGs include 17 goals, 169 targets, and 244 indicators [2,3]. The goal number 7 in SDGs aims to enclose the affordable and adoption of clean energy, which has become the priority of many countries worldwide.

Saudi Arabia is the top country among the Gulf cooperation countries GCC in electricity consumption. Figure 1 illustrates the growth consumption of electricity for the six GCC countries from 1990 to 2018 [4,5]. The carbon emissions from energy consumption have a long-term effect on the economic development in GCC [6–8]. Saudi Arabia is the top among the GCC countries in CO_2 emissions where it produces 471.82 M tonnes of CO_2 (18 tonnes of CO_2/person). Figure 2 shows the forecast of CO_2 emissions from 2011 to 2050 in GCC countries [9,10]. Therefore, the GCC countries, via new policies, measures, and legislative instruments, are promoting sustainable urban development and clean energy efficiency aspects, e.g., PV systems technology [10–14].

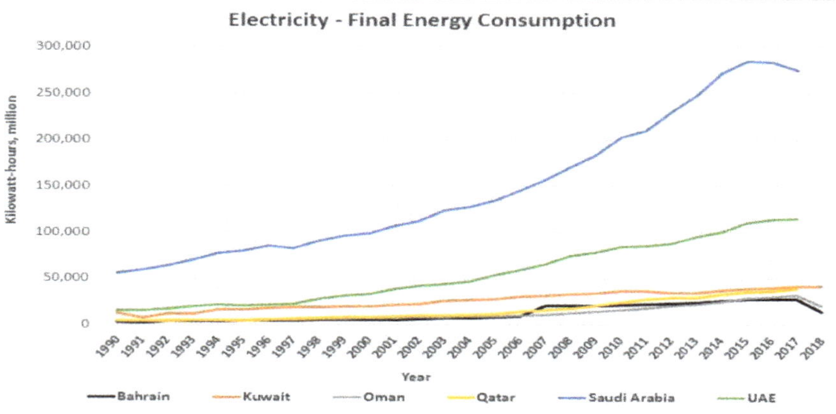

Figure 1. Gulf Cooperation Council (GCC) countries electric power consumption 1990–2018.

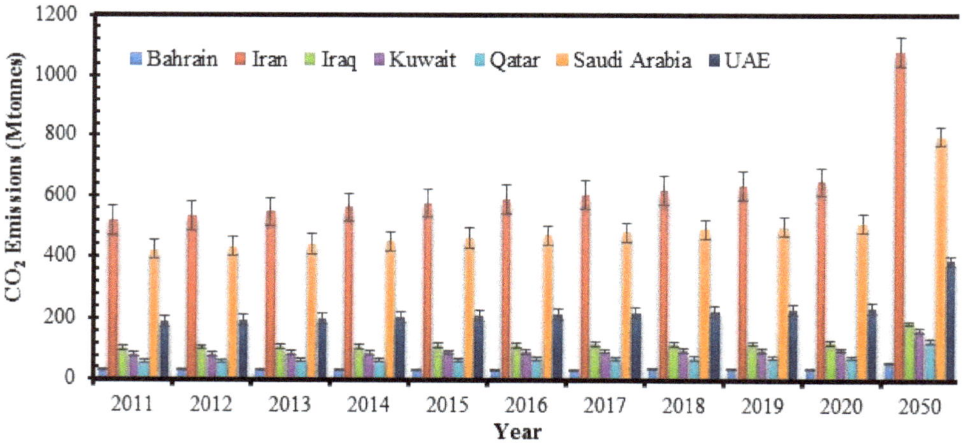

Figure 2. Forecast of CO_2 emissions from 2011 to 2050 in GCC countries.

The Saudi Arabia Kingdom (KSA) with 2030 Vision, is aiming to regulate the newly constructed buildings after the country was labeled with the highest ecological footprint in 2007 [15]. The major goals of KSA 2030 Vision include a good life for society, a flourishing economic status, and an aspirant Saudi nation [16]. The three main goals comprise nine sub-goals with broad line projects and implementation measures, especially in energy aspects, 13 implementation programs to achieve the 96 strategic objectives [17–19]. The report of the renewable energy projects development office at the Ministry of Energy, will reduce the domestic energy consumption which is expected to exceed 120 gigawatts by the year 2032 [20].

The national renewable energy program under the prosperous economy goal aims to establish the largest solar energy project plant in the world (located between 35° north and 35° south of the Kingdom) to generate 200 gigawatts at the cost of USD 200 billion in 2030 via solar plants. The first stage costs USD 5 billion with a capacity of 7.2 gigawatts [21,22]. The second stage is in Sakaka city—Al-Jouf region with 300 megawatts of clean energy for 45,000 housing units. The third stage offers 12 projects in phases with a total value of about USD 4 billion and produces 6.77 MW [21,22]. The solar system in construction projects in Saudi Arabia is still limited, where the use of renewable energy production in building roofs is about 1.6% of the existing renewable energy in Saudi Arabia. The KSA Vision plans to make the Kingdom a global logistics hub, promoting mining and energy industries, and solar systems in construction projects in line with the gulf standards for green building construction projects, which became mandatory procedures for all local municipalities in the major gulf countries [23,24].

1.1. Sustainable Rating Systems

Green and sustainable buildings contribute to a better environment, sustainability processes, and benefits to building owners and users throughout the project lifecycle [25]. The regulations, which were addressed in green buildings, include a coalition of more than 80 countries around the world become mandatory in most of these countries in the building code for all public and private sector construction projects [26]. Qatar has incorporated QSAS and GSAS certification in green building, comprising 140 sustainability assessment mechanisms, divided into eight sections [27,28]. Abu Dhabi green building regulations is running under the name of the pearl rating system in UAE and is the sustainability rating system for UAE to support sustainability from design to implementation to operation, including communities, buildings, and villas [29]. Lebanon has a Lebanese Green Building Association LGBC as a cedar system for green building evaluation and assessment [30]. Saudi Arabia established a sustainable building program and launched a building sustainability assessment "Mostadam" aims to raise the quality of life in residential buildings besides reducing water and electricity consumption, which will positively affect family health, the building internal environment, and reduce the operational cost [22,31].

LEED is an American sustainability rate system that supports buildings to consider a triple bottom-line approach to achieve returns for people, planet, and profit. LEED 2009 consists of rating systems for new design and construction, building operation, houses, and residential neighborhoods. Five overarching categories correspond to the specialties available under the LEED program called LEED rating systems, consisting of credit and prerequisites for the green building certification program [32,33]. The LEED 2009 system established points of potential environmental impacts and human benefits for each credit. LEED v4.1 version includes efficiency selections of energy, water, site, material, daylight, and waste management. The LEED rating system consists of perquisites, credits, and points that could be managed in a points system or a scorecard in eight categories with a total of 110 points. The four certification levels start from certified (40–49 points), then silver (50–59 points), gold (60–79 points), and platinum (80+ points) [32–34].

BREEAM is the British Research Establishment Environmental Assessment Method, which is the sustainability assessment for buildings, master planning, infrastructure, and asset conservation [35,36]. The Australian Green Star rating system uses a robust assessment process. [37]. The common international sustainable building certification and rating systems are LEED, BREEAM, and DGNB—"German Association for Sustainable Building". A comparison between these international well-known rating systems is illustrated in Table 1. The BREEAM and LEED have several advantages, a strong system and large market use with a score of more than 75 points [38,39].

Table 1. Review comparison of well-known sustainable rating systems.

	BREEAM	LEED	CASBEE	Green Star	HK-BEAM
Popularity and influence	10	10	6	5	5
Availability	7	7	7	8	8
Methodology	11	10	13	9	11
Applicability	13	13	11.5	10	9
Data collecting process	7	7	6	9	8
Accuracy and verification	8	7	9	5	5
User-friendliness	8	10	6	8	8
Development	8	8	7	8	8
Results presentation	3	3	4	3	4
Final Score	75	75	69.5	65	66

1.2. Solar BIPV Modules in Building Envelopes

Different types of Solar PV panels serve different needs and purposes, while the classification by generation focuses on the materials and efficiency of different types of solar PV panels. The PV panels in the first-group solar are the traditional types of solar PV panels made of monocrystalline silicon or polysilicon, which are most commonly used with efficiency 21–23%. The PV panels cells in the second-group solar are thin-film solar PV cells (TFSC) such as silicon, cadmium, amorphous silicon solar PV cell, or copper onto a substrate, primarily used for to integrate buildings with photovoltaic power stations or smaller solar PV systems with efficiency 15–41%. The PV panels in the third generation solar include a variety of thin-film technologies; most of them are still in the research or development phase using organic materials. and some using inorganic substances, e.g., CdTe, concentrated PV cell curved mirror surfaces, CVP, and HCVP [40,41] with efficiency 15–18%. These different renewable resources and energy storage systems can reduce CO_2 emissions and costs by 50% [27,42] and affect financial returns [35,38].

The application of photovoltaic PV as a construction element in architectural structures and buildings is an abbreviation of the building-integrated photovoltaic BIPV. The key market driver for building integrated photovoltaics (BIPV) was the European Directive 2010/31/EU [43]. The BIPV facades consider the energy road map for several countries in the construction industry [44]. The advantages of using BIPV façades are the production of renewable electric clean energy, contributing to increase the degree of buildings sustainability towards net-zero energy construction [45], producing more renewable energy on-site or close to the building, and support for CO_2 and heat island reduction [46]. Planning buildings with multifunctional BIPV systems is an essential for architectural design and environmental concern [47].

BIPV module surfaces are manufactured as flat or flexible type to be integrated in the building envelope. BIPV efficiency and productivity, which can be installed on roofs and façades, are affected by orientation, shading, and surrounding surfaces reflections [48,49]. Figure 3 shows the component of BIPV panels and Figure 4 shows different international examples of BIPV facade design with the production rate [50].

The idea of integrating PV panels with the building elements increases the prospects of renewable energy systems, and the assessment of BIPV potential is considered as a preliminary fundamental step towards supporting public decision-makers to achieve energy transition goals [51]. The global BIPV market experienced fast growth, and the annual worldwide BIPV market was predicted to be more than 11,500 MWp in 2019 with high investment in the solar energy market. Table 2 shows the global installation forecast of the BIPV growth from 2014 to 2020 [51,52].

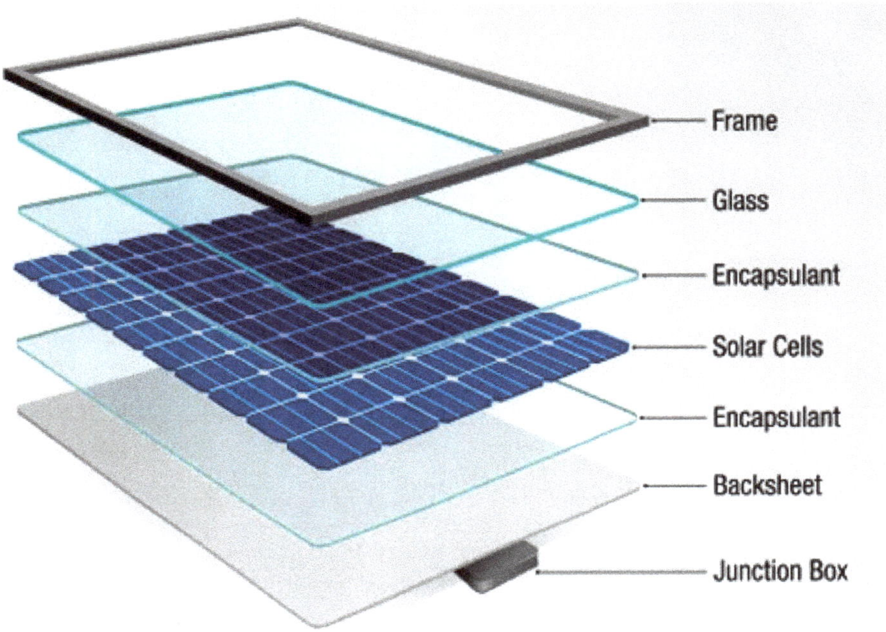

Figure 3. Component of BIPV panels.

Dubai Electricity and Water Authority R&D (United Emirates) BIPV façade produce 134 kW

Opera Theater (France) BIPV façade produce 1.1 MW

Train station (Italy) BIPV façade produce 1.0 MW

Figure 4. Different international examples of BIPV facade design.

Table 2. The global installation forecast of the BIPV growth from 2014 to 2020 (MW).

Region/Country	2014	2015	2016	2017	2018	2019	2020	CAGR %
Asia/Pacific	300	492	722	1159	1672	2329	3.134	47.8
Europe	650	967	1441	2103	2929	3807	4838	39.7
USA	319	476	675	917	1200	1491	1766	33.0
Canada	42	61	86	119	157	190	228	32.6
Japan	143	201	268	349	434	520	612	27.5
Rest of world	81	125	184	263	355	451	561	37.9
Total (GW)	1.5	2.3	3.4	4.9	6.7	8.8	11.1	

This research discusses a holistic approach, which presents a comprehensive guideline of measuring and calculating the specifications of sustainable clean energy in the construction industry. The research adopted this approach in a pioneer case study to achieve the triple bottom sustainability benefits in energy consumption for an ongoing construction project inside an existing public campus in Saudi Arabia. Therefore, significant goals could be accomplished. First, at the project level, complete analysis and simulation were conducted to improve the specifications of the external envelope. It contributed to shifting the building towards a nearly zero energy building by covering more than 68% of the needed energy from renewable energy resources. It also contributed to the sustainability ranking of the case study in the sustainable rating system without any contractual conflicts. Second, at the campus level, a complete actual sustainable guideline approach was presented to the decision maker to apply the results to the remaining 76 campuses construction project. This contributed to reduction of the energy consumption, CO_2 emission, and heat island loads, and enhancing the skyline looking at the whole campus. Third, at the Saudi Arabian level, these guideline specifications submitted an actual updating of applying renewable energy regulations and specifications. The significant contribution of the building facades and rooftops is to contribute to the required clean energy resource as a part of the Saudi vision 2030.

The study analysis for the case study built its approach based on two phases. The first phase is a numerical feasibility comparison in energy performance between the tender façade composite design and the new façade composite design. The second phase is the calculation of the clean energy production value and potential quantity of the sustainability points in energy performance. It can be applied in the case study to upgrade the existing sustainability ranking. The case study which was selected is registered in the LEED NC v3 sustainable rating system and awarded 37 points in the design phase and possibly pending 31 points in the construction phase. The holistic approach focuses on upgrading the design system of the building envelope from a passive energy envelope to an active energy envelope based on technical feasibility assessment and numerical comparison analysis, giving due consideration to project execution status, project cost, contractual situation, environmental impacts, and excellency needs.

The study opens the gate for various studies for improving contractual construction project documents with sustainability for enhancing building energy performance, under the sustainability rating system focusing on adapting active and passive envelope energy applications. Therefore, the next main questions arose. What is the practical approach to modifying contractual construction project documents to improve the building energy performance? Is it worth making a cost analysis to convert the passive envelope to an active envelope? What are the contractual risks encountered in the project in this case ? To answer these questions, the paper highlights the main research problem and explores the methods used. Then, it presents the results of the evaluation and assessment for the numerical calculations of the design of a new solar module cladding to the case study shell. The paper then concludes the research and discusses its limitation and future search venues.

2. Methodology

The energy efficiency enhancing process using the composite active facade based renewable energy technologies system to adapt contractual construction documents to the sustainability requirements in public building is an increasing concern. It aims to maximize the environmental benefits. The case study project was registered in Leadership in Energy and Environmental Design LEED organization (LEED NC v3) to obtain golden certification with 68 points (37 point in design document—31 points expected after project handover) according to first design document review. The study used PVsyst V7.1.0 software for data analysis, the design process, and the sizing system for solar systems, performing a simulation run system, and a comparison analysis. PVsyst V7.1.0 software specifies parameter details, and analyzes fine effects such as thermal behavior, wiring, module quality, mismatch and incidence angle losses, horizon (far shading), or partial shadings of near objects on the array. Results include several dozen simulation variables displayed in monthly, daily, or hourly values [53]. The outcomes are the specific PV production (kWh/kWp year), annual PV production (MW), and the performance factor. The research used HAP software for energy analysis to make energy consumption comparisons, to operate design costs, and support green building design alternatives in buildings [54].

The holistic evaluation used technical feasibility assessment and numerical comparison analysis. It aims to explore the quantity of the upgrading process for the sustainable points earned in energy and atmosphere EA criteria of the sustainable rating system in LEED NCv3 in a unique project in Saudi Arabia as a case study between 2019–2022. Therefore, the holistic evaluation was conducted based on two stages. The first stage was designed based on PVsyst V7.1.0 software for energy simulation, focusing on the main shell skeleton envelope structure. The study conducted the numerical feasibility comparison between the tender façade design document, consisting of composite aluminum with tempered glass. It also redesigned the facade with BIPV modules to calculate energy improvement quantity, considering the unbalance in bidding value status.

The second stage involves adjusting the results according to energy improvement quantity in the first stage. Therefore, the study explores points obtained in an energy sustainable rating system score based on using the available points in two credits from energy and atmosphere EA criteria. The first credit is EAc2: on-site renewable energy credit, while the second credit is EAc1: optimized energy performance credit. The study used the data from three essential sustainable ranking tables from LEED NC v3 to explore and illustrate the study results. Table 3 illustrates the sustainable 7 criteria with a total of 110 points in LEED NC v3. Table 4 illustrates sustainability credit and points in US LEED NC v3. Energy and Atmosphere (EA) criteria. It consists of 3 perquisites and 6 credits with a total of 36 points. Table 5 illustrates sustainable credit and points details in Energy and Atmosphere criteria including sustainable points in credit EAc1: optimized energy performance and credit EAc2: on-site renewable energy. It is used as an achievement parameter in energy generation [31–33]. The other parameters to evaluate the feasibility of the results include a contractual impact study of this comparison and financial issue in variation order. The building shell skeleton was executed by the main contractor and the Chinese subcontractor. To make the final envelope from solar BIPV modules, high technical coordination of all technical teams was required [32,40].

Table 3. Sustainability credit and points in US LEED NC v3.

Section to Cover	Total No. of Criteria	No. of Prerequisites	No. of Credits	No. of Points
Sustainable sites	15	1	14	26
Water efficiency	4	1	3	10
Energy & atmosphere	12	3	9	35
Material & Resources	8	1	9	14
Indoor environmental quality	17	2	15	15
Innovation & design	2	0	2	6
Regional priority	4	0	4	4
Total	65	8	73	110

2.1. Case Study

The case study project is one of the important projects at the King Faisal university campus. The design of the project was started in 2015 in accordance with the strategic plan of the university under the name of the project of the center for talent and the center for research and consultation. It aimed to support the university's strategic objectives in developing talent and research and experimental studies for students and faculty members. It also aimed to become one of the centers for research and development as well as to achieve the Kingdom's vision in developing human energy [54]. The building was designed with a unique design on a building area of about 15,000 square meters, with a basement and four recurring floors. Figure 5 illustrates the project tender ground floor and perspective. Figure 6 illustrates the tender shell steel structure with length 117 m width, 71 m, maximum height 35 m, and gate height 12 m in the conceptual design of the envelope from composite aluminum and structure glazing, which in its philosophy represents the human mind as a center of talent, sense, and development [54,55]. Both buildings under the shell form the left and right lobes of the mind. The building consists of 66 classes and training halls, seminar exhibition and discussion halls, halls for visiting researchers, and a modernized hall that can accommodate about 280 students. It also contains a large hall for students that can accommodate about 280 students. The building's exterior envelope dimension is 117 m in length, 71 m in width, and 35 m in maximum height, with a total area of about 9520 m^2 [54,55]. The envelope is a steel structure with 1200 tonnes, and the cover for the steel structure is designed from composite aluminum with 6200 m^2 and double structure glazing with 3300 m^2. Figure 7 illustrates the calculated electrical and mechanical energy load consumption in the tender design after execution which include the total demand loads for the building reaching about 1275 kW, including 355 kW for mechanical loads (air handling units, elevators, fan coil units, fountains, water pumps), 852 kW for power loads, and 426 kW for lighting loads [55]. The project in the design stage accomplished an energy cost saving of 23.23% which helps the project to obtain 6 points and raise the sustainability rating system points [54,55].

Table 4. Energy and atmosphere criteria in LEED NC V3 (35 points).

Credit	Credit Title	Criteria Points
EA Prereq 1	Fundamental commissioning of the building energy systems	
EA Prereq 2	Minimum energy performance	
EA Prereq 3	Fundamental refrigerant management	
EA Credit 1	Optimize energy performance	19
EA Credit2	On-site renewable energy	7
EA Credit 3	Enhanced commissioning	2
EA Credit 4	Enhanced refrigerant management	2
EA Credit 5	Measurement & verification	3
EA Credit 6	Green power	2
Total		35

2.2. Case study: Energy Simulation Document

The study used PVsyst V7.1.0 software for energy simulation, focusing on the main shell skeleton envelope structure which structurally is separate from the building structure. The study divided the envelope into five main areas based on project longitude and latitude, solar zone radiation, and sun movement. Figure 8 shows the five division for the shell on the satellite image of the actual constructed case study project location with primary simulation for each area. The study conducted energy simulation for 4750 BIPV module design cover for the shell skeleton envelope. The results from using the PVsyst V7.1.0 software include simulation parameters, grid-connected system, near shading definition, main results, special graphs, loss diagram, cost of the system, financial analysis, CO_2 balance. Figure 9 illustrates the simulation results for the final study, the solar BIPV module project using PVsyst V7.1.0 software.

Table 5. Sustainability points in Credit EAc2: on site renewable energy, and Credit EAc1: optimize energy.

Credit EAc1: Optimize Energy Performance			Credit EAc2: on Site Renewable	
New Building	**Existing Building Renovation**	**LEED Points**	**Percentage of Renewable Energy**	**LEED Points**
12%	8%	1	1%	1
14%	10%	2	3%	2
16%	12%	3	5%	3
18%	14%	4	7%	4
20%	16%	5	9%	5
22%	18%	6	11%	6
24%	20%	7	13%	7
26%	22%	8		
28%	24%	9		
30%	26%	10		
32%	28%	11		
34%	30%	12		
36%	32%	13		
38%	34%	14		
40%	365	15		
42%	38%	16		
44%	40%	17		
46%	42%	18		
48%	44%	19		

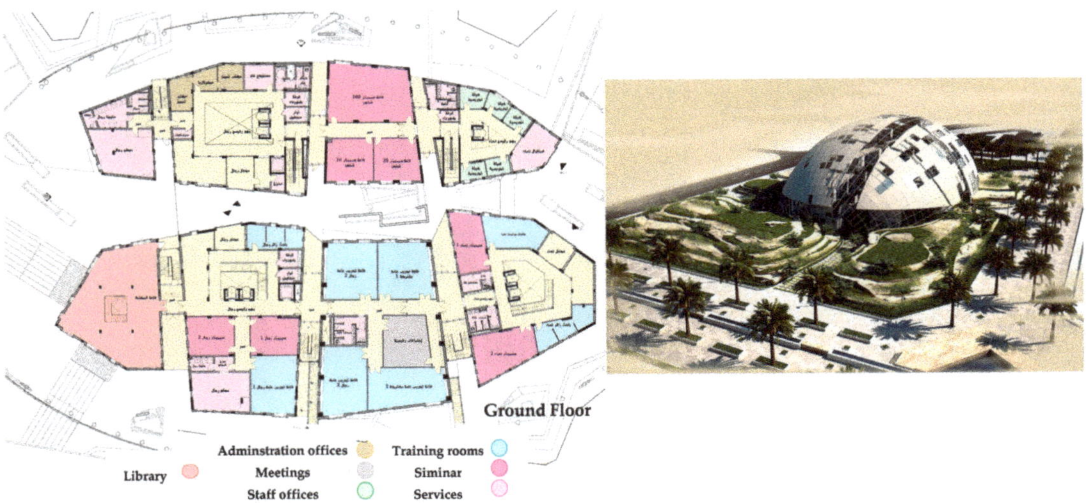

Figure 5. The project tender ground floor and perspective.

2.3. *Case Study: Sustainable Rating System Document*

The challenge for the building professionals and building designers in the sustainable design process is how a building meets all sustainable requirements. The project was registered in the LEED organization under LEED NC v3 rating system for golden certification. According to the design document review, the project obtained in total 68 points, including 37 points awarded in the design phase and potential expected 31 points that could be achieved in the construction processes phase. Tables 6 and 7 illustrate awarded points distribution in the design phase of the LEED NC v3 sustainability rating system checklist for the case study project. It includes the Talent & Research Project which awarded points distribution in all criteria of the LEED NC v3 sustainability rating system checklist, the points distribution in the case study project design phase and that expected in the construction phase in all criteria of the LEED NC v3 sustainability rating system checklist—the sustainability score reached 37 points for the design document and expected 31 in the construction execution processes phase—and the points distribution expected in the construction phase after applying the case study results in energy and atmosphere EA criteria (EA credit1+EA credit2) in the LEED NC v3 sustainability rating system. The authors and sustainability team started to make full details of all points gained in the design phase. They focused on the attempted points to classify the potential to achieve more points in ongoing construction based on Energy and Atmosphere (EA) criteria [33–35]. The total criteria goal points in Energy and Atmosphere EA in LEED NC are 35 points. The project's total awarded points in Energy and Atmosphere EA criteria were 8 points (6 points awarded in the credit optimized energy performance, and 2 points awarded in the enhanced refrigerant management) in the design phase review [33,34].

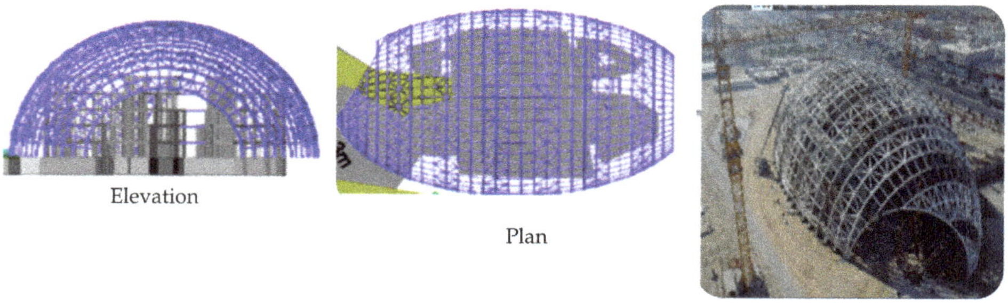

Figure 6. Tender shell steel structure with length 117 m, width 71 m, and maximum height 35 m.

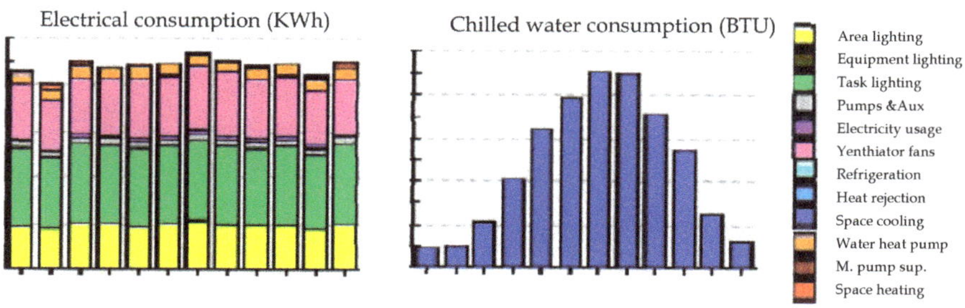

Figure 7. The calculated electrical and mechanical energy loads consumption in tender design.

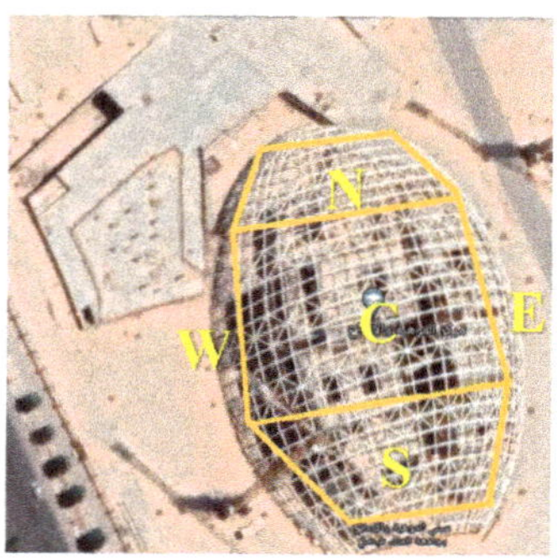

Array	M Len m	M Wid m	M2 Area m²	N Panels	kWp pmax	Deg Tilt	Deg Azi	Sheet	MWh/y MWh/y
1N	40	25	1000	504	181.4	30		180 T30_A180	202.3
2C	40	35	1400	706	254.2	0		0 T0_A0	397.8
3S	40	25	1000	504	181.4	30		0 T30_A0	304.9
4E	85	10	850	428	154.1	45		-90 T45_A_90	232.9
5W	85	10	850	428	154.1	45		90 T45_A90	182.7
	SUM		5100	2570	925.2				1320.6

Figure 8. Shows the five division for the shell in the case study actual satellite image with primary simulation for each area.

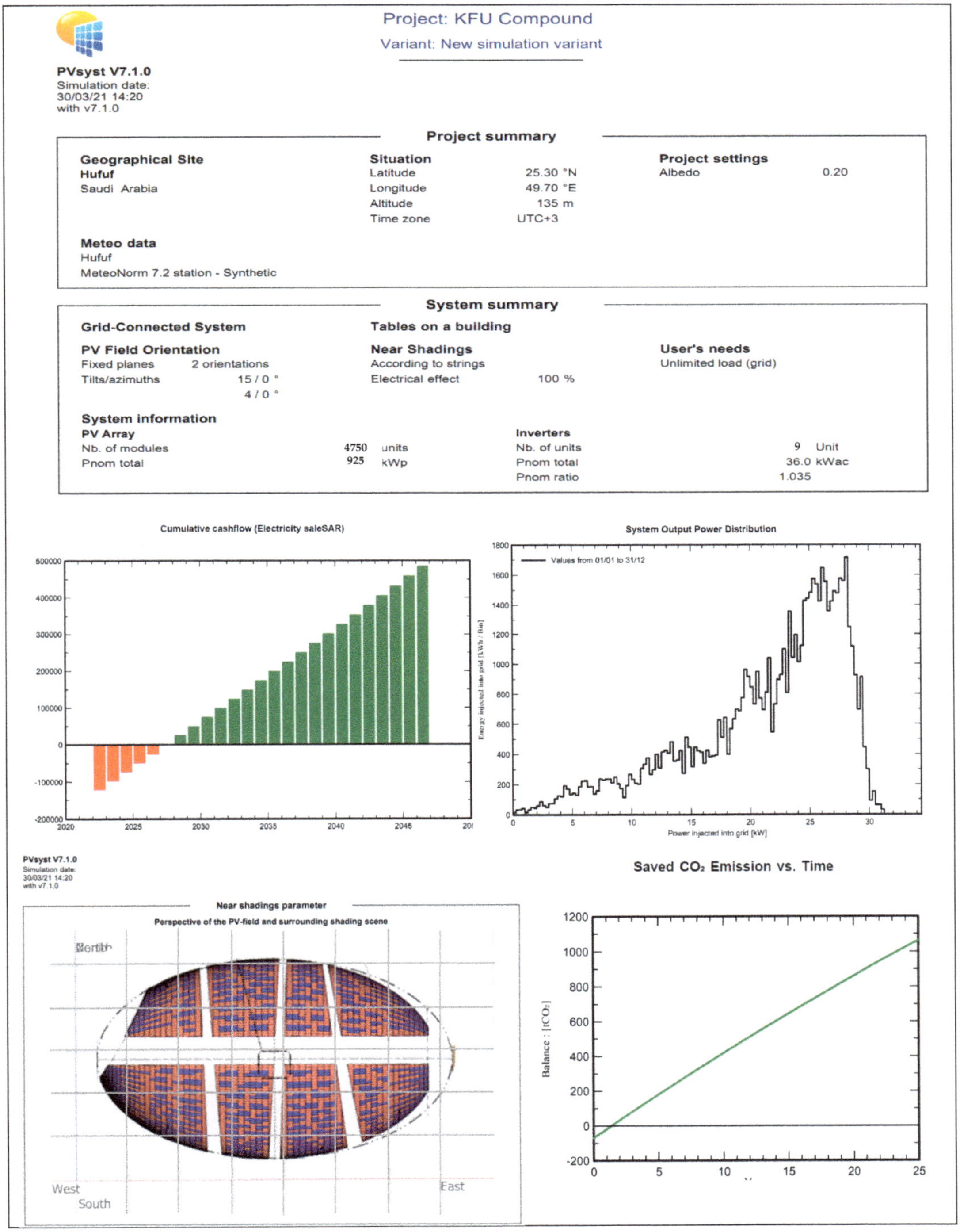

Figure 9. Simulation results for final study Solar BIPV module project using PVsyst V7.1.0 software.

Table 6. Awarded Points distribution in design phase of LEED NC v3 rating system checklist for the case study project.

	Project Checklist			Talented and Research Center		
	Project ID 1000036424	**Rating System & Version LEED-NC v2009**		**Project Registration Date 10/22/2013**		
	Sustainable Sites	**Possible Points:**	**9 of 26**	**Indoor Environmental Quality**	**Possible Points:**	**6/15**
Prereq 1	Construction Activity Pollution Prevention		N	Prereq. 1	Minimum Indoor Air Quality Performance	Y
Credit 1	Site Selection		1/1	Prereq 2	Environmental Tobacco Smoke ETS Control	Y
Credit 2	Development Density and Community Connectivity		0/5	Credit 1	Outdoor Air Delivery Monitoring	1/1
Credit 3	Brownfield Redevelopment		0/1	Credit 2	Increased Ventilation	0/1
Credit 4.1	Alternative Transportation—Public Transportation Access		0/6	Credit 3.1	Construction IAQ Management Plan—During Construction	0/1
Credit 4.2	Alternative Transportation—Bicycle Storage and Changing Rooms		0/1	Credit 3.2	Construction IAQ Management Plan—Before Occupancy	0/1
Credit 4.3	Alternative Transportation—Low-Emitting and Fuel-Efficient Vehicles		3/3	Credit 4.1	Low-Emitting Materials—Adhesives and Sealants	0/1
Credit 4.4	Alternative Transportation—Parking Capacity		2/2	Credit 4.2	Low-Emitting Materials—Paints and Coatings	0/1
Credit 5.1	Site Development—Protect or Restore Habitat		0/1	Credit 4.3	Low-Emitting Materials—Flooring Systems	0/1
Credit 5.2	Site Development—Maximize Open Space		1/1	Credit 4.4	Low-Emitting Materials—Composite Wood and Agri fiber Products	0/1
Credit 6.1	Stormwater Design—Quantity Control		1/1	Credit 5	Indoor Chemical and Pollutant Source Control	1/1
Credit 6.2	Stormwater Design—Quality Control		0/1	Credit 6.1	Controllability of Systems—Lighting	1/1
Credit 7.1	Heat Island Effect—Non-roof		1/1	Credit 6.2	Controllability of Systems—Thermal Comfort	1/1
Credit 7.2	Heat Island Effect—Roof		1/1	Credit 7.1	Thermal Comfort—Design	1/1
Credit 8	Light Pollution Reduction		0/1	Credit 7.2	Thermal Comfort—Verification	1/1
				Credit 8.1	Daylight and Views—Daylight	0/1
	Water Efficiency	Possible Points:	**8 of 10**	Credit 8.2	Daylight and Views—Views	0/1
				Innovation and Design Process	Possible Points:	**2/6**
Prereq 1	Water Use Reduction—20% Reduction		Y			
Credit 1	Water Efficient Landscaping		2/4	Credit 1.1	Innovation in Design: Specific Title	0/1
Credit 2	Innovative Wastewater Technologies		2/2			
Credit 3	Water Use Reduction		4/4	Credit 1.2	Innovation in Design: Specific Title	0/1

Table 6. *Cont.*

	Project Checklist			**Talented and Research Center**	
	Project ID 1000036424	**Rating System & Version LEED-NC v2009**		**Project Registration Date 10/22/2013**	
	Sustainable Sites	**Possible Points: 9 of 26**	**Indoor Environmental Quality**	**Possible Points:**	**6/15**
			Credit 1.3	Innovation in Design: Specific Title	1/1
	Energy and Atmosphere	Possible Points: **8 of 35**	Credit 1.4	Innovation in Design: Specific Title	1/1
			Credit 1.5	Innovation in Design: Specific Title	0/1
			Credit 2	LEED Accredited Professional	0/1
Prereq 1	Fundamental Commissioning of Building Energy Systems	N			
Prereq 2	Minimum Energy Performance	Y	**Regional Priority Credits**	Possible Points:	**4/4**
Prereq 3	Fundamental Refrigerant Management	Y			
Credit 1	Optimize Energy Performance	6/19			
Credit 2	On-Site Renewable Energy	0/7	Credit 1.1	Regional Priority: Specific Credit	1/1
Credit 3	Enhanced Commissioning	0/2	Credit 1.2	Regional Priority: Specific Credit	1/1
Credit 4	Enhanced Refrigerant Management	2/2	Credit 1.3	Regional Priority: Specific Credit	1/1
Credit 5	Measurement and Verification	0/3	Credit 1.4	Regional Priority: Specific Credit	1/1
Credit 6	Green Power	0/2			
			Total	Possible Points:	**37/110**
	Materials and Resources	Possible Points: **0/14**			
Prereq 1	Storage and Collection of Recyclables	Y			
Credit 1.1	Building Reuse—Maintain Existing Walls, Floors, and Roof	0/3			
Credit 1.2	Building Reuse—Maintain 50% of Interior Non-Structural Elements	0/1			
Credit 2	Construction Waste Management	0/2			
Credit 3	Materials Reuse	0/2			
Credit 4	Recycled Content	0/2			
Credit 5	Regional Materials	0/2			
Credit 6	Rapidly Renewable Materials	0/1			
Credit 7	Certified Wood	0/1			

Table 7. The classification of points in energy and atmosphere checklist in LEED NC v3 sustainability rating system checklist in the construction phase for the case study project.

Points				Credit	Credit Title	Type of Credit	Status after Final Design Review
Attempted	Awarded	Pending	Denied				
16	9	0	0		Sustainable Sites		
				Prereq 1	Construction Activity Pollution Prevention	construction	Awarded
1	1			Credit 1	Site Selection	design	Awarded
5				Credit 2	Development Density &Community Connectivity	design	Awarded
3	3			Credit 4.3	Alternative transportation, Low Emittance & Fuel efficiency vehicles	design	Awarded
2	2			Credit 4.4	Alternative transportation & Parking capacity	design	Awarded
1				Credit 5.1	Site development, protect or restore habitat	construction	
1	1			Credit 5.2	Site development, maximize open space	design	Awarded
1	1			Credit 6.1	Stream water design, quantity control	design	Awarded
1				Credit 7.1	Heat island effect, non roof	construction	
1	1			Credit 7.2	Heat island effect, roof	design	Awarded
13	11	0	0		Water Efficiency		
				Prereq	Water Use Reduction	design	Awarded
5	3			Prereq	Water efficient landscaping	design	Awarded
2	3			Prereq	Innovative wastewater technologies	design	Awarded
5	5			Credit	Outdoor Water Use Reduction	design	Awarded
19	9	0	0		Energy and Atmosphere		
7				Prereq 1	Fundamental Commissioning of the building energy systems	construction	
				Prereq 2	Minimum Energy Performance	design	Awarded
				Prereq 3	Fundamental Refrigerant Management	design	Awarded
12	7		3	Credit 1	Optimize Energy Performance	design	Awarded
3				Credit 3	Enhanced comissioning	construction	
2	2			Credit 4	Enhanced Refrigerant Management	design	Awarded
2				Credit 5	Measurement & verfication	construction	
4	0	0	0		Materials and Resources		
				Prereq 1	Storage and Collection of Recyclables	design	Awarded
1				Credit 2	Construction and Demolition Waste Management Planning	construction	

Table 7. *Cont.*

Points				Credit	Credit Title	Type of Credit	Status after Final Design Review
Attempted	Awarded	Pending	Denied				
				Credit 3	Material reuse	construction	
2				Credit 5	Regional material		
1				Credit 4	recycled content	construction	
11	6	0	0		Indoor Environmental Quality		
				Prereq 1	Minimum IAQ performance	design	Awarded
				Prereq. 2	Environmental Tobacco Smoke (ETS) Control	design	Awarded
1	1			Credit 1	Outdoor Air delivery monitoring	design	Awarded
1				Credit 3.1	Construction Indoor Air Quality Management Plan during construction	construction	
1				Credit 3.2	Construction Indoor Air Quality Management Plan before occupancy	construction	
1				Credit 4.1	Low-Emitting Materials, adhesive & sealant	construction	
1				Credit 4.2	Low-Emitting Materials, plants &coatings	construction	
1				Credit 4.3	Low-Emitting Materials, flooring system	construction	
1	1			Credit 5	Indoor chemical & pollution source control	design	Awarded
1	1			Credit 6.1	construability of systems, lighting	design	Awarded
1	1			Credit 6.2	construability of systems, thermal comfort	design	Awarded
1	1			Credit 7.1	Thermal Comfort, design	design	Awarded
1	1			Credit 7.2	Thermal Comfort, verification	design	Awarded
6	2	0	0		Innovation		
1			1	Credit 1.1	Green Foundation program		
1				Credit 1.2	Innovation in design		
1	1			Credit 1.3	Innovation wastewater technologies		
1	1			Credit 1.4	Green cleaning policy		
1				Credit 1.5	Innovation in design		
1				Credit 2	LEED accredited professional		
					Total Points		
68	37	31	4				

Therefore, the official sustainability team held more than 14 workshops to discuss upgrading the building tools, materials, and systems in a technical and financial study [53]. One important alternative was to convert the shell envelope from composite aluminum with 6000 m^2 and tempered double glazing roof with 3500 m^2 to solar BIPV modules as on-site renewable energy with an area of 9500 m^2 to achieve five significant goals; first: enhance energy performance to build a pioneer project in the whole gulf countries in the public campus sector to achieve nearly net zero energy by retrofitting for the ongoing or existing project, second: avoiding the risk to the project of not be gold certification and achieve the potential to upgrade the certification to the platinum certificate, third: to achieve the modern architectural shape, fourth: maximize energy saving in total demand loads for the building, which reaches about 1.2 MWp, and fifth: to maximize the sustainability impacts, and extend the life span of the building while providing a healthy and safe living environment in cities as well as promoting a culture of green buildings based on international sustainability standards with actual application in hot areas [53,54].

Solar BIPV (building integrated photovoltaic) modules as renewable energy can significantly contribute to LEED certification. The solar BIPV contributes in the Energy and Atmosphere category (EA), e.g., on-site renewable energy credit, which offers up to 7 LEED points, demonstrating over 17% of the points for certification. Ventures chasing for certification through LEED-NC V3009 use the benefits of on-site renewable energy, which give up to 7 points for providing up to 13% of the building's energy with on-site renewables as illustrated in Table 5. The performance of the venture was calculated according to the energy produced by the renewable systems as annual energy cost percentage of the building and the number of points achieved according to Table 5. Electricity and heat generated on-site were sold to the local grid connections at a premium stage. Nonetheless, this relatively humble delineation of what constitutes "renewable energy" has become more complex by integrating technologies. Hence, LEED-NC V3009 has attempted to define renewable energy more comprehensively [53–55].

3. Results and Discussion

3.1. Sustainable Rating System Impact

The study team members with external experts reviewed every detail related to the efforts made to draw the maximum benefits of building a shell structure envelope. It was carried out to enhance the sustainability team's effort to raise the ability to be an active shell as well as to obtain the golden certificate from the LEED organization. In contrast, the project could submit another 20 points for energy and atmosphere credit in the construction phase process to have 51 points instead of 31 points. It means that the total points in the design and construction phases will be 88 points. This, the project will be under a platinum certificate instead of a gold certificate. These 20 points are explained in the succeeding paragraphs.

The credit EAc1: Optimize Energy Performance intends to increase energy performance levels behind the prerequisite standard to make the environmental and economic impact reduction associated with extensive energy use, by using option 1 mentioned in the credit for whole building energy simulation. The committee with experts demonstrated a percentage improvement in the proposed building performance rating as compared to the baseline building performance rating. The committee, with support from experts, also calculated the baseline building performance mentioned in ANSI/ASHRAE/IESNA Standard [56] (see Appendix G).

The sustainable team used solar BIPV energy PVsyst V7.1.0 software by applying 4750 bifacial monocrystalline solar BIPV modules with 310 W: 360 W power in different efficiency and transparent for the whole Solar BIPV shell envelope. 90.1-2007. The team used a software simulation for the case study building. According to the tender design and after electromechanical system selection by the site technical team using HAPP software, the total load connecting TLC was 1267 kWh (0.355 MW for mechanical equipment and 0.896 MW for power and lighting loads). It means that the total load demand TLD was

1367 kWh, resulting in 3700 MW/year with total cost (according to tariff cost 0.33 SAR/kW) of 1221.000 SAR/year. The project with an active shell can generate about 925 kWp with 1320 MW/year, with total cost (according to tariff cost 0.33 SAR/kW) of 436.000 SAR/year as a cost-saving. It presents about 36% of the total demand of energy in the project which can give 13 points in the rating system LEED NC v3.

The credit EAc2: requirements use on-site renewable energy systems to offset building energy cost intents to increase on-site renewable energy self-supply to reduce negative environmental and economic consequences. According to it, the team calculated venture performance by checking the energy produced by the active shell as a renewable energy system as a percentage of annual energy cost and use in the building. The active shell produces 925 kW/p and 1320 MW/year. The total demand energy TDL in the project is 1367 kWh and 3600 MW/year. Therefore, according to production hours, the percentage of energy produced from the active shell as a renewable energy source as kW/p is 68% of the total energy needed for the case study building. It enables the project to obtain 7 points in the rating system. Table 7 illustrates classification of points in the energy and atmosphere checklist in the LEED NC v3 sustainability rating system checklist in the construction phase for the case study project. Table 8 shows the classification of points in energy and atmosphere in the case study checklist in credit EAc2: on site renewable energy, and Credit EAc1: optimize energy performance in the construction phase.

Table 8. The classification of points in energy and atmosphere in case study checklist in credit EAc2: on site renewable energy, and Credit EAc1: optimize energy performance in construction phase.

Criteria Goal	Points		Credit	Credit Title	Type of Credit	Status after Final Design Review
	Awarded in Design Phase	Awarded in Construction Phase				
			EA Prereq 1	Fundamental Commissioning of the building energy systems	construction	on going
			EA Prereq 2	Minimum Energy Performance	design	Awarded
			EA Prereq 3	Fundamental Refrigerant Management	design	Awarded
19	6	13	EA Credit 1	Optimize Energy Performance	design	Awarded
7		**7**	EA Credit2	On-site renewable energy	construction	on going
2			EA Credit 3	Enhanced commissioning	construction	on going
2	2		EA Credit 4	Enhanced Refrigerant Management	design	Awarded
3			EA Credit 5	Measurement & verification	construction	on going
2			EA Credit 6	Green Power	construction	on going
35	**8**	**20**				**28**

3.2. *Contractual Document Conflict*

The new item cost is less than 17% of the main contractor tender item price and could be contractually approved. Therefore, the authors reviewed and compared all tender bidders' documents related to the shell structure component with the solar BIPV energy system cost as a new contractual item. They rechecked if there was any contractual conflict between the costs for all bidders in this item, so that any kind of contractual objection did not occur from any related authorized reviewers.

3.3. *Execution Process Impact*

The sustainable team coordinated with solar BIPV energy experts to submit design drawings between the existing shell structure and the active shell envelope from solar BIPV energy modules. The team also undertook the technical procedures to connect the solar BIPV system inverter. The main project electrical board was connected to the main building switchgear according to the distribution of the solar BIPV envelope and the remaining solar BIPV system components such as cables, junctions, and combiners. In addition, the execution time matched the project's approved baseline schedule time. Table 9 shows the multi-benefits of using BIPV in the shell envelop compared with the tender envelope that affects reduction of 7 tonnes of CO_2 annually and saving of oil burning consumption to about 7 Barrels/m^2. The cost of the proposed solar BIPV modules is less than the cost of the corresponding item. Additionally, all building roofs are designed as outdoor areas. Thus, maintaining the financial balance of the project and increasing the size of the glass block with the use of solar cells in solar panels with different degrees of transparency increase the positive visual interaction between the outdoor and the indoor and the psychological comfort of the building's occupants.

Table 9. The multi-benefits of using BIPV in shell envelop compared with tender envelope.

Project Talent & Research	Total Lighting Points	CO_2 Emission	Barrels of Oil Saved	No. of Modules	Electricity Generated	Total Saving	Total Cost	Total Area
	Numbers.	Ton	m^2	No	MW/year	Thous. SAR/year	Thous SAR	m^2
 The Study	57,000	7	7	4750	1320	436	15,000	9500
The Tender 	N/A	N/A	N/A	-	N/A	N/A	18,500	9500

Figure 10 illustrates the comparison study of upgrading results in a sustainability rating system for the case study. It adds 20 sustainability points to shift the project to the platinum certificate as well as to prove all the advantages and goals mentioned in this holistic study. Applying the study proposal of using BIPV modules instead of composite aluminum with tempered glass in the construction phase processes assures that the project earned 37 points in the design document phase and 31 points expected in the construction phase. However, after applying the study proposal, the project earned 57 points in construction phase processes with 20 sustainability points extra.

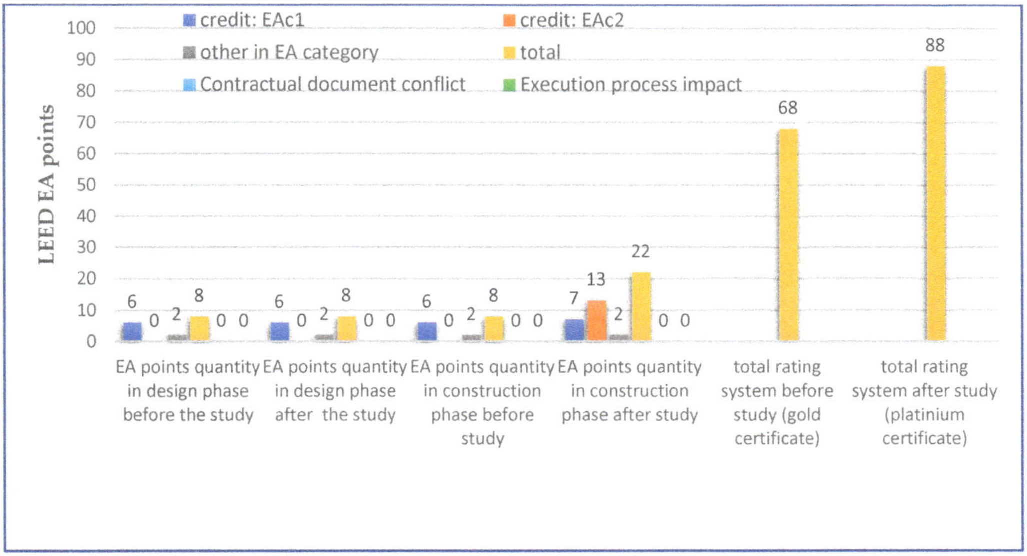

Figure 10. Comparison study of results in the LEED sustainability rating system for the project.

4. Conclusions

The Saudi Arabia Vision 2030 has three main axes. One of them is a prosperous economy, which aims to achieve production of about 50% of renewable energy from the total energy that Saudi Arabia needs based on solar plants. The construction projects are still not clearly considered in the policies and regulations. The construction projects contribute only 1.6% based on the official renewable energy 2020 statistic. Therefore, a holistic study as an assessment and evaluation approach for the selected case study inside a university campus as a public construction project in Saudi Arabia was conducted. The aim of this study was to provide a pioneering example for achieving a retrofitting process and a near zero energy campus to support the Saudi Arabia 2030 Vision. The holistic study explored the quantity of contributions in improving the energy performance and updating the bidding documents to achieve advanced ranking in the construction sustainability standards. The selected case study from the campus projects was the shell envelope of the talent and research center project. This project was under construction and was registered in the LEED to obtain the golden certificate in sustainability with a total of 68 points (37 points in the design phase and potential 31 points in the construction phase). By applying the holistic approach and analysis for the entire contract documents for the shell envelope, significant goals were achieved. PVsyst V7.1.0 software was used for solar analysis to redesign and implement the steel dome covered with 30% double structured glass and 70% composite aluminum on a surface of 9500 m^2 to give 4750 active BIPV solar panel panels on the entire surface of the case study building. It involved two comparative analysis stages. The significant goals included the upgrading of the building energy performance by generating electricity with approx. 925 kWp and about 1,320,000 kWh/year, which covers 68% of the energy building needs, achieving an annual saving of about 436 thousand riyals annually. The significant goals also include the building's upgrading in the sustainability rating system from gold certificate to platinum certificate based on energy performance by obtaining another 20 points in credit EAc1: optimize energy performance and credit EAc2: on-site renewable energy. Other significant results include reducing 7 tonnes of CO_2 annually and saving oil burning consumption to about 7 Barrels/m^2. In addition, the cost of the proposed solar BIPV modules is less than the cost of the design bidding item, and transparency increased. Since all building roofs were designed as outdoor areas, this enhances the positive visual interaction between the outdoor and the indoor and the psychological comfort of the building's occupants.

5. Limitations and Future Research Opportunities

This research focused on enhancing energy performance, upgrading the sustainability rating certificate, and adopting a nearly-zero energy project based on an active envelope to asset clean energy for a public campus project. Contractual assessment and technical evaluation with the latest PVsyst V7.1.0 software were applied in the construction project case study. The results could be limited for other public construction projects in the same context, but it has opened the door for future studies on public organization construction projects concerning the use of active envelope aspects such as commercial and healthcare projects. Additionally, design processes for economic and environmental impacts can be another area of future research opportunity.

Author Contributions: Conceptualization, E.M.H.I. and A.E.E.S.; methodology, E.M.H.I.; software, E.M.H.I. validation, E.M.H.I. and A.E.E.S.; formal analysis, E.M.H.I.; investigation, E.M.H.I. and A.E.E.S.; resources, E.M.H.I. and A.E.E.S.; data curation, E.M.H.I.; writing—original draft preparation, E.M.H.I. and A.E.E.S.; writing—review and editing, E.M.H.I. and A.E.E.S.; visualization, E.M.H.I.; supervision E.M.H.I.; project administration, E.M.H.I.; funding acquisition, E.M.H.I. and A.E.E.S. All authors have read and agreed to the published version of the manuscript.

Funding: This work was supported by the Deanship of Scientific Research, Vice Presidency for Graduate Studies and Scientific Research, King Faisal University, Saudi Arabia [Project No. GRANT3283].

Institutional Review Board Statement: Not applicable.

Informed Consent Statement: Not applicable.

Data Availability Statement: Data available upon request from the first author.

Conflicts of Interest: The authors declare no conflict of interest.

References

1. Griffiths, T.G. Education to transform the world: Limits and possibilities in and against the SDGs and ESD. *Int. Stud. Sociol. Educ.* **2021**, *30*, 73–92. [CrossRef]
2. Abbott, K.W.; Bernstein, S. High-Level Political Forum on Sustainable Development. In *Essential Concepts of Global Environmental Governance*; Routledge: Oxford, UK; pp. 120–121.
3. Leal Filho, W. Accelerating the implementation of the SDGs. *Int. J. Sustain. High. Educ.* **2020**, *21*, 507–511. [CrossRef]
4. Yonehara, A.; Saito, O.; Hayashi, K.; Nagao, M.; Yanagisawa, R.; Matsuyama, K. The role of evaluation in achieving the SDGs. *Sustain. Sci.* **2017**, *12*, 969–973. [CrossRef]
5. AlKhars, M.; Miah, F.; Qudrat-Ullah, H.; Kayal, A. A systematic review of the relationship between energy consumption and economic growth in GCC countries. *Sustainability* **2020**, *12*, 3845. [CrossRef]
6. Bekhet, H.A.; Matar, A.; Yasmin, T. CO_2 emissions, energy consumption, economic growth, and financial development in GCC countries: Dynamic simultaneous equation models. *Renew. Sustain. Energy Rev.* **2017**, *70*, 117–132. [CrossRef]
7. Menegaki, A.N. On Energy Consumption and GDP Studies, a Meta-Analysis of the Last Two Decades. *Renew. Sustain. Energy Rev.* **2014**, *29*, 31–36. [CrossRef]
8. Richardson, C.; Bithas, K. A Meta-Analysis Investigation of the Direction of the Energy-GDP Causal Relationship: Implications for the Growth-Degrowth Dialogue. *J. Clean. Prod.* **2014**, *67*, 1–13.
9. Olabemiwo, F.A.; Danmaliki, G.I.; Oyehan, T.A.; Tawabini, B.S. Forecasting CO_2 emissions in the Persian gulf states. *Glob. J. Environ. Sci. Manag.* **2017**, *3*, 1–10.
10. Mehrjerdi, H.; Iqbal, A.; Rakhshani, E.; Torres, J.R. Daily-seasonal operation in net-zero energy building powered by hybrid renewable energies and hydrogen storage systems. *Energy Convers. Manag.* **2019**, *201*, 112156. [CrossRef]
11. Reiche, D. Energy policies of Gulf Cooperation Council (GCC) countriesd possibilities and limitations of ecological modernization in rentier states. *Energy Policy* **2010**, *38*, 2395–2403. [CrossRef]
12. Alzubaidi, S. Cost Impact of Green Building in Qatar–A case study CCTC 2013 Paper Number (1569693237). In Proceedings of the EIC Climate Change Technology Conference 2013, Montreal, QC, Canada, 27–29 May 2013.
13. Koch, N. Green laboratories: University campuses as sustainability "exemplars" in the Arabian Peninsula. *Soc. Nat. Resour.* **2018**, *31*, 525–540. [CrossRef]
14. Sharples, S.; Radhi, H. Assessing the technical and economic performance of building integrated photovoltaics and their value to the GCC society. *Renew. Energy* **2013**, *55*, 150–159. [CrossRef]

15. Yousif, J.H.; Kazem, H.A.; Boland, J. Predictive models for photovoltaic electricity production in hot weather conditions. *Energies* **2017**, *10*, 971. [CrossRef]
16. AlArjani, A.; Modibbo, U.M.; Ali, I.; Sarkar, B. A new framework for the sustainable development goals of Saudi Arabia. *J. King Saud Univ.-Sci.* **2021**, *33*, 101477. [CrossRef]
17. Amran, Y.A.; Alyousef, R.; Alabduljabbar, H. Renewable and sustainable energy production in Saudi Arabia according to Saudi Vision 2030; Current status and future prospects. *J. Clean. Prod.* **2020**, *247*, 119602. [CrossRef]
18. Salam, M.A.; Khan, S.A. Transition towards sustainable energy production–A review of the progress for solar energy in Saudi Arabia. *Energy Explor. Exploit.* **2018**, *36*, 3–27. [CrossRef]
19. Alnaser, W.E.; Alnaser, N.W. The status of renewable energy in the GCC countries. *Renew. Sustain. Energy Rev.* **2011**, *15*, 3074–3098. [CrossRef]
20. Gonand, F.; Hasanov, F.J.; Hunt, L.C. Estimating the impact of energy price reform on Saudi Arabian intergenerational welfare using the MEGIR-SA model. *Energy J.* **2019**, *40*, 101–123. [CrossRef]
21. Bdour, M.; Dalala, Z.; Al-Addous, M.; Radaideh, A.; Al-Sadi, A. A comprehensive evaluation on types of microcracks and possible effects on power degradation in photovoltaic solar panels. *Sustainability* **2020**, *12*, 6416. [CrossRef]
22. Chang, H.S. Power Output in Various Types of Solar Panels in the Central Region of Korea. *J. Korean Sol. Energy Soc.* **2018**, *38*, 37–44.
23. Zhang, T.; Wang, M.; Yang, H. A Review of the Energy Performance and Life-Cycle Assessment of Building-Integrated Photovoltaic (BIPV) Systems. *Energies* **2018**, *11*, 3157. [CrossRef]
24. Ananthakumar, S.; Kumar, J.R.; Babu, S.M. Third-Generation Solar Cells: Concept, Materials and Performance-An Overview. In *Emerging Nanostructured Materials for Energy and Environmental Science*; Springer: Cham, Switzerland, 2019; pp. 305–339.
25. Soman, A.; Antony, A. Colored solar cells with spectrally selective photonic crystal reflectors for application in building integrated photovoltaics. *Sol. Energy* **2019**, *181*, 1–8. [CrossRef]
26. Saifullah, M.; Ahn, S.; Gwak, J.; Ahn, S.; Kim, K.; Cho, J.; Park, J.H.; Eo, Y.J.; Cho, A.; Yoo, J.S.; et al. Development of semitransparent CIGS thin-film solar cells modified with a sulfurized-AgGa layer for building applications. *J. Mater. Chem. A* **2016**, *4*, 10542–10551. [CrossRef]
27. Mas'ud, A.A.; Wirba, A.V.; Alshammari, S.J.; Muhammad-Sukki, F.; Abdullahi, M.M.; Albarracín, R.; Hoq, M.Z. Solar energy potentials and benefits in the gulf cooperation council countries: A review of substantial issues. *Energies* **2018**, *11*, 372. [CrossRef]
28. Alobaidi, K.A.; Rahim, A.B.A.; Mohammed, A.; Baqutayan, S. Sustainability Achievement and Estidama Green Building Regulations in Abu Dhabi Vision 2030. *Mediterr. J. Soc. Sci.* **2015**, *6*, 509–518. [CrossRef]
29. Zhang, Y.; Wang, W.; Wang, Z.; Gao, M.; Zhu, L.; Song, J. Green building design based on solar energy utilization: Take a kindergarten competition design as an example. *Energy Rep.* **2021**, *7*, 1297–1307. [CrossRef]
30. Al-Surf, M.; Balabel, A.; Alwetaishi, M.; Abdelhafiz, A.; Issa, U.; Sharaky, I.; Shamseldin, A.; Al-Harthi, M. Stakeholder's perspective on green building rating systems in Saudi Arabia: The case of LEED, Mostadam, and the SDGs. *Sustainability* **2021**, *13*, 8463. [CrossRef]
31. U.S. Green Building Council. *LEED Reference Guide for Building Design and Construction*; U.S. Green Building Council: Washington, DC, USA, 2019; ISBN 978-1-932444-18-6.
32. Wu, S.R.; Fan, P.; Chen, J. Incorporating culture into sustainable development: A cultural sustainability index framework for green buildings. *Sustain. Dev.* **2016**, *24*, 64–76. [CrossRef]
33. Babaei, M.; Azizi, E.; Beheshti, M.T.; Hadian, M. Data-Driven load management of stand-alone residential buildings including renewable resources, energy storage system, and electric vehicle. *J. Energy Storage* **2020**, *28*, 101221. [CrossRef]
34. Felseghi, R.A.; Bolboacă, A.; Răboaca, M.S.; Aşchilean, I. Hybrid energy systems for power of sustainable buildings. Case study: A renewable energy based on-site green electricity production. In *Reference Module in Earth Systems and Environmental Sciences*; Elsevier: London, UK, 2021; ISBN 9780124095489.
35. Ferwati, M.S.; Al Saeed, M.; Shafaghat, A.; Keyvanfar, A. Qatar sustainability assessment system (QSAS)-neighborhood development (ND) assessment model: Coupling green urban planning and green building design. *J. Build. Eng.* **2019**, *22*, 171–180. [CrossRef]
36. Ramani, A.; García de Soto, B. Estidama and the pearl rating system: A comprehensive review and alignment with LCA. *Sustainability* **2021**, *13*, 5041. [CrossRef]
37. Sabbagh, M.J.; Mansour, O.E.; Banawi, A.A. Grease the Green Wheels: A Framework for Expediting the Green Building Movement in the Arab World. *Sustainability* **2019**, *11*, 5545. [CrossRef]
38. Mohamed, S.M. Greening Existing Residential Buildings in Saudi Arabia with Mostadam as an Objective. Ph.D. Thesis, Arizona State University, Tempe, AZ, USA, 2022.
39. Alawneh, R.; Ghazali, F.; Ali, H.; Asif, M. A new index for assessing the contribution of energy efficiency in LEED 2009 certified green buildings to achieving UN sustainable development goals in Jordan. *Int. J. Green Energy* **2019**, *16*, 490–499. [CrossRef]
40. Awadh, O. Sustainability and green building rating systems: LEED, BREEAM, GSAS and Estidama critical analysis. *J. Build. Eng.* **2017**, *11*, 25–29. [CrossRef]
41. Agha, A.; Shibani, A.; Hassan, D.; Salmon, A. Building research establishment environmental assessment methodology on the UK residential projects. *Int. J. Constr. Eng. Manag.* **2020**, *9*, 183–189.

42. Illankoon, I.C.S.; Tam, V.W.; Le, K.N.; Wang, J. Life cycle costing for obtaining concrete credits in green star rating system in Australia. *J. Clean. Prod.* **2018**, *172*, 4212–4219. [CrossRef]
43. Gui, X.; Gou, Z. Association between green building certification level and post-occupancy performance: Database analysis of the National Australian Built Environment Rating System. *Build. Environ.* **2020**, *179*, 106971. [CrossRef]
44. Nguyen, B.K.; Altan, H. Comparative review of five sustainable rating systems. *Procedia Eng.* **2011**, *21*, 376–386. [CrossRef]
45. Doan, D.T.; Ghaffarianhoseini, A.; Naismith, N.; Zhang, T.; Ghaffarianhoseini, A.; Tookey, J. A critical comparison of green building rating systems. *Build. Environ.* **2017**, *123*, 243–260. [CrossRef]
46. Agathokleous, R.; Kalogirou, S.A. Double skin facades (DSF) and building integrated photovoltaics (BIPV): A review of configurations and heat transfer characteristics. *Renew. Energy* **2016**, *89*, 743–756. [CrossRef]
47. Verberne, G.; Bonomo, P.; Frontini, F.; Van Den Donker, M.N.; Chatzipanagi, A.; Sinapis, K.; Folkerts, W. BIPV products for facades and roofs: A market analysis. In Proceedings of the 29th European Photovoltaic Solar Energy Conference and Exhibition, Amsterdam, The Netherlands, 22–26 September 2014; pp. 3630–3636.
48. Boccalatte, A.; Fossa, M.; Ménézo, C. Best arrangement of BIPV surfaces for future NZEB districts while considering urban heat island effects and the reduction of reflected radiation from solar façades. *Renew. Energy* **2020**, *160*, 686–697. [CrossRef]
49. Schuetze, T.; Willkomm, W.; Roos, M. Development of a holistic evaluation system for BIPV façades. *Energies* **2015**, *8*, 6135–6152. [CrossRef]
50. Sustainable Development Goals, BIPV Projects Allover World, Solar Energy, Spain. Available online: https://onyxsolar.com/projects (accessed on 30 March 2023).
51. Transparency Market Research. *Building Integrated Photovoltaics Market: Global Industry Analysis, Size, Share, Growth, Trends and Forecast, 2013–2019*; Transparency Market Research: Albany, NY, USA, 2015.
52. Nanomarkets. *BIPV Technologies and Market, 2015–2022*; Nano-839; N-Tech Research: Glen Allen, VA, USA, 2015.
53. Kumar, R.; Rajoria, C.; Sharma, A.; Suhag, S. Design and simulation of standalone solar PV system using PVsyst Software: A case study. *Mater. Today Proc.* **2021**, *46*, 5322–5328. [CrossRef]
54. King Faisal University. Building Tender Document Archive. 2016. Available online: https://www.kfu.edu.sa/ar/Departments/Campus/Pages/Home-new.aspx (accessed on 2 January 2023).
55. Dar Alomran for Engineering and Consultant, Designer and Consultant of the Project, Jordan. 2016. Available online: http://www.daralomran.com/index.html (accessed on 2 January 2023).
56. Goel, S.; Rosenberg, M.; Athalye, R.; Xie, Y.; Wang, W.; Hart, R.; Zhang, J.; Mendon, V. *Enhancements to ASHRAE Standard 90.1 Prototype Building Models*; Pacific Northwest National Lab. (PNNL): Richland, WA, USA, 2014.

buildings MDPI

Article

Cold Housing in Central Mexico: Environmental Dissatisfaction and Underheating Lowers Self-Perceived Health in Central Mexico

Carlos Zepeda-Gil [1,*] and Augusto Jacobo Montiel-Castro [2]

1 Escuela de Arquitectura y Ciencias del Habitat, Universidad de Monterrey, Av. Ignacio Morones Prieto 4500, San Pedro Garza García 66238, Mexico
2 Departamento de Ciencias de la Salud, División de Ciencias Biológicas y de la Salud, Universidad Autónoma Metropolitana, Unidad Lerma, Lerma de Villada 52005, Mexico
* Correspondence: carlos.zepeda@udem.edu

Abstract: Despite being perceived as a warm country, winters in the Central Mexican Plateau frequently reach temperatures below zero Celsius. Prolonged exposures to low temperatures resulting in heart and respiratory morbidities are estimated to be responsible for 50% of the reported illness in the plateau, attributable primarily to the design of homes ill-suited to extreme temperatures. Consequently, there is a growing need to ensure that dwellings provide adequate indoor thermal conditions in the region. Hence, on-site sensors were used to collect temperature and relative humidity data every five minutes in 26 living rooms in the Plateau for 11 months. From these data, a subsample was determined, resulting in dwelling-level thermal comfort and health surveys on 15 homes. Computer simulations were used to investigate whether the building itself could provide thermal comfort under different retrofitting scenarios. Multiple linear regression relating the Predicted Percentage Dissatisfaction (PPD) index to self-perceived health was undertaken. Both monitored and simulated results were matched against our underheating model, finding that 92% of the homes had cold indoor environments, some even during summer. High PPD and intense levels of underheating were positive predictors of higher self-reported health problems. More self-reported health problems were correlated with both lower life satisfaction and self-worth, and with subjects' use of more adaptive strategies against environmental dissatisfaction. Dynamic computer simulations suggested that indoor thermal environments could be improved by enforcing the non-utilised standard NOM-ENER-020, which recommends the addition of insulation on walls and roofs. These findings suggest that the cold environments within homes of the plateau influence the self-perceived physical and mental health of its population. Hence, the application of adequate measures, such as retrofitting homes with stronger standards than the existing NOM-ENER-020 are needed in place.

Keywords: underheating; self-perceived health; indoor temperatures; Mexico; NOM-ENER-020; thermal comfort

Citation: Zepeda-Gil, C.; Montiel-Castro, A.J. Cold Housing in Central Mexico: Environmental Dissatisfaction and Underheating Lowers Self-Perceived Health in Central Mexico. *Buildings* **2023**, *13*, 814. https://doi.org/10.3390/buildings13030814

Academic Editors: Paulo Santos and Mark Bomberg

Received: 15 February 2023
Revised: 12 March 2023
Accepted: 12 March 2023
Published: 20 March 2023

1. Introduction

Climate change is expected to convert conventional season patterns to hotter summers and colder winters, threatening human health [1]. Estimations indicate a global temperature rise of more than 5 °C by 2070 [2] making, among other consequences, seasonal cold waves worldwide (cold fronts and cold spells) stronger and longer [3], homemakers and children spend about 80% of their day at home (and likely to increase with the COVID-19 pandemic) [4], homes without adequate protection will be more exposed, endangering the health of their occupants. To date, Mild Climate Countries (MCC) show higher Excess Winter Deaths (EWD) than countries with severe winters [5]. This may be because there are no adequate building standards in MCCs that address this issue, or they exist but are not enforced [6]. The latter is the case in areas such as Latin America [7], Greece [8], Australia [9], Spain [10], and New Zealand [11].

In this context, Mexico fits the definition of a "hot climate" country with an unrecognised "cold housing" problem. Although to date there are no studies that examine this issue in the country, this can be confirmed by its high percentage of EWD, as the 11% presented in 2020 [12] is similar to countries with more severe winters, such as Hungary (11.3%), Germany (11%), and even higher than Poland (10.2%) or Finland (10%) [13]. In fact, it is common for MCCs to have higher EWDs than countries with severe winters [14]. Unfortunately, vulnerable social groups such as the elderly [15], those of lower-income [16] and people with chronic diseases [17] are more likely to be affected by this issue. Hence, it is important to examine this unattended issue affecting thousands of people every year, as mortalities and morbidities related to respiratory and cardiovascular diseases rank first and second, respectively, as the most common in the country.

This paper examines different retrofitting solutions that can potentially solve this issue. In this sense, its relevance is underlined by being one of the first field studies of winter cold discomfort and indoor temperatures of houses in the Mexican Plateau, coupled with strategies for homes whose building fabric coincides with 27.4 K homes in Mexico, corresponding to 78% of the country's housing stock. In this research, we aimed to address the following two questions:

Q1. Is the current building fabric of the majority of the housing stock in Mexico putting at risk the health of the inhabitants in Central Mexico?

Q1.1. What effective solutions can be implemented to improve the indoor thermal comfort of houses without relying on inefficient electric or gas heaters that contribute to increased energy consumption?

Q2. Is there any relationship between the current internal environments in the Central Mexican Plateau and householders' self-reported perception of their physical and mental health?

1.1. Thermal Comfort

Thermal comfort models are the most common means of assessing the balance between indoor environmental conditions and the personal factors that make a person feel thermally comfortable [18]. There are two approaches to evaluating thermal comfort that emerged in the last century. The steady-state approach (also known as Predicted Mean Vote PMV) developed by Fanger [18], currently used in the international standard ISO 7730 [19], and the adaptive method, proposed by Nicol and Humphreys [20] used in the American Society of Heating, Refrigerating and Air-Conditioning Engineers [21]. The Mexican government adopted the first, published as its voluntary standard NMX-C-7730-ONNCCE-2018 [22]. Both are based on building users' votes on a scale ranging from cold (−3) to neutral (0) and from neutral to hot (+3). The vote is called Thermal Sensation Vote (TSV). Both standards, ISO7730 [19] and the ASHRAE 55 [23], comply if the vote range is TSV is within the range of [−1, +1].

The steady-state approach assumes that any effort to adjust the internal thermal environment makes it undesirable [24]. This method was developed in climate chambers and for the air conditioning industry, and it should be used in conditioned spaces. The PMV value is calculated with four environmental variables (air velocity (A_v), external air temperature (T_{air}), internal operative temperature (T_{op}), and relative humidity (RH), as well as two personal variables metabolic rate (Met), and clothing insulation (Clo). The Predicted percent dissatisfied (PPD) term indicates the percentage of people that are not thermally comfortable in a space. The ASHRAE 55 standard limits this percentage to 10%.

The adaptive method considers that the building occupants adapt to their indoor environment through three general mechanisms: behavioural, physiological, and psychological. This method should be used in naturally ventilated buildings, where usually, people are thermally satisfied at a more extensive range of temperatures, compared to those in a more mechanically ventilated space [25]. Hence, this model does not aim to find a fixed temperature, but a temperature band at which the occupant, with sufficient adaptive opportunities (i.e., wearing an extra layer of clothing, drinking a hot beverage, etc.), can experience thermal comfort.

1.2. Housing and Health

The experience when arriving at a heated home during a cold day depends on temperature-balanced regulation between the internal environment and external conditions. Our sympathetic nervous system uses physiological activation to control the body's temperature, allowing us to adjust our behaviour, contributing to faster and more efficient response and adjustment [26]. In this context, thermal comfort refers to balancing environmental and personal control elements leading an individual to feel satisfied in their thermal environment [25]. When temperature variations in the external environment are more difficult to control (i.e., when extreme temperatures become environmental stressors), thermal comfort is harder to achieve [27]. Thermal comfort is highly associated with distinct health indicators. Many social determinants of health, including economic resources, housing conditions, and social resources, can modify an individual's thermal sensations [28]. For instance, housing quality has been associated with infectious diseases. Cold, damp, and mouldy housing are often associated with respiratory problems such as asthma, and even mental disorders, including anxiety and depression [29]. The World Health Organization (WHO) has set the minimum recommended indoor temperature to 18 °C for healthy environments. In addition, the WHO quantified the environmental burden of disease-associated inadequate housing in Europe, describing the variety of maladies that can be associated with lower quality home environments. Table 1 summarises the relationships found by that extensive report [30].

Table 1. Diseases caused by poor quality indoor environments according to the WHO.

Housing Characteristic	Health Impact
Indoor dampness/mould	Asthma onset in children
Physical conditions	Home injury
Crowding	Tuberculosis
Cold	Mortality
Noise exposure	Ischemic heart disease
Indoor radon	Lung cancer
Second-hand smoke	Respiratory infection; asthma; heart disease; lung cancer
Lead exposure (e.g., paints)	Anaemia decreased renal function, cognitive, developmental, neurological, behavioural, and cardiovascular effects.
Carbon monoxide	CO_2 intoxication, tissue hypoxia.
Formaldehyde exposure	Respiratory symptoms in children.
Indoor smoke from solid fuel use	Pneumonia in children, and chronic obstructive pulmonary disease and lung cancer in adults.
Housing quality	*Mental health:* anxiety, behaviour conduct disorders in children, depression, feelings of inadequacy, social isolation, stigmatisation.

In this sense, self-rated health is a widely used index of actual health status in research on neighbourhood environments and health, often measured through Likert-type scale answers [31] including self-assessments of mental health [32].

Ormandy et al. [16] found that even after controlling for age, gender, socio-economic status and smoking, poor health (i.e., self-reported) had a significant association with perceptions of poor thermal comfort. Moreover, they found that asthma attacks in the past year, allergies, hypertension, colds and sore throats, migraines, and gastric and duodenal ulcers were associated with poor thermal comfort. It seems then that the perception of cold in the internal living environment can affect essential health indicators. Moreover, blood pressure and viscosity may be fundamental causes of higher winter morbidity in MCCs, associated with strokes and heart attacks. Investigating the relationship between indoor temperature and the risk of high blood pressure in Scotland, Shiue et al. [33] determined that an indoor temperature below 8 °C could account for 9% of the population with a risk of high blood pressure. Indeed, people with lower income and poor access to various resources are less likely to live in decent housing, where exposure to environmental factors detrimental to health is more likely [28].

Regarding a relationship between low air quality and poor health in social vs. non-social housing, Patino et al. [34] found no evidence that social housing residents were significantly more exposed to pollutants such as formaldehyde or dampness. Instead, they found that poor thermal comfort was highly prevalent and associated with adverse health effects. The above suggests that low-income households may accept lower indoor temperatures and reach lower thermal comfort status due to budgetary constraints [35].

1.3. NOM ENER 020

The NOM-020-ENER-2011 (NOM-020) [36] is a mandatory Mexican standard that defines acceptable heat transfers for walls and roofs (among others, i.e., shading coefficients) in residential dwellings. This standard does not include criteria for floors or windows. Despite its obligatory nature, it is not enforced for various reasons, e.g., it is not requested to process a building's construction permit as well as the lack of technical knowledge of building inspectors [37].

Appendix A of the NOM-020 provides an allowed U value ($Wm^{-2}\,K^{-1}$) for residential buildings up to three floors high of 0.909 $Wm^{-2}\,K^{-1}$ for walls and roofs in Toluca. However, this varies depending on the region or state. To date, 88% of the total Mexican housing stock is built with 150 mm solid walls made from brick, block or concrete that provides an average U value of 2.88 $Wm^{-2}\,K^{-1}$ (sd = 0.52 $Wm^{-2}\,K^{-1}$) Further, 77% of the homes in the country have a roof composed of 100 mm reinforced concrete slab, plus 38 mm of lightweight stone (locally known as "Tepojal") and a 12 mm waterproof membrane, providing together a U value of 3.06 $Wm^{-2}\,K^{-1}$. Both values are outside the parameters of this standard.

Given that only 2% of Mexican houses comply the NOM-020 standards [38] it is practically impossible to study the effectiveness of the NOM-020 standard. Hence, alternative methods should be used to study its effectiveness. Ruiz Torres et al. [39] evaluated the NOM-020 standard with three houses located in the city of Tuxtla (hot and humid climate, Köppen Aw) in southwest Mexico, with dynamic simulations in Energy-Plus [40]. Their results show that the homes could not provide environmental conditions within the acceptable ranges of the ISO-7730 Standard [19]. Alpuche Cruz et al. [41] calculated the thermal balance of three houses located in different cities (Hermosillo—hot and arid; Colima-hot and humid, and Mexico City—temperate, within the Central Plateau) as built (without insulation), and with the NOM-020 parameters. They found that even after manually calculating the thermal balance of the homes under the NOM-020, houses in hot climates showed thermal imbalances, contrary to the one in Mexico City. Romero-Moreno et al. [42] conducted a similar study assessing the thermal balance of three homes in Mexicali (hot and humid climate) against the NOM-020 with a web-based tool provided by the government [43], and reached similar conclusions to those of Alpuche Cruz et al. However, we found no peer-reviewed studies in which a dynamic simulation tool was used to evaluate the parameters of the standard in the Central Mexican Plateau. Therefore, it is necessary to study the effectiveness of the NOM-020 for temperate climates such as those found in the Central Plateau, ideally with larger samples and with calibrated dynamic simulations.

2. Methodology

2.1. Field Study and Participants

This study aimed to examine the internal environments of the type of house with constructive characteristics predominating in central Mexico. An adaptive cluster [44] coupled with a targeted sampling [45] process was used. The study was promoted through posters and Facebook in local markets, schools, and churches in Toluca, Mexico. Thirty householders volunteered to participate and, after four dropouts, we had a final sample of 26 householders/homes. Every home visit took approximately 1 to 1.5 h, as surveys included 122 questions. Due to time restrictions, sometimes respondents only answered the thermal comfort survey, leaving the health questions out. As the temperature starts to drop in August, partly due to high levels of precipitation and relative humidity, the thermal sensation in the region becomes colder. Therefore, surveys applied in Autumn and Winter were pooled in what we defined as a "cold season". Then, the data were reduced to the houses that responded to the questionnaires at least once per season. This process left us with a final sample of 15 homes. While we understand that this is a small sample, it can be considered representative of the region since the characteristics of all the homes match at least 71% of the total housing stock in the country [46]. Additionally, the occupancy levels of our sample match the country's housing occupancy of 3.4 occupants per home [46]. The characteristics of the full sample are described in Table 2. However, we should mention that this sample is reduced only for the health section, as applying surveys took much longer.

Table 2. Sample characteristics.

Characteristics of the Sample		
Sex	Female	9
	Male	6
Age	20–30	6
	30–40	4
	40–50	3
	50 or more	2
Qualifications	Undergraduate	9
	Postgraduate	4
	Preferred not to answer	2
Socio-Economic Characteristics		
Room numbers	0–5	10
	5–10	5
House age	0–5	7
	5–10	5
	10 or more	3
Income	Less than 9000 MXN	10
	More than 9000 MXN	5

The houses selected in this study are built with solid brick walls rendered with cement on the external face, and with plaster on the internal, resulting in a wall thickness of 150 mm. The roof is built with a 100 mm reinforced concrete slab, with plaster on the internal face and a waterproof membrane on the external. None of the homes had any insulation, double glazing or any passive design strategies to avoid solar gains.

On average, the homes in this study had 1.6 storeys (sd = 0.5), with an average occupancy of 3.15 inhabitants (sd = 1.5). Four homes were oriented to the north, seven to the east, nine to the south and six to the west. Figure 1 provides an overview of key characteristics provided by the household. It shows that the age of the homes ranged between 1 and 50 years, that the age of the respondents was between 20 and 65 years old, and that heating was used only in isolated cases.

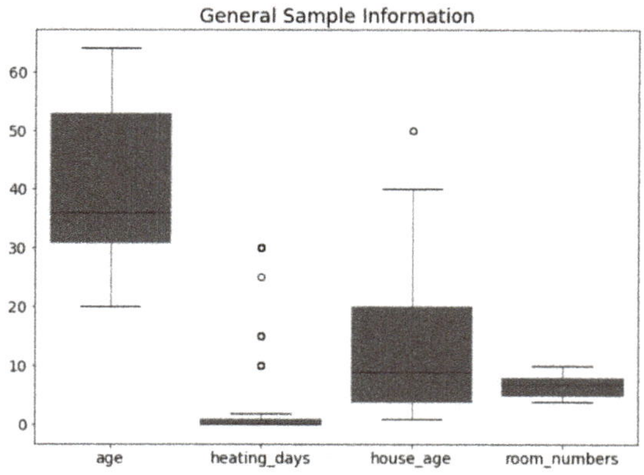

Figure 1. General information of the selected sample. The figure shows box plots of the information gathered during the fieldwork stage from the sample. The black line within the box represents the mean. The circles represent the outliers.

Figure 2 shows the location of the homes and the weather station *"CODAGEM, METEPEC*, 15266 (Clave OMM-76675, altitude 19°17′28″, longitude 99°42′51″)", where the weather data were obtained. We always ensured that the same householder responded to the questionnaire for every house. Lastly, the values used as responses in the questionnaires were transformed from qualitative responses (i.e., never, rarely, sometimes) to scalar data (i.e., 1, 2, 3) for our statistical model. Further information on this transformation is found in Appendix A.

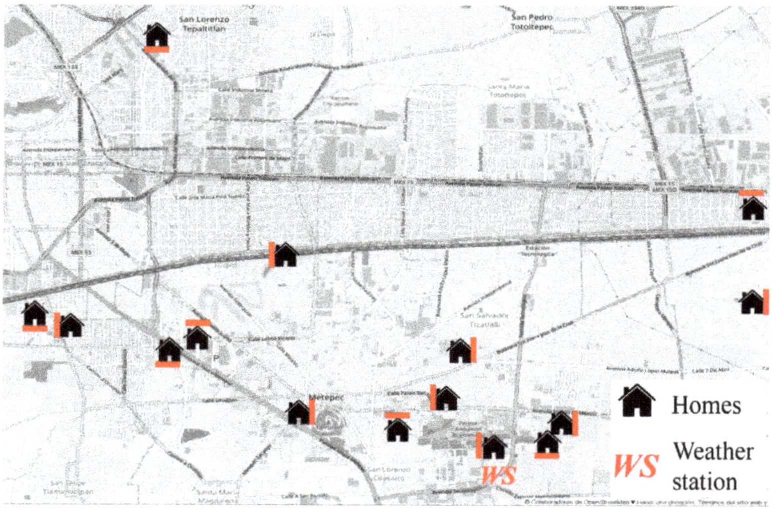

Figure 2. Location of the participating homes in Toluca and the weather station from which the external climatic data were obtained. Source of the background image was OpenStreetMap® [47]. Source of the house icon was. The red line next to the house represents the orientation of the room in which the sensor was positioned.

2.2. Site

Toluca is a city with approximately 900,000 people. However, fifteen cities are attached to it, resulting in an urban area with 2.3 M inhabitants, making it the fifth most populated urban area of Mexico, with the second largest area [48]. The city is located on the Meseta Central Mexicana (Central Mexican Plateau). We should underline that, at 2600 m of altitude, it is the highest city in North America. It has a Köppen climate classification of Cwb or "dry winter subtropical highland climate", an average temperature of 12.5 °C (sd = 1.5 °C), and an average daily temperature swing of 15.1 °C (sd = 4.3 °C), as seen in Figure 3. It has an average yearly precipitation of 113.4 mm (sd = 96.57 mm), the period from June to October being the most humid and rainy with an average rainfall of 209.4 mm (sd = 63 mm), and average relative humidity of 78% (sd = 4.2%). These climatic conditions are not limited to this urban area, characterising others such as Mexico City and Puebla, which have up to 30 M people.

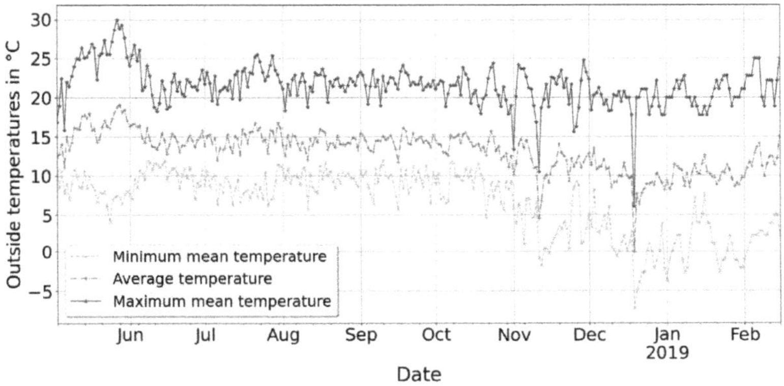

Figure 3. Daily outdoor temperatures in Toluca from May 2018 to February 2019. Light grey represents the minimum, mid-grey the average, and dark grey the maximum.

2.3. *Indices*

2.3.1. Clothing and Metabolic Activity

The Metabolic Rate (*MET*) was calculated according to the methods suggested in the ASHRAE 55 standard. Our calculations were made based on the activities we observed 15 min prior to the application of the questionnaire. Particular references were made to Table A1 of the ASHRAE 55 [49]. Participants described their clothing in a given table, where they would tick the piece of clothing they were wearing. After, these garments were matched against the values listed in Table 5-3 "Garment insulation" of ASHRAE 55 [23] and computed following the "Method 2" guidelines of Appendix B of the same standard.

2.3.2. Predicted Mean Vote and Predicted Percentage Dissatisfied Model

The predicted mean vote (PMV) method aims to predict the thermal sensation vote (TSV) a person would choose from a seven-point scale survey. The PMV model uses the heat balance of the human body and considers the conditions of the internal environment and the effect these have on the clothes worn by the occupant (*CLO*), as well as their metabolic activity (*MET*).

The PPD index aims to quantitatively predict what percentage of people do not feel thermally comfortable in their environment, i.e., too cold or hot. The ISO 7730 standard defines this group as voting +2, +3, −2 and −3 on its 7-point scale.

The PMV and PPD were computed with an "R" script [50], based on equations from the ISO-7730 [19] (PMV Equations (1)–(4); PPD Equation (5)) found in Section 4 of the standard.

The PPD index was used as a predictor in our multiple regression analysis, as described in Section 2.6 This is because (1) the PPD index already includes the PMV, which itself includes both environmental and personal variables described in Section 1.1, and (2) it only considers "cool (−2)" or "cold (−3)" votes, i.e., those who were thermally dissatisfied towards cold.

2.3.3. Adaptive Model

The adaptive method is based on the principle that the building occupant will adapt to a range of temperatures. This adaptation is made via physiological (e.g., body performs actions to return to the temperature of 37 °C), psychological (acclimatisation), and behavioural (those performed by the individual to be in thermal balance) processes. The thermal comfort band for naturally ventilated buildings is 7 °C wide for 80% of acceptability, centred around the comfortable temperature T_{conf} defined as:

$$T_{conf} = 0.31\ T_{a,out} + 17.8 \quad (1)$$

where $T_{a,out}$ is the average external dry bulb temperature [51]. Further, the adaptive graph requires the operative temperature T_{op} on the Y-axis. Our sensors only recorded dry bulb air temperature Ta. As in indoor spaces there are minor differences between radiant and air temperatures [52], and many studies have assumed that Ta is a good proxy to Top [50,53]. Hence, the results in this article should be read as $Top = Ta$.

2.4. Sensors

The fieldwork stage included the measurement of environmental conditions such as air temperature t_a and relative humidity Rh. The ISO-7730 standard requires operative temperature t_{op} in the adaptive thermal comfort model. However, it has been recognized [53,54] that in small environments with low wind speed, air temperature may be a good proxy for operative temperature. Hence, the findings in this paper should be read under the assumption that $t_a = t_{op}$.

The characteristics of the sensors selected are shown in Table 3. These sensors were attached to a Raspberry Pi computer and left in the living room of our participants for eleven months. The sensors recorded relative humidity and temperature every five minutes, and data were averaged hourly as established in the ISO-7730 standard. To measure windspeed, an anemometer ATP AVM-8880 was used. This is slightly out of standards, but has been used in other, similar field studies [53].

Table 3. Characteristics of the sensors.

Parameter	Sensor Model	Range	Accuracy
Temperature	DS18B20	−10 to +85 °C	>±0.5 °C
Relative Humidity	RHT03	0–100%	>±2%

2.5. Questionnaires

The fieldwork stage of the study took place from March 2018 to February 2019. Houses were visited every month based on availability. Two types of surveys were applied, a "short" one that aimed to capture information about aspects such as the participant's home (aspect, orientation, windows and doors), personal information (gender, age, income, occupation, weight and height), thermal comfort (7-point ASHRAE scale) and the necessary information for the PMV calculations as described in the ISO-7730 standard [19]. The "long" survey included data regarding participants' self-perceived mental and physical health. Table 4 shows the types of questions included in the health survey.

Questionnaire Data Conversion Methods

The answer options available in the health questionnaires were written in a way that they could later be translated into a scalar form for regression analysis. This subsection describes how the translation was made for all the question types.

Table 4. Types of questions and answers included in the "long" surveys deployed during the fieldwork stage.

Self-Reported Physical Health	
Participants responded: "Last month I had __ problems"	**Responses available**
Vision **Ear** **Arthritis** **Cardiovascular** **Hypertension** **Respiratory** **Neurodegenerative** **Depression** **Circulatory**	Yes No Rather not to say
Life Satisfaction and Self-worth **Participants responded: "Last month I felt __"**	**Responses available**
Happy **Confident** **Excited** **Loved** **Content** **Joyful** **Healthy**	Never Rarely Sometimes Often Very often Rather not to say
Life anxiety **Participants responded: "Last month I felt __"**	**Responses available**
Nervous **Hopeless** **Worthless** **Restless** **Powerless** **Lonely**	Never Rarely Sometimes Often Very often Rather not to say
Adaptation Style **Participants responded: "Last month when I felt too cold, I __"**	**Responses available**
Used extra blankets **Wore extra layers** **Drank a hot drink** **Stayed longer in bed** **Took a hot shower** **Left the cold room**	Never Rarely Sometimes Often Very often Rather not to say
Self-perceived physical capacity **Participants responded: "Last month I could __"**	**Responses available**
Walk half kilometre **Climb ten steps** **Stand for two hours** **Bend over kneel** **Reach something above head level** **Carry a supermarket bag** **Pull furniture** **Go out to socialize**	Extremely difficult Moderately difficult Easy Manageable Very easy Rather not to say

(a) Self-reported health problems

This section included nine "yes or no" questions. Every "yes" was later transformed into a value "1". A single home/respondent could obtain a maximum of none "points" for their answers (i.e., when its answers were all "Yes" in the nine questions). Thus, we obtained the total score per home/respondent by adding the values of each question into a single score per home or respondent per season. Furthermore, we calculated a single mean for each home or respondent across the three seasons and obtained an "average total health problems score".

(b) Life satisfaction and self-worth, life anxiety, and adaptation style

The answers in these sections were scalar. They ranged from never to very often, as seen in Table 5. We obtained a total score per home or respondent across these seven questions per season by adding the values of each into a single score per home/respondent. Thus, a single response could obtain a maximum of 35 points in seven questions (i.e., if the answers were all "very often" in all seven questions). Afterwards, we calculated the median value for each home or respondent across the three seasons, obtaining a single "median life satisfaction and worth self-reported score", "mean life anxiety self-reported score" and "median added adaptation score".

Table 5. Scores assigned to the different answer options in the group of questions regarding life-satisfaction and worth.

Option	Score Assigned
Never	0
Rarely	2
Sometimes	3
Often	4
Very often	5
Rather not say	1

(c) Self-perceived physical capacity

The results in this section were also scalar, ranging from extremely difficult to very easy, as seen in Table 6. A single home/respondent could obtain a maximum of 45 points for their answers to these nine questions (i.e., if its answers were all "very easy" in the nine questions). Thus, first we obtained the total score per home or respondent in each season by adding the values of each of these questions into a single score per home or respondent. Afterwards, we calculated the mean of these values for each home or respondent, obtaining the "average physical self-perception score".

Table 6. Scores assigned to the different answer options in the group of questions regarding physical capacity.

Option	Score Assigned
Extremely difficult	1
Moderately difficult	2
Easy	3
Manageable	4
Very easy	5

2.6. Regression Methods

Similar field studies have used multiple linear regression for modelling subjects' thermal sensation in a range of environments [55]. Therefore, we also used the multiple regression method for predicting a quantitative dependent variable based on data from other quantitative predictors (i.e., independent variables), given a series of assumptions. In this case, once the fieldwork stage finished and the survey data were compiled into a single database (as seen in Appendix A), a multiple linear regression was used to determine whether any of the self-perception variables collected in the surveys, or any of the data from the homes' inner or outer environments predicted the subjects' average total self-reported health problems score. The particular regression method chosen was a stepwise, backwards multiple linear regression. This approach differs from the traditional regression method. Its utility lies in initially including all predictors saturating the model, and subsequently excluding each predictor based on its contribution to the final model. All predictors are first included but are then excluded based on their statistical significance, testing each predictor's importance against the overall result. The significance value for a t-test for each predictor is compared against a removal criterion (in this case set to $p \leq 0.050$): if the predictor does not make a significant contribution, then it is removed, and the model is re-estimated with the remaining predictors (i.e., reassessing the contribution of each remaining predictor). The stepwise backward method is considered preferable to the forward method due to suppressor effects and a lower risk of making a Type II error [56]. This statistical test was carried out using SPSS 21.0.

2.7. Simulations

After the eleven months of temperature monitoring were completed, the information about house geometry was used in Design Builder® models. The Design builder® (DB) software is a Graphical User Interphase (GUI) that runs an 'Energy plus®' simulation engine. This software is widely used in energy and thermal simulation in buildings, and is one of the few considered to be standard practice software in dynamic simulations [57]. The weather file was created with the open-source software LADDER® created by the Rocky Mountain Institute® with weather data provided by the Mexican Water National Commission. The files were provided in XLS format in hourly intervals of wet and dry bulb temperature, relative humidity, vapour tension, cloud levels, wind speed, and wind direction. The files were provided in MS Excel® format, where one file corresponded to one month, and each day was in a different Table. Then, the data were saved into a single file for conversion through a Macro written in M.S. Visual basic®.

Figure 4 shows the simulation process divided into three stages. The calibration process against the temperatures monitored is described in Section 2.7.1. Figure 5 illustrates the cross-section details of the different parameters inputted to Design Builder®. The first was the final calibrated model of the homes "as built". This includes a layer of a solid brick of 13 cm, rendered on both sides by gypsum and cement on the inner and outer faces, respectively. This layout corresponds to all of the homes used in this study, and at least 73% of the total housing stock in Mexico [58] as mentioned earlier. The second was made with similar characteristics to the first but added double glazing. The third included the same 15 cm solid wall, with the addition of 3 cm fibreglass insulation on the inner face of the walls and roofs with single glazing, as determined in the Mexican Standard NOM-ENER-020. The fourth was similar to the previous one, but with 5 cm of insulation, as this is the standard industry size for a layer of glass fibre insulation. The fifth had 5 cm of insulation and double glazing. All the simulations were undertaken with a 100 mm concrete slab for the floor. It was impossible to determine the percentage of homes from the existing housing stock in the country for each simulation layout, as official information only states that 4% of the total housing stock has insulation on the roofs and 1% on the walls [58]. The U-values of the different building envelope typologies are shown in Table 7.

Table 7. U-Values in $Wm^{-2}K^{-1}$ for each building fabric configuration used in dynamic simulations. The letters at the top of the table correspond to the sections shown in Figure 5.

Building Element	(a)	(b)	(c)	(d)	(e)
Wall	2.97	2.97	0.91	0.56	0.56
Roof	3.06	3.06	0.89	0.57	0.57
Window	6.12	3.15	6.12	6.12	3.15

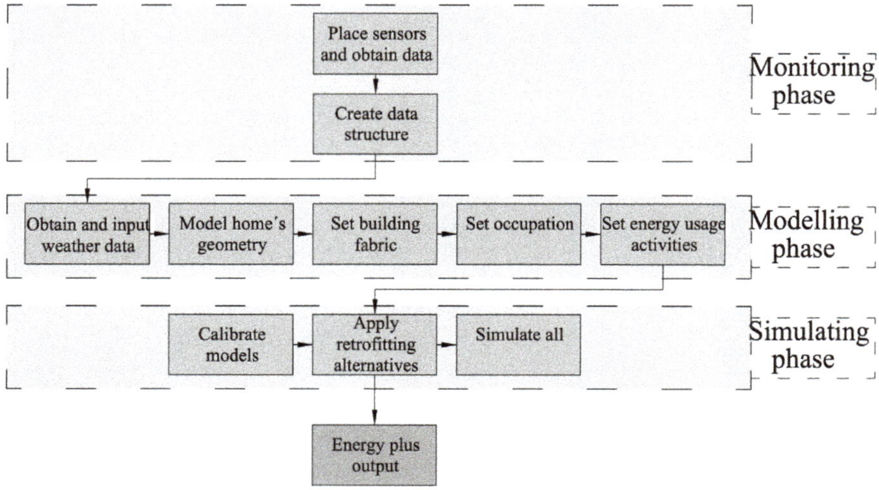

Figure 4. Stages of the simulation process. Stage 1 (monitoring phase) consisted of obtaining the sensors, ensuring their correct operation, and placement in the different houses. Stage 2 (modelling phase) surveyed the homes (measuring the perimeters and taking photographs). Later, these data were used in D.B. to create the simulation models. Finally, Stage 3 (simulation stage) consisted of simulations with the different building envelopes.

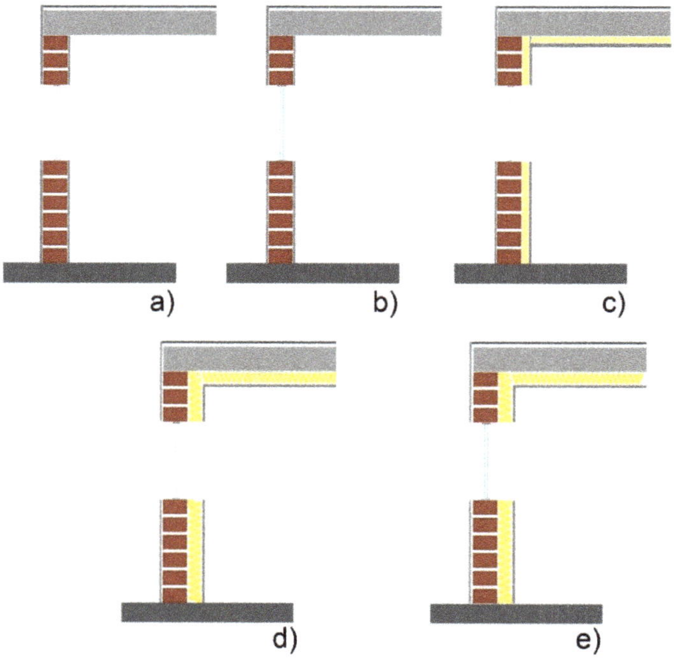

Figure 5. Section drawings of the different building fabric typologies used in the dynamic simulations. Each cross-section is explained below identified by its letter: (**a**) as built; (**b**) double glazing, no insulation; (**c**) single glazing and 3 cm of insulation (as established by the NOM-ENER-020); (**d**) single glazing and 5 cm of insulation (standard practice); (**e**) double glazing and 5 cm of insulation.

2.7.1. Model Calibration

The calibration of a computer model to real data is one of the most complex steps of a simulation study [59]. It compares the real data collected to the simulation results, proving the validity of the computer model for further studies and applications related to improving house thermal behaviour and energy conservation measures [60]. One of the most accepted calibration techniques is to compare real monitored hourly data against the simulated data [61]. The magnitude of the difference is evaluated with mean bias error (MBE) and the coefficient of variation of the root mean square error (RMSE). A model is considered calibrated if these values are not above 25% (MBE) and 35% (RMSE), according to ASHRAE guideline 14 [21]. Table 8 shows the MBE and RMSE values for the different calibrations undertaken. The mean value for MBE is 12.5 (sd = 3.6), and for RMSE is 15.2 (sd = 3.5). None exceeds the threshold values set by ASHRAE. A graphical check shows the comparisons in line and box graphs per hour as seen in Appendix A.

Table 8. Mean Bias Error (MBE) and Root Mean Square Error (RMSE) comparing the calibrated temperatures provided by our sensors against those provided by the Design-Builder model.

	ID_2	ID_3	ID_4	ID_6	ID_7	ID_14	ID_16	ID_17	ID_18	ID_20	ID_22	
MBE	10.08	12.88	12.8	7.79	12.43	20.32	16.47	12.34	9.65	11.16	9.69	
RMSE	12.69	19.32	13.7	9.96	14.08	20.05	20.61	18.44	12.81	12.76	12.22	
	ID_24	**ID_26**	**ID_27**	**ID_28**	**ID_29**	**ID_30**	**ID_31**	**ID_33**	**ID_37**	**ID_38**	**ID_39**	**ID_40**
MBE	14.22	12.64	11	10.66	12.31	7.11	18.29	12.35	21.48	10.98	13.03	9.02
RMSE	15.81	16.41	12.78	13.32	16.37	9.03	19.5	15.53	22.34	13.09	16.39	11.72

2.8. *Underheating*

The WHO has established that a temperature range between 18 °C and 24 °C presents no risk to human health during sedentary activities in indoor spaces [17]. Although existing literature [53] and standards [62] confirm that the upper threshold may vary depending on other variables, such as the outside temperature, many studies have agreed on the lower threshold of 18 °C [9,63,64], and even 20 °C for spaces occupied by vulnerable groups [50], for naturally ventilated buildings.

The CIBSE TM59 Standard [62] was first used as a benchmark to develop an underheating model. This standard considers that the building must provide thermal comfort by without external air conditioning or heating devices. This principle was a key consideration, as our sample homes in the Toluca Valley do not use central heating. The TM59 defines ΔT as the difference between T_{op} and T_{max}. This difference should not exceed 1 K for more than 3% of the occupied hours. This means that a space can be considered overheated if T_{op} = 26 °C and T_{max} = 24.93 °C. In contrast, the lower threshold must not contain this flexibility, as 18 °C is accepted as a "healthy temperature threshold". Hence, we define ΔT_{min} as the difference between 18 °C and T_{op}, when T_{op} < 18 °C being:

$$\Delta T_{min} = 18 - T_{op} \tag{2}$$

Our Criterion 1 considers the hours of exceedance of the lower threshold ($H_{e,low}$), where the total amount of hours at $\Delta T_{min} > 0$ should not exceed 3% of the year. This 3% maximum percentage is based on the standard TM52.

Criterion 2 assesses the severity of underheating. It is well established that exposure to temperatures below 18 °C damages the health of the occupant; however, the lower the operating temperature, the more damage. For example, Saeki et al. [65] found that the blood pressure of the adult population in their study worsened when the temperature dropped by one Centigrade degree. In addition, Shiue et al. [66] found a risk of high blood pressure rises when subjects spent time at temperatures below 16 °C. In addition, a house exposed to temperatures below 14 °C encourages mould growth, affecting the health of its occupants [67]. Hence, we deemed it necessary to evaluate the severity of underheating in the homes in this study. Criterion 2 uses a metric based on the heating degree day (HDD) [68]. This metric is used as a proxy to calculate the energy needed to heat a space. The HDD formula is:

$$D = 18.3 - t, D \geq 0 \tag{3}$$

where D is the degree day, and t is the average daily temperature in °C. Disregarding the fractional part, we propose that our model measures the severity of underheating as

$$Uh_{day} = 18 - t, Uh_{day} \geq 0 \tag{4}$$

where Uh_{day} is the underheated day, and t is the average temperature of any day in °C of all the monitored (or simulated) hours equal to or below 18 °C. The Kelvin hours (Kh) above 18 °C are not considered in this metric. Hence, if a sensor captures readings as described in Table 9:

Table 9. Example of monitored hourly temperature data.

Hour	**00:00**	**01:00**	**02:00**	**03:00**	**04:00**	**05:00**	**06:00**	**07:00**
Temperature in °C	16	16	16	16	15	16	17	18
Hour	08:00	09:00	10:00	11:00	12:00	13:00	14:00	15:00
Temperature in °C	19	20	20	21	23	23	21	21
Hour	16:00	17:00	18:00	19:00	20:00	21:00	22:00	23:00
Temperature in °C	21	20	18	18	18	17	17	16

(i) there will be ten data points (Kelvin hours—Kh) with temperatures higher than 18 °C (in the period from 8:00 h–17:00 h), and (ii) the remaining 14 Kh are then equal or less than 18 °C (0:00 h–7:00, & 18:00–23:00). The mean, in this case, will be calculated over the fourteen Kh, as in (iii) where $T_{op} \leq 18$ °C (0:00 h–7:00, & 18:00–23:00). Hence, t = 16.7 °C.

If the $Uh_{day} \leq 1$, and the daily temperature average $T_{op} \leq 18$°C, the space is considered as "mild underheated. If $Uh > 1$ and ≤ 2, the space is medium underheated. When $Uh \geq 3$, the space is considered severely underheated.

3. Results

3.1. Home Characteristics

The homes studied had, on average, three occupants per house (sd = 1.1), the average age of the householders was 40 years old (sd = 12.8 years), and the average age of the homes was 23.5 years (sd = 17 years). Only 5% of households used any kind of external heating systems (including electric heaters and ethanol chimneys) to raise the temperature of their homes, and these were used rarely. The most common strategies to achieve a personal thermal balance were related to the use of clothing, since 66.2% of the participants affirmed that they used more than two layers of clothing when they felt "too cold" at home. During the winter site visits, the interviewees affirmed having various layers of clothing, equivalent to 1.2 to 1.6 *CLO* according to Figure A7 from Appendix B of the ASHRAE 55 standard. In these cases, the building would not meet the criteria set in ASHRAE 55 since they are no longer valid when *CLO* values are above 1.5

To illustrate the contrast of internal temperatures across the seasons, Figure 6 presents a heatmap of the monitored bedroom temperatures averaged per hour for the different seasons. For the hot seasons of spring and summer, the living rooms' mean temperature was 21.4 °C (sd = 2.5 °C) and 20.4 °C (sd = 3.38 °C), respectively. However, in Autumn, the average temperature drops to 18.7 °C (sd = 3 °C), slightly above the WHO recommended threshold. Worryingly, the winter average living room temperature was 17.7 °C (sd = 3.2) below the WHO threshold, and heating systems were rarely used. For this period, most homes (i.e., except for id11, id14, and id16) averaged temperatures between 15 °C and 20 °C, especially during the night hours. Figure 7 shows a cumulative graph of the hours monitored during the whole study, compared to the "cold season". It shows that 50% of the monitored hours recorded temperatures below 18 °C for 60% of the houses. This problem was exacerbated during the cold season, when only 20% of the houses had more than 3% of the total recorded hours above 18 °C. This suggests a serious underheating problem in more than half of the studied houses.

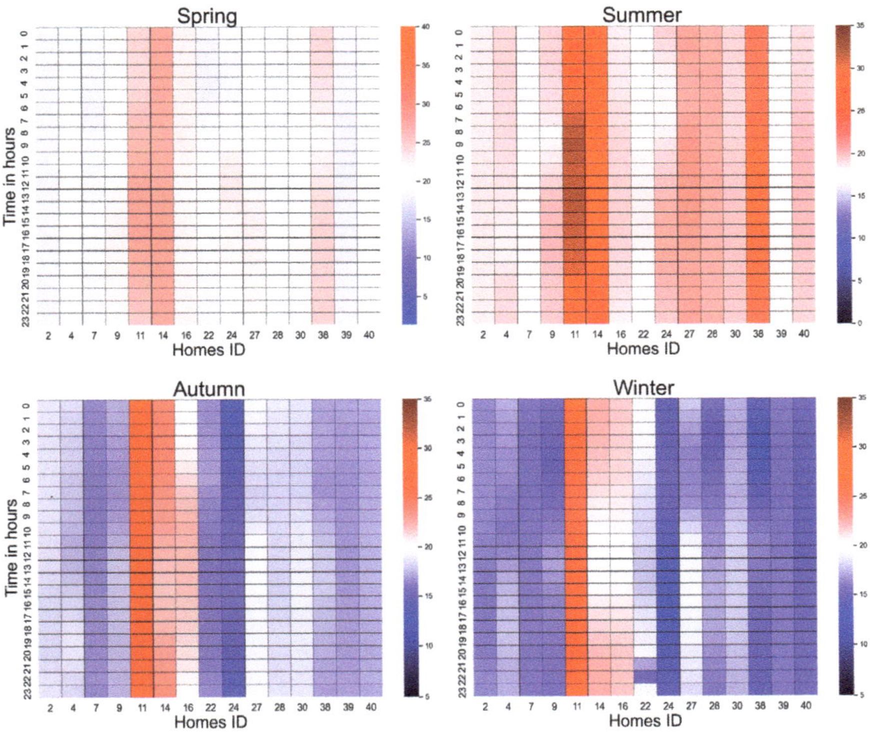

Figure 6. Heatmap of the average of each house's seasonal living room temperatures across 2018 (spring, summer and autumn) and 2019 (winter).

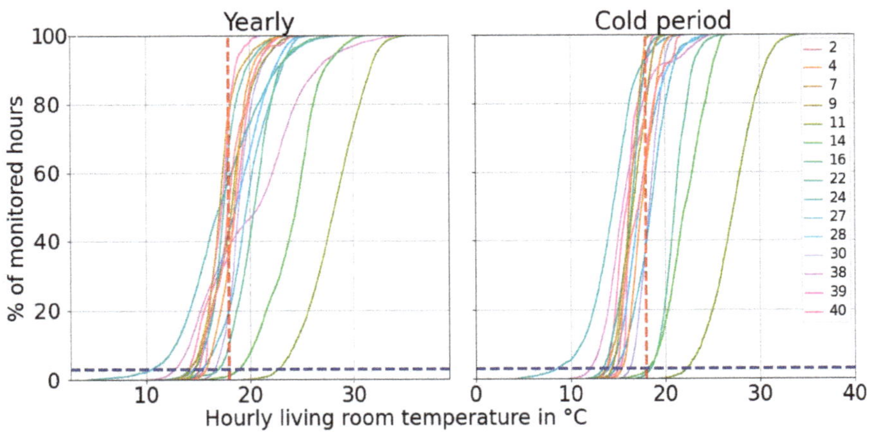

Figure 7. Hourly living room temperatures for the length of the study (left) and those in the cold season (right). The cold season comprises the period from 1 October to 28 February. The dashed horizontal blue line shows the limit of hours (3%) that the homes must not exceed before meeting the dashed red vertical line, which represents the 18 °C temperature threshold recommended by WHO.

Figure 8 shows seasonal box plots of the monitored living room relative humidity measured across the spring, summer and autumn of 2018, and the winter of 2018–2019. The average relative humidity throughout the whole monitoring period was 35% (sd = 7%). This remained consistent throughout the year (spring: 32.5%, sd = 6.7; summer: 37.4%, sd = 6.4%; autumn: 37.5%, sd = 8.1%; winter: 31.9%, sd = 8%).

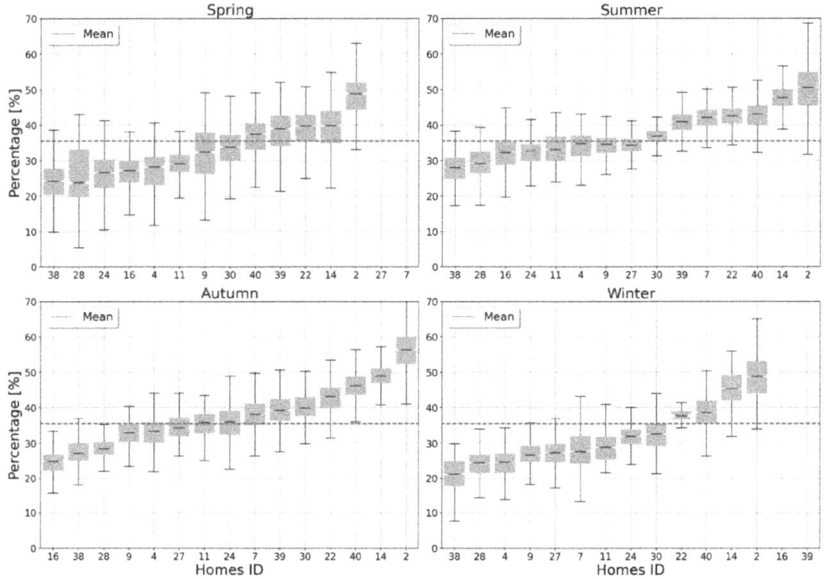

Figure 8. Ranked living room relative humidity by season. The red dotted line represents the mean.

3.2. *Underheating*

Criterion 1 indicates that only two houses did not experience underheating for the annual period, as shown in Figure 9. For the annual period, the average of the underheated hours accounted for 32% (sd = 20%). However, this figure increased to 47% on average (sd = 26.5%) for the cold season between August and January. The high standard deviation observed in both the annual and cold seasons suggests a significant disparity in the number of underheated hours in both periods within our sample. We identified three patterns of behaviour that if divided into groups, showed very low standard deviations, implying a high degree of proximity to each other.

The first group consisted of id30 and id28, averaging 26% of underheated hours with a standard deviation of 0.2%. The results of both health and thermal comfort surveys did not reveal any significant patterns or commonalities among these houses. The next group, which comprised id2, id4, id22, and id28, had an average of 45% underheated hours with a standard deviation of 4.1%. The results of our surveys (further discussed in Section 3.4) showed that this group of houses had high levels of adaptation coupled with an average PMV of −0.7 (sd = 0.35), still within the comfort zone. This means that despite the low temperatures recorded by our sensors, the house occupants were able to be in comfort ranges after taking appropriate adaptive measures. Finally, it is concerning that the majority of houses averaged 59% underheated hours, indicating longer cold periods. The results of the thermal comfort surveys for this group showed a similar pattern to the previous, in that PMV and PPD average values were, in general, within comfort ranges, indicating a consistent trend similar to the previous group.

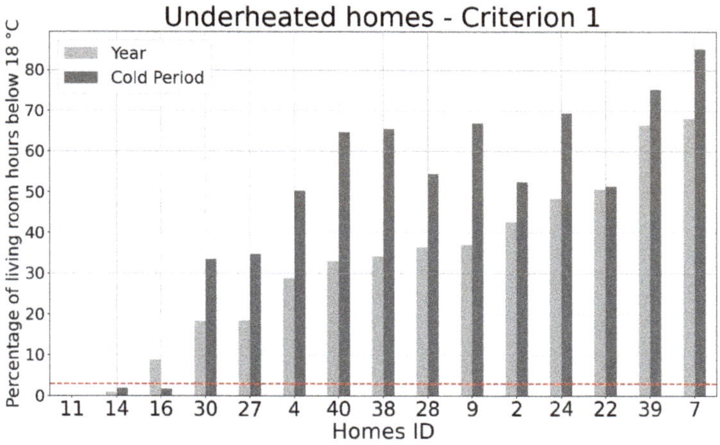

Figure 9. Percentage of underheated hours according to criterion 1.

Figure 10 shows the frequency of underheated days according to criterion 2. The summer months show long diurnal temperature swings resulting in 66% of the houses with light underheating throughout the year. Only one house (id11) did not experience any underheated days. Houses in May and June averaged 4.2 underheated days (sd = 6 days). This number increased to 11.5 days on average (sd = 1.2 days) from July to October. However, it decreased to an average of 6.4 days (sd = 2 days) from October 2018 to February 2019.

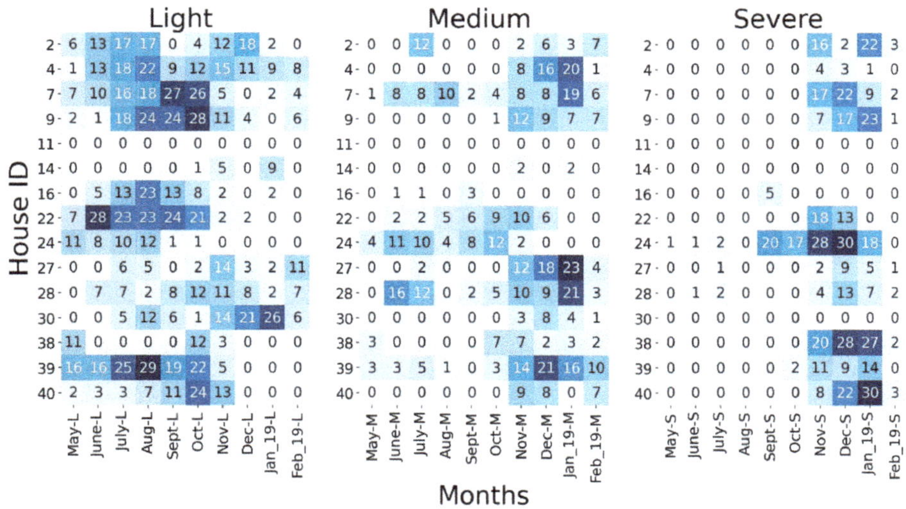

Figure 10. Number of underheating days per home split monthly according to criterion 2.

The medium underheated houses during the months of May to October were reported only on 1.7 days on average (sd = 1 day). However, from October 2018 to January 2019, the number of reported cases increased to an average of 6.2 days (sd = 1.3 days). In terms of severe underheating, from May to October 2018 there was an average of 0.5 days (sd = 1 day). However, from November 2018 to January 2019, the number of severely underheated days increased significantly to an average of 10.2 days (sd = 0.9 days) across the sample. In addition, extreme cases of severe underheating were observed in three instances: id24 and id40 with 30 days of severe underheating during December, and ID38 which had 28 days.

3.3. *Dynamic Simulations*

3.3.1. Home Characteristics

After identifying that the majority of homes in this study (93%) were underheated, various retrofitting options were simulated. The addition of double glazing resulted in an average T_{op} increase of 20.6 °C (sd = 7.3 °C). However, the significant standard deviation indicates that some houses (i.e., id7, id14, id28, and id38) had hours below 18 °C. Adding insulation made a substantial difference compared to the previous two simulations. Living room temperatures under the NOM-020 parameters (with 3 cm of insulation) averaged 22 °C (sd = 2.5 °C), and homes with double glazing and insulation averaged 23 °C (sd = 2.4 °C). For all simulations using the insulation parameter, the yearly average was always above the 18 °C threshold, as seen in Figure 11.

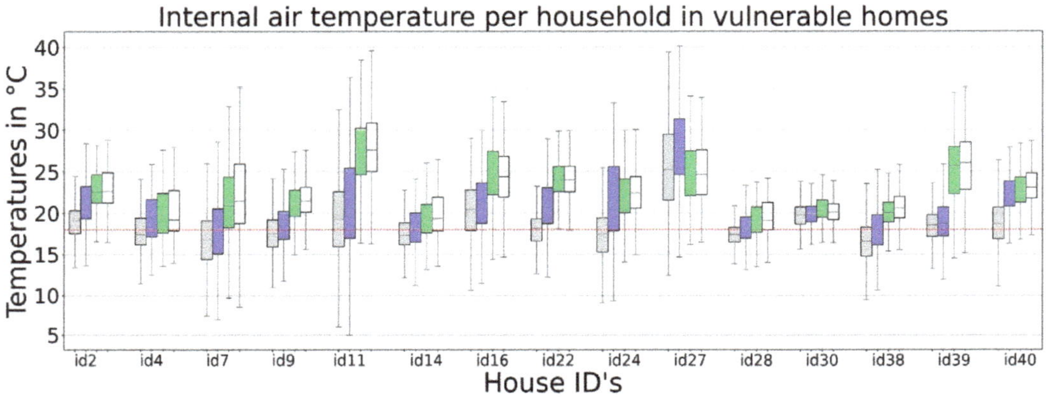

Figure 11. Boxplots of internal temperatures, where the light grey (first from left to right) corresponds to the temperatures monitored with sensors. The blue boxes (second from left to right) correspond to the simulations with double glazing, the green (third from left to right) to the homes only with insulation, and the white (fourth from left to right) correspond to the ones under the NOM-ENER-021 parameters.

3.3.2. Thermal Comfort

This sub section presents the results of both monitored and simulated environmental data against the adaptive model from the ASHRAE 55 Standard, as shown in Figure 12. Although we found data points below the lower threshold of the adaptive graph, there was an increase in data-points within the ASHRAE 55 comfort bands in the retrofitting cases. For the monitored data, 30% of the living room hours were within the 80% acceptability limits. This percentage was increased to 40% with double glazing only. There was a noticeable change when insulation was adhered to ceilings and walls, as 50 mm of fibreglass increased comfort hours to 67%. The comfort hours increased to 65% when using the parameters established under the NOM-020.

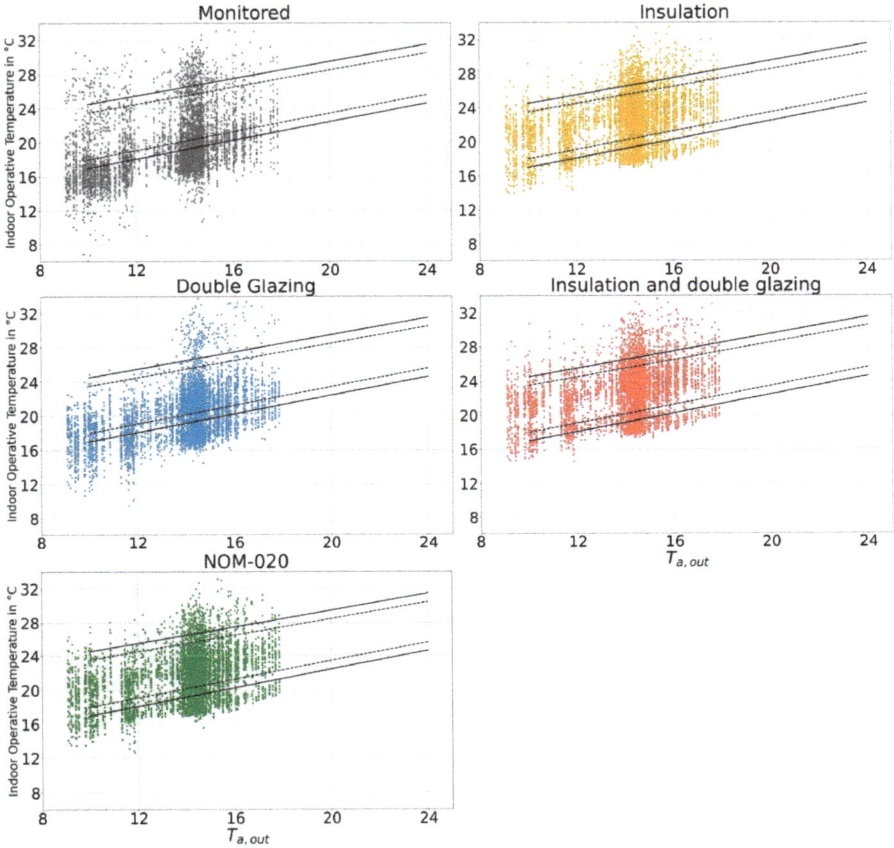

Figure 12. Adaptive approach graphs showing the outdoor running mean temperatures (°C) (*x*-axis) and the indoor operative temperature (°C) (*y*-axis) of the homes included in this study against the parameters established in the ASHRAE 55 standard. The segmented line (inner) represents 90% acceptability. The continuous (external) line represents 80% acceptability ASHRAE thresholds. Finally, the spots represent one hour (monitored or simulated). The top left (grey) shows the monitored (actual) temperatures. The different simulations provide the remaining results, as stated in each graph's title.

3.3.3. Underheating

Figure 13 shows the results of the underheating analysis under criterion 1. The results show a significant decrease in underheated hours for the retrofitted options compared to the calibrated models. The addition of double glazing lowered the number of underheated hours compared to the calibrated models to an average of 15.4 °C (sd = 9.5 °C) for the annual period, and 18.5 °C (sd = 11 °C) for the cold season. Insulation alone considerably reduced the number of underheating hours, since these were reduced on average by 34.2 °C (sd = 13.5 °C) and 40 °C (sd = 12.8 °C) for the annual and cold periods, respectively compared to the calibrated models. A combination of insulation and double glazing resulted in a slight increase of 2 °C (in both periods) concerning the batch that only included insulation. Finally, simulations based on the NOM-020 standard resulted in an average reduction of 28.5 °C (sd = 13.5 °C) for the annual period, and of 31.3 °C (sd = 12.7 °C) for the cold period.

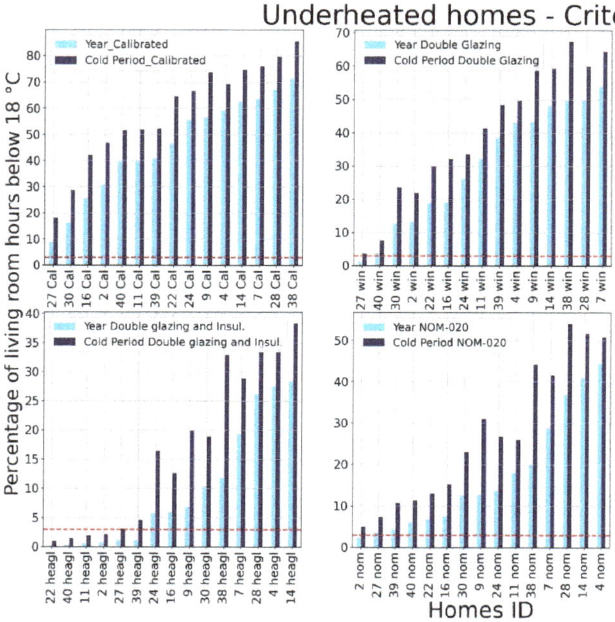

Figure 13. Percentage of underheated hours according to our criterion 1 after simulations.

Table 10 presents the results of the analysis with underheating criterion 2. The calibrated models showed an average of 6.5 days per month of light underheating (sd = 1.8 days). On average, the number of medium underheated days per month was 3.2 (sd = 1.2 days). While the average number of severe underheated days observed was 1.6 days (sd = 2.2 days), it is concerning that this value increased to 8.6 days on average (sd = 4.5 days) in the cold period. The data indicate that the installation of double glazing resulted in a reduction of 52% in the average number of light underheated days. Despite this improvement, the cold period still had 3.8 days on average (sd = 2.6 days) of light underheating. As expected, significant reductions in underheating were observed after insulation was added, mainly in medium and severe underheating, decreasing from 3.5 days per month to 0.3 days. The application of the NOM-020 standard parameters resulted in a reduction of severe underheating in most of the houses (all but three: id7, id24 and id38). Despite the addition of insulation, some homes still had light and medium underheating days during November and December, averaging 3 days in both months (sd = 2.6 days). Additional information on the evaluations conducted under criterion 2 can be found in Appendix B.

Table 10. Average days per month with light, medium and severe underheating. D G stands for double glazing.

		Calibrated	Double Glazing	Insulation	D G and Insul	NOM-020
Light	mean	6.5	3.1	0.7	0.6	1.3
	sd	1.8	1.1	1	1	1.2
Medium	mean	3.5	1.3	0.3	0.3	0.7
	sd	1.2	1.2	0.6	0.5	1.2
Severe	mean	4.1	1.6	0.3	0.1	0.6
	sd	4.1	2.2	0.6	0.3	1

3.4. Health Surveys

Table 11 below presents the summary of the results for each house ID, after the data reduction methods were applied. It shows the results for the variables where people ranked their self-perceived thermal comfort (mean predicted percentage dissatisfied, the mean added adaptation score and mean PMV), the variables of self-reported physical health (the mean total health problems score, and the median life satisfaction and worth), self-reported mental health (median life satisfaction and worth, mean life anxiety), and variables related to their personal characteristics (days with underheating, house age, and respondents age). It is important to mention that these are the variables that resulted in the step-backwards regression model after iterations.

Table 11. Summary of survey results per house ID, for each of the variables used in multiple regression analyses.

House ID	Mean Predicted Percentage Dissatisfied (PPD)	Median Added Adaptation Score	Mean Total Health Problems Score	Median Life Satisfaction and Worth Self-Reported Score	Days w/ Underheating	House Age	Respondent's Age	Mean Physical Self-Perception Score	Mean Life Anxiety Self-Reported Score	Mean Predicted Mean Vote (PMV)
2	31.37	15	2.67	7	134	10	60	27.66	7	−0.76
4	27.02	20	2	7	92	50	55	30.33	6	−0.85
7	23.06	15	3	6.5	220	20	32	30.33	3.67	−0.65
9	12.48	19	1.75	6.25	108	15	44	34.75	3	−0.6
11	12.88	16	1.8	5	0	30	31	39.4	2.2	−0.11
14	27.89	14.5	4	3.25	2	6	31	33.5	4.5	−0.41
16	11.11	12	2	7	20	10	24	37	3.5	−0.43
22	10.25	15	3.6	3.5	176	20	52	37.4	5	−1.03
24	6.78	10	2.5	5	156	30	33	29.5	6	−0.08
27	48.89	18	4.33	6	71	20	60	45	5	−1.45
28	23.8	14	2.5	7	115	15	32	30.25	2.75	−0.62
30	5.17	6	1.66	6.5	61	30	49	41.33	5.67	−0.05
38	9.73	8	2.8	4.5	114	3	30	33.4	7	−1.03
39	14.7	4	3	6	217	15	27	30.33	5.67	−0.35
40	12.7	4	3.66	4	103	25	63	36.33	9	−0.23
Mean ± SD	18.52 ± 11.74	12.70 ± 5.20	2.75 ± 0.85	5.63 ± 1.32	105.93 ± 68.95	19.93 ± 11.90	41.53 ± 13.69	34.43 ± 4.95	5.06 ± 1.85	−0.58 ± −0.40

n = 15.

3.5. Multiple Linear Regression

Results for the best-fit final multiple linear regression model (R^2 = 0.91, F (4,14) = 28.4, p > 0.01, n = 15) suggest that the PPD index was a first, significant, positive, and strong predictor of the average total self-reported health problems score. The second-best, negative predictor was the median life satisfaction and worth score; the median added adaptation score was also a significant and negative predictor. Total days with underheating was the last significant but positive predictor (Table 12).

Table 12. Best-fit final stepwise linear regression (backward) model with average total self-reported health problems score as the dependent variable for 15 homes in the city of Toluca, State of Mexico, Mexico.

Predictor	B	SE	StB	t	p	Tolerance	VIF
(Constant)	4.47	0.37		12.87	>0.01		
Median Life satisfaction and worth self-reported score	−0.45	0.061	−0.71	−7.51	>0.01	0.91	1.10
Median added adaptation score	−0.053	0.018	−0.32	−2.94	0.015	0.66	1.51
Average PPD	0.06	0.008	0.89	8.21	>0.01	0.69	1.44
Total days w/Underheating	0.003	0.001	0.25	2.68	0.023	0.93	1.07

R^2 = 0.91; Adjusted R^2 = 0.88; F(4, 14) = 28.40, p > 0.01, n = 15.

In contrast, variables such as "House age", "Respondent's Age", "Average Physical self-perception score", "Mean Life anxiety score", and "PMV" were all non-significant and were thus excluded (Table 13). PPD was positively correlated with the median added adaptation score (r = 0.54; p = 0.37) and the median life satisfaction and worth score (r = 0.23; p = 0.41), but negatively correlated with total days with underheating (−0.09; p = 0.74). The median added adaptation score and the median life satisfaction and worth score were positively correlated (r = 0.24; p = 0.39). The correlation between Total days with Underheating and Median life satisfaction and worth score (r = 0.1; p = 0.71) and that between Total days with Underheating and Median added adaptation score (r = −0.21; p = 0.45) were also non-significant, as seen on Figure 14.

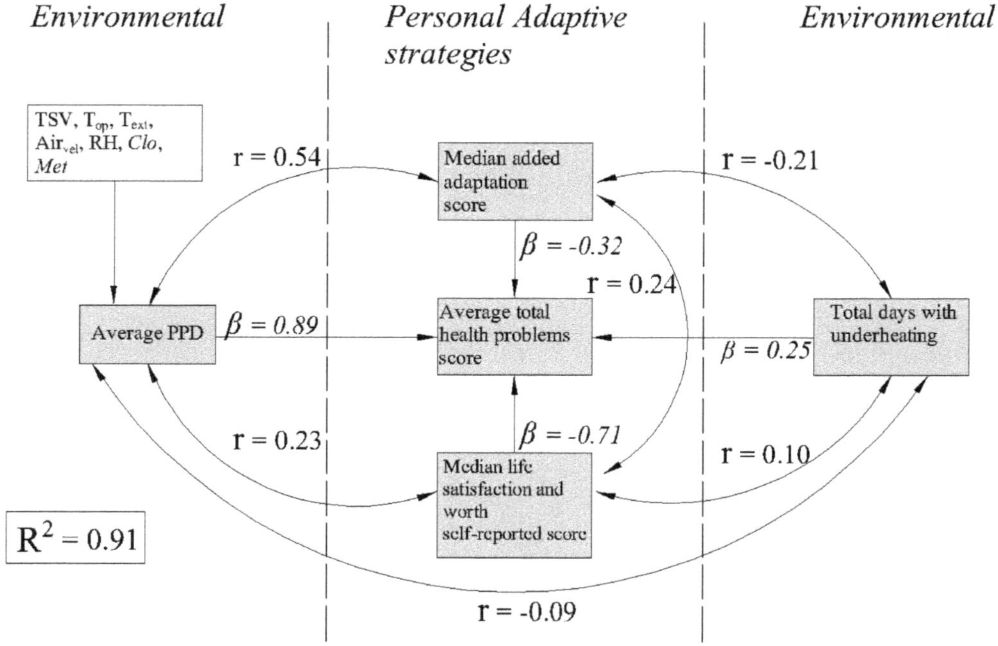

Figure 14. Relationships among predictors of average total self-reported health score in the best-fit multiple linear regression model (backward method; R^2 = 0.91; Adjusted R^2 = 0.88); distinguished by whether predictors are of the environmental origin or personal adaptive strategies. Direct effects are displayed by straight arrows, whereas correlations among predictors are described as curved arrows. Variables composing the PPD index are pictured within a lighter-shaded box (upper-left corner) for informative purposes only.

Table 13. Excluded variables for the best-fit final stepwise linear regression (backward) model with average total self-reported health problems score as the dependent variable for the data of 15 homes in the city of Toluca, State of Mexico, Mexico.

	Beta in	t	*p*	Partial Correlation	Tolerance	VIF
House Age	−0.91	−0.96	0.36	−0.30	0.905	1.106
Respondent's Age	−0.018	−0.18	0.86	−0.059	0.847	1.181
Average Physical self-perception score	0.131	1.25	0.24	0.384	0.699	1.431
Mean Life anxiety self-reported score	−0.087	−0.71	0.498	−0.229	0.563	1.777
Average (PMV)	−0.137	−1.03	0.328	−0.326	0.461	2.172

4. Discussion

4.1. Underheating

The results of this study suggest that many houses in the Central Mexican Plateau do not provide adequate internal environments to their inhabitants because of issues with the building envelope. The underheating analysis indicates that while applying the NOM-020 standard parameters lead to a decrease of underheated hours, it was not sufficient to fully resolve this issue. It is important to state that we observed significant disparities in underheating percentages among the monitored houses without any evident explanation, as their parameters did not show any relevant differences. This suggests that external factors may be at play, potentially related to variations in airtightness that may be contributing to the observed disparities. Unfortunately, no permeability study ($m^3/hr/m^2$@50Pa) has been conducted in Mexico to date, indicating a further need to investigate this area.

The different simulations suggest that the single addition of double glazing does not significantly reduce the number of underheated hours. Further, the average temperatures between the models with insulation and single glazing, against those with insulation and double glazing only, provide an average temperature rise of 2 °C. This does not seem worth the investment, as double-glazed windows are still new in Mexico, making their manufacture and installation costly. Hence, they are not recommended as a retrofitting strategy for the homes in our sample or in the whole plateau. Although the models with insulation did not meet our 3% target, considerable amounts of underheated hours were reduced. Therefore, this suggests a need to enforce the NOM-020 standard in the region, coupled with adequate air leakage measures.

4.2. Underheating and Health

It is clear that improving the internal environment of the houses on the Central Mexican Plateau may also positively impact the self-perceived mental health of the population. However, based on our analysis, we found that self-reported indicators of poor health are significantly associated with perceptions of poor thermal discomfort (i.e., PPD), consistent with other studies. Additionally, our findings indicate that a combination of environmental factors and personal adaptive strategies can be used as predictors of self-reported health problems, at least within the limited but representative sample of houses examined in this study.

Our results suggest that an interaction of both environmental factors and personal adaptive strategies can be used as predictors of self-reported health problems, at least for this small sample of houses in Mexico. Our results concur with the multidimensional characteristics of the adaptive approach to human thermal comfort, based on physiological acclimatisation, behavioural adjustment and psychological adaptation [25,55] against environmental (i.e., thermal) insult [69]. Conversely, the substantial predictive influence of both PPD and total days w/underheating upon average total self-reported health problems scores suggests that subjects perceive environmental insults (i.e., temperature drops, dampness, cold winds) that elicit physiological responses, yet still influence their psychological welfare. In this regard, it is interesting to observe that the significant negative relationship between self-reported health problems and life-satisfaction and self-worth perception suggests that subjects with better self-regard tend to report fewer health problems. This factor has been reported before, indicating that life satisfaction, self-esteem and perceived health are related [70]. In contrast, however, the same results suggest that when subjects have a worse self-perception, they report more health problems associated with greater dissatisfaction with their environments, perhaps even increasing the possibility that resulting psychological stress worsens any previous health condition. On the other hand, the behavioural aspect of the model is underlined by the possibility that, despite the environmental insult, subjects with higher adaptive capabilities can reduce the order of magnitude of their self-perceived health problems. The inverse interpretation is also of interest. As subjects exert worse adaptative strategies (one of the possible reasons could be related to lower incomes), they may consider that they have more health problems or perceive their health as worse. This confirms previous results related to the health locus of control and suggests that as subjects perceive that their control upon their lives is higher and their self-esteem better, their behavior and cognitive processes are greater promoters of their mental and physical health [71]. The correlations between predictors complete this interesting scenario. Dissatisfaction with the environment and house underheating increase subjects' display of adaptive responses; those with higher self-regard and higher life satisfaction showing more (or more intense) adaptive responses. However, such responses could be discouraged if subjects have insufficient financial means to exert a compensatory response, or lack the sufficient psychological drive to display a resolute behavioural response. Overall, would be interesting to investigate if the interaction between the perception of health and self-esteem is one of the factors that contribute to the unfortunate statistic of the high amount of EWD reported for Mexico. This should be a question for study with a larger sample of homes across different locations in Mexico.

5. Conclusions and Limitations

The houses in our study were below established thermal comfort standards and below the parameters set by our underheating model. Both a higher Predicted Percentage of Dissatisfaction (PPD) and more days with underheated homes led to poorer self-perception of health. As subjects felt worse in their internal environments, the worse their self-reported physical health became, the lower their life satisfaction, and the poorer their self-worth, i.e., the less satisfied they felt at home. This was associated with greater use of adaptive strategies against environmental dissatisfaction.

The results of this study of cold comfort in Mexico align with the vast body of literature on cold comfort in hot countries. Paradoxically, the less extreme the winter, the lower the internal temperature and the greater the cold discomfort. In addition, we found that these events also correlate with worse self-perceived health. In the case of the Central Mexican Plateau, the external air temperatures are not close to what is considered an extreme environment. Nevertheless, the vast majority of homes were underheated. Furthermore, a crucial result provided by dynamic simulations is that would be possible to reduce heat losses if the unenforced, but mandatory, standard NOM-ENER-020 was strictly applied. The findings of our PPD self-reported health relationship model suggest that raising internal temperatures in winter could also contribute to raising subjects self-reported health perceptions.

Despite the focus of this research being centered on the NOM-ENER-020 standard, it is noteworthy that the results have relevance to other cold cities in hot countries, particularly in cities with high altitudes in Latin America. This is also due to the similarity in housing typologies between Mexican houses and those located in the region. In fact, a comparative analysis of green building regulations in Latin America [7] determined that the NOM-ENER-020 ranks among the most stringent regulations in the area with regards to the building envelope. Therefore, the findings presented in this paper may offer valuable insights for policy makers and stakeholders seeking to enforce building codes so homes in other Latin American regions provide adequate environments.

This study's limitations include a relatively restricted sample size. While it is acknowledged that a larger sample size is generally necessary to draw valid conclusions to ensure representativeness, a qualitative approach was adopted, given that 71% of the homes in Mexico share the characteristics of the homes examined in this sample. Therefore, while the sample size may be considered small, the homes studied are representative and the findings may still be valuable to policymakers and stakeholders seeking to enhance the quality of houses in the region.

Three aspects should be considered for future research. First, a study with a larger sample size should be conducted to confirm and expand upon the current findings of this study. Another crucial aspect would be to conduct a pilot study involving the retrofitting of a home using a "before and after" approach, where thermal comfort surveys and temperature sensors are installed prior to and following the intervention in several homes. This methodology would enable the assessment of the effectiveness of retrofitting in enhancing thermal comfort and reducing energy consumption. Finally, a cost-benefit analysis is required to investigate the expense of retrofitting the housing stock of the most vulnerable, while examining the benefits of such investment in terms of reduced healthcare costs in the region, and improved health outcomes. Addressing these aspects will be critical in building upon the findings of this study and developing effective strategies for enhancing the quality of housing in the region.

Author Contributions: First Author C.Z.-G.: Data curation, Funding acquisition, Methodology, Software, Visualization, and writing original draft. Second Author A.J.M.-C.: Formal analysis, Validation, and Writing original draft. All authors have read and agreed to the published version of the manuscript.

Funding: This research was funded by the Mexican Council of Science and Technology (CONACyT) under its scheme "Becas para postgrado al Extranjero". The fieldwork stage was funded by the "International Mobility Fund 2019" of the University of Bath. PRODEP-SEP México (PTC-915038) and UAM, Department of Health Sciences (institutional funds) provided financial support to AJMC. The open access fee has been convered by the Universidad de Monterrey, under its scheme "Fondo de publicaciones 2023".

Data Availability Statement: The data supporting the findings of this study are available upon reasonable request from the authors.

Acknowledgments: The authors are deeply thankful to all the volunteers who willingly decided to participate in this study. We wish to thank Héctor Saldaña Márquez for his valuable inputs on the paper.

Conflicts of Interest: The authors declare no conflict of interest.

Appendix A. Results of Visual Calibrations

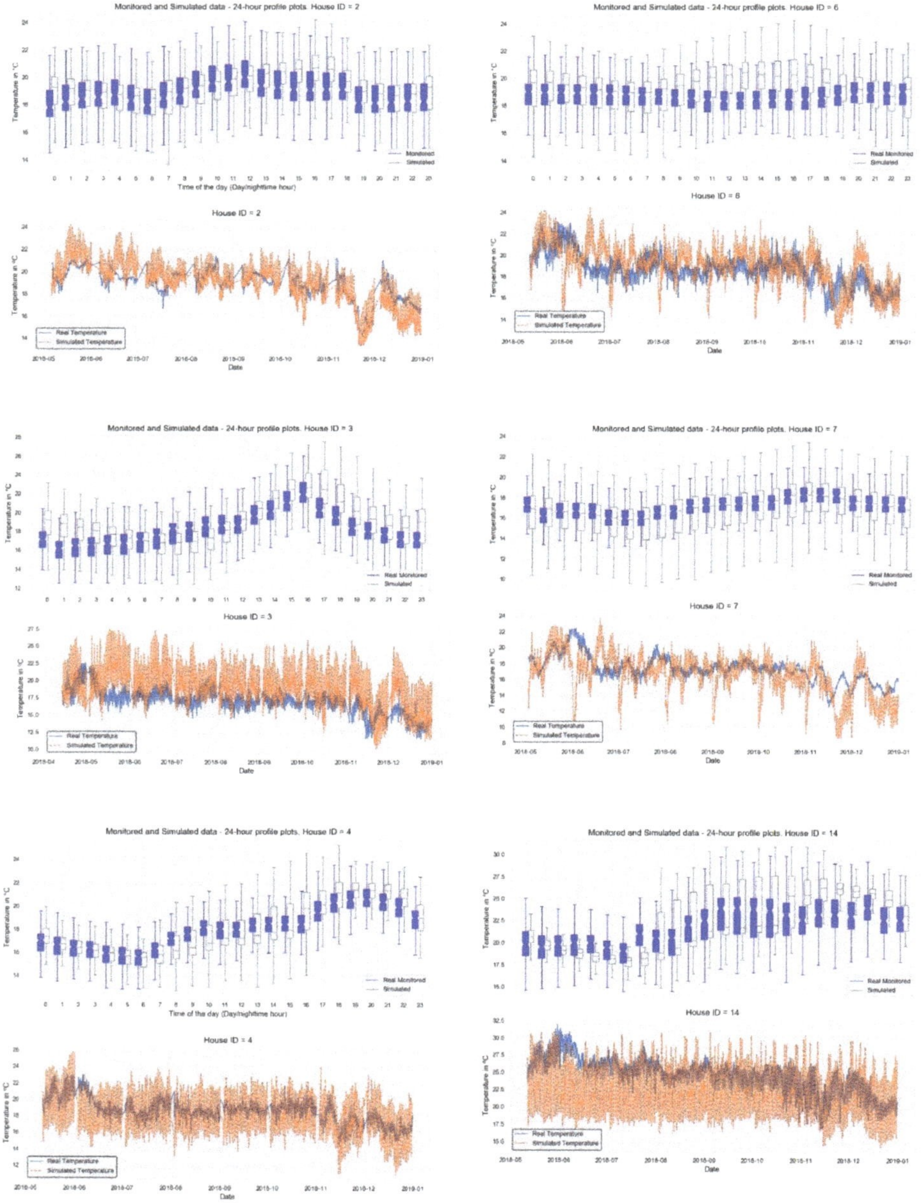

Figure A1. Results of the calibration of the models used for simulation—page 1.

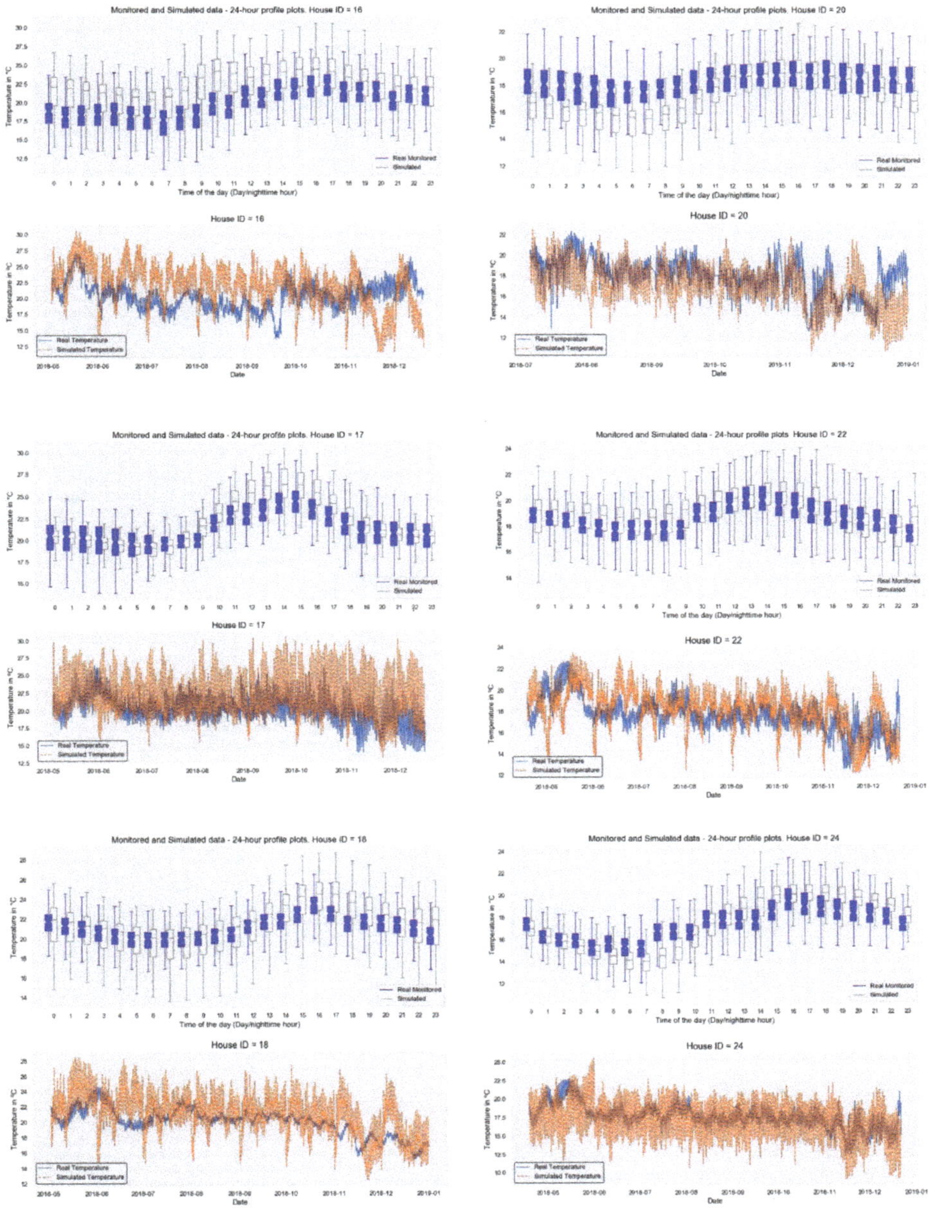

Figure A2. Results of the calibration of the models used for simulation—page 2.

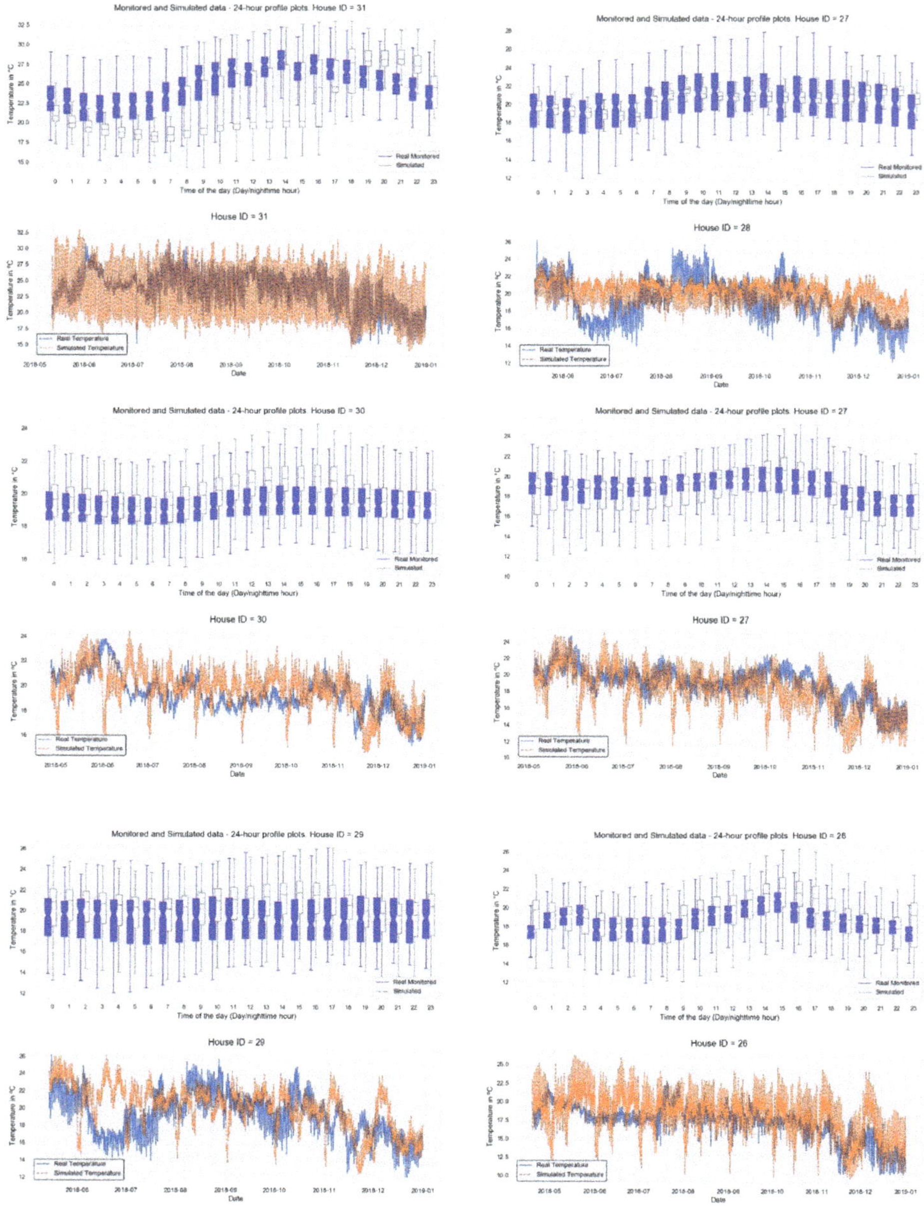

Figure A3. Results of the calibration of the models used for simulation—page 3.

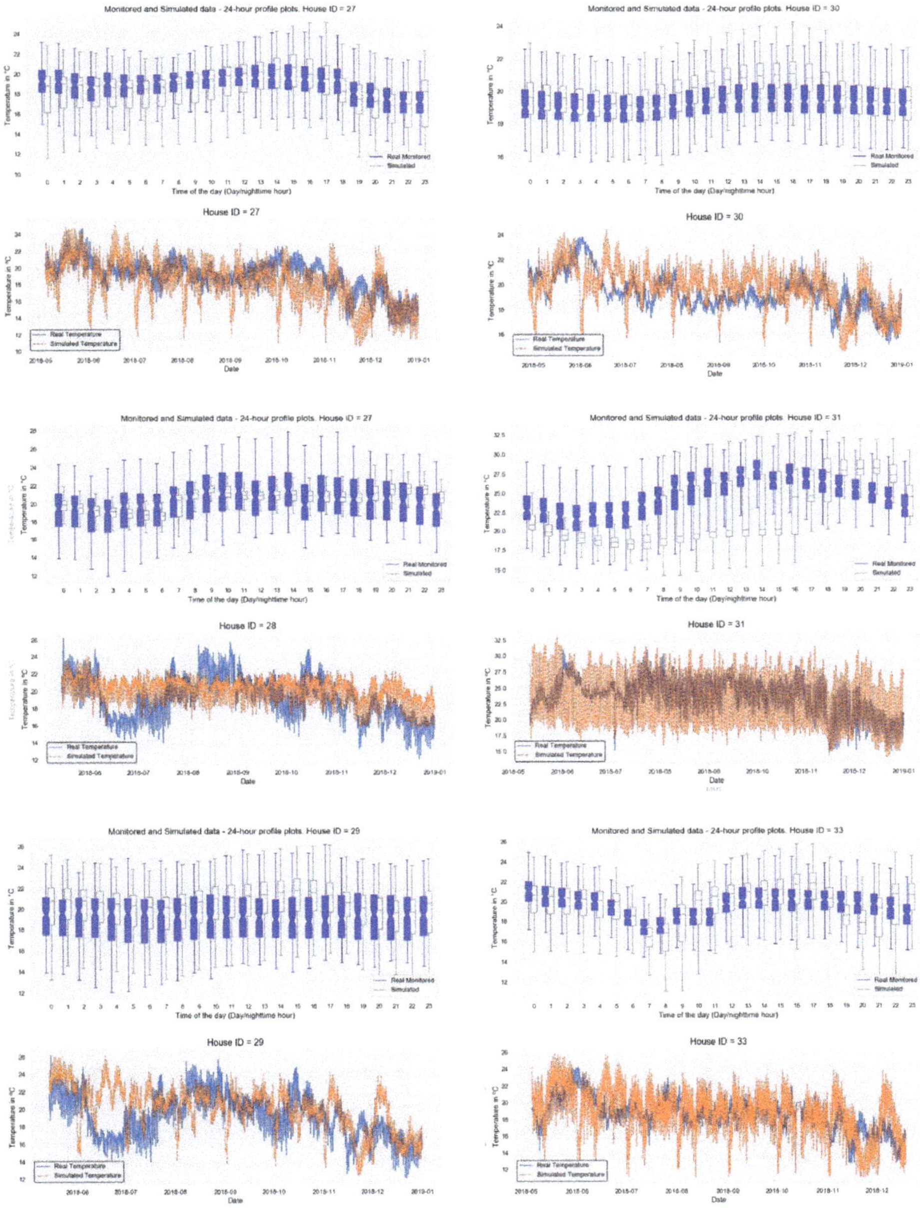

Figure A4. Results of the calibration of the models used for simulation—page 4.

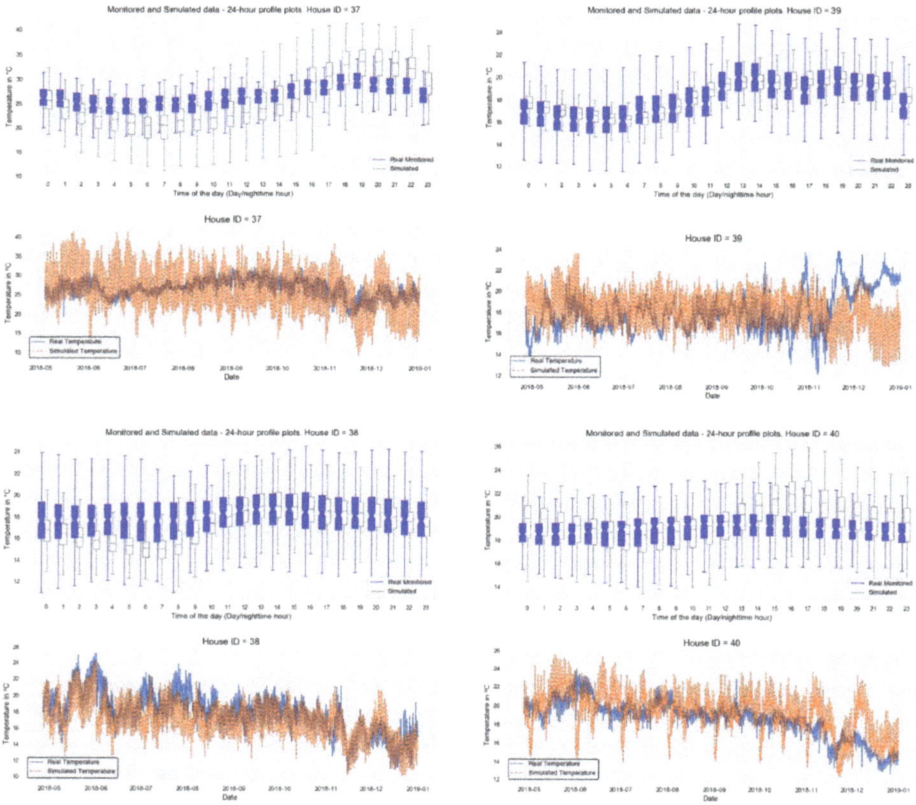

Figure A5. Visual results of the calibration of the models used for simulation—page 5.

Appendix B. Results of the Underheating Criteria 2 Applied to the Simulations

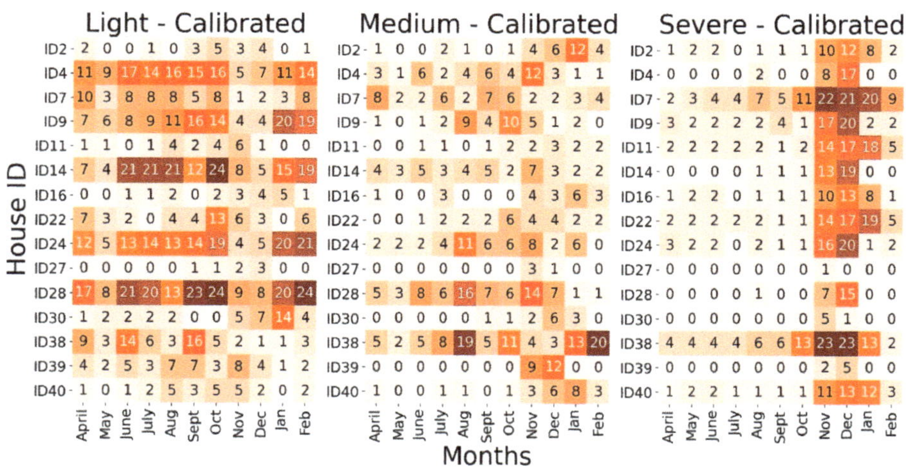

Figure A6. Results of the underheating criteria 2 applied to the calibrated models. The *X* axis shows the months, and *Y* axis shows the house ID. The intersection of both shows the number of underheated light (left), medium (middle), and severe (right) days in that month.

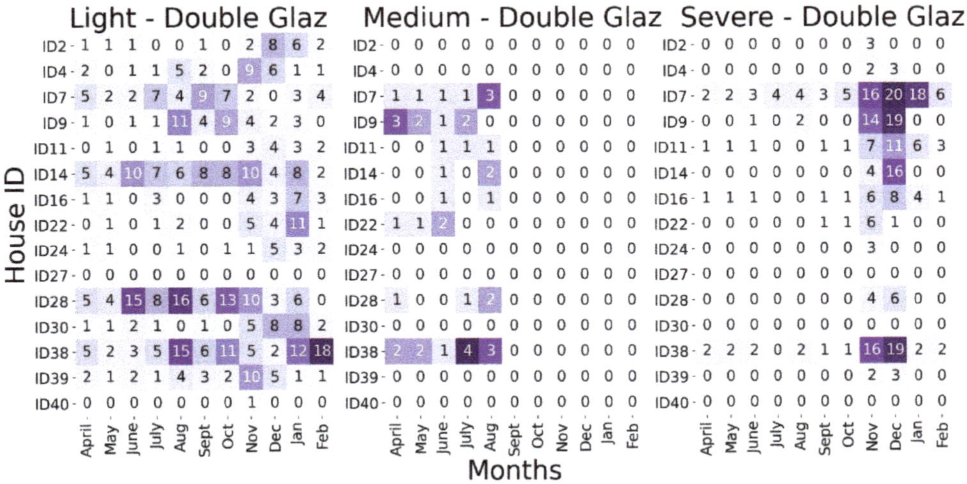

Figure A7. Results of the underheating criteria 2 applied to the models with double glazing. The *X* axis shows the months, and *Y* axis shows the house ID. The intersection of both shows the number of underheated light (left), medium (middle), and severe (right) days in that month.

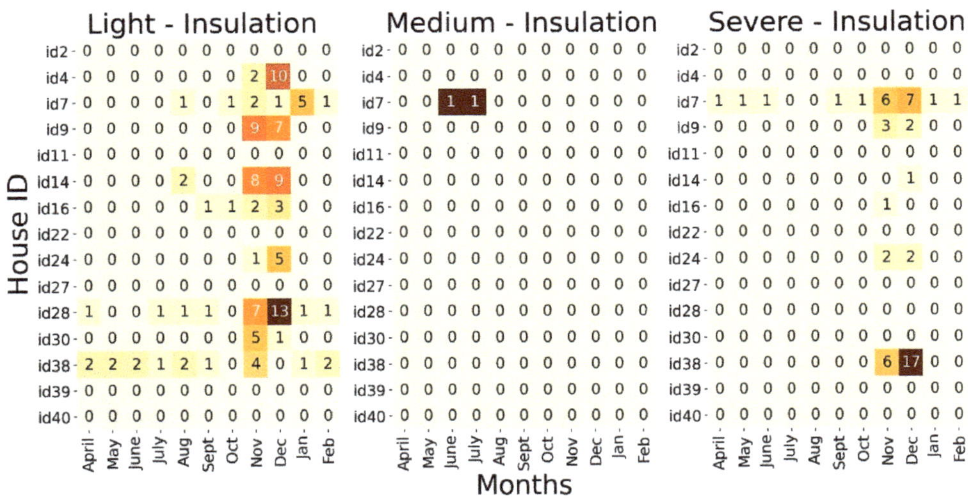

Figure A8. Results of the underheating criteria 2 applied to the models with only insulation. The *X* axis shows the months, and *Y* axis shows the house ID. The intersection of both shows the number of underheated light (left), medium (middle), and severe (right) days in that month.

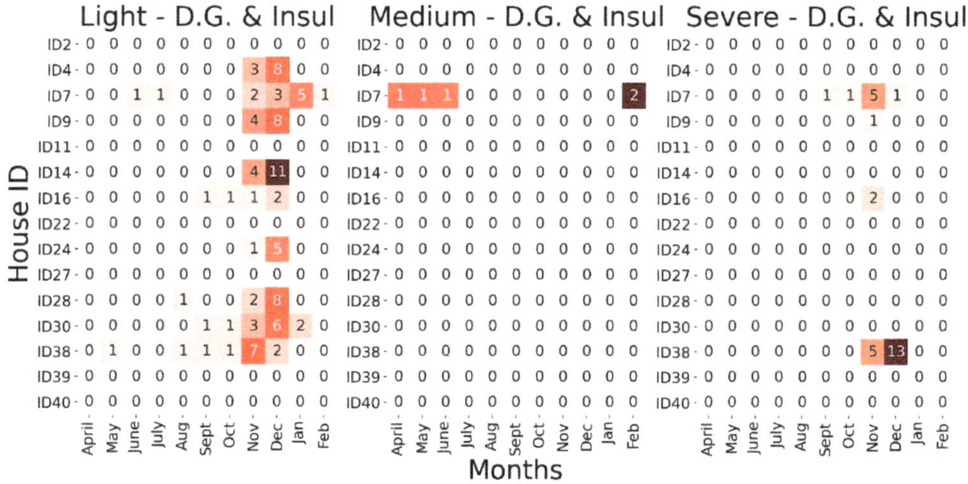

Figure A9. Results of the underheating criteria 2 applied to the models with insulation and double glazing. The *X* axis shows the months, and *Y* axis shows the house ID. The intersection of both shows the number of underheated light (left), medium (middle), and severe (right) days in that month.

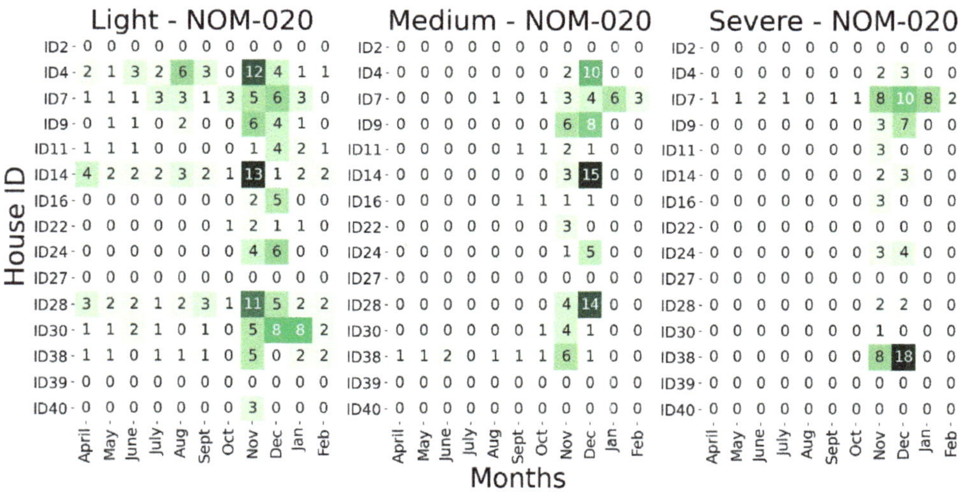

Figure A10. Results of the underheating criteria 2 applied to the models simulated under the NOM-020 standard. The *X* axis shows the months, and *Y* axis shows the house ID. The intersection of both shows the number of underheated light (left), medium (middle), and severe (right) days in that month.

References

1. Kjellstrom, T.; McMichael, A.J. Climate change threats to population health and well-being: The imperative of protective solutions that will last. *Glob. Health Action* **2013**, *6*, 20816. [CrossRef] [PubMed]
2. Murphy, J.M.; Sexton, D.; Jenkins, G.; Boorman, P.; Booth, B.; Brown, C.; Clark, R.; Collins, M.; Harris, G.; Kendon, E. *UK Climate Projections Science Report: Climate Change Projections*; Met Office Hadley Centre: Exeter, UK, 2009.
3. Perkins-Kirkpatrick, S.E.; Gibson, P.B. Changes in regional heatwave characteristics as a function of increasing global temperature. *Sci. Rep.* **2017**, *7*, 12256. [CrossRef] [PubMed]
4. Farrow, A.; Taylor, H.; Golding, J. Time Spent in the Home by Different Family Members. *Environ. Technol.* **1997**, *18*, 605–613. [CrossRef]

5. Healy, J.D. Excess winter mortality in Europe: A cross country analysis identifying key risk factors. *J. Epidemiol. Community Health* **2003**, *57*, 784–789. [CrossRef] [PubMed]
6. Clinch, J.P.; Healy, J.D. Housing standards and excess winter mortality. *J. Epidemiol. Community Health* **2000**, *54*, 719–720. [CrossRef]
7. Zepeda-Gil, C.; Natarajan, S. A Review of "Green Building" Regulations, Laws, and Standards in Latin America. *Buildings* **2020**, *10*, 188. [CrossRef]
8. Santamouris, M.; Alevizos, S.; Aslanoglou, L.; Mantzios, D.; Milonas, P.; Sarelli, I.; Karatasou, S.; Cartalis, K.; Paravantis, J. Freezing the poor—Indoor environmental quality in low and very low income households during the winter period in Athens. *Energy Build.* **2014**, *70*, 61–70. [CrossRef]
9. Daniel, L.; Baker, E.; Williamson, T. Cold housing in mild-climate countries: A study of indoor environmental quality and comfort preferences in homes, Adelaide, Australia. *Build. Environ.* **2019**, *151*, 207–218. [CrossRef]
10. Escandón, R.; Suárez, R.; Sendra, J.J. On the assessment of the energy performance and environmental behaviour of social housing stock for the adjustment between simulated and measured data: The case of mild winters in the Mediterranean climate of southern Europe. *Energy Build.* **2017**, *152*, 418–433. [CrossRef]
11. Butler, S.; Williams, M.; Tukuitonga, C.; Paterson, J. Problems with damp and cold housing among Pacific families in New Zealand. *N. Z. Med. J.* **2003**, *116*.
12. INEGI. Estadísticad de Mortalidad en México. 2020. Available online: https://www.inegi.org.mx/temas/mortalidad/ (accessed on 11 March 2023).
13. Fowler, T.; Southgate, R.J.; Waite, T.; Harrell, R.; Kovats, S.; Bone, A.; Doyle, Y.; Murray, V. Excess Winter Deaths in Europe: A multi-country descriptive analysis. *Eur. J. Public Health* **2014**, *25*, 339–345. [CrossRef]
14. Gasparrini, A.; Guo, Y.; Hashizume, M.; Lavigne, E.; Zanobetti, A.; Schwartz, J.; Tobias, A.; Tong, S.; Rocklöv, J.; Forsberg, B.; et al. Mortality risk attributable to high and low ambient temperature: A multicountry observational study. *Lancet* **2015**, *386*, 369–375. [CrossRef]
15. Hajat, S.; Kovats, R.; Lachowycz, K. Heat-related and cold-related deaths in England and Wales: Who is at risk? *Occupat. Environ. Med.* **2007**, *64*, 93–100. [CrossRef]
16. Ormandy, D.; Ezratty, V. Health and thermal comfort: From WHO guidance to housing strategies. *Energy Policy* **2011**, *49*, 116–121. [CrossRef]
17. WHO. *WHO Housing and Health Guidelines*; World Health Organization: Geneva, Switzerland, 2018.
18. Fanger, P.O. *Thermal Comfort. Analysis and Applications in Environmental Engineering*; Danish Technical Press: Copenhagen, Denmark, 1970.
19. *ISO 7730:2005*; Ergonomics of the Thermal Environment—Analytical Determination and Interpretation of Thermal Comfort Using Calculation of the PMV and PPD Indices and Local Thermal Comfort Criteria. ISO: Geneva, Switzerland, 2005.
20. Nicol, J.F.; Humphreys, M.A. Thermal comfort as part of a self-regulating system. *Build. Res. Pract.* **1973**, *1*, 174–179. [CrossRef]
21. *Guideline 14-2002*; Measurement of Energy and Demand Savings. American Society of Heating, Ventilating, and Air Conditioning Engineers: Atlanta, Georgia, 2002.
22. El Director General de Normas y Secretariado Técnico de la Comisión Nacional de Normalización. *NMX-C-7730-ONNCCE-2018—Industria de la Construcción-Ergonomía del Ambiente Térmico-Determinación Analítica e Inter-Pretación del Confort Térmico Mediante el Cálculo de Los Índices VME y PEI y Los Criterios de Confort Térmico Local*; Diario Oficial de la Nación: Mexico City, Mexico, 2018.
23. ASHRAE. *STANDARD 55—Thermal Environmental Conditions for Human Occupancy*; ASHRAE: Atlanta, GA, USA, 2022.
24. Heschong, L. *Thermal Delight in Architecture*; MIT Press: Cambridge, MA, USA, 1979.
25. Brager, G.S.; de Dear, R.J. Thermal adaptation in the built environment: A literature review. *Energy Build.* **1998**, *27*, 83–96. [CrossRef]
26. Kandel, E. Essentials of Neural Science and Behavior. *Appleton&Lange Norwalk CT* **1995**, *6*, 425–451.
27. Ortiz, M.A.; Kurvers, S.R.; Bluyssen, P.M. A review of comfort, health, and energy use: Understanding daily energy use and wellbeing for the development of a new approach to study comfort. *Energy Build.* **2017**, *152*, 323–335. [CrossRef]
28. Marmot, M.; Allen, J.; Bell, R.; Bloomer, E.; Goldblatt, P. WHO European review of social determinants of health and the health divide. *Lancet* **2012**, *380*, 1011–1029. [CrossRef]
29. Krieger, J.; Higgins, D.L. Housing and Health: Time Again for Public Health Action. *Am. J. Public Health* **2002**, *92*, 758–768. [CrossRef]
30. Braubach, M.; World Health Organization. *Environmental Burden of Disease Associated with Inadequate Housing: A Method Guide to the Quantification of Health Effects of Selected Housing Risks in the WHO European Region*; World Health Organization: Geneva, Switzerland, 2011.
31. Lyu, Y.; Forsyth, A.; Worthington, S. Built Environment and Self-Rated Health: Comparing Young, Middle-Aged, and Older People in Chengdu, China. *HERD Health Environ. Res. Des. J.* **2021**, *14*, 229–246. [CrossRef] [PubMed]
32. Cheung, E.S.L.; Mui, A.C. Do Home and Community Environments Explain Self-Rated Health Among Older Canadians? Evidence From the 2018 Canadian Housing Survey. *HERD Health Environ. Res. Des. J.* **2022**, *15*, 112–125. [CrossRef]
33. Shiue, I. Duration of daily TV/screen watching with cardiovascular, respiratory, mental and psychiatric health: Scottish Health Survey, 2012–2013. *Int. J. Cardiol.* **2015**, *186*, 241–246. [CrossRef] [PubMed]

34. Diaz Lozano Patino, E.; Siegel, J.A. Indoor environmental quality in social housing: A literature review. *Build. Environ.* **2018**, *131*, 231–241. [CrossRef]
35. Fairburn, J.; Braubach, M. Social inequalities in environmental risks associated with housing and residential location. Environment and health risks: A review of the influence and effects of social inequalities. *Eur. J. Public Health* **2010**, *20*, 36–43.
36. Secretaría de Energía. *NORMA Oficial Mexicana NOM-020-ENER-2011, Eficiencia Energética en Edificaciones.—Envolvente de Edificios para uso Habitacional, in Apéndice A, Tabla 1*; SEGOB: Mexico City, Mexico, 2011.
37. Martin-Dominguez, I.R.; Rodriguez-Muñoz, N.A.; Romero-Perez, C.K.; Najera-Trejo, M.; Ortega-Avila, N. Analysis of the Methodologic Assumptions of the NOM-020-ENER-2011—Mexican Residential Building Standard. *Environments* **2018**, *5*, 118. [CrossRef]
38. INEGI. *Inventario Nacional de Vivienda*; INEGI: Mexico City, Mexico, 2016; Available online: https://www.inegi.org.mx/app/mapa/inv/ (accessed on 11 March 2023).
39. Torres, R.P.R.; Nolasco, G.C.; Méndez, T.A.; Morillón, D.; Reséndiz, O. Comparación del comportamiento térmico de una vivienda en clima cálido subhúmedo con la NOM-020-ENER. In Proceedings of the Memorias del XXXV Congreso Nacional de Energía Solar. ANES, Chihuahua, Mexico, 5–7 October 2011.
40. Crawley, D.B.; Lawrie, L.K.; Pedersen, C.O.; Winkelmann, F.C. Energy plus: Energy simulation program. *ASHRAE J.* **2000**, *42*, 49–56.
41. Cruz, M.G.A.; Aguilar, E.A.D. La NOM-020-ENER-2011 en viviendas económicas ubicadas en diferentes regiones climáticas de México. *Vivienda Y Comunidades Sustentables* **2017**, 75–90. [CrossRef]
42. Romero-Moreno, R.; Bojórquez-Morales, G.; Luna-León, A.; Reyes-Barajas, K. Energy efficiency in mass-built housing and demonstrative bioclimatic models in a hot-dry climate zone Eficiencia energética en vivienda de construcción en serie y en modelos demostrativos bioclimáticos en zona de clima cálido seco. *Archit. Des.* **2020**, *4*, 1–8.
43. Secretaría de Energía. Herramienta Cálculo NOM-020-ENER-2002. 2002. Available online: https://www.gob.mx/conuee/acciones-y-programas/herramienta-calculo-nom-020-ener-2001 (accessed on 11 March 2023).
44. Thompson, S.K. Adaptive cluster sampling. *J. Am. Stat. Assoc.* **1990**, *85*, 1050–1059. [CrossRef]
45. Watters, J.K.; Biernacki, P. Targeted sampling: Options for the study of hidden populations. *Soc. Probl.* **1989**, *36*, 416–430. [CrossRef]
46. INEGI. *México—Encuesta Nacional Sobre Consumo de Energéticos en Viviendas Particulares 2018*; ESD3.04-ENCEVI-2018; INEGI: Mexico City, Mexico, 2018.
47. OpenStreetMap. *Map of Toluca. Open Data Commons Open Database License (ODbL) of the OpenStreetMap (OSMF) Foundation*; OpenStreetMap Foundation: Cambridge, UK, 2022.
48. Consejo Estatal de Población. *Zonas Metropolitanas del Estado de México*; Government of "Estado de México": Toluca, Mexico, 2021.
49. *ANSI/ASHRAE Standard 55-2013*; Thermal Environmental Conditions for Human Occupancy. ASHRAE: Atlanta, GA, USA, 2013.
50. Hughes, C.; Natarajan, S.; Liu, C.; Chung, W.J.; Herrera, M. Winter thermal comfort and health in the elderly. *Energy Policy* **2019**, *134*, 110954. [CrossRef]
51. de Dear, R.J.; Brager, G.S. Thermal comfort in naturally ventilated buildings: Revisions to ASHRAE Standard 55. *Energy Build.* **2002**, *34*, 549–561. [CrossRef]
52. Walikewitz, N.; Jänicke, B.; Langner, M.; Meier, F.; Endlicher, W. The difference between the mean radiant temperature and the air temperature within indoor environments: A case study during summer conditions. *Build. Environ.* **2015**, *84*, 151–161. [CrossRef]
53. Hughes, C.; Natarajan, S. Summer thermal comfort and overheating in the elderly. *Build. Serv. Eng. Res. Technol.* **2019**, *40*, 426–445. [CrossRef]
54. Zepeda-Gil, C.; Natarajan, S. Thermal comfort in naturally ventilated dwellings in the central Mexican plateau. *Build. Environ.* **2022**, *211*, 108713. [CrossRef]
55. Albadra, D.; Vellei, M.; Coley, D.; Hart, J. Thermal comfort in desert refugee camps: An interdisciplinary approach. *Build. Environ.* **2017**, *124*, 460–477. [CrossRef]
56. Field, A. *Multilevel Linear Models, in Discovering Statistics Using SPSS (and Sex and Drugs and Rock "n" Roll)*; SAGE: London UK, 2009.
57. Wasilowski, H.A.; Reinhart, C.F. Modelling an existing building in DesignBuilder/EnergyPlus: Custom versus default inputs. In Proceedings of the Eleventh International IBPSA Conference, Glasgow, Scotland, 27–30 July 2009.
58. INEGI. Características de Los Hogares. 2017. Available online: http://www.beta.inegi.org.mx/temas/hogares/ (accessed on 11 March 2023).
59. Oberkampf, W.L.; Trucano, T.G. Verification and validation in computational fluid dynamics. *Prog. Aerosp. Sci.* **2002**, *38*, 209–272. [CrossRef]
60. Yoon, J.; Lee, E.J.; Claridge, D.E. Calibration Procedure for Energy Performance Simulation of a Commercial Building. *J. Sol. Energy Eng.* **2003**, *125*, 251–257. [CrossRef]
61. Kaplan, M.; Caner, P. *Guidelines for Energy Simulation of Commercial Buildings: Final*; Kaplan Engineering: Portland, OR, USA, 1992. [CrossRef]
62. CIBSE. *TM59 Design Methodology for the Assessment of Overheating Risk in Homes*; U. Chartered Institution of Building and Service Engineers, Ed.; CIBSE: London, UK, 2017.

63. Rudge, J.; Gilchrist, R. Measuring the health impact of temperatures in dwellings: Investigating excess winter morbidity and cold homes in the London Borough of Newham. *Energy Build.* **2007**, *39*, 847–858. [CrossRef]
64. Critchley, R.; Gilbertson, J.; Grimsley, M.; Green, G. Living in cold homes after heating improvements: Evidence from Warm-Front, England's Home Energy Efficiency Scheme. *Appl. Energy* **2007**, *84*, 147–158. [CrossRef]
65. Saeki, K.; Obayashi, K.; Iwamoto, J.; Tone, N.; Okamoto, N.; Tomioka, K.; Kurumatani, N. Stronger association of indoor temperature than outdoor temperature with blood pressure in colder months. *J. Hypertens.* **2014**, *32*, 1582–1589. [CrossRef]
66. Shiue, I.; Shiue, M. Indoor temperature below 18 °C accounts for 9% population attributable risk for high blood pressure in Scotland. *Int. J. Cardiol.* **2014**, *171*, e1–e2. [CrossRef]
67. Zock, J.-P.; Jarvis, D.; Luczynska, C.; Sunyer, J.; Burney, P. Housing characteristics, reported mold exposure, and asthma in the European Community Respiratory Health Survey. *J. Allergy Clin. Immunol.* **2002**, *110*, 285–292. [CrossRef]
68. Thom, H.C.S. The rational relationship between heating degree days and temperature. *Mon. Weather. Rev.* **1954**, *82*, 1–6. [CrossRef]
69. De Dear, R.J.; Brager, G.S.; Reardon, J.; Nicol, F. Developing an adaptive model of thermal comfort and preference/discussion. *ASHRAE Trans.* **1998**, *104*, 145.
70. An, J.-Y.; An, K.; O'Connor, L.; Wexler, S. Life Satisfaction, Self-Esteem, and Perceived Health Status Among Elder Korean Women: Focus on Living Arrangements. *J. Transcult. Nurs.* **2008**, *19*, 151–160. [CrossRef]
71. Moshki, M.; Ashtarian, H. Perceived Health Locus of Control, Self-Esteem, and Its Relations to Psychological Well-Being Status in Iranian Students. *Iran. J. Public Heal.* **2010**, *39*, 70–77.

buildings

MDPI

Review

Review of Parameters Measured to Characterize Classrooms' Indoor Environmental Quality

Minh Tien Tran [1,2,*], Wenjuan Wei [1], Claire Dassonville [1], Christophe Martinsons [1], Pascal Ducruet [1], Corinne Mandin [1], Valérie Héquet [2] and Pawel Wargocki [3]

1 Scientific and Technical Center for Building (CSTB), Health and Comfort Department, French Indoor Air Quality Observatory (OQAI), 77420 Champs sur Marne, France
2 IMT Atlantique, CNRS, GEPEA, UMR 6144, CEDEX 3, 44307 Nantes, France
3 International Centre for Indoor Environment and Energy, Department of Environmental and Resource Engineering (DTU SUSTAIN), Technical University of Denmark (DTU), 2800 Lyngby, Denmark
* Correspondence: minh-tien.tran@cstb.fr

Abstract: As attention to indoor environmental quality (IEQ) grows, a systematic strategy for assessing IEQ in schools needs to be developed. For this purpose, this paper presents a summary of parameters measured in school classrooms to characterize the quality of thermal, acoustic, and visual environments and indoor air quality (IAQ). The summary is based on a review of published literature reporting measurements in schools in Europe and North America in the past ten years. It also summarizes the measurement protocols and measured concentrations. Eighty-eight papers describing measurements in schools were identified and analyzed. No unique standardized measuring method was used in the reviewed studies and different parameters were measured. The most often measured parameters were those describing the thermal environment and IAQ. The former mainly comprised air temperature and relative humidity. The latter mainly comprised concentrations of carbon dioxide, particulate matter, radon, formaldehyde, and some volatile organic compounds. The measured parameters describing acoustic and visual environments mainly comprised noise level, reverberation time, and illuminance. A few studies reported additional measurements of radiant temperature, operative temperature, and speech intelligibility. Measurement protocols from different studies show inconsistency in sampling duration and location and expressed results. Measured concentrations also show high variation between studies, with some pollutants exceeding the threshold values proposed by local and/or international organizations such as the World Health Organization (WHO). This review provides the reference for developing a rating scheme and protocols for uniform characterization of classroom IEQ.

Keywords: IEQ; thermal; acoustic; IAQ; visual; measurement

Citation: Tran, M.T.; Wei, W.; Dassonville, C.; Martinsons, C.; Ducruet, P.; Mandin, C.; Héquet, V.; Wargocki, P. Review of Parameters Measured to Characterize Classrooms' Indoor Environmental Quality. *Buildings* **2023**, *13*, 433. https://doi.org/10.3390/buildings13020433

Academic Editor: Ricardo M. S. F. Almeida

Received: 14 December 2022
Revised: 20 January 2023
Accepted: 31 January 2023
Published: 3 February 2023

1. Introduction

Indoor environmental quality (IEQ) depends on the quality of the thermal environment, acoustic environment, indoor air quality (IAQ), and visual environment [1]. IEQ is a primary concern because people spend a significant portion of their time in buildings [2]. A growing body of studies has shown the influence of IEQ on occupants' health, comfort, and well-being, at homes, offices, and schools [3,4].

In the last twenty years, it has been shown that poor schools' IEQ can affect children's health. The study of Gaffin et al. [5] showed that exposure to a concentration of NO_2 greater than 8 ppb in urban American schools was associated with respiratory airflow obstruction in children. Meanwhile, a growing body of studies showed that low indoor environmental quality in schools can impact children's school performance. The study of Wargocki et al. [6] showed that improved classroom ventilation, as indicated by reduced carbon dioxide (CO_2) concentrations from 2100 to 900 ppm, resulted in improved children's academic performance by 12%, while the performance was increased by 20% when classroom temperatures were reduced from 30 to 20 °C [7]. Most of the work published to date examined the effects of one parameter, and no relationships were created between the quality of the indoor environment and health effects and children's school performance. One reason is the complexity and cost of performing measurements. The other is the lack of a rating scheme for IEQ. The latter is needed to take into account interactions between different parameters and their influence on children.

Recently, Wei et al. [8] reviewed parameters measured to characterize IEQ in offices and hotels. Nearly 100 parameters were identified, but no common rating scheme was identified for IEQ. Consequently, the TAIL (Thermal, Acoustic, IAQ, Lighting) rating scheme was developed including twelve IEQ parameters [9]. Moreover, the method to predict the parameters included in TAIL through simulation was developed and is called predicTAIL [10]. No similar rating exists for schools. In this context, the development of this approach to assess IEQ in school buildings is a relevant issue.

Green Building Certification (GBC) schemes identify some parameters to be monitored to describe IEQ. These are Beam plus [11], BREEAM [12], DGNB [13], Green Globes [14], Green Mark [15], Green Star [16], GREENSHIP [17], HQE [18], KLIMA [19], LEED [20], Lotus [21], Trees [22]. They, however, do not share similar methods or a homogeneous approach. Furthermore, only DGNB, HQE and LEED have a specific section focusing on classrooms and schools. GBC schemes are voluntary, therefore criteria applied in these schemes are not systematically used during measurements in schools. In an attempt to develop an IEQ rating scheme for schools, the present work aimed to review existing measurements of IEQ parameters in schools. Three specific research questions were examined: (1) which IEQ parameters were often measured in schools; (2) what measuring methods were used; and (3) what are the main findings from these measurements?

2. Materials and Methods

The Scopus database was used to search for relevant literature. The following combination of keywords was used: ("indoor environmental quality" OR "IEQ" OR "thermal" OR "acoustic" OR "indoor air quality" OR "IAQ" OR "luminous" OR "visual" OR "lighting") AND ("school" OR "daycare center" OR "nursery" OR "university") AND ("measurement"). To ensure a collection of studies with the latest and up-to-date measurement protocols, the search only covered papers published from 2010 with the following topics: environmental sciences, engineering, social sciences, energy, and multidisciplinary.

A total of 573 papers were identified. A geo-localization filter was applied as only studies in Europe and North America were included to ensure comparability across studies regarding climate conditions, surrounding environment, and building characteristics. Consequently, 324 articles were retained and used for screening. Non-relevant to IEQ measurements in schools, simulation studies, and studies based only on questionnaires were removed. This screening resulted in 79 papers. Nine additional papers were identified and added manually although they did not appear in the initial search. Finally, 88 papers were analyzed in this review. Figure 1 presents a PRISMA flowchart describing the paper selection. It should be noted that 11 studies in university classrooms were included in the review because universities are school environments, even though the students are young adults and no longer children.

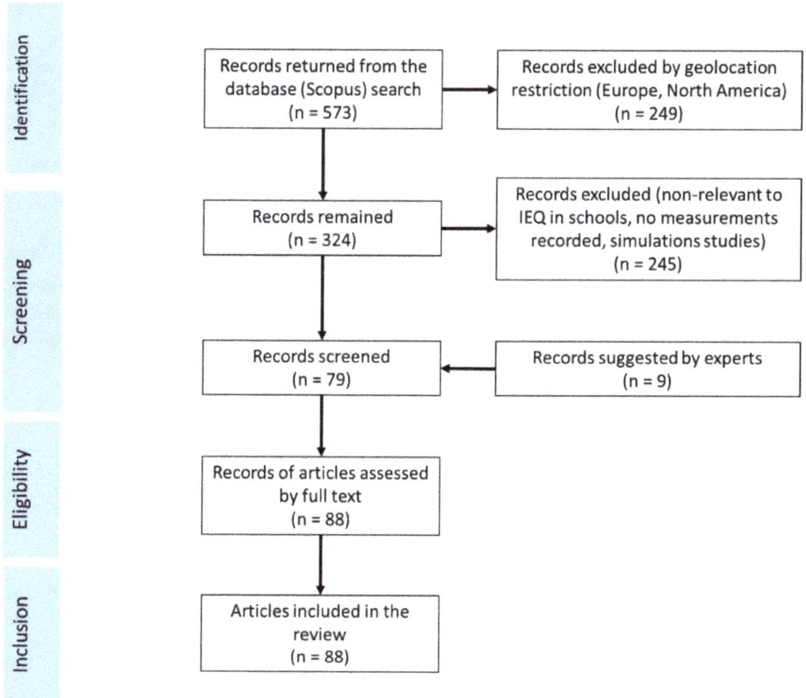

Figure 1. Review flowchart.

3. Results

Among the 88 articles analyzed in this review, some were issued from the same study. Two articles by Branco et al. [23,24] reported results from a measuring campaign in four nurseries in Porto, Portugal; one focused on CO_2 and comfort, the second one on IAQ pollutants. The report by Csobod et al. [25] on the thermal environment and IAQ and the article by Baloch et al. [26] on the results of the visual environment both reported findings from the SINPHONIE pan-European study. Among the 88 articles, IAQ parameters were measured in 73 studies, thermal parameters in 43 studies, visual parameters in 15 studies, and acoustic parameters in 13 studies. In 40 studies, the parameters describing at least two of the four IEQ components were measured. Only four studies [27–30] measured parameters of the four IEQ components. These studies were carried out in 28 countries (Figure S1 in Supplementary Materials (SM)).

Nine studies were conducted in day-care centers, nine in nursery schools, 40 in elementary schools, 24 in secondary schools, 12 in high schools, and 11 in universities. In 59 studies, schools had no mechanical ventilation system. In eight studies, an HVAC (Heating, ventilation, and air conditioning) system with heat recovery was installed. In nine studies, measurements were performed in both classrooms with no ventilation system and classrooms equipped with a mechanical ventilation system. In three studies, no information on ventilation type was available, and only acoustic and visual measurements were conducted [31–33]. Table 1 summarizes all studies presented in this review. Table 2 summarizes the main findings with the range of results from the 88 articles.

Table 1. Summary of the investigated studies (*n* = 88 articles).

Study [Reference]	Location	Season	Type of School	IEQ Component
Aguilar et al., 2022 [34]	Spain	Winter	University	Thermal, IAQ
Ahmed et al., 2019 [35]	Finland, Estonia	Winter	Daycare, elementary	Thermal, IAQ
Alves et al., 2013 [36]	Portugal	Winter	Kindergarten, elementary	Thermal, IAQ
Annesi-Maesano et al., 2012 [37]	France	N/A	Elementary	IAQ
Azara et al., 2018	Italy	Spring Summer	Elementary, secondary, high school	IAQ
Baloch et al., 2021 [38]	Europe	Spring Summer	Kindergarten, elementary	Visual
Barmparesos et al., 2018 [39]	Greece	Summer	Elementary	Thermal, IAQ
Becerra et al., 2020 [40]	Spain	Spring	Kindergarten, elementary, secondary, high school	IAQ
Branco et al., 2015 [23]	Portugal	Spring Autumn	Daycare	Thermal, IAQ
Branco et al., 2015 [24]	Portugal	Spring Autumn	Daycare	IAQ
Branco et al., 2016 [41]	Portugal	Year long	Daycare, elementary	IAQ
Brdaric et al., 2019 [42]	Croatia	Spring	Elementary	Thermal, IAQ
Buratti et al., 2018 [43]	Italy	Spring Autumn	University	Thermal, Acoustic, Visual
Canha et al., 2016 [44]	France	Winter	Daycare, elementary	Thermal, IAQ
Cequier et al., 2014 [45]	Norway	Winter Spring	Elementary	IAQ
Chetoni et al., 2016 [46]	Italy	N/A	Secondary, high school	Acoustic
Csobod et al., 2014 [25]	Europe	Spring Summer	Kindergarten, elementary	Thermal, IAQ, Visual
de Gennaro et al., 2013 [47]	Italy	N/A	Elementary	IAQ
De Giuli et al., 2012 [48]	Italy	Spring	Elementary	Thermal, IAQ, Visual
De Giuli et al., 2014 [49]	Italy	Spring	Elementary	Thermal, IAQ, Visual
De Giuli et al., 2015 [50]	Italy	Spring	Elementary	Thermal, IAQ, Visual
de la Hoz –Torres et al., 2022 [29]	Portugal, Spain	Fall	University	Thermal, Acoustic, IAQ, Visual
Dhoqina et al., 2019 [51]	Albania	Spring	Elementary, secondary, high school	IAQ
Erlandson et al., 2019 [52]	United States		University	Thermal, IAQ
Fabbri 2013 [53]	Italy	Fall	Kindergarten	Thermal
Franci et al., 2014 [54]	Italy	Winter	Elementary, secondary, high school	IAQ
Gaffin et al., 2018 [5]	United States	Fall Spring	Elementary, secondary	IAQ
Harcarova et al., 2020 [28]	Slovakia	N/A	Elementary	Thermal, Acoustic, IAQ, Visual
Heracleous et al., 2019 [55]	Cyprus	Winter	Secondary	Thermal, IAQ
Irulegi et al., 2017 [56]	Spain	Spring	University	Thermal
Istrate et al., 2016 [57]	Romania	Summer	High school	Thermal, IAQ
Ivanova et al., 2014 [58]	Bulgaria	Spring	Kindergarten	IAQ
Ivanova et al., 2021 [59]	Bulgaria	Fall Winter	Elementary	IAQ
Jovanovic et al., 2014 [60]	Serbia	Spring	Elementary	Thermal, IAQ
Klatte et al., 2010 [61]	Germany	N/A	Elementary	Acoustic
Kojo et al., 2020 [62]	Finland	Winter Spring	Daycare, Elementary	IAQ
Korsavi et al., 2019 [63]	England	Summer Fall Winter Spring	Elementary	Thermal, IAQ, Visual

Table 1. *Cont.*

Study [Reference]	Location	Season	Type of School	IEQ Component
Kristiansen et al., 2011 [64]	Denmark	Fall Winter Spring	Secondary	Acoustic
Krugly et al., 2014 [65]	Lithuania	Winter	Elementary	IAQ
Laborda et al., 2020 [66]	Spain	Winter	Secondary	Thermal, IAQ, Visual
Larsson et al., 2017 [67]	Sweden	Spring Fall	Kindergarten	IAQ
Leccese et al., 2020 [31]	Italy	Spring Fall	University	Visual
Liaud et al., 2021 [68]	France	Spring	High school	IAQ
Loreti et al., 2016 [30]	Italy	N/A	Secondary	Thermal, Acoustic, IAQ, Visual
Madudeira et al., 2015 [69]	Portugal	Fall Winter	Elementary	Thermal, IAQ
Mainka et al., 2015 [70]	Poland	Winter	Daycare	IAQ
Mikulski et al., 2011 [71]	Poland	N/A	Elementary	Acoustic
Müllerova et al., 2017 [72]	Hungary, Poland, Slovakia	Fall Winter	Kindergarten	IAQ
Nunes et al., 2016 [73]	Portugal	Spring	Nursery	IAQ
Oldham et al., 2020 [74]	United States	Fall Spring	Elementary	Thermal, IAQ, Visual
Oliveira et al., 2016 [75]	Portugal	Spring	Kindergarten	Thermal, IAQ
Oliveira et al., 2017 [76]	Portugal	Winter Spring	Kindergarten	IAQ
Oliveira et al., 2017 [77]	Portugal	Spring	Kindergarten	Thermal, IAQ
Onishchenko et al., 2017 [78]	Russia	N/A	Kindergarten	IAQ
Papadopoulos et al., 2020 [79]	Greece	Winter	University	Thermal, IAQ
Papazoglou et al., 2019 [80]	Greece	Summer	University	Thermal
Pereira et al., 2014 [81]	Portugal	Spring	Secondary	Thermal, IAQ
Pereira et al., 2015 [82]	Portugal	Spring	Secondary	Thermal, IAQ
Persson et al., 2018 [83]	Sweden	Year long	Kindergarten	IAQ
Poulin et al., 2012 [84]	Canada	Winter	Elementary, secondary, high school	IAQ
Raffy et al., 2017 [85]	France	N/A	Nursery, elementary	IAQ
Ramalho et al., 2015 [86]	France	N/A	Nursery, elementary	IAQ
Rivas et al., 2014 [87]	Spain	Winter Spring Summer	Elementary, secondary	IAQ
Romagnoli et al., 2014 [88]	Italy	Winter Spring Summer	Elementary, secondary, high school	IAQ
Rovelli et al., 2014 [89]	Italy	Winter	Elementary, secondary	IAQ
Rucinska et al., 2020 [33]	Poland	Winter	University	Visual
Russo et al., 2019 [32]	Italy	N/A	Elementary	Acoustic
Sarantopoulos et al., 2014 [90]	Greece	Spring	Elementary	Acoustic
Sarka Langer et al., 2020 [91]	Sweden	Fall Winter Spring	Elementary	Thermal, IAQ
Senitkova et al., 2017 [92]	Czech Republic	N/A	Daycare	Thermal, IAQ
Shield et al., 2015 [93]	England	N/A	Secondary	Acoustic
Simanic et al., 2019 [94]	Sweden	Fall Winter Spring	Elementary	Thermal, IAQ
Sivanantham et al., 2021 [95]	France	Fall Winter Spring	Daycare, Elementary	Thermal, IAQ
Slezakova et al., 2019 [96]	Portugal	Winter Spring	Elementary	IAQ

Table 1. *Cont.*

Study [Reference]	Location	Season	Type of School	IEQ Component
Smith et al., 2019 [97]	United States	N/A	Elementary, secondary	Acoustic
Stamp et al., 2020 [98]	United Kingdom	N/A	Secondary	Thermal, IAQ
Toftum et al., 2015 [99]	Denmark	N/A	Elementary	Thermal, IAQ
Trevisi et al., 2012 [100]	Italy	Year long	Daycare, elementary, secondary	IAQ
Ulla Haverinen-Shaughnessy et al., 2015 [101]	United States	Fall Winter Spring	Elementary	IAQ
Verriele et al., 2016 [102]	France	N/A	Elementary, secondary	Thermal, IAQ
Vilcekova et al., 2017 [27]	Slovakia	Fall	Elementary	Thermal, Acoustic, IAQ, Visual
Villanueva et al., 2018 [103]	Spain	Spring	Elementary	IAQ
Vornanen Winqvist et al., 2018 [104]	Finland	Spring	Secondary	Thermal, IAQ
Vornanen Winqvist et al., 2020 [105]	Finland	Winter	Secondary	Thermal, IAQ
Z.Curguz et al., 2020 [106]	Bosnia and Herzegovina	N/A	Elementary, secondary, high school	IAQ
Zecevic et al., 2018 [107]	Bosnia and Herzegovina	Winter Summer	University	Thermal, IAQ
Zhong et al., 2017 [108]	United States	Winter Spring	Elementary	Thermal, IAQ
Živković et al., 2015 [109]	Serbia	Winter Spring	Elementary, secondary, high school	IAQ

Table 2. Summary of the main results (n = 88 articles).

Parameters	Number of Studies	Main Findings	Reference Values
		Thermal environment	
Air temperature (°C)	43	Western Europe: Range: 13 °C to 38 °C with a mean of 22 °C Northern Europe: Range: 12 °C to 26 °C with a mean of 21 °C	22 ± 1 °C (EN 16798-1)
Relative humidity (%)	43	Naturally ventilated classrooms range 22% with a mean air temperature of 23 °C to 78% with a mean air temperature of 25 °C. Mechanically ventilated classrooms, range: 30 to 72%	30–50% (EN 16798-1)
PMV/PPD (derivative)	8	Mean result: ±0.5 from 0 °C	±0.2 °C (EN 16798-1)
Mean radiant temperature (°C)	7	Range: 13 to 24 °C	N/A
Air speed (m/s)	7	All reported results are under 0.1 m/s	N/A
Operative temperature (°C)	5	Range: 19 to 22 °C	N/A
		Acoustic environment	
Background noise level (db(A))	8	Range: 41 to 82 db(A)	<30 db(A) (EN 16798-1)
Reverberation time (s)	8	Range: 0.9 to 1.1 s	0.5 s for small spaces 0.8 for large spaces (EN 16798-1)
Speech intelligibility (%)	7	SNR range: 12 ± 3.6 db STI range: 41–76% C50 range: −6.3 to 5.6 db	N/A

Table 2. *Cont.*

Parameters	Number of Studies	Main Findings	Reference Values
		IAQ	
CO_2 (ppm)	42	Naturally ventilated classrooms, range: 591 to 3494 ppm Mechanically ventilated classroom, all under 1000 ppm	≤550 ppm (concentration above outdoor) (EN 16798-1)
PM (μg/m^3)	22	PM_{10} range: 34 to 2061 μg/m^3 PM_5 range: 31 to 206 μg/m^3 $PM_{2.5}$ range: 1.3 to 106 μg/m^3 PM_1 range: 6.0 to 33 μg/m^3 $PM_{0.5}$ range: 2.1 to 22 μg/m^3	$PM_{2.5} \leq 5$ μg/m^3 (WHO)
Radon (Bq/m^3)	16	Range: 56 to 579 Bq/m^3	100 Bq/m^3 (WHO)
BTEX (μg/m^3)	14	Benzene range: 0.5 to 3.2 μg/m^3 Toluene range: 0.2 to 17 μg/m^3 Ethylbenzene range: <Limit of detection to 9.0 μg/m^3 Xylene range: 1 to 12 μg/m^3	Benzene: <2 μg/m^3
SVOCs (ng/m^3)		Tables S14–S16	N/A
Aldehydes (μg/m^3)	15	Formaldehyde range: 1.4 to 89 μg/m^3	Formaldehyde: <30 μg/m^3
ACR/ VR (h^{-1} or l/s/p)	11	ACR range: 0.1 to 0.4 h^{-1} VR range: 0.8 l/s per person to 3.4 l/s per person	≥10 L/s per person + 2.0 L/s/m^2 floor
VOCs (μg/m^3)	10	Table S12	N/A
NO_2 (μg/m^3)	11	Range: 4.9 to 125 μg/m^3	<10 μg/m^3 (WHO)
Mold inspection (cm^2 or CFU/m^3)	4	Range: 22 to 260 CFU/m^3	<400 cm^2 (Nordic classification and Levels)
		Visual environment	
Artificial illuminance (lx)	4	Range: 241 to 748 lx	500 lx Can be drop to 300 lx for younger children (EN 12464-1)
Total lighting (natural + artificial)	4	Table S19	N/A
Natural lighting	3	Table S19	>5% (EN 17037)

3.1. Thermal Environment

Thermal environment was assessed in 43 studies. The air temperature, radiant temperature, operative temperature, humidity or air speed were measured. Some studies additionally estimated thermal comfort using the model developed by Fanger [110].

3.1.1. Air Temperature

The air temperature was measured in 43 studies; it was often measured simultaneously with relative humidity. Measurements were performed during winter (heating season) in eight studies, during non-heating season, i.e., spring and autumn in 29 studies, during summer in six studies, and seven studies did not provide any information.

Temperature was mainly recorded continuously for the periods of over thirty minutes to three months with a time step ranging from two to ten minutes. In three studies, only spot measurements were carried out. In one study, a combination of continuous measurements over three months and spot measurements four times in each classroom were carried out.

Different numbers of sensors, at various locations in each classroom, were used. Thirty-eight studies used only one sensor for each classroom, and three studies used three sensors per classroom. Laborda et al. [66] measured air temperature in 12 locations in the classroom: six at 0.6 m and six at 1.7 m, divided into two arrays close to the window and next to the entrance doors. Papadopoulos et al. [79] measured air temperature in 11 locations in the classrooms, with their height situated from 0.5 to 2.5 m above the ground to be compliant with the students' breathing zone. Nine studies reported that their sensors were placed at 0.6 m above the ground, and seven other studies placed their sensors from 0.7 to 2 m above the ground. Fabbri et al. [53] measured air temperature at two different heights: 0.6 and 1.3 m. The detailed protocols and results can be found in Table S1 in Supplementary Materials.

Studies from Western Europe (Italy, Spain, Portugal, France, and Greece) measured temperatures ranging from 13 to 38 °C with a mean of 22 °C. Studies from Northern Europe (Sweden and Finland) measured temperatures ranging from 12 to 26 °C with a mean of 21 °C.

3.1.2. Humidity

Relative humidity (RH) was measured simultaneously with air temperature using hygrometer sensor. In one study, absolute humidity was also determined along with relative humidity. Humidity was reported in 43 studies. Findings on the number of sensors, their location, and the duration of measurements are identical to air temperature measurements since the two parameters were always measured together in the reviewed studies. RH ranged between 22% with a mean air temperature of 23 °C and 96% with a mean air temperature of 25 °C. Among the classrooms with mechanical ventilation, the relative humidity varied between 30 and 72%, while the air temperature varied between 21 and 25 °C.The detailed protocols and results can be found in Table S2 in Supplementary Materials.

3.1.3. Mean Radiant Temperature

Mean radiant temperature was measured in 11 studies. In eight studies, the measurements were made using one globe thermometer, placed in the center of the room in each study, with a measurement duration ranging from one to five days. Other studies made spot measurements of the radiant plane temperature to estimate the mean radiant temperature but the monitoring protocol was not detailed and in two studies, air and mean radiant temperature were measured over 24 h. Papadopoulos et al. [79] used a thermal imaging camera to estimate the mean radiant temperature from surface temperatures. The mean radiant temperature measured in the 11 studies ranged between 13 and 24 °C. Table S3 in Supplementary Materials shows the details concerning measuring protocols and results.

3.1.4. Operative Temperature

Operative temperature was reported in six studies based on the measurement of globe temperature, and air velocity. In these studies, a small sample of classrooms (less than ten) was monitored, but the measurements were made at different classroom locations and repeated. In their study in 145 classrooms in Sweden, Simanic et al. [94] assumed that the operative temperature was similar to air temperature, as these schools were well insulated [15].Reported operative temperature from five studies ranged from 19 to 22 °C. Table S3 in Supplementary Materials provides details regarding measuring protocols and results.

3.1.5. Airspeed

Airspeed was measured in 12 studies in schools. These measurements were made to assess thermal comfort. Spot measurements were always made using an anemometer, but its position was never clearly mentioned. The reported results were all under 0.1 m/s.

3.2. Acoustic Environment

Thirteen studies reported measurements of parameters characterizing acoustic environment in schools. Parameters measured included noise level, reverberation time (RT) and speech intelligibility.

3.2.1. Noise Level

The background noise level was the most frequently measured acoustic parameter. It is also a parameter commonly used to assess occupants' long-term noise exposure [111]. Background noise level can be estimated using the variation of the pressure in the air caused by sound waves (sound pressure level-SPL) or the equivalent continuous sound level, e.g., LA_{eq}, defined as the total energy from the sound pressure level during the measurement period [112].

LA_{eq} was measured in nine studies, with one or two sound meters per classroom. The location of the sound meter was mentioned in three studies: Chetoni et al. [46] measured in the center of the room and one meter from the window, Shield et al. [93] also measured at two positions in the classroom, and de la Hoz-Torres et al. [29] only used one sound-meter placed in the center of the room.

Measurements were a spot measurement of one minute or continuous from two hours to two days. In some studies, measurements were made during day without students in the classrooms, while in other studies, measurements were made when the students were occupying the classrooms.

The average measured LA_{eq} in studied classrooms varied between 29 and 82 dB(A). Chetoni et al. [46] used the L_{DAY} indicator, defined as the daily LA_{eq} over the 12 h diurnal period from 7 a.m. to 7 p.m. since this parameter is used in Italy's national regulation for outdoor acoustic. The L_{DAY} ranged between 23 and 63 dB(A). Smith et al. [97] in their study quantified the influence of the different mechanical ventilation systems (single- and multi-zone HVAC systems) on the non-speech noise during occupied periods. The average LA_{eq} was 66 dB(A) for unit ventilators, 67 dB(A) for centralized systems, and 66 dB(A) for systems with decentralized heat pumps. Details on measurement protocol and results can be found in Table S4 in Supplementary Materials.

3.2.2. Reverberation Time

A room reverberation time (RT) expresses the time required for the sound to decay after the sound source has stopped; for example, T_{20} is the time it takes for sound to decay by 20 dB, and T_{60} is the time for a decay of 60 dB, T_{20} as the time for a decay by 20 dB, respectively [112]. There is a linear relationship between T_{20} and T_{60} in the same environment, as the measurement of T_{20} can be used to evaluate T_{60}, which is the case in the study by Loreti et al. [30].

RT was measured in in four studies in Italy [30,32,43,46], one study in Poland [71], one in Denmark [64], one in Germany [61], and one in England [93]; all referred to the measurement methods defined in the ISO 3382 standard [112], using an impulsive response method for a controlled and continuous (white noise) generated by an omnidirectional loudspeaker, blank gun noise, or maximum-length sequence signals. Two to twelve microphones were placed in children's seat positions, i.e., at least 1.1 m of height, and the results were expressed as the average of all these measurements. Mikulski et al. [71] measured the T_{mf} which is the arithmetic mean of RT for 500 Hz, 1000 Hz, and 2000 Hz, as well as the T_{wf} which is the arithmetic mean of RT for 250 Hz, 500 Hz, 1000 Hz, 2000 Hz, and 4000 Hz. The authors specified that measuring T_{mf} alone does not always correspond with the subjective evaluation of the acoustical properties of the room. Other studies, such as the Klatte et al. [61] study, measured the T_{20} averaged from results of octave bands from 250 to 2000 Hz. The measured RT ranged from 0.9 to 1.1 s for T_{60} and 1.1 to 1.4 s for T_{20}. Details on measurement protocol and results can be found in Table S6 in Supplementary Materials.

3.2.3. Speech Intelligibility

Speech intelligibility depends on the spoken language familiarity of the listeners and is limited in children due to a lack of vocabulary and grammar skills [113]. Speech transmission is the physical measurement of the speech intelligibility, which depends on classroom acoustic characteristics such as the RT and the background noise level.

One indicator of speech intelligibility is the signal-to-noise (SNR) ratio. It is defined as the ratio of the signal power to the background noise power and is expressed in decibels. A review on speech intelligibility in school has shown that high background noise levels can mask speech sounds, as the authors stated that a recommended SNR should be greater than +15 dB, and ideally at +25 dB [114]. Sarantopoulos et al. [90] measured the SNR using LA_{90} in an occupied classroom with a teacher talking. LA_{90}, defined as the 90th percentile of LA_{eq} in a one-minute measurement period, was considered as a proxy to background noise during the active teaching period. The SNR was calculated by subtracting the measured teacher's speech noise to the LA_{90}. For 41 teachers in 15 classrooms, the average SNR for teaching was 12.0 ± 3.6 dB(A), ranging from +6.8 dB(A) to + 21.6 dB(A).

Another indicator of speech intelligibility is the speech transmission index (STI), which ranges between 0 and 100% and represents the transmission quality of speech concerning intelligibility by a speech transmission channel, according to the standard EN 60268-16, 2011 [115]. STI was assessed in four studies in Italy [30,32,43,46], one study in Poland [71], and one in England [93] by emitting a speech-like sound signal and measuring its transmission quantity at another point in the room, referring to the methods described in the standard EN 60268-16, 2011 [115]. Measurements were performed in unoccupied classrooms as students' presence can alter the STI results, and the results ranged from 41 to 76%.

The clarity index is also used to assess speech intelligibility. It is the difference between the emitted sound energy and the later arriving sound energy after a time limit [112], expressed in dB. C_{50} is the clarity index in case of a time limit of 50 ms (ISO 3392-1 standard, 2010). C_{50} was measured simultaneously with STI in the same three studies in Italy [30,32,43], and in one study in England [93]. The results ranged from −6.28 to 5.55 dB in 66% of classrooms. Details on measurement protocol and results of measured SNR, STI, and C_{50} can be found in Table S5 in Supplementary Materials.

3.2.4. Sound Insulation

Sound insulation is the ability of buildings' components to reduce sound transmission through the envelope and the internal walls and floors. Façade insulation measurement can be determined by measuring the airborne sound reduction index between outside and inside the buildings. Different methods exist (ISO 16283-3 standard, 2016 [116]) and aim at assessing either the sound reduction index of an element of the building façade, such as windows, or the reduction of indoor noise levels due to building façade with actual traffic conditions. In some countries, minimum sound insulation level requirements exist in building regulations: at least 38 dB(A) façade insulation in the Italian technical regulation and at least 30 dB(A) façade insulation in the French regulation [116].

Chetoni et al. [46] in Italy determined the total insulation of the school building façade exposed to road traffic and calculated the weighted standardized level insulation ($D_{2m,nT,w}$), following ISO 16283-3 [116]. The study found that the façade insulation index was below the regulatory value of 38 dB in 23 out of 24 classrooms. Different classrooms in the same school had different results, as there are various conditions of poor quality or even damaged windows and doors.

In the same study, airborne sound insulation between two classrooms or between classrooms and corridors was assessed by measuring wall insulation between interior spaces (R'_w). Results showed that in 11 out of 24 classrooms, the R'_w between two classrooms [46] was below the Italian regulatory value of 41 dB [54]. A large variability of R'_w (at a maximum of 18 dB) between the classrooms was observed. It was explained by the different construction technologies, either with load bearing or only with a partition wall.

3.3. *Indoor Air Quality (IAQ)*

IAQ depends on the concentrations of pollutants having outdoor or indoor origin. Sixty-two studies performed IAQ measurements, including the following parameters: carbon dioxide, different ventilation parameters, formaldehyde and other aldehydes, volatile organic compounds (VOC), semi-volatile organic compounds (SVOC), particulate matter, nitrogen dioxide, bio-contaminants and radon.

3.3.1. Carbon Dioxide (CO_2), Ventilation Rate (VR), and Air Change Rate (ACR)

CO_2 is a marker of ventilation adequacy in the presence of people indoors and is the most prevalently measured parameter in connection with IAQ monitoring [117]. CO_2 was measured mainly with a non-dispersive infrared (NDIR) sensor. The concentrations were determined using spot or continuous measurements, the latter for a period of 30 min to three months with a time interval ranging from two to ten minutes. The number of measurement locations was either one in the center of the room or in multiple locations.

Across 32 studies that measured CO_2 in a naturally ventilated classroom, average mean concentrations ranged between 591 and 3494 ppm. In all naturally ventilated classrooms, CO_2 concentration varied throughout the day depending on children's presence and the frequency of window opening. Studies in classrooms equipped with mechanical ventilation systems showed that CO_2 concentrations did not exceed 1000 ppm.

Details on measurement protocols of CO_2 and results from studies can be found in Table S7 in Supplementary Materials.

The air change rate (ACR) and ventilation rate (VR) can be calculated using the measured CO_2 concentrations. Six studies calculated the ACR using the CO_2 decay rate during the non-occupied period, and one study calculated VR using the CO_2 production rate during the occupied period. The estimated ACR ranged between 0.11 and 0.39 h^{-1} in naturally ventilated classrooms, and between 1.4 and 3.2 h^{-1} in classrooms with mechanical ventilation systems.

In 70 classrooms equipped with either air handling units (17%), fan coil units (21%), or individual unit ventilators (62%) in the USA, VR was calculated using the peak level of measured CO_2 and mean VR was estimated to be 3.6 ± 2.3 L/s per person. The VR was calculated using measured CO_2 concentrations in 51 classrooms in France (14 classrooms had mechanical ventilation systems and 37 had natural ventilation). Mean VR was estimated to be 4.2 ± 1.7 L/s per person in mechanically ventilated classrooms and 2.4 ± 1.4 L/s per person in naturally-ventilated classroom. In the pan-European SINPHONIE study, the mean VR ranged from 0.87 L/s per child in Western Europe to 3.4 L/s per child in Northern Europe.

Detailed protocols regarding ACR/VR can be found in Table S8 in Supplementary Materials.

3.3.2. Formaldehyde and Other Aldehydes

Formaldehyde was measured in 15 studies, while other aldehydes were measured in six studies. A summary of formaldehyde measurements is provided in Table S9 in Supplementary Materials while Table S10 presents the measurements of other aldehydes.

Formaldehyde can be measured using passive (12 studies) or active (three studies) methods. It is sampled on a cartridge containing an organic reagent, such as 2,4-dinitrophenylhydrazine (DNPH), then analyzed with high-performance liquid chromatography (HPLC) and ultraviolet (UV) detection, as recommended in the ISO 16000-4 standard [118] for lightweight aldehydes. One or two passive samplers were deployed per classroom, and their locations were not always reported. The sampling duration for passive samplers ranged from two days to two weeks. The three studies that used active sampling had measured in either spot measurements from one minute [24] and 30 min [108], or 2.5 h [119]. The mean concentration of formaldehyde ranged between 1.4 and 89 $\mu g/m^3$. Summaries of measurement protocols and results of formaldehyde and other aldehydes are provided in Tables S9 and S10 in Supplementary Materials.

3.3.3. Volatile Organic Compounds (VOCs)

The most frequently measured VOCs were BTEX: benzene, toluene, ethylbenzene, and xylenes. Fourteen studies measured BTEX, among which three also measured concentration of total volatile organic compounds (TVOC), and eleven measured a larger number of VOCs. Three studies used a portable analyzer including a photoionization detector with UV to measure organic compounds. The other studies used diffusive passive samplers for a duration of two to four weeks. The location of the samplers was usually in the center of the room with other measurement devices, but in most studies, it was not reported. The mean indoor concentrations ranged between 0.5 and 3.2 $\mu g/m^3$ for benzene, 0.2 and 17 $\mu g/m^3$ for toluene, less than the limit of detection and 9.0 $\mu g/m^3$ for ethylbenzene, and 0.6 to 12 $\mu g/m^3$ for xylenes. A summary of all the measured BTEX compounds and other VOCs, their protocols and results can be found in Table S11 and Table S12 in Supplementary Materials, respectively.

3.3.4. Semi-Volatile Organic Compounds (SVOCs)

SVOCs are less volatile than VOCs and can also be present in the particulate phase in addition to the gas phase. Four groups of SVOCs were measured and reported in eleven studies. They were polycyclic aromatic hydrocarbons, PAH (acenaphthene, anthracene, benzo(a)pyrene), flame retardants (tributylphosphate, polybrominated diphenyl ethers), phthalates (BBP, DBP, DEHP, DEP, DiBP), and synthetic musks (tonalide, galaxolide). Phthalates were the most frequently detected SVOCs in the air. Active sampling on polyurethane foam (PUF) was often used to trap the SVOC gas phase. The samples were then analyzed using GC-MS. The sampling duration ranged from 24 h to one week at one point in the room, with the accurate location not specified. SVOC measurement protocols and concentrations can be found in Tables S14–S16 in Supplementary Materials.

3.3.5. Particulate Matter (PM)

Particulate matter (PM) can originate from indoor (e.g., cooking and heating) and outdoor (e.g., traffic) sources. Twenty-two studies measured PM in classrooms, among which ten measured PM_{10}, 21 measured $PM_{2.5}$, one measured PM_4, two measured PM_1, and one measured $PM_{0.1}$.

PM concentrations can be measured using optical and gravimetric methods. Optical PM counters use a laser to count the particles passing a small volume, with results expressed in the total particle count adjusted to the volume. The gravimetric method uses an air pump to drive air through an impactor that collects PM according to their size. Eight studies used an optical counter, and twelve used the gravimetric method to measure PM concentrations. Two studies used condensation particle counters (CPC) that can measure small particles, such as $PM_{0.1}$ and $PM_{0.5}$. The sampling duration ranged from eight hours to ten months, with a time interval from one to ten minutes. The sampling locations in the classrooms were not specified.

The measured concentrations ranged from 34 to 2061 $\mu g/m^3$ for PM_{10}, and from 1.3 to 106 $\mu g/m^3$ for $PM_{2.5}$. PM_1 was measured in one study with a mean concentration of 19.2 ± 7.2 $\mu g/m^3$ in two classrooms during school hours. One study measured $PM_{0.5}$ in two classrooms and reported concentrations ranging from 2.1 to 22 $\mu g/m^3$. One study measured the ultrafine particles, with concentrations ranging from 1560 to 16,780 particles/cm^3. The measurement protocols and results can be found in Table S13 in Supplementary Materials.

3.3.6. Nitrogen Dioxide (NO_2)

Nitrogen dioxide (NO_2) is primarily emitted by combustion and mainly comes from outdoors, particularly from traffic. Among eleven studies that measured NO_2, nine conducted long-term (five to fourteen days) passive measurements, and two used a chemiluminescence continuous analyzer for a period of thirty minutes or 24 h. The location of the samplers was usually in the center of the room with other measurement devices, but in most studies, it was not reported. The mean NO_2 concentration ranged from 4.9 to 125 $\mu g/m^3$, with a maximum concentration of 292 $\mu g/m^3$. Details on the measurement protocols and results can be found in Table S17 in Supplementary Materials.

3.3.7. Bio-Contaminants

Dampness and high relative humidity in buildings lead to microbial growth, dust mites and their allergens [120]. Bio-contaminants assessment was reported in three studies in Finland, one in Portugal, and in one pan-European study.

Mold exposure can be assessed either by visual inspection or by measurements of airborne spores. Visual inspection has been reported in one study in Finland, with two out of seventeen schools showing visible mold. For measurements of airborne spores, two studies in Finland measured indoor airborne cultivable microorganisms, and reported an average concentration of 22 to 260 CFU/m^3 of airborne cultivable microorganisms. Finally, the SINPHONIE pan-European study reported that 7% of classrooms had visible signs of mold.

3.3.8. Radon

Radon is a chemically inert gas emitted naturally from underground. It was measured in sixteen studies identified in the present review. Measurements can be made with a passive dosimeter, which was the case in fifteen studies, for a duration ranging from three months to one year. Passive dosimeters were placed on the ground floor or on the lowest floor of the buildings. Only one study used an active measurement device that provided a value every sixty minutes over 24 h. The mean radon concentration ranged from 56 to 579 Bq/m^3. The measurement protocol and the results showing measured radon concentrations are presented in Table S18 in Supplementary Materials.

3.4. *Visual Environment*

Lighting conditions determine the quality of the visual environment. They include the contributions of both daylight and artificial light emitted by the installed luminaires. Visual environment was investigated in 15 studies; most of them were spot measurement using an illuminance meter placed at students' desks. The number of measurement points varied from one per class to 319. Leccese et al. [31] measured numerous parameters for the purpose of determining the most influential parameters on students' visual comfort.

3.4.1. Daylighting

Daylighting measurements can be made by assessing the daylight factor via simulation or by simply measuring desk illuminance in classrooms with artificial light turned off. Daylight factor (%) is the ratio between the indoor horizontal illuminance at a given location and the outdoor horizontal illuminance measured under the unobstructed sky vault in overcast conditions. One study reported the daylight factor, with a result of 2.2%. Leccese et al. [31] also reported that daylight glare is the most important factor contributing to students' visual comfort.

Five studies reported daylighting by measuring the total illuminance with artificial lighting on students' desk. Rucinska et al. [33] measured the average illuminance in a classroom during two periods: one period with a clear sky and the other period with an overcast sky. This study included the highest number of measurement points per classroom, with 319 points grouped in three different longitudinal rows of tables, i.e., near the windows, in the middle of the room, and next to the entrance door away from the windows. Results showed a 9-fold decrease of mean illuminance in the middle of the room and an 18-fold decrease next to the entrance door, compared to the illuminance on the desks near the windows. In overcast conditions, these ratios were about 3 and 8, respectively. In the other studies, mean measured total illuminance ranged from 303 lx to 1255 lx.

3.4.2. Artificial lighting

Artificial lighting is provided by luminaires installed inside the classroom. Artificial lighting should be able to compensate for an insufficient level of daylight indoors.

The measurement of artificial lighting follows the same principle with the use of an illuminance-meter. To correctly evaluate only artificial lighting, measurements should be done with controlled daylighting, by selecting low daylight timeframe, or simply by closing solar protections if any in the classroom. All performed measurements were spot measurements. Artificial lighting was measured in eight studies. Six out of the eight studies measured only at the center of the classroom, one study reported at least four measurements points and one study measured at six points in the room, additionally reporting the values of the lighting uniformity factors. The illuminance values reported were in the range of 241 lx to 748 lx. Table S19 combines all measurement protocols and results of all studies on lighting conditions.

4. Discussion

This review highlighted that out of the 88 reviewed articles, all four components of IEQ were measured only in four studies [27–30]. Identical parameters were measured in these four studies (air temperature, RH, CO_2, illuminance, noise level), with a sample size of less than five classrooms in three studies. The study by de la Hoz–Torres et al. [29] measured IEQ parameters in 15 classrooms in six buildings in Portugal and Spain. As shown in Figure 2, eight studies measured at least three out of the four IEQ components—seven studies measured IAQ, visual and thermal parameters [25,48–50,63,66,74], and one study measured visual, thermal and acoustic parameters [43].

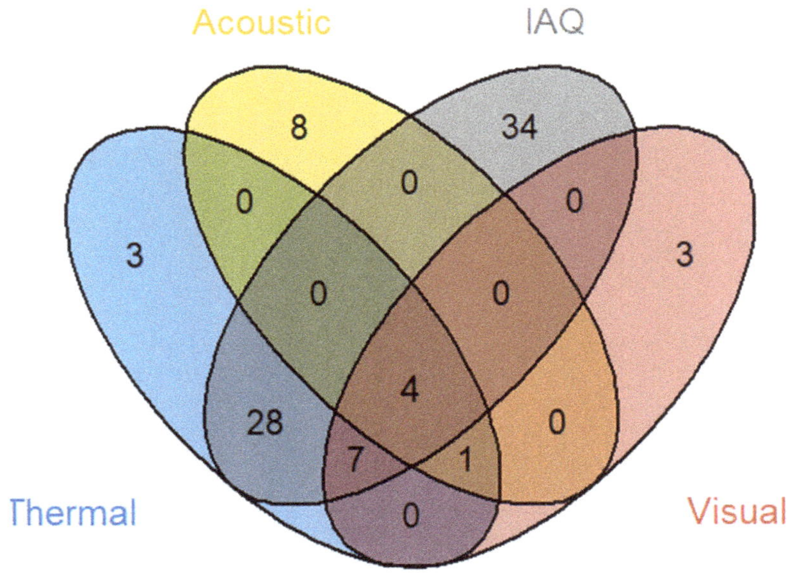

Figure 2. Number of studies according to the IEQ components measured (n = 88 articles).

As shown in Figure 3, most studies had a small sample size; the median was six schools and 17 classrooms. Twenty-six studies only measured IEQ parameters in one school, and thirteen of them only instrumented one classroom. In 19 out of these 26 studies, there were at least two IEQ components measured including three studies that measured all four IEQ components [27,28,30]. In 11 studies that targeted more than 100 schools, only three performed measurements of more than one IEQ component, including the pan-European SINPHONIE study, which measured thermal, IAQ, and visual parameters in 114 schools and 342 classrooms across 23 countries [25,38]. One study had the largest sample size of 1000 classrooms in 438 schools but only radon was measured with passive dosimeters [100]. Overall, it was seldom that many parameters were measured when the study included many classrooms and schools.

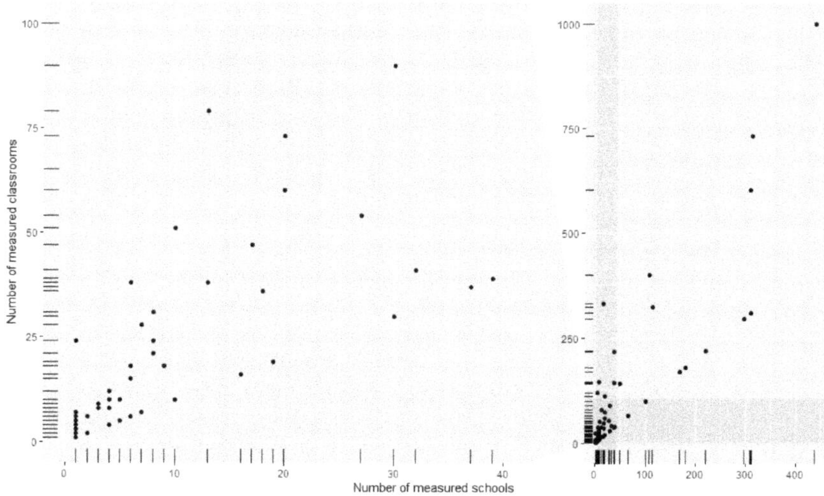

Figure 3. Number of schools and classrooms per study where the measurements were made. Each dot represents a study with the number of measured schools on the x-axis and the number of measured classrooms on the y-axis.

Among the different aspects retrieved from the reviewed studies (sampling location, type of sampling, duration, etc.), the measured parameters were the first point of interest of the review. Targeted parameters varied between studies depending on their objective and their capabilities in terms of equipment, safety considerations, and cost. Whilst most parameters could be used to characterize IEQ or evaluate children's pollutant exposure and comfort, some parameters show limitations. Indeed, to characterize thermal environment, air temperature was measured in all studies. However, this measurement may not be sufficient to characterize children's thermal sensations which depends also on other parameters such as relative humidity, airspeed, and mean radiant temperature [53,121]. Operative temperature is based on the measurement of air temperature with a temperature thermocouple sensor, air velocity with an anemometer, and mean air radiant temperature with a globe thermometer. It requires advanced measurement systems which are complex and expensive to implement in a large sample of classrooms over a long period [94]. We stipulate that it was the reason why it was measured only in five studies and in two of them as spot measurements before the monitoring week due to the safety considerations [27,81]. In these studies, the differences between mean radiant temperature and air temperature were always below 3 °C, suggesting that the choice of measuring air temperature is somewhat justified. In 145 classrooms of six schools in Sweden, air temperature was measured from May to October and the difference between operative temperature and air temperature was below 1 °C because the schools were well insulated; this provides additional justification for using only air temperature measurements [94].

The second discussion point deals with the measurement protocols. Measuring protocols for some parameters have shown significant differences in sampling strategies and statistical indicators. The sampling strategies were inhomogeneous concerning the position, the height, the duration of the measurements in the classrooms, the sampling frequency, and the devices 'accuracy' were not systematically mentioned in many studies including calibration. This variability was observed for the four IEQ components, especially for the numerous IAQ parameters. For example, particles were monitored with optical or gravimetric methods, with various sampling strategies during occupancy or non-occupancy periods. These variations were also observed for the background noise level, which was measured either when children were in the classrooms or not in the classrooms. These differences led to inhomogeneity in results, metrics, and difficulty to interpret the results. Many measuring methods and indicators were complex and trained field technicians were needed to perform measurements. This mainly concerns parameters of acoustic and visual components. For example, speech intelligibility, essential for providing an adequate learning environment for children, was measured in five studies using three different indicators: SNR, STI, and C_{50}, RT, which is also an essential factor for acoustic quality in classrooms [122], was measured with different protocols, especially for the generation of the noise, but always followed the standard ISO 3382 [112]. For the visual environment, many parameters, including illuminance, uniformity, luminance distribution, glare, effects of temporal light modulation such as flicker and the stroboscopic effect, and color temperature, should be considered to have well-balanced lighting in classrooms essential for health, well-being, and learning. Illuminance on desks was measured in all the studies dealing with visual environment. It is the less complex parameter to assess using a lux-meter. Other indices, such as glare indices, have been shown to have more impact on students' perception but are based on equations correlating luminance values in the occupant's field of view and human' sensation. These parameters were only measured one time, in one lighting's specific study in one classroom, demonstrating its difficulty to implement them in a large sample of schools [31]. The factors that primarily influence occupant visual comfort were daylight glare and luminance, which are not commonly measured as previously mentioned.

Finally, main results from the reviewed studies provide an overall picture of IEQ in schools within the scope of our review. Most IEQ parameters identified in this review have reference values in the regulations, standards, and guidelines from WHO and the governments or GBC. Table 2 presents the summary results for each IEQ parameter from 88 retrieved papers and their reference values. IAQ parameters have fewer guideline values compared to other parameters, especially for VOCs and SVOCs. While a growing body of studies on IAQ parameters in schools has shown associated risks to children's health, well-being, and cognitive performance, further development and future studies are still needed to highlight the relationship between children' exposure to air pollutants and health outcomes, for the purpose of establishing their reference values. For acoustic and visual components, some parameters are regulated in European countries and the USA for school buildings, such as façade insulation, sound insulation between two rooms, and RT. The illuminance-regulated value has been mentioned in the labor code in France. This review highlighted that many parameters in the four components exceeded threshold values [123].

This review creates the background for developing an IEQ rating scheme in schools by listing the measured parameters and assessing their respective prominence, limitation, and applicability to the school environment. The most measured parameters for the four IEQ components are air temperature and relative humidity for thermal, concentrations of carbon dioxide, particulate matter, radon, formaldehyde, and some volatile organic compounds for IAQ, noise level, and reverberation time for acoustic and illuminance for visual environments. The popularity of these parameters in studies indicates their pertinence in characterizing the IEQ parameters in schools. However, a consensus measurement protocol of all IEQ parameters must be set to facilitate inter-comparison between measurements at different schools. Another question can be raised on whether and how occupant perception can be used to assess the IEQ, as studies have shown limitations of some parameters when evaluating children's comfort. Future works should propose harmonized protocols and define the threshold values for each parameter to ensure children can spend time at school without harming their health, well-being, and cognitive performance.

5. Conclusions

This review provides an overview of IEQ parameters measured in schools in Europe and North America since 2010 and compares different monitoring protocols and results in 88 articles. This review aims to provide an overview of IEQ parameters often measured in schools, their measuring methods, and the main findings from these measurements. These studies mainly focused on individual IEQ components in small samples of schools (median of six schools). Twenty-two parameters or families of parameters were reviewed. Measurement protocols, including the number and position of samples and the sampling duration, did not present a consensus between studies, which leads to the difficulty in comparing the results. Some parameters also present limitations in the current schools' environment and need further developments to be adapted to the school's environment and children's exposure. Air temperature and relative humidity for thermal, concentrations of carbon dioxide, particulate matter, radon, formaldehyde, and some volatile organic compounds for IAQ, noise level, and reverberation time for acoustic and illuminance for visual environments can be defined as adequate parameters for the purpose of characterizing IEQ in schools.

Overall, a holistic approach to quantifying IEQ in schools is needed with a set of measurable parameters for the four components with consensus on measurement protocols, and threshold values that reflect children's pollutant exposure and comfort perception.

Supplementary Materials: The following supporting information can be downloaded at: https://www.mdpi.com/article/10.3390/buildings13020433/s1, Figure S1: Distribution of studies per country (n = 88); Figure S2: Types of ventilation in reviewed studies; Table S1: Measurements of air temperature (°C) in schools in Europe and North America since 2010; Table S2: Measurement of relative humidity in schools across Europe and North America from 2010; Table S3: Thermal comfort assessment in schools in Europe and North America since 2010; Table S4: Measurements of background noise level (dB) in schools in Europe and North America since 2010; Table S5: Measurement of clarity index in schools across Europe and North America from 2010; Table S6: Measurements of reverberation time in schools across Europe and North America from 2010; Table S7: Measurements of carbon dioxide (CO_2) in schools in Europe and North America since 2010; Table S8: Calculation of air exchange rate/ ventilation in schools across Europe and North America from 2010; Table S9: Measurements of formaldehyde in schools in Europe and North America since 2010; Table S10: Measurements of aldehydes in schools across Europe and North America from 2010; Table S11: Measurements of BTEX in schools across Europe and North America from 2010; Table S12: Measurements of volatile organic compounds in schools across Europe and North America from 2010; Table S13: Measurements of particulate matter (PM) in schools in Europe and North America since 2010; Table S14: Measurements of flame retardants (SVOCs) in schools across Europe and North America since 2010; Table S15: Measurements of polycyclic aromatic hydrocarbons (SVOCs) in schools across Europe and North America since 2010; Table S16: Measurements of phthalates and musks (SVOCs) in schools across Europe and North America since 2010; Table S17: Measurements of NO_2 in schools across Europe and North America since 2010; Table S18: Measurements of radon in schools across Europe and North America since 2010; Table S19: Measurements of illuminance (Lux) in schools in Europe and North America since 2010.

Author Contributions: Conceptualization, M.T.T., W.W., C.D., P.W., V.H. and C.M. (Corinne Mandin); methodology, M.T.T., C.D., P.D. and C.M. (Christophe Martinsons); investigation, M.T.T.; writing—original draft preparation, M.T.T. and C.D.; writing—review and editing, W.W., P.W., V.H., C.M. (Christophe Martinsons), P.D. and C.M. (Corinne Mandin); supervision, P.W. and V.H.; project administration, W.W. and C.D. All authors have read and agreed to the published version of the manuscript.

Funding: This research was co-funded by the French Agency for Ecological Transition (ADEME), contract number TEZ20-047 and the Scientific and Technical Centre for Building (CSTB).

Institutional Review Board Statement: Not applicable.

Informed Consent Statement: Not applicable.

Data Availability Statement: The data presented in this study are available in Supplementary Materials.

Conflicts of Interest: The authors declare no conflict of interest.

Nomenclature

Abbreviation	Signification
IEQ	Indoor environmental quality
IAQ	Indoor air quality
CO2	Carbon dioxide
TAIL	Thermal, Acoustic, IAQ, Lighting
GBC	Green building certification
SM	Supplementary material
HVAC	Heating, ventilation, and air conditioning
PMV	Perceived mean vote
PPD	Percentage person dissatisfied
VOCs	Volatile organic compounds
BTEX	Benzene, toluene, ethyl-benzene, xylene
SVOCs	Semi-volatile organic compounds
PM	Particulate matter
ACR	Air change rate
VR	Ventilation rate
NO2	Nitrogen dioxide
CFU	Colony forming unit
RH	Relative humidity
LAeq	Background noise equivalent level
RT	Reverberation time
SNR	Speech to noise ratio
STI	Speech transmission index
HPLC	High-performance liquid chromatography
UV	Ultra violet
TVOC	Total volatile organic compounds
DNPH	2,4-dinitrophenylhydrazine
PAH	Polycyclic aromatic hydrocarbons
BBP, DBP, DEHP, DEP, DiBP	Phthalates
PUF	Polyurethane foam

References

1. Altomonte, S.; Allen, J.; Bluyssen, P.M.; Brager, G.; Heschong, L.; Loder, A.; Schiavon, S.; Veitch, J.A.; Wang, L.; Wargocki, P. Ten Questions Concerning Well-Being in the Built Environment. *Build. Environ.* **2020**, *180*, 106949. [CrossRef]
2. Klepeis, N.E.; Nelson, W.C.; Ott, W.R.; Robinson, J.P.; Tsang, A.M.; Switzer, P.; Behar, J.V.; Hern, S.C.; Engelmann, W.H. The National Human Activity Pattern Survey (NHAPS): A Resource for Assessing Exposure to Environmental Pollutants. *J. Expo. Anal. Environ. Epidemiol.* **2001**, *11*, 231–252. [CrossRef] [PubMed]
3. Nimlyat, P.S. Indoor Environmental Quality Performance and Occupants' Satisfaction [IEQPOS] as Assessment Criteria for Green Healthcare Building Rating. *Build. Environ.* **2018**, *144*, 598–610. [CrossRef]
4. Wu, H.; Wu, Y.; Sun, X.; Liu, J. Combined Effects of Acoustic, Thermal, and Illumination on Human Perception and Performance: A Review. *Build. Environ.* **2020**, *169*, 106593. [CrossRef]
5. Gaffin, J.M.; Hauptman, M.; Petty, C.R.; Sheehan, W.J.; Lai, P.S.; Wolfson, J.M.; Gold, D.R.; Coull, B.A.; Koutrakis, P.; Phipatanakul, W. Nitrogen Dioxide Exposure in School Classrooms of Inner-City Children with Asthma. *J. Allergy Clin. Immunol.* **2018**, *141*, 2249–2255.e2. [CrossRef]
6. Wargocki, P.; Porras-Salazar, J.A.; Contreras-Espinoza, S.; Bahnfleth, W. The Relationships between Classroom Air Quality and Children's Performance in School. *Build. Environ.* **2020**, *173*, 106749. [CrossRef]
7. Wargocki, P.; Porras-Salazar, J.A.; Contreras-Espinoza, S. The Relationship between Classroom Temperature and Children's Performance in School. *Build. Environ.* **2019**, *157*, 197–204. [CrossRef]
8. Wei, W.; Wargocki, P.; Zirngibl, J.; Bendžalová, J.; Mandin, C. Review of Parameters Used to Assess the Quality of the Indoor Environment in Green Building Certification Schemes for Offices and Hotels. *Energy Build.* **2020**, *209*, 109683. [CrossRef]
9. Wargocki, P.; Wei, W.; Bendžalová, J.; Espigares-Correa, C.; Gerard, C.; Greslou, O.; Rivallain, M.; Sesana, M.M.; Olesen, B.W.; Zirngibl, J.; et al. TAIL, a New Scheme for Rating Indoor Environmental Quality in Offices and Hotels Undergoing Deep Energy Renovation (EU ALDREN Project). *Energy Build.* **2021**, *244*, 111029. [CrossRef]
10. Wei, W.; Wargocki, P.; Ke, Y.; Bailhache, S.; Diallo, T.; Carré, S.; Ducruet, P.; Maria Sesana, M.; Salvalai, G.; Espigares-Correa, C.; et al. PredicTAIL, a Prediction Method for Indoor Environmental Quality in Buildings Undergoing Deep Energy Renovation Based on the TAIL Rating Scheme. *Energy Build.* **2022**, *258*, 111839. [CrossRef]

11. B.S Limited BEAM Plus New Buildings Version 2.0. 2019. Available online: https://www.hkgbc.org.hk/eng/beam-plus/file/BEAMPlus_New_Buildings_v2_0.pdf (accessed on 30 January 2023).
12. BRE Global Breeam UK New Construction. 2018. Available online: https://files.bregroup.com/breeam/technicalmanuals/NC2018/ (accessed on 30 January 2023).
13. DGNB GmbH. DGNB System Buildings in Use Criteria Set. 2020. Available online: https://static.dgnb.de/fileadmin/dgnb-system/downloads/criteria/DGNB-Criteria-Set-Buildings-In-Use-Version-2020.pdf (accessed on 30 January 2023).
14. Green Building Initiative. Green Globes Technical Reference Manual. 2018. Available online: http://www.greenglobes.com/v3/newconstruction/Green_Globes_for_New_Construction_2021_Technical_Reference_Manual.pdf (accessed on 30 January 2023).
15. Ministry of National Development. Green Mark for Non-Residential Buildings NRB:2015. 2015. Available online: https://www.bca.gov.sg/greenmark/others/GM_NREB_V3.pdf (accessed on 30 January 2023).
16. GBCA. Green Star—Design & As Built **2017**, *1.2*, 1–48. Available online: https://new.gbca.org.au/green-star/rating-system/design-and-built/ (accessed on 30 January 2023).
17. Green Building Council Indonesia. GREENSHIP Rating Tool for INTERIOR SPACE. **2012**, *22*. Available online: https://www.gbcindonesia.org/greens/existing (accessed on 30 January 2023).
18. Certivea Certification NF HQE Bâtiments Tertiaires En Exploitation. 2014. Available online: https://certivea.fr/solutions/hqe-batiment/ (accessed on 30 January 2023).
19. Ministry Republic of Austria for Sustainability and Tourism. Klimaaktiv Quality and Criteria New Buildings and Renovations. 2019; ISBN 9783990910054. Available online: https://www.klimaaktiv.at/bauen-sanieren.html (accessed on 30 January 2023).
20. USGBC. Leed v4.1 Building Design and Construction. 2021. Available online: https://www.usgbc.org/leed/v41 (accessed on 30 January 2023).
21. USGBC. LOTUS-New Construction V3. 2019. Available online: https://vgbc.vn/en/lotus-new-construction-lotus-nc/ (accessed on 30 January 2023).
22. TGBI. TREES—NC Thai' s Rating of Energy and Environmental Sustainability; Volume 75. Available online: https://tgbi.or.th/uploads/trees/2017_03_TREES-EB-Eng.pdf (accessed on 30 January 2023).
23. Branco, P.T.B.S.; Alvim-Ferraz, M.C.M.; Martins, F.G.; Sousa, S.I.V. Children's Exposure to Indoor Air in Urban Nurseries-Part I: CO2 and Comfort Assessment. *Environ. Res.* **2015**, *140*, 1–9. [CrossRef]
24. Branco, P.T.B.S.; Nunes, R.A.O.; Alvim-Ferraz, M.C.M.; Martins, F.G.; Sousa, S.I.V. Children's Exposure to Indoor Air in Urban Nurseries—Part II: Gaseous Pollutants' Assessment. *Environ. Res.* **2015**, *142*, 662–670. [CrossRef]
25. Csobod, É.; Annesi-Maesano, I.; Carrer, P.; Kephalopoulos, S.; Madureira, J.; Rudnai, P.; de Oliveira Fernandes, E. *Schools Indoor Pollution & Health Observatory Network in Europe SINPHONIE*; Publications Office of the European Union: Luxembourg, 2014. [CrossRef]
26. Baloch, R.M.; Maesano, C.N.; Christoffersen, J.; Banerjee, S.; Gabriel, M.; Csobod, É.; de Oliveira Fernandes, E.; Annesi-Maesano, I.; Szuppinger, P.; Prokai, R.; et al. Indoor Air Pollution, Physical and Comfort Parameters Related to Schoolchildren's Health: Data from the European SINPHONIE Study. *Sci. Total Environ.* **2020**, *739*, 139870. [CrossRef]
27. Vilcekova, S.; Meciarova, L.; Burdova, E.K.; Katunska, J.; Kosicanova, D.; Doroudiani, S. Indoor Environmental Quality of Classrooms and Occupants' Comfort in a Special Education School in Slovak Republic. *Build. Environ.* **2017**, *120*, 29–40. [CrossRef]
28. Harčárová, K. Indoor Air Quality in Classrooms of a Newly Built School. *IOP Conf. Ser. Mater. Sci. Eng.* **2020**, *867*, 012008. [CrossRef]
29. De la Hoz-Torres, M.L.; Aguilar, A.J.; Costa, N.; Arezes, P.; Ruiz, D.P.; Martínez-Aires, M.D. Reopening Higher Education Buildings in Post-Epidemic COVID-19 Scenario: Monitoring and Assessment of Indoor Environmental Quality after Implementing Ventilation Protocols in Spain and Portugal. *Indoor Air* **2022**, *32*, e13040. [CrossRef]
30. Loreti, L.; Barbaresi, L.; De Cesaris, S.; Garai, M. Overall Indoor Quality of a Non-Renewed Secondary-School Building. *Build. Acoust.* **2016**, *23*, 47–58. [CrossRef]
31. Leccese, F.; Salvadori, G.; Rocca, M.; Buratti, C.; Belloni, E. A Method to Assess Lighting Quality in Educational Rooms Using Analytic Hierarchy Process. *Build. Environ.* **2020**, *168*, 106501. [CrossRef]
32. Russo, D.; Ruggiero, A. Choice of the Optimal Acoustic Design of a School Classroom and Experimental Verification. *Appl. Acoust.* **2019**, *146*, 280–287. [CrossRef]
33. Rucińska, J.; Trzaski, A. Measurements and Simulation Study of Daylight Availability and Its Impact on the Heating, Cooling and Lighting Energy Demand in an Educational Building. *Energies* **2020**, *13*, 2555. [CrossRef]
34. Aguilar, A.J.; de la Hoz-Torres, M.L.; Martínez-Aires, M.D.; Ruiz, D.P. Thermal Perception in Naturally Ventilated University Buildings in Spain during the Cold Season. *Buildings* **2022**, *12*, 890. [CrossRef]
35. Ahmed, K.; Kuusk, K.; Heininen, H.; Arumägi, E.; Kalamees, T.; Hasu, T.; Lolli, N.; Kurnitski, J. Indoor Climate and Energy Performance in Nearly Zero Energy Day Care Centers and School Buildings. *E3S Web Conf.* **2019**, *111*, 02003. [CrossRef]
36. Alves, C.; Nunes, T.; Silva, J.; Duarte, M. Comfort Parameters and Particulate Matter (PM10 and PM2.5) in School Classrooms and Outdoor Air. *Aerosol Air Qual. Res.* **2013**, *13*, 1521–1535. [CrossRef]
37. Annesi-Maesano, I.; Hulin, M.; Lavaud, F.; Raherison, C.; Kopferschmitt, C.; De Blay, F.; Charpin, D.A.; Denis, C. Poor Air Quality in Classrooms Related to Asthma and Rhinitis in Primary Schoolchildren of the French 6 Cities Study. *Thorax* **2012**, *67*, 682–688. [CrossRef]

38. Baloch, R.M.; Maesano, C.N.; Christoffersen, J.; Mandin, C.; Csobod, E.; De, E.; Fernandes, O.; Annesi-Maesano, I. Daylight and School Performance in European Schoolchildren. *Public Health* **2021**, *18*, 258. [CrossRef]
39. Barmparesos, N.; Assimakopoulos, M.N.; Assimakopoulos, V.D.; Loumos, N.; Sotiriou, M.A.; Koukoumtzis, A. Indoor Air Quality and Thermal Conditions in a Primary School with a Green Roof System. *Atmosphere* **2018**, *9*, 75. [CrossRef]
40. Becerra, J.A.; Lizana, J.; Gil, M.; Barrios-Padura, A.; Blondeau, P.; Chacartegui, R. Identification of Potential Indoor Air Pollutants in Schools. *J. Clean. Prod.* **2020**, *242*, 118420. [CrossRef]
41. Branco, P.T.B.S.; Nunes, R.A.O.; Alvim-Ferraz, M.C.M.; Martins, F.G.; Sousa, S.I.V. Children's Exposure to Radon in Nursery and Primary Schools. *Int. J. Environ. Res. Public Health* **2016**, *13*, 386. [CrossRef]
42. Brdarić, D.; Capak, K.; Gvozdić, V.; Barišin, A.; Jelinić, J.D.; Egorov, A.; Šapina, M.; Kalambura, S.; Kramarić, K. Indoor Carbon Dioxide Concentrations in Croatian Elementary School Classrooms during the Heating Season. *Arh. Hig. Rada Toksikol.* **2019**, *70*, 296–302. [CrossRef]
43. Buratti, C.; Belloni, E.; Merli, F.; Ricciardi, P. A New Index Combining Thermal, Acoustic, and Visual Comfort of Moderate Environments in Temperate Climates. *Build. Environ.* **2018**, *139*, 27–37. [CrossRef]
44. Canha, N.; Mandin, C.; Ramalho, O.; Wyart, G.; Ribéron, J.; Dassonville, C.; Hänninen, O.; Almeida, S.M.; Derbez, M. Assessment of Ventilation and Indoor Air Pollutants in Nursery and Elementary Schools in France. *Indoor Air* **2016**, *26*, 350–365. [CrossRef]
45. Cequier, E.; Ionas, A.C.; Covaci, A.; Marcé, R.M.; Becher, G.; Thomsen, C. Occurrence of a Broad Range of Legacy and Emerging Flame Retardants in Indoor Environments in Norway. *Environ. Sci. Technol.* **2014**, *48*, 6827–6835. [CrossRef]
46. Chetoni, M.; Ascari, E.; Bianco, F.; Fredianelli, L.; Licitra, G.; Cori, L. Global Noise Score Indicator for Classroom Evaluation of Acoustic Performances in LIFE GIOCONDA Project. *Noise Mapp.* **2016**, *3*, 157–171. [CrossRef]
47. De Gennaro, G.; Farella, G.; Marzocca, A.; Mazzone, A.; Tutino, M. Indoor and Outdoor Monitoring of Volatile Organic Compounds in School Buildings: Indicators Based on Health Risk Assessment to Single out Critical Issues. *Int. J. Environ. Res. Public Health* **2013**, *10*, 6273–6291. [CrossRef]
48. De Giuli, V.; Da Pos, O.; De Carli, M. Indoor Environmental Quality and Pupil Perception in Italian Primary Schools. *Build. Environ.* **2012**, *56*, 335–345. [CrossRef]
49. De Giuli, V.; Zecchin, R.; Corain, L.; Salmaso, L. Measured and Perceived Environmental Comfort: Field Monitoring in an Italian School. *Appl. Ergon.* **2014**, *45*, 1035–1047. [CrossRef]
50. De Giuli, V.; Zecchin, R.; Corain, L.; Salmaso, L. Measurements of Indoor Environmental Conditions in Italian Classrooms and Their Impact on Childrens Comfort. *Indoor Built Environ.* **2015**, *24*, 689–712. [CrossRef]
51. Dhoqina, P.; Tushe, K.; Xhixha, G.; Daci, B.; Bylyku, E. Measurements of Indoor Radon Concentrations in Schools in Some Cities of North Albania. AIP Conf. Proc. 2019; 2075, p. 170003.
52. Erlandson, G.; Magzamen, S.; Carter, E.; Sharp, J.L.; Reynolds, S.J.; Schaeffer, J.W. Characterization of Indoor Air Quality on a College Campus: A Pilot Study. *Int. J. Environ. Res. Public Health* **2019**, *16*, 17–26. [CrossRef]
53. Fabbri, K. Thermal Comfort Evaluation in Kindergarten: PMV and PPD Measurement through Datalogger and Questionnaire. *Build. Environ.* **2013**, *68*, 202–214. [CrossRef]
54. Franci, D.; Aureli, T. Long- and Short-Term Indoor Radon Survey in the Ardea Municipality, South Rome. *Radiat. Prot. Dosimetry* **2014**, *162*, 625–629. [CrossRef]
55. Heracleous, C.; Michael, A. Experimental Assessment of the Impact of Natural Ventilation on Indoor Air Quality and Thermal Comfort Conditions of Educational Buildings in the Eastern Mediterranean Region during the Heating Period. *J. Build. Eng.* **2019**, *26*, 100917. [CrossRef]
56. Irulegi, O.; Serra, A.; Hernández, R. Data on Records of Indoor Temperature and Relative Humidity in a University Building. *Data Br.* **2017**, *13*, 248–252. [CrossRef] [PubMed]
57. Istrate, M.A.; Catalina, T.; Cucos, A.; Dicu, T. Experimental Measurements of VOC and Radon in Two Romanian Classrooms. *Energy Procedia* **2016**, *85*, 288–294. [CrossRef]
58. Ivanova, K.; Stojanovska, Z.; Tsenova, M.; Badulin, V.; Kunovska, B. Measurement of Indoor Radon Concentration in Kindergartens in Sofia, Bulgaria. *Radiat. Prot. Dosim.* **2014**, *162*, 163–166. [CrossRef]
59. Ivanova, K.; Stojanovska, Z.; Djunakova, D.; Djounova, J. Analysis of the Spatial Distribution of the Indoor Radon Concentration in School's Buildings in Plovdiv Province, Bulgaria. *Build. Environ.* **2021**, *204*, 108122. [CrossRef]
60. Jovanović, M.; Vučićević, B.; Turanjanin, V.; Živković, M.; Spasojević, V. Investigation of Indoor and Outdoor Air Quality of the Classrooms at a School in Serbia. *Energy* **2014**, *77*, 42–48. [CrossRef]
61. Klatte, M.; Hellbroock, J. Effects of Classroom Acoustics on Performance and Wellbeing in Elementary School Children: A Field Study. *Environ. Behav.* **2010**, *42*, 659–692. [CrossRef]
62. Kojo, K.; Kurttio, P. Indoor Radon Measurements in Finnish Daycare Centers and Schools-Enforcement of the Radiation Act. *Int. J. Environ. Res. Public Health* **2020**, *17*, 2877. [CrossRef]
63. Korsavi, S.S.; Montazami, A. Developing a Valid Method to Study Adaptive Behaviours with Regard to IEQ in Primary Schools. *Build. Environ.* **2019**, *153*, 1–16. [CrossRef]
64. Kristiansen, J.; Lund, S.P.; Nielsen, P.M.; Persson, R.; Shibuya, H. Determinants of Noise Annoyance in Teachers from Schools with Different Classroom Reverberation Times. *J. Environ. Psychol.* **2011**, *31*, 383–392. [CrossRef]

65. Krugly, E.; Martuzevicius, D.; Sidaraviciute, R.; Ciuzas, D.; Prasauskas, T.; Kauneliene, V.; Stasiulaitiene, I.; Kliucininkas, L. Characterization of Particulate and Vapor Phase Polycyclic Aromatic Hydrocarbons in Indoor and Outdoor Air of Primary Schools. *Atmos. Environ.* **2014**, *82*, 298–306. [CrossRef]
66. Campano-Laborda, M.Á.; Domínguez-Amarillo, S.; Fernández-Agüera, J.; Acosta, I. Indoor Comfort and Symptomatology in Non-University Educational Buildings: Occupants' Perception. *Atmosphere* **2020**, *11*, 357. [CrossRef]
67. Larsson, K.; Lindh, C.H.; Jönsson, B.A.; Giovanoulis, G.; Bibi, M.; Bottai, M.; Bergström, A.; Berglund, M. Phthalates, Non-Phthalate Plasticizers and Bisphenols in Swedish Preschool Dust in Relation to Children's Exposure. *Environ. Int.* **2017**, *102*, 114–124. [CrossRef]
68. Liaud, C.; Chouvenc, S.; Le Calvé, S. Simultaneous Monitoring of Particle-Bound PAHs inside a Low-Energy School Building and Outdoors over Two Weeks in France. *Atmosphere* **2021**, *12*, 108. [CrossRef]
69. Madureira, J.; Paciência, I.; Rufo, J.; Ramos, E.; Barros, H.; Teixeira, J.P.; de Oliveira Fernandes, E. Indoor Air Quality in Schools and Its Relationship with Children's Respiratory Symptoms. *Atmos. Environ.* **2015**, *118*, 145–156. [CrossRef]
70. Mainka, A.; Zajusz-Zubek, E. Indoor Air Quality in Urban and Rural Preschools in Upper Silesia, Poland: Particulate Matter and Carbon Dioxide. *Int. J. Environ. Res. Public Health* **2015**, *12*, 7697–7711. [CrossRef]
71. Mikulski, W.; Radosz, J. Acoustics of Classrooms in Primary Schools—Results of the Reverberation Time and the Speech Transmission Index Assessments in Selected Buildings. *Arch. Acoust.* **2011**, *36*, 777–793. [CrossRef]
72. Müllerová, M.; Mazur, J.; Csordás, A.; Grzadziel, D.; Holý, K.; Kovács, T.; Kozak, K.; Kureková, P.; Nagy, E.; Neznal, M.; et al. Preliminary Results of Radon Survey in the Kindergartens of V4 Countries. *Radiat. Prot. Dosim.* **2017**, *177*, 95–98. [CrossRef]
73. Nunes, R.A.O.; Branco, P.T.B.S.; Alvim-Ferraz, M.C.M.; Martins, F.G.; Sousa, S.I.V. Gaseous Pollutants on Rural and Urban Nursery Schools in Northern Portugal. *Environ. Pollut.* **2016**, *208*, 2–15. [CrossRef]
74. Oldham, E.; Kim, H. IEQ Field Investigation in High-Performance, Urban Elementary Schools. *Atmosphere* **2020**, *11*, 81. [CrossRef]
75. Oliveira, M.; Slezakova, K.; Delerue-Matos, C.; Pereira, M.D.C.; Morais, S. Assessment of Polycyclic Aromatic Hydrocarbons in Indoor and Outdoor Air of Preschool Environments (3-5 Years Old Children). *Environ. Pollut.* **2016**, *208*, 382–394. [CrossRef] [PubMed]
76. Oliveira, M.; Slezakova, K.; Madureira, J.; de Oliveira Fernandes, E.; Delerue-Matos, C.; Morais, S.; do Carmo Pereira, M. Polycyclic Aromatic Hydrocarbons in Primary School Environments: Levels and Potential Risks. *Sci. Total Environ.* **2017**, *575*, 1156–1167. [CrossRef] [PubMed]
77. Oliveira, M.; Slezakova, K.; Delerue-Matos, C.; do Pereira, M.C.; Morais, S. Indoor Air Quality in Preschools (3- to 5-Year-Old Children) in the Northeast of Portugal during Spring–Summer Season: Pollutants and Comfort Parameters. *J. Toxicol. Environ. Health-Part A Curr. Issues* **2017**, *80*, 740–755. [CrossRef] [PubMed]
78. Onishchenko, A.; Malinovsky, G.; Vasilyev, A.; Zhukovsky, M. Radon Measurements in Kindergartens in Ural Region (Russia). *Radiat. Prot. Dosim.* **2017**, *177*, 112–115. [CrossRef] [PubMed]
79. Papadopoulos, G.; Panaras, G.; Tolis, E. Thermal Comfort and Indoor Air Quality Assessment in University Classrooms. *IOP Conf. Ser. Earth Environ. Sci.* **2020**, *410*, 012095. [CrossRef]
80. Papazoglou, E.; Moustris, K.P.; Nikas, K.S.P.; Nastos, P.T.; Statharas, J.C. Assessment of Human Thermal Comfort Perception in a Non-Air-Conditioned School Building in Athens, Greece. *Energy Procedia* **2019**, *157*, 1343–1352. [CrossRef]
81. Dias Pereira, L.; Raimondo, D.; Corgnati, S.P.; Gameiro da Silva, M. Assessment of Indoor Air Quality and Thermal Comfort in Portuguese Secondary Classrooms: Methodology and Results. *Build. Environ.* **2014**, *81*, 69–80. [CrossRef]
82. Pereira, L.D.; Cardoso, E.; Da Silva, M.G. Indoor Air Quality Audit and Evaluation on Thermal Comfort in a School in Portugal. *Indoor Built Environ.* **2015**, *24*, 256–268. [CrossRef]
83. Persson, J.; Wang, T.; Hagberg, J. Organophosphate Flame Retardants and Plasticizers in Indoor Dust, Air and Window Wipes in Newly Built Low-Energy Preschools. *Sci. Total Environ.* **2018**, *628–629*, 159–168. [CrossRef]
84. Poulin, P.; Leclerc, J.M.; Dessau, J.C.; Deck, W.; Gagnon, F. Radon Measurement in Schools Located in Three Priority Investigation Areas in the Province of Quebec, Canada. *Radiat. Prot. Dosim.* **2012**, *151*, 278–289. [CrossRef]
85. Raffy, G.; Mercier, F.; Blanchard, O.; Derbez, M.; Dassonville, C.; Bonvallot, N.; Glorennec, P.; Le Bot, B. Semi-Volatile Organic Compounds in the Air and Dust of 30 French Schools: A Pilot Study. *Indoor Air* **2017**, *27*, 114–127. [CrossRef]
86. Ramalho, O.; Wyart, G.; Mandin, C.; Blondeau, P.; Cabanes, P.A.; Leclerc, N.; Mullot, J.U.; Boulanger, G.; Redaelli, M. Association of Carbon Dioxide with Indoor Air Pollutants and Exceedance of Health Guideline Values. *Build. Environ.* **2015**, *93*, 115–124. [CrossRef]
87. Rivas, I.; Viana, M.; Moreno, T.; Pandolfi, M.; Amato, F.; Reche, C.; Bouso, L.; Àlvarez-Pedrerol, M.; Alastuey, A.; Sunyer, J.; et al. Child Exposure to Indoor and Outdoor Air Pollutants in Schools in Barcelona, Spain. *Environ. Int.* **2014**, *69*, 200–212. [CrossRef]
88. Romagnoli, P.; Balducci, C.; Perilli, M.; Gherardi, M.; Gordiani, A.; Gariazzo, C.; Gatto, M.P.; Cecinato, A. Indoor PAHs at Schools, Homes and Offices in Rome, Italy. *Atmos. Environ.* **2014**, *92*, 51–59. [CrossRef]
89. Rovelli, S.; Cattaneo, A.; Nuzzi, C.P.; Spinazzè, A.; Piazza, S.; Carrer, P.; Cavallo, D.M. Airborne Particulate Matter in School Classrooms of Northern Italy. *Int. J. Environ. Res. Public Health* **2014**, *11*, 1398–1421. [CrossRef]
90. Sarantopoulos, G.; Lykoudis, S.; Kassomenos, P. Noise Levels in Primary Schools of Medium Sized City in Greece. *Sci. Total Environ.* **2014**, *482–483*, 493–500. [CrossRef]
91. Langer, S.; Ekberg, L.; Teli, D.; Cabovska, B.; Bekö, G.; Wargocki, P. Study of the Measured and Perceived Indoor Air Quality in Swedish School Classrooms. *IOP Conf. Ser. Earth Environ. Sci.* **2020**, *588*, 032070. [CrossRef]

92. Senitkova, I.J. Occurrence of Indoor VOCs in Nursery School—Case Study. IOP Conf. Ser. Mater. Sci. Eng. 2017; 245, p. 082027.
93. Shield, B.; Conetta, R.; Dockrell, J.; Connolly, D.; Cox, T.; Mydlarz, C. A Survey of Acoustic Conditions and Noise Levels in Secondary School Classrooms in England. *J. Acoust. Soc. Am.* **2015**, *137*, 177–188. [CrossRef]
94. Simanic, B.; Nordquist, B.; Bagge, H.; Johansson, D. Indoor Air Temperatures, CO2 Concentrations and Ventilation Rates: Long-Term Measurements in Newly Built Low-Energy Schools in Sweden. *J. Build. Eng.* **2019**, *25*, 100827. [CrossRef]
95. Sivanantham, S.; Dassonville, C.; Grégoire, A.; Malingre, L.; Ramalho, O.; Mandin, C. Coexposure to Indoor Pollutants in French Schools and Associations with Building Characteristics. *Energy Build.* **2021**, *252*, 111424. [CrossRef]
96. Slezakova, K.; de Oliveira Fernandes, E.; do Pereira, M.C. Assessment of Ultrafine Particles in Primary Schools: Emphasis on Different Indoor Microenvironments. *Environ. Pollut.* **2019**, *246*, 885–895. [CrossRef] [PubMed]
97. Smith, K. Explorations of Acoustic Trends in Data Logged in K-12 Classrooms; Architetural Engineering and Construction, Durham School. 2019. Available online: https://digitalcommons.unl.edu/cgi/viewcontent.cgi?article=1058&context=archengdiss (accessed on 30 January 2023).
98. Stamp, S.; Burman, E.; Shrubsole, C.; Chatzidiakou, L.; Mumovic, D.; Davies, M. Long-Term, Continuous Air Quality Monitoring in a Cross-Sectional Study of Three UK Non-Domestic Buildings. *Build. Environ.* **2020**, *180*, 107071. [CrossRef]
99. Toftum, J.; Kjeldsen, B.U.; Wargocki, P.; Menå, H.R.; Hansen, E.M.N.; Clausen, G. Association between Classroom Ventilation Mode and Learning Outcome in Danish Schools. *Build. Environ.* **2015**, *92*, 494–503. [CrossRef]
100. Trevisi, R.; Leonardi, F.; Simeoni, C.; Tonnarini, S.; Veschetti, M. Indoor Radon Levels in Schools of South-East Italy. *J. Environ. Radioact.* **2012**, *112*, 160–164. [CrossRef]
101. Haverinen-Shaughnessy, U.; Shaughnessy, R.J.; Cole, E.C.; Toyinbo, O.; Moschandreas, D.J. An Assessment of Indoor Environmental Quality in Schools and Its Association with Health and Performance. *Build. Environ.* **2015**, *93*, 35–40. [CrossRef]
102. Verriele, M.; Schoemaecker, C.; Hanoune, B.; Leclerc, N.; Germain, S.; Gaudion, V.; Locoge, N. The MERMAID Study: Indoor and Outdoor Average Pollutant Concentrations in 10 Low-Energy School Buildings in France. *Indoor Air* **2016**, *26*, 702–713. [CrossRef]
103. Villanueva, F.; Tapia, A.; Lara, S.; Amo-Salas, M. Indoor and Outdoor Air Concentrations of Volatile Organic Compounds and NO2 in Schools of Urban, Industrial and Rural Areas in Central-Southern Spain. *Sci. Total Environ.* **2018**, *622–623*, 222–235. [CrossRef]
104. Vornanen-Winqvist, C.; Salonen, H.; Järvi, K.; Andersson, M.A.; Mikkola, R.; Marik, T.; Kredics, L.; Kurnitski, J. Effects of Ventilation Improvement on Measured and Perceived Indoor Air Quality in a School Building with a Hybrid Ventilation System. *Int. J. Environ. Res. Public Health* **2018**, *15*, 1414. [CrossRef]
105. Vornanen-Winqvist, C.; Järvi, K.; Andersson, M.A.; Duchaine, C.; Létourneau, V.; Kedves, O.; Kredics, L.; Mikkola, R.; Kurnitski, J.; Salonen, H. Exposure to Indoor Air Contaminants in School Buildings with and without Reported Indoor Air Quality Problems. *Environ. Int.* **2020**, *141*, 105781. [CrossRef]
106. Curguz, Z.; Venoso, G.; Zunic, Z.S.; Mirjanic, D.; Ampollini, M.; Carpentieri, C.; Di Carlo, C.; Caprio, M.; Alavantic, D.; Kolarz, P.; et al. Spatial Variability of Indoor Radon Concentration in Schools: Implications on Radon Measurement Protocols. *Radiat. Prot. Dosim.* **2020**, *191*, 133–137. [CrossRef]
107. Zečević, N.; Husika, A.; Džaferović, E. Impact of Energy Efficiency Measures on Indoor Air Quality in Building of Mechanical Engineering Faculty Sarajevo. *Ann. DAAAM Proc. Int. DAAAM Symp.* **2018**, *29*, 0197–0201. [CrossRef]
108. Zhong, L.; Su, F.C.; Batterman, S. Volatile Organic Compounds (VOCs) in Conventional and High Performance School Buildings in the U.S. *Int. J. Environ. Res. Public Health* **2017**, *14*, 100. [CrossRef]
109. Živković, M.; Jovašević-Stojanović, M.; Cvetković, A.; Lazović, I.; Tasić, V.; Stevanović, Ž.; Gržetić, I. Pahs Levels in Gas and Particle-Bound Phase in Schools at Different Locations in Serbia. *Chem. Ind. Chem. Eng. Q.* **2015**, *21*, 159–167. [CrossRef]
110. Fanger, P.O. *Thermal Comfort. Analysis and Applications in Environmental Engineering*; Danish Technical Press: Copenhagen, Denmark, 1970.
111. Basner, M.; Babisch, W.; Davis, A.; Brink, M.; Clark, C.; Janssen, S.; Stansfeld, S. Auditory and Non-Auditory Effects of Noise on Health. *Lancet* **2014**, *383*, 1325–1332. [CrossRef]
112. *ISO 3382:1997*; Acoustics—Measurement of Room Acoustic Parameters—Part 1: Performance Spaces. International Organization for Standardization: Geneva, Switzerland, 1997. Available online: https://www.iso.org/standard/40979.html (accessed on 30 January 2023).
113. Bradley, J.S.; Sato, H. The Intelligibility of Speech in Elementary School Classrooms. *J. Acoust. Soc. Am.* **2008**, *123*, 2078–2086. [CrossRef]
114. Subramaniam, N.; Ramachandraiah, A. Speech Intelligibility Issues in Classroom Acoustics- A Review. *J. Inst. Eng. Archit. Eng. Div.* **2006**, *87*, 29–33.
115. *IEC 60268-16*; Sound System Equipment, Part 16: Objective Rating of Speech Intelligibility by Speech Transmission Index. International Electrotechnical Commission: London, UK, 2011; pp. 1–28.
116. *ISO 16283-3-EN-Façade*; Acoustics — Field measurement of sound insulation in buildings and of building elements. ISO: Geneva, Switzerland, 2016.
117. Ragazzi, M.; Albatici, R.; Schiavon, M.; Ferronato, N.; Torretta, V. CO2 Measurements for Unconventional Management of Indoor Air Quality. *WIT Trans. Ecol. Environ.* **2019**, *236*, 277–286. [CrossRef]
118. Trocquet, C.; Bernhardt, P.; Guglielmino, M.; Malandain, I.; Liaud, C.; Englaro, S.; Le Calvé, S. Near Real-Time Monitoring of Formaldehyde in a Low-Energy School Building. *Atmosphere* **2019**, *10*, 763. [CrossRef]

119. Madureira, J.; Paciência, I.; Rufo, J.; Severo, M.; Ramos, E.; Barros, H.; de Oliveira Fernandes, E. Source Apportionment of CO2, PM10 and VOCs Levels and Health Risk Assessment in Naturally Ventilated Primary Schools in Porto, Portugal. *Build. Environ.* **2016**, *96*, 198–205. [CrossRef]
120. Arlian, L.G.; Neal, J.S.; Morgan, M.S.; Vyszenski-Moher, D.A.L.; Rapp, C.M.; Alexander, A.K. Reducing Relative Humidity Is a Practical Way to Control Dust Mites and Their Allergens in Homes Temperate Climates. *J. Allergy Clin. Immunol.* **2001**, *107*, 99–104. [CrossRef]
121. Lamberti, G.; Salvadori, G.; Leccese, F.; Fantozzi, F.; Bluyssen, P.M. Advancement on Thermal Comfort in Educational Buildings: Current Issues and Way Forward. *Sustainability* **2021**, *13*, 10315. [CrossRef]
122. Shield, B.M.; Dockrell, J.E. The Effects of Noise on Children at School: A Review. *Build. Acoust.* **2003**, *10*, 97–116. [CrossRef]
123. ASHRAE Board of Directors. ASHRAE Position Document on Indoor Carbon Dioxide; 2022; Volume 21. Available online: https://www.ashrae.org/file%20library/about/position%20documents/pd_indoorcarbondioxide_2022.pdf (accessed on 30 January 2023).

buildings

MDPI

Article

Uncertainty Assessment of Mean Radiant Temperature Estimation for Indoor Thermal Comfort Based on Clustering Analysis of Reduced-Input Surfaces

Eunho Kang [1], Ruda Lee [1], Jongho Yoon [1], Heejin Cho [2] and Dongsu Kim [1,*]

1 Department of Architectural Engineering, Hanbat National University, Daejeon 34158, Republic of Korea
2 Department of Mechanical Engineering, Mississippi State University, 210 Carpenter Engineering Building, Starkville, MS 39762, USA
* Correspondence: dongsu.kim@hanbat.ac.kr; Tel.: +82-042-821-1122

Abstract: Mean radiant temperature (MRT) is important for indoor thermal comfort determination. Several good ways to practically obtain accurate MRT include measuring all indoor surface temperatures for MRT calculation or using a black globe thermometer. Still, it can be hard to apply in practice because using such experimental measurements increases the efforts of data management times and acquisition costs. In this regard, there is a practical advantage in reducing the number of measured surfaces by grouping similar surfaces rather than measuring all indoor surface temperatures individually to obtain MRT. However, since even those similar surfaces are not the same, it can lead to erroneous MRT estimation, which needs to be investigated. This study analyzes the uncertainty of MRT estimates by categorizing the surfaces with similar temperature behaviors to examine the risk of such inaccuracy. In this study, the input data required for the MRT calculation are generated using a measurement data-based simulation model, and the uncertainty of the MRT is quantified using the Monte Carlo method. As a result of the study, it is observed that excluding surfaces with similar temperatures for MRT estimation does not significantly affect the uncertainty. When the appropriate number of input surfaces is satisfied, its MRT shows a difference of less than 1% compared to the results calculated with all surfaces.

Keywords: mean radiant temperature; plane radiant temperature; uncertainty analysis; thermal comfort

Citation: Kang, E.; Lee, R.; Yoon, J.; Cho, H.; Kim, D. Uncertainty Assessment of Mean Radiant Temperature Estimation for Indoor Thermal Comfort Based on Clustering Analysis of Reduced-Input Surfaces. *Buildings* **2023**, *13*, 342. https://doi.org/10.3390/buildings13020342

Academic Editors: Paulo Santos and Mark Bomberg

Received: 11 January 2023
Revised: 18 January 2023
Accepted: 19 January 2023
Published: 26 January 2023

1. Introduction

Indoor environment quality can be affected by many related factors, such as heat, acoustic and visual conditions, and indoor air quality. In particular, indoor thermal comfort is a very important factor related to the health of occupants [1]. According to the definition of ASHRAE 55 [2], thermal comfort is achieved by maintaining a thermal balance between the human body and the surrounding environment, so it is determined by the physical parameters of the surrounding environment and the occupant's condition.

Therefore, since various variables determine thermal comfort, indicators such as the predicted mean vote (PMV) are widely used to represent it. PMV represents thermal comfort using two parameters (metabolism, clothing) related to occupant's condition and four parameters (air temperature, mean radiant temperature, air velocity, and humidity) associated with the indoor environment. PMV is widely used to indicate indoor thermal comfort. PMV represents thermal comfort by considering heat transfer between the human body and the surrounding environment using the occupant's state and indoor physical variables. This indicator was developed through experimental research by Fanger [3], and it is introduced as an indicator of thermal comfort in ISO 7730 [4] and ASHRAE 55 [2]. Because estimating all input parameters for PMV calculation requires much effort and resources in a field [5], it is relatively hard to obtain accurate PMV estimates practically.

In particular, the mean radiation temperature (MRT), one of the parameters required for PMV calculation, is a particular variable and a physical parameter that greatly affects PMV significantly. MRT is defined as the surface temperature of a blackbody that radiates from its surrounding surface a radiative heat flux equal to the radiative heat flux incident on that point [4]. Therefore, in a room surrounded by warm surfaces, which is an environment with a high MRT, occupants can feel the warmth even when the ambient air temperature is low. Similarly, if there are cold walls or windows around the occupant, the occupant may feel cold even if the air temperature is within comfortable ranges [6,7]. Because the calculation requires a professional measuring instrument to obtain physical and personal variables [8,9], MRT is often assumed to be simplified, which may increase uncertainty about indoor thermal comfort [10].

Many studies have shown that inaccurate MRT observations can be a source of PMV uncertainty. Ekici (2016) [11] reported that the ambient air temperature and MRT among the input variables of PMV have the most direct influence on the uncertainty of thermal comfort. Chaudhuri et al. (2016) [12] also showed that assumptions considering MRT equal to ambient air temperature could cause large errors in thermal comfort estimation, which means that error propagation of MRT can result in inaccurate indoor comfort controls. Accordingly, although measuring MRT is difficult in an accurate manner, the procedure for observing MRT needs to be considered for PMV calculation because it significantly affects thermal comfort behaviors [13–15]. ISO 7726, an international standard, introduces the method of using the globe temperature and the angle factor method using the ambient surface temperature as a method of observing the mean radiant temperature [13,16].

Measurement of MRT with the globe temperature is a commonly used method [17] because the price of the measuring instrument is relatively low and easy to use. This method can calculate the MRT using three parameters: the globe temperature, air temperature, and air velocity around the globe thermometer. However, the response time (20–30 min) is high, and it may cause inconvenience to occupant activities because it must be installed in all locations where measurement is required [9,15].

Since the angle factor method calculates the MRT by taking the surface temperature and angle factor between the human body and the surrounding surfaces as inputs, the measuring sensor of surface temperatures, such as a contact thermometer or an infrared thermometer, is used. Contact thermometers are inexpensive and relatively handy for measuring temperature. However, since the reliability of the measured value may be reduced due to contact resistance, steady management is required. In the case of an infrared thermometer, there is an advantage in usability because it can measure remotely. The measurement value is greatly affected by the physical properties of the surface and the measurement environment. Therefore, the target surface's historical emissivity and reflectance values are required. Additionally, the user's expertise is necessary to correct the error caused by the measurement environment [18,19].

In the case of the measurement of MRT with a global thermometer, the accuracy is high, but since it provides only the MRT for the measurement location, there is a disadvantage that it may cause inconvenience to occupants' activities due to the installation of the measuring instrument. On the other hand, in the case of the angle factor method, since the physical relationship of the surrounding surfaces is used, it is possible to calculate the MRT of not only a single point but also all points in the room. It is also possible to evaluate the radiation asymmetry using the plane radiant temperature. It has been reported that the MRT derived by the calculation method using the surface temperature is reliable with an error of about 1 °C [8].

Data observation and management costs may increase if the angle factor method is used because most of the surrounding surface temperature must be measured. However, when calculating the room's surface temperature, since the surfaces exist in the same space, a similar pattern is shown except for the surface strongly influenced by external conditions or heating equipment. Therefore, if the MRT is derived by representing similar surfaces with one surface after grouping similar surfaces in advance, the number of surface temperatures, which are input variables for calculation, is reduced, so there may be practical advantages such as data measurement cost reduction.

There have been various previous studies to reduce the resources required for MRT calculation. Vorre et al. (2015) [20] geometrically simplified the human body shape to reduce the amount of computation necessary for MRT calculation. They compared it with the results of complex methods such as ray tracing. Dogan et al. (2021) [21] predicted outdoor MRT with an accuracy of about ±2 °C by clustering surrounding surfaces. However, few studies have simplified indoor MRT prediction methods by grouping indoor surfaces. In addition, no study has explored uncertainty analysis due to the simplification method of MRT calculation. Even though indoor surfaces are on similar boundary conditions, they are not the same, so if some surfaces are considered equally and the average radiant temperature is calculated, an error will inevitably occur. When the error for MRT observation becomes large, thermal comfort may be incorrectly evaluated, so it is necessary to investigate the uncertainty in advance for reliable indoor MRT observation [22,23]. In this study, to quantitatively analyze the reliability of the MRT calculated according to the grouping of similar surface temperatures, the uncertainty of MRT due to the reduction of surface temperature input is investigated through a case study.

2. Methodology

This study investigates the uncertainty of indoor MRT derived according to the indoor surface temperature grouping for a specific case. This study also analyzes the effect of the uncertainty of indoor thermal comfort. The input data for the MRT calculation are generated through a simulation model, and the Monte Carlo method is used to investigate the uncertainty. The overall research flow proceeds, as shown in Figure 1.

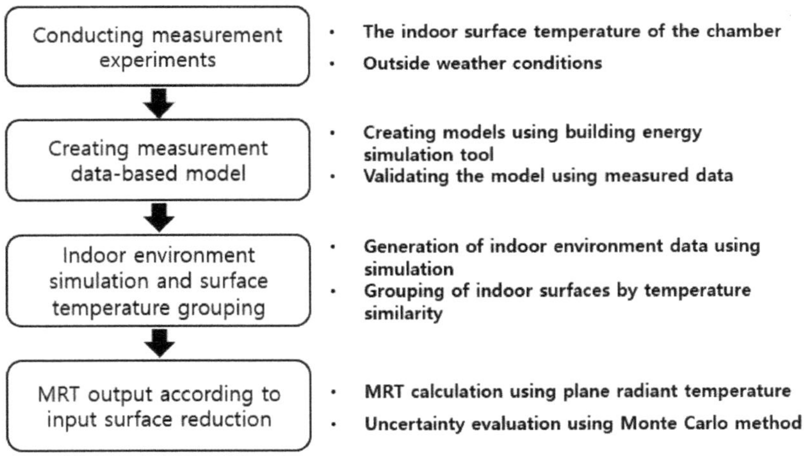

Figure 1. Flowchart of the research.

In steps 1 through 2, measurement experiments are conducted for simulation modeling for data generation. For the measurement data-based modeling, the chamber, in which the surface temperature measurement experiment is performed, is modeled using the building energy simulation program. The model is then validated by comparing the measured data with the model's output data. After the model validation, the second step is carried out. In step 3, as a data generation step, simulated results for MRT calculation are used as outputs, and then grouping is performed for surface temperatures. The main output variables are the indoor surface temperatures and the solar radiation transmitted into the room, simulated under the input conditions of a typical meteorological year. Then, clustering analysis is performed on the simulated surface temperature data to categorize the surfaces, including similar temperature behaviors. The final step is to calculate the MRT according to the input surface temperature reduction and examine the uncertainty of the MRT. The MRT is calculated using the simulation outputs of the preceding steps, and the Monte Carlo method is used because uncertainty must also be taken into account. The Monte Carlo method is a technique used for estimating results, such as uncertainty values. Since it proceeds with sufficiently many iterations with input values randomly extracted from a set of input variables, almost all possible consequences can be derived. At this time, the variable input set is the grouped surface temperature, which is randomly selected from the group. The extracted surface temperature is an input to the MRT calculation by replacing the remaining surfaces in the group. After sufficiently repeating the trial, the results of all trials are aggregated to calculate the uncertainty and statistical indicators.

3. Models and Generation Datasets

3.1. Experimental Setup for Model Validation

The experiment to measure the indoor surface temperature was conducted for 8 days from May 4 to May 12 in a test facility located at Hanbat National University (Republic of Korea, Daejeon, latitude: 36.35, longitude: 127.30). Figure 2 shows a photograph and plan view of the test facility where the measurement experiment is conducted. The interior size of the test facility is 5.5 m × 2.4 m × 2.3 m (length × width × height), and a window (2.0 m × 1.0 m) is located on the south wall at 0.8 m from the floor. The south and roof surfaces are exposed to the outside air, and the remaining surfaces are in contact with the adjacent zone. The roof has a 5-degree slope. The facility envelope is composed of SIP (Structural Insulated Panels) with a thickness of 220 mm, and the thermal transmittance is 0.17 W/m^{2}K. The window on the south side is a fixed window made of triple glazing (6 mm Clear, 6 mm LowE, 6 mm LowE) and the thermal transmittance is 1.44 W/m^{2}K.

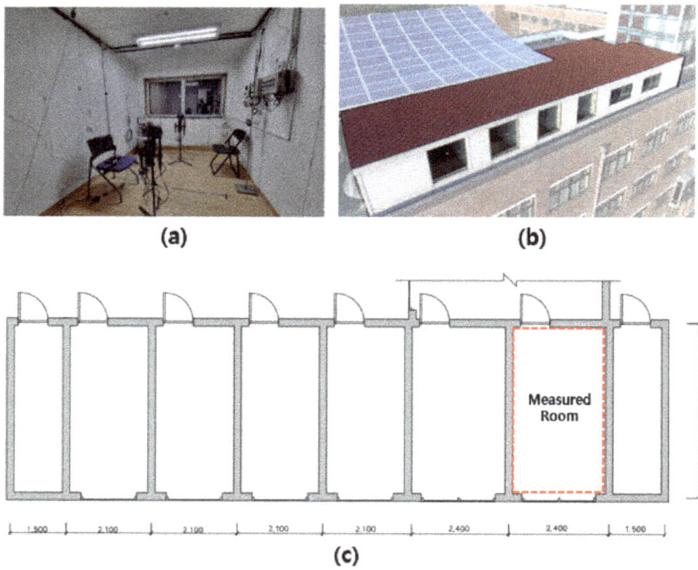

Figure 2. (**a**) Interior space of the chamber; (**b**) exterior view of the chamber; and (**c**) floor plan of the chamber.

To minimize heat transfer with the adjacent zone, the internal wall of the test chamber has the same insulation (220 mm, SIP) as the outer wall. The indoor surface temperature is measured using a thermocouple wire attached to each surface center point. The area-weighted average temperature of the frame, glass, and edge determines the window surface temperature. The temperature of the remaining surface is calculated by averaging the temperatures measured at three points in the center of each surface, and the average surface temperature of each surface is assumed to be constant over the entire area of each surface. The measurement data was logged at 10 min intervals.

External meteorological data (e.g., external temperature and global horizontal irradiance) is also measured with 10 min of time-step. The outside temperature varied from 8.9 °C to 30.5 °C, the global horizontal irradiance was up to 1011 W/m^2, and the outside weather was mostly sunny during the experimental period. During the measurement period, the indoor thermal condition of the chamber was kept without the heating and cooling controls. As a result of the measurement, the indoor air temperature ranged from a minimum of 21.1 °C to a maximum of 28.6 °C due to the change in outdoor temperature and the influence of solar radiation, and the average daily temperature difference was 4.2 °C. In the case of indoor surface temperature, the average daily temperature difference of opaque surfaces such as walls and floors was 3.3 °C, but the average daily temperature difference between windows and doors was 7.1 °C, showing a relatively large range of changes. Figure 3 is a graph showing the measured indoor surface temperature and external weather conditions. The measuring instruments used in the measurement experiment are shown in Table 1.

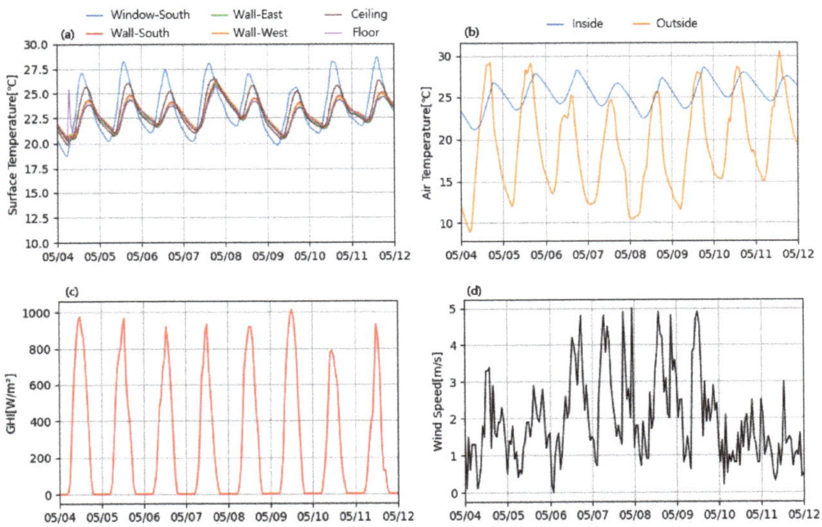

Figure 3. (**a**) Measured inside surface temperature; (**b**) measured internal and external air temperatures; (**c**) measured GHI (global horizontal irradiance); and (**d**) measured wind speed.

Table 1. Measurement equipment specifications used in the measurement experiment.

Variable	Instrument	Measuring Range	Accuracy
Dry Bulb Temperature	Thermocouple (T type)	−250 °C to 350 °C	±1.0 °C
Wind velocity	Hot Wire Anemometer	0 to 25 m/s	±5.0%
Global Solar Radiation	Pyranometer	0 to 2000 W/m^2	±15 W/m^2
Surface Temperature	Thermocouple (T type)	−250 °C to 350 °C	±1.0 °C

A simulation model of the test chamber is developed based on the geometry information from the experiment facility. The simulation model is created using the EnergyPlus [24] program, and its validity is checked by simulating the internal surface temperature under the same weather conditions as the measurement experiment. At this time, direct and diffuse components of solar radiation input to the simulation were determined using the Perez model [25], and parameter identification was performed using MCMC sampling to improve data simulation accuracy [26]. The simulated chamber internal surface temperature and R2 of the measured data showed good similarity to 0.8 or more. Figure 4 is a graph comparing simulation results and measured data for surface temperature.

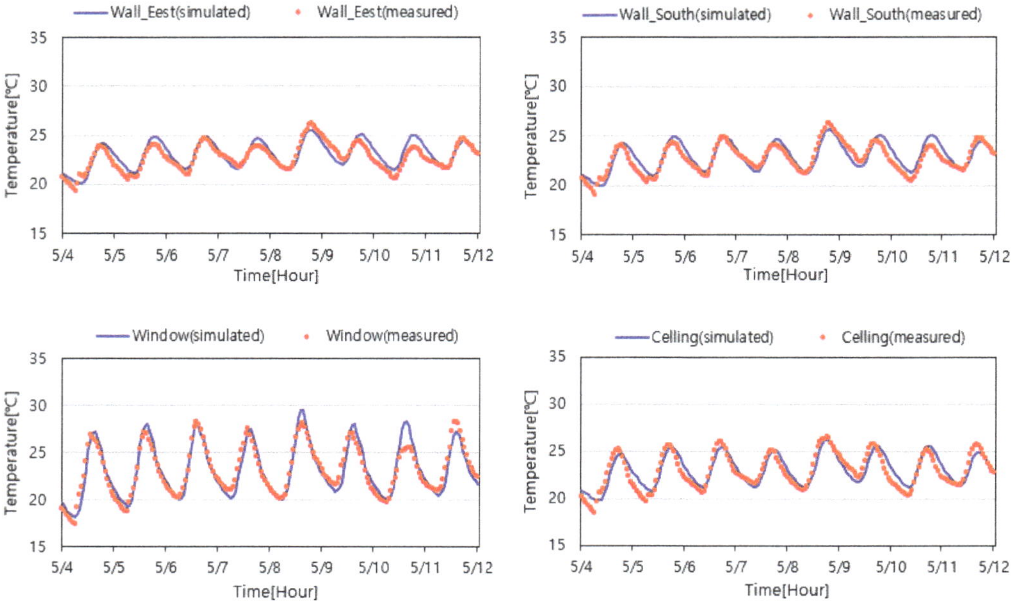

Figure 4. Comparison of measured and simulated surface temperature.

3.2. Generation of the Input Dataset

Input data required for MRT calculation is generated through the simulation model. The selected representative month of the typical meteorological year (TMY) is an input as the meteorological condition of the simulation. The TMY used for this analysis is the standard meteorological data of Daejeon, located in South Korea. The data is IWEC2 type TMY provided by ASHRAE [27]. The month with the lowest outdoor temperature of the TMY is in January, and the month with the highest is in August. In addition, the relative humidity in summer is higher than in other seasons. Accordingly, input data were generated in winter (January) and summer (August) weather conditions. The simulation period was 2 weeks from the 1st to the 14th of each month, and all result values were output at every 1-h step. Assuming that the room is controlled at a constant air temperature, the set temperature for cooling and heating is set at 25 °C for the entire simulation period.

Figures 5 and 6 show the simulation results under the set weather conditions. Figure 5 shows the indoor surface temperature, and Figure 6 shows the direct and diffusion components of irradiance transmitted into the room. The window's surface temperature clearly indicates a large difference compared to other surfaces. Especially in January, it shows a big difference under the influence of irradiance and low outside temperature. The surface temperature of the ceiling in August is higher than that of other surfaces except for windows due to the influence of irradiance during the day. In the case of irradiance transmitted indoors, the diffuse irradiance transmitted indoors was slightly higher in August than in January. In contrast, the direct irradiance transferred indoors was much higher in January than in August due to the influence of the solar's incident angle.

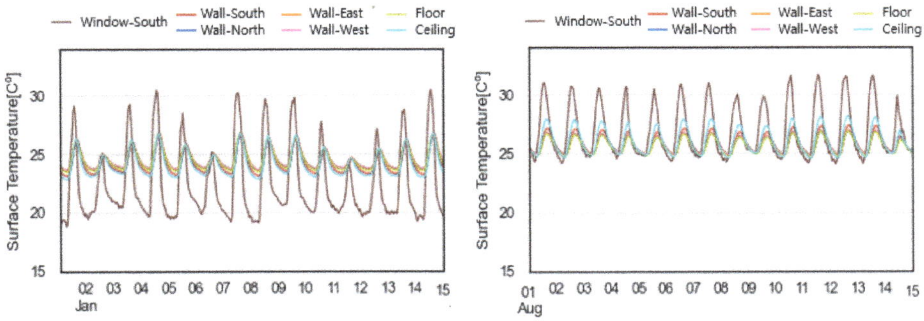

Figure 5. Simulation results: Indoor surface temperature: Winter (**left**), summer (**right**).

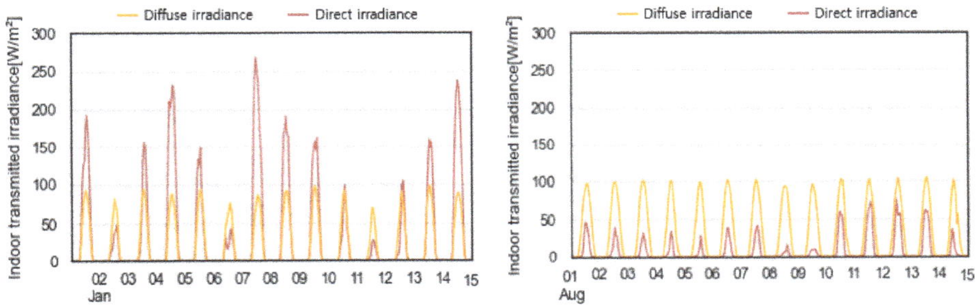

Figure 6. Simulation results: Indoor transmitted irradiance: winter (**left**), summer (**right**).

3.3. Grouping Indoor Surfaces Using Clustering

Clustering analysis is performed on the indoor surface temperature data to group surfaces exhibiting similar temperatures. Clustering is an unsupervised method that defines the similarity between data and clusters each data based on the similarity. Since MRT depends on the indoor conditions at the time, clustering is performed by calculating the Euclidean distance between data observed simultaneously. The k-mean algorithm was used as the cluster analysis algorithm to perform this analysis.

K-means was first proposed by M. Queen (1967) [28] as a clustering method that groups the given data into k number of clusters. The K-means algorithm classifies data based on the assumption that similar data are distributed around the centroid of each cluster. When the number (k) of clusters set by the user is given, a grouping of similar data is performed by continuously updating the centroid so that the distance between each data and the centroid of the cluster is minimized. K-mean clustering was performed according to the procedure below and was performed using Scikit-learn, a Python open-source library [29].

- Step (1) Randomly select the initial centroid of each cluster;
- Step (2) Calculate the similarity between the centroid and each data using Euclidean distance, and assign the data to the nearest cluster;
- Step (3) Update a new centroid for each cluster using the expectation-maximization algorithm;
- Step (4) If the selected center value satisfies the convergence condition, the center point update is stopped, and the finally calculated center value is adopted. If the convergence condition is not met, repeat steps 1–3.

Since clustering results may vary depending on randomly set central values, the expected results were used after performing clustering 30 times. As all 7 surfaces were classified into 6 groups from 2 groups, 5 types of clustering were performed, and the surface temperature for each time was used as the property for grouping the surfaces. The results of clustering are shown in the following Table 2.

Table 2. Clustering results for interior surfaces.

Count of Cluster		2	3	4	5	6
Window	South	2	2	2	2	2
Exterior Wall	South	1	3	4	4	4
Interior Wall	North	1	1	1	1	6
	West	1	1	1	1	1
	East	1	1	1	1	1
Floor		1	1	1	5	5
Ceiling		1	3	3	3	3

As a result of the clustering, when the surfaces are classified into two groups, the first group includes only the window surface, and the second group has all other surfaces. Since the window surface is most affected by the outside air, it is clearly differentiated from different surfaces. When the surfaces are classified into three groups, the surfaces in contact with the outside are organized into a new group. The surfaces classified into this group are the ceiling surface and the south wall surface. Then, as the number of clusters increases, a group, including a single surface, is created.

4. Calculation Method of Output

This section describes the mean radiant temperature and the PMV calculation procedure. To calculate these, the procedure presented in ISO7726 is mainly referred to, and since MRT may be affected by solar radiation transmitted through windows, the calculation method is modified by considering these effects [30]. In addition, these indicators are calculated for points with a distance of 1.0 m (A), 2.0 m (B), and 3.0 m (C) from the window to examine the trend according to the indoor location. The height of all points is 0.6 m, which is the height of the occupant's sitting position [15]. Figure 7 is a schematic diagram of the location of the target points.

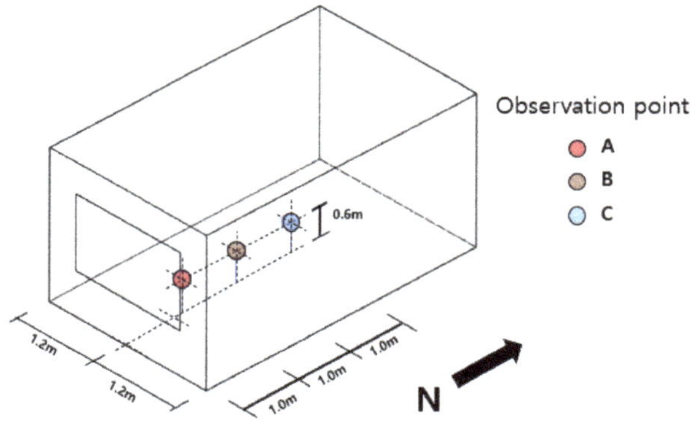

Figure 7. Defined calculation points in the room.

4.1. The Mean Radiant Temperature Algorithm

Mean radiant temperature (MRT) T_r is calculated using the plane radiant temperature presented in the international standard ISO 7726. According to the standard, the MRT is calculated using the plane radiant temperature for each surface and the human body projected area factor for the surface direction, assuming that the occupant is a regular hexahedron. Therefore, the MRT is derived by Equation (1) using the plane radiant temperatures for the six infinitesimal surfaces of the target point.

$$T_r = c_z\left(T_{pr,Top} + T_{pr,Bottom}\right) + c_y\left(T_{pr,Right} + T_{pr,Left}\right) + c_x\left(T_{pr,Front} + T_{pr,Back}\right) \quad (1)$$

where c_z, c_y, c_x are the projected area factors in the *z*, *y*, and *x* directions on the human body surface, and $T_{pr,D}$ is the plane radiation temperature in the *D*-direction. The projected area factor was entered as the value of the sitting position [15].

For an environment where only long-wave radiation from surrounding surfaces is considered, and other radiation effects are small enough to be negligible, the plane radiation temperature can be calculated as Equation (2) below. However, in a location close to a surface that can transmit solar radiation, such as a window, the plane radiation temperature may rise due to the radiation transmitted into the room. Since Equation (2) does not include solar radiation transmitted into the room, the MRT may be underestimated during the daytime when solar radiation is present. Several results have been reported in the literature that the MRT can be increased by more than 20 °C under the influence of transmitted solar radiation and can cause radiation asymmetry [30].

$$T_{pr,D} = \sqrt[4]{\sum_{i=1}^{N} F_{D\to i} T_{surf,i}^4} \tag{2}$$

where $T_{surf,i}$ means the *i*-th indoor surface temperature, and $F_{d\to i}$ means the view factor between the *D*-direction surface of the target point and the surrounding *i*-th surface.

Since the indoor environment modeled in this study can be affected by irradiance, the plane radiation temperature is calculated by Equation (3) considering the direct and diffuse irradiance transmitted into the room [31].

$$T_{pr,D} = \sqrt[4]{\frac{\alpha_S}{\varepsilon_S}\sum_{i=1}^{N} F_{D\to surf,i} T_{surf,i}^4 + \frac{\alpha_d}{\sigma\varepsilon_S}\sum_{j=1}^{M} F_{D\to surf,j} I_d + C_S\frac{\alpha_b}{\sigma\varepsilon_S} I_b cos\theta_b} \tag{3}$$

where T_{surf} is the surrounding surface temperature, ε_S is the emissivity of the target surface, and σ is the Stefan-Boltzmann constant ($\sigma = 5.67 \times 10^{-8}\ W/m^2K^4$). I_d and α_d indicate the radiation intensity and absorption coefficient of the diffuse solar radiation transmitted into the room. And I_b, α_b, and θ_b mean the radiation intensity, absorption coefficient, and incident angle for the direct solar radiation transmitted into the room, respectively. C_S is a shading coefficient for direct solar radiation.

The absorption coefficient α_s and the emissivity ε_S of the target surface are set to 1 from the black body assumption, but the absorption coefficient α_b and α_d for irradiation can vary depending on the type of clothing worn as well as the skin color of the occupants. These parameters are assumed to be 0.67 for average clothing and skin color (white) [32].

The solar radiation incident through the window can be partially blocked due to the shading effect of the envelope. Therefore, it is necessary to verify whether the target point is affected by the shading effect. This is checked hourly because it changes with the sun's position. The shading effect of the envelope can partially block the solar radiation incident through the window. So it needs to be verified that the target point is affected by the shading effect and this is checked every timestep as it depends on the sun's position.

The area where the direct sunlight Is incident was calculated to confirm the shading effect [33]. This area is determined by the coordinates A, B, C and D points shown in Figure 8. These coordinates can be determined using the su"s position and the geometric parameters of the envelope. (Equations (4)–(11)) When the target point is located within the calculated area, the shading coefficient C_S is 1, otherwise, it is 0.

$$x_A = x_W - \left(\frac{z_W - z_b}{tan\beta}cos\gamma + \frac{s}{2}\right)tan\gamma + |\frac{s}{2}tan\gamma| \tag{4}$$

$$y_A = \frac{z_W - z_b}{tan\beta}cos\gamma \tag{5}$$

$$x_B = x_W - \left(\frac{H_W + z_W - z_b}{tan\beta}cos\gamma - \frac{s}{2}\right)tan\gamma + |\frac{s}{2}tan\gamma| \tag{6}$$

$$y_B = \frac{H_W + z_W - z_b}{tan\beta} cos\gamma - s \quad (7)$$

$$x_C = x_W + L_W + \left(\frac{H_W + z_W - z_b}{tan\beta} cos\gamma - \frac{s}{2}\right) tan\gamma - |\frac{s}{2} tan\gamma| \quad (8)$$

$$y_C = \frac{H_W + z_W - z_b}{tan\beta} cos\gamma - s \quad (9)$$

$$x_D = x_W + L_W + \left(\frac{z_W - z_b}{tan\beta} cos\gamma + \frac{s}{2}\right) tan\gamma - |\frac{s}{2} tan\gamma| \quad (10)$$

$$y_D = \frac{z_W - z_b}{tan\beta} cos\gamma \quad (11)$$

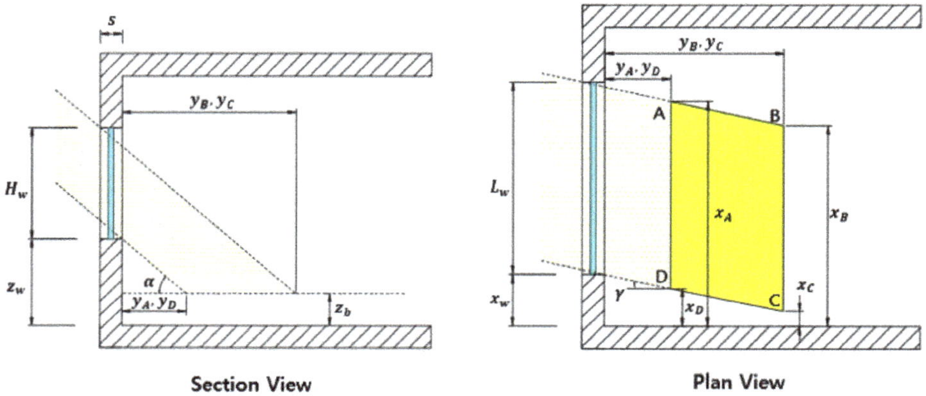

Figure 8. Definition of the geometric parameters involved in the delimitation of the irradiated zone modified with permission from Ref. [33].

The view factor can be calculated by Equation (12). Still, since it is often difficult to apply in actual geometric boundary conditions, the view factor is calculated numerically after being discretized as in Equation (13).

$$F_{d \to j} = \frac{1}{A_i} \iint_{A_i A_d} \frac{cos\theta_i cos\theta_d}{\pi R^2} dA_d dA_i \quad (12)$$

$$F_{d \to j} \approx \frac{1}{A_i} \sum \sum \frac{cos\theta_i cos\theta_d}{\pi R^2} dA_d dA_i \quad (13)$$

where R is the distance between dA_d and dA_i, and θ is the angle between the normal vector of dA_i and the vector connecting the centers of dA_d and dA_i.

4.2. Predicted Mean Vote (PMV)

Predicted Mean Vote (PMV) is used as the thermal comfort index used in this study. PMV represents people's average thermal sensation with the same activity and clothing for a given environment. To comply with ASHRAE 55 [2], the recommended thermal limit on PMV is between −0.5 and 0.5. In the case of ISO 7730 [4], the recommended PMV is between −0.7 and 0.7 for existing buildings, and new buildings range between −0.5 and +0.5. PMV is widely used to evaluate the thermal comfort level of indoor environments such as residential buildings, offices, and hospitals.

According to the procedure presented in ISO-7730 [4], PMV is calculated using 5 physical parameters (i.e., air temperature, relative humidity, air velocity, mean radiant temperature, water vapor pressure) and 2 occupant's parameters (i.e., clo and metabolic rate). Indoor physical parameters, such as relative humidity, air velocity, and air temperature, are assumed to be uniform throughout the indoor space. The values simulated by EnergyPlus are used.

As shown in Table 3, For the clo value, different values are applied depending on the winter and summer seasons, and the occupant's activity (metabolic rate) is assumed to be a sedentary activity (office, dwelling, school, laboratory). The parameters dependent on the occupant are assumed to be constant during the simulation period.

Table 3. Input parameters for occupant's state.

Parameter Description [Unit]	Winter	Summer
Metabolic rate [W/m^2]	70	70
External work [W/m^2]	0	0
Clo [-]	1.0	0.5

4.3. *Uncertainty Analysis Method*

In this study, the Monte Carlo method (MCM), a sampling-based method, was used to derive results, including uncertainty. MCM is a widely used uncertainty analysis method in the building energy field, such as GUM (Guide to the expression of uncertainty in measurement), TSM (Taylor Series Method) [34–36]. According to the literature, researchers obtained reliable uncertainty results and objective estimates of mean and variance for thermal comfort using MCM [37,38]. The following process is carried out to derive the MRT uncertainty using MCM.

Considering each clustered surface group G_k as a sample space, one surface is randomly sampled from each group. In this step, all surfaces in the group have the same probability of being sampled. (Equation (14))

$$T_{surf,k} \sim U_k(G_k), G_k = \left\{T_{surf,a}, T_{surf,b}, \ldots\right\} \tag{14}$$

The temperature of the sampled surface is input to the MRT calculation model f_{MRT}, and the surface is representative of all surfaces in the sampled group. For each sampled surface temperature, 0.5 °C of Gaussian noise is added to reflect the uncertainty of the surface temperature sensor. (Equation (15))

$$T_{r,k} = f_{MRT}(N\left(T_{surf,1}, 0.5\right), \ldots, N\left(T_{surf,k}, 0.5\right)) \tag{15}$$

Therefore, the output is calculated from the MRT model using the temperature sampled on the surface group, as shown in Figure 9. Repeating this process, an MRT sample is collected, and when the number of iterations achieves a set value, an uncertainty indicator is derived from the data generated in all trials.

The uncertainty indicator is expressed as an expanded uncertainty with 95% confidence interval based on the GUM method. When there are sufficient observed samples, and the systematic error is zero, an expanded uncertainty of 95% confidence level $U_{95,k}$ can be obtained by multiplying the standard deviation by two. (Equation (16)) [35] Using MCM, standard deviations are calculated for every time step, and finally these indicators are expressed as expected values for the set simulation period.

$$U_{95,k} = 2\overline{s_k} \tag{16}$$

$$s_k = \sqrt{\frac{\sum_{i=1}^{n}\left(T_{mrt,k} - \overline{T}_{mrt,k}\right)^2}{n}} \tag{17}$$

where k means the number of surfaces used for MRT calculation, and n is the number of MRT data for each time step (n = 100).

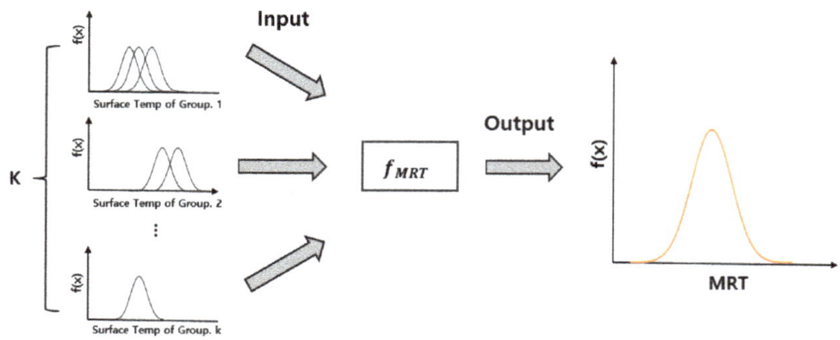

Figure 9. Schematic view of Monte Carlo simulation to estimate the uncertainty of MRT.

5. Results and Discussion

5.1. Uncertainty Analysis of Mean Radiant Temperature

Based on MCM, the uncertainty distribution of MRT was derived. Figure 10 shows a histogram of the MRT uncertainty over the entire simulation period. Except for the MRT calculated with two surfaces, the uncertainty distribution of the other MRTs appeared similar to the normal distribution. The mean and standard deviation of each distribution are shown in Table 4. The uncertainty of the MRT calculated with only two surfaces shows the largest value with an average of about 0.62 °C and a maximum of about 1.1 °C. The uncertainty of MRT calculated with 5 or 6 surfaces is relatively small, with an average of about 0.48 °C and a maximum of 0.59 °C.

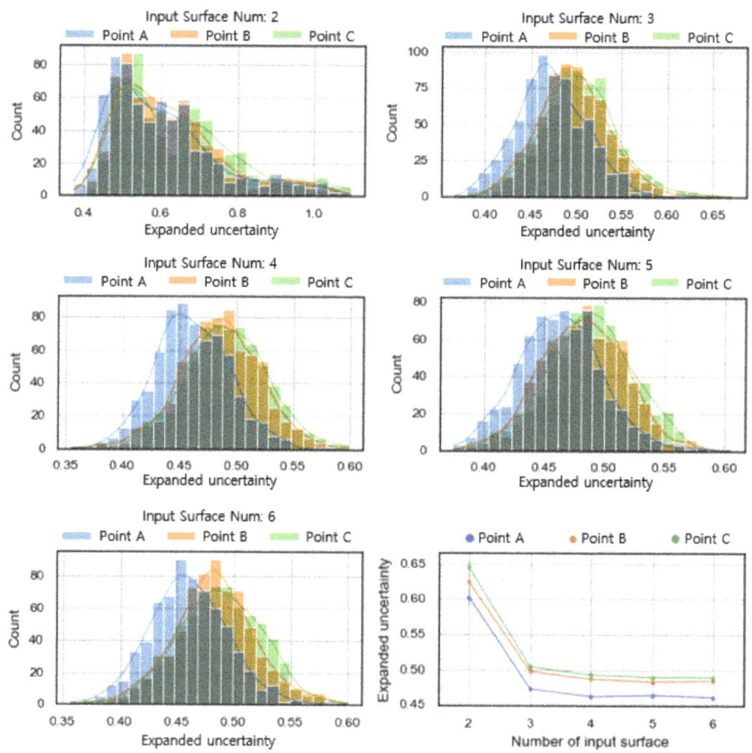

Figure 10. Histogram of expanded uncertainty of MRT calculation results.

Table 4. Summary of the expanded uncertainty of 95% confidence level.

Number of Inputs Surface	Point	Mean	Max	Standard Deviation	Coefficient of Variation
2	A	0.603	1.082	0.144	23.8%
	B	0.623	1.080	0.140	22.4%
	C	0.644	1.097	0.151	23.5%
3	A	0.473	0.627	0.039	8.2%
	B	0.497	0.641	0.039	7.8%
	C	0.504	0.672	0.042	8.3%
4	A	0.461	0.571	0.032	7.0%
	B	0.486	0.600	0.035	7.2%
	C	0.492	0.595	0.035	7.1%
5	A	0.463	0.581	0.034	7.3%
	B	0.483	0.605	0.034	7.1%
	C	0.489	0.586	0.034	7.1%
6	A	0.460	0.587	0.034	7.4%
	B	0.484	0.600	0.035	7.3%
	C	0.489	0.600	0.036	7.4%

The uncertainty tends to decrease as the number of surfaces used in the calculation increases. In particular, when increasing the input surface temperature from 2 to 3, the uncertainty reduction rate was 20% to 22%, showing a great improvement. When the number of input surfaces is increased from 2 to 3, the surfaces facing the outside air (ceiling, south wall) are classified into a new group, and MRT is calculated. In this case, it is judged that individually considering the surfaces facing the outside significantly influences the improvement of the uncertainty.

The uncertainty of $T_{r,4}$ was on average about 2.5% smaller than the uncertainty of $T_{r,3}$. In the case of more than 5 surfaces, the uncertainty change by increasing the input surface was small, within $\pm$1%. The newly added surfaces in $T_{r,5}$ and $T_{r,6}$ are surfaces classified from the surface group facing the adjacent zone. This indicates that the uncertainty change with the addition of the input surface is negligible because the surfaces facing the adjacent zone have similar temperatures.

Comparing the expected uncertainty values at each observation point, the smallest value is shown at point A, which is close to the window, and the largest at point C, which is close to the center of the room. (Figure 10) Uncertainty deviation, according to the observation point, varies from 3% to 10% depending on the number of input surface temperatures. The uncertainty increases as the distance from the window increases. This shows that the uncertainty propagation of the surface temperature can appear variously depending on the target location.

In addition, increasing the distance of the target location from the surfaces considered as individual surfaces (window and south wall) means that the influence of the grouped surfaces increases, which causes additional uncertainty. To investigate the effect of each surface, sensitivity analysis is performed for each surface according to the observation position. For sensitivity analysis, the Sobol method, which can quantitatively express the influence of input variables, is used using SALib, a Python library [39,40]. The Sobol method is a sensitivity analysis method based on variance decomposition. The sensitivity index is defined by dividing the expected value of the variance of the output variable for the input variable by the total variance of the output variable. (Equation (18)).

$$S_i = \frac{E(V(Y|X_{\sim i}))}{V(Y)} \quad (18)$$

where, S is the sensitivity index, X is the input variable, and Y is the model's output.

Figure 11 shows the sensitivity index of each surface to the MRT according to the distance from the south wall. As the distance increases, the sensitivity of the windows and the south wall decreases, and the sensitivity of other surfaces increases. It means that as the target point gets closer from A to C, the weight of the grouped surface increases so that the MRT uncertainty may increase. The increase was up to 0.1 °C in all cases.

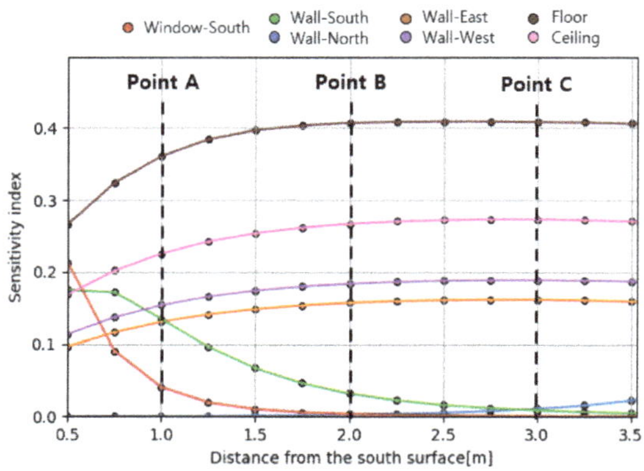

Figure 11. Results of sensitivity analysis.

The uncertainty of MRT according to the number of input surfaces is classified into the winter and summer periods and investigated. Figure 12 shows the expanded uncertainty for winter (left) and summer (right) over 24 h a day. It shows that the time-dependent change of the summer uncertainty and the winter uncertainty is different in $T_{r,2}$. $T_{r,2}$ is MRT calculated with two surface temperatures, and the input surface temperature is randomly selected from a group including only the window surface and a group including all other surfaces.

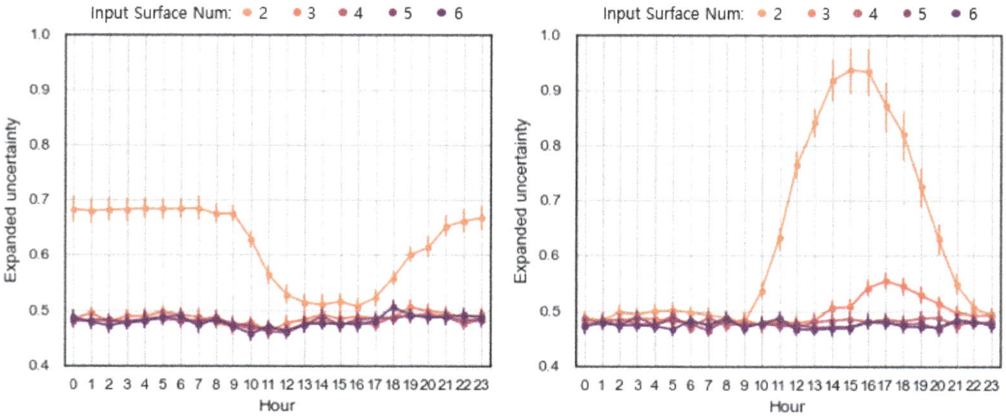

Figure 12. Expanded uncertainty of MRT over 24 h: Winter (**left**), summer (**right**).

In the winter period, the temperature difference between the surfaces in the group increases because the temperature of the surface in contact with the outdoor air is relatively low at night. Therefore, since there are cases where the surface sampled from the group is a surface in contact with the outside air, the MRT may be underestimated, which is a major cause of its uncertainty increase. On the other hand, the uncertainty is higher during the day than at night for the summer period. During the daytime, the temperature difference between the surface in contact with the outside and the rest of the surface increases due to the influence of solar radiation incidents on the envelope. This is a major factor in reducing the temperature similarity of surfaces within a group and can increase the uncertainty of MRT.

In the case of $T_{r,(k\leq3)}$, the uncertainty shows a different trend depending on the weather conditions. Figure 13 shows the MRT uncertainty in winter (left) and summer (right) as boxplots according to the number of input surface temperatures. The deviation of the expected value is less than 0.1 °C, showing a slight difference, but the standard deviation of the uncertainty distribution in $T_{r,2}$ is relatively large in summer. The uncertainty of $T_{r,2}$ shows relatively greater fluctuations in summer than in winter. In the case of $T_{r,(k\leq3)}$, the uncertainty expected value deviation between the two periods is less than 0.1 °C, showing little difference in uncertainty depending on the season.

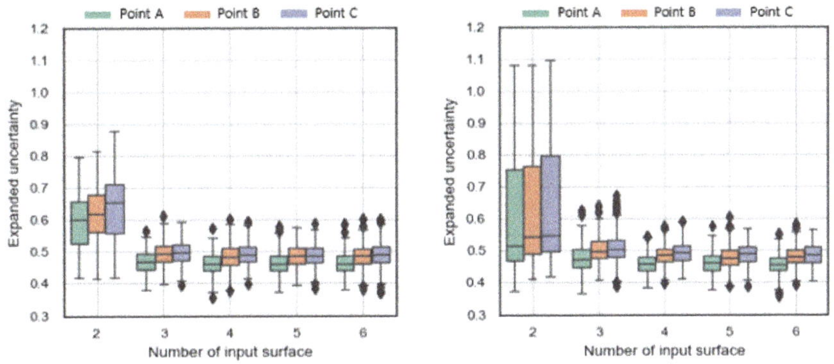

Figure 13. Expanded uncertainty: winter (**left**), summer (**right**).

5.2. Propagation of Uncertainty to Thermal Comfort

The PMV uncertainty is derived using the MRT calculated by the MCM. Figure 14 shows the PMV uncertainty distribution for each season according to the number of input surface temperatures. The uncertainty of PMV shows a similar trend to that of MRT shown in the previous section due to its linear relationship with MRT.

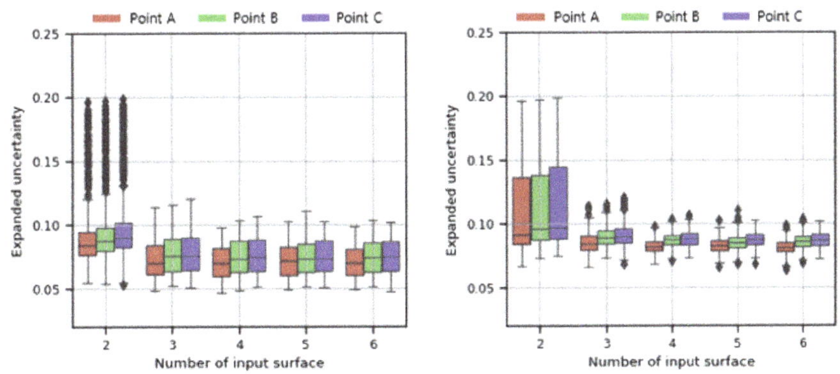

Figure 14. Expanded uncertainty of PMV: winter (**left**), summer (**right**).

As the number of input surfaces increased, the expected value converged to about 0.062 for PMV uncertainty in winter and about 0.084 for PMV uncertainty in summer. In both seasons, the expected value and standard deviation of PMV uncertainty are high at $T_{r,2}$, but the PMV uncertainties calculated from $T_{r,(k>3)}$ show similar values without large deviations. Additionally, the range of expected PMV uncertainty at $T_{r,(k>3)}$ is 0.060–0.088, which are not exceeding 0.1 for most of the observation period.

The uncertainty propagation of PMV shows different characteristics in summer and winter. The expected value of PMV uncertainty was 27–30% higher in summer than in winter. The main reason for the difference was the propagation of uncertainty according to the input clo, since PMV reflects the heat transfer between the occupant and the surrounding environment according to the clo value. As a result, the lower the clo value, the greater the effect of radiation and convection by the surrounding environment, so the uncertainty propagation of MRT becomes relatively large under the condition of wearing light clothing. Figure 15 is a graph showing the uncertainty of PMV propagated by MRT uncertainty by clo value. The PMV distributions are calculated using Monte Carlo simulations for six clothing conditions. It shows that the uncertainty of PMV decreases as the clo value increases.

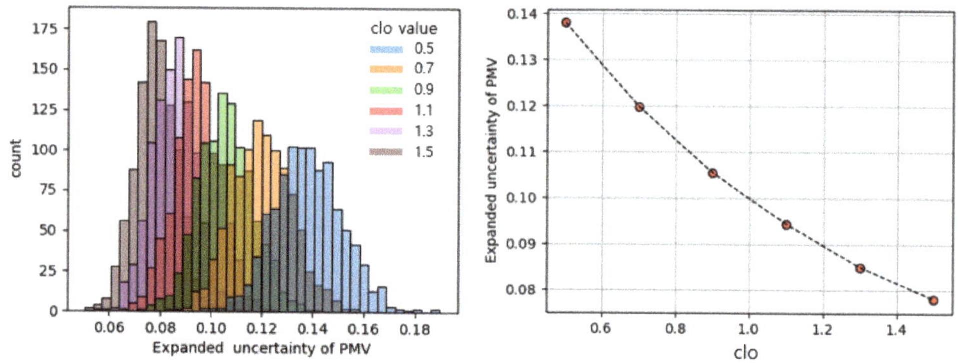

Figure 15. Propagation of uncertainty in PMV by clo value.

6. Conclusions

In this study, MRT was calculated using grouped surface temperatures to investigate the possibility of data monitoring cost savings for MRT observation, and uncertainty due to input surface temperature reduction was analyzed using MCM. The results are summarized as follows:

1. As the number of input surface temperatures used for MRT calculation decreased, the uncertainty increased up to 0.64 °C. This uncertainty change showed a significant difference between $T_{r,2}$ and $T_{r,3}$, and in $T_{r,(k>4)}$, the uncertainty change due to the addition of the input surface was less than 1%. In the case of $T_{r,(k>4)}$, since the surfaces not in contact with the outside show similar temperatures, it was shown that measuring these surfaces individually does not significantly affect the results;
2. When examining the MRT uncertainty according to the observation point, the expected value of the uncertainty increased at the point farther from the southern surface. Since the parameters of the MRT model (view factors for the surrounding surfaces) can vary depending on the observation point, the uncertainty propagated at each surface can also vary. In this case, the uncertainty increased by up to 0.1 °C as the observation point approached the grouped surface;
3. In $T_{r,(k\leq3)}$, when only the surface in contact with the outside was selected, the MRT uncertainty tended to fluctuate according to the weather conditions. For stable MRT observation, it is necessary to appropriately select a surface that can be representative of the other surfaces in the group;
4. As a result of investigating the effect of MRT uncertainty on PMV uncertainty, the expected value of PMV uncertainty was about 30% higher in summer than in winter. This is a result of the input clothing value difference, and the lighter the occupant's clothing, the greater the MRT's effect on PMV.

As the results indicate, it was found that excluding surfaces with similar temperatures for MRT estimation did not significantly affect the uncertainty. Observation of MRT using only representative surfaces through preliminary research shows the possibility of effectively reducing the cost of data measurement. As to limitations, since this study was conducted under limited indoor conditions, the number of surface temperatures required, and the uncertainty may vary depending on the characteristics of the applied zone. Hence, more geometry and zone types with different indoor conditions should be investigated to validate our results further.

Author Contributions: Conceptualization, D.K.; methodology, H.C.; software, E.K.; validation, R.L.; formal analysis, H.C.; investigation, E.K.; resources J.Y.; data curation, R.L.; writing—original draft preparation, E.K.; writing—review and editing, D.K.; visualization, R.L.; supervision, D.K.; project administration, J.Y.; funding acquisition, J.Y. All authors have read and agreed to the published version of the manuscript.

Funding: This work is supported by the Korea Agency for Infrastructure Technology Advancement (KAIA) grant funded by the Ministry of Land, Infrastructure and Transport (Grant 21CTAP-C163698-01).

Data Availability Statement: The data presented in this study are available on request from the corresponding author.

Acknowledgments: The authors wish to acknowledge supports from the Korea Agency for Infrastructure Technology Advancement (21CTAP-C163698-01).

Conflicts of Interest: The authors declare no conflict of interest.

References

1. Jiang, Y.; Wang, Z.; Lin, B.; Mumovic, D. Development of a Health Data-Driven Model for a Thermal Comfort Study. *Build. Environ.* **2020**, *177*, 106874. [CrossRef]
2. *ASHRAE Standard 55*; Thermal Environmental Conditions for Human Occupancy. The American Society of Heating, Refrigerating and Air-Conditioning Engineers, Inc.: Atlanta, GA, USA, 2013.
3. Fanger, P.O. *Thermal Comfort: Analysis and Applications in Environmental Engineering*; Mcgraw-Hill: New York, NY, USA, 1970.
4. *ISO 7730:2005 Ergonomics of the Thermal Environment—Analytical Determination and Interpretation of Thermal Comfort Using Calculation of the PMV and PPD Indices and Local Thermal Comfort Criteria*, 3rd ed.; International Organization for Standardization: Geneva, Switzerland, 2005.
5. Kim, J.; Schiavon, S.; Brager, G. Personal Comfort Models–A New Paradigm in Thermal Comfort for Occupant-Centric Environmental Control. *Build. Environ.* **2018**, *132*, 114–124. [CrossRef]
6. Fanger, P.O. Local Discomfort to the Human Body Caused by Non-Uniform Thermal Environments. *Ann. Occup. Hyg.* **1977**, *20*, 285–291. [CrossRef] [PubMed]
7. McNall, P.E.; Biddison, R.E. Thermal and Comfort Sensations of Sedentary Persons Exposed to Asymmetric Radiant Fields. *ASHRAE Trans.* **1970**, *76*, 123–136.
8. Özbey, M.F.; Turhan, C. A Comprehensive Comparison and Accuracy of Different Methods to Obtain Mean Radiant Temperature in Indoor Environment. *Therm. Sci. Eng. Prog.* **2022**, *31*, 101295. [CrossRef]
9. Alfano, F.R.D.A.; Dell'Isola, M.; Palella, B.I.; Riccio, G.; Russi, A. On the Measurement of the Mean Radiant Temperature and Its Influence on the Indoor Thermal Environment Assessment. *Build. Environ.* **2013**, *63*, 79–88. [CrossRef]
10. Guo, H.; Ferrara, M.; Coleman, J.; Loyola, M.; Meggers, F. Simulation and Measurement of Air Temperatures and Mean Radiant Temperatures in a Radiantly Heated Indoor Space. *Energy* **2020**, *193*, 116369. [CrossRef]
11. Ekici, C. Measurement Uncertainty Budget of the PMV Thermal Comfort Equation. *Int. J. Thermophys.* **2016**, *37*, 48. [CrossRef]
12. Chaudhuri, T.; Soh, Y.C.; Bose, S.; Xie, L.; Li, H. On Assuming Mean Radiant Temperature Equal to Air Temperature during PMV-Based Thermal Comfort Study in Air-Conditioned Buildings. In Proceedings of the IECON 2016-42nd Annual Conference of the IEEE Industrial Electronics Society, Florence, Italy, 23–26 October 2016; pp. 7065–7070.
13. Halawa, E.; van Hoof, J.; Soebarto, V. The Impacts of the Thermal Radiation Field on Thermal Comfort, Energy Consumption and Control—A Critical Overview. *Renew. Sustain. Energy Rev.* **2014**, *37*, 907–918. [CrossRef]
14. Wang, D.; Chen, G.; Song, C.; Liu, Y.; He, W.; Zeng, T.; Liu, J. Experimental Study on Coupling Effect of Indoor Air Temperature and Radiant Temperature on Human Thermal Comfort in Non-Uniform Thermal Environment. *Build. Environ.* **2019**, *165*, 106387. [CrossRef]
15. Atmaca, İ.; Kaynaklı, Ö.; Yiğit, A. Effects of Radiant Temperature on Thermal Comfort. *Build. Environ.* **2007**, *42*, 3210–3220. [CrossRef]
16. *Standard ISO 7726*; Ergonomics of the Thermal Environment-Instruments for Measuring Physical Quantities. International Organization for Standardization: Geneva, Switzerland, 1998.
17. Da Silva, M.G.; Santana, M.M.; e Sousa, J.A. Uncertainty Analysis of the Mean Radiant Temperature Measurement Based on Globe Temperature Probes. *J. Phys. Conf. Ser.* **2018**, *1065*, 072036. [CrossRef]

18. Lee, D.-S.; Kim, E.-J.; Cho, Y.-H.; Kang, J.-W.; Jo, J.-H. A Field Study on Application of Infrared Thermography for Estimating Mean Radiant Temperatures in Large Stadiums. *Energy Build.* **2019**, *202*, 109360. [CrossRef]
19. ISO. *ISO 9869-2: 2018—Thermal Insulation—Building Elements—In-Situ Measurement of Thermal Resistance and Thermal Transmittance—Part 2: Infrared* Method for Frame Structure Dwelling; International Organization for Standardization: Geneva, Switzerland, 2018.
20. Vorre, M.H.; Jensen, R.L.; le Dréau, J. Radiation Exchange between Persons and Surfaces for Building Energy Simulations. *Energy Build.* **2015**, *101*, 110–121. [CrossRef]
21. Dogan, T.; Kastner, P.; Mermelstein, R. Surfer: A Fast Simulation Algorithm to Predict Surface Temperatures and Mean Radiant Temperatures in Large Urban Models. *Build. Environ.* **2021**, *196*, 107762. [CrossRef]
22. Alfano, F.R.D.A.; Palella, B.I.; Riccio, G. The Role of Measurement Accuracy on the Thermal Environment Assessment by Means of PMV Index. *Build. Environ.* **2011**, *46*, 1361–1369. [CrossRef]
23. Moutela, R.; Carrilho, J.D.; da Silva, M.G. Sensitivity of the PMV Index to the Thermal Comfort Parameters. In Proceedings of the 2nd Energy for Sustainability Multidisciplinary Conference, Coimbra, Portugal, 14–15 May 2015.
24. Crawley, D.B.; Lawrie, L.K.; Pedersen, C.O.; Winkelmann, F.C. Energy plus: Energy Simulation Program. *ASHRAE J.* **2000**, *42*, 49–56.
25. Loutzenhiser, P.G.; Manz, H.; Felsmann, C.; Strachan, P.A.; Frank, T.H.; Maxwell, G.M. Empirical Validation of Models to Compute Solar Irradiance on Inclined Surfaces for Building Energy Simulation. *Sol. Energy* **2007**, *81*, 254–267. [CrossRef]
26. Chong, A.; Menberg, K. Guidelines for the Bayesian Calibration of Building Energy Models. *Energy Build.* **2018**, *174*, 527–547. [CrossRef]
27. IWEC2 Weather Files. Available online: https://www.Ashrae.Org/Technical-Resources/Bookstore/Ashrae-International-Weather-Files-for-Energy-Calculations-2-0-Iwec2 (accessed on 10 January 2023).
28. McQueen, J.B. Some Methods of Classification and Analysis of Multivariate Observations. In Proceedings of the 5th Berkeley Symposium on Mathematical Statistics and Probability, Berkeley, CA, USA, 21 June–18 July 1965 and 27 December 1965–7 January 1966; pp. 281–297.
29. Pedregosa, F.; Varoquaux, G.; Gramfort, A.; Michel, V.; Thirion, B.; Grisel, O.; Blondel, M.; Prettenhofer, P.; Weiss, R.; Dubourg, V. Scikit-Learn: Machine Learning in Python. *J. Mach. Learn. Res.* **2011**, *12*, 2825–2830.
30. Marino, C.; Nucara, A.; Pietrafesa, M. Thermal Comfort in Indoor Environment: Effect of the Solar Radiation on the Radiant Temperature Asymmetry. *Sol. Energy* **2017**, *144*, 295–309. [CrossRef]
31. La Gennusa, M.; Nucara, A.; Rizzo, G.; Scaccianoce, G. The Calculation of the Mean Radiant Temperature of a Subject Exposed to the Solar Radiation—A Generalised Algorithm. *Build. Environ.* **2005**, *40*, 367–375. [CrossRef]
32. Blum, H.F. The Solar Heat Load: Its Relationship to Total Heat Load and Its Relative Importance in the Design of Clothing. *J. Clin. Investig.* **1945**, *24*, 712–721. [CrossRef] [PubMed]
33. La Gennusa, M.; Nucara, A.; Pietrafesa, M.; Rizzo, G. A Model for Managing and Evaluating Solar Radiation for Indoor Thermal Comfort. *Sol. Energy* **2007**, *81*, 594–606. [CrossRef]
34. Joint Committee for Guides in Metrology (JCGM), Supplement 1 to the 'Guide to the Expression of Uncertainty in Measurement'—Propagation of Distributions Using a Monte Carlo Method JCGM 101:2008. 2008. Available online: http://www.bipm.org/utils/common/documents/jcgm/JCGM_100_2008_E.pdf (accessed on 10 January 2023).
35. Tian, W.; Heo, Y.; de Wilde, P.; Li, Z.; Yan, D.; Park, C.S.; Feng, X.; Augenbroe, G. A Review of Uncertainty Analysis in Building Energy Assessment. *Renew. Sustain. Energy Rev.* **2018**, *93*, 285–301. [CrossRef]
36. Cho, H.; Smith, A.; Luck, R.; Mago, P.J. Transient Uncertainty Analysis in Solar Thermal System Modeling. *J. Uncertain. Anal. Appl.* **2017**, *5*, 1–15. [CrossRef]
37. Broday, E.E.; Ruivo, C.R.; da Silva, M.G. The Use of Monte Carlo Method to Assess the Uncertainty of Thermal Comfort Indices PMV and PPD: Benefits of Using a Measuring Set with an Operative Temperature Probe. *J. Build. Eng.* **2021**, *35*, 101961. [CrossRef]
38. Ricciu, R.; Galatioto, A.; Desogus, G.; Besalduch, L.A. Uncertainty in the Evaluation of the Predicted Mean Vote Index Using Monte Carlo Analysis. *J. Environ. Manag.* **2018**, *223*, 16–22. [CrossRef] [PubMed]
39. Herman, J.; Usher, W. SALib: An Open-Source Python Library for Sensitivity Analysis. *J. Open Source Softw.* **2017**, *2*, 97. [CrossRef]
40. Saltelli, A.; Annoni, P.; Azzini, I.; Campolongo, F.; Ratto, M.; Tarantola, S. Variance Based Sensitivity Analysis of Model Output. Design and Estimator for the Total Sensitivity Index. *Comput. Phys. Commun.* **2010**, *181*, 259–270. [CrossRef]

buildings

MDPI

Article

Assessment of Energy, Environmental and Economic Costs of Buildings' Thermal Insulation–Influence of Type of Use and Climate

António M. Raimundo [1,*], Afonso M. Sousa [1] and A. Virgílio M. Oliveira [2]

[1] Department of Mechanical Engineering, University of Coimbra, Pólo II, Rua Luís Reis Santos, 3030-788 Coimbra, Portugal
[2] Polytechnic Institute of Coimbra, ISEC, Department of Mechanical Engineering, Rua Pedro Nunes, Quinta da Nora, 3030-199 Coimbra, Portugal
* Correspondence: antonio.raimundo@dem.uc.pt

Abstract: Among the aspects with major impacts on the energy and environmental performance of a building, the thermal insulation of the opaque elements of its envelope stands out. This work assesses the influence of the application of thermal insulation to the opaque elements of the building's envelope on the thermal comfort conditions indoors; moreover, the influence of the thermal insulation on the energy, environmental, and economic costs over the building's complete life cycle is evaluated. For this purpose, the three most commonly used thermal insulating materials (expanded polystyrene—EPS, extruded polystyrene—XPS, and mineral wool—MW), thicknesses between 0 (without insulation) and 40 cm, five climates (hot, warm, moderate, cold, and very cold), and six types of use (apartment, housing, clinic, school, bank branch, and supermarket) were considered. EPS reveals itself to be the most promising thermal insulation material, both in economic and environmental terms, so it was selected for this study. The EPS' optimal thickness depends on the building's type of use, the climate, and the perspective from which the assessment is carried out (energy, environmental, or economic). The results show that the economically optimal thicknesses of thermal insulation are significantly lower than the corresponding ones in environmental terms. Furthermore, the application of thermal insulation to the opaque building's envelope is more beneficial in energy and environmental terms than from an economic perspective.

Keywords: optimal buildings' thermal insulation; influence of type of use; influence of climate; life cycle cost analysis (LCCA); life cycle energy analysis (LCEA); life cycle impact analysis (LCIA)

Citation: Raimundo, A.M.; Sousa, A.M.; Oliveira, A.V.M. Assessment of Energy, Environmental and Economic Costs of Buildings' Thermal Insulation–Influence of Type of Use and Climate. *Buildings* **2023**, *13*, 279. https://doi.org/10.3390/buildings13020279

Academic Editor: Elena Lucchi

Received: 15 December 2022
Revised: 13 January 2023
Accepted: 14 January 2023
Published: 18 January 2023

1. Introduction

Good indoor environmental quality is essential to achieving well-being and ensuring work efficiency. Among the aspects relevant to indoor environmental quality, thermal comfort is usually pointed out as being more important than visual and acoustic comfort and indoor air quality [1]. To assure a low environmental impact, thermal comfort must be ensured with low primary energy consumption. The preference of building holders is that thermal comfort conditions should be achieved at a low economic cost [2].

1.1. Overview

According to the Intergovernmental Panel on Climate Change (IPCC) report on climate change mitigation from 2022 [3], total greenhouse gas (GHG) emissions have increased between 1970 and 2019, with a larger absolute increase after year 2000. Despite a growing number of climate change mitigation policies, annual GHG emissions grew on average by 2.2% per year from 2000 to 2019, compared with 1.3% per year from 1970 to 2000. Slightly different values for the GHG emissions are reported on the Emissions Gap Report 2022 of the United Nations Environment Programme [4], where an average annual growth rate of

2.6% per year from 2000 to 2009 and 1.1% per year from 2010 to 2019 is reported. According to both reports, a high was reached in 2019, followed by a decrease in 2020 (associated with the COVID-19 confinement); it is also suggested that total global GHG emissions in 2021 will be similar to, or even surpass, 2019 levels. According to the IPCC report [3], the building sector was responsible for 32% of the final energy consumption and 19% of the global equivalent CO_2 emissions.

The actual European Union (EU) "energy performance of buildings directive" (EPBD) [5] demands that new buildings be "near-zero-energy buildings" (nZEB). To progress towards more sustainable buildings, a new EPBD directive is being developed [6], in which it is required that new buildings be "zero-emission buildings" of greenhouse gases (GHG), defined as "buildings with a very high energy performance in line with the energy efficiency first principle, and where the very low amount of energy still required is fully covered by energy from renewable sources at the building or district or community level where technically feasible".

1.2. *State of the Art*

This section provides a summary of some previous research about the energy, environmental, and economic costs of thermal comfort, presents the methodologies of analysis, details the studies more related to the present one, and highlights the aspects that require a further assessment.

The environmental impacts of a building are essentially divided into three parts: those caused by the construction, those due to the building's operation, and those resulting from the building's demolition at the end of its useful life. Between them, the impacts caused by a building's end-of-life are negligible or have a small share [7–10], so studies related to this area normally consider only the environmental impacts due to the construction and operation phases.

The energy consumption of a building during its use is directly related to the quality of its constructive solutions (passive and active) and the efficiency of its heating, ventilation, and air conditioning (HVAC) system. Among its passive constructive solutions, the opaque elements of the envelope stand out, whose thermal transmission coefficient and thermal mass strongly influence the thermal and energy behavior of the building; these properties depend on the materials included, namely the thermal insulator (material, thickness, and location in the constructive element) [11]. Furthermore, opaque building elements generally represent the largest share of the construction expenditures [11].

The economic costs of a building are always supported by whoever is responsible for its use, be it the owner or the tenant [11,12]. To carry out an economic assessment of a building, it is necessary to know the corresponding expenses during its complete life cycle [13,14], whose value depends on the financial framework of its holder [11,12,15]. This assessment can be performed using the concept of "equivalent annual cost" [€/(m^2 year)] [2,11,15], which is an improvement of the concept of "equivalent global cost" (€/m^2) [13,14].

The environmental impacts caused by the construction phase arise from the energy used for the extraction, manufacture, transport, and application of buildings' constituent elements [16], which is recognized as embodied energy. This energy is one of the most commonly used indicators to represent the sustainability of materials and equipment. Therefore, a lower embodied energy indicates a lower environmental impact.

The methodologies used to identify the best thermal insulation material, its better position in the building opaque envelope, and its optimal thickness are normally based on a life cycle energy analysis (LCEA), a life cycle impact analysis (LCIA), and a life cycle cost analysis (LCCA), depending on whether the focus is on minimizing the energy consumption, the environmental impacts, or the economic costs, respectively [7–11,17–21]. As the relationship between the economic cost of buildings' thermal insulation and its energy efficiency and environmental impacts is not linear, there may be large differences between the solutions identified as the best, depending on the assessment methodology used [11,12,17–21].

Anastaselos et al. [17] proposed a methodology to classify the economic, the energy, and the environmental qualities of thermal insulation solutions, which assigns a performance category to each perspective. According to its proponents, this "simplistic" approach allows an easy comparison of thermal insulation solutions. They tested a fixed thickness of 6 cm of expanded polystyrene (EPS), extruded polystyrene (XPS), and mineral wool (MW) insulating materials. The selection of the best solution requires a ranking of priorities between economic cost, energy performance, and environmental impact.

Bastos et al. [9] carried out an analysis of the energy consumed and of the greenhouse gas (GHG) emissions of three types of buildings located in Lisbon (Portugal). Considering 75 years of buildings' lifetime, they modeled their life cycle focusing on the construction, restoration, and maintenance phases. They noticed that the use phase is the one with the highest primary energy demands (69–83%) and the greatest emissions of GHG. They also observe that, per square meter, larger buildings present lower energy demands, which is reflected in lower GHG emissions.

Using energy and economic assessments, Raimundo et al. [11] focused their study on identifying the most advantageous thermal insulation solutions for opaque elements for buildings located in the Portuguese temperate climate. For this purpose, they considered buildings of five types of use (apartment, detached house, clinic, school, and supermarket) and three types of thermal insulation materials (EPS—expanded polystyrene, XPS—extruded polystyrene, and MW—mineral wool) applied in three alternative positions (on the inner surface, in the middle of the air gap, and on the outer surface). They identified EPS as the most economically advantageous thermal insulation material. They also found that, in economic terms, it is more advantageous to apply thermal insulation in the intermediate position (in the air gap); the ETICS solution is more advantageous from the energy perspective; and the optimal thickness of thermal insulation depends on which perspective the analysis is performed. Even so, they concluded that, according to both perspectives, buildings located in regions with more intense climates require substantially greater thermal insulation thicknesses than those located in milder climates, and that residential buildings are the ones that need the highest thickness values. They identified the detached house as the building with the highest optimal thermal insulation thickness and the school (a service building with intermittent use only during the daytime) with the lowest.

Dylewski and Adamczyk [18] addressed the economic and environmental benefits of using thermal insulation on the external walls of buildings located in the various Polish climate regions. Variants in terms of different thermal insulation materials, thermal insulation systems, climatic zones, and heat sources were taken into consideration. They concluded that the environmental impact of a building strongly decreases with the decrease in energy demands for HVAC equipment operation, with the use of thermal insulation being a highly effective way to reduce these needs. They stated that the ideal insulation thickness, from both perspectives, increases with the number of heating degree-days.

Totland et al. [22] state that, in the case of buildings with insufficient thermal insulation thickness, 90% of their environmental impacts occur during the use phase; the increase in thermal insulation thickness leads to a decrease in this impact during the use phase and to its growth during the construction phase.

There is a great variability of climates, which can be identified using a wide range of classification types [23,24]. However, a simplistic categorization system is normally used on energy and environmental assessments, where the climates are classified as: hot, warm, moderate, cold, and very cold. As each type of climate leads to a different level of indoor-outdoor thermal exchanges, the climate of the building site is undoubtedly of vital importance for the identification of the optimum thermal insulation solution [11,23–26].

In terms of type of use, there are published studies focused on the optimal insulation of various types of buildings [9,17,25,27]. However, these studies considered only the buildings located in a specific place and did not involve a range of buildings sufficiently representative of the various types of use. Therefore, a comparative analysis in which various typologies of use are included is helpful and must be undertaken at sites representative of the various possibilities of climatic conditions.

1.3. Objectives and Scope

The present work aims to analyze the influence of thermal insulation on the economic, energy, and environmental parameters of buildings located in different climate types, namely the identification of the most suitable insulating material and its optimal thickness according to the economic, energy, and environmental perspectives. These assessments were performed using life cycle energy analysis (LCEA), life cycle impact analysis (LCIA), and life cycle cost analysis (LCCA) methodologies.

A unified comparison between various types of buildings, chosen to be a good representation of the building stock and supposedly located in the different climate alternatives, from very cold to hot, was used to perform an original assessment. In this analysis, several parameters frequently missing in the literature are considered, namely the building holder's financial framework and tax burden, both of which have a large impact on the economic cost associated with the use of a building. Additionally, the economic and environmental costs of applying the thermal insulation system and of other necessary materials are also taken into account. The usefulness of the results of this research is that they can be used to select the recommended thermal insulation thickness to apply in a given building located in a specific climate. Clearly, the value of the optimal thickness depends on the perspective considered.

The choice of locations considered to represent the various climate types was based on the number of heating degree days (HDD) and cooling degree days (CDD). Five cities with distinct climatic conditions were selected, which, taken together, are thought to be a good representation of a wide range of climate types: hot, warm, moderate, cold, and very cold.

The building stock is represented by buildings of six types of use: an apartment located in the middle of a multifamily building and a detached house (residential buildings), a clinic (a service building with permanent occupancy), a school and a bank branch (service buildings with intermittent use), and a supermarket (a commercial building with intermittent utilization).

It was assumed that all buildings are built using the same type of passive construction solutions (opaque and glazed), and the opaque elements of the buildings' envelope are equipped with a traditional External Thermal Insulation Composite System (ETICS) [2,11,24,25,28–32]. The three most common thermal insulation materials used in building construction are expanded polystyrene (EPS), extruded polystyrene (XPS), and mineral wool (MW) [20,21,30,33,34]. Thus, these are the insulating materials considered.

2. Methodology

SEnergEd software [2,11,15,35], a validated in-home tool developed for research purposes, was used to perform the life cycle energy analysis (LCEA), the life cycle impact analysis (LCIA), and the life cycle cost analysis (LCCA) of the buildings. It is a user-friendly software integrating algorithms for the dynamic simulation of the thermal and energy behavior of buildings (residential, commercial, and service), including thermal comfort evaluation, environmental impact analysis, and economic assessment of the building life cycle (Figure 1).

2.1. Energy Demands and Consumption

SEnergEd software predicts the thermal behavior and energy needs of buildings using a reformulated version of the dynamic hourly model 5R1C (5 thermal resistances and one thermal capacitance) described in ISO 13,790 [36]. Other energy demands are also obtained by a dynamic hourly calculation [35]. Using the energy performance of equipment and systems (HVAC, DHW, lighting, appliances, etc.), the energy demands are converted into consumption. More details on the energy model are described elsewhere [2,11,15,35].

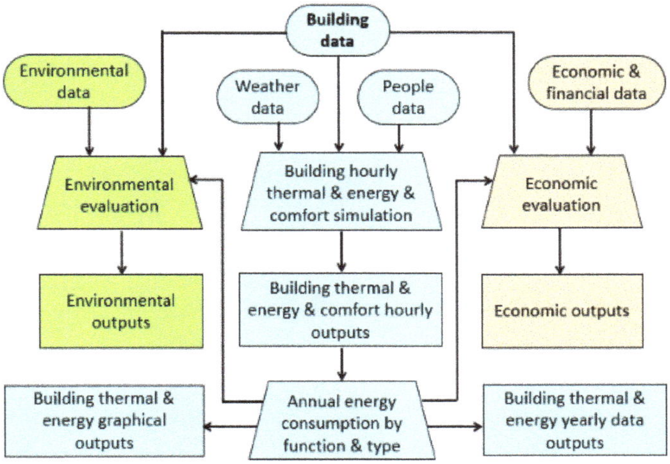

Figure 1. Global block diagram of the SEnergEd calculation methodology.

The energy component of this software was validated by comparison of its predictions with the real monthly energy consumption (during 2014, full year) of a high school [35] (Figure 2). This school was divided into 7 different thermal zones and has a global net floor area of 11,246 m^2, a gross floor area of 14,148 m^2, and a weighted average ceiling height of 3.84 m. Classrooms, rooms for teachers, offices, and secretaries occupy a net floor area of 7669 m^2, common spaces make up 3171 m^2, and showers and toilets make up 406 m^2. Each thermal zone includes acclimatized and non-acclimatized spaces (archives, storage rooms, warehouses, and technical spaces). There are no constructive elements that promote significant shading of the glazing. As this building is considered in this study, the occupancy and operating profiles are described ahead. As revealed by Figure 2, a reliable predictive capacity of SEnergEd software was observed, with a difference of 2.9% for the annual energy consumption and a maximum difference of 8.4% for the monthly consumption. These differences were justified by the difficulty in accurately establishing the utilization profile of each thermal zone, by the climatic data measured by a meteorological station located about 4 km away from the building, and by the impossibility of considering all the exact details of the building, such as constructive aspects, equipment performance, and occupation profiles.

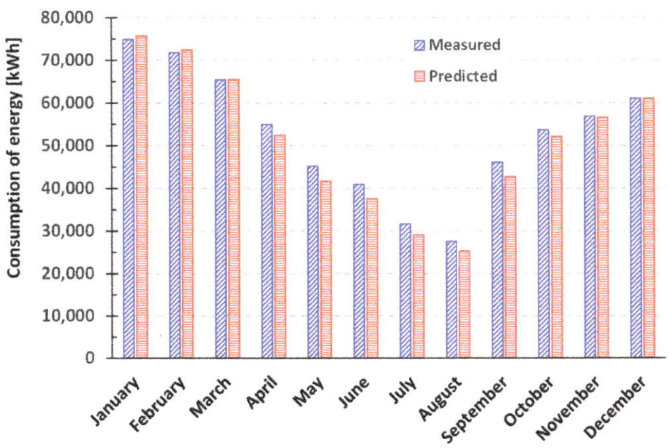

Figure 2. Measured versus predicted energy consumption of the high school during 2014.

2.2. Energy and Environmental Relations

The energy resources found in nature are known as primary energy and can be renewable (hydro, solar, wind, plant biomass, etc.) or non-renewable (coal, oil, natural gas, etc.). Final (or secondary) energy is energy in a state that allows its use in powering equipment, which may correspond to primary energy (as is the case with natural gas, among others) or result from refining (gasoline, diesel, etc.) or transformation (electricity, liquefied petroleum gas, thermal, etc.) processes. Useful (or tertiary) energy is energy capable of serving people directly (heat introduced or removed from a space to keep it comfortable, heat contained in domestic hot water, energy in the form of light, etc.), being produced by equipment powered with final energy.

The overall energy consumption of a building is equal to the sum of its energy consumption associated with its construction (represented by the energy absorbed in construction solutions), occupation (HVAC equipment, lighting, etc.), and demolition phases. However, energy consumption during the demolition phase is much lower than in the other two, and, as such, it can be neglected [7–10]. Thus, in the determination of buildings' overall energy consumption during their full life cycle, the end-of-life phase was not considered (Figure 3).

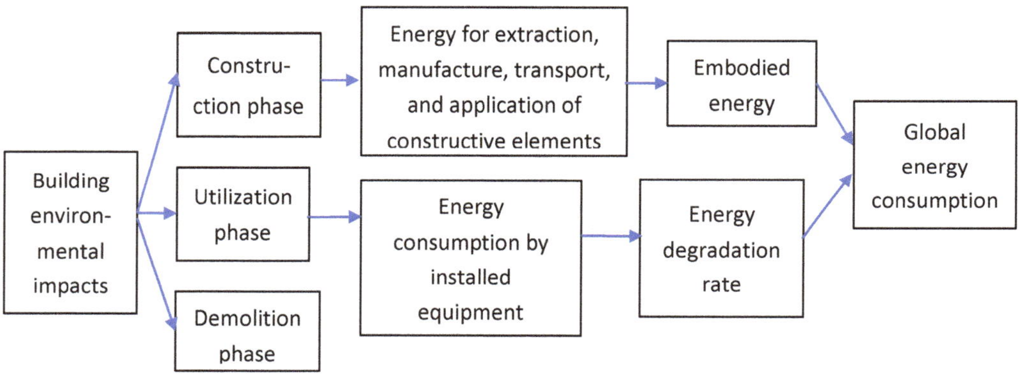

Figure 3. Block diagram of the building's global energy consumption calculation.

SEnergEd software predicts, for the first year of the use phase, the energy demands and the final energy consumption, disaggregated for: heating, cooling, ventilation, and pumping; non-HVAC equipment; indoor lighting; outdoor lighting; domestic hot water; specific equipment; kitchens; laundry; parking lots; warehouses; and complementary spaces. The final energies consumed can be of various types, so they are standardized by their conversion to a reference base in terms of primary energy. Annual primary energy consumption during the remaining years of the building's lifetime is determined assuming a rate of increase, which is representative of the loss of energy performance of the building's passive and active systems.

In this work, the kWh_{OE} (kilowatt-hour oil equivalent) is assumed as a reference for primary energy, and it is considered for the conversion factors of final energy to primary reference energy (F_{FP} [kWh_{OE}/kWh]). The values of 2.5 for electricity, 1.0 for natural gas (a non-renewable fuel), and 0 for energy from renewable sources were used [37]. To carry out an environmental impact assessment, it is necessary to link primary energy consumption with equivalent CO_2 emissions. For this, conversion factors of primary energy to emissions of equivalent kilograms of CO_2 (F_{PC} [$kgCO_2$ eq/kWh_{OE}]) were used, assuming the values of 0.144 for electricity, 0.202 for natural gas, and 0 for energy from renewable sources [37].

2.3. Equivalent Annual Cost of Thermal Insulation

The equivalent annual cost (EAC) of a building includes the costs of acquisition, installation/construction, maintenance, replacement, occupation, and end-of-life [2,11,35]. The cost of energy must consider the tariffs to be paid for each type of final energy.

The global cost of a building, as defined in EN 15,459 [13] and in European Regulation [14], can be represented by the net present value (NPV), which corresponds to the "overall economic loss" of the building when discounted to the instant when its use starts and can be expressed by [11]

$$NPV = \frac{RV}{(1+r)^n} - \sum_{k=0}^{n} \frac{I_k - CF_k}{(1+r)^k} \quad (1)$$

where k is the index representing each one of the n years of the period under analysis, r is the real interest rate (discount rate), RV is the residual value of the building at the end of the period under analysis, and I_k is the required amount of investment in year k [11]. The cash flow (CF_k) represents the net economic flow related to the building in each year, k.

$$CF_k = R_k + ST_k - C_k \quad (2)$$

where R_k, ST_k, and C_k represent the annual value of revenues, tax relief, and costs, respectively [11]. Annual revenues (R_k) may arise from selling energy or from renting. The existence of any income directly generated by the buildings was not considered in this study ($R_k = 0$). The annual tax relief (ST_k) is the reduction of the payable tax amount the building holder can benefit from by including the costs related to the building as expenses of his economic activity. This tax relief depends on the building holder's tax framework, which is null if the holder is an individual. In contrast, a significant monetary amount of tax relief can be present if the building holder is a company. The costs, given by

$$C_k = CT_k + CR_k + CM_k + CW_k + CE_k + CO_k \quad (3)$$

include taxes due to building ownership (CT_k), equipment replacement (CR_k), building and systems maintenance (CM_k), water consumption, sewage treatment, waste collection (CW_k), energy consumption (CE_k), and other costs (CO_k), such as, for example, condominium costs (in the case of the apartment) [11]. As they have no influence on the cost of thermal comfort, the expenses CT_k, CW_k, and CO_k are not relevant for the purposes of the present analysis.

The equivalent annual cost (*EAC*) of the building's use during its useful life is obtained from the building's net present value using the following equation [11]

$$EAC = -\frac{NPV}{A_{cl}} \cdot \frac{r\,(1+r)^n}{(1+r)^n - 1} \quad (4)$$

where A_{cl} represents the floor area of the building's acclimatized spaces.

However, the focus of this study is not the *EAC* of the building but only the equivalent annual economic cost of the thermal insulation (*ECTI*); therefore,

$$ECTI = EAC_b - EAC_0 \quad (5)$$

where EAC_b is the *EAC* of the building under assessment equipped with a HVAC system operating when necessary to maintain thermal comfort conditions, and EAC_0 represents the *EAC* of that building on the condition that its opaque elements do not include thermal insulation and the HVAC system only assures the renovation of indoor air (thus, no heating or cooling of indoor air).

To obtain the required amount of investment, it is necessary to know in advance the economic cost of implementation of each thermal insulation solution (*ECS*, €/m^2), which depends on the thermal insulating material, the thickness applied, the type of solution (ETICS in this case), the type of construction element (exterior wall, floor, roof, etc.), and the country in question. Thus, it is reasonable to consider that *ECS* is basically composed of three parts: (i) the cost of the thermal insulating material; (ii) the cost of the extra material required to apply the solution; and (iii) the cost of labor. Accordingly, for each type of constructive element, the three parcels are: (i) the cost of the thermal insulating material, which depends on the unit cost of EPS (*Ctim*, €/m^2 per cm of thickness) and the thickness applied (*thick*, cm), not depending on the country; (ii) the cost of the extra material (*Cextra*, €/m^2), which varies with the type of building element and is the same in all countries; and (iii) the cost of labor (*CL*, €/m^2), which depends on the type of building element and varies from country to country in proportion to the respective minimum wage (*MW*). To carry out this study, the values obtained by Raimundo et al. [11] for *Ctim*, *Cextra*, and *CL*, which are valid for Portugal, were assumed. Thus, the *ECS* value of each opaque constructive element was obtained using the expression

$$ECS = Ctim \cdot thick + Cextra + CL_{PT} \cdot MW_C / MW_{PT} \quad (6)$$

where $Ctim$ = 0.75 €/m^2 per centimeter of EPS thickness, CL_{PT} is the labor cost in Portugal of applying the thermal insulation solution, MW_C is the annual minimum wage in the country in question, and MW_{PT} is the annual minimum wage in Portugal, which, in the first semester of 2022, were 9870 €/year in Portugal, 13,510 €/year in Spain, 7858 €/year in Poland, and 24,840 €/year in Iceland [38].

3. Research Objects and Conditions

3.1. Climate Conditions

Even when only considering areas with a significant human population density, terrestrial climatic conditions are very diverse. Among the different methodologies used to classify the various types of climates, the one based on the number of heating degree-days (*HHD* [°C.day/year]) and the number of cooling degree-days (*CDD* [°C.day/year]) is the one that allows a more direct relationship between outdoor weather conditions and energy demands for heating and cooling, respectively [18,19,24–26,34]. Then, to perform this study, the selection process of the representative locations was based on the *HDD* and *CDD* values, defined in relation to base values of 20 °C and 24 °C, respectively. The selected five cities with distinct climate conditions are shown in Table 1, where also their Köppen–Geiger climate classification is indicated [23,24].

Table 1. Climatic data of the selected 5 cities with distinct climate conditions.

Climate Type	City, Country	Latitude [North]	HDD_{20} [°C.Day/Year]	CDD_{24} [°C.Day/Year]	Köppen-Geiger Class
Hot	Almeria, Spain	36.84°	1069	281	BSk
Warm	Lisbon, Portugal	38.73°	1601	174	Csa
Moderate	Salamanca, Spain	40.96°	3040	120	Csb
Cold	Warsaw, Poland	53.23°	4367	26	Cfb
Very Cold	Reykjavik, Iceland	64.14°	5670	0	Cfc

3.2. Buildings' Layout and Occupancy

Six buildings with different acclimatized areas, occupancy, internal thermal gains, and types of use were considered: (i) an apartment at the midlevel of a multi-story building (residential with permanent use); (ii) a detached house (residential with permanent use); (iii) a private clinic with hospitalization (services with permanent use); (iv) a private high school (services with intermittent use); (v) a bank branch (services with intermittent use); and (vi) a medium-sized supermarket (commerce with intermittent use). Table 2 shows a summary of the main characteristics of these buildings. The net and gross areas do not include non-acclimatized spaces. Additional information about the layout and the characteristics of these buildings can be found in Raimundo et al. [2,11,15].

Table 2. Summary of the characteristics of the 6 buildings considered: *Np*—maximum number of occupants; *Nf*—number of floors; A_{cl}—acclimatized floor area; A_{gf}—gross floor area; *Ch*—ceiling height; *Vol*—acclimatized volume; A_{opc}—opaque area of external envelope; A_{glz}—glazed area of external envelope; *AR*—aspect ratio = $(A_{opc} + A_{glz})/Vol$; *EA*—envelope area ratio = $(A_{opc} + A_{glz})/A_{cl}$); *GA*—glazed area ratio = A_{glz}/A_{cl}.

	Apartment	Detached House	Private Clinic	Private School	Bank Branch	Super-Market
Np [–]	4	4	151	1100	12	194
Nf [–]	1	3	2	4	1	1
A_{cl} [m^2]	109.4	167.1	926.7	11,246.0	111.4	1035.3
A_{gf} [m^2]	141.6	212.6	1161.2	14,147.5	134.7	1176.1
Ch [m]	2.62	2.96	3.72	3.84	2.60	3.60
Vol [m^3]	286.6	494.6	3447.3	43,184.6	316.2	3727.1
A_{opc} [m^2]	58.6	343.4	743.4	22,703.8	181.0	2830.6
A_{glz} [m^2]	21.3	49.7	192.8	2975.3	37.2	96.6
AR [m^{-1}]	0.28	0.79	0.27	0.59	0.69	0.79
EA [–]	0.73	2.35	1.01	2.28	1.96	2.83
GA [–]	0.19	0.30	0.21	0.26	0.33	0.09

In general terms, occupancy and operating profiles are characterized by:

- In all buildings, the occupancy and the operating profiles vary according to the time of the day, the day of the week, and the week of the year;
- When the building is closed, the HVAC system is off, and the lighting systems are either off or operate at very low power;
- Residential buildings are assumed to be unoccupied during the first fifteen days of August and permanently occupied during the remaining days of the year, by four people on Saturdays and Sundays, and between 6 p.m. and 8 a.m. on weekdays (Mondays to Fridays), and by one person the rest of the time;
- The clinic runs continuously during all days of the year, but with higher occupation intensity between 8 a.m. and 8 p.m. on weekdays and on Saturdays;
- The high school is only occupied between 8 a.m. and 6 p.m. on weekdays, is closed on Saturdays and Sundays, and its operation follows a common school calendar, so it works at 100% during school periods, at 50% during the 1st examination phase (15–30 June), at 25% during the 2nd examination phase (1–15 July), at 25% during the admission phase (16–31 July), and is closed during school holidays (the first 15 days of April, 1 to 31 August, and the last 15 days of December);
- The bank branch operates every weekday of the year, is occupied between 8 a.m. and 6 p.m., and is closed on Saturdays and Sundays;
- The supermarket operates every day of the year and is occupied between 8 a.m. and 10 p.m., but with more intense activity on Saturdays and Sundays.

3.3. *Air-Conditioning Systems*

In temperate climates, electric air-source heat pumps have reasonably good energy performances in heating mode, and so systems based on air-source chillers/heat pumps are the recurrent option [2,11,15,39]. Alternatively, in cold climates, air-source heat pumps have very low performance in heating mode, so in these climates, heating systems are usually based on a boiler (using gas or another type of fuel) [18,33,34]. Therefore, two different HVAC systems are considered: (A) for buildings located in hot, warm, and moderate climates; and (B) for buildings located in cold and very cold climates.

HVAC systems type A are assisted by a single chiller/heat pump device, and HVAC systems type B are supported by a natural-gas boiler and a chiller machine. The chillers and heat pumps considered are based on a compression-expansion cycle and are of European Class A+ [40], as this class is the one that better represents the equipment currently being installed and it leads to a lower economic cost to obtain thermal comfort conditions [2].

The chiller has a seasonal energy efficiency ratio *SEER* = 5.85 in cooling mode [40], the heat pump has a seasonal coefficient of performance *SCOP* = 4.30 in heating mode [40], and the natural-gas boiler has a seasonal thermal efficiency *STE* = 0.95 [18,33,34].

The results of the present study were obtained assuming that HVAC systems only work when the corresponding space is expected to be occupied. During these periods, the indoor air temperature is maintained between setpoints T_{min} = 21 °C and T_{max} = 24 °C. As stated by Raimundo and Oliveira [2], this range of setpoints is the one that most probably guarantees category II of thermal comfort, as endorsed by the European Standard EN 16798-1:2019 [41] for new buildings and renovations. The air renewal is ensured by air handling units (AHU) and air extraction fans, both with an efficiency of 70% [37]. Heat recovery from the rejected air and free-cooling strategies were not considered.

3.4. Glazing Elements

The windows identified by Raimundo et al. [15] as the most economically advantageous glazing for buildings located in Portugal were selected. The windows are composed of an aluminum frame with a thermal barrier and double glass (colorless 6 mm + 11 mm of air layer + colorless 4 mm). Electric blinds composed of horizontal plastic strips were considered external protection. This glazing system has a thermal transmission coefficient (U) and a solar factor ($g_{\perp}$) of U_w = 3.05 W/(m^2 K) and $g_{\perp w}$ = 0.79, respectively, and when the external protection is active, U_{wp} = 1.56 W/(m^2 K) and $g_{\perp wp}$ = 0.05.

3.5. Economic and Environmental Costs of Thermal Insulation

The thermal insulation materials most commonly used in building construction are expanded polystyrene (EPS), extruded polystyrene (XPS), and mineral wool (MW) [11,16,17,20,21,30,33]. The option for these materials is a result of their economic advantages, which is the first criterion used by the constructors and the owners of buildings. Thus, our focus will be on these three thermal insulation materials.

By consulting six Portuguese suppliers, Raimundo et al. [11] obtained an average price (VAT not included) of 0.75, 2.61, and 1.15 €/m^2 per centimeter of thickness for sheets of EPS, XPS, and MW, respectively. The embodied energy was obtained from Anastaselos et al. [17], where values of 80.76 for EPS, 87.10 for XPS, and 24.61 MJ/kg for MW are suggested, which correspond to 11.31, 27.87, and 16.00 MJ/m^2 per each centimeter of thickness of EPS, XPS, and MW, respectively.

The economic and environmental costs associated with a building depend substantially on the material used to insulate the opaque elements. Thus, an important issue is the identification of the most advantageous thermal insulating material from economic and environmental perspectives. For this, the environmental and economic costs of thermal insulating materials were obtained for the functional unit R = 1 m^2 K/W, which are shown in Table 3. Considering the values obtained, it can be observed that the most promising thermal insulation material is EPS (expanded polystyrene). Therefore, EPS was the thermal insulation material selected for this study.

Table 3. Environmental and economic costs of the thermal insulation materials for a functional unit R = 1 m^2 K/W.

Material	Environmental Cost [MJ/m^2]	Economic Cost [€/m^2]
EPS	40.29	2.66
XPS	86.09	8.05
MW	53.98	3.82

Several other works had identified EPS (expanded polystyrene) as among the most advantageous thermal insulation materials for application to building elements [11,16,17,20,21,30,33]. In addition, it can be used in almost all opaque building elements (façades, roofs, floors, etc.) [11,42] and has a durability of at least 50 years [16,33,39]. Alternatively, there are other thermal insulation materials with better environmental performance [43,44], namely natural pumice, cellulose flocks, and kenaf fibers [43]. However, there is not a regular market for this kind of thermal insulator, so their prices make it economically unviable.

3.6. Opaque Elements of Buildings' Envelope

Each type of opaque construction element relies on a base structure, which is the same for all buildings and for all climates. Table 4 describes the base structure of the opaque elements in contact with the exterior and the corresponding values of thickness, useful thermal mass (*Mt*), and thermal transmission coefficient (*U*).

Table 4. Base structure of the opaque elements of the external envelope.

Element	Description (from Outside to Inside)	Values
Wall	Traditional plaster with 2 cm, bored brick of 22 cm, not-ventilated air space with 1 cm, bored brick of 11 cm, traditional plaster with 2 cm	*Thickness* = 38 cm Mt = 150 kg/m^2 U = 0.88 W/(m^2 K)
Pillar/Beam	Traditional plaster with 2 cm, reinforced concrete (iron volume less than 1%) of inerts with 22 cm, not-ventilated air space of 1 cm, bored brick of 11 cm, traditional plaster with 2 cm	*Thickness* = 38 cm Mt = 150 kg/m^2 U = 1.36 W/(m^2 K)
Floor above outside	Traditional plaster with 2 cm, lightened slab of 38 cm, light-sand concrete of 7.5 cm, screed (mortar) of 5.5 cm, oak wood with 2 cm	*Thickness* = 55 cm Mt = 150 kg/m^2 U = 1.17 W/(m^2 K)
Ground floor	Waterproofing layer, lightened slab of 38 cm, light-sand concrete of 7.5 cm, screed (mortar) of 5.5 cm, oak wood with 2 cm	*Thickness* = 54 cm Mt = 150 kg/m^2 U = 1.23 W/(m^2 K)
Accessible roof	Mosaic tile with 1 cm, screed (mortar) of 5.5 cm, waterproofing of 3 mm, light-sand concrete of 7.5 cm, lightened slab of 38 cm, traditional plaster with 2 cm	*Thickness* = 55 cm Mt = 150 kg/m^2 U = 1.39 W/(m^2 K)
Not accessible roof	Sandstone (inert) with 4 cm (or ceramic tile), waterproofing of 3 mm, screed (mortar) of 4 cm, lightened slab of 23 cm, traditional plaster with 2 cm	*Thickness* = 33 cm Mt = 150 kg/m^2 U = 2.40 W/(m^2 K)

The basic structure is complemented with the application of EPS thermal insulation on the outer face through an ETICS-type solution, identified by Raimundo et al. [11] as the most efficient in terms of energy demands. Another advantage is that it can be applied both in new buildings and in refurbishments. Thermal insulation thicknesses between 0 cm (without insulation) and 40 cm were tested.

Bearing in mind that the objective of this work is not the analysis of the environmental impact of buildings but the application of thermal insulation to their opaque elements, it is reasonable to consider the embodied energy in buildings null when their opaque elements are not thermally insulated. The embodied energy in the thermal insulation solution (*EES*, kWh/m^2) mainly depends on the thermal insulating material, its thickness, the type of solution, and the type of construction element.

Tables with the useful thermal mass (*Mt*) and the thermal transmission coefficient (*U*) of each type of construction element and with the economic cost of each thermal insulation solution (*ECS*), obtained using Equation (6), and the embodied energy (*EES*) of the respective thermal insulation solution were built. Table 5 shows the values obtained for the external walls.

3.7. Economic Assessment

The tax burden is high in almost all countries, so any economic assessment must consider them. Each country has its own tax system. Usually, the differences are associated with the applicable rates and not with the types of taxes. The tax system considered is based on EU practice, where the more relevant fees for this study are: (i) the value-added tax (VAT), which is a consumption tax; and (ii) the annual tax relief, i.e., the reduction of the amount paid in each year, related to the professional activity of the holder due to the annual expenses attributable to the building. The value of these taxes is strongly influenced by the taxable person's fiscal framework, which can be: (i) a natural person (individual or family), which, as the final consumer, is unable to recover the VAT paid; (ii) a legal person (company or organization), which is unable to recover the VAT paid; or (iii) a legal person capable of recovering the VAT paid.

Table 5. Useful thermal mass (*Mt*) and thermal transmission coefficient (*U*) of the exterior walls and the respective economic cost (*ECS*) and embodied energy (*EES*) of the thermal insulation solution.

Thickness [cm]	*Mt* [kg/m^2]	*U* [W/(m^2 K)]	*ECS* [€/m^2] Portugal	Spain	Poland	Iceland	*EES* [kWh/m^2]
0	150	0.88	0.00	0.00	0.00	0.00	0.00
3	150	0.54	20.34	26.05	17.19	43.80	12.56
4	150	0.48	21.09	26.80	17.94	44.55	15.70
5	150	0.43	21.84	27.55	18.69	45.30	18.84
6	150	0.39	22.59	28.30	19.44	46.05	21.98
8	150	0.33	24.09	29.80	20.94	47.55	28.26
10	150	0.28	25.59	31.30	22.44	49.05	34.55
12	150	0.25	27.09	32.80	23.94	50.55	40.83
14	150	0.22	28.59	34.30	25.44	52.05	47.11
16	150	0.20	30.09	35.80	26.94	53.55	53.39
20	150	0.17	33.09	38.80	29.94	56.55	65.95
25	150	0.14	36.84	42.55	33.69	60.30	81.66
30	150	0.12	40.59	46.30	37.44	64.05	97.36
35	150	0.11	44.34	50.05	41.19	67.80	113.06
40	150	0.009	48.09	53.80	44.94	71.55	128.77

It was assumed that the holders of the apartment and the detached house are natural persons, while the ones of the clinic and the school are corporations without the ability to recover the VAT paid; on the contrary, the bank branch and the supermarket are companies capable of recovering the VAT paid. Therefore, to all investments and all expenses (including energy costs) related to the apartment and the detached house, it is necessary to add the VAT; to all investments and expenses related to the clinic and the school, the VAT must be added; and both the investments and the expenses related to the supermarket and the bank branch do not include the VAT. Investments and disbursements on buildings are normally subject to the maximum VAT rate, which is currently 21% in Spain, 23% in Portugal and Poland, and 24% in Iceland.

In most countries, the annual tax paid due to the professional activity of individuals is determined by applying a rate to the total amount of revenues. Therefore, if the building holders are individuals, tax relief due to annual expenses or investments in the building does not exist, and in Equation (2), $ST_k = 0$. This happens in the case of residential buildings. The business activity is normally taxed in proportion to the total profit (revenues minus expenses). When the building is associated with the activity of its holder and their investments and expenditures decrease the holders' profit, then the amount of activity fees to pay also decreases. Consequently, if the building holder is a corporation, there is tax relief due to the annual expenses and the investments within the building. In these cases, to obtain the corresponding value of ST_k to be used in Equation (2), a profit tax rate of 25% was considered, which is what is supposed to happen in the case of the four non-residential buildings.

For buildings constructed using passive solutions of the type considered, a lifespan of at least 50 years is often reported in the literature [11,16,33,39]. Thus, it is appropriate to assume $n = 50$ years in the economic analysis. Besides usual maintenance work, any major intervention in the passive elements was assumed to occur during the first 50 years of a building's lifespan. The maintenance costs (CM_k in Equation (3)) are related to the corresponding investment value, considering a rate of 1%/year for all passive elements [2,11].

A large portion of annual expenditure is related to energy consumption (CE_k in Equation (3)). The tariffs of energy (electric and natural gas) were obtained from the European Union energy price statistics website [45,46], where the consumers are classified as household or non-household. The most recent data refer to the first semester of 2022, which was considered (Table 6).

Table 6. Electric energy and natural gas tariffs in the first semester of 2022 for household or non-household consumers [45,46] (VAT not included).

Type of Consumer	Country	Electric Energy [€/kWh]	Natural Gas [€/kWh]
Household (Residential buildings)	Spain	0.2822	0.3290
	Poland	0.1379	0.1511
	Portugal	0.1831	0.2067
	Iceland	0.1234	0.1478
Non-household (Commerce and Services buildings)	Spain	0.1879	0.2111
	Poland	0.1555	0.1825
	Portugal	0.1202	0.1415
	Iceland	0.0756	0.0843

In research papers involving buildings' energy consumption [2,11,33,39,42], rates ranging from 0 to 10%/year are presumed for the increase of average prices, between 0 and 15%/year for the increase of energy tariffs, and between 0 and 10%/year for the financial discount rate. As they are considered to be highly probable, a global inflation rate of i_{fg} = 5%/year and a financial discount rate of r = 5%/year were assumed. The energy market is perhaps one of the most unstable, so it is not possible to recognize a value as the most likely for the energy price evolution rate. Thus, it was considered that the energy price inflation rate (i_{fe}) is equal to the global inflation rate, $i_{fe} = i_{fg}$ = 5%/year. The economic assessment was carried out using a current price analysis.

4. Results and Discussion

To facilitate comparisons between the different buildings, it was assumed that they were all built using the same type of passive construction solutions (opaque and glazed). The combination of opaque element materials with an External Thermal Insulation Composite System (ETICS) leads to buildings with high thermal inertia, which is known as an effective strategy to mitigate both overheating and cooling load peaks [2,11,24,25,28–32].

As stated elsewhere [28,31,32], the buildings' energy performance is very sensitive to several deterioration factors, particularly the degradation with time of the HVAC and the thermal insulation systems. Thus, this loss of efficiency was considered assuming an increase of 1%/year in primary energy consumption for heating and cooling.

Energy, environmental, and economic perspectives are used to relate the thickness of the thermal insulation with the climate and the type of building. For this, five climate types (hot, warm, moderate, cold, and very cold) and six different buildings (an apartment, a detached house, a clinic, a school, a bank branch, and a supermarket) were considered. The results shown in the following sections are normalized per m^2 of acclimatized spaces' floor area. The net (A_{cl}) and gross (A_{gf}) floor areas are presented in Table 2.

4.1. Energy Cost of Thermal Comfort

The primary energy consumption for the heating and cooling functions is used to represent the energy cost of thermal comfort. This cost depends on the building type of use, on the climate characteristics, on the energy efficiency of the acclimatization systems, and on the thickness of thermal insulation.

As shown in Figure 4, the buildings with day and night uses (apartment, dwelling, and clinic) have a higher primary energy consumption for climatization than those with only daytime occupation (school and bank branch) or occupied during daytime and part of the night (supermarket). As it has a much more favorable aspect ratio (AR), the apartment's energy requirements are lower than those of the dwelling. The school is the building with the lowest energy consumption for air conditioning per m^2 of acclimatized floor area due to the fact that it is occupied only during the daytime and is closed, or is partially closed, during school holidays.

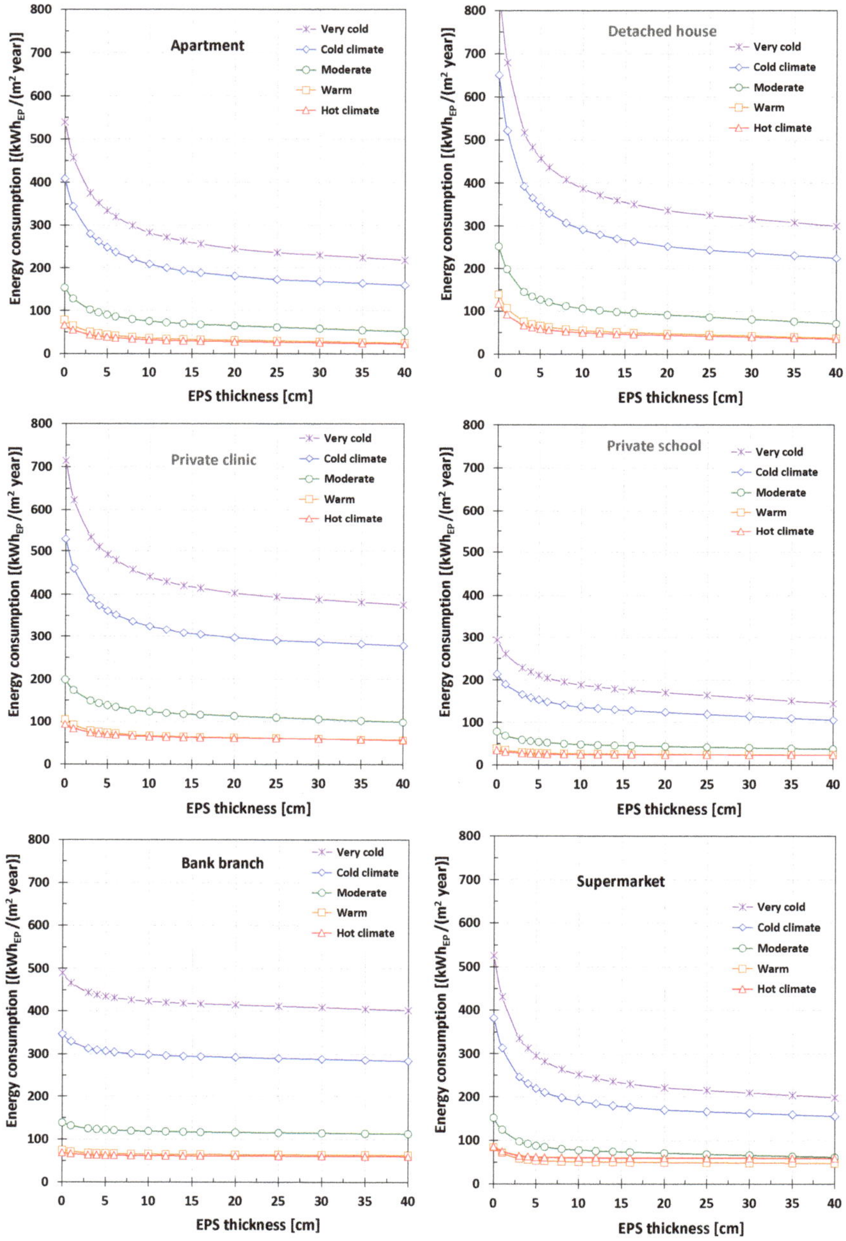

Figure 4. Primary energy consumption for heating and cooling, to maintain thermal comfort.

Figure 4 shows that the energy cost of thermal comfort is substantially higher in the case of the two cold climates (cold and very cold) than in the remaining three (hot, warm, and moderate). This is due to the high energy requirements for heating in cold climates and the fact that the heating system considered for these has a significantly lower efficiency than that assumed for the remaining three.

The increase in the applied thickness of EPS leads to a continuous decrease in the consumption of primary energy by the air-conditioning equipment, promoting a decrease in the thermal comfort energy cost. This decrease has an inverse exponential behavior, showing a declining rate of decrease as the EPS thickness increases, and, within the range of tested EPS thicknesses (0–40 cm), the minimum value occurs for the thickness of 40 cm, which leads to the projection that this consumption will continue to decline with the growth in the thickness of thermal insulation. Therefore, it was not possible to identify, in concrete, the optimal thickness of EPS from an energy perspective.

For the same building, the decrease in energy consumption for air conditioning with the increase in EPS thickness is much more significant in the cases of cold and very cold climates. Otherwise, the decrease in energy consumption with the increase in the applied EPS thickness is small in the case of buildings located in hot and warm climates. Furthermore, primary energy consumption is very similar in these two climates. Except for the supermarket, primary energy consumption in the warm climate is slightly higher than in the hot climate. Due to its high internal thermal loads, the supermarket has high energy needs for cooling, which leads to a slightly higher consumption of primary energy in a hot climate compared with a warm climate.

4.2. Environmental Cost of Thermal Comfort

The environmental impact of ensuring thermal comfort conditions inside buildings is assessed using the concept of the "annual environmental cost of thermal comfort," which is represented here by equivalent CO_2 emissions. These emissions include those associated with the building's construction phase and those related to its operation throughout its useful life cycle, ignoring the emissions related to the end-of-life of the materials. Equivalent CO_2 emissions associated with the construction phase include those caused by energy consumption for the extraction, manufacture, transport, and application of the thermal insulation solution. Since they are not related to the thickness of thermal insulation, the environmental impacts associated with other construction materials were ignored. CO_2 equivalent emissions associated with the building use phase are due to energy consumption by HVAC equipment to meet heating and cooling demands.

Figure 5 shows the equivalent annual CO_2 emissions of the six buildings considered, according to the type of climate and the thickness of thermal insulation. As expected, these emissions depend on the building's type of use, the characteristics of the climate, and the EPS thickness.

The results presented in Figure 5 reveal that annual CO_2 equivalent emissions depend mostly on the annual consumption of primary energy for heating and cooling, with a minor influence from the emissions associated with building construction. Thus, with few exceptions, the statements for the energy cost of thermal comfort are also valid for the environmental cost of thermal comfort. Namely, the environmental cost of thermal comfort is substantially higher in the case of the two cold climates (cold and very cold) than in the remaining three (hot, warm, and moderate). For the same typology of use, buildings located in hot and warm climates have very close environmental costs. The buildings with day and night uses (apartment, dwelling, and clinic) have higher environmental costs than those with only daytime occupation (school and bank branch) or occupied during daytime and part of the night (supermarket). Per m^2 of acclimatized floor area, the school is the building with the lowest environmental cost of thermal comfort.

Figure 5 reveals that the increasing EPS thickness leads to a continuous decrease in CO_2 equivalent emissions, but only up to a certain thickness, after which emissions increase. Therefore, from an environmental perspective, there is an EPS thickness for which the environmental impact is minimal.

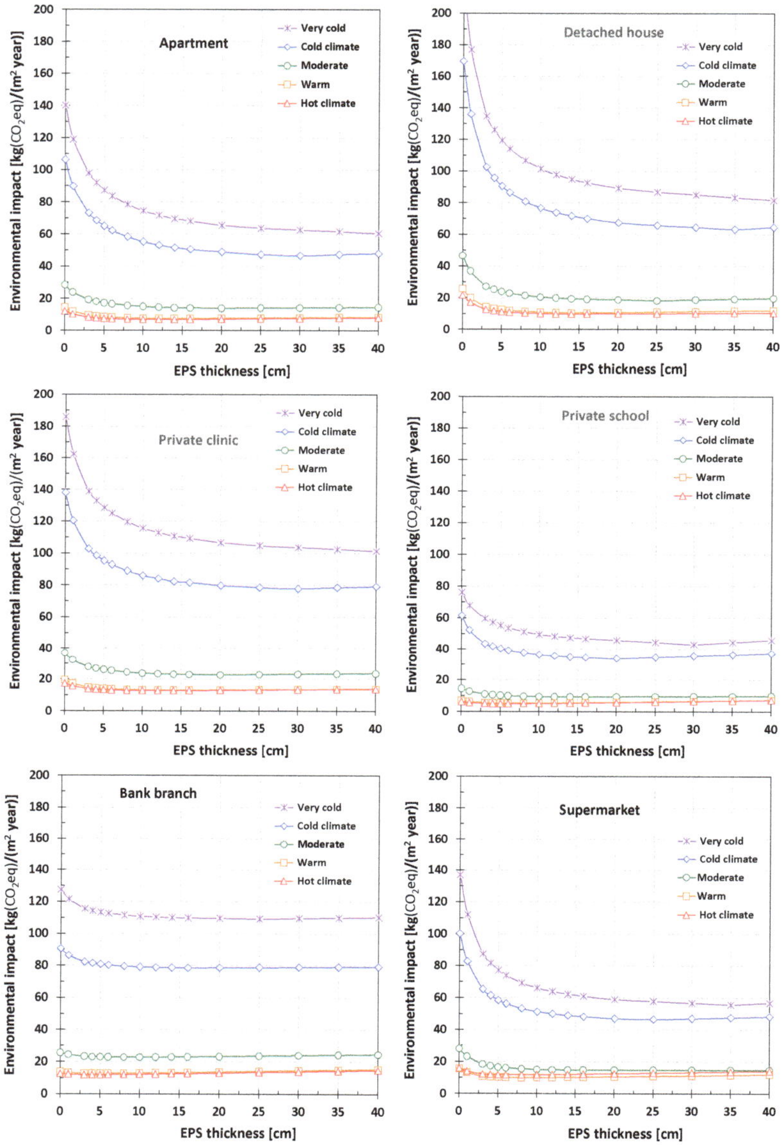

Figure 5. CO_2 equivalent emissions to maintain thermal comfort conditions.

4.3. Economic Cost of Thermal Comfort

The economic cost of thermal comfort is represented by the "equivalent annual economic cost of the thermal insulation" (*ECTI*), obtained considering the investment costs in the thermal insulation system, the expenses in its maintenance, and the costs due to the energy consumed to maintain thermal comfort conditions. Taxes were also considered and, if they exist, tax savings.

For a specific country and a particular HVAC system, the *ECTI* value depends on the type of building, the climate characteristics, and the thickness of the thermal insulation [2,11]. However, if the analysis involves different countries and distinct HVAC systems, the *ECTI* value also depends on the labor cost, the energy efficiency of heating and cooling devices, and the price of energy purchase. As can be seen in Section 3.3 and Tables 5 and 6, these parameters can show very substantial differences between distinct countries. Therefore, the *ECTI* value is only valid for the HVAC system, the type of climate, and the country for which it is determined. Thus, the type of climate must appear to be associated with the country of the place considered to represent it.

Figure 6 shows the equivalent annual economic cost of the thermal insulation (*ECTI*) of the six buildings considered as a function of the climate type (and country) and of the thermal insulation thickness. The optimal thermal insulation thickness is the one for which the *ECTI* value is the lowest. Both the *ECTI* value and its evolution with the increase in EPS thickness are strongly related to the type of building. This relationship, however, is substantially more evident in the case of cold and very cold climates than in the others (hot, warm, and moderate). In general terms, for the same type of climate (and country), buildings with permanent occupancy (apartment, detached house, and clinic) have higher *ECTI* values than buildings with intermittent use (school, bank branch, and supermarket). The school is the building with the lowest *ECTI* values, followed by the supermarket.

The highest *ECTI* values always occur in very cold climates (Iceland). This is due to the high demand for thermal energy for heating, the low performance of the heating system, and the high value of labor costs in Iceland. The difference compared with other types of climates (countries) is very substantial in the case of the two residential buildings and very small in the case of the supermarket.

The consumption of primary energy for air conditioning is always higher in the cold climate (Poland) than in the moderate climate (Spain), but the *ECTI* value is higher in the case of the moderate climate (Spain) than in the cold climate (Poland) for the two residential buildings. This is a result of the substantially higher values of labor costs and energy prices in Spain than in Poland. These causes are not enough to lead to this "inversion of order" in the case of non-residential buildings, but even so, it leads to a greater approximation of values between *ECTI* and energy consumption.

The difference in primary energy consumption for air conditioning between hot and warm climates is very small, being slightly higher in warm climates (except in the case of the supermarket, where it is the opposite, due to the demands of thermal energy for cooling associated with this type of building). Even so, the difference in *ECTI* values between hot (Spain) and warm (Portugal) climates is not small, and the value of this economic cost is always higher in the case of a hot climate (Spain) than in a warm climate (Portugal). This results from substantially higher values in Spain than in Portugal for labor costs and energy purchase prices.

Previous outcomes lead to the conclusion that the optimal value of the thermal insulation thickness from an economic perspective is only valid for the HVAC system, the type of climate, and the country for which it is determined.

4.4. Optimal Thickness of Thermal Insulation

The optimal thermal insulation thickness of buildings' envelopes depends on the perspective of analysis, which can be energetic, environmental, or economic [11,12,17–21]. In energy terms, it corresponds to the one that leads to the lowest primary energy consumption for climatization; from an environmental perspective, it is the one where the emission of CO_2 equivalent reaches a minimum; and in economic terms, it is the one that leads to the lowest value of the economic cost of thermal comfort. The application of these criteria to the values represented in Figures 4–6 allowed the elaboration of a summary table (Table 7), in which it is evident that each perspective leads to different values of the optimal EPS thickness.

The results shown in Table 7 reveal that, within the range of tested EPS thicknesses (0 to 40 cm), from an energy point of view, the optimal thermal insulation thickness is greater than 40 cm for all types of buildings and climates. This is a consequence of the fact that increasing thermal insulation thickness always leads to a decrease in energy consumption for heating and an increase in energy consumption for cooling, as shown by the results of this study (figures not shown) and what is reported in the bibliography [2,11,24,25,33].

Additionally, the rate of decrease in energy consumption for heating is greater than the rate of increase in consumption for cooling, which is reflected in a continuous decrease in energy consumption for air conditioning with the increase of thermal insulation thickness. Clearly, this statement is valid only for the situations analyzed here. In the authors' opinion, a reversal of this trend can occur in buildings with high internal thermal loads and low thermal mass that are located in regions with hot or warm climates. The buildings in this study have a high thermal mass, so eventually, a thermal insulation thickness that leads to a minimum of energy consumption does not exist.

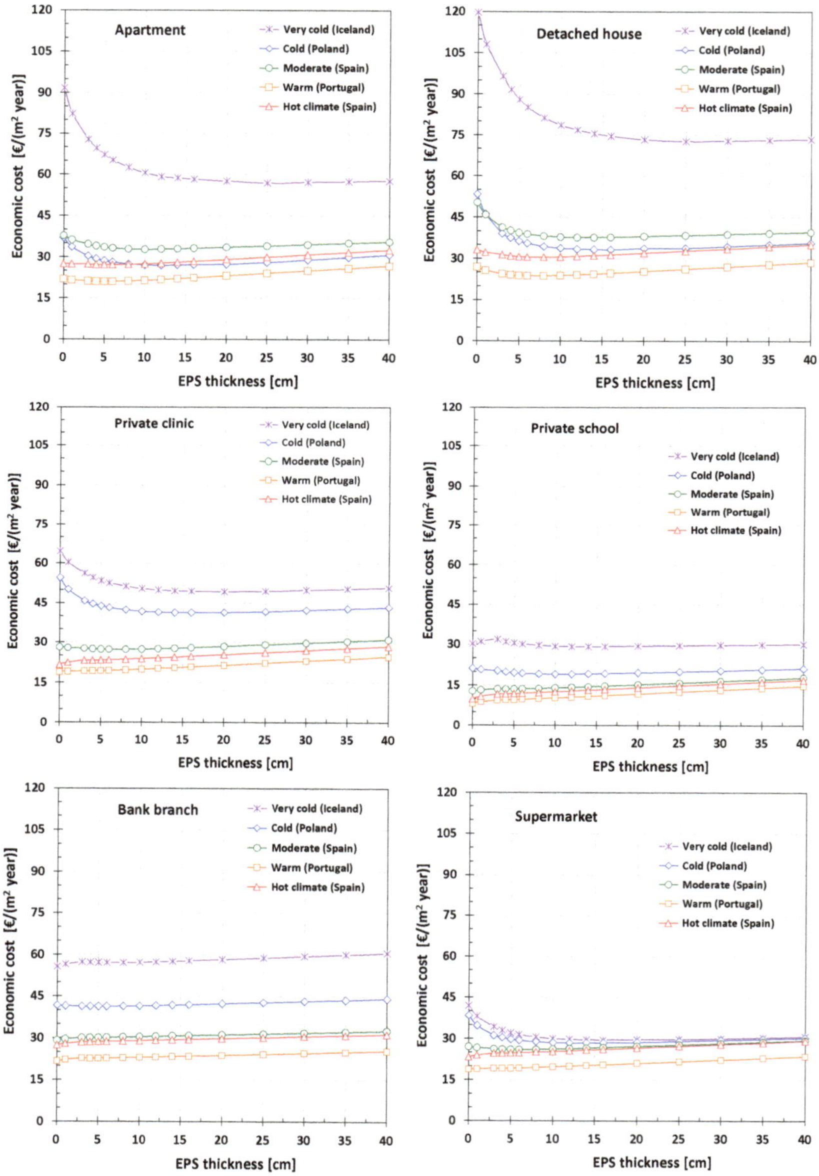

Figure 6. Economic equivalent annual cost per unit area to maintain thermal comfort conditions.

Table 7. Optimal EPS thickness [cm] from energy, environmental, and economic perspectives.

Building	Perspective	Hot	Warm	Climate Moderate	Cold	Very Cold
Apartment	Energetic	>40	>40	>40	>40	>40
	Environmental	12	14	20	30	>40
	Economic	5	5	10	12	25
Detached house	Energetic	>40	>40	>40	>40	>40
	Environmental	14	16	25	35	>40
	Economic	8	8	14	16	25
Private clinic	Energetic	>40	>40	>40	>40	>40
	Environmental	12	14	20	30	>40
	Economic	0	0	6	14	20
Private school	Energetic	>40	>40	>40	>40	>40
	Environmental	5	8	14	20	30
	Economic	0	0	0	10	14
Bank branch	Energetic	>40	>40	>40	>40	>40
	Environmental	3	5	8	16	25
	Economic	0	0	0	6	0
Super-market	Energetic	>40	>40	>40	>40	>40
	Environmental	6	8	16	25	35
	Economic	0	0	6	14	16

As Table 7 highlights, there is an EPS thickness that minimizes equivalent CO_2 emissions, so it can be stated that there always exists an environmentally optimal thermal insulation thickness. This optimal thickness increases with the cooling of the climate in all types of buildings, always presenting the lowest value for a hot climate and the highest for a very cold climate. In the situation of very cold weather, it was not possible to identify the environmentally optimal EPS thickness for buildings with permanent occupancy (apartments, detached houses, and clinics). Even so, the evolution of emissions of CO_2 equivalent clearly suggests that its value lies between 40 and 50 cm.

In global terms and for the same type of climate, the value of the environmentally optimal thickness of buildings with permanent occupancy is substantially higher than that of buildings with intermittent use (school, bank branch, and supermarket). According to the environmental perspective, the detached house requires the greatest EPS thickness, and the bank branch the smallest.

In very general terms and as expected, it can be outlined that the optimal EPS thickness according to the economic perspective increases with the cooling of the climate; thus, the lowest corresponds to a hot climate and the highest to a very cold climate. In terms of type of use, buildings can be ordered from the least economically optimal thickness to the greatest: bank branch (0 to 6 cm), school (0 to 14 cm), supermarket (0 to 16 cm), clinic (0 to 20 cm), apartment (5 to 25 cm), and detached house (8 to 25 cm). Among the situations analyzed in the present study, it is not economically advantageous to isolate the opaque elements of the four non-residential buildings when located in hot (Spain) and warm (Portugal) climates. In the case of the moderate climate (Spain), this also occurs in the cases of the school and the bank branch.

In addition to the energy demands for air conditioning, the energy efficiency of HVAC systems, the cost of acquiring EPS, the cost of labor, and the price of purchasing energy have also been considered in the economic analysis, which differs from country to country. As such, the economic optimal value of thermal insulation thickness is only valid for the type of climate and the country for which it was determined. Furthermore, it was assumed that, for the next 50 years, energy prices would increase at a rate of 5%/year. However, as demonstrated by Raimundo and Oliveira [2], the economic advantages of thermally insulating the opaque elements of buildings grow with the rate of inflation of energy prices.

Figure 7 shows the relation between the environmental impact and the economic cost of the six buildings considered for the five types of climates. As revealed by these graphs and by Table 7, the EPS thickness that leads to the lower environmental impacts does not correspond to the optimal economic cost. However, as shown in Figure 7, the optimal EPS thickness in environmental terms shows an economic cost relatively close to its optimal value. So, in the authors' point of view, the best option for thermal insulation thickness lies between the optimal values of the economic and environmental costs.

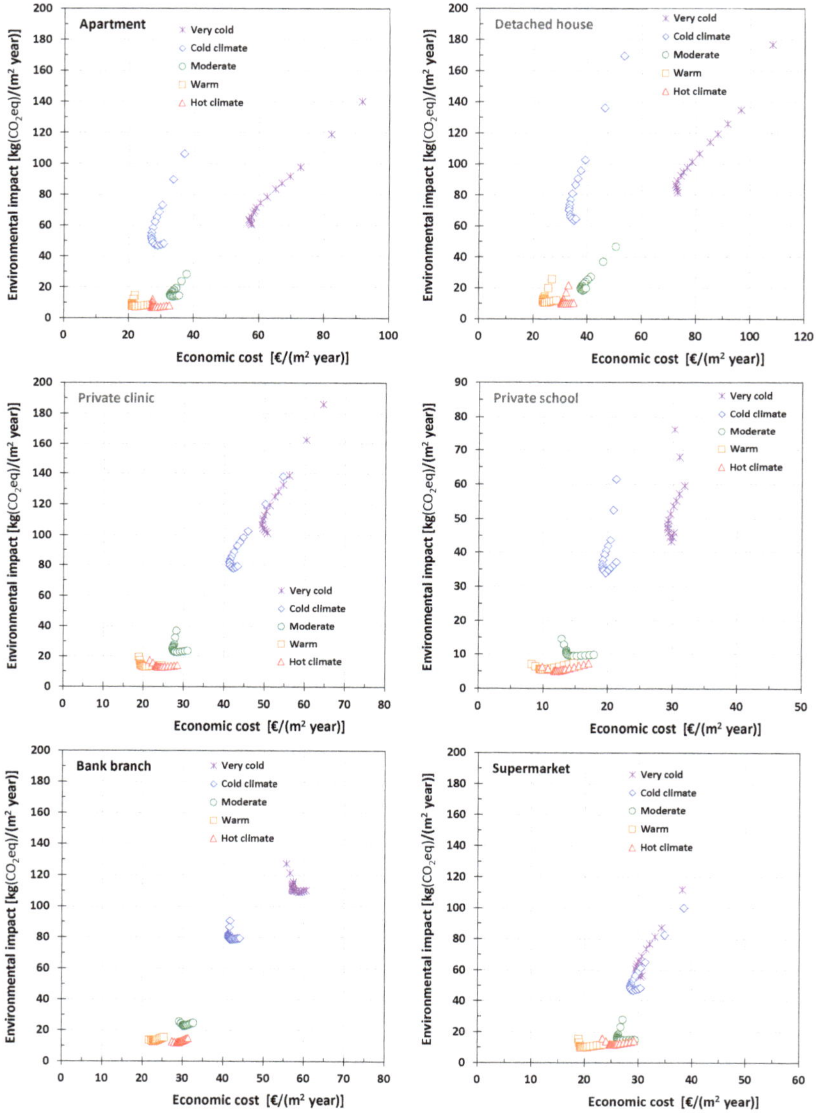

Figure 7. Relationship between the environmental impact and the economic cost.

In summary, for the cases considered in this study, the greater the thickness of EPS, the lower the energy consumption for air conditioning. From an environmental perspective, there is always an optimal EPS thickness below which emissions of CO_2 equivalent decrease with increasing EPS thickness. For thicknesses above this ideal value, the energy embodied in constructive solutions is not offset by the reduction in energy consumption for air conditioning. The economically optimal thicknesses of EPS are significantly lower than the corresponding ones in environmental terms. From a strictly economic perspective, the additional investment in a thermal insulation system is only clearly offset by the reduction in energy expenditure for air conditioning when buildings are in a cold or very cold climate. However, the higher the price of purchasing energy, the greater the economic interest in increasing the thickness of thermal insulation [2], so it is highly likely that in the future it will be economically interesting to apply generous thicknesses of thermal insulation, even in moderate climates.

5. Conclusions

An assessment of the relationship between the cost of thermal comfort in residential and non-residential buildings and the type of building use, the type of climate, and the thermal insulation of opaque elements was carried out using energy, environmental, and economic analyses. Three buildings with permanent use (an apartment, a detached house, and a clinic) and three buildings with intermittent utilization (a school, a bank branch, and a supermarket) were selected, which were supposedly located in places with five different climates (hot, warm, moderate, cold, and very cold).

The concepts of "economic cost of thermal insulation efficiency" and "environmental cost of thermal insulation efficiency" were applied to the three thermal insulating materials most widely used (EPS—expanded polystyrene, XPS—extruded polystyrene, and MW—mineral wool). Both perspectives reveal the EPS as the most promising. Therefore, this was the thermal insulation material selected for this study.

Regardless of the climatic region and the type of the building, the predicted values show a continuous decrease in the consumption of primary energy for climatization with the increase in the EPS thickness applied to opaque elements of the building's envelope. This decrease has an inverse exponential behavior, showing a lessening of the decreasing rate as the EPS thickness increases. The energy cost of thermal comfort is substantially higher in the case of the two cold climates than in the other three, and it is also in the cold climates that there is a more significant decrease in energy consumption for air conditioning with the increase in EPS thickness. The buildings with day and night uses have higher energy consumption for climate control than those with only daytime occupation or occupied during the day and part of the night.

The environmental cost of thermal comfort mainly depends on the annual primary energy consumption for heating and cooling, with emissions associated with building construction having a smaller influence. The buildings with permanent use have a higher environmental cost than those with intermittent occupation, and this cost is substantially higher for the two cold climates than for the remaining three. Clearly, it is more environmentally advantageous to thermally insulate opaque elements of buildings located in cold climates than in moderate, warm, or hot climates. Thus, there is a great sensitivity to this parameter in relation to climate and building types, with a substantially stronger relationship with the type of climate.

In general terms, the economic optimal EPS thickness increases with the cooling of the climate, with the lowest value for hot climates and the highest for very cold ones. Furthermore, it varies with the building's type of use, with higher values for residential buildings than for commercial and service buildings. Economically, the investment in a thermal insulation system is only clearly offset by the reduction in energy expenditure for climatization when buildings are in cold or very cold climates and/or if they are of the residential type.

In short, it has been demonstrated that the application of thermal insulation in the opaque constructive solutions of the building envelope is a highly efficient way of reducing energy consumption and environmental impacts, and it is economically advantageous in the case of residential buildings and/or all buildings located in cold climates. The optimal economic thicknesses of EPS are significantly lower than the corresponding ones in environmental terms.

The optimal thermal insulation thicknesses of the opaque elements of six types of buildings, supposedly located in different types of climates, were obtained using a unified assessment. The results can be used to select the recommended thermal insulation thickness to apply in a given building located in a specific climate; this fact represents the main added value of this work. Clearly, the optimal thickness depends on the perspective considered.

It should be noted that the influence of installing electrical energy production systems from renewable sources was not considered, which is noteworthy in environmental terms and might become economically significant due to the probable energy price escalation. These limitations deserve to be addressed in future work. Finally, the results of this research provide useful insights for building professionals and policymakers.

Author Contributions: Conceptualization, A.M.R.; methodology, A.M.R.; software, A.M.R. and A.M.S.; validation, A.M.R.; investigation, A.M.S.; writing-first draft, A.M.S.; formal analysis, A.M.R. and A.V.M.O.; writing-second draft, review, and editing, A.M.R. and A.V.M.O. All authors have read and agreed to the published version of the manuscript.

Funding: This research received no external funding.

Institutional Review Board Statement: Not applicable.

Informed Consent Statement: Not applicable.

Data Availability Statement: Not applicable.

Conflicts of Interest: The authors declare no conflict of interest.

References

1. Frontczak, M.; Wargocki, P. Literature survey on how different factors influence human comfort in indoor environments. *Build. Environ.* **2011**, *46*, 922–937. [CrossRef]
2. Raimundo, A.M.; Oliveira, A.V.M. Analyzing thermal comfort and related costs in buildings under Portuguese temperate climate. *Build. Environ.* **2022**, *219*, 109238. [CrossRef]
3. Intergovernmental Panel on Climate Change. Technical Summary. In *Climate Change 2022: Mitigation of Climate Change. Contribution of Working Group III to the Sixth Assessment Report of the Intergovernmental Panel on Climate Change*; Cambridge University Press: Cambridge, UK; New York, NY, USA, 2022. Available online: https://www.ipcc.ch/report/sixth-assessment-report-working-group-3/ (accessed on 2 September 2022). [CrossRef]
4. United Nations Environment Programme. *Emissions Gap Report 2022: The Closing Window—Climate Crisis Calls for Rapid Transformation of Societies*; United Nations Environment Programme: Nairobi, Kenya, 2022. Available online: https://www.unep.org/resources/emissions-gap-report-2022 (accessed on 20 September 2022).
5. European Union. Directive (EU) 2018/844 of the European Parliament and of the Council, of 30 May 2018, amending Directive 2010/31/EU on the energy performance of buildings and Directive 2012/27/EU on energy efficiency. *Offic. J. Eur. Union* **2018**, *156*, 75–91. Available online: https://eur-lex.europa.eu/legal-content/EN/TXT/?uri=uriserv%3AOJ.L_.2018.156.01.0075.01.ENG (accessed on 2 September 2022).
6. *2021/0426 (COD)*; Proposal for a Directive of the European Parliament and of the Council on the Energy Performance of Buildings (Recast). European Union: Brussels, Belgium, 2021. Available online: https://eur-lex.europa.eu/procedure/EN/2021_426 (accessed on 2 September 2022).
7. Scheuer, C.; Keoleian, G.A.; Reppe, P. Life cycle energy and environmental performance of a new university building: Modeling challenges and design implications. *Energy Build.* **2003**, *35*, 1049–1064. [CrossRef]
8. Beccali, M.; Cellura, M.; Fontana, M.; Longo, S.; Mistretta, M. Energy retrofit of a single-family house: Life cycle net energy saving and environmental benefits. *Renew. Sustain. Energy Rev.* **2013**, *27*, 283–293. [CrossRef]
9. Bastos, J.; Batterman, S.A.; Freire, F. Life-cycle energy and greenhouse gas analysis of three building types in a residential area in Lisbon. *Energy Build.* **2014**, *69*, 344–353. [CrossRef]
10. Atmaca, A.; Atmaca, N. Life cycle energy (LCEA) and carbon dioxide emissions ($LCCO_2A$) assessment of two residential buildings in Gaziantep, Turkey. *Energy Build.* **2015**, *102*, 417–431. [CrossRef]
11. Raimundo, A.M.; Saraiva, N.B.; Oliveira, A.V.M. Thermal insulation cost optimality of opaque constructive solutions of buildings under Portuguese temperate climate. *Build. Environ.* **2020**, *182*, 107107. [CrossRef]
12. Jafari, A.; Valentin, V. Selection of optimization objectives for decision-making in building energy retrofits. *Build. Environ.* **2018**, *130*, 94–103. [CrossRef]

13. *EN 15459:2007*; Energy Performance of Buildings—Economic Evaluation Procedure for Energy Systems in Buildings. CEN (European Committee for Standardization): Brussels, Belgium, 2007.
14. European Commission. Commission Delegated Regulation (EU) 244/2012 of 16 January 2012 supplementing Directive 2010/31/EU on the energy performance of buildings by establishing a methodology framework for calculating cost optimal levels of minimum energy performance. *Offic. J. Eur. Union* **2012**, *81*, 18–36. Available online: https://eur-lex.europa.eu/legal-content/EN/TXT/?uri=CELEX%3A32012R0244 (accessed on 2 September 2022).
15. Raimundo, A.M.; Saraiva, N.B.; Dias, L.P.; Rebelo, A.C. Market-oriented cost-effectiveness and energy analysis of windows in Portugal. *Energies* **2021**, *14*, 3720. [CrossRef]
16. Grazieschi, G.; Asdrubali, F.; Thomas, G. Embodied energy and carbon of building insulating materials: A critical rewiew. *Clean. Environ. Syst.* **2021**, *2*, 100032. [CrossRef]
17. Anastaselos, D.; Giama, E.; Papadopoulos, A. An assessment tool for the energy, economic and environmental evaluation of thermal insulation solutions. *Energy Build.* **2009**, *41*, 1165–1171. [CrossRef]
18. Dylewski, R.; Adamczyk, J. Optimum thickness of thermal insulation with both economic and ecological costs of heating and cooling. *Energies* **2021**, *14*, 3835. [CrossRef]
19. Las-Heras-Casas, J.; López-Ochoa, L.M.; López-González, L.M.; Olasolo-Alonso, P. Energy renovation of residential buildings in hot and temperate Mediterranean zones using optimized thermal envelope insulation thicknesses: The case of Spain. *Appl. Sci.* **2021**, *11*, 370. [CrossRef]
20. Wang, H.; Huang, Y.; Yang, L. Integrated economic and environmental assessment-based optimization design method of building roof thermal insulation. *Buildings* **2022**, *12*, 916. [CrossRef]
21. Yang, Q.; Kong, L.; Tong, H.; Wang, X. Evaluation model of environmental impacts of insulation building envelopes. *Sustainability* **2020**, *12*, 2258. [CrossRef]
22. Totland, M.; Kvande, T.; Bohne, R.A. The effect of insulation thickness on lifetime CO_2 emissions. *IOP Conf. Ser. Earth Environ. Sci.* **2019**, *323*, 012033. [CrossRef]
23. Congedo, P.M.; Baglivo, C.; Seyhan, A.K.; Marchetti, R. Worldwide dynamic predictive analysis of building performance under long-term climate change conditions. *J. Build. Engin.* **2021**, *42*, 103057. [CrossRef]
24. Ounis, S.; Aste, N.; Butera, F.M.; Pero, C.D.; Leonforte, F.; Adhikari, R.S. Optimal balance between heating, cooling and environmental impacts: A method for appropriate assessment of building envelope's U-value. *Energies* **2022**, *15*, 3570. [CrossRef]
25. Bo, R.; Shao, Y.; Xu, Y.; Yu, Y.; Guo, H.; Chang, W.S. Research on the relationship between thermal insulation thickness and summer overheating risk: A case study in severe cold and cold regions of China. *Buildings* **2022**, *12*, 1032. [CrossRef]
26. Chen, S.; Zhang, G.; Xia, X.; Setunge, S.; Shi, L. A review of internal and external influencing factors on energy efficiency design of buildings. *Energy Build.* **2020**, *216*, 109944. [CrossRef]
27. Andreotti, M.; Bottino-Leone, D.; Calzolari, M.; Davoli, P.; Pereira, L.D.; Lucchi, E.; Troi, A. Applied research of the hygrothermal behaviour of an internally insulated historic wall without vapour barrier: In situ measurements and dynamic simulations. *Energies* **2020**, *13*, 3362. [CrossRef]
28. Kvande, T.; Bakken, N.; Bergheim, E.; Thue, J.V. Durability of ETICS with rendering in Norway—Experimental and field investigations. *Buildings* **2018**, *8*, 93. [CrossRef]
29. Brambilla, A.; Bonvin, J.; Flourentzou, F.; Jusselme, T. On the influence of thermal mass and natural ventilation on overheating risk in offices. *Buildings* **2018**, *8*, 47. [CrossRef]
30. Michałowski, B.; Marcinek, M.; Tomaszewska, J.; Czernik, S.; Piasecki, M.; Geryło, R.; Michalak, J. Influence of rendering type on the environmental characteristics of expanded polystyrene-based external thermal insulation composite system. *Buildings* **2020**, *10*, 47. [CrossRef]
31. D'Agostino, D.; Landolfi, R.; Nicolella, M.; Minichiello, F. Experimental study on the performance decay of thermal insulation and related influence on heating energy consumption in buildings. *Sustainability* **2022**, *14*, 2947. [CrossRef]
32. Landolfi, R.; Nicolella, M. Durability assessment of ETICS: Comparative evaluation of different insulating materials. *Sustainability* **2022**, *14*, 980. [CrossRef]
33. Dylewski, R.; Adamczyk, J. Economic and environmental benefits of thermal insulation of building external walls. *Build. Environ.* **2011**, *46*, 2615–2623. [CrossRef]
34. Adamczyk, J.; Dylewski, R.; Sobierajewicz, P. Economic and ecological benefits of thermal insulation of external partitions depending on the temperature in residential premises. *J. Clean. Product.* **2023**, *384*, 135622. [CrossRef]
35. Raimundo, A.M. *SEnergEd—Software for Buildings' Dynamic Energy Simulation and Calculation of Their Environmental and Economic Life-Cycle Costs, User Guide 2019*; DEM, FCTUC, University of Coimbra: Coimbra, Portugal. (In Portuguese)
36. *ISO 13790:2008*; Energy Performance of Buildings—Calculation of Energy Use for Space Heating and Cooling. International Organization for Standardization: Geneva, Switzerland, 2008. Available online: https://www.iso.org/obp/ui/#iso:std:iso:13790:ed-2:v1:en (accessed on 2 September 2022).
37. General Directorate of Energy and Geology of Portugal. *Technical Manual for the Assessment of the Energy Performance of Buildings. Approved by Dispatch n° 6476-H/2021 of July 1*; General Directorate of Energy and Geology of Portugal: Lisbon, Portugal, 1 July 2021. Available online: https://www.sce.pt/legislacao/ (accessed on 2 September 2022). (In Portuguese)
38. Eurostat. Monthly Minimum Wages—Bi-annual Data, First Semester of 2022. Available online: https://ec.europa.eu/eurostat/en/web/products-datasets/-/EARN_MW_CUR (accessed on 2 September 2022).

39. Baglivo, C.; Congedo, P.M.; D'Agostino, D.; Zacá, I. Cost-optimal analysis and technical comparison between standard and high efficient mono-residential buildings in a warm climate. *Energy* **2015**, *83*, 560–575. [CrossRef]
40. European Commission. Consolidated text of Commission Delegated Regulation (EU) 626/2011 of 4 May 2011 supplementing Directive 2010/30/EU with regard to energy labelling of air conditioners. *Offic. J. Eur. Union* **2011**, *L178*, 1–78. Available online: https://eur-lex.europa.eu/legal-content/EN/TXT/?uri=CELEX%3A02011R0626-20200809 (accessed on 16 September 2022).
41. *EN 16798-1:2019*; Energy Performance of Buildings—Part 1: Indoor Environmental Input Parameters for Design and Assessment of Energy Performance of Buildings Addressing Indoor Air Quality, Thermal Environment, Lighting and Acoustics, CEN. European Committee for Standardization: Brussels, Belgium, 2019.
42. Guardigli, L.; Bragadin, M.A.; Fornace, F.D.; Mazzoli, C.; Prati, D. Energy retrofit alternatives and cost-optimal analysis for large public housing stocks. *Energy Build.* **2018**, *166*, 48–59. [CrossRef]
43. Schiavoni, S.; D'Alessandro, F.; Bianchi, F.; Asdrubali, F. Insulation materials for the building sector: A review and comparative analysis. *Renew. Sustain. Energy Rev.* **2016**, *62*, 988–1011. [CrossRef]
44. Braulio-Gonzalo, M.; Bovea, M.D. Environmental and cost performance of building's envelope insulation materials to reduce energy demand: Thickness optimization. *Energy Build.* **2017**, *150*, 527–545. [CrossRef]
45. Eurostat. Electricity Price Statistics, First Semester of 2022. Available online: http://ec.europa.eu/eurostat/statistics-explained/index.php/Electricity_price_statistics (accessed on 16 September 2022).
46. Eurostat. Natural Gas Price Statistics, First Semester of 2022. Available online: http://ec.europa.eu/eurostat/statistics-explained/index.php/Natural_gas_price_statistics (accessed on 16 September 2022).

buildings

MDPI

Article

Assessment of the Retrofit Strategies on Thermal Insulation Applied in Buildings Located on the Southern Border of the EU: The Case of the Canary Islands

Eduardo González-Díaz [1], José Miguel Márquez-Martinón [1], Ana Pérez-García [1], Norena Martín-Dorta [1] and Benjamín González-Díaz [2,*]

[1] Departamento de Técnicas y Proyectos en Ingeniería y Arquitectura, Universidad de La Laguna (ULL), 38200 San Cristóbal de La Laguna, Spain
[2] Departamento de Ingeniería Industrial, Escuela Superior de Ingeniería y Tecnología, Universidad de La Laguna (ULL), 38206 San Cristóbal de La Laguna, Spain
* Correspondence: bgdiaz@ull.edu.es; Tel.: +34-922-316-502 (ext. 6252)

Citation: González-Díaz, E.; Márquez-Martinón, J.M.; Pérez-García, A.; Martín-Dorta, N.; González-Díaz, B. Assessment of the Retrofit Strategies on Thermal Insulation Applied in Buildings Located on the Southern Border of the EU: The Case of the Canary Islands. *Buildings* **2022**, *12*, 1994. https://doi.org/10.3390/buildings12111994

Academic Editors: Paulo Santos, Mark Bomberg and Xing Jin

Received: 3 October 2022
Accepted: 11 November 2022
Published: 16 November 2022

Abstract: Nowadays, a large amount of the total primary energy is consumed by buildings, accounting for about 40% of the total energy demand. Aligned with the EU objectives and the strategies to reduce the demand, cooling and heating are stated as the most energy consuming processes and the building envelope plays an important role to reduce the energy consumption. In this work, the energy demand related to heating and cooling in a typical building has been evaluated, which has been simulated in 35 cities located in different climatic zones, using the DesignBuilder v.6.1.7.007 software. Although the increase in insulation and the replacement of windows lead to a reduction in energy demand, in the case of the cities of Santa Cruz de Tenerife and Las Palmas de Gran Canaria without insulation, the demands are lower than 1.7 kWh/m^2/year and 5 kWh/m^2/year, respectively, and these results indicate that energy saving strategies, driven by policies and economic support, based on the renovation and improvement of the thermal insulation of the building envelope, are not the most appropriate due to the need for an additional energy load for cooling and to maintain comfort within the regulatory limits.

Keywords: energy transition; energy consumption; building envelope; retrofitting strategies

1. Introduction

The global energy demand has been increasing during the last years mainly due to the economic development and globalization processes. The building sector is one of the principal energy-consuming sectors in the world, representing around 40% of the total demand [1] and contributing over 30% of the CO_2 emissions [2], where heating and cooling processes are responsible for more than 33% of the total energy consumed in buildings [3]. Additionally, in some trend scenarios proposed for 2050, the energy demand for cooling will reach 150% of the actual values and the cooling energy demand will rise by 300% to 600% in buildings [4].

In this sense, EU has established clear objectives on energy saving in the construction sector, defining broad major objectives and assumptions aimed at lowering the energy consumption of buildings [5], complemented by the Directive 2010/31//UE which introduced the concept of nearly zero-energy building (nZEB) [6,7]. In the nZEB concept, some authors have defined two approaches for the retrofitting strategy to achieve the targets: reducing the power consumption and perform an energy transition, replacing the actual energy mix and introducing renewable energies [1].

Therefore, the improvements to reduce the heating and cooling demand are the most relevant retrofitting actions in the buildings to produce an effective reduction of the energy demand, where the building envelope [8] plays a key role, together with other construction elements.

In this aforementioned sense, the most used renovation proposals to reduce the energy consumption are to increase the insulation thickness of the thermal envelope as well as installing double-glazed windows. Several authors have been focused on the optimization of the insulation thickness of the exterior walls of buildings to minimize the heating demand required in cold regions [9] and to reduce the temperature [10–12], where the retrofitting actions in continental climates have reported reduction over 30% of the energy demand [13]. Similar studies referred to hot climates [14] have reported the optimization of insulation thickness for the reduction of heat transfer in buildings during the summer, considering solar radiation and the feasibility of retrofit actions, in combination with Solar Passive Technologies [15]. Additionally, the thermal performance of the building considering the percentage of window surface and type of glazing [16] has been analyzed for different climate conditions.

The aforementioned strategies have different results depending on the climatic zone [17] and the specific boundary conditions of the environment, in order to provide results in context-specific solutions [18]. For example, in Algeria, reducing cooling demand by replacing single glass with double glass contributes, on average, to a reduction in energy demand of 8%. However, the application of other strategies, such as solar protection, window glazing, air tightness and insulation, and reductions in energy consumption can achieve energy savings of 33% [19].

The analysis of the comfort for buildings without installed heating or cooling systems reveals that it is possible to have winter thermal comfort without additional thermal insulation on façades in climates similar to the cities of southern Europe [20].

In general, the results of the energy evaluation in different locations show the reduction in energy demand associated with retrofit strategies depends, to a large extent, on the climate zone where the building is located. Therefore, the criteria for achieving the objectives set by the EU in terms of improving the energy performance of the building stock require adaptation to the climatic singularities of each region [18,21].

The scientific literature has extensively reported the effect of insulation in buildings located in areas with a remarkable heating demand where the implications of passive design measures on heating and cooling energy have been properly addressed [22]. In particular, the Spanish standard, which is the transposition of European directives, promotes and partially finances the placement of building insulation and the replacement of windows.

In this sense, following the aforementioned studies published in the literature, we present in this work a novel study of the energy evaluation of a building on the southern border of the EU, to evaluate the behavior of these strategies for the Canary Islands, in comparison with other EU cities.

The present work analyzes the energy demand of a reference building located in 33 cities in different climate zones, following the classification reported in [23], and compared with the two Canary Islands capital cities.

The paper is structured as follows. In Section 2, the building model and simulation variables are listed. In Section 3, the obtained results and discussion are shown. This section is followed by the conclusions in Section 4.

2. Building Model and Methodology

2.1. Description of the Building Used in This Work

We have selected one of the twin buildings used as a reference for the European ENCORE H2020 Project (ENergy aware BIM Cloud Platform in a COst-effective Building REnovation Context), to carry out the simulations. They are two single-family dwellings built at full scale: a standard dwelling, used as a reference and the twin dwelling, used to carry out experimental innovations to reduce the energy consumption. These buildings were built in the framework of the Experimental Demonstrators in Energy and Architecture (EDEA) project co-funded by the LIFE program of the European Commission [24].

The experimental building used in the simulations has a façade of 140.5 m^2 (north and south), dividing walls of 160.7 m^2 (east and west), a roof of 70.4 m^2, and 22.0 m^2 of openings in the façade walls (Figure 1). In terms of the window-wall ratio (WWR), the building presents 11% of openings to the north and 20% of openings to the south (see Table 1). The selected building model does not present window opening areas in the east and west façades, because this model represents a rowhouse. The window-wall areas and WWR are the standard values in Spain for the typology of the selected building.

Figure 1. EDEA characteristic building BIM model (**left**) used in the simulations and demonstration houses of the EDEA project built in Cáceres, Spain (**right**).

Table 1. Wall areas, window opening areas, window wall ratios, and gross wall areas for the different façades of the simulated building.

Walls and Areas	Total	North (315 to 45 deg)	East (45 to 135 deg)	South (135 to 225 deg)	West (225 to 315 deg)
Gross Wall Area (m^2)	325.11	75.3	74.89	75.3	99.61
Above Ground Wall Area (m^2)	301.33	75.3	74.89	65.25	85.88
Window Opening Area (m^2)	21.98	8.43	0	13.55	0
Gross Window-Wall Ratio (%)	6.76	11.2	0	17.99	0
Above Ground Window-Wall Ratio (%)	7.29	11.2	0	20.76	0
Gross Wall Area (m^2)	325.11	75.3	74.89	75.3	99.61

The building materials and construction solutions used in the simulation have been chosen to represent a single-family house type of construction system. Table 2 shows the description and thermal properties of the building materials included in the construction solutions of the building envelope. The orientation of the EDEA building has been unaltered in the different emplacements to minimize a crossed effect due to the orientation changes.

Table 2. Building envelope construction solutions and material properties: t: Thickness; λ: Thermal conductivity; ρ: Density; C_p: Specific heat; R: Thermal resistivity; i(t): Insulation thickness where i = 0, 40, 80 or 120 mm. (1) Depending on the thickness R = i(t)/λ. (2) Spanish Technical Building Code [25].

Parameters	U (W/m²K)	Materials	t (10^{-2} m)	λ (W/m K)	ρ (kg/m³)	C_p (J/kg K)	R (m²K/W)
Roof	2.40 $^{i(0)}$, 0.62 $^{i(40)}$, 0.36 $^{i(80)}$, 0.25 $^{i(120)}$	Crushed stone	10	2.0	1450	1050	0.05
		Concrete with lightweight aggregates (1600 < ρ < 1800)	8	1.15	1700	1000	0.07
		XPS expanded with CO_2	i(t)	0.034	38	1000	(1)
		Reinforced concrete (2300 < ρ < 2500)	30	2.3	2400	1000	0.13
		Gypsum plaster (1000 < ρ < 1300)	1.5	0.57	1150	1000	0.03
Floor slab	1.87 $^{i(0)}$, 0.59 $^{i(40)}$, 0.35 $^{i(80)}$, 0.25 $^{i(120)}$	Stoneware tile	2	2.3	2500	1000	0.01
		Cement mortar for plastering (1600 < ρ < 1800)	3	1.0	1525	1000	0.03
		XPS expanded with CO_2	i(t)	0.034	38	1000	(1)
		Reinforced concrete slab (2300 < ρ < 2500)	15	2.3	2400	1000	0.07
		Hardcore (stone)	40	2.0	1450	1050	0.20
Outer wall (North and South)	1.69 $^{i(0)}$, 0.61 $^{i(40)}$, 0.37 $^{i(80)}$, 0.27 $^{i(120)}$	Ceramic perforated brick	11.5	0.667	1140	1000	0.17
		Unvented air chamber insulation	5	-	-	-	0.18 (2)
		Mineral wool insulation	i(t)	0.04	40	1000	(1)
		Gypsum board (750 < ρ < 900)	1.5	0.25	825	1000	0.06
Outer wall (East and West)	1.13 $^{i(0)}$, 0.52 $^{i(40)}$, 0.33 $^{i(80)}$, 0.25 $^{i(120)}$	Viroc® Cement Bonded Particle Board (CBPB)	2	0.22	1350	1500	0.09
		Unvented air chamber insulation	10	-	-	-	0.19 (2)
		Mineral wool insulation	i(t)	0.04	40	1000	(1)
		Cement mortar for plastering (1600 < ρ < 1800)	1.5	1	1525	1000	0.02
		Ceramic perforated brick	11.5	0.667	1140	1000	0.17
		Cement mortar for plastering (1600 < ρ < 1800)	1.5	1	1525	1000	0.02
		Unvented air chamber insulation	5	-	-	-	0.18 (2)
		Gypsum board (750 < ρ < 900)	1.5	0.25	825	1000	0.06

To implement a systematic comparison, we have selected commercial insulation thickness of 0, 40, 80, and 120 mm, in accordance with what has been reported in the literature [26], to evaluate the insulation influence in the chosen cities. Despite the selected values not being the most appropriate for northern and central countries according to EU country policies, the aim of this work is focused on the analysis and comparison of the insulation performance and constraints for the southern cities under study.

The characteristics of the construction materials of the building's interior construction solutions are shown in Table 3.

2.2. Methodology of Calculation and Selection of Cities

The proposed methodology evaluates the energy demand of the building, when the insulation thicknesses and the type of glazing in windows are modified, aligned with European retrofitting guidelines. The selection of cities has been performed in two groups: (i) cities with heating demand and (ii) cities without heating demand. The energy analysis was performed with DesignBuilder (Calculation Engine EnergyPlus) v.6.1.7.007 software, whose calculation engine is Energy Plus v.8.9.0.001. The EPW (Energy Plus Weather) files from the Energy Plus database were used as climate files for each of the selected locations.

Table 3. Interior construction solutions and material properties: t: Thickness; λ: Thermal conductivity; ρ: Density; C_p: Specific heat.

Interior Parameters	Materials	t (10^{-2} m)	λ (W/m K)	ρ (kg/m³)	C_p (J/kg K)
Interior slab 1.66 W/m²K	Ceramic tile	2	2.3	2500	1000
	Cement mortar for plastering 1600 < ρ < 1800	3	1	1525	1000
	Concrete with lightweight aggregates 1600 < ρ < 1800	5	1.15	1700	1000
	Reinforced concrete slab 2300 < ρ < 2500	37	2.3	2400	1000
	Gypsum plaster 1000 < ρ < 1300	2	0.57	1150	1000
Interior wall 2.09 W/m²K	Gypsum plaster 1000 < ρ < 1300	1.5	0.57	1150	1000
	Ceramic perforated brick	11.5	0.667	1140	1000
	Gypsum plaster 1000 < ρ < 1300	1.5	0.57	1150	1000

The Engine EnergyPlus uses finite difference calculation method, the Conduction Transfer Function (CTF) algorithm, TARP method for the internal convection algorithm, and DOE-2 for the external convection algorithm.

To carry out the simulations, four thermal zones have been considered in the building, one zone per floor. The interior partition walls have not been considered for the energy evaluation. Additionally, the building has been simulated considering the slab in contact with the first floor for all the locations.

The energy demand of the test building, located in 35 cities in different locations of the EU, has been calculated (Figure 2). The results have been considered appropriate when the insulation thickness reduces the cooling energy demand. In the opposite case, when the insulation thickness increases the cooling energy demand, the previous valid insulation thickness has been considered. We have chosen different locations to consider the different climates zones in the EU, and the cities have been grouped by latitude, forming different groups according to the classification shown in [23].

Moreover, in this study, we have included European cities located outside of the continental climates, to evaluate the impact of the political guidelines of retrofitting from the EU.

2.3. *Transmittances of the Envelope and Windows*

The transmittances of the different elements of the thermal envelope according to the insulation level are shown in Table 4. Regarding the glazing type, we have carried out simulations for two window types and frame combinations: 6 mm glass with aluminum frame (the most widely used window in buildings in the Canary Island cities under study) and 4 + 12 + 6 mm glass with PVC frame, to assess the energy demand changes due to the replacement of the windows.

Table 4. Transmittance (W/m²K) of envelope elements (U_{env}), window types (U_W), and g/SHGC values: t = Insulation thickness (mm); N = North; S = South; E = East; W = West.

Elements	U_{env}				Window Type	U_W	g/SHGC
	t = 0	t = 40	t = 80	t = 120			
Roof	2.56	0.63	0.36	0.25	6 mm glass with aluminum frame	5.8 5.7	0.85
External wall (N and S)	1.69	0.61	0.37	0.27			
Internal wall (E andW)	1.12	0.51	0.33	0.25			
Interior slab	1.66	-	-	-	4 + 12 + 6 mm glass with PVC frame	2.7 1.8	0.77
Floor slab	1.83	-	-	-			
Interior wall	2.09	-	-	-			

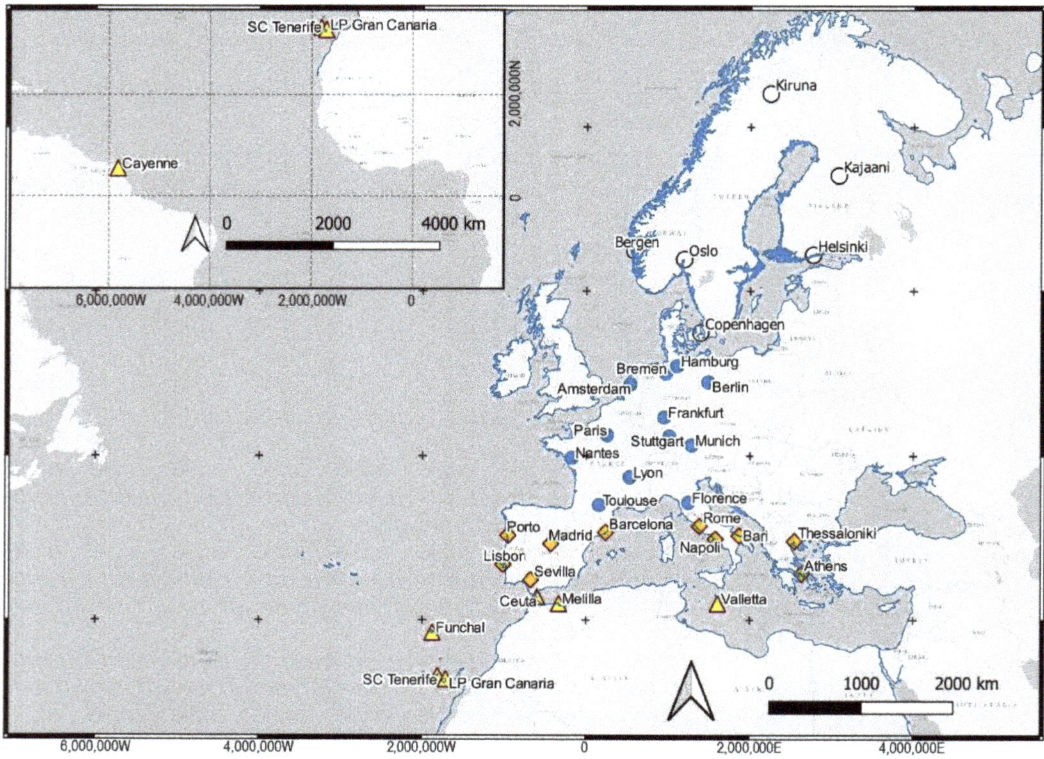

Figure 2. Location of the 35 cities studied grouped into 4 zones: north (white circle), center (blue circle), south on the continent (yellow rhombus), and south (yellow triangle).

2.4. Calculation of Infiltrations in the Building

The calculation of infiltration was performed according to the Energy Performance of Buildings Directive 2010/31/EU, the Energy Efficiency Directive 2012/27/EU, and subsequent amendments contained in the Spanish transposition of the aforementioned Directives [25].

According to study [27], the infiltration value obtained was 0.186 ACH (Air Changes per Hour) with a permeability value equal to 9 $m^3/h{\cdot}m^2$ (at 100 Pa) and a mechanical ventilation equal to 0.63 ACH for the volume of the dwelling, façade and roof areas, and percentage of openings indicated in the building description.

The year-round mechanical ventilation of 0.63 ACH ensures healthiness through proper aeration of the living spaces. However, a natural ventilation of 4 ACH has also been included during the summer months (June, July, August, and September) from 0:00 to 07:59 to cool the interior spaces in summer and improve the thermal comfort of the occupants to minimize the use of active cooling systems.

Night ventilation has been used exclusively in the summer months (June, July, August, and September, between 1 and 7 a.m.). In the case of northern European cities, night ventilation produces indoor temperatures of between 17 °C and 20 °C.

2.5. Internal Loads, Usage Profiles, Metabolic Rate, and Set-Point Temperatures

The internal loads define the heat generated inside the building due to internal sources: occupancy, lighting, equipment, etc. These loads are involved in the calculation of the energy demand of the analyzed models. The internal loads and associated operating hours used in the simulations are described in Table 5.

Table 5. Internal loads and schedules used in the simulations [25]. WD: Working days; NWD: Weekend and public holidays.

Internal Load (W/m^2)		Schedule (Typical Week)					
		0:00–6:59	7:00–14:59	15:00–17:59	18:00–18:59	19:00–22:59	23:00–23:59
Occupation (Sensitive)	WD	2.15	0.54	1.08	1.08	1.08	2.15
	NWD	2.15	2.15	2.15	2.15	2.15	2.15
Occupation (Latent)	WD	1.36	0.34	0.68	0.68	0.68	1.36
	NWD	1.36	1.36	1.36	1.36	1.36	1.36
Lighting	Both	0.44	1.32	1.32	2.20	4.40	2.20
Equipment	Both	0.44	1.32	1.32	2.20	4.40	2.20

The metabolic rate was estimated as the sum of the sensible occupation (O_{sen} = 2.15 W/m^2) and latent occupation (O_{lat} = 1.36 W/m^2) for an occupancy density (ρ_d) equal to 33.33 m^2/person. Applying Equation (1), we obtain a metabolic rate (M_{rate}) equal to 117 W/m^2.

$$M_{rate} = O_{sen} \cdot \rho_d + O_{lat} \cdot \rho_d \tag{1}$$

The set-point temperatures used for the winter months were 20 °C and 17 °C (heating temperatures) and for the summer months were 25 °C and 27 °C (cooling temperatures). These four set-point temperatures were used with the times indicated in Table 6 which correspond to those established in [28].

Table 6. Set-point temperatures and times used in the simulations [25].

Set Point Temperatures	Period	Schedule (Typical Week)			
		0:00–6:59	7:00–14:59	15:00–22:59	23:00–23:59
Winter set-point (C) temperatures (heating)	January–May	17	20	20	17
	June–September	-	-	-	-
	October–December	17	20	20	17
Summer set-point temperatures (C) (cooling)	January–May	-	-	-	-
	June–September	27	-	25	27
	October–December	-	-	-	-

2.6. Inverse Distance Weighting Interpolation

Inverse distance weighting interpolation (IDW) is a deterministic interpolation method that assumes that the interpolated value will be more similar to nearby data than to remote data. IDW interpolation uses distance as the weight, so sample points that are close will have greater weight and the amount of weight will decrease as the distance from the sample point increases. The equation used for IDW interpolation is as follows:

$$hd_j = k_j \sum_{i=1}^{n} \frac{1}{d_{ij}^p} hd_i \tag{2}$$

$$k_j = \sum_{i=1}^{n} \frac{1}{d_{ij}} \tag{3}$$

where:

hd_j estimated heating demand (kWh/m^2/year) at point j;

hd_i experimental heating demand (kWh/m^2/year) at point i;

d_{ij} distance from point i to j;

p power, in this case p = 2 (weighting with the square of the distance);

n number of cases.

The effect of decreasing the heating demand can be obtained continuously by interpolating the experimental data. The interpolation of the inverse square of the distance is an easy method to apply to estimate the parameters required in the calculation of the energy consumption of buildings [29]. Additionally, this interpolation can be used for the prediction of the buildings' operation through analysis of energy consumption [30]. This technique allows the estimation of the values at the unknown point from a weighted sum of the values of N known points.

3. Results and Discussion

3.1. Heating Demand as a Function of Insulation Thickness

In this work, the total energy demand has been defined as the required energy for cooling and heating systems, to maintain comfort temperature conditions inside the building. Figure 3 shows the heating demand values (kWh/m^2/year) of the house with double-glazed windows 4 + 12 + 6 mm located in northern, central, and southern EU cities, as a function of insulation thickness.

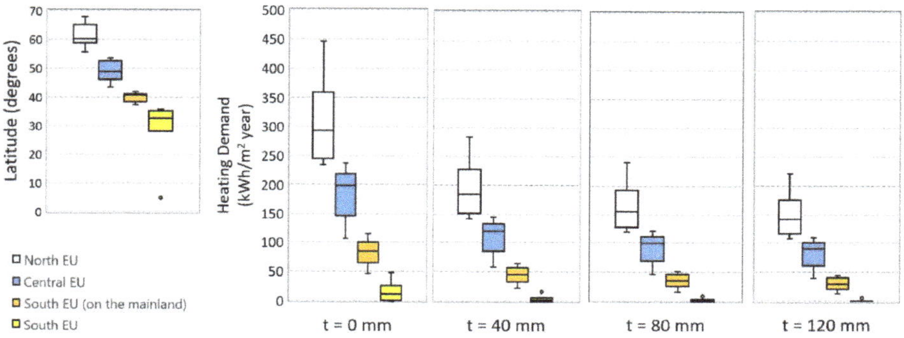

Figure 3. Heating demand (kWh/m^2/year) in cities belonging to the northern, central, southern (on the continent), and southern EU for different insulation thicknesses.

In the northern cities of the EU, heating demand without insulation ranges between 446 and 142 kWh/m^2/year, achieving values of 240 and 118 kWh/m^2/year with 80 mm insulation. For the cities located in the central part of the EU, heating demand varies between 105 and 237 kWh/m^2/year (without insulation) and 45–120 kWh/m^2/year (with 80 mm insulation).

This decrease in heating demand is almost 4 times lower in southern cities of the EU, with the highest demand in cities located within the continent with a heating demand ranging between 114 and 46 kWh/m^2/year (without insulation) and between 50 and 16 kWh/m^2/year with 80 mm thick insulation, despite the fact that the thicknesses selected in this comparative study are not the optimal thicknesses that could be found in those countries.

Specifically, the cities of Ceuta (Spain), Melilla (Spain), Funchal (Portugal), and Valleta (Malta) have a heating demand ranging between 50 and 12 kWh/m^2/year for the building without insulation. This heating demand decreases to values of 15 and 2 kWh/m^2/year, respectively, when 40 mm insulation is included. These values also reach 9 and 0.5 kWh/m^2/year increasing the insulation thickness to 80 mm.

The determination of the interpolated points allows an adequate visualization of the heating energy demand using the values obtained for the cities depicted in Figure 2. The values obtained from the interpolation method compared with the precise values, show an error below 10% on average. The objective of Figure 4 is to show graphically how the heating demand is in the Canary Islands compared to the rest of the EU. This demand remains below the threshold of 15 kWh/m^2/year regardless of the thickness of the insulation.

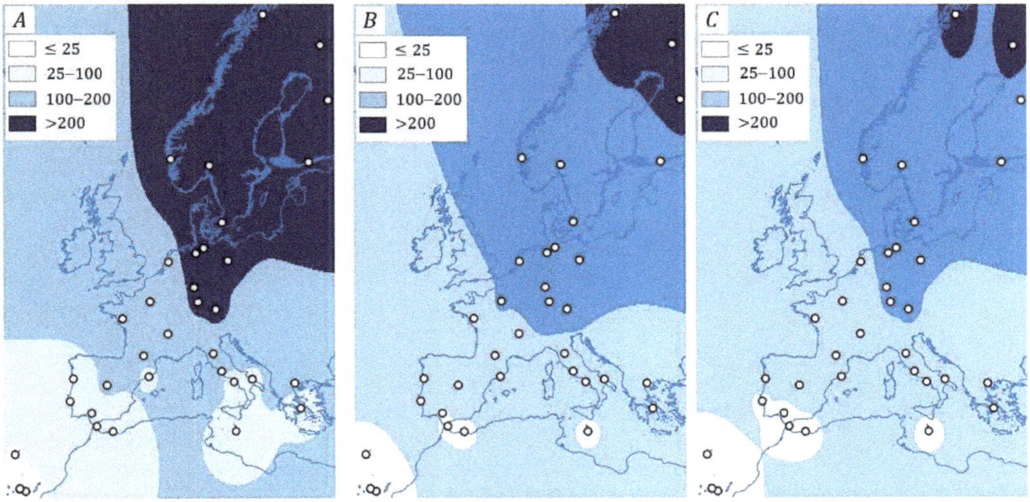

Figure 4. Heating demand intervals (kWh/m^2/year) interpolated by IDW for the reference building with double-glazed windows of 4 + 12 + 6 mm and without insulation in the envelope (**A**), with insulation thicknesses of 40 mm (**B**) and with thicknesses of 80 mm insulation (**C**).

In general, this improvement of heating demand with insulation in northern, central, and southern EU cities can also be observed graphically in the heating demand map by applying IDW, shown in Figure 4.

However, the general behavior observed in northern, central, and southern European cities contrasts with those observed in cities even further south in the EU: Cayenne, Santa Cruz de Tenerife, and Las Palmas de Gran Canaria. In these cases, the installation of insulation does not result in significant savings in heating demand. In these cities, the heating demand is less than 1.7 kWh/m^2/year without insulation (Figure 5A). Therefore, the installation of insulation in Santa Cruz de Tenerife, Las Palmas de Gran Canaria, and Cayenne does not significantly reduce the heating demand in the buildings constructed.

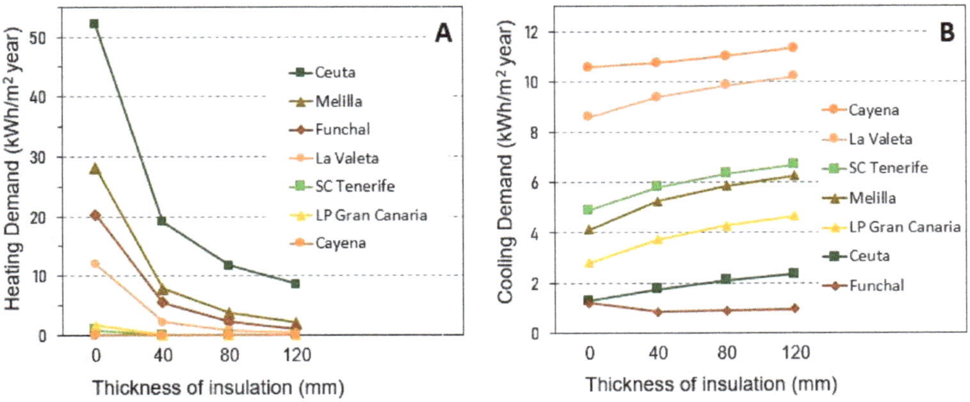

Figure 5. Heating demand (**A**) and cooling demand (**B**) in southern EU cities for different insulation thicknesses and with single-glazed windows (6 mm).

On the other hand, it is remarkable that there was different behavior shown in the Spanish cities of North Africa (Ceuta and Melilla), compared to the results obtained in the Canary Islands capitals, for the same reference building. In the cities of the Canary Islands, unlike the cities of North Africa and the rest of the cities of continental Europe, the installation of insulation does not improve the demand for heating (Figure 5A) and even produces an increase in the demand for cooling (Figure 5B). These results obtained for the reference building show the need to evaluate the rehabilitation strategies that are currently applied in the Canary Islands.

3.2. Cooling Demand as a Function of Insulation Thickness

In a previous work, the cooling demand in Helsinki, Berlin, and Madrid, which are located in the northern and central areas of the EU defined in this work [23], were calculated. The results of this study show the cooling demand in these cities is very low compared to the heating demand. Specifically, the values range between 12 and 0.3 kWh/m^2/year in uninsulated envelopes, and between 9 and 0.3 kWh/m^2/year with 120 mm insulation, being higher in Madrid and lower in Helsinki. In general, northern and central EU cities do not require active systems to control cooling demand.

On the other hand, in the cities of Santa Cruz de Tenerife, Las Palmas de Gran Canaria, Funchal, Ceuta, Melilla, and Cayenne, it is observed (Figure 5B) that the cooling demand is practically constant with insulation thickness, the value always being less than 12 kWh/m^2/year, corresponding to Cayenne. In the case of the Canary Island cities, the cooling demand is less than 5 kWh/m^2/year without insulation, increasing slightly to a value of 7 kWh/m^2/year when the insulation thickness is increased.

This effect of increased cooling demand observed in many cities in the south of the EU as insulation thickness increases is due precisely to the fact that, in hot weather, heat cannot be dissipated through the envelope due to the incorporation of that insulation. This fact could cause overheating and therefore, active cooling systems could be required. Therefore, it can be deduced from the results of these simulations that, in these Canary Island cities, the use of insulation in buildings does not lead to an improvement in energy demand, and even causes a slight increase in cooling demand. In these cities, the reduction of energy demand should be focused on strategies such as orientation, use of shading, and incorporation of renewable energies.

3.3. Single-Glazed vs. Double-Glazed Windows

The installation of double-glazed windows is another frequently used intervention to reduce the energy demand in the dwelling. Figure 6 shows the improvement by replacing single-glazed windows with double-glazed windows in the EDEA building used as a reference in the simulations.

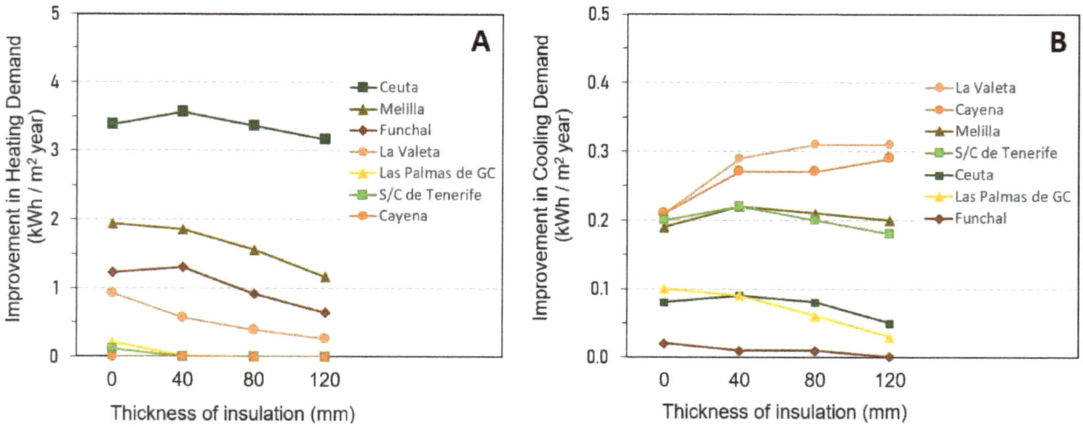

Figure 6. Improvement in heating (**A**) and cooling (**B**) demand replacing a single-glazed window (6 mm) with a double-glazed window (4 + 12 + 6 mm).

Regarding heating demand, it is observed that in Ceuta and Melilla the improvement in heating demand is in the order of 3.5 and 2 $kWh/m^2/year$, respectively. In Funchal and Valletta, the improvement is of the order of 1 $kWh/m^2/year$, and in the Canary Island cities, it is less than 0.2 $kWh/m^2/year$. Therefore, there are not significant changes observed increasing insulation thickness (Figure 6A). In general, these changes do not present significative improvements and therefore, the payback period is dramatically increased.

Moreover, cooling demand in all cases is always less than 0.5 $kWh/m^2/year$ and the cooling demand with insulation thickness is lower than 0.1 $kWh/m^2/year$ (Figure 6B).

The results show that the retrofitting actions consisting in the change of glazing do not lead to significant savings in either heating or cooling demand. In the cities Santa Cruz de Tenerife and Las Palmas de Gran Canaria, the improvement in heating and cooling demand does not exceed 0.2 and 0.25 $kWh/m^2/year$, respectively. In this work, the influence of the WWR on the energy demand has not been studied. The modification of the WWR values implies important changes on the energy demand and must be studied for each case [31]. Even when the WWR values of the simulated building are not among the typical values for current buildings in the northern and central areas of the EU, these values have been used in order to compare with the obtained results for the Canary Island cities. Moreover, it is expected that in the near future, new buildings in the north and central EU will present WWR values close to the ones considered in this study [22].

4. Conclusions

In this article, an energy analysis of a reference building placed in 35 EU cities has been presented where the effectiveness of retrofit strategies for energy savings in buildings in EU has been evaluated, considering different cities in different climate zones.

The behavior observed in northern, central, and southern European cities contrasts with those observed in cities even further south in the EU: Santa Cruz de Tenerife and Las Palmas de Gran Canaria. In these cases, the installation of insulation does not represent a significant saving in the heating demand since the heating demand without insulation in the building is already less than 1.7 $kWh/m^2/year$.

In relation to the cooling demand, in the case of these cities located on the Canary Islands, the cooling demand without insulation is also low (<5 $kWh/m^2/year$). This demand increases slightly to a value of 7 $kWh/m^2/year$ when the thickness of the insulation increases, due to the heat which cannot be dissipated through the enclosure due to the insulation improvement. In these cities, the reduction of energy demand should be focused on other strategies, such as orientation, use of shading, and incorporation of renewable energies.

On the other hand, the rehabilitation actions consisting of changing the glazing do not represent a significant saving in either the demand for heating or cooling. In the cities of Santa Cruz de Tenerife and Las Palmas de Gran Canaria, the improvement in heating and cooling demand is less than 0.25 $kWh/m^2/year$, which implies an excessively long amortization period.

The results reveal the energy-saving strategies, and therefore, the European subsidies for energy rehabilitation of buildings, based on the renovation and improvement of the thermal insulation of the building envelope are suitable for the cities of the north and center of the EU. However, in the case of the southern evaluated cities, the placement of insulation in the walls, which is required by law to comply with thermal transmittance requirements, does not imply an improvement in energy demand for the Canary Island cities studied.

Therefore, the general retrofit strategies for energy savings in buildings in the EU are not suitable for the southern cities under study, revealing the need for new strategies and policies to save energy in buildings placed in the studied cities from the Canary Islands.

Although the typologies of buildings can be very diverse in each of the cities studied, the results obtained on the reference building highlight the need to evaluate the rehabilitation strategies currently applied in the Canary Islands territory.

Author Contributions: Conceptualization, J.M.M.-M. and E.G.-D.; methodology, J.M.M.-M., E.G.-D., N.M.-D., A.P.-G. and B.G.-D.; software, J.M.M.-M.; validation, E.G.-D., N.M.-D., A.P.-G. and B.G.-D.; formal analysis, J.M.M.-M. and E.G.-D.; investigation, A.P.-G., E.G.-D. and B.G.-D.; resources, J.M.M.-M.; data curation, N.M.-D., E.G.-D. and A.P.-G.; writing—original draft preparation, J.M.M.-M. and E.G.-D.; writing—review and editing, N.M.-D., B.G.-D. and A.P.-G.; visualization, B.G.-D. and A.P.-G.; supervision, E.G.-D., B.G.-D. and N.M.-D.; project administration, N.M.-D.; funding acquisition, N.M.-D. All authors have read and agreed to the published version of the manuscript.

Funding: This work was funded by the EU H2020 ENCORE Project—"Energy aware BIM Cloud Platform in a Cost-effective Building Renovation Context", European Union's Horizon 2020 research and innovation program under grant agreement number 820434. Also this work has received co-funding support by the Agencia Canaria de Investigación, Innovación y Sociedad de la Información (ACIISI) de la Consejería de Economía, Conocimiento y Empleo, and by the European Social Fund (ESF)—Canary Islands Integrated Operational Programme 2014–2020, Axis 3—Priority Theme 74 (85%).

Data Availability Statement: Not applicable.

Conflicts of Interest: The authors declare no conflict of interest.

References

1. Li, S.Y.; Han, J.Y. The Impact of Shadow Covering on the Rooftop Solar Photovoltaic System for Evaluating Self-Sufficiency Rate in the Concept of Nearly Zero Energy Building. *Sustain. Cities Soc.* **2022**, *80*, 103821. [CrossRef]
2. Yang, L.; Yan, H.; Lam, J.C. Thermal Comfort and Building Energy Consumption Implications—A Review. *Appl. Energy* **2014**, *115*, 164–173. [CrossRef]
3. Rathore, P.K.S.; Gupta, N.K.; Yadav, D.; Shukla, S.K.; Kaul, S. Thermal Performance of the Building Envelope Integrated with Phase Change Material for Thermal Energy Storage: An Updated Review. *Sustain. Cities Soc.* **2022**, *79*, 103690. [CrossRef]
4. International Energy Agency (IEA). Technology Roadmap—Energy Efficient Building Envelopes. Available online: https://www.iea.org/reports/technology-roadmap-energy-efficient-building-envelopes (accessed on 20 July 2022).
5. Fedorczak-Cisak, M.; Radziszewska-Zielina, E.; Orlik-Kozdoń, B.; Steidl, T.; Tatara, T. Analysis of the Thermal Retrofitting Potential of the External Walls of Podhale's Historical Timber Buildings in the Aspect of the Non-Deterioration of Their Technical Condition. *Energies* **2020**, *13*, 4610. [CrossRef]
6. EU. Parliament Directive 2010/31/EU of the European Parliament and of the Council of 19 May 2010 on the Energy Performance of Buildings. Available online: https://eur-lex.europa.eu/LexUriServ/LexUriServ.do?uri=OJ:L:2010:153:0013:0035:en:PDF (accessed on 20 July 2022).
7. European Parliament Directive (EU). 2018/844 of the European Parliament and of the Council of 30 May 2018 Amending Directive 2010/31/EU on the Energy Performance of Buildings and Directive 2012/27/EU on Energy Efficiency. Available online: https://eur-lex.europa.eu/legal-content/EN/TXT/?uri=celex%3A32018L0844 (accessed on 20 July 2022).
8. Sadineni, S.B.; Madala, S.; Boehm, R.F. Passive Building Energy Savings: A Review of Building Envelope Components. *Renew. Sustain. Energy Rev.* **2011**, *15*, 3617–3631. [CrossRef]
9. Çomakli, K.; Yüksel, B. Optimum Insulation Thickness of External Walls for Energy Saving. *Appl. Therm. Eng.* **2003**, *23*, 473–479. [CrossRef]
10. de la Flor, F.S.; Jara, E.R.; Pardo, .R.; Lissén, J.S.; Kolokotroni, M. Energy-Efficient Envelope Design for Apartment Blocks—Case Study of a Residential Building in Spain. *Appl. Sci.* **2021**, *11*, 433. [CrossRef]
11. Nyers, J.; Kajtar, L.; Tomić, S.; Nyers, A. Investment-Savings Method for Energy-Economic Optimization of External Wall Thermal Insulation Thickness. *Energy Build* **2015**, *86*, 268–274. [CrossRef]
12. Dylewski, R.; Adamczyk, J. Optimum Thickness of Thermal Insulation with Both Economic and Ecological Costs of Heating and Cooling. *Energies* **2021**, *14*, 3835. [CrossRef]
13. Zyczyńska, A.; Suchorab, Z.; Majerek, D. Influence of Thermal Retrofitting on Annual Energy Demand for Heating in Multi-Family Buildings. *Energies* **2020**, *13*, 4625. [CrossRef]
14. Al-Khawaja, M.J. Determination and Selecting the Optimum Thickness of Insulation for Buildings in Hot Countries by Accounting for Solar Radiation. *Appl. Therm. Eng.* **2004**, *24*, 2601–2610. [CrossRef]
15. Brito-Coimbra, S.; Aelenei, D.; Gomes, M.G.; Rodrigues, A.M.; Gomes, G.; Rodrigues, M.; Façade, A.B. Building Façade Retrofit with Solar Passive Technologies: A Literature Review. *Energies* **2021**, *14*, 1774. [CrossRef]
16. Derradji, L.; Imessad, K.; Amara, M.; Boudali Errebai, F. A Study on Residential Energy Requirement and the Effect of the Glazing on the Optimum Insulation Thickness. *Appl. Therm. Eng.* **2017**, *112*, 975–985. [CrossRef]
17. Yang, Y.; Javanroodi, K.; Nik, V.M. Climate Change and Energy Performance of European Residential Building Stocks—A Comprehensive Impact Assessment Using Climate Big Data from the Coordinated Regional Climate Downscaling Experiment. *Appl. Energy* **2021**, *298*. [CrossRef]
18. Walker, L.; Hischier, I.; Schlueter, A. Does Context Matter? Robust Building Retrofit Decision-Making for Decarbonization across Europe. *Build. Environ.* **2022**, *226*. [CrossRef]

19. El-Darwish, I.; Gomaa, M. Retrofitting Strategy for Building Envelopes to Achieve Energy Efficiency. *Alex. Eng. J.* **2017**, *56*, 579–589. [CrossRef]
20. Curado, A.; de Freitas, V.P. Influence of Thermal Insulation of Facades on the Performance of Retrofitted Social Housing Buildings in Southern European Countries. *Sustain. Cities Soc.* **2019**, *48*, 101534. [CrossRef]
21. European Parliament Maximising the Energy Efficiency Potential of the EU Building Stock. Available online: https://www.europarl.europa.eu/doceo/document/A-9-2020-0134_EN.html (accessed on 20 July 2022).
22. Pajek, L.; Košir, M. Strategy for Achieving Long-Term Energy Efficiency of European Single-Family Buildings through Passive Climate Adaptation. *Appl. Energy* **2021**, *297*, 117116. [CrossRef]
23. Márquez-Martinón, J.M.; Martín-Dorta, N.; González-Díaz, E.; González-Díaz, B. Influence of Thermal Enclosures on Energy Saving Simulations of Residential Building Typologies in European Climatic Zones. *Sustainability* **2021**, *13*, 8646. [CrossRef]
24. Gamero, E.; Vizcaíno, A.M.; Méndez, L.V.; Conejero, S. Proyecto EDEA: Desarrollo de la Eficiencia Energética en la Arquitectura. In Proceedings of the 13th International Conference on Project Engineering (AEIPRO), Badajoz, Spain, 8–10 July 2009.
25. Spanish Technical Building Code, C. Royal Decree 732/2019, 314/2006 Which Modifies the CTE, Approved by Royal Decree 314/2006. Available online: https://www.boe.es/buscar/act.php?id=BOE-A-2006-5515 (accessed on 20 July 2022).
26. Kaynakli, O. A Review of the Economical and Optimum Thermal Insulation Thickness for Building Applications. *Renew. Sustain. Energy Rev.* **2012**, *16*, 415–425. [CrossRef]
27. Aurea Consulting-DesignBuilder España Cálculo de Infiltraciones Conforme al DB-HE y Certificación Energética. Available online: https://ecoeficiente.es/infiltracioneshe/ (accessed on 20 March 2022).
28. Ministerio de Transportes Movilidad y Agenda Urbana Basic Document on Energy Saving: Annex D. Available online: https://www.codigotecnico.org/pdf/Documentos/HE/DcmHE.pdf (accessed on 26 August 2022).
29. Qiao, R.; Liu, T. Impact of Building Greening on Building Energy Consumption: A Quantitative Computational Approach. *J. Clean. Prod.* **2020**, *246*, 119020. [CrossRef]
30. Belussi, L.; Danza, L. Method for the Prediction of Malfunctions of Buildings through Real Energy Consumption Analysis: Holistic and Multidisciplinary Approach of Energy Signature. *Energy Build.* **2012**, *55*, 715–720. [CrossRef]
31. Veillette, D.; Rouleau, J.; Gosselin, L. Impact of Window-to-Wall Ratio on Heating Demand and Thermal Comfort When Considering a Variety of Occupant Behavior Profiles. *Front. Sustain. Cities* **2021**, *3*, 700794. [CrossRef]

buildings

MDPI

Article

Estimating Heating Load in Residential Buildings Using Multi-Verse Optimizer, Self-Organizing Self-Adaptive, and Vortex Search Neural-Evolutionary Techniques

Fatemeh Nejati [1], Nayer Tahoori [2], Mohammad Amin Sharifian [3], Alireza Ghafari [4,*] and Moncef L. Nehdi [5,*]

1 Department of Art and Architecture, Faculty of Architecture, Khatam University, Tehran 1417466191, Iran
2 Department of Art, Science and Research Branch, Islamic Azad University, Tehran 1477893855, Iran
3 Department of Architecture, Science and Research Branch, Islamic Azad University, Tehran 1477893855, Iran
4 Department of Architecture, Rafsanjan Branch, Islamic Azad University, Rafsanjan 1477893855, Iran
5 Department of Civil Engineering, McMaster University, Hamilton, ON L8S 4M6, Canada
* Correspondence: alireza.ghafari4@gmail.com (A.G.); nehdim@mcmaster.ca (M.L.N.); Tel.: +1-905-525-9140 (ext. 23824) (M.L.N.)

Abstract: Using ANN algorithms to address optimization problems has substantially benefited recent research. This study assessed the heating load (HL) of residential buildings' heating, ventilating, and air conditioning (HVAC) systems. Multi-layer perceptron (MLP) neural network is utilized in association with the MVO (multi-verse optimizer), VSA (vortex search algorithm), and SOSA (self-organizing self-adaptive) algorithms to solve the computational challenges compounded by the model's complexity. In a dataset that includes independent factors like overall height and glazing area, orientation, wall area, compactness, and the distribution of glazing area, HL is a goal factor. It was revealed that metaheuristic ensembles based on the MVOMLP and VSAMLP metaheuristics had a solid ability to recognize non-linear relationships between these variables. In terms of performance, the MVO-MLP model was considered superior to the VSA-MLP and SOSA-MLP models.

Keywords: self-organizing self-adaptive; vortex search algorithm; multi-verse optimizer; heating load; residential

Citation: Nejati, F.; Tahoori, N.; Sharifian, M.A.; Ghafari, A.; Nehdi, M.L. Estimating Heating Load in Residential Buildings Using Multi-Verse Optimizer, Self-Organizing Self-Adaptive, and Vortex Search Neural-Evolutionary Techniques. *Buildings* **2022**, *12*, 1328. https://doi.org/10.3390/buildings12091328

Academic Editors: Paulo Santos and Mark Bomberg

Received: 1 July 2022
Accepted: 8 August 2022
Published: 30 August 2022

1. Introduction

The heating, ventilation, and air conditioning (HVAC) systems of a freshly constructed building regulate indoor air quality [1]. On the other hand, the rising trend of individuals living in energy-efficient buildings necessitates a thorough comprehension of the entire thermal loads necessary to choose appropriate HVAC systems. Several mathematical and analytic techniques [2–4] have optimized HVAC systems. According to a recent study, machine learning techniques (i.e., inverse modeling) can be used to predict and evaluate the buildings' energy performance [5]. Due to developments in programming sciences and computation, various innovative approaches have been created over the past several years [6–8]. The main goal of these simulations is to make simulations of actual events more practical [8–10]. Using a range of methods (e.g., numerical [11,12], experimental [13,14], empirical [15,16]), scientists have been able to select the most suitable technique for the unsolved problem. Several more conventional processes could be supplanted by machine learning, which has shown promising outcomes. Using various machine-learning programs, it is feasible to solve intricate problems with high accuracy.

The artificial neural network (ANN) [17,18] is a powerful processor capable of simulating a variety of scientific objectives and tasks [19–24]. Due to its neural processors and several layers, the multi-layer perceptron (MLP) [25] is a characteristic form of ANN. The utilization of these processors in simulations involving energy has been effective [26–28]. Using an MLP, researchers can study the relationship between a dependent parameter and other independent factors. Each dependent parameter is assigned a weight in the neurons of the MLP, which are its processors. The resultant value will then be used to activate a function by combining it with a bias term. The subsequent development of neurons uses this strategy to have a unique mathematically forward progress [29].

Consequently, the MLP has become a "feed-forward instrument" [30]. Analytical approaches were congruent with Ren et al.'s hypothesized ANN results heat loss prediction [31–33]. This model surpasses all others in calculating the strain in a concrete beam's tie section, as Mohammadhassani et al. [34]. Sadeghi et al. [35] utilized an MLP to predict a residential structure's cooling and heating demands. A sensitivity analysis also indicated the ideal network response. Sholahudin and Han [36] employed the Taguchi method to develop a simplified dynamic ANN to accurately predict heat loss (HL) in an HVAC system. Several prior studies [37–39] have proved the efficacy of ANNs in energy modeling. In addressing energy-related issues, fuzzy networks [40], random forests, and support vector machines have all been useful [41–43].

In energy analysis, metaheuristic scholars have been increasingly interested in HVAC systems and energy analysis [44–49]. Martin et al. [50] calibrated the HVAC subsystem component via a metaheuristic and sensitivity analysis. Bamdad Masouleh [51] implemented ant colony optimization to optimize energy. Moreover, several benchmarks revealed that the proposed models excelled in traditional methods. Numerous research studies have demonstrated that machine learning models can benefit from various techniques [52–55]. As part of their research, Zhou et al. [56] investigated how to best estimate the HL and CL by ANN, utilizing ABC and PSO applied to an ANN [57]. The PSO outperformed the other algorithm by approximately 22 to 24 percent, demonstrating that both approaches are effective. Bui et al. [58] used a firefly technique based on electromagnetism to optimize the ANN for calculating energy use. Researchers discovered that hybrid approaches were more exact than a conventional ANN technique. In this sense, Moayedi et al. [59] assessed the performance of grasshopper optimization algorithm (GOA) and grey wolf optimization (GWO) optimizers in conjunction with an ANN, for estimating the heating load of green residential construction [60]. As a result of these tactics, the prediction error dropped from 2.9859 to 2.4460 and 2.2898, respectively.

Metaheuristic approaches have developed to overcome common computing restrictions, including local minima [61–70]. Employing these methods to find the intelligent models' training would result in very accurate predicting models for various goals [71,72]. Because there are so many optimization methods, comparative research on the next generation of metaheuristics is necessary.

Environmentally and economically, finding a realistic model for thermal load modeling could be advantageous. The main goal of this article is to forecast the heating and cooling load via metaheuristic algorithms and check whether these algorithms can predict the heating and cooling load precisely. Metaheuristic optimizers, such as the multi-verse optimizer (MVO), self-organizing and self-adaptive (SOSA), and vortex search algorithm (VSA), are being evaluated to discover whether they can aid in estimating the HL. Also, these three methods are compared, and the best one is presented at the end of the task.

2. Established Database

The connection between these influencing factors and parameters must be investigated to estimate a parameter. Hence, the supplied data must be accurate. A total of 768 thermal load scenarios are employed to train, test, and validate the models in this work. The data was initially developed by Tsanas and Xifara, who analyzed the heating load and cooling load of various residential buildings [73]. Due to their work, a valuable dataset was compiled and made accessible for download at https://archive.ics.uci.edu/ml/datasets/Energy+efficiency (accessed on 15 July 2022). Overall height (OH), roof area (RA), glazing area (GA), wall area (WA), relative compactness (RC), orientation (OR), surface area (SA), and glazing area distribution (GAD) are independent factors identified to affect the HL output parameters. A box plot of the heating load and input components is displayed in Figure 1.

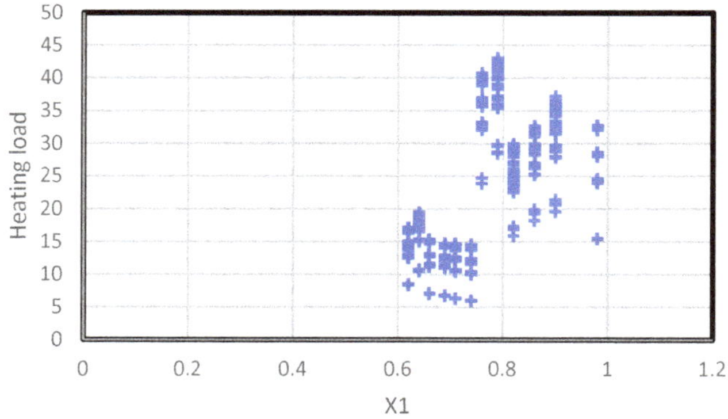

(**a**)

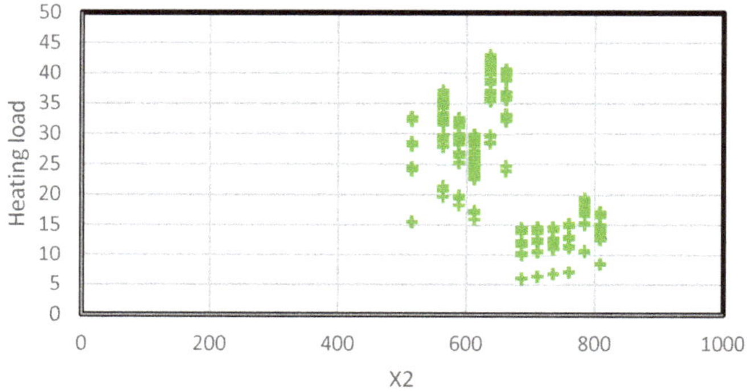

(**b**)

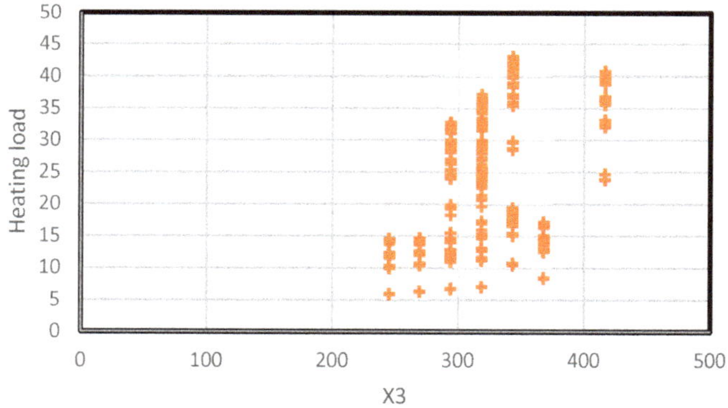

(**c**)

Figure 1. *Cont.*

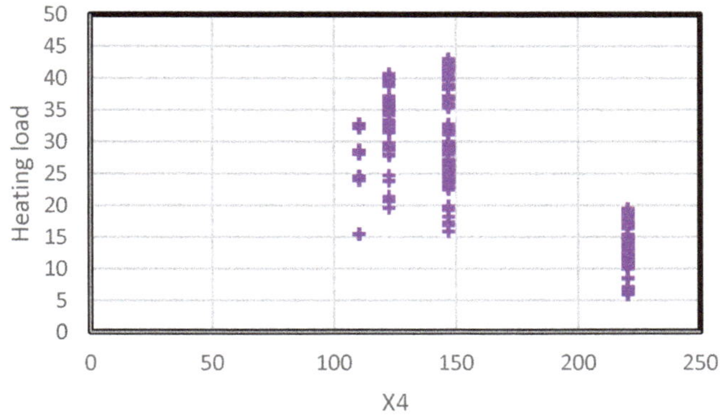

(**d**)

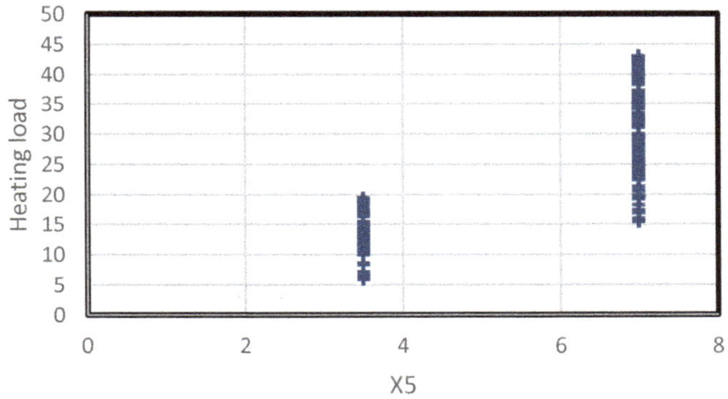

(**e**)

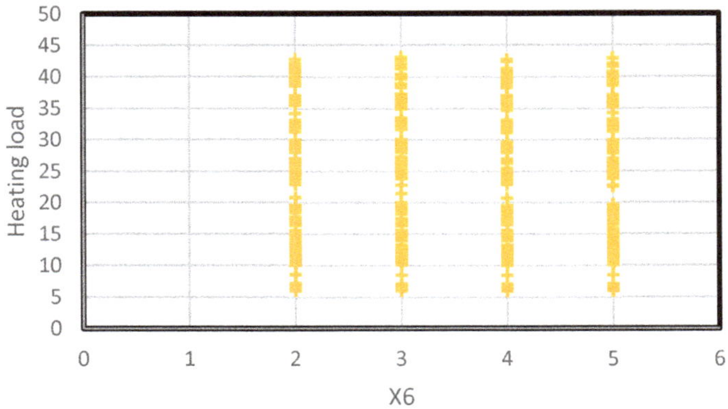

(**f**)

Figure 1. *Cont.*

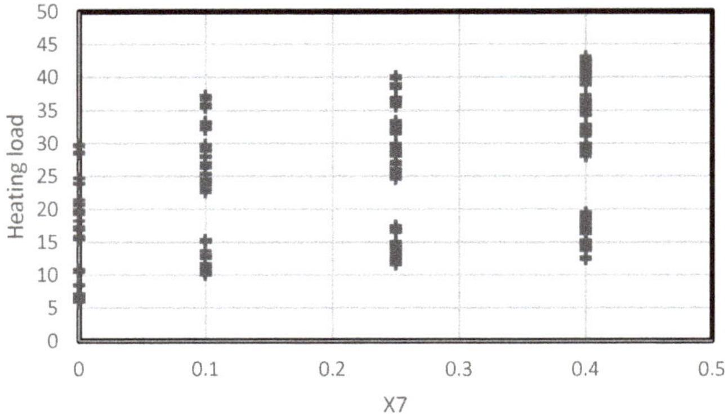

(g)

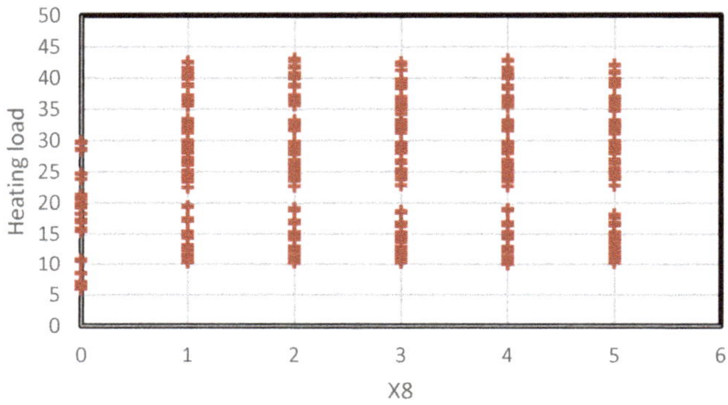

(h)

Figure 1. Box plot of used dataset variations with the heating load. (**a**) Relative compactness (RC), (**b**) surface area, (**c**) wall area, (**d**) roof area, (**e**) overall height, (**f**) orientation, (**g**) glazing area, (**h**) glazing area distribution, with the heating load.

3. Methodology

This research examines an ANN with three novel optimizers, MVO, SOSA, and VSA, to test their investigation of how they affect the limits of a typical neural network. These algorithms seek better hyperparameters than those proposed by more conventional learning methods (backpropagation and Levenberg–Marquardt).

3.1. Multilayer Perceptron

Multilayer perceptrons, a type of neural network, have recently been demonstrated to be a viable alternative to conventional statistical methods [74]. Hornik et al. (1989) [75] demonstrated that the MLP could simulate any smooth and measurable function. Despite other methods, the MLP method does not consider data processing. This method can model and teach complex nonlinear functions to generalize correctly using previously unexplored new data. These properties make it a possible alternative to statistical and numerical modeling techniques. The multilayer perceptron has several atmospheric scientific uses, as will be demonstrated.

Figure 2 depicts the predefined connection between the main inputs and output(s) vectors for the multilayer perceptron, a network of fundamentally interconnected neurons or nodes. Each network node's output signals and weights are derived from a primary activation function or nonlinear transfer. The MLP can only model linear functions if the transfer function is linear. The node's output can serve as an input for other network-connected nodes for each network-connected node. In light of this, the multilayer perceptron is a feed-forward neural network. There are a variety of structural configurations for multilayer perceptrons, but they all contain layers of neurons. The input layer serves as a conduit for data transfer from the input layer to other network layers. A multilayer perceptron's input and output vectors can be expressed as single vectors (Figure 2). An MLP structure consists of multiple hidden layers and one output layer. Multilayer perceptron refers to a network in which each node is interconnected in the layers above and below with every other node.

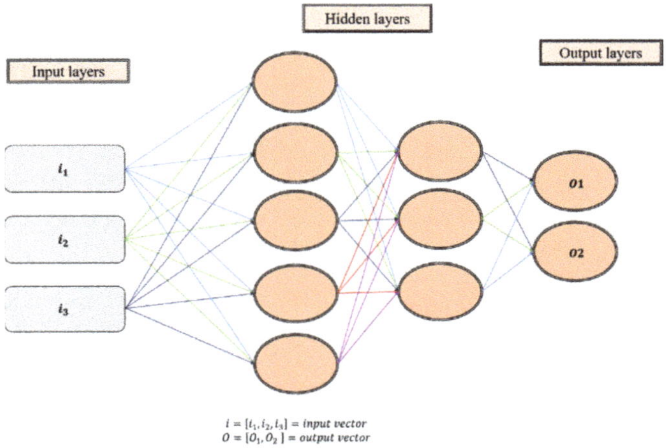

Figure 2. A two hidden layers multilayer perceptron.

As proven, multilayer perceptrons can estimate any computable function between two sets of input and output vectors by selecting a fine collection of linking weights and transfer functions [75]. A multilayer perceptron is capable of learning new abilities by training. You will require input and output vector-based training data to learn a new algorithm. A multilayer perceptron decides on the network's weights until the required input-output mapping is reached. It can only acquire knowledge in the presence of an observer. When training an MLP, it is possible that its output for a given input vector may not match the anticipated output. The difference between the actual and desired outputs characterizes error signals. Adjusting the direct networks depending on this error signal during training can help lower the total error of the MLP. A multilayer perceptron can be trained in various methods with several different algorithms. Once trained with adequate training data, the multilayer perceptron can generalize to new, unknown inputs.

3.2. Multi-Verse Optimizer (MVO)

The multi-verse optimizer [76] is known as a growing metaheuristic algorithm that tries to mimic the laws of a multi-verse theory. It is a relatively recent development. Parallel universe theories, including the presence of black, white, and wormholes, were the primary source of inspiration for the design of this optimizer. A population-based stochastic method is employed to determine the global optimum for optimization problems [77]. To update the answers using this method, the probability of wormhole existence (*WEP*) and the rate of travel (*TDR*) must first be computed. These parameters determine the frequency and magnitude of solution changes during the optimization process and are formulated as:

$$WEP = a + t \times \left(\frac{b-a}{T}\right) \quad (1)$$

The total iterations' number is T, corresponding to the minimum, b to the maximum, and t to the current iteration.

$$TDR = 1 - \frac{t^{1/P}}{T^{1/P}} \quad (2)$$

p indicates the exploitation accuracy. P is the most essential TDR measure. The emphasis on exploitation increases as the value of this choice rises.

The following equation can be used to update the solution positions when WEP and TDR have been calculated:

$$x_i^j \begin{cases} \begin{cases} x_j + TDR + ((ub_j - lb_j) * r_4 + lb_j) & if\ r_3 < 0.5 \\ x_j - TDR + ((ub_j - lb_j) * r_4 + lb_j) & if\ r_3 \geq 0.5 \end{cases} & if\ r_2 < WEP \\ x^j_{roulette\ Wheel} & if\ r_2 \geq WEP \end{cases} \quad (3)$$

where x_j is set to be the jth element from the best predefined individual, WEP, TDR are coefficients, lb_i and ub_i are the lower and upper bounds of the jth element, r_2; r_3; r_4 are randomly generated numbers drawn from the interval of [0, 1], x_i^j represents the jth parameter in ith individual, and $x^j_{roulette\ Wheel}$. does the roulette wheel selection mechanism to pick the jth element of a solution.

This equation can be used to compute a new solution position and compare it to the most recent best-in-class participant in the WEP. If r_3, a random number in the interval [0, 1], is less than 0.5, then an optimal solution value for the jth dimension requires a solution. By increasing WEP during optimization, MVO increases the use of the most proper solution so far.

3.3. Self-Organizing and Self-Adaptive (SOSA)

Self-organization (SO) parallels the biologically inspired notions of emergence and swarm intelligence very closely. Frequently, in this technique, SO and emergence are conflated. De and Holvoet (2005) [78] examine the phrase's origins and the difference between the two conceptions. This is known as SO:

SO is an adaptive and dynamic computational process through which systems retain their structure independently of external stimuli [78,79]. However, SO can also refer to the emergence-causing process [80,81]. In addition, ref. [82] differentiates between the terms called strong SO schemes with no explicit central internal or external control and weak SO systems with some central internal control. SO and emergent systems are separate concepts, although they share one characteristic: the absence of direct exterior control. Although the external effect on self-organized systems is studied more thoroughly in directed SO, less attention has been paid to it in the context of unguided SO [83]. In this text, external effect is characterized as either specific or non-specific, with specific influence suggesting straight control on the functional structure or temporal, spatial, or other non-specific impacts indicating that the system determines its response to an external stimulus. Consequently, Prokopenko (2009) [83] defines SO guidance as the potential limiting of the domain or extent of functions/structures, or selecting a subset of the multiple alternatives that the dynamics might take.

According to ref. [78], the main distinction between SO and emergence is that individual entities are informed of the systems planned by global behavior in the former scenario. Consequently, self-organization may be considered a weak kind of emergence. Utilizing feedback loops is a common and straightforward method for achieving SO. Components of the system monitor the state, interpret it according to the expected behavior, and initiate the required actions. This method is also employed by "single entity systems." This notion is referred to as self-adaptation [84,85]. Self-adaptation happens when a decentralized system composed of several entities adapts to external changes. Self-adaptation within the context of software is set as follows: SA software modifies its behavior in response to modifications within its operating environment. The operating environment refers to everything the software system may see, including human input, sensors and external hardware devices, and programmed instrumentation [86].

3.4. Vortex Search Algorithm (VSA)

Ölmez and Doğan [87] initially developed the vortex pattern generated by the vertical flow of stirred fluids to design the VSA algorithm. As with countless other methods, the algorithm seeks to balance exploratory and exploitative actions. The VSA uses an adaptive step-size-adjustment method to determine the optimal response. Consequently, exploratory behavior is accounted for in the early phases of the VSA, resulting in a better global search capability. In the following, the optimal response is achieved by employing an exploitative strategy around the suggested replies [88].

The vortex is depicted by stacked circles, assuming a set in two dimensions. Given U and L as the current space's boundaries, Equation (4) produces the starting point λ_0. of the outer circle:

$$\lambda_0 = \frac{U+L}{2} \tag{4}$$

Then, several neighbor solutions *Ct*(*s*) are generated at random. This production makes use of a Gaussian distribution technique.

$$C_0(s) = \{S_1, S_2, \ldots, S_g\} \qquad g = 1, 2, \ldots, z, \tag{5}$$

where t is the number of cycles and z represents the total number of potential solutions. Let x and Σ be the vector and covariance matrix of the random variable. The multivariate Gaussian distribution is denoted by Equation (6):

$$P(x|\lambda, \Sigma) = \frac{1}{\sqrt{(2\pi)^D \Sigma}} \exp\left\{\frac{-1}{2}(x-\lambda)^T \Sigma (x-\lambda)\right\}, \tag{6}$$

where D is the magnitude of the issue and is the mean vector introduced as sample.

The main distribution will be spherical if the off-diagonal elements are uncorrelated and the co-variance matrix values have similar variances (circular for two-dimensional concerns). I, where I is a $D \times D$ identity matrix and σ^2 is the distribution's variance, Σ may be written as follows:

$$\Sigma = \sigma^2 \times [I]_{D \times D}. \tag{7}$$

Using Equation (8), the initial standard deviation of the distribution is computed (σ_0). This parameter may correspond to r_0. (which requires significant values) [89]:

$$\sigma_0 = \frac{\text{maximum}\,(U) - \text{minimum}\,(L)}{2} \tag{8}$$

As is well known, the essential concept of metaheuristic algorithms for enhancing the final result is to update the obtained answers. During the VSA selection phase, the current λ_0 is replaced with the most promising alternative. This requires the proposed solution to exist inside the given space. This item is assessed using the Equation (9).

$$\begin{cases} s_g^i = rand \cdot (U^i - L^i) + L^i, & if\ s_g^i < L^i \\ s_g^i = rand \cdot (U^i - L^i) + L^i, & if\ s_g^i > U^i \end{cases} \tag{9}$$

where rand is a random integer with uniform distribution.

The best answer discovered thus far is then applied to the second (or inner) circle's center. After successively decreasing the effective radius of the current solution, a new group of solutions (C1(s)) is produced close to it. Repetitioning the same approach might yield a more viable answer [89]. Other researches have also described the VSA well [90,91].

4. Results and Discussion

This study analyzes the HL approximation capabilities of three unique neural network upgrades described in Section 1. The algorithms are synthesized using an MLP neural network to accomplish this objective. Each approach uses a unique search strategy to get the optimal computational weights for the MLP (and biases).

As is commonly known, the size and number of neurons contained inside a hidden layer define the MLP's structure. Therefore, these parameters must initially be modified. Numerous studies have demonstrated that a single hidden layer is excellent at simulating complicated processes [92,93]. However, the hidden neurons' optimal number was established by trial and error. Among the designs studied, $8 \times 6 \times 1$ demonstrated the most promising performance (where the middling layer contained $1, 2, 3, \ldots, 10$ neurons). Figure 2 illustrates the used MLP.

4.1. Accuracy Indicators

Mean absolute error (MAE) as the first used statistical index and root mean square error (RMSE) as the second index was specified for assessing the potential errors in proposed structures. Equations (10) and (11) produce are used for RMSE and MAE. Additionally, Equation (12) defines the coefficient of determination (R^2) required to compute the compatibility between the measured and predicted HLs:

$$\text{MAE} = \frac{1}{U}\sum_{i=1}^{U}\left|S_{i_{observed}} - S_{i_{predicted}}\right| \tag{10}$$

$$\text{RMSE} = \sqrt{\frac{1}{U}\sum_{i=1}^{U}\left[\left(S_{i_{observed}} - S_{i_{predicted}}\right)\right]^2} \tag{11}$$

$$R^2 = 1 - \frac{\sum_{i=1}^{U}\left(S_{i_{predicted}} - S_{i_{observed}}\right)^2}{\sum_{i=1}^{U}\left(S_{i_{observed}} - \overline{S}_{observed}\right)^2} \tag{12}$$

$S_{i_{observed}}$ and $S_{i_{anticipate}}$ represent the measured and expected HLs, respectively, in these equations. In addition, U represents the number of recordings, whereas $S_{observed}$ is the average of the observed HLs.

4.2. Combining the MLP with Hybrid Optimizers

After combining hybrid algorithms with the MLP, three ensembles of MVO-MLP, SOSA-MLP, and SOSA-MLP are constructed. Each costume is supplied with training data to determine the relationship between associated parameters and heating load. One thousand repetitions are assessed for each model's optimization behavior in order to carry out the optimization. The objective function is represented using the RMSE of each iteration's findings. In swarm-based optimization algorithms, the population size is a critical variable. Ten distinct population sizes (50, 100, 150, 200, 250, 300, 350, 400, 450, and 500) are evaluated for each proposed model, and the population size results in the lowest MSE chosen as the optimal population size. The MSEs for all calculated iterations are shown in Figure 3. The populations with the lowest RMSE values (0.3540, 8.8064, and 0.2887, respectively) are 300, 4500, and 100 for MVO-MLP, SOSA-MLP, and VSA-MLP, respectively. The SOSA-MLP method, on the other hand, is less sensitive than the other two; the explanation for this may be found in the optimization approaches' characteristics. Figure 4 also displays the RMSE values achieved for different levels of complexity over all rounds.

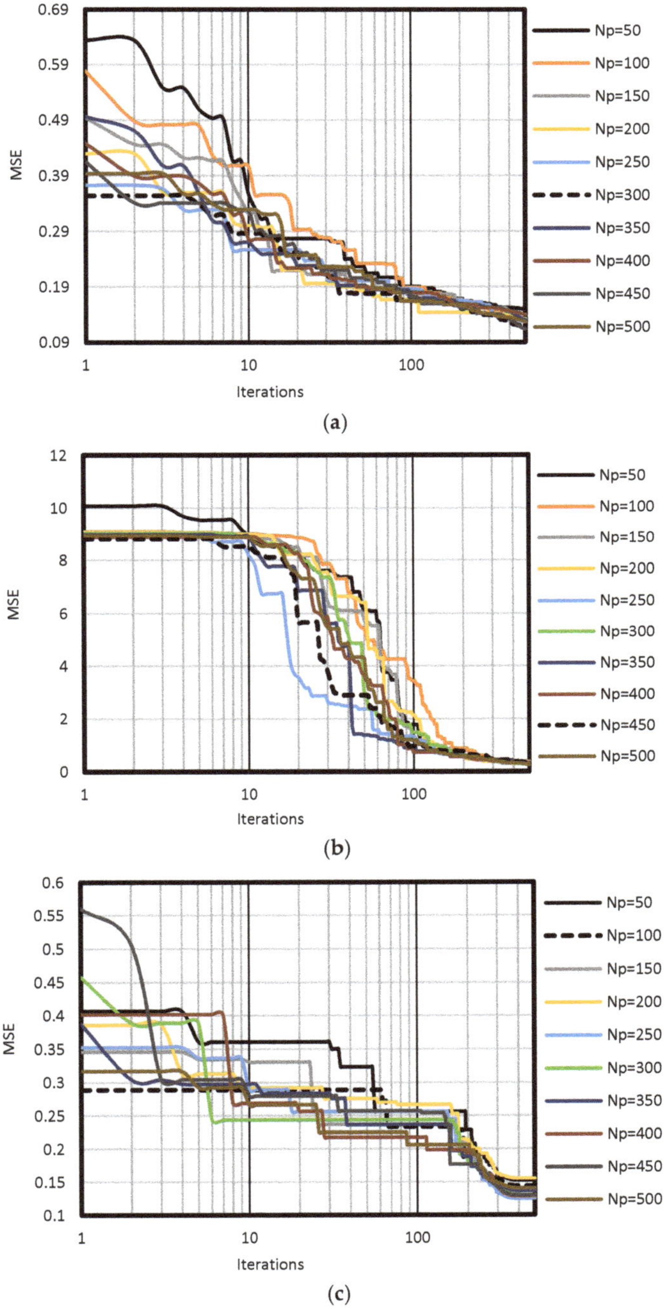

Figure 3. Model iterations versus the variation of MSE; (**a**) MVO-MLP, (**b**) SOSA-MLP, (**c**) VSA-MLP.

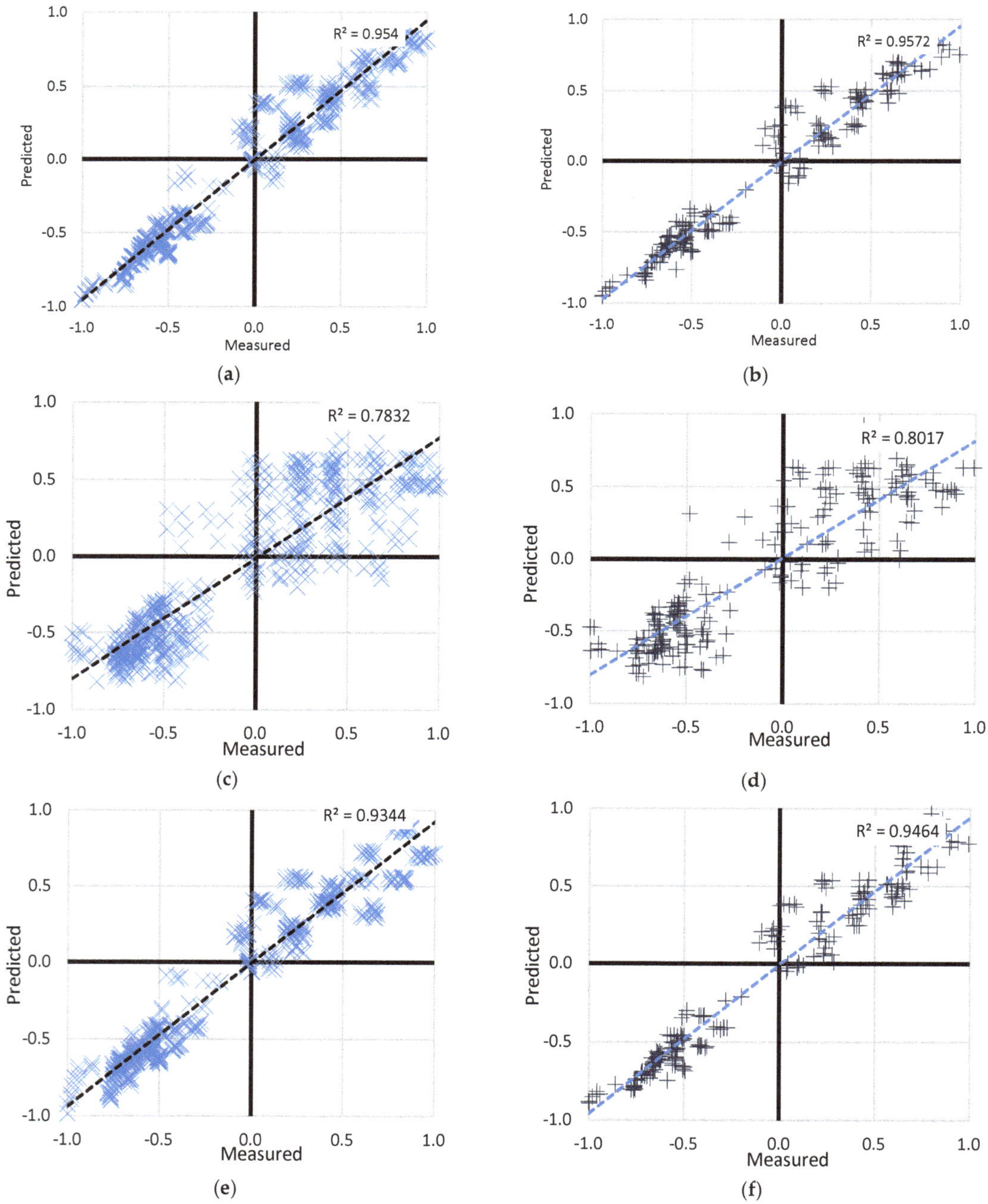

Figure 4. The accuracy of the best-fit proposed model for the (**a**) MVO-MLP training dataset, (**b**) MVO-MLP testing dataset, (**c**) SOSA-MLP training dataset, (**d**) SOSA-MLP testing dataset, (**e**) VSA-MLP training dataset, and (**f**) VSA-MLP testing dataset.

The value of R^2 for three methods of MVO, SOSA, and VSA is (0.977 and 0.978), (0.885 and 0.895), and (0.974 and 0.975) for testing and training phases, respectively. Also, in the case of RMSE, MVO, SOSA, and VSA have the value of (0.117 and 0.110), (0.255 and

0.239), and (0.124 and 0.112) in the training and testing phases, respectively. These results show that the lowest value of RMSE and the highest value of R^2 are related to the MVO technique, indicating the best performance of MVO-MLP. According to R^2 and RMSE values (Tables 1–4), the second technique for predicting HL and CL is VSA-MLP, and the last is SOSA-MLP.

Table 1. The network results for the MVO-MLP.

Population Size	Network Result				Scoring				Total Score	RANK
	Train		Test		Train		Test			
	R^2	RMSE	R^2	RMSE	R^2	RMSE	R^2	RMSE		
50	0.962	0.149	0.964	0.143	1	1	1	1	4	10
100	0.972	0.130	0.974	0.120	5	5	4	5	19	6
150	0.972	0.129	0.975	0.117	6	6	6	7	25	5
200	0.971	0.132	0.975	0.119	3	3	5	6	17	7
250	0.973	0.127	0.976	0.115	8	8	7	8	31	3
300	0.977	0.117	0.978	0.110	9	9	9	10	37	1
350	0.971	0.130	0.973	0.123	4	4	3	4	15	8
400	0.967	0.140	0.966	0.137	2	2	2	2	8	9
450	0.978	0.113	0.980	0.127	10	10	10	3	33	2
500	0.973	0.127	0.976	0.115	7	7	8	9	31	3

Table 2. The network results for the SOSA-MLP.

Population Size	Network Result				Scoring				Total Score	RANK
	Train		Test		Train		Test			
	R^2	RMSE	R^2	RMSE	R^2	RMSE	R^2	RMSE		
50	0.810	0.343	0.806	0.355	2	2	2	2	8	9
100	0.776	0.378	0.781	0.384	1	1	1	1	4	10
150	0.874	0.289	0.893	0.274	6	7	6	7	26	4
200	0.889	0.289	0.894	0.278	10	6	7	6	29	3
250	0.881	0.307	0.898	0.302	7	4	9	5	25	5
300	0.871	0.278	0.836	0.304	4	9	4	4	21	7
350	0.832	0.327	0.807	0.342	3	3	3	3	12	8
400	0.884	0.285	0.899	0.270	8	8	10	8	34	2
450	0.871	0.293	0.880	0.255	5	5	5	9	24	6
500	0.885	0.255	0.895	0.239	9	10	8	10	37	1

Table 3. The network results for the VSA-MLP.

Population Size	Network Result				Scoring				Total Score	RANK
	Train		Test		Train		Test			
	R^2	RMSE	R^2	RMSE	R^2	RMSE	R^2	RMSE		
50	0.961	0.152	0.965	0.140	2	2	2	3	9	9
100	0.965	0.143	0.967	0.135	3	3	3	4	13	8
150	0.968	0.138	0.968	0.133	5	5	4	5	19	6
200	0.959	0.155	0.964	0.141	1	1	1	1	4	10
250	0.974	0.124	0.977	0.112	10	10	10	10	40	1
300	0.969	0.136	0.973	0.122	7	7	8	8	30	3
350	0.968	0.136	0.970	0.128	6	6	6	7	25	4
400	0.972	0.128	0.974	0.120	9	9	9	9	36	2
450	0.972	0.130	0.973	0.140	8	8	7	2	25	4
500	0.967	0.140	0.970	0.129	4	4	5	6	19	6

Table 4. Selection of the best fit structures among the most accurate items of each model.

	Swarm Size	Training Dataset		Testing Dataset		Scoring				Total Score	Rank
		RMSE	R2	RMSE	R2	Training		Testing			
MVOMLP	300	0.977	0.117	0.978	0.11	3	3	3	3	12	1
SOSAMLP	500	0.885	0.255	0.895	0.239	3	3	1	1	8	2
VSAMLP	250	0.974	0.124	0.977	0.112	2	2	2	2	8	2

According to Figure 3, the MVO method has a little more constrained convergence curve than the other methods. This shows that this approach decreases error rates when ANN parameters are altered. As a result, the algorithm's findings are given to develop a prediction model. Referring to Figure 2, the output of the most recent neuron consists of seven parameters (one bias and six weights). This neuron is nourished by six layers of neurons, each responsible for nine parameters (one bias and eight weights). The network consists of 61 optimized variables with metaheuristic methods.

4.3. Prediction Results

In this section, the reliability of the applied models is assessed by considering both the outputs (i.e., the predicted HLs) to the target values (i.e., the measured HLs). Figure 5 illustrates the outcomes of the training phase by displaying the difference between each pair of output and HL goals. During this phase, the error rate for the MVO-MLP, SOSA-MLP, and VSA-MLP range between [−0.000034913 and 0.11776], [−0.011611 and 0.25559], and [-5.4249×10^{-5} and 0.12416], respectively. The preceding section indicates that the RMSE values are 0.3540, 8.8064, and 0.2887. In addition, the estimated MAEs of the three models (0.08499, 0.19662, and 0.088861) demonstrate a small degree of training error. Moreover, the computed R2 values indicate that greater than 93% of the objective and output HLs are consistent.

4.4. Efficiency Comparison

The models with the lowest RMSE (or MAE) and the highest R^2 are chosen as the most exact HL predictors, considering the learning and prediction stages. Table 4 displays the accuracy standards that must be satisfied to attain this objective. As demonstrated, the MLP constructed utilizing the MVO's weights and biases provide the most accurate knowledge of the HL and predicting it. The VSA appears as the second possible optimizer after the MVO. This study's MVO and VSA algorithms appear to outperform previously proposed models in the training and testing phases. For example, six different MLP network's hybrids (for instance, based on other hybrid techniques, such as whale optimization algorithm (WOA) [94], ABC [95], PSO [96], the salp swarm algorithm (SSA) [97], wind-driven optimization (WDO) [98], the spotted hyena optimization (SHO) [99], the imperialist competitive algorithm (ICA) [100], GOA [101], the genetic algorithm (GA) [102], and GWO [103]) were utilized to estimate the HL by using the same dataset. This suggests that the objective of developing more effective HL assessment tools has been met.

4.5. Discussion

In several engineering applications, the superiority of intelligent computational techniques over conventional and even solid experimental methods is well acknowledged. In addition to appropriate accuracy, the simplicity of applying these models is a determining factor in their application. In energy-efficiency studies, for instance, forward modeling methodologies (low capabilities for inhabited buildings [104]) and prevalent simulation software may have limitations (low capabilities for occupied buildings [104]). (Different accuracy of simulation [105]). Consequently, like the models reported in this study, indirect evaluative models outperform destructive and expensive methods. This is emphasized further when an optimal strategy is created using metaheuristic methods [106]. In other words, these optimization techniques yield competent ensembles that function optimally.

Realistic applications for the offered approaches may be developed in terms of applicability. Here are two illustrations:

The developed technique can provide an accurate estimate of the needed heating thermal load for an upcoming construction project based on the size and features of the structure [26,107,108]. Engineers and property owners might benefit from the models when developing HVAC systems. Another early-stage support for reconstruction projects is modifying structural design and architecture based on input parameters. Consequently, it is also feasible to examine the effect of each input parameter separately to comprehend the thermal load behavior. Although the trend is not predictable nor regular, the MVO-MLP predicts it precisely. Consequently, this approach may yield approximations of real-world structures that are correct.

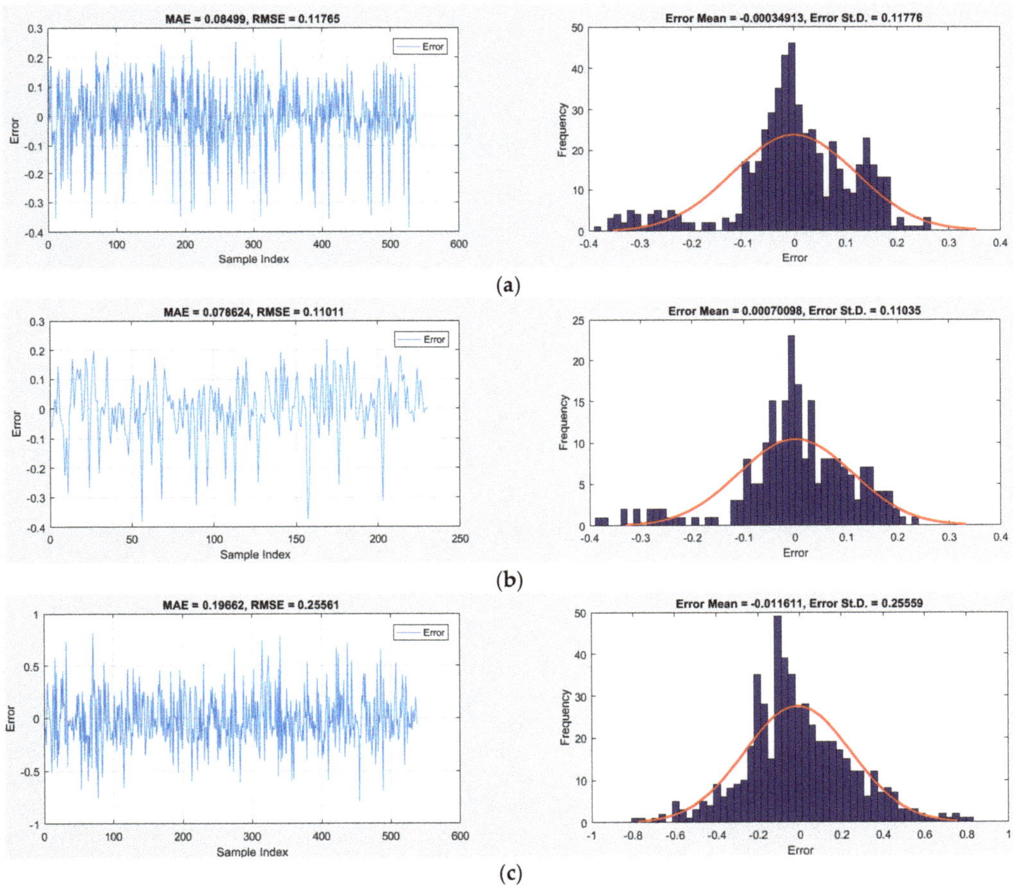

Figure 5. *Cont.*

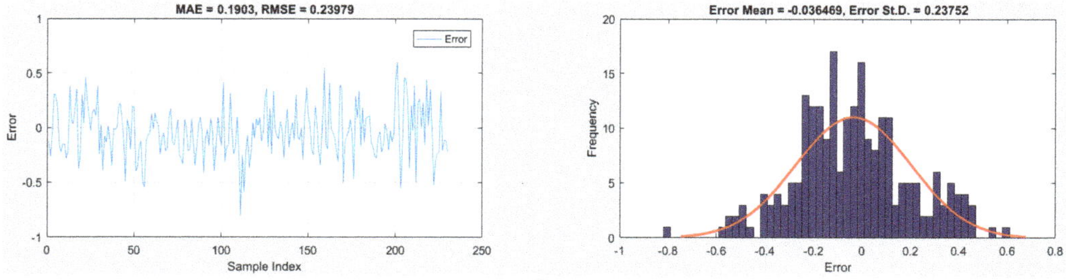

(d)

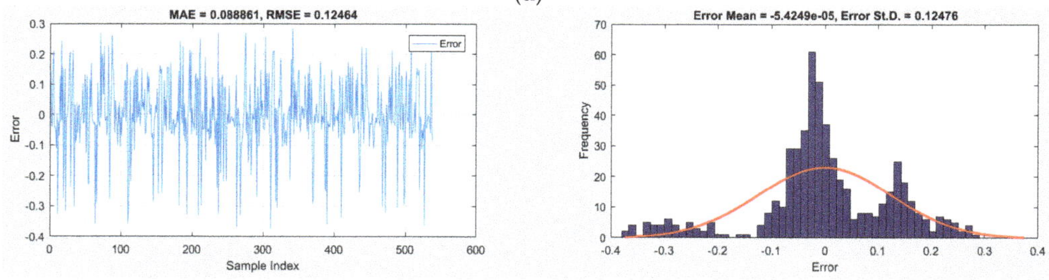

(e)

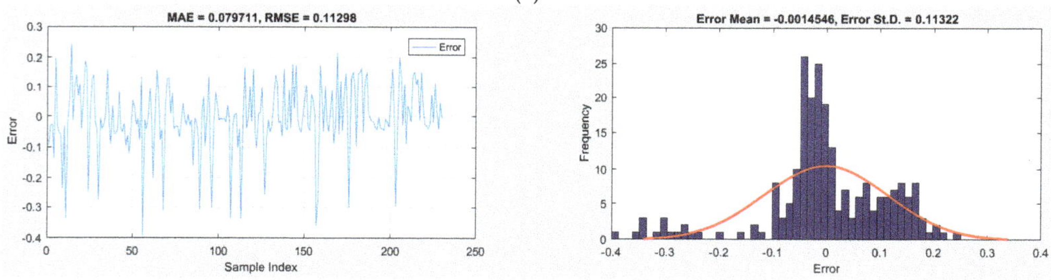

(f)

Figure 5. The error analysis for the best-fit proposed model for the (**a**) MVO-MLP training dataset, (**b**) MVO-MLP testing dataset, (**c**) SOSA-MLP training dataset, (**d**) SOSA-MLP testing dataset, (**e**) VSA-MLP training dataset, and (**f**) VSA-MLP testing dataset.

Even if there are several benefits to addressing an optimization problem, it is essential to commit the time necessary to discover a global solution. Consequently, achieving a balance between model time economy and precision may impact selecting the most efficient model. Nevertheless, according to the authors, lowering the complexity of the problem space and locating better solutions may be as simple as configuring the hyper-parameters of optimizers correctly and doing feature validity analysis. In contrast, the MVO model was the most precise; this requires establishing the optimal time and accuracy-based method. In projects in which time is not a factor, for instance, it makes sense to choose the most precise technique (regardless of how time-consuming), but in time-sensitive applications, a tolerance for accuracy may be considered in order to find a speedier solution. However, the models' overall performance was comparable, and it should be emphasized that all versions would be adequate for real-world applications. Table 5 indicates the previous research focused on heating load prediction. Noting that the outcomes were less accurate, either using R^2 or RMSE, as those were the hybrid techniques that we employed in the current study.

Table 5. Studies focused on research on heating load prediction.

References	Article Title	Scope
Refs. [26,60]	Comprehensive preference learning and predicting heating load in residential buildings using machine learning techniques	Using traditional machine learning in predicting heating and cooling load
Refs. [57,107]	Proposing a novel predicting technique using M5Rules-PSO and M5Rules-GA model in estimating CL and HL in residential building system	Estimating cooling and heating load via a novel predictive technique using M5Rules
Ref. [108]	Predicting heating and cooling loads in residential buildings using two hybrid intelligent models	Hybrid intelligent models in predicting heating and cooling load
Ref. [109]	Optimal modification of HVAC system performances in energy-efficient buildings using the integration of metaheuristic optimization and neural computing	Using neural networks and metaheuristic optimization in modifying HVAC systems
Ref. [56]	Employing ABC and PSO techniques for optimizing a neural network in prediction of HL and CL of residential buildings	Using neural network algorithms in predicting cooling and heating load in residential green buildings
Ref. [39]	A teaching-learning based optimization Neural Processor for Predicting HL in Residential Buildings	Predicting heating load using a novel neural network algorithm of TLBO

5. Conclusions

This study evaluates the MVO, SOSA, and VSA metaheuristic algorithms for analyzing and determining the HL. These methods served as the optimizer for a common neural predictive network simulation. The models predicted the HL based on a total of 768 design scenarios of the heating load. The following conclusions can be drawn from this work:

According to the sensitivity analysis, the MVA-MLP, SOSA-MLP, and VSA-MLP ensembles achieved optimal complexity at corresponding swarm sizes of 300, 500, and 250, respectively. The optimal MVO design required more calculation time than alternative MLP optimization algorithms. In terms of precision (MAEs of 0.08499, 0.19662, and 0.088861), all three ensembles profoundly understood the link between the HL and essential factors. During the testing phase, the measured value for the R^2 was 0.978, 0.895, and 0.977 demonstrating that the developed models were successful and had minimal prediction error. The most powerful model was the MVO-MLP, followed by the VSA-MLP and the SOSA-MLP. The MVO-MLP methodology was presented for use in real-world situations, but potential ideas for future projects were also presented in light of the shortcomings of the research, such as data enhancement and future selection, optimizing building characteristics using the model, and comparing the model to improved time-saving methods.

Author Contributions: F.N.: methodology, software, data curation. N.T.: writing—original draft preparation, investigation, validation. M.A.S.: conceptualization, methodology. A.G.: writing—original draft preparation, resources, final draft preparation. M.L.N.: supervision, project administration, funding acquisition. All authors have read and agreed to the published version of the manuscript.

Funding: This research received no external funding.

Data Availability Statement: The data used in this study is freely available on http://archive.ics.uci.edu/ml/datasets/Energy+efficiency (accessed on 15 July 2022).

Conflicts of Interest: The authors declare no conflict of interest.

References

1. McQuiston, F.C.; Parker, J.D.; Spitler, J.D. *Heating, Ventilating, and Air Conditioning: Analysis and Design*; John Wiley & Sons: Hoboken, NJ, USA, 2004.
2. Ihara, T.; Gustavsen, A.; Jelle, B.P. Effect of facade components on energy efficiency in office buildings. *Appl. Energy* **2015**, *158*, 422–432. [CrossRef]
3. Knight, D.; Roth, S.; Rosen, S.L. Using BIM in HVAC design. *Ashrae J.* **2010**, *52*, 24–29.
4. Ikeda, S.; Ooka, R. Metaheuristic optimization methods for a comprehensive operating schedule of battery, thermal energy storage, and heat source in a building energy system. *Appl. Energy* **2015**, *151*, 192–205. [CrossRef]
5. Sonmez, Y.; Guvenc, U.; Kahraman, H.T.; Yilmaz, C. A comperative study on novel machine learning algorithms for estimation of energy performance of residential buildings. In Proceedings of the 2015 3rd International Istanbul Smart Grid Congress and Fair (ICSG), Istanbul, Turkey, 29–30 April 2015; pp. 1–7.
6. Lu, N.; Wang, H.; Wang, K.; Liu, Y. Maximum probabilistic and dynamic traffic load effects on short-to-medium span bridges. *Comput. Model. Eng. Sci.* **2021**, *127*, 345–360. [CrossRef]
7. Chen, Y.; Lin, H.; Cao, R.; Zhang, C. Slope stability analysis considering different contributions of shear strength parameters. *Int. J. Geomech.* **2021**, *21*, 04020265. [CrossRef]
8. Zhang, S.-W.; Shang, L.-Y.; Zhou, L.; Lv, Z.-B. Hydrate Deposition Model and Flow Assurance Technology in Gas-Dominant Pipeline Transportation Systems: A Review. *Energy Fuels* **2022**, *36*, 1747–1775. [CrossRef]
9. Liu, E.; Li, D.; Li, W.; Liao, Y.; Qiao, W.; Liu, W.; Azimi, M. Erosion simulation and improvement scheme of separator blowdown system—A case study of Changning national shale gas demonstration area. *J. Nat. Gas Sci. Eng.* **2021**, *88*, 103856. [CrossRef]
10. Peng, S.; Zhang, Y.; Zhao, W.; Liu, E. Analysis of the influence of rectifier blockage on the metering performance during shale gas extraction. *Energy Fuels* **2021**, *35*, 2134–2143. [CrossRef]
11. Zhang, W.; Tang, Z. Numerical modeling of response of CFRP–Concrete interfaces subjected to fatigue loading. *J. Compos. Constr.* **2021**, *25*, 04021043. [CrossRef]
12. Peng, S.; Chen, Q.; Liu, E. The role of computational fluid dynamics tools on investigation of pathogen transmission: Prevention and control. *Sci. Total Environ.* **2020**, *746*, 142090. [CrossRef]
13. Wei, J.; Xie, Z.; Zhang, W.; Luo, X.; Yang, Y.; Chen, B. Experimental study on circular steel tube-confined reinforced UHPC columns under axial loading. *Eng. Struct.* **2021**, *230*, 111599. [CrossRef]
14. Mou, B.; Bai, Y. Experimental investigation on shear behavior of steel beam-to-CFST column connections with irregular panel zone. *Eng. Struct.* **2018**, *168*, 487–504. [CrossRef]
15. Xie, S.-J.; Lin, H.; Chen, Y.-F.; Wang, Y.-X. A new nonlinear empirical strength criterion for rocks under conventional triaxial compression. *J. Cent. South Univ.* **2021**, *28*, 1448–1458. [CrossRef]
16. Ju, B.-K.; Yoo, S.-H.; Baek, C. Economies of Scale in City Gas Sector in Seoul, South Korea: Evidence from an Empirical Investigation. *Sustainability* **2022**, *14*, 5371. [CrossRef]
17. Braspenning, P.J.; Thuijsman, F.; Weijters, A.J.M.M. *Artificial Neural Networks: An Introduction to ANN Theory and Practice*; Springer Science & Business Media: Berlin, Germany, 1995; Volume 931.
18. Liu, F.; Zhang, G.; Lu, J. Multisource heterogeneous unsupervised domain adaptation via fuzzy relation neural networks. *IEEE Trans. Fuzzy Syst.* **2020**, *29*, 3308–3322. [CrossRef]
19. Yahya, S.I.; Rezaei, A.; Aghel, B. Forecasting of water thermal conductivity enhancement by adding nano-sized alumina particles. *J. Therm. Anal. Calorim.* **2021**, *145*, 1791–1800. [CrossRef]
20. Peng, S.; Chen, R.; Yu, B.; Xiang, M.; Lin, X.; Liu, E. Daily natural gas load forecasting based on the combination of long short term memory, local mean decomposition, and wavelet threshold denoising algorithm. *J. Nat. Gas Sci. Eng.* **2021**, *95*, 104175. [CrossRef]
21. Seyedashraf, O.; Mehrabi, M.; Akhtari, A.A. Novel approach for dam break flow modeling using computational intelligence. *J. Hydrol.* **2018**, *559*, 1028–1038. [CrossRef]
22. Zhao, Y.; Foong, L.K. Predicting Electrical Power Output of Combined Cycle Power Plants Using a Novel Artificial Neural Network Optimized by Electrostatic Discharge Algorithm. *Measurement* **2022**, *198*, 111405. [CrossRef]
23. Khajehzadeh, M.; Taha, M.R.; Eslami, M. Multi-objective optimisation of retaining walls using hybrid adaptive gravitational search algorithm. *Civ. Eng. Environ. Syst.* **2014**, *31*, 229–242. [CrossRef]
24. Eslami, M.; Neshat, M.; Khalid, S.A. A novel hybrid sine cosine algorithm and pattern search for optimal coordination of power system damping controllers. *Sustainability* **2022**, *14*, 541. [CrossRef]
25. Pinkus, A. Approximation theory of the MLP model in neural networks. *Acta Numer.* **1999**, *8*, 143–195. [CrossRef]
26. Gao, W.; Alsarraf, J.; Moayedi, H.; Shahsavar, A.; Nguyen, H. Comprehensive preference learning and feature validity for designing energy-efficient residential buildings using machine learning paradigms. *Appl. Soft Comput.* **2019**, *84*, 105748. [CrossRef]
27. Ahmad, A.; Ghritlahre, H.K.; Chandrakar, P. Implementation of ANN technique for performance prediction of solar thermal systems: A Comprehensive Review. *Trends Renew. Energy* **2020**, *6*, 12–36. [CrossRef]
28. Liu, T.; Tan, Z.; Xu, C.; Chen, H.; Li, Z. Study on deep reinforcement learning techniques for building energy consumption forecasting. *Energy Build.* **2020**, *208*, 109675. [CrossRef]
29. Guo, Y.; Yang, Y.; Kong, Z.; He, J. Development of Similar Materials for Liquid-Solid Coupling and Its Application in Water Outburst and Mud Outburst Model Test of Deep Tunnel. *Geofluids* **2022**, *2022*, 8784398. [CrossRef]

30. Hornik, K. Approximation capabilities of multilayer feedforward networks. *Neural Netw.* **1991**, *4*, 251–257. [CrossRef]
31. Ren, Z.; Motlagh, O.; Chen, D. A correlation-based model for building ground-coupled heat loss calculation using Artificial Neural Network techniques. *J. Build. Perform. Simul.* **2020**, *13*, 48–58. [CrossRef]
32. Wei, G.; Fan, X.; Xiong, Y.; Lv, C.; Li, S.; Lin, X. Highly disordered VO_2 films: Appearance of electronic glass transition and potential for device-level overheat protection. *Appl. Phys. Express* **2022**, *15*, 043002. [CrossRef]
33. Fan, X.; Wei, G.; Lin, X.; Wang, X.; Si, Z.; Zhang, X.; Shao, Q.; Mangin, S.; Fullerton, E.; Jiang, L. Reversible switching of interlayer exchange coupling through atomically thin VO2 via electronic state modulation. *Matter* **2020**, *2*, 1582–1593. [CrossRef]
34. Mohammadhassani, M.; Nezamabadi-pour, H.; Suhatril, M.; Shariati, M. Identification of a suitable ANN architecture in predicting strain in tie section of concrete deep beams. *Struct. Eng. Mech.* **2013**, *46*, 853–868. [CrossRef]
35. Sadeghi, A.; Younes Sinaki, R.; Young, W.A.; Weckman, G.R. An Intelligent Model to Predict Energy Performances of Residential Buildings Based on Deep Neural Networks. *Energies* **2020**, *13*, 571. [CrossRef]
36. Sholahudin, S.; Han, H. Simplified dynamic neural network model to predict heating load of a building using Taguchi method. *Energy* **2016**, *115*, 1672–1678. [CrossRef]
37. Ryu, J.-A.; Chang, S. Data Driven Heating Energy Load Forecast Modeling Enhanced by Nonlinear Autoregressive Exogenous Neural Networks. *Int. J. Struct. Civ. Eng. Res.* **2019**, *8*, 246–252. [CrossRef]
38. Khalil, A.J.; Barhoom, A.M.; Abu-Nasser, B.S.; Musleh, M.M.; Abu-Naser, S.S. Energy efficiency prediction using artificial neural network. *Int. J. Acad. Pedagog. Res.* **2019**, *3*, 1–7.
39. Almutairi, K.; Algarni, S.; Alqahtani, T.; Moayedi, H.; Mosavi, A. A TLBO-Tuned Neural Processor for Predicting Heating Load in Residential Buildings. *Sustainability* **2022**, *14*, 5924. [CrossRef]
40. Cao, B.; Zhao, J.; Liu, X.; Arabas, J.; Tanveer, M.; Singh, A.K.; Lv, Z. Multiobjective evolution of the explainable fuzzy rough neural network with gene expression programming. *IEEE Trans. Fuzzy Syst.* **2022**. [CrossRef]
41. Adedeji, P.A.; Akinlabi, S.; Madushele, N.; Olatunji, O.O. Hybrid adaptive neuro-fuzzy inference system (ANFIS) for a multi-campus university energy consumption forecast. *Int. J. Ambient Energy* **2022**, *43*, 1685–1694. [CrossRef]
42. Ahmad, T.; Chen, H. Nonlinear autoregressive and random forest approaches to forecasting electricity load for utility energy management systems. *Sustain. Cities Soc.* **2019**, *45*, 460–473. [CrossRef]
43. Namlı, E.; Erdal, H.; Erdal, H.I. Artificial intelligence-based prediction models for energy performance of residential buildings. In *Recycling and Reuse Approaches for Better Sustainability*; Springer: Berlin/Heidelberg, Germany, 2019; pp. 141–149.
44. Yepes, V.; Martí, J.V.; García, J. Black hole algorithm for sustainable design of counterfort retaining walls. *Sustainability* **2020**, *12*, 2767. [CrossRef]
45. Jamal, A.; Tauhidur Rahman, M.; Al-Ahmadi, H.M.; Ullah, I.; Zahid, M. Intelligent intersection control for delay optimization: Using meta-heuristic search algorithms. *Sustainability* **2020**, *12*, 1896. [CrossRef]
46. Jitkongchuen, D.; Pacharawongsakda, E. Prediction Heating and cooling loads of building using evolutionary grey wolf algorithms. In Proceedings of the 2019 Joint International Conference on Digital Arts, Media and Technology with ECTI Northern Section Conference on Electrical, Electronics, Computer and Telecommunications Engineering (ECTI DAMT-NCON), Nan, Thailand, 30 January–2 February 2019; pp. 93–97.
47. Ghahramani, A.; Karvigh, S.A.; Becerik-Gerber, B. HVAC system energy optimization using an adaptive hybrid metaheuristic. *Energy Build.* **2017**, *152*, 149–161. [CrossRef]
48. Foong, L.K.; Zhao, Y.; Bai, C.; Xu, C. Efficient metaheuristic-retrofitted techniques for concrete slump simulation. *Smart Struct. Syst. Int. J.* **2021**, *27*, 745–759.
49. Eslami, M.; Shareef, H.; Mohamed, A. Optimization and coordination of damping controls for optimal oscillations damping in multi-machine power system. *Int. Rev. Electr. Eng.* **2011**, *6*, 1984–1993.
50. Martin, G.L.; Monfet, D.; Nouanegue, H.F.; Lavigne, K.; Sansregret, S. Energy calibration of HVAC sub-system model using sensitivity analysis and meta-heuristic optimization. *Energy Build.* **2019**, *202*, 109382. [CrossRef]
51. Bamdad Masouleh, K. Building Energy Optimisation Using Machine Learning and Metaheuristic Algorithms. Ph.D. Thesis, Queensland University of Technology, Brisbane City, QLD, Australia, 2018.
52. Moayedi, H.; Mu'azu, M.A.; Foong, L.K. Novel swarm-based approach for predicting the cooling load of residential buildings based on social behavior of elephant herds. *Energy Build.* **2020**, *206*, 109579. [CrossRef]
53. Moayedi, H.; Mosavi, A. Electrical power prediction through a combination of multilayer perceptron with water cycle ant lion and satin bowerbird searching optimizers. *Sustainability* **2021**, *13*, 2336. [CrossRef]
54. Yang, F.; Moayedi, H.; Mosavi, A. Predicting the degree of dissolved oxygen using three types of multi-layer perceptron-based artificial neural networks. *Sustainability* **2021**, *13*, 9898. [CrossRef]
55. Zhao, Y.; Wang, Z. Subset simulation with adaptable intermediate failure probability for robust reliability analysis: An unsupervised learning-based approach. *Struct. Multidiscip. Optim.* **2022**, *65*, 172. [CrossRef]
56. Zhou, G.; Moayedi, H.; Bahiraei, M.; Lyu, Z. Employing artificial bee colony and particle swarm techniques for optimizing a neural network in prediction of heating and cooling loads of residential buildings. *J. Clean. Prod.* **2020**, *254*, 120082. [CrossRef]
57. Nguyen, H.; Moayedi, H.; Jusoh, W.A.W.; Sharifi, A. Proposing a novel predictive technique using M5Rules-PSO model estimating cooling load in energy-efficient building system. *Eng. Comput.* **2020**, *36*, 857–866. [CrossRef]

58. Bui, D.-K.; Nguyen, T.N.; Ngo, T.D.; Nguyen-Xuan, H. An artificial neural network (ANN) expert system enhanced with the electromagnetism-based firefly algorithm (EFA) for predicting the energy consumption in buildings. *Energy* **2020**, *190*, 116370. [CrossRef]
59. Moayedi, H.; Nguyen, H.; Kok Foong, L. Nonlinear evolutionary swarm intelligence of grasshopper optimization algorithm and gray wolf optimization for weight adjustment of neural network. *Eng. Comput.* **2021**, *37*, 1265–1275. [CrossRef]
60. Moayedi, H.; Bui, D.T.; Dounis, A.; Lyu, Z.; Foong, L.K. Predicting heating load in energy-efficient buildings through machine learning techniques. *Appl. Sci.* **2019**, *9*, 4338. [CrossRef]
61. Moayedi, H.; Mehrabi, M.; Bui, D.T.; Pradhan, B.; Foong, L.K. Fuzzy-metaheuristic ensembles for spatial assessment of forest fire susceptibility. *J. Environ. Manag.* **2020**, *260*, 109867. [CrossRef] [PubMed]
62. Zhao, Y.; Hu, H.; Song, C.; Wang, Z. Predicting compressive strength of manufactured-sand concrete using conventional and metaheuristic-tuned artificial neural network. *Measurement* **2022**, *194*, 110993. [CrossRef]
63. Zhao, Y.; Yan, Q.; Yang, Z. A Novel Artificial Bee Colony Algorithm for Structural Damage Detection. *Adv. Civ. Eng.* **2020**, *2020*, 3743089. [CrossRef]
64. Khajehzadeh, M.; Taha, M.R.; Eslami, M. A new hybrid firefly algorithm for foundation optimization. *Natl. Acad. Sci. Lett.* **2013**, *36*, 279–288. [CrossRef]
65. Zhao, Y.H.; Joseph, A.; Zhang, Z.W. Deterministic Snap-Through Buckling and Energy Trapping in Axially-Loaded Notched Strips for Compliant Building Blocks. *Smart Mater. Struct.* **2020**, *29*, 02LT03. [CrossRef]
66. Khajehzadeh, M.; Keawsawasvong, S.; Nehdi, M.L. Effective hybrid soft computing approach for optimum design of shallow foundations. *Sustainability* **2022**, *14*, 1847. [CrossRef]
67. Khajehzadeh, M.; Taha, M.R.; Eslami, M. Multi-objective optimization of foundation using global-local gravitational search algorithm. *Struct. Eng. Mech.* **2014**, *50*, 257–273. [CrossRef]
68. Zhao, Y.; Hu, H.; Bai, L. Fragility Analyses of Bridge Structures Using the Logarithmic Piecewise Function-Based Probabilistic Seismic Demand Model. *Sustainability* **2021**, *13*, 7814. [CrossRef]
69. Eslami, M.; Shareef, H.; Mohamed, A.; Khajehzadeh, M. Damping controller design for power system oscillations using hybrid GA-SQP. *Int. Rev. Electr. Eng.* **2011**, *6*, 888–896.
70. Khajehzadeh, M.; Taha, M.R.; Keawsawasvong, S.; Mirzaei, H.; Jebeli, M. An effective artificial intelligence approach for slope stability evaluation. *IEEE Access* **2022**, *10*, 5660–5671. [CrossRef]
71. Mehrabi, M.; Moayedi, H. Landslide susceptibility mapping using artificial neural network tuned by metaheuristic algorithms. *Environ. Earth Sci.* **2021**, *80*, 804. [CrossRef]
72. Zhao, Y.; Moayedi, H.; Bahiraei, M.; Kok Foong, L. Employing TLBO and SCE for optimal prediction of the compressive strength of concrete. *Smart Struct. Syst.* **2020**, *26*, 753–763. [CrossRef]
73. Tsanas, A.; Xifara, A. Accurate quantitative estimation of energy performance of residential buildings using statistical machine learning tools. *Energy Build.* **2012**, *49*, 560–567. [CrossRef]
74. Robert, J.S. *Pattern Recognition: Statistical, Structural and Neural Approaches*; Wiley India Pvt. Limited: Noida, India, 1992; ISBN 8126513705.
75. Hornik, K.; Stinchcombe, M.; White, H. Multilayer feedforward networks are universal approximators. *Neural Netw.* **1989**, *2*, 359–366. [CrossRef]
76. Mirjalili, S.; Mirjalili, S.M.; Hatamlou, A. Multi-verse optimizer: A nature-inspired algorithm for global optimization. *Neural Comput. Appl.* **2016**, *27*, 495–513. [CrossRef]
77. Zhao, Y.; Zhong, X.; Foong, L.K. Predicting the splitting tensile strength of concrete using an equilibrium optimization model. *Steel Compos. Struct. Int. J.* **2021**, *39*, 81–93.
78. Wolf, T.D.; Holvoet, T. Emergence versus self-organisation: Different concepts but promising when combined. In *Proceedings of the International Workshop on Engineering Self-Organising Applications*; ESOA: Berlin/Heidelberg, Germany, 2005; pp. 1–15.
79. Wang, J.; Tian, J.; Zhang, X.; Yang, B.; Liu, S.; Yin, L.; Zheng, W. Control of Time Delay Force Feedback Teleoperation System With Finite Time Convergence. *Front. Neurorobot.* **2022**, *16*, 877069. [CrossRef] [PubMed]
80. Goldstein, J. Emergence as a construct: History and issues. *Emergence* **1999**, *1*, 49–72. [CrossRef]
81. Noël, V.; Zambonelli, F. Methodological guidelines for engineering self-organization and emergence. In *Software Engineering for Collective Autonomic Systems*; Springer: Berlin/Heidelberg, Germany, 2015; pp. 355–378.
82. Serugendo, G.D.M.; Gleizes, M.-P.; Karageorgos, A. Self-organization in multi-agent systems. *Knowl. Eng. Rev.* **2005**, *20*, 165–189. [CrossRef]
83. Prokopenko, M. *Guided Self-Organization*; Taylor & Francis: London, UK, 2009.
84. Brun, Y.; Marzo Serugendo, G.D.; Gacek, C.; Giese, H.; Kienle, H.; Litoiu, M.; Müller, H.; Pezzè, M.; Shaw, M. Engineering self-adaptive systems through feedback loops. In *Software Engineering for Self-Adaptive Systems*; Springer: Berlin/Heidelberg, Germany, 2009; pp. 48–70.
85. Lemos, R.d.; Giese, H.; Müller, H.A.; Shaw, M.; Andersson, J.; Litoiu, M.; Schmerl, B.; Tamura, G.; Villegas, N.M.; Vogel, T. Software engineering for self-adaptive systems: A second research roadmap. In *Software Engineering for Self-Adaptive Systems II*; Springer: Berlin/Heidelberg, Germany, 2013; pp. 1–32.
86. Oreizy, P.; Gorlick, M.M.; Taylor, R.N.; Heimhigner, D.; Johnson, G.; Medvidovic, N.; Quilici, A.; Rosenblum, D.S.; Wolf, A.L. An architecture-based approach to self-adaptive software. *IEEE Intell. Syst. Appl.* **1999**, *14*, 54–62. [CrossRef]

87. Doğan, B.; Ölmez, T. A new metaheuristic for numerical function optimization: Vortex Search algorithm. *Inf. Sci.* **2015**, *293*, 125–145. [CrossRef]
88. Doğan, B.; Ölmez, T. Vortex search algorithm for the analog active filter component selection problem. *AEU-Int. J. Electron. Commun.* **2015**, *69*, 1243–1253. [CrossRef]
89. Dogan, B.; Ölmez, T. Modified off-lattice AB model for protein folding problem using the vortex search algorithm. *Int. J. Mach. Learn. Comput.* **2015**, *5*, 329. [CrossRef]
90. Altintasi, C.; Aydin, O.; Taplamacioglu, M.C.; Salor, O. Power system harmonic and interharmonic estimation using Vortex Search Algorithm. *Electr. Power Syst. Res.* **2020**, *182*, 106187. [CrossRef]
91. Qyyum, M.A.; Yasin, M.; Nawaz, A.; He, T.; Ali, W.; Haider, J.; Qadeer, K.; Nizami, A.-S.; Moustakas, K.; Lee, M. Single-solution-based vortex search strategy for optimal design of offshore and onshore natural gas liquefaction processes. *Energies* **2020**, *13*, 1732. [CrossRef]
92. Nguyen, H.; Mehrabi, M.; Kalantar, B.; Moayedi, H.; Abdullahi, M.a.M. Potential of hybrid evolutionary approaches for assessment of geo-hazard landslide susceptibility mapping. *Geomat. Nat. Hazards Risk* **2019**, *10*, 1667–1693. [CrossRef]
93. Mehrabi, M. Landslide susceptibility zonation using statistical and machine learning approaches in Northern Lecco, Italy. *Nat. Hazards* **2022**, *111*, 901–937. [CrossRef]
94. Mirjalili, S.; Lewis, A. The whale optimization algorithm. *Adv. Eng. Softw.* **2016**, *95*, 51–67. [CrossRef]
95. Karaboga, D. *An Idea Based on Honey Bee Swarm for Numerical Optimization*; Technical Report-tr06; Erciyes University, Engineering Faculty, Computer: Kayseri, Turkey, 2005.
96. Kennedy, J.; Eberhart, R. Particle swarm optimization. In Proceedings of the ICNN'5-International Conference on Neural Networks, Perth, Australia, 27 November 1995; pp. 1942–1948.
97. Mirjalili, S.; Gandomi, A.H.; Mirjalili, S.Z.; Saremi, S.; Faris, H.; Mirjalili, S.M. Salp Swarm Algorithm: A bio-inspired optimizer for engineering design problems. *Adv. Eng. Softw.* **2017**, *114*, 163–191. [CrossRef]
98. Bayraktar, Z.; Komurcu, M.; Werner, D.H. Wind Driven Optimization (WDO): A novel nature-inspired optimization algorithm and its application to electromagnetics. In Proceedings of the 2010 IEEE Antennas and Propagation Society International Symposium, Toronto, ON, Canada, 11–17 July 2010; pp. 1–4.
99. Dhiman, G.; Kumar, V. Multi-objective spotted hyena optimizer: A multi-objective optimization algorithm for engineering problems. *Knowl.-Based Syst.* **2018**, *150*, 175–197. [CrossRef]
100. Atashpaz-Gargari, E.; Lucas, C. Imperialist competitive algorithm: An algorithm for optimization inspired by imperialistic competition. In Proceedings of the 2007 IEEE Congress on Evolutionary Computation, Singapore, 25–28 September 2007; pp. 4661–4667.
101. Saremi, S.; Mirjalili, S.; Lewis, A. Grasshopper optimisation algorithm: Theory and application. *Adv. Eng. Softw.* **2017**, *105*, 30–47. [CrossRef]
102. Holland, J.H. Genetic algorithms. *Sci. Am.* **1992**, *267*, 66–73. [CrossRef]
103. Mirjalili, S.; Mirjalili, S.M.; Lewis, A. Grey wolf optimizer. *Adv. Eng. Softw.* **2014**, *69*, 46–61. [CrossRef]
104. Park, J.; Lee, S.J.; Kim, K.H.; Kwon, K.W.; Jeong, J.-W. Estimating thermal performance and energy saving potential of residential buildings using utility bills. *Energy Build.* **2016**, *110*, 23–30. [CrossRef]
105. Yezioro, A.; Dong, B.; Leite, F. An applied artificial intelligence approach towards assessing building performance simulation tools. *Energy Build.* **2008**, *40*, 612–620. [CrossRef]
106. Gong, X.; Wang, L.; Mou, Y.; Wang, H.; Wei, X.; Zheng, W.; Yin, L. Improved Four-channel PBTDPA control strategy using force feedback bilateral teleoperation system. *Int. J. Control Autom. Syst.* **2022**, *20*, 1002–1017. [CrossRef]
107. Bui, X.-N.; Moayedi, H.; Rashid, A.S.A. Developing a predictive method based on optimized M5Rules–GA predicting heating load of an energy-efficient building system. *Eng. Comput.* **2020**, *36*, 931–940. [CrossRef]
108. Fang, J.; Kong, G.; Yang, Q. Group Performance of Energy Piles under Cyclic and Variable Thermal Loading. *J. Geotech. Geoenviron. Eng.* **2022**, *148*, 04022060. [CrossRef]
109. Tien Bui, D.; Moayedi, H.; Anastasios, D.; Kok Foong, L. Predicting heating and cooling loads in energy-efficient buildings using two hybrid intelligent models. *Appl. Sci.* **2019**, *9*, 3543. [CrossRef]

buildings

MDPI

Perspective

Position Paper Introducing a Sustainable, Universal Approach to Retrofitting Residential Buildings

Małgorzata Fedorczak-Cisak [1,*], Mark Bomberg [2,3,*], David W. Yarbrough [4], Lowell E. Lingo [3] and Anna Romanska-Zapala [5]

1 Faculty of Civil Engineering, Cracow University of Technology, 31-155 Kraków, Poland
2 Mechanical Engineering, Clarkson University, Potsdam, NY 13699, USA
3 DFI Enterprises, Inc, Morrisville, NY 13408, USA; lelingo@syr.edu
4 R&D Services, Inc., Watertown, TN 37184, USA; dave@rdservices.com
5 Department of Automation and Information Technology, Faculty of Electrical and Computer Engineering, Cracow University of Technology, 31-155 Kraków, Poland; a.romanska@pk.edu.pl
* Correspondence: mfedorczak-cisak@pk.edu.pl (M.F.-C.); mark.bomberg@gmail.com (M.B.)

Citation: Fedorczak-Cisak, M.; Bomberg, M.; Yarbrough, D.W.; Lingo, L.E.; Romanska-Zapala, A. Position Paper Introducing a Sustainable, Universal Approach to Retrofitting Residential Buildings. *Buildings* **2022**, *12*, 846. https://doi.org/10.3390/buildings12060846

Academic Editor: Gerardo Maria Mauro

Received: 13 March 2022
Accepted: 10 May 2022
Published: 17 June 2022

Abstract: Protests during the 2021 Climate Conference in Glasgow exemplified our dilemma. The establishment perpetuates old thinking, while young people demand a new approach to mitigate the impact of climate change. The authors agree with the young people, and as a solution we propose to replace the current fragmentary approach with a new holistic one. The passive house approach that was conceptualized by the University of Illinois and built in Canada in 1977 showed us that energy consumption can be reduced about half of that used in the traditional design. Seventeen years later, a European passive house was built in Darmstadt. In 2008, a demonstration house in Syracuse, NY, showed that integrated passive measures produced energy use by about half of the NY state code for 2004. At the same time, some advanced houses in the USA showed total energy use of about 70 kWh/(m^2·y). In 2008, at the first Building Enclosure Science and Technology Conference, two equally important objectives for 2030 were proposed by the Lawrence Berkeley National Laboratory: (1) a 90% reduction of energy use in new buildings and (2) 50% for the retrofitting of existing buildings, i.e., to the level achieved in the 1980s. The first objective has recently been achieved in small buildings while the large residential buildings remain on the level obtained in the 2000s. Yet, the retrofitting of existing buildings (the second objective) has been a dismal failure. This paper acknowledges progress in hydronic heating and cooling involving electric heat pumps and hybrid solar panels, building automatics used for operation of HVAC, and modification of air distribution systems that comes from experience with the SARS-CoV-2 pandemic. Furthermore, it highlights that to accelerate energy efficiency and carbon emission reductions, there must be broad public-private educational programs with demonstrations of a new generation of retrofitting. Economically and ecologically retrofitted buildings will create a new approach to real estate investment.

Keywords: energy efficiency; residential retrofits; building automatics control; integrated HVAC; retrofitting technology

1. Foreword

In science, a discovery or invention precedes application. This is not the case for building science. Years ago, Confucius (Lao Tse) highlighted the bond between buildings and culture. Modern building science was created by an analysis of traditions. The mighty cathedrals built in the Middle Ages in Europe were based more on scientific intuition than knowledge. Yet, since the development of the natural sciences, it is the tradition, and its failures demonstrated when trying to modify some concepts, that defined what became the mainstream of building technology. Therefore, today, when trying to develop a rapid solution for climate change, we cannot analyze the construction technology alone, but we must do it in the context of the socio-economic conditions of the country. Several behavioral studies showed that occupants want to actively participate in shaping their environment and several economic studies highlighted that the well-being and productivity of the building occupants are more important than the initial cost of building. Yet, the traditional structure of investments does not include these aspects of design. As we propose a market disruption with new technology, we must provide a cost-benefit analysis of all critical aspects of housing, not only the life-cycle cost analysis. In this context, the indoor environment and, related to them, the hygro-thermal considerations or indoor air quality are critical. We do not discuss other aspects of building science as they do not change during the proposed market disruption. We do not neglect their importance. Yet, dealing with a limit of 21 pages of text, we decided to examine only one approach. Furthermore, in a time of economic globalization, using examples from North America is not a problem and restricting the selection of technology to a potential winner is also justified. The environmental quality management (EQM) technology selected for this paper provides accessibility of environmental controls allowing occupants to correct unsatisfactory environments and is wide enough to include several different practical strategies. In contrast to strategies that eliminated other options, e.g., passive houses or solar houses, the EQM includes thermo-active buildings from Japan, or thermo-active thermal insulations from Hungary, or multi-stage construction methods applied in Canada. In a nutshell, the EQM is an aggregate of the best individual developments in different cases.

2. Limitations of Current Technology in Shaping the Indoor Environment

The following article considers history and forces the shaping of environmental progress in residential construction with the focus on retrofits. It proposes changes to residential buildings and formulates a universal retrofitting technology. Yet, to achieve progress one must first understand the limitations of the current technology.

2.1. Indoor Environment as Defined by an Architect

In 1970 Flynn and Segil [1] wrote,

But rather than a simple correction of climatic deficiencies, the environmental control function of building must be oriented toward the more extensive sensory demands of various occupant activities and experiences. This occupant perceives light as the surface brightness and color; he absorbs heat from warmer surfaces and warmer air; and he himself emits heat to the cooler surfaces and cooler air. He responds physiologically to humidity, to air motion, to radiation and to air freshness. He also responds to sound. A major function of the building, then, is to provide for all the sensory responses concurrently-to establish and maintain order and harmony in the sensory environment.

Why does this statement not represent construction practice? Because the design, construction, and commissioning professionals are thinking only about their individual fields. These experts do not have tests to ensure field performance of materials. The existing tests measure material characteristics and only after collecting enough field performance information may one correlate them with field performance to gain enough judgement value allowing a designer to select the material. Structural, fire, acoustic, lighting, and material experts are all qualified in their disciplines and have little understanding for "how to establish and maintain order and harmony in the sensory environment" when developing designs for a "durable and cost-effective building shell". The current design process does not provide the facility for predicting the future performance of new systems, nor the means for quality assurance for the building. As we do not have a unified approach to design low-energy buildings, many demonstrations of monitored whole-building performance, e.g., the high-environmental-performance house in Syracuse NY [2–4] or the net-zero-energy equilibrium house [5], despite analyzing the monitored discussion and the demonstrated field performance that have limited impact in the marketplace.

2.2. The Role of the Architect Is Changing

While in the past, architects had a holistic view of occupants and the building, this is not the case today. There are two major reasons for this, namely (a) an increase in the number of materials, and (b) a lack of a two-way street between building science and building practice.

(a) Speaking about Swedish statistics, in 1900, there were about 500 different construction products to choose from and today we can find 55,000–60,000 different products. This highlights the growth of specialized expertise and the fragmentation of the design process that erased the capability of an architect to control all stages of the design and construction process. Today, more than in the past, the architect must be able to produce an integrated product satisfying all occupants and all aspects of building performance.

(b) The lack of real exchange between science and practice is demonstrated below. Previously, moisture was not a serious consideration because masonry is resilient to moisture (unless exposed to freezing and thawing). The masonry wall could wet and slowly dry and thus temper large changes in moisture level introduced by climate or people. Scientists knew about diffusion theory and the calculation of condensation as early as 1938 and five different books were published in Russian on calculating water vapor transport through walls before 1958, when Glaser described a simple method to calculate water condensation in layman's terms. But in the 1950s, commercially manufactured glass fiber insulation was typically placed in the wood-frame wall cavities reducing the cavity surface temperature and causing condensation. So, having visible problems in practice, the codes and standards embraced an easy-to-understand solution. Moisture transport by diffusion became a worldwide accepted concept. Despite computer modeling for simultaneous heat and moisture (water and vapor) being available in the early 1970s and showing shortcomings of the Glaser's method, the widespread models came only in the 2000s. The lead author in 2015, while giving a course on building physics at one of the EU Universities, was told, "Please include the Glaser theory because it is included in this course and approved by our government." He replied, "Of course, already 40 years ago I published the paper explaining why using it is wrong".

2.3. The Paths from Materials to an Exterior Wall Assembly

For the sake of discussion, we distinguish between four functions typically associated in standards for a material layer, namely (1) exterior cladding (façade), (2) exterior continuous insulation, (3) structure or structural layer for load transmission layer, and (4) interior trim and finish; see discussion in reference [6]. The façade layer (1) controls fire, rain, air and water vapor entry, light, sound, solar radiation, and vermin; the external insulation (2) controls heat, but may also control air, water vapor, and sound; the structural layer (3) provides strength and rigidity but may also control air, water, and vapor transports; and the interior finish layer (4) controls fire, air, water and vapor movements, and sound.

(1) The façade layer can be either directly attached or be a rain screen with an air gap behind it (e.g., brick veneer) to provide rain control. In this case, the layer on the interior side of the air gap should fulfill all other façade requirements.

(2) The thermal insulating layer should also control acoustics. Note that popular materials such as mineral fiber with wind protection or polystyrene boards with taped joints do not fulfill the many requirements for air, water vapor, and vermin entry.

(3) The structural design is not discussed here.

(4) The requirements for airtightness and fire resistance of interior finishes are fulfilled by gypsum board, which is water-vapor permeable but lacks the moisture buffer capability. Thus, in the next generation of technology we need to reintroduce materials with moisture buffering ability.

The summer overheating of rooms associated with airtight, well-insulated buildings is caused by a large area of glazing and a high precision of heating controls. One may re-distribute the energy gains caused by solar radiation with increased thermal mass, circulation of indoor air, night ventilation systems, or cooling.

2.4. Energy Use in the Building Systems of 1978 and 2005

Figure 1 from the U.S. Energy Information Administration shows total residential energy use in 1978 and 2005. The total energy consumption in the building sector did not change from 1978 to 2005; while the fraction of the space heating was reduced from 66% to 41%, the comfort components took all the savings.

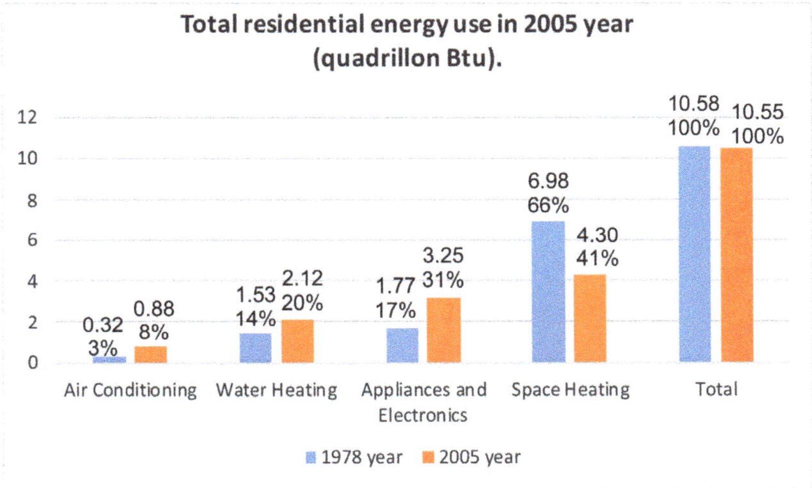

Figure 1. A comparison of energy use in 2005 with use in 1978.

Similar observations were reported in different countries and even by the authors of this paper [7]. An event that we call an "energy conundrum" was a discovery that in the years 1929 and 2002, large residential buildings in the city of Vancouver used the same amount of total energy, namely 250 kWh/(m^2·y). The fact that heavy masonry buildings without insulation could provide reasonable indoor climate highlighted the role of thermal mass that was later eliminated by use of automatic temperature controls. Yet, research on an adaptable comfort approach shows that one can modify the indoor air temperature by at least 6 °C, over a 6 h period and the change in human comfort does not exceed 4% of the comfort index [8]. This means that we should re-examine the approach to the design of low-energy buildings. If we want to reduce the use of air conditioning and use electrical energy during designated hours in a night, then add utilization of thermal mass and switch to the adaptable indoor comfort approach.

The second conclusion that one can draw from these comparisons is that only the total energy gives us a realistic measure for energy; measuring one of its components using metrics such as U-value or R-value may only introduce confusion and misjudgment. We recall the problem of log houses that did not meet the standard requirements, while providing a good indoor climate. This tradition was valid up to the 1950s when small but continuous changes in material technology started improving the performance of the whole system. Those changes became strongly accelerated in the 1970s when the concept of sustainability appeared. Sustainable building includes aspects related to reducing the consumption of natural resources [9–11], improving the external environment and ensuring the comfort of users [12–14].

2.5. Case Studies in Retrofitted Buildings

The examples below relate to summer overheating of masonry buildings with natural ventilation. In a renovated, four-star hotel, with high-ceiling rooms, located in Brussels, a room on the fifth floor in the summer afternoon showed a temperature of 26 °C when the day was cloudy and the outside temperature was 22 °C. After a window was fully opened, the temperature went down to 25 °C and when a huge fan (provided by the hotel concierge) was activated, the temperature remained at 25 °C, but the occupants felt comfortable.

The explanation is simple. Retrofitting included adding insulation and air-tight windows. In this way, retrofitting increased the impact of the stack effect (hot air rising because of buoyancy). The multi-unit masonry was built in the 1930s and as burning coal at that time required one chimney for each dwelling, the European tradition was to connect the exhaust from a ceramic stove to the same chimney stack at every second floor. Newly installed electrical heaters and natural ventilation all worked well, but the forgotten, hidden connections though an old chimney moved hot air from the first floor to the third floor and subsequently to the fifth floor. Note that these rooms had no air mixing and high ceilings facilitated a significant vertical gradient of temperature. Since level five was near the neutral pressure zone, opening or closing windows had no significant impact on air buoyancy and the additional heating delivered from the two lower floors kept the room temperature three degrees above the outdoor temperature.

A second example is from Canada, from a passive-solar design, an 18 story, concrete building with air pressurized corridors to compensate for the stack effect. The living room has a large heating and cooling unit located under the window and a leaky entry door (gaskets are worn). With closed windows, cigarette smoke coming from outside was smelt in the bathroom. During the summertime, this two-bedroom apartment had four different temperature zones. Of course, the occupants opened windows and used extra ventilators in the kitchen and living room.

With 192 apartments in the building and a total of 384 ventilators, the building used much more energy than was saved by the retrofitting measures. The design concept, however, worked well, keeping the average indoor temperature about two degrees above the outdoors in summer and on sunny days in winter, although in summer it caused excessive use of cooling. In effect, this design would be good if an indoor air circulation system was properly designed.

Another example is a case in Portland, Oregon where the best technical solutions were applied, and the post-retrofitting survey gave only a passing level of satisfaction with the indoor environment [15]. This reinforces the observation from the Canadian example that any indoor space without air flows between 0.1 and 0.3 m/s appear to American people as having a poor environment. Why American? Because the tradition here was air-borne heating that always included air flows in the space.

Elsewhere, we discussed an example where books became covered with mold while being kept in a closed desk during the winter in the warm and humid climate of Nanjing, China. Yet, books also became covered with mold in the corner of a basement room on an open bookshelf in the cold climate in Ottawa, Canada in a house with good mechanical ventilation when an adaptable comfort was used for temperature control [8]. This is a reminder that the universal approach to indoor environment during retrofitting must also *include air humidity management*.

Now, after repeated waves of the SARS-CoV-2 pandemic, one comes to the realization that the next generation of new or renovated dwellings must provide indoor space suitable for quarantining people infected with an air-borne virus, be it SARS-CoV-2 or influenza. A bedroom likely to be used for sick people should be underpressurized to guard against the spread of illness. A designated outdoor air system (DOAS) should be used [16]. Therefore, the requirements for retrofitting must be broadened to include an adequate handling of DOAS ventilation with interior air circulation and air humidity controls.

2.6. The Quest for a Sustainable Built Environment

The quest for sustainability resulted in dramatic changes in the process of residential construction. The new concepts of an integrated design team, building information modeling, commissioning of the building enclosure, and passive house standards have reached maturity. Global work on the development of new construction materials has not changed, but their evaluation is different from in the past when each material was considered on its own merits. Today, we look at the performance of a building as a system and on the material as a contributor to this system. Sustainability involves harmony between different aspects of the environment, society, and economy. Figure 2 shows the scope of considerations that broadens the field of building science in North America (building physics in Europe). Now, in the quest of improving building performance, building science merges concepts of passive houses with solar engineering and integrates building shells with mechanical services, but is still missing an overall vision. Physics does not tell us how to integrate people with their environment.

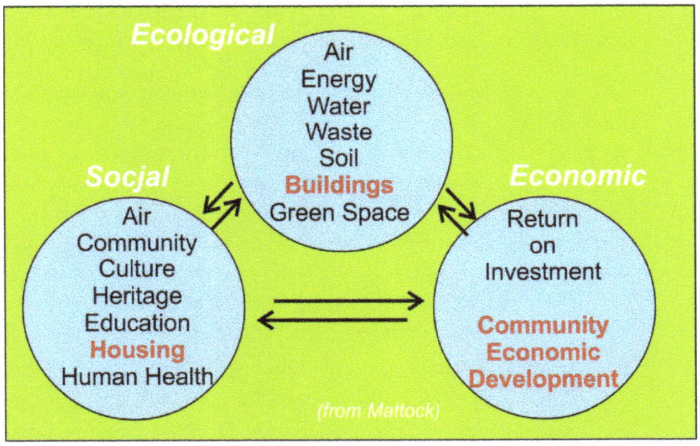

Figure 2. All three components must be satisfied for the sustainable built environment.

The authors introduced a new term, namely environmental quality management (EQM) for buildings, because the vision of building physics must be redirected toward occupants. In doing so, building physics will include the indoor environment with aspects such as energy efficiency, ventilation, indoor air quality, and thermal comfort on one side and added durability of the shell, affordability, and building resilience.

2.7. Case of Market Thinking Replacing System Thinking

Scientists demonstrated the system approach, highlighting the significance of passive measures in the design of energy-efficient buildings [17]. One may be surprised that 44 years ago one built with almost the current level of technology. This fact then begs the question: why was this technology not used in the marketplace? This occurred primarily because there was no real contact between the building science community and the socioeconomic forces driving the evolution of building construction. The design and construction shown in Figure 3 represented the ideas of building science leaders in North America, but the construction industry was not prepared to follow because the gap between building science (building physics) and builders was then, and is today, a critical issue that slows progress in climate-change reduction [18,19]. In a nutshell, 44 years ago we had a concept for a new technology, but not the understanding of the building energy performance. The current definition of energy performance is based on the 2005 U.S. Energy Act that talks about all factors affecting energy efficiency and the well-being of the occupants (specifically an occupant's productivity if this is a nonresidential construction).

Figure 3. Saskatchewan Energy Conservation house designed by the Illinois U demonstrated passive technology in 1978 (Regina). Solar-exposed surfaces with large windows are inclined. Evacuated solar pipes are placed on the attic level. It was provided with an air-sourced heat pump and polyethylene-based heat recovery ventilator. Reprinted with the permission of Harold Orr.

In 1978, builders liked the idea of airtight, well-insulated houses as shown in Figure 3 and, to make them affordable, they applied an economic trade-off. With much smaller heating loads, one could eliminate furnaces and replace them with electrical heating. Yet, the air-borne heating system had more functions than heating. The builders did not realize that eliminating the chimney also changes the indoor environment. Sick buildings (not enough fresh ventilation air) and wet attics (increased humidity and condensation on top floors and in attics) were problems introduced by the elimination of the chimneys [20].

Now we understand better the interactions between the different functions of a building. Yet, this understanding came in steps:

- In 1985, mechanical ventilation became mandatory in all residential buildings of Canada.
- Any residential building must be considered as a system with interacting subsystems.
- The design process was changed to the integrated design process (or protocol, IDP), where a whole team started collective work in conceptual stages [21]. Incidentally, bringing critical decisions to the initial stage of work reduced the cost of late changes in the design and therefore IDP was readily accepted in construction.

The following observations summarize the difference between then and now:

1. Heat, air, and moisture transports are inseparable and cannot be separately assessed. Today, we talk about "environmental control".
2. Often, when we modify a material or a construction detail, we find that the cost of repairs following these small changes is significant. This happens each time we analyze only one function without interaction with the other elements of the system. In other words, we fail when we lose track of the integrated approach.
3. The modification of building enclosures is slow but continuing. While old, leaky, and poorly insulated walls dry quickly, airtight walls that are insulated on the exterior dry slowly. If water enters around the window frames, then it will stay inside the wall.
4. Traditionally, building science followed the evolution of practice and lessons from any solution of the encountered problems enhanced understanding of construction performance. In the 1980s there was no practice of low-energy housing construction and Timusk (see [15]), when reviewing moisture control issues of the previous decade, stated:

"At the moment we are in a position where the traditional approach of learning from failures and copying what worked, has broken down. ... it is extremely difficult to accommodate all of the new information in view of the rapid changes in materials, details and performance expectations."

The Canadian R-2000 and Build America programs as guides for housing design, highlighted that formal education must be complemented by seminars for consultants and building practitioners. Indeed, this became the widespread practice in the USA. Furthermore, one needed to expand the environmental control in two dimensions:

- Improve the tools of field monitoring and field diagnostics and integrate them with user-oriented, computer-based design tools.
- Stress the objective-based design process, much in the same manner as it is done in structural engineering.

One should highlight that the impact of the interstitial pressure field became significant only when the building enclosure became more airtight [22–24]. Figure 4 shows mold growth on an interior wall in Florida [22] as the effect of an interstitial air pressure field.

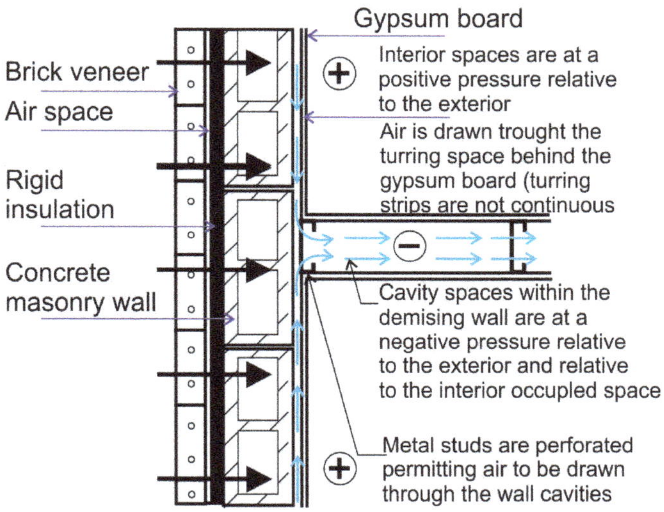

Figure 4. Interstitial airflow results in mold growth on the interior in a Florida hotel. With permission of J. Lstiburek.

The new performance expectations emerged slowly and mainly because of an accelerated pace of building science developments. Here are a few critical performance issues:

1. Discovery of interstitial air pressure fields and observations that several airflow paths are possible between any two points in a building. Furthermore, the total airflow resistance is a sum of all local airflow resistances and each of them varies with the frequency and direction of the wind gusts.
2. Introduction of air-barrier systems [25,26] affected heat and moisture flows.
3. The need for exterior insulating sheathing [27] to avoid moisture condensation within exterior walls.
4. Ecological considerations modified material choice and the wall-field performance [28].
5. Introduction of an integrated design process permits a system approach but also forces the making of critical decisions in the beginning of the design process.
6. With air-tight and highly insulted houses, one needs to modify humidity in indoor air [29] and the capillary active technology is developed.
7. Finally, in retrofitting, one may introduce a second air gap [30] and use it for several different functions, such as control of water content in the outer walls, as a means to modify air pressure in the indoor spaces, or even air recirculation in the dwelling.

The methods of building evaluation are different today than they were three decades ago because today we understand the interactions between the different functions of the building. The architectural design process includes four steps:

1. First, all passive energy measures and factors affecting indoor environment such as temperature, indoor air quality, acoustics, daylight, illumination, hot and sewer water management, aesthetics, and building resilience in disaster situations are addressed.
2. Second, the building's automatic control systems to integrate heating, cooling, ventilation, and other indoor climate controls including use of geothermal and solar means for energy generation and storage are addressed.
3. Next, an economic analysis to determine the level of investment for the initial building design or the initial stage of retrofitting. For example, one must decide to what extent photovoltaics should be included in stage 1 of the construction or retrofitting process.
4. Finally, one develops a comprehensive operational manual for the building and provides the design for stage 2 of new construction or retrofitting. This step also estimates costs for stage 2 of a project (see later text about two-stage design).

New elements in the above list are step 2 on building automatics and step 4 on the operational manual. The latter must be requested in the contract documents. Observe that there is no standard for a passenger car manual but all buyers expect to receive one. So, buyers should start expecting a manual for a smart house.

2.8. Review of the Deep Energy Retrofits (DER) in the USA

Published in a 2021 report from Lawrence Berkeley National Laboratory written by Iain Walker, Brennan Less, and Nuria Casquero-Modrego, entitled Emerging Pathways to Upgrade the US Housing Stock: A Review of the Home Energy Upgrade Literature [27] provides a detailed analysis of the state of the art.

With permission of the authors, we quote some of the key elements of the review.

This review addresses whole home energy upgrades targeting deep energy reductions (i.e., Deep Energy Retrofits, or DERs), from 30 to >50% site energy savings. The intent of this work is to characterize how energy upgrade projects and programs have evolved and improved over the past decade, and to identify what changes are needed to drive expansion of the U.S. retrofit market in such a way that addresses carbon emission from buildings, improves resilience and upgrades the housing stock for the 21st century. The topics covered in this review are wide-ranging, including trends in U.S. and European retrofit programs, measure costs (e.g., ductless heat pumps, heat pump water heaters, exterior wall insulation), emerging technologies, advancements in simulation tools, surveys of energy upgrade homeowners and practitioners, business economics (e.g., soft costs, gross margins), and health effects. Key changes in project design noted in this review include the: (1) electrification of dwellings with rapidly improving heat pump systems and low-cost PV technology, (2) shift away from high-cost super-insulation strategies and towards more traditional home performance/weatherization envelope upgrades, (3) recognition of the importance of when energy is used and from what fuel sources in terms of both energy cost and carbon emissions, and (4) emerging smart home technologies, such as batteries or thermal storage, smart ventilation and HVAC controls, and energy feedback devices. Promising program design strategies covered in this review include: (1) end-use electrification programs, (2) novel financing approaches (e.g., Pay-As-You-Save and local lender networks), (3) Pay-for-Performance incentive structures, (4) securitization of portfolios of upgraded homes as investment products, and (5) One-Stop Shop programs that integrate financing, project management, design, and support services. In addition to these project- and program-innovations, the industry should adopt new project performance metrics, namely those for carbon, peak demand, and energy storage, along with metrics characterizing resilience and health. Market drivers are needed to spur widespread energy upgrades in the U.S. housing stock, which will require valuation of DERs by the real estate industry, reduced project costs (in part by cutting soft costs), and projects designed to appeal to homeowners' while being enjoyable and profitable for contractors.

Key barriers that provide opportunities for modification:

- Projects focused solely on energy savings are not appealing.

The past focus on energy efficiency or annual site energy savings is not enough. There is a need to emphasize other metrics including health, resilience, affordability, maintenance, and environmental aspects. Energy upgrade projects must address the actual needs and goals of homeowners, and these projects must be profitable and enjoyable for the contractors and trade workers implementing them.

- The workforce remains inadequate.

Despite market development efforts over the past decades and the presence of many dedicated and very skillful companies, the general workforce is inadequate to implement complex projects at scale. The emergence of new technologies, metrics and processes make this inadequacy even more evident, as no centralized databases exist of contractors who have experience with electrification or low-carbon projects, for example.

- The costs remain too high.

Finding the lowest-cost way to save energy and reduce carbon emissions is likely to include PV, thermal storage, simple weatherization, and electrification, rather than high-cost envelope upgrades. Other novel approaches may include leaving existing heating systems in place and augmenting with higher-performance systems to save the cost of existing system decommissioning. Another aspect of cost control is to invest in technologies that can more reliably reduce energy use (or CO_2 emissions) in homes to reduce financial risks for homeowners and post-retrofit home performance risks for contractors. Across the industry, soft costs are a substantial fraction of the total, and efforts are needed to reduce these soft costs to levels equivalent to or less than the general remodeling industry.

- Economic justifications are challenging and possibly inadequate.

Due to low energy costs and the failure to appropriately price carbon emissions, the direct financial benefits of home-energy upgrades are difficult to prove using simplistic methods, such as the number of years it takes to pay back an investment. Other approaches, such as net-monthly cost and pay-as-you-save programs are making progress in this area, but more work is needed to incorporate health and environmental costs that are typically ignored. We also need to recognize that for many homeowners, their motivation and decisions regarding home energy upgrades are not purely based on simplistic financial analyses.

As our article does not deal with the financial side of retrofitting, we want to highlight that the LBNL report discuss strategy such as the electrification programs and methods of financing (pay-as-you-save, pay-for-performance, and retrofit as an investment product) and perhaps the most important from the system integration point of view is the one-stop shop program that integrates project financing and design management and support services.

2.9. *Decarbonization of Buildings Is Critical for Tomorrow's Built Environment*

We distinguish between three different areas of concern [31,32]:

(1) *Direct emissions* are the greenhouse gas (GHG) emissions from controlled sources such as combustion in boilers, furnaces, and equipment (including fugitive GHG emissions such as refrigerant leaks).
(2) *Indirect emissions* are indirect GHG emissions associated with the purchased media such as electricity, district heating, or cooling used in buildings.
(3) *Activity-related emissions* include embodied emissions within materials and resources used or consumed by the organization—paper used, waste produced, coffee consumed, and emissions of any suppliers. Embodied carbon includes mining, processing, manufacturing, transportation, and installation of materials.

The atmospheric GHGs typically associated with buildings are carbon dioxide, methane, and a few refrigerants used in cooling. An index, called the global warming potential (GWP), which compares the global warming impact of a mass unit of a given material to the same mass unit of carbon dioxide, is used to quantify the GWP effect.

We define a zero-carbon building (ZCB) as: A zero-carbon building is a highly energy-efficient building in which carbon-free renewable energy or high-quality carbon offsets are used to counterbalance the annual carbon emissions from building materials and operations so that, with time, it offsets the carbon emissions embodied in the original construction process. One needs to use the definition of ZCB to compare different construction strategies. Without an accepted ZCB definition, one cannot analyze the building as a system and spend time and effort on useless talk. A good example is what many green people consider as an option, which is to ban the use of concrete. While concrete has huge, embodied energy, when used with the controlled effect of thermal mass it may be a material of choice in many specific applications. The highest ASHRAE green award in 2020 went to a concrete building in Tokyo using thermo-active technology (an equivalent of environmental quality management). Therefore, the definition of a zero-carbon building includes a comparison of initial and operational carbon emissions.

We realize that analysis of a building as a system is not easy; it requires technical proficiency higher than used today in construction. Furthermore, the modeling of building performance and cost-benefit analysis are much broader than the current energy modeling. To optimize for net-zero gas emissions, one must use modeling of the building operation under actual weather and occupancy conditions, as well as consideration of carbon emissions from the electrical grid.

Including electric grid interaction with the building complicates modeling in a dramatic way because the GHG emissions depend on both location and time of the year. To minimize the GHG emissions from the generation of electricity requires re-evaluation of the way we do modeling and the way we design the building operation. As today decarbonization is not included in the cost calculations, one must be prepared for the inclusion of decarbonization in the next generation of buildings. This paper highlights the need for rethinking the fundamentals of the retrofitting technology.

2.10. Shaping the Internal Environment in Thermomodernized Buildings, Conclusion

Perhaps the single most visible limitation of the current design is fragmentation. The fragmented nature of the current approach often leads to results different than expected. As we started systematic work on energy conservation in the mid-1970s, it is interesting to examine Figure 1, which shows that between the years 1978 and 2005 the space heating component was dramatically reduced, but there were no changes in the total energy use. In effect, fragmentation often improves one aspect of field performance but may destroy others.

Figure 2 reminds us that sustainability includes three components and ecology alone does not make buildings sustainable. In effect, by including economic and social aspects, the concept of sustainability is opposite to fragmentation and brings another perspective to the construction process.

A failure of the integrated passive house system (Figure 3), when applied by builders in a simplified approach, demonstrates the role of interaction. Without understanding the interaction between heating and ventilation, builders changed the heating system, worsening the indoor environment. This example emphasized that reducing the gap between science and practice and creating a new generation of technology must proceed in parallel and that one must create *broad public–private educational programs.* In 2008, two equally important objectives were proposed by the Lawrence Berkeley National Laboratory, namely a 90% reduction of energy use in new buildings and the retrofitting of all existing buildings. The second objective, retrofitting, failed because it was not supported by a broad public–private educational program such as R2000 (Canada) or Building America (USA) that were introduced in the 1980s to support the sustainability approach.

Therefore, *if one wants to succeed with the climate-change* aspect of economically and ecologically retrofitted buildings, one must institute such programs again, as well as create a new approach to investment in building and city renovation.

3. An Example of the Next Generation of Low Energy Buildings

This section presents and discusses principles of *environmental quality management* (EQM), a generic public domain technology that we consider a next-generation technology. Several different practical examples are given of its application.

3.1. Thermal Storage Requires Transient Operation of Buildings

EQM technology includes thermal storage from geothermal sources and contribution from thermal mass that is controlled by the building automated system. Geothermal storage system has been discussed [33–37]. EQM integrates HVAC with the building structure, uses transient but controlled changes of indoor climate (adaptable indoor climate) to control the input of thermal mass and moisture buffers. To control the operation, one develops algorithms based on monitored information from field performance and uses them for system optimization.

3.2. Heating and Cooling Systems in the EQM Technology

Bomberg et al. [6] recommended using water-based heat-pump technology with coupling between thermal mass and a large surface of a hydronic heating or cooling system. This type of heating/cooling was found to be more efficient than air-borne systems (Brennan et al. [3]) or other air-based heat pumps because of the large thermal mass of water in the system. The most important aspect of water-to-water heat-pump use is, however, summer cooling. We know that highly insulated, air-tight buildings even with a window-to-wall percentage as low as 20 percent [29] will result in summer overheating in a typical mixed climate and we typically use 40 to 60 percent of glazing on the solar side of the building.

A reinforced polyethylene (PEX) tubing can be used in heating on the wall (Hu et al. [28], Fadiejev et al. [37]) and cooling on the floor. To improve control of the thermal mass, a hydronic heating on the surface of the interior walls is used. The impact of radiant panel location is significant (Table 1).

Table 1. Effect of location of the radiant panel on energy demand in dynamic operation mode.

Panel Location	Heating Demand (GJ)	Cooling Demand (GJ)
Wall vs. floor	58 vs. 98	24 vs. 31

These values were calculated using energy-plus with film coefficients typical for horizontal and vertical orientations. Hu also found that to achieve more than 90% efficiency in the desired heating, a thermal resistance of the insulation layer of at least 1 (m^2K)/W) must be used.

Figure 5 shows that the integrated design process (IDP) may start with an initial lowering of utility bills without a cost increase and that all passive measures create only a small increase in the ownership cost that is here expressed as a mortgage. With a further increase of the ownership cost, this passes through a minimum. Another characteristic point on the curve shown in Figure 5 is the point of equilibrium where the use of photovoltaic (PV) panels is approximately the same as traditional passive measures. One may continue using PV panels until reaching zero energy at a substantial mortgage investment, typically about a 50–70% increase of the minimum cost. Currently, the typical investor stops after placing a few solar panels.

Thus, the rational design of low-energy buildings hinges now on the capability of selecting the reference point for the photovoltaic (PV) technology. In line with this need, the PHIUS selected reference buildings based on the ASHRAE/DOE climate zones [38,39] and considered 115 locations for cost optimization that included air tightness, window upgrades with a 15 °C minimum interior surface temperature, heating and cooling demands, and peak heating and cooling loads. Statistical models were fit so that the cost of the target properties can be generated for any location from parameters such as degree-days and design temperatures. In this manner, both the German and U.S. PH developments moved housing toward the goal of sustainable development. As we have selected the maximum period of return on investment to be 15 years, the energy level attained may be below the zero-energy building.

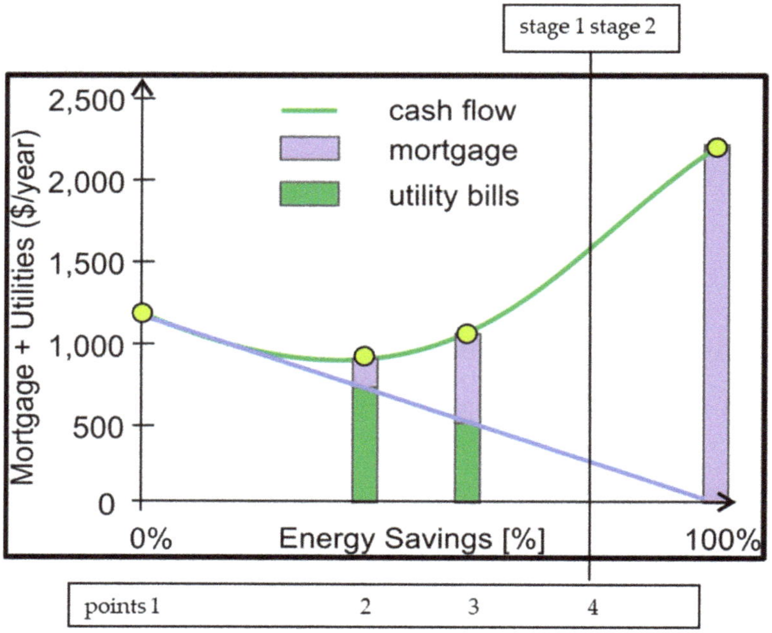

Figure 5. Costs of utilities (green) and mortgage (blue) versus energy savings from zero savings to 100% savings. Point 1 is the reference cost, point 2 is an optimum with passive measures alone, and point 3 is the price of photovoltaic (PV) when equal to passive measures, point 4 represent the end of the stage 1. Modified graph from PHIUS (own archives).

The proposed way to alleviate the conflict between the investor (with limited finds) and society (which wants to reach a net-zero energy level) is to introduce *a two-stage construction process*. In such a process, stage one is designed to achieve a performance level limited by the available funding, while stage two continues to the selected performance level. In the first stage, the building is completed at a minimum performance level that is acceptable to both the building code and the investor, yet the design predicts continuation of construction to the zero-energy level.

Generally, the second stage would start a few years later and often will also include the charging of vehicles used by the household. For the two-stage solution to be successful, both stages must be designed at the same time, and only the construction process is divided to secure funds after the basic building exists.

3.3. *Hybrid Ventilation Modified for SARS-CoV-2 Containment*

The pandemic showed shortcomings of both the EU and U.S. approaches to indoor air handling in residential buildings. Therefore, one must capability to change ventilation rate between 0. and 3.0 air change per hour and either add filtration to the existing system or use pressure gradient within the habitable space. and used DOAS with exterior filtration In the latter a case one must resolve the management of air humidity, see the next section.

3.4. Over-Pressure Indoor Air Requires New Moisture Management Technology

Introduction of air pressure higher than outdoors is useful for controlling ventilation rates, but it also requires modifying the indoor humidity control and provision of moisture buffer capabilities that were lost with the elimination of lime from the interior plasters. To this end, an air gap between the old wall (exterior or interior) and panelized interior finish system is introduced. The panelized interior system has surfaces covered with a hygroscopic and capillary active layer called *eco-wrap* that will act as a moisture buffer.

This concept is based on research on dynamic walls in CRIR (Centre Recherché Idustrielle de Rantigny), France, in the 1980s that showed the difference between static and dynamic performance of the wall is negligible. So, the wall acts as the heat exchanger and the energy loss is minimal. In this manner the ventilation and moisture management are integrated with the heating/cooling system.

There is a need to use a material with the capability to catch and release water, i.e., wetting and drying though an interaction with the passing air, this material was labeled eco-wrap. This material may also be a new subcategory of plaster/mortar because it must be continuous, strongly hygroscopic, and capillary active, as well as show field performance like an elastic plaster. Hygroscopic means attracting and holding water molecules from the surrounding environment by either absorption or adsorption, which is usually at room temperature. As the Thompson (Kelvin) law explains, a pore size lower than 10^{-7} m reduces the saturation value of the water vapor concentration above water meniscus. With the pore curvature corresponding to a size smaller than 10^{-8} m the condensation of water vapor takes place at 90% RH and at 10^{-9} m it would happen at 30% RH. Thus, wood is more hygroscopic than inorganic materials and using mortars that include organic fibers is a method to increase the hygroscopicity of the composite material.

The hygroscopic layer may also be drawing moisture from other materials with larger pores and this phenomenon was introduced under the name of the capillary active layer in the work of Grunewald and Häupl [40]. Figure 6 shows a capillary active layer that brings the condensed water vapor to the surface of the material, allowing evaporation to the interior air.

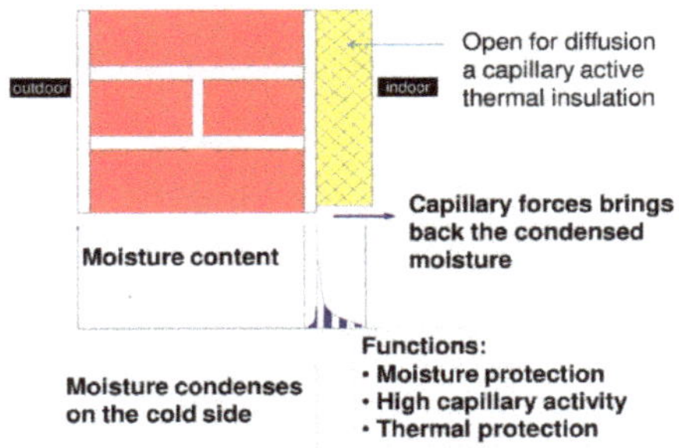

Figure 6. Capillary active layer brings the condensed water vapor to the surface of the material, allowing evaporation to the interior air (Reprinted with permission from Ref. [40]).

The first capillary active layer was developed in the application of calcium silicate by Grunewald [38] using a try-and-error sequence that involved a pilot manufacturing enterprise, determination of the moisture transport characteristics at a university, and a second round of pilot manufacturing. The observation that tobermorite, one of the two basic minerals forming the calcium silicate, had a higher sorption curve (i.e., contained more micropores with nominal diameter lower than 10^{-8} m) led to the optimization of the mixture of these two components to improve the overall hygric performance of the material.

A clay-based plaster is another capillary active material used as the interior finish in historic caves in Japan. This plaster was developed to replace the mechanical control system that could not cope with opening and closing the entrance door by a highly variable number of visitors and changing weather conditions. In principle, the capillary active material technology is an extension of traditional materials such as lime plasters and wood-based cement boards used on the interior of buildings. A low-density cement-bonded wood fiber (CBWF) was used in masonry breathing walls that were highly water-vapor permeable and have high sorption. CBWF had a WV permeance like that of a lime plaster (i.e., about 10 times higher than wood). The sorption of CBWF, i.e., water in equilibrium with very humid air (for instance 50% RH), is about six times greater than lime plaster. While under a stepwise change in humidity, this material is much slower than wood, but after ten hours of exposure to new conditions, the amount of water stored by CBWF exceeds that of an unpainted log wood house. In conclusion, CBWF and strawbale walls were found to provide good solutions for a breathing wall.

In effect, cement-bonded wood fiber material represents a purely historic development, while the Dresden Technical University team added a next step in knowledge to improve the process of moisture management and complimentary work in Japan expanded the concept of capillary active technology. The next step was motivated not with the material but with the system level. Exterior thermal insulation and composite systems typically that use expanded polystyrene foam and cementitious lamina should not be used above the 11th floor, i.e., to the limit of evacuation ladders.

For higher buildings, the only suitable exterior insulation is high-density mineral fiber insulation (MFI). Generally, it is a self-draining material except for the layer at the bottom. For the fresh material, the water-retention strip is about 100 mm, but with aging (weathering) of the material this is increased to about 180–200 mm. Furthermore, the MFI does not have the required impact resistance. Both these shortcomings are alleviated by the work of Fort et al. [41], by providing a water-vapor diffusion to the surface of MFI, a capillary active layer is applied. While the main point of this research relates to a combined experimental-computational analysis of hygrothermal performance, the engineering solution addresses a known market need. This paper modifies the hygrothermal model of Kuenzel [42] closer to that used in soil science, e.g., de Vries, and see the analysis in Ph.D. thesis of Bomberg [38]. A laboratory experiment to determine temperature and moisture fields served as a basis for model calibration and the identification of unknown parameters. The calibrated model is subsequently used for a long-term hygrothermal assessment of the studied detail.

Hygrothermal properties of insulating materials, connecting layers, and their mutual interactions therefore belong to the main topics of recent studies aimed at interior thermal insulation systems (Fořt et al. [41], Vereecken and Roles [43]). Both computer modeling and testing are used in their work. Measurements were made in a real-time mode, but modeling is versatile. The results, however, are strongly dependent on the quality of input data and material characteristics.

We have three different methods for model calibration:

(1) During the experimental work as used by Fořt et al. [41]
(2) Using one of the material properties verification methods (see: Bomberg and Pazera [32]) or a new, emerging approach of the system-monitoring application of statistical analysis of data [44] and performance evaluation (MAPE modeling, see: Romanska-Zapala et al. [37]).

3.5. Next Generation of Capillary Active Materials

During pilot manufacturing in the USA, recycled wood with 100-micron diameter and unspecified cellulose fibers were used. In China, fibers from rice plants and rice hulls were used. The choice of different components in the fiber mixture is made with the view to: (a) ratio between strong and weak fibers, (b) ratio between inert and electrostatically charged, and (c) ratio between the micropores and the skeleton. Furthermore, to reduce shrinkage, the fraction of Portland cement, replacing it with pozzolanic material such as fly ash, or a mixture of fly ash with silica were limited. The choice of fillers in addition to the preselected size of sand is unlimited. To obtain the combability of different materials we use re-dispersing and bonding polymers, as well as polymers regulating the water retention of the mix by eco-wrap.

Generally, the eco-wrap mix is designed for demolding after 24 h and 48 h water retention to apply the next layer without prewetting the surface. Eco-wrap can be designed for application in either one or two layers. The base layer would have more of the coarse sand and bonding polymer than the interior finish layer. This also improves the control of the wetting and drying performance.

The mix design process takes a few stages. In the first stage we are looking for wet ting, drying, shrinkage, and workability performance and this is done purely on an experimental basis. One tries to develop a mix that behaves as a typical mortar with a reasonable workability, consistency, and critical physical properties. In stage 2 (Figure 7), one evaluates the basic hygric properties by comparison to the reference mortar/plaster, in our case a standard Portland cement mortar.

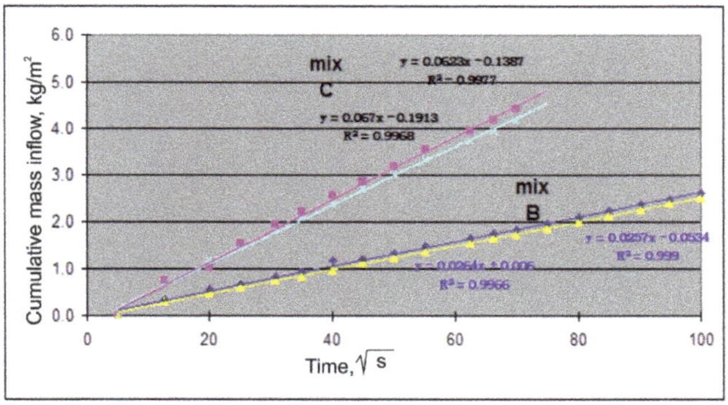

Figure 7. Lime-pozzolanic mix C shows wetting from free water surface faster than mix B, which is based on standard Portland cement. Note that an initial stage of the test is purposely discarded.

The outcome of the water absorption test is easily predictable. But the test is needed to determine the maximum liquid water conductivity of the material. Conversely, the drying rate permits better evaluation. Figure 8 shows clear differences between different eco-wrap mixes. The mix C shows a much better drying rate than mixes D and E. At this stage one may decide that mix C3 is more suitable for a one-coat application, while a combination of D2 and C3 will be better suited for the finishing interior surfaces.

Stage 3 is a repeat of stage 1 but with a rigorous framework of special mortar design and this includes review of the cost–benefit relation, early strength development, and use of a chemical addition to retain water for 48 h. The fraction of this admixture is different in hot or cold climates.

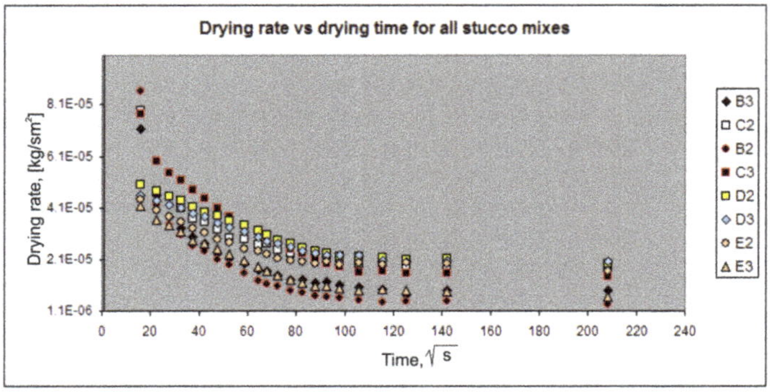

Figure 8. Drying rate versus drying time. Mix B is based on standard Portland Cement; mixes (C, D, E) represent eco-wrap based on lime with varying fractions of fibers (e.g., mix D has a density of 800 kg/m^3).

3.6. The Need for Building Automatic Control Network (BACnet)

EQM technology includes the development of an integrated control system that performs optimization of different subsystems such as the water-to-water heat pumps, solar thermal and photovoltaic panels, air intake ventilators, heat exchangers for pre-heat or pre-cooling of the ventilation and modification of its humidity level, heat exchangers for the exhaust air, instantaneous hot-water delivery systems (or hot-water tanks), cold-water tanks and rain- and/or gray-water tanks, illumination, and control of temperature, air flow and humidity in several spaces depending on their functions. Furthermore, it controls the use of surface heating/cooling that are integrated with building partitions to modify the contribution of thermal mass to the energy consumption. With an adequate steering algorithm based on the history of use and predicted values in relation to the outdoor climate, one can optimize the indoor climate for any type of use and climate (except for the extreme).

The above list explains why the building of an automatic control network (BACnet) is included and optimization during the post-occupancy stage. The system includes two subsystems: (a) a metering system to monitor the selected parameters with a view to discovering malfunctions of the separate units and provide information on their performance, and (b) the building management system to ensure the required quality of indoor environment. To this end, one may introduce the self-learning functions of the steering algorithms to permit optimization of these subsystems in the post-occupancy period. Yet, to optimize HVAC performance we need to have information collected over four different seasons, i.e., a minimum of one year of post-occupancy.

Figure 9 shows the concept of building automatic control in the EQM system.

Note that the charging of motor vehicles and PV systems are now common practices for residential properties; any combination of AC and low-voltage DC on the same grid may also affect the frequency distribution of the electrical current and associate with its loss of quality in the energy systems. Thus another, yet not discussed, function of the BACnet is to ensure the quality of the electrical current.

3.7. Rehabilitation (Retrofitting) of Old Buildings—A Continuum in Time

Another dimension of the EQM technology is that it allows use of the same approach for new low-energy buildings and for rehabilitation of the old. A project in Montreal, Canada demonstrates the power of planning the construction process for a longer period. Atelier Rosemont in Montreal, Canada is a cluster of buildings designed to be upgraded over a period of 10 years to demonstrate the power of integrated planning and execution from the standard neighborhood to near-zero-energy buildings that includes community areas. The Atelier Rosemont project highlights how vague the boundary between new constructions and retrofitting of old buildings can be and that the system integration proposed in dynamic EQM projects is not a dream. EQM technology is to make a double integration, in time and space, much easier to achieve.

Figure 10 shows the buildings to which different stages of energy reductions that were applied from 2008 (0% of total energy reduction) to 2018 (92% of total energy reduction).

Stages of improvements from 2008 to 2018 in Atelier Rosemount, Montreal,

- High-performance enclosure, common water loops, solar wall—36 percent reduction in the total energy use per square meter and year;
- Gray-water power pipe—42% reduction in energy use;
- Heat-pump heating (with horizontal heat exchanger)—60% reduction;
- Renewable 1: evacuated solar panels for hot water—74% reduction;
- Renewable 2—photovoltaic bring the total reduction to 92% (in 2018).

Thus, in the span of 10 years, these building reduced energy use by 92 percent of the original use.

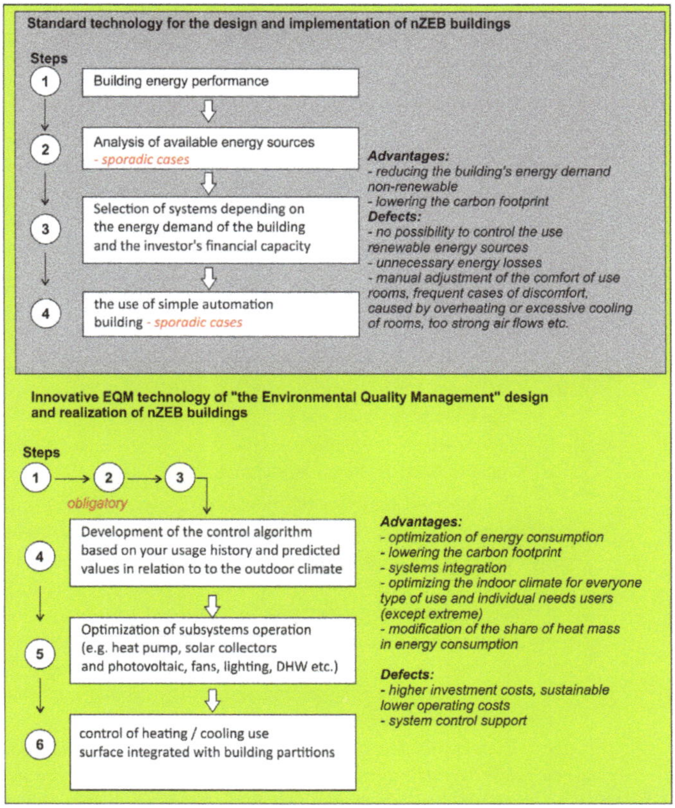

Figure 9. The concept of building automatic control in the EQM system. Source (MFC own study).

Figure 10. An affordable, low rise, energy efficient multi-unit residential building "Atelier Rosemount" in Montreal; rain retention basin in the bottom right (credit Nikkol Rot, reprinted with permission).

3.8. Thermal Storage Is Located in Ground Next to Building Foundations

Elsewhere, we analyzed earth–air heat exchangers for pre-conditioning air, postulating that efficient use of EAHX (earth–air heat exchanger) requires use of both: a fresh air inlet and EAHX. This paper explained that when outdoor air reaches a specific temperature, e.g., 19 °C, one should use outdoor air and switch back to the EAHX when the air temperature cools below 19 °C. To switch between two sources of air we need to know the following:

- Temperature of outdoor air;
- Temperature on exit from EAHX;
- Temperature required on entry to the indoor space;
- The need for heating or cooling of air to the temperature of entry to the indoor space.
- Electric energy needed for operating EAHX;
- Dynamics of temperature changes in the soil surrounding the EAHX pipes;
- Temperature of the air being removed from the indoor space and recuperation of energy.

The highest efficiency of interaction between EAHX and the ventilation center (mechanical room) is achieved when fully integrated control/steering systems for the low-energy building is used. In such a case, separate needs in different indoor spaces (different rooms), e.g., presence of people in the room are considered. Such a system will address optimization of both comfort and energy use.

In discussing this holistic approach to design, one must also consider resiliency of the building, i.e., what happens when the electrical supply is interrupted. Buildings must be airtight but not too tight because continuous use of mechanical ventilation requires a significant supply of electricity.

Experience from Finland indicates that the uncontrolled air leakage that may provide 60–70% of the minimum ventilation appears to be good guidance. We also suggest using 10 Pa overpressure of buildings if the walls are designed for adequate moisture management and the delivery of fresh air is restricted to 60% of time.

While use of a direct air pre-conditioning is restricted to small houses and in larger residential units water-based heat exchangers, the above discussed considerations are still valid. The preferred location for a water tank used as ground for the thermal storage is under the house or in the house's perimeter.

3.9. An Example of Hungarian Demonstration House

An extension of passive house technology (proposed by Bomberg [38] and Krecké, see [45]) was an introduction of pre-conditioned air or water tubing to produce an 'active insulation' layer. This would allow using energy from other sources and reduce heat transmission through external enclosure. The system requires only an ordinary pump to transport the heating/cooling medium that connects an active insulation layer with the earth heat exchanger. In a cold climate, the medium temperature is lower than the indoor space but higher than the outdoor air.

A patent introducing the concept for an active thermal insulation layer for buildings as a contrast to passive house technology was issued in 2012 and the demonstration building in the town of Nyiregyhaza in Hungary that has a direct coupling between ground and wall heat exchangers is described elsewhere [46]. Data recorded for eight years demonstrate that active thermal insulation regularly improves thermal performance of external walls. The equivalent thermal transmittance U_{eq} of the analyzed wall (dependent upon climatic conditions) varied from 0.047 W/(m^{2}K) in November to 0.11 W/(m^{2}K) in March, while the steady-state value without ground coupling was 0.282 W/(m^{2}K). In the tested building, the average heat loss reduction was 63% in relation to standard insulation.

In the demonstration house, a continuous circulation of the medium is used for the purpose of analysis. Indeed, the medium temperature varied from 12 to 23 °C with an average of 17.9 °C, while the average outdoor temperature was 11.9 °C, indicating the summer recharging the ground with energy from the walls. Figure 11 presents the daily temperature differences between the coil medium and the outside air. As the positive temperature difference (5843 degree-days) are much higher than the negative values (−493 degree-days), one sees large heating potential.

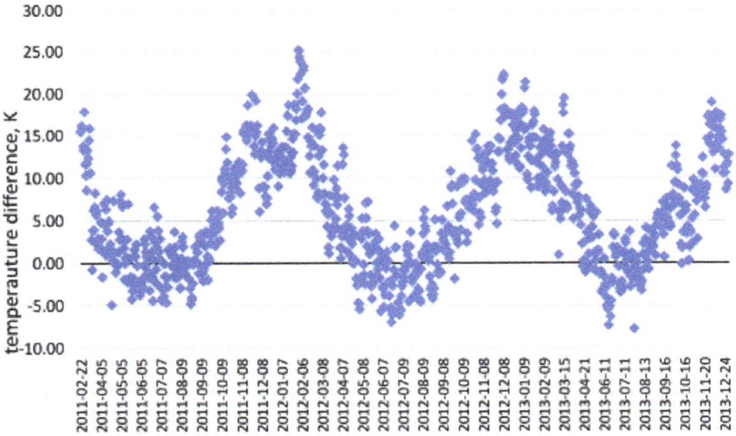

Figure 11. Temperature difference between fluid in coil and external air (Adapted from Ref. [46]).

In effect, if the system operated only in the heating fashion, one would see that the heat-loss reduction in the cold period was higher than 50%.

3.10. Next Generation of Low Energy Buildings, Conclusion

A comparison of residential buildings from 1929 to 2002 in Vancouver highlighted the need for a *transient operation* of buildings. In the next generation of buildings, large surface of hydronic heating or a cooling system coupled with a thermal mass and operated with a water-to-water heat pump will be used. The Hungarian demonstration house shows the possibility of 40–50% energy reduction.

To alleviate the conflict between the investor (limited finds) and society (wants net-zero energy level), we introduce *a two-stage construction process, where stage one is designed to achieve a limited performance level, though acceptable to* the building code, and a limited cost acceptable to the investor, while the second stage continues to the selected performance level. The Canadian (*Atelier Rosemount" in Montreal*) shows 92% energy reduction in the social housing project.

As the indoor environment should not be worse than the outdoor, the lesson from SARS-CoV-2 leads us to a variable rate of over-pressurized ventilation in all rooms except for a selected bedroom. Overpressure of indoor air requires a fundamental change of the humidity-control strategy and the development of new moisture-buffer technology.

4. Conclusions from the Work, Further Action Plan

In this article, the authors touched upon many important topics related to the energy efficiency of buildings., namely optimization of the HVAC and implementation of new nearly zero-energy retrofitting technology. The authors also stress the need for improving the ventilation by using filtration with M13 device or designated outdoor air system (DOAS) with exterior filtration.

The currently used concept of a passive house includes many improvements but also uses many arbitrary criteria that restrict the applicability of this technology. Furthermore, when an occupant opens windows during inclement weather, he/she may destroy "energy efficiency". The history of building science from its beginnings almost 100 years ago up to the development of net-zero-energy buildings gave us an understanding that the next generation of buildings will be designed with the indoor environment as the starting point. If we satisfy the occupant's need for large windows, individual ventilation on demand, and hydronic heating/cooling systems built in walls or floors, we are making progress toward sustainable buildings. Furthermore, when our heating/cooling system operates at low temperature without noise or visible heaters or ventilators we are on a solid basis for the next generation of buildings.

This work was started about 12 years ago. Yet, while we agreed with the basis for a change in the approach, our previous attempts to synthesize the environmental aspects of building science were shallow. The indoor environment is more than thermal comfort. While we knew that electrical heat-pump technology will be a key in the next generation of buildings, we also realized that the focus of building science must be on renovation. The review showed that the next generation of technology must be active and involve use of air cavities. The most significant contribution to our research came from the NY test house [38] and Hungarian demonstration house [46], where a direct coupling between hydronic heating and a ground heat exchanger was used. Yet, as we will incorporate hybrid solar panels and water-sourced heat pumps in the next-generation technology, we will use a building's automatic control system instead of a direct connection.

The critical issue today is the necessity for rapid action. There are a few components of housing obsolescence that, combined with an urgent need to reduce climate change, provide us with the perfect opportunity to change the challenge into a win-win situation. Construction today is facing four crisis elements: (1) affordability, (2) lack of skilled trades, (3) productivity (no change for the last 10 years while the manufacturing sector in the U.S. increased productivity by 70%), and (4) the digitalization of industry (again less than 20% production in the U.S. is not on the building site; in Sweden, for instance, more than 80% is manufactured off the site). If we add the need for zero energy (zero carbon emissions) and health concerns highlighted by the spread of COVID-19, we get a situation calling.

Yet, until recently, all these pieces of knowledge did not have linkage to the occupant. The breakthrough came with the analysis of SARS-CoV-2 in the low-energy buildings. In the recovery bedroom, an occupant needs air pressure to be lower than in the rest of the indoor space and also a strongly variable ventilation rate. Yet the whole dwelling needs to have some air movement. This implies that living space itself must have air pressure gradients and effectively one must use much more carefully designed humidity control. In North America, we had air circulation imposed by the air-borne heating systems, but we did not appreciate its significance. Today we know that neither opening widows (Europe) nor having air conditioner in the window (U.S.) make economic sense. Today we can design buildings that are energy efficient, inexpensive, and produce more energy than they use. We can do in the harsh climates of Canada or humid climates like Florida. The design principle is the same, but the details are different.

The methods of building evaluation are different today than they were four decades ago because today we understand the interactions between the different functions of the building. Yet, in this editorial overview, we highlighted that it is not the technology itself, but the manner of how we deal with technology that makes the socio-economic impact. Overdoing the intellectual part of technology in academic research or overdoing the commercial impact in the technology transfer stage (application of the technology) does not help. On the contrary, it often reduces the impact of the technology. Furthermore, if for the sake of the administration policy, one aspect, e.g., energy efficiency, is stressed while others are neglected, the whole technology becomes simplified and unbalanced. Our current system of supporting applied research requires a visible fragmentation, while the socio-economic progress requires integration. As an example, if we want to develop an energy model based on the data collected from the actual building to optimize the system and later develop a more advanced model for HVAC operation, such a project does not fit into academic research as it is too practical; it does not fit into applied research because one cannot estimate the monetary savings of this approach. Of course, those who realize the significance of the continuum of data and control models for the optimization of the HVAC will continue doing this type of research, but the whole world of academia is excluded as they rely on the government support.

In contrast to structural engineering where new materials can come with an impact on construction, the management of environmental quality will require many small details to be thought through and built with care. While the change in environmental control is urgently needed, this change requires an understanding of the building as a system and must come from the scientific community. For this to happen, the scientific community must have a vision of the next generation of buildings. Two critical issues in this vision are: (1) the occupant must be able to control the indoor environment and the building automatics must enable the occupant to do this, and (2) the design must be focused on the level of the components or assembly, while materials will be judged upon whether they fulfill the requirements for assembly and the subsystems.

Meadows produced a list of the most important factors that modify people's motivation. In the first place he lists transcending paradigms and in the last (11th place) numbers, parameters such as subsidies, taxes, or standards. He contradicts *paradigms*, saying:

> *"The shared ideas in the minds of society, the great big unstated assumptions, constitute that society's paradigm, or deepest set of beliefs about how the world works. These beliefs are unstated because it is unnecessary to state them—everyone already knows them."*

To the sustainability research:

> *"Notice, however, that most of the current sustainability research ... is focused on the least effective leverage points like the economic aspects ... politicians believe that sustainability is mainly an economic problem. So, "Numbers" ... and parameters such as subsidies, taxes, and standards become the main focus."*

We fully understand the concerns of Meadows who stressed that sustainability research has more in common with the change in the transcendental paradigm than the economics of construction. Kuhn [47] highlighted that a scientific revolution that comes with a small step in the socio-economic situation is close to the change. The authors claim that the change-of-thinking-paradigm needed for the next generation of building technology is just this last step and slowing or reversing climate change is *the transcending paradigm of our time*. Understanding this paradigm should motivate all involved in building science (physics) to mobilize the public and explain to politicians that the important approach in the post-COVID world is to invest in the renovation of our buildings. In effect, this review shows that it is not gas emission or energy saving, but an *emerging holistic vision* that must be communicated to the broad public. The only way to accelerate the green revolution is not through green materials but through broad public–private programs of education and demonstrations of the need for reinvesting in the next generation of retrofitted buildings.

Now, after repeated waves of pandemic, comes the realization that the next generation of new or renovated dwellings must provide an indoor space suitable for the quarantine of people infected with an air-borne virus, be it SARS-CoV-2 or influenza. A bedroom likely to be used by sick people should be underpressurized to guard against the spread of illness. A direct outdoor air system (DOAS) should be used. Therefore, the requirements for retrofitting must be broadened to include an adequate handling of DOAS ventilation with interior air circulation. In effect, one may summarize the above discussion as follows:

"The next generation of environmental quality management technology should have a dedicated, hybrid ventilation system with an adequate management of air humidity."

Author Contributions: Conceptualization, M.B., A.R.-Z. and M.F.-C.; methodology, M.B., A.R.-Z. and M.F.-C.; validation, M.B., A.R.-Z. and M.F.-C.; formal analysis, M.F.-C.; investigation, D.W.Y. and L.E.L.; resources, M.B.; data curation, M.F.-C.; writing—original draft preparation, M.B., A.R.-Z. and M.F.-C.; writing—review and editing, M.F.-C., D.W.Y. and L.E.L.; visualization, M.F.-C.; supervision, M.B. and M.F.-C.; project administration, D.W.Y. and L.E.L.; funding acquisition, M.B. All authors have read and agreed to the published version of the manuscript.

Funding: This research received no external funding.

Institutional Review Board Statement: Not applicable.

Informed Consent Statement: Not applicable.

Data Availability Statement: No extra data is produced.

Conflicts of Interest: The authors declare no conflict of interest.

References

1. Flynn, J.E.; Segil, A.W. *Architectural Interior Systems: Lighting, Air Conditioning, Acoustics*; U.S. Department of Energy Office of Scientific and Technical Information: Oak Ridge, TN, USA, 1970.
2. Bomberg, M.; Brennan, T.; Henderson, H.; Stack, K.; Walburger, A.; Zhang, J. *High Environmental Performance (HEP)—Residential Housing for NY State*; New York State Energy Research and Development Authority: New York, NY, USA; National Center for Energy Management and Building Technologies: Alexandria, VA, USA, 2009.
3. Brennan, T.; Henderson, H.; Stack, K.; Bomberg, M. Quality Assurance and Commissioning Process in High Environmental Performance (HEP) House in NY State. Available online: https://www.semanticscholar.org/paper/QUALITY-ASSURANCE-AND-COMMISSIONING-PROCESS-IN-HIGH-Brennan-Bomberg/6093cda3698953a834b37a7b12ced95ad1ef9636 (accessed on 7 May 2022).
4. Walburger, A.; Brennan, T.; Bomberg, M.; Henderson, H. Energy Prediction and Monitoring in a High—Performance Syracuse House. 2010. Available online: http://thebestconference.org/BEST2 (accessed on 2 October 2019).
5. Aelenei, L.; Frattari, A.; Riscala, L.; Altan, H.; Hashemi, A.; Aoul, K.A.T.; Noguchi, M. Zero energy homes. In *ZEMCH: Toward the Delivery of Zero Energy Mass Custom Homes*; Noguchi, M., Ed.; Springer: Cham, Switzerland, 2016; pp. 275–309.
6. Bomberg, M.; Kisilewicz, T.; Mattock, C. *Methods of Building Physics*; Cracow University of Technology: Kraków, Poland, 2015.
7. Fedorczak-Cisak, M.; Bomberg, M. Dlaczego wielorodzinne budynki mieszkalne w Vancouver zużywały tyle samo energii w 2002 co w 1929 roku? *Mater. Bud.* **2022**, *1*, 58–60. [CrossRef]
8. Bomberg, M.; Onysko, D. Heat, Air and Moisture Control in Walls of Canadian Houses: A Review of the Historic Basis for Current Practices. *J. Build. Phys.* **2002**, *26*, 3–31. [CrossRef]
9. Radziszewska-Zielina, E.; Kania, E. Problems in Carrying Out Construction Projects in Large Urban Agglomerations on the Example of the Construction of the Axis and High5ive Office Buildings in Krakow. *MATEC Web Conf.* **2017**, *117*, 00144. [CrossRef]
10. Bielski, A.; Zielina, M.; Młyńska, A. Analysis of heavy metals leaching from internal pipe cement coating into potable water. *J. Clean. Prod.* **2020**, *265*, 121425. [CrossRef]
11. Młyńska, A.; Zielina, M.; Bielski, A. Contamination of drinking water soon after cement mortar lining renovation depending on the disinfectant doses. *SN Appl. Sci.* **2019**, *1*, 516. [CrossRef]
12. Piasecki, M.; Radziszewska-Zielina, E.; Czerski, P.; Fedorczak-Cisak, M.; Zielina, M.; Krzyściak, P.; Kwaśniewska-Sip, P.; Grześkowiak, W. Implementation of the Indoor Environmental Quality (IEQ) Model for the Assessment of a Retrofitted Historical Masonry Building. *Energies* **2020**, *13*, 6051. [CrossRef]
13. Fedorczak-Cisak, M.; Kowalska-Koczwara, A.; Pachla, F.; Radziszewska-Zielina, E.; Szewczyk, B.; Śladowski, G.; Tatara, T. Fuzzy Model for Selecting a Form of Use Alternative for a Historic Building to be Subjected to Adaptive Reuse. *Energies* **2020**, *13*, 2809. [CrossRef]

14. Dudzik, M. Towards characterization of indoor environment in smart buildings: Modelling PMV index using neural network with one hidden layer. *Sustainability* **2020**, *12*, 6749. [CrossRef]
15. Lstiburek, J.W.; Pressnail, K.; Timusk, J. Transient Interaction of Buildings with HVAC Systems—Updating the State of the Art. *J. Therm. Envel. Build. Sci.* **2016**, *24*, 111–131. [CrossRef]
16. Romanska-Zapala, A.; Bomberg, M.; Dechnik, M.; Fedorczak-Cisak, M.; Furtak, M. On Preheating of the Outdoor Ventilation Air. *Energies* **2019**, *13*, 15. [CrossRef]
17. Kisilewicz, T. Passive control of indoor climate conditions in low energy buildings. *Energy Procedia* **2015**, *78*, 49–54. [CrossRef]
18. Bomberg, M.; Kisilewicz, T.; Nowak, K. Is there an optimum range of airtightness for a building? *J. Build. Phys.* **2015**, *39*, 395–421. [CrossRef]
19. Bomberg, M.; Romanska-Zapala, A.; Yarbrough, D. Journey of American Building Physics: Steps Leading to the Current Scientific Revolution. *Energies* **2020**, *13*, 1027. [CrossRef]
20. The Saskatchewan Conservation House—Passive House Canada; Maison Passive Canada. Available online: https://www.passivehousecanada.com/policy-series-2-operating-efficiency-what-is-the-target/the-saskatchewan-conservation-house/?gclid=Cj0KCQjwsdiTBhD5ARIsAIpW8CIG5VAnKScXvx1ukIH26-VUe49Wy29QrP9lwPXvS0v54UIxF4RVGy0aAucoEALw_wcB (accessed on 7 May 2022).
21. Romanska-Zapala, A.; Bomberg, M.; Fedorczak-Cisak, M.; Furtak, M.; Yarbrough, D.; Dechnik, M. Buildings with environmental quality management (EQM), part 2: Integration of hydronic heating/cooling with thermal mass. *J. Build. Phys.* **2018**, *41*, 397–417. [CrossRef]
22. Lstiburek, J.; Pressnail, K.; Timusk, J. Evaluating the Air Pressure Response of Multizonal Buildings. *J. Therm. Envel. Build. Sci.* **2016**, *25*, 299–319. [CrossRef]
23. Yarbrough, D.W.; Bomberg, M.; Romanska-Zapala, A. Buildings with environmental quality management, part 3: From log houses to environmental quality management zero-energy buildings. *J. Build. Phys.* **2019**, *42*, 672–691. [CrossRef]
24. Fedorczak-Cisak, M.; Furtak, M.; Radziszewska-Zielina, E. Certification of "Nearly Zero-Energy Buildings" as a Part of Sustainability. In Proceedings of the International Scientific Conference "People, Buildings and Environment 2018 (PBE), Brno, Czech Republic, 17–19 October 2018; Volume 222.
25. Air Barrier Systems in Buildings; WBDG—Whole Building Design Guide. Available online: https://www.wbdg.org/resources/air-barrier-systems-buildings (accessed on 7 May 2022).
26. Perrault, J.C. *Air barrier systems: Construction applications. An Air Barrier for the Building Envelope;* National Research Council of Canada: Saskatoon, SK, Canada, January 1989.
27. Kisilewicz, T. Wpływ izolacji termicznej ścian na mikroklimat w budynkach w okresie letnim. *Mater. Bud.* **2015**, *1*, 112–115. [CrossRef]
28. Hu, X.; Shi, X.; Bomberg, M. Radiant heating/cooling on interior walls for thermal upgrade of existing residential buildings in China. In Proceedings of the In-Build Conference, Cracow TU, Cracow, Poland, 17 July 2013.
29. Fadejev, J.; Simson, R.; Kurnitski, J.; Bomberg, M. Thermal mass and energy recovery utilization for peak load reduction. *Energy Procedia* **2017**, *132*, 39–44. [CrossRef]
30. Bomberg, M.; Onysko, D. (Eds.) Energy Efficiency and Durability of Buildings at the Crossroads. Available online: http://thebestconference.org/BEST1 (accessed on 25 February 2022).
31. Less, B.D.; Walker, I.S.; Casquero-Modrego, N. *Emerging Pathways to Upgrade the US Housing Stock: A Review of the Home Energy Upgrade Literature;* Lawrence Berkeley National Laboratory: Berkeley, CA, USA, 2021.
32. Klingenberg, K.; Kernagis, M.; Knezovich, M. Zero energy and carbon buildings based on climate-specific passive building standards for North America. *J. Build. Phys.* **2016**, *39*, 503–521. [CrossRef]
33. Bomberg, M.; Pazera, M. Methods to check reliability of material characteristics for use of models in real time hygrothermal analysis. In Proceedings of the First Central European Symposium on Building Physics, Cracow–Lodz, Poland, 13–15 September 2010.
34. Bomberg, M.; Gibson, M.; Zhang, J. A concept of integrated environmental approach for building upgrades and new construction: Part 1—Setting the stage. *J. Build. Phys.* **2014**, *38*, 360–385. [CrossRef]
35. Yarbrough, D.W.; Bomberg, M.; Romanska-Zapala, A. On the next generation of low energy buildings. *Adv. Build. Energy Res.* **2021**, *15*, 223–230. [CrossRef]
36. Romanska-Zapala, A.; Furtak, M.; Fedorczak-Cisak, M.; Dechnik, M. Need for Automatic Bypass Control to Improve the Energy Efficiency of a Building Through the Cooperation of a Horizontal Ground Heat Exchanger with a Ventilation Unit during Transitional Seasons: A Case Study. In Proceedings of the IOP Conference Series: Materials Science and Engineering, Polzunov Altai State Technical University, Barnaul, Russia, 27–28 June 2019; Volume 471.
37. Dudzik, M.; Romanska-Zapala, A.; Bomberg, M. A Neural Network for Monitoring and Characterization of Buildings with Environmental Quality Management, Part 1: Verification under Steady State Conditions. *Energies* **2020**, *13*, 3469. [CrossRef]
38. Bomberg, M. A concept of capillary active, dynamic insulation integrated with heating, cooling and ventilation, air cond. system. *Front. Archit. Civ. Eng. China* **2010**, *4*, 431–437. [CrossRef]
39. Wright, G.; Klingenberg, K. *Climate-Specific Passive Building Standards;* The National Renewable Energy Laboratory: Golden, CO, USA, July 2015.

40. Häupl, P.; Grunewald, J.; Fechner, H. Moisture behavior of a "Gründerzeit"—house by means of a capillary active int. insulation. In Proceedings of the 5th Building Physics Symposium in the Nordic Countries, Gothenburg, Sweden, 24–26 August 1999; pp. 225–232.
41. Fořt, J.; Kočí, J.; Pokorný, J.; Podolka, L.; Kraus, M.; Černý, R. Characterization of Responsive Plasters for Passive Moisture and Temperature Control. *Appl. Sci.* **2020**, *10*, 9116. [CrossRef]
42. Künzel, H.M. *Simultaneous Heat and Moisture Transport in Building Components, One- and Two- Dimensional Calculations Using Simple Parameters*; Fraunhofer Institute of Building Physics: Holzkirchen, Germany, 1995.
43. Vereecken, E.; Roels, S. Capillary active interior insulation: Do the advantages really offset potential disadvantages? *Mater. Struct. Constr.* **2015**, *48*, 3009–3021. [CrossRef]
44. Romanska-Zapala, A.; Dudek, P.; Górny, M.; Dudzik, M. Modular statistical system for an integrated environmental control. *E3S Web Conf.* **2020**, *172*, 19006. [CrossRef]
45. Krecké, E.D. Passive House Building Technology. Available online: http://www.isomax-terrasol.eu/fileadmin/_migrated/content_uploads/ISOMAX-TERRA-SOL-engl.pdf (accessed on 1 May 2022).
46. Kisilewicz, T.; Fedorczak-Cisak, M.; Barkanyi, T. Active thermal insulation as an element limiting heat loss through external walls. *Energy Build.* **2019**, *205*, 109541. [CrossRef]
47. Kuhn, J.; Ebert, H.P.; Arduini-Schuster, M.C.; Büttner, D.; Fricke, J. Thermal transport in polystyrene and polyurethane foam insulations. *Int. J. Heat Mass Transf.* **1992**, *35*, 1795–1801. [CrossRef]

buildings

MDPI

Article

Determination of Optimum Building Envelope Parameters of a Room concerning Window-to-Wall Ratio, Orientation, Insulation Thickness and Window Type

Ayşe Fidan Altun

Orhangazi Yenikoy Asil Celik Vocational School, Bursa Uludag University, Bursa 16850, Turkey; aysealtun@uludag.edu.tr

Abstract: The building envelope includes all materials (glazing, external walls, doors, etc.) that separate the conditioned space from the outside environment. Building envelope characteristics significantly influence the energy consumption of buildings. In this study, research was carried out to find optimum building envelope design parameters, such as insulation thickness, orientation, glazing type, and the window-to-wall ratio of a room, using actual climatic data of two cities with different characteristics according to the Köppen climatic classification. The insulation thickness and the window type that minimizes the net present worth of the building façade over 20 years of a lifetime gave the optimum values. In addition, the effect of the various parameters, such as the infiltration rate through the envelope, room set-point temperature, and the fuel type, on the net present cost was also analyzed. It was found that appropriate selection of windows, orientation, and insulation thickness would lead to a significant reduction in the annual energy consumption. Despite having the lowest initial investment cost, the room with single glazed windows had the highest energy requirement and the net present cost. The building façade with double glazed windows, oriented towards the south-west, yielded the minimum net present cost in both locations. Results showed that the optimum external wall thickness is 9 cm in Hakkari (Dsa—Continental Climate) and 6 cm in Istanbul (Csa—Mild Climate).

Keywords: insulation thickness; energy efficiency; windows

Citation: Altun, A.F. Determination of Optimum Building Envelope Parameters of a Room concerning Window-to-Wall Ratio, Orientation, Insulation Thickness and Window Type. *Buildings* **2022**, *12*, 383. https://doi.org/10.3390/buildings12030383

Academic Editors: Paulo Santos and Mark Bomberg

Received: 3 February 2022
Accepted: 16 March 2022
Published: 20 March 2022

1. Introduction

The energy required for buildings constitutes 40% of the total energy consumption globally [1]. A significant share of the energy consumed globally can be related to the heating demand, with a considerable amount associated with heat losses through the building envelope [2]. Recently, there has been an increasing interest in analyzing the thermal characteristics of the building envelope to decrease energy consumption. The building envelope is a crucial design subject to achieve indoor comfort levels with minimum energy requirements. As a result, it is imperative to consider all aspects, such as shape, orientation, climate, the envelope structure, and glazing type, from the beginning of a project. Proper design of the building envelope significantly decreases carbon emissions and energy consumption, and provides occupants with a healthy and comfortable indoor environment.

The appropriate design of the external walls can help achieve energy savings of around 50–60% [3]. Finding the optimum insulation thickness is part of this process. Due to the strong interest in reducing the energy use of the buildings, many researchers have focused on the optimization of insulation thickness. Many researchers have used a simplified equation by following the degree days method to minimize the net present value of the total costs, including of insulation and energy consumption. In a previous study, Kaynaklı presented a procedure to optimize thermal insulation thickness for external walls by considering costs of energy, insulation material, and installation. The study showed that the payback periods for the optimum insulation thickness vary between 3.85 and 16.25 years [4]. In another study,

Kaynaklı [5] investigated the influence of parameters such as building lifetime, inflation rate, cost of insulation material, thermal conductivity of the insulation material, and installation cost, on the total life cycle cost, energy savings rate and payback period. The results of the study showed that project lifetime, fuel cost, inflation rate, and thermal conductivity increase the optimum insulation thickness. In contrast, insulation material cost and Coefficient of Performance (COP) decrease the optimum insulation thickness. Kurekci [6] investigated the optimum insulation thickness for buildings with only heating, only cooling, and both heating and cooling energy requirements. Various insulation materials (extruded polystyrene, rock wool, expanded polystyrene, glass wool, and polyurethane) and different fuel types were used in the study. The calculations were undertaken for all city centers in Turkey. Ekici et al. conducted a study to investigate optimum insulation thickness for selected cities in the different climate regions by considering different wall types, different insulation materials, and different fuel types [7]. Based on the calculated optimum insulation thicknesses, energy savings and payback periods were presented for each selected location. Yuan et al. [8] investigated the optimum insulation thickness of external walls for 32 regions of China using the degree days method and life cycle cost analysis. Results of the study showed that 63% of the CO_2 emissions could be reduced when the optimum insulation thickness is applied. Canbolat et al. [9] investigated the optimum insulation thickness and payback period for two different climates. Using the Taguchi method, the importance order of the examined parameters was found, and the heating degree days was found to be the most efficient parameter based on the results. Alsayed and Tayeh [10] analyzed the optimum insulation thickness for Palestinian buildings considering weather data, insulation types, energy prices, and wall construction. Results of the study highlight the influence of degree days base temperature and insulation type on the optimum insulation thickness. Ozel et al. [11] investigated the optimum insulation thickness according to degree days, life cycle cost, and entransy loss methods. The calculations were undertaken for two different insulation materials. Acikkalp and Kandemir [12] presented a technique that combines economic and environmental effects to determine optimum insulation thickness. The proposed method is based on the degree day approach. A case study was carried out for the Bilecik province of Turkey. The optimum insulation thickness values were found to be between 0.13 and 0.47 m, depending on the environmental and economic priority. Barrau et al. [13] calculated the optimum insulation thickness considering different lifetimes of building and insulation materials. Results of the study show that changing the building lifetime from 20 to 50 years increases the optimum insulation thickness. Moreover, changing the optimization criteria from economic to environmental priority highly affects the results. Some researchers also combined life cycle assessment and exergy analysis to find the optimum insulation thickness [14–16]. To enhance the accuracy of predictions for the optimal thermal performance of the buildings, effective software programs were recommended to be used [17]. Simulation programs such as EnergyPlus [18–23] and TRNSYS [24,25] have been used by some researchers.

Windows also have a dominant role in energy consumption and the visual comfort of buildings. Finding the adequate window characteristics (such as orientation, area, and type of window) is part of the initial design decision, and is hard to change later [26]. Windows affects the energy requirement in different ways, such as heat conduction, solar radiation, and daily light transmission [27]. Windows can either reduce or increase the energy loads through solar heat gains or conduction heat losses. The energy transfer through windows depends on many parameters such as climatic conditions, shading levels, orientation, frame material, glazing type, area of the glazing, and many other factors. Detailed analysis must be undertaken to select the glazing types based on their impact according to the geographical location and respective climate conditions. There are only a few studies that have investigated the thermal performance of windows. Altun and Kılıç [26] presented a study to examine the influence of windows and shading device characteristics on the energy consumption of the buildings. Parameters such as the window-to-wall ratio of the façade, total solar energy transmittance value of the glazing, and shading levels regarding orientations were studied. Gasparella [28] studied the impact of the glazing type, window size, and internal gains on the energy need of a residential building. Climatic data of Paris, Milan, Nice, and Rome were used in the research. Tsikaloudaki et al. [29,30] investigated the influence of the window configuration in terms of geometrical characteristics, thermo-physical properties, orientation, and shading levels on its energy performance. Kon [31] conducted a study to investigate optimum insulation thickness and glass number for different climatic conditions, fuel types, and insulation materials. Ozel [32] conducted a study to examine the effect of the window-to-wall ratio on the optimum insulation thickness considering wall orientations. Results of the study showed that the window area and orientation have a significant influence on the optimum insulation thickness. Karabay and Arıcı [33] optimized multi-pane windows for different locations using the degree day method. The optimum number of the glazing for each selected location was found by considering the life cycle cost. Derradji et al. [34] conducted a study to investigate the optimum insulation thickness of a prototype building both numerically and experimentally. Various parameters such as wall structure, window area, and fuel type were studied.

In the literature, previous studies primarily focus on optimizing the insulation thickness based on heating degree days or cooling degree days. Although it is straightforward and fast to estimate, the degree days concept has been criticized for not considering the solar radiation and thermal mass effect [35]. Various researchers have questioned the accuracy of the method. It is essential to consider the building envelope as a whole when conducting optimization studies. Only a handful of studies have considered the building façade as a whole (windows and external walls together). Previous studies have mainly been conducted separately for exterior walls or windows. To the best of the author's knowledge, no study has simultaneously focused on insulation thickness, window-to-wall ratio, orientation, different climatic conditions, and window type of the building envelope.

In this study, a thermo-economic analysis of a designed zone was conducted for two selected cities in Turkey, which belong to different climate zones according to the Köppen climatic classification. The dynamic energy behavior of the zone was simulated for a year. Net present cost analyses were used for economic calculations. A comprehensive parametric evaluation was undertaken by varying the insulation thickness of the external wall, glazing type, glazing area, infiltration rate, room set-point temperature, fuel type, and wall orientation. The dynamic simulation results yielded essential insights for both energy savings and minimizing the net present cost. The methodology used in the study can be applied by other researchers, engineers, and architects around the world to design both energy-efficient and cost-effective buildings.

2. Materials and Methods

2.1. Description of the Reference Zone

Parametric evaluation was carried out for a reference room. The plan of the room is shown in Figure 1.

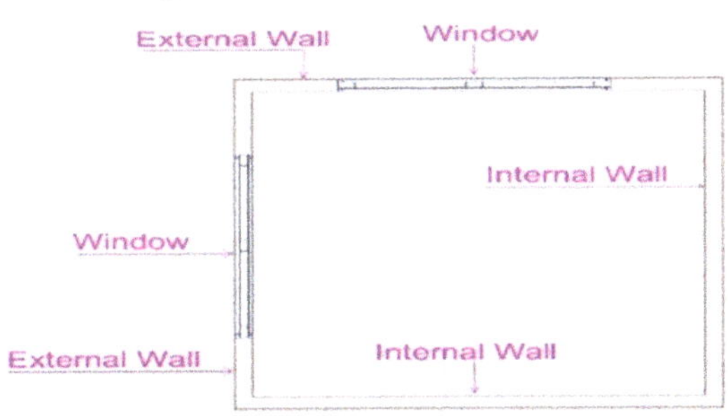

Figure 1. The layout of the reference room.

The designed zone has two external walls facing two different orientations. The room has a 100 m^2 floor area. The windows are considered to be located at the center of each exterior wall of the room. The dimensions of the room are 10 m (l), 10 m (w), 3 m (h). External walls are insulated, and insulation thickness is an investigation parameter. The properties of the building components are presented in Table 1. Three different glazing types (single, double, and triple glazed) were used to determine the optimum one. The heating temperature set-points were assumed to equal to 24 °C for the base case; however, this was also investigated parametrically. Occupant density was considered equivalent to 0.1 occupants/m^2, and specific lighting gains were determined as 10 W/m^2 during occupied hours if the total horizontal radiation level was lower than 120 W/m^2. For the base case, the infiltration rate was considered to be 0.2 ACH (Air Change per Hour); this was also investigated parametrically. In Table 1, the thermal properties of the external wall are given. The main design parameters of the zone are shown in Table 2. Three types of windows were investigated: (1) single glazed window, (2) double glazed window, (3) triple glazed window with argon filling. Four different façade orientations were analyzed as (1) south-east, (2) south-west, (3) north-east, (4) north-west.

Table 1. Thermal properties of the external wall.

Material	Thickness (m)	Conductivity ($kJ \cdot h^{-1} m^{-1} K^{-1}$)	Density ($kg \cdot m^{-3}$)
Plaster	0.020	5	2000
Brick	0.210	3.2	1800
Plaster	0.030	5	2000
Insulation	0.03–0.15	0.144	40

Table 2. Main design parameters of the room.

Parameter	Value
Area	100 m^2
Height	3 m
Window Type	Single glazed (5.69 W/$m^2 \cdot$K)
	Double glazed (1.1 W/$m^2 \cdot$K)
	Triple glazed (0.61 W/$m^2 \cdot$K)
Infiltration rate	0.2–0.4–0.6–0.8–1 ACH
Orientation	North-east, north-west
	South-east, south-west
Heating set-point temperature	18–20–22–24–26 °C

2.2. Model Design and Calculation Methods

The degree day method is a widespread method for assessing and classifying climate regions with common climatic characteristics. In the literature, most of the studies have used this method to calculate optimum insulation thickness [9,33,36–38]. However, energy consumption in buildings depends on many parameters, such as building occupancy, solar radiation, equipment usage, and infiltration rate. Therefore, to find optimum insulation thickness, all these aspects should be considered to obtain realistic results.

In this study, dynamic and transient analysis was conducted using a simulation tool. TRNSYS 18 Simulation software was used for the thermal model. The TRNSYS program is a widely-known energy analysis program primarily used by researchers [39]. The TRNBuild interface of the program was used to create the building. TRNBuild allows the user to define building construction layers and create various infiltration and ventilation types and occupancy schedules. For the meteorological data, the Meteonorm database of the program library was used. Typical meteorological year (TMY) weather data for the selected locations were used.

Space heating and cooling loads have four major components; solar heat gains through apertures, heat conduction, ventilation/infiltration, and internal loads [40]. The heat transfer between the building envelope, and outside and inside environments, can be described by conduction, convection, and radiation mechanisms. Convection heat flux Q_i to a zone due to the difference between the indoor and outdoor temperatures can be expressed as:

$$Q_i = Q_{surf} + Q_{inf} + Q_{ven} + Q_g + Q_{cpl} + Q_{sol} + Q_{ISH} \tag{1}$$

The infiltration gains/losses (Q_{inf}) are expressed as below:

$$Q_{inf} = \mathrm{V} \cdot \rho \cdot c_p \cdot (\mathrm{T}_{out} - \mathrm{T}_{\mathrm{air}}) \tag{2}$$

In Equation (1), Q_{ven} is the ventilation gains/losses, Q_g and Q_{cpl} are the internal convective gains (by people, equipment, illumination etc.) and gains due to connective air from the boundary condition, respectively.

$$Q_{ven} = \mathrm{V} \cdot \rho \cdot c_p \cdot (\mathrm{T}_{ven} - \mathrm{T}_{\mathrm{air}}) \tag{3}$$

where ρ is the air density (kg/m^3), c_p is the air specific heat (kJ/kg·K), V is the airflow rate (m^3/s). Q_{sol} is the fraction of solar radiation entering a building zone through external windows that transfer as a convective gain to the inside air. Q_{ISH} is the absorbed solar radiation on all internal shading devices that is directly transferred to the inside air.

2.3. System Performance Evaluation Parameters

It is apparent that as the insulation thickness increases, the cost of insulation also increases, and that the energy cost decreases. Similarly, single pane windows have a lower investment cost than double glazed and triple glazed windows. However, the application of windows with lower energy transmittance values decreases the annual energy cost. Therefore, it is essential to consider the net present worth of the building envelope that considers the initial investment cost and annual energy cost over the lifetime. In Equation (4), the investment cost of the external wall insulation is given:

$$C_{ins} = C_i \times x \times A_w \tag{4}$$

In Equation (4), C_i is the cost of the insulation material per volume (\$/m^3), x is the insulation thickness (m), and A_w is the wall area without the glazing (m^2). Karabay and Arıcı [33] obtained manufacturer prices and correlated the cost of the multiple panes. The investment cost of the multi-pane window (C_I) per unit is given below [33]:

$$C_I = (19.25 \times n) + 49 \tag{5}$$

In Equation (5), n is the number representing the glazing. The total investment cost of a multi-pane window C_w can be calculated by multiplying C_I with the glazing area A_g as below:

$$C_w = C_I \times A_g \tag{6}$$

Assuming C_F is the unit price of the fuel, LHV is the lower heating value of the fuel, and η_b is the efficiency of the heating equipment, the annual heating cost of the designed zone (C_h) can be calculated as below:

$$C_h = \frac{Q_h}{LHV \times \eta} \times C_F \tag{7}$$

Q_h is the annual heating load, which is integrated for a year. For N years of the project lifetime, the total heating cost can be determined by multiplying Equation (7) with the present worth factor (*PWF*), which is given as:

$$PWF = \frac{(1+r)^N - 1}{r(1+r)^N} \quad (8)$$

In Equation (8), i is the interest rate, and g is the inflation rate; for $i > g$, r can be written as:

$$r = \frac{i-g}{1+g} \quad (9)$$

For $i < g$:

$$r = \frac{g-i}{1+i} \quad (10)$$

For $i = g$:

$$PWF = \frac{N}{1+i} \quad (11)$$

To find the optimum thickness of thermal insulation, the insulation of the external walls of the façade and the window were considered to be an investment [37]. The total net present cost of the building envelope can be obtained as below:

$$C_t = C_h \times PWF + C_{ins} + C_w \quad (12)$$

To investigate energy savings due to insulation, the energy savings rate parameter is used. Energy savings rate can be defined as below:

$$E = \frac{Q_{h_{un}} - Q_{h_{ins}}}{Q_{h_{un}}} \quad (13)$$

In Equation (13), $Q_{h_{un}}$ is the annual heating energy requirement of the zone with uninsulated external walls. $Q_{h_{ins}}$ is the annual heating energy requirement with external wall insulation. Energy savings rate changes between 0 and 1, and greater values show better energy efficiency.

The economic predictions were made using published data from the Central Bank of Turkey [41] and the National Institute of Statistics [42]. All of the parameters selected in this study are given in Table 3. The flow chart of the simulation process that was used for all simulation cases is depicted in Figure 2.

Table 3. Parameters for the financial analysis.

Parameter	Value [1]
Natural Gas	Unit price: 0.18 $/m^3 [43]
	LHV: 9.59 kWh/m^3
	η : 98%
Coal	Unit price: 0.13 $/m^3 [43]
	LHV: 5.76 kWh/kg
	η : 65% [7]
Liquid Petrol Gas (LPG)	Unit price: 1.60 $/kg [43]
	LHV: 12.9 kWh/kg
	η : 90% [7]
Electricity	Unit price: 0.15$/kWh [43]
	LHV: 1 kWh/kWh
	η : 99%

Table 3. *Cont.*

Parameter	Value [1]
Fuel-Oil	Unit price: 1.03 $/kg [43]
	LHV: 11.28 kWh/kg
	η : 80% [44]
Interest rate (i)	15%
Inflation rate (g)	14.6%
Project lifetime	20 years [45,46]
The unit price of the selected insulation material	100 $/m^3$ [12]

[1] TL/USD currency conversion set at 01.03.2022 1 $ = 13.93 TL.

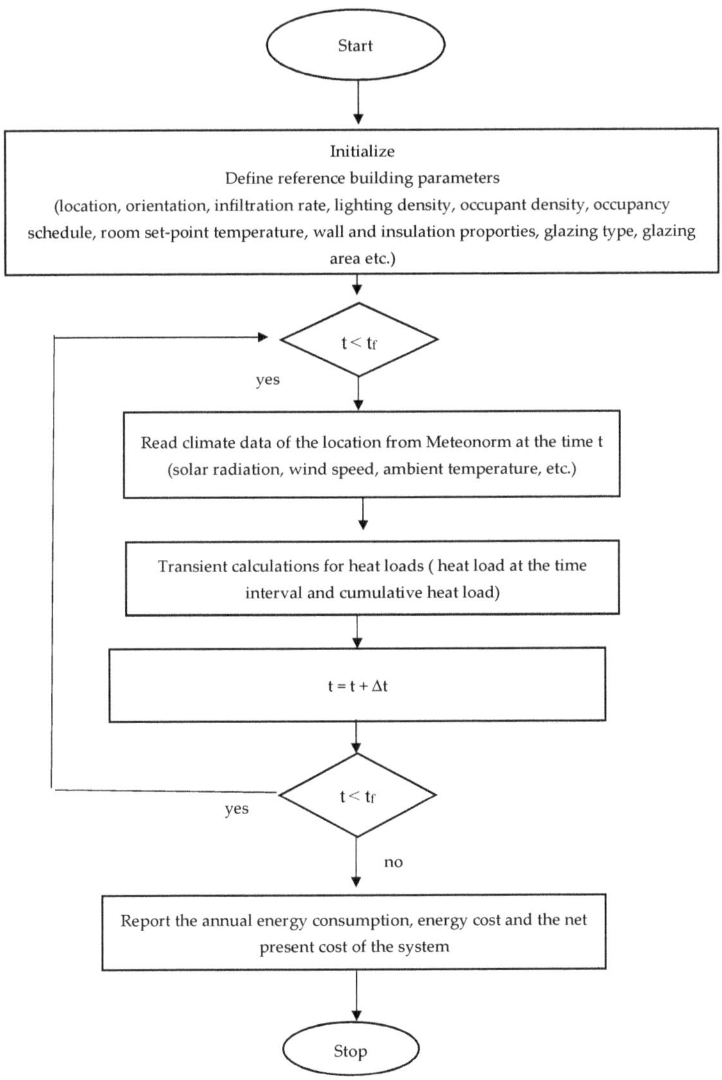

Figure 2. The flow chart of the simulation process.

3. Analysis

Turkey is positioned at the intersection of Europe, Asia, and the Middle East. It experiences different climatic conditions [47] and is divided into four climate zones. Two representative cities from different zones were selected to show the influence of the climatic conditions on optimum building envelope parameters. In Table 4, geographic information is presented for the selected cities. The heating degree days of Istanbul and Hakkari are 1860 and 3470, respectively [4], whereas cooling degree days are only 6 and 18. Both locations are heating-dominated; therefore, thermo-economic optimization of the building envelope was conducted based on the heating load. The Köppen climate classification system is widely used to classify the climate of a region. This classification has five major climate groups, which are A (tropical), B (dry), C (mild), D (continental), and E (polar). Each major group is divided into sub-groups. As shown in Table 4, based on this classification, Istanbul is an example of the Mild Climate (Csa), and Hakkari is an example of the Continental Climate (Dsa). In Figures 3 and 4, the monthly minimum, maximum, and mean ambient temperatures of the selected locations are presented.

Table 4. Climatic zones, topographic features, and degree days of the selected locations.

Selected City	Istanbul	Hakkari
TS 825 Climate Zone	2	4
Latitude	41°00′ N	37°44′ N
Longitude	28°97′ E	43°74′ E
Altitude (Elevation)	40 m	1728 m
HDD (Heating Degree Days)	1865	3470
CDD (Cooling Degree Days)	6	18
Köppen Classification Major group	C (Mild)	D (Continental)
Köppen Classification Sub group	Csa	Dsa

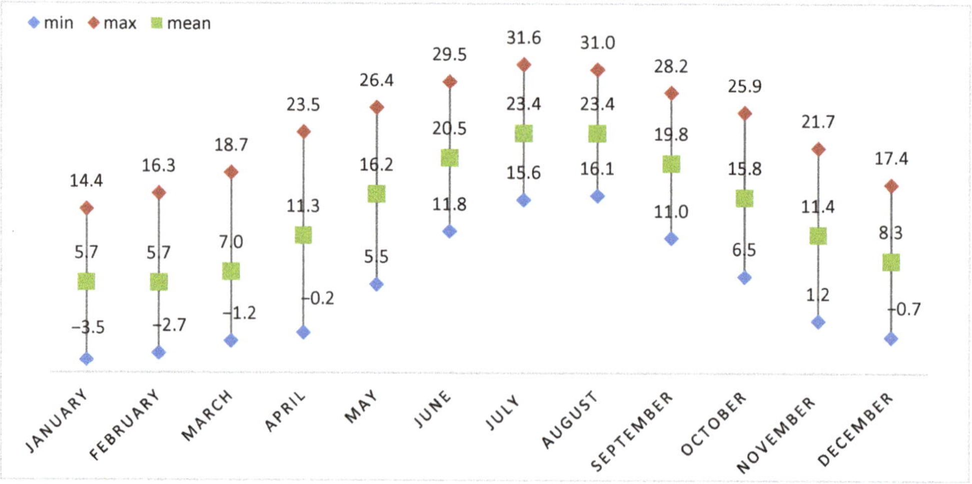

Figure 3. Mean, minimum, and maximum monthly temperatures for Istanbul.

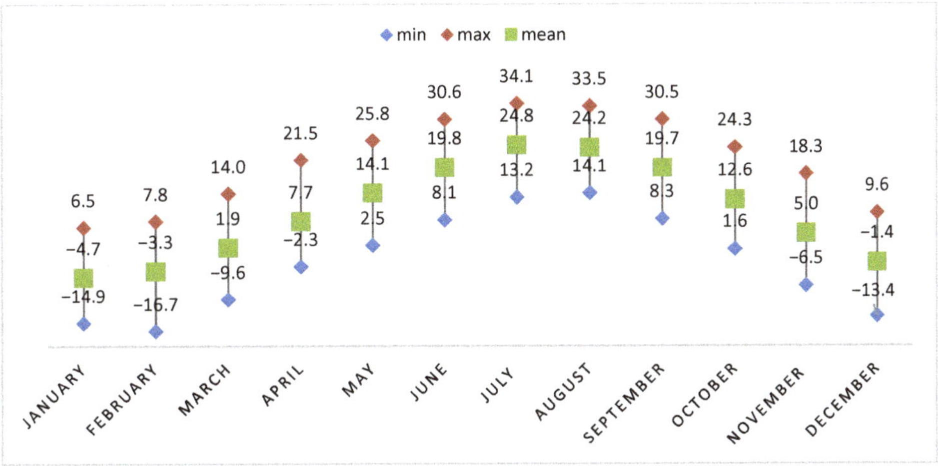

Figure 4. Mean, minimum, and maximum monthly temperatures for Hakkari.

In this study, the influence of insulation thickness, glazing type, infiltration rate, fuel type, room set-point temperature, window-to-wall ratio, and wall orientation on annual heating energy requirement, investment cost, and the net present worth was investigated. Results revealed information about design of a building façade with the minimum energy requirement. Table 5 presents all of the studied cases.

Table 5. Investigated cases.

Case	Insulation Thickness (cm)	Glazing Type	Window to Wall Ratio	Orientation	Infiltration Rate	Heating Set-Point	Fuel Type
1	3	Single/Double/Triple	40%	NW	0.2 ACH	24 °C	Natural Gas
2	6	Single/Double/Triple	40%	NW	0.2 ACH	24 °C	Natural Gas
3	9	Single/Double/Triple	40%	NW	0.2 ACH	24 °C	Natural Gas
4	12	Single/Double/Triple	40%	NW	0.2 ACH	24 °C	Natural Gas
5	15	Single/Double/Triple	40%	NW	0.2 ACH	24 °C	Natural Gas
6	3	Single/Double/Triple	40%	NE	0.2 ACH	24 °C	Natural Gas
7	3	Single/Double/Triple	40%	NW	0.2 ACH	24 °C	Natural Gas
8	3	Single/Double/Triple	40%	SE	0.2 ACH	24 °C	Natural Gas
9	3	Single/Double/Triple	40%	SW	0.2 ACH	24 °C	Natural Gas
10	3	Single/Double/Triple	20%	NW	0.2 ACH	24 °C	Natural Gas
11	3	Single/Double/Triple	40%	NW	0.2 ACH	24 °C	Natural Gas
12	3	Single/Double/Triple	60%	NW	0.2 ACH	24 °C	Natural Gas
13	3	Single/Double/Triple	80%	NW	0.2 ACH	24 °C	Natural Gas
14	3	Single/Double/Triple	100%	NW	0.2 ACH	24 °C	Natural Gas
15	3	Single/Double/Triple	40%	NW	0.2 ACH	24 °C	Natural Gas
16	3	Single/Double/Triple	40%	NW	0.4 ACH	24 °C	Natural Gas
17	3	Single/Double/Triple	40%	NW	0.6 ACH	24 °C	Natural Gas
18	3	Single/Double/Triple	40%	NW	0.8 ACH	24 °C	Natural Gas

Table 5. *Cont.*

Case	Insulation Thickness (cm)	Glazing Type	Window to Wall Ratio	Orientation	Infiltration Rate	Heating Set-Point	Fuel Type
19	3	Single/Double/Triple	40%	NW	1.0 ACH	24 °C	Natural Gas
20	3	Single/Double/Triple	40%	NW	0.2 ACH	18 °C	Natural Gas
21	3	Single/Double/Triple	40%	NW	0.2 ACH	20 °C	Natural Gas
22	3	Single/Double/Triple	40%	NW	0.2 ACH	22 °C	Natural Gas
23	3	Single/Double/Triple	40%	NW	0.2 ACH	24 °C	Natural Gas
24	3	Single/Double/Triple	40%	NW	0.2 ACH	26 °C	Natural Gas
25	3	Single/Double/Triple	40%	NW	0.2 ACH	24 °C	Natural Gas
26	3	Single/Double/Triple	40%	NW	0.2 ACH	24 °C	Fuel-Oil
27	3	Single/Double/Triple	40%	NW	0.2 ACH	24 °C	Coal
28	3	Single/Double/Triple	40%	NW	0.2 ACH	24 °C	LPG
29	3	Single/Double/Triple	40%	NW	0.2 ACH	24 °C	Electricity

4. Results and Discussion

4.1. Orientation

In this part of the study, the reference room was rotated in such a way that exterior walls faced one of the four main orientations. The two outer walls were changed to meet north-east (NE), south-east (SE), north-west (NW), and south-west (SW) directions. In Figure 5, the effect of façade orientation on the net present cost is given. Results revealed that the net present cost of the system is minimized for the façade with double glazed windows, oriented towards the south-west. Despite having the lowest investment cost, the façade with single glazed windows has the highest net present cost values due to the high annual heating energy requirement. The net present cost can be cut by 35% if the façade is oriented towards the south-west instead of the north-east, and a double glazed window is selected instead of the single glazed window.

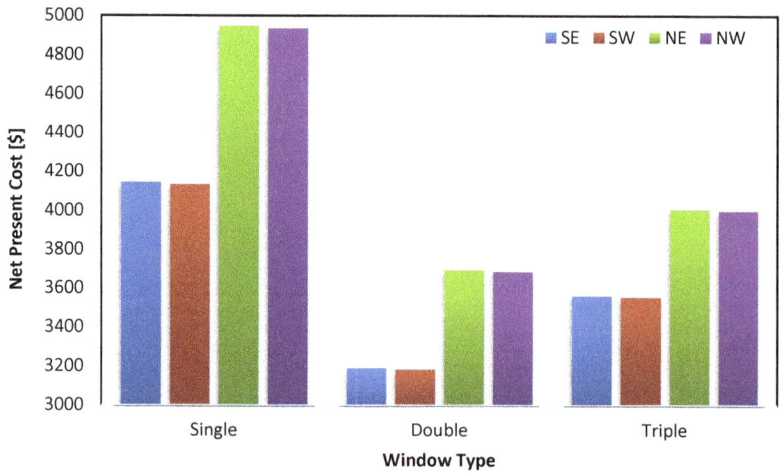

Figure 5. The influence of orientation on the net present cost of the building façade in Istanbul.

4.2. Window-to-Wall Ratio (WWR)

Windows are key design elements in architectural applications because they improve the appearance of buildings and enable daylight penetration [48]. Therefore, the effect of the glazing area on the thermal performance of the buildings is an important research goal. A parametric evaluation was conducted to investigate various window-to-wall ratios (20%, 40%, 60%, 80%, 100%) according to be single glazed, double glazed or triple glazed windows. In Figures 6 and 7, the effect of the WWR on the annual heating cost is presented for Istanbul and Hakkari, respectively. Results revealed that the annual heating energy consumption of the room façade with a single glazed window increases with increasing window area. Due to lower energy transmittance values, the annual heating energy cost decreases with growing window area for double glazed and triple glazed façades. Moreover, the influence of the WWR on the heating energy cost is more dramatic in Hakkari than in Istanbul due to colder climatic conditions throughout the winter.

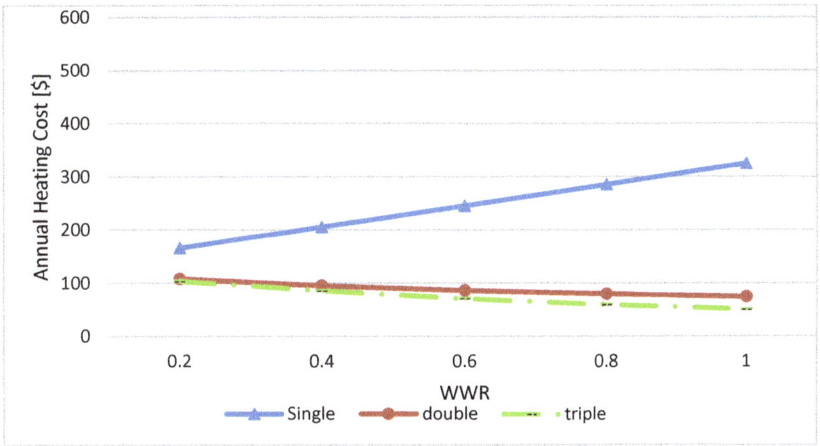

Figure 6. The influence of the window-to-wall ratio on annual heating cost in Istanbul (3 cm insulation, NW oriented).

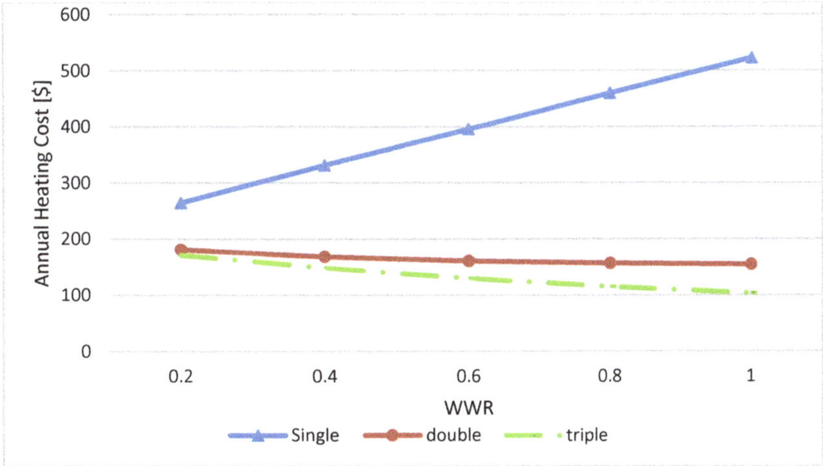

Figure 7. The influence of the window-to-wall ratio on the annual heating cost in Hakkari (3 cm insulation, NW oriented).

In Figures 8 and 9, the effect of the window-to-wall ratio and the glazing type on the net present cost of the system is presented. Results show that the net present cost increases with increasing window area. A fully glazed façade with a single pane window has the highest net present cost, of USD 9134 and 12,189 in Istanbul and Hakkari, respectively. A double glazed façade with a 20% window-to-wall ratio has the lowest net present cost value in Istanbul and Hakkari of USD 2870 and 4000, respectively. Although increasing the window surface area increases the net present cost, window sizes cannot be reduced without considering visual comfort.

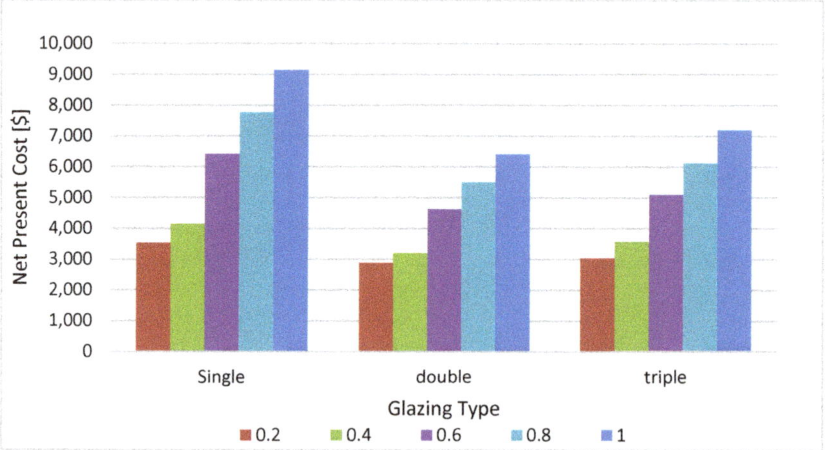

Figure 8. The influence of the window-to-wall ratio and the glazing type on the net present cost of the system in Istanbul (3 cm insulation, NW oriented).

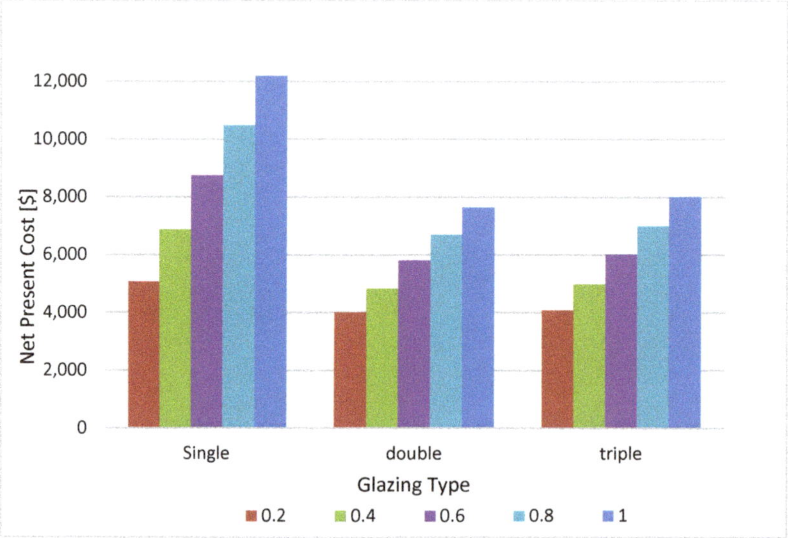

Figure 9. The influence of the window-to-wall ratio and the glazing type on the net present cost of the system in Hakkari (3 cm insulation, NW oriented).

In Figure 10, the influence of the window-to-wall ratio, orientation, and the glazing type on the net present cost of the façade are shown. Increasing the window area increases the net present value of the façade for both south-west and north-west directions. However, due to passive heating and higher solar radiation levels in south-oriented façades, the net present cost is always lower than that of the north-oriented façades.

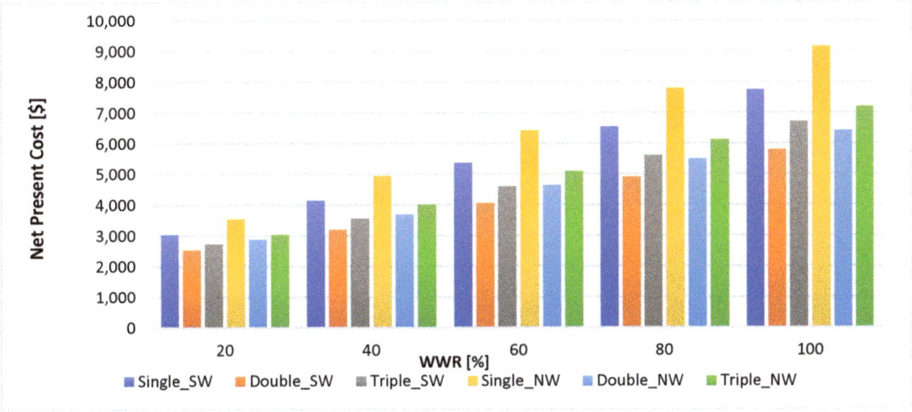

Figure 10. The influence of the window-to-wall ratio, orientation, and the glazing type on the net present cost of the system in Istanbul (3 cm insulation).

4.3. Insulation Thickness

In this part of the study, energy consumption and the net present cost were calculated by applying various insulation thicknesses of the external walls and window types. Optimum insulation thickness was obtained for which the net present cost value was minimized. In Figure 11, the influence of the insulation thickness on the net present cost of the system in Hakkari and Istanbul is shown. Results show that from 3 to 9 cm insulation thickness, the net present cost of the zone decreases dramatically for the façades with both double glazed and triple glazed windows. It can be seen that choosing a thickness value greater than 9 cm will increase the net present cost in Hakkari; therefore, it is unnecessary. The minimum net present cost was found to be USD 4584 for the façade with double glazed windows and 9 cm external wall insulation thickness in Hakkari. The optimum insulation thickness was found to be 6 cm for single, double, and triple glazed windows in Istanbul. The minimum net present cost was found to be USD 3583 for the façade with double glazed windows and 6 cm external wall insulation thickness.

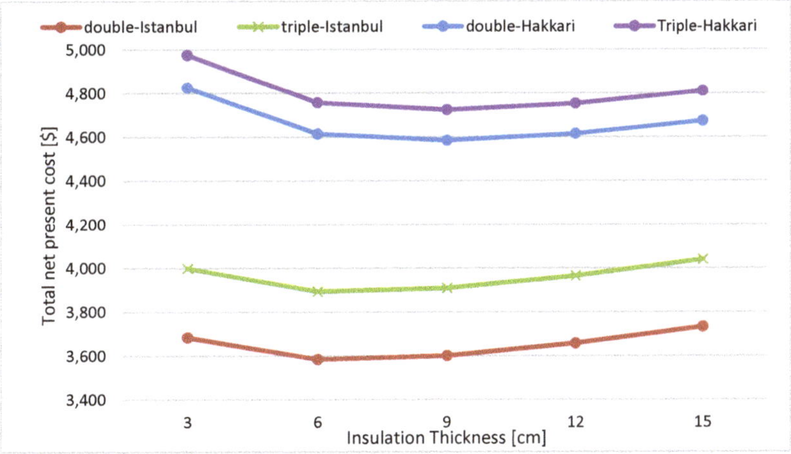

Figure 11. The influence of insulation thickness on the net present cost (40% WWR, NW oriented façade) for the selected locations.

Figure 12 demonstrates the investment cost of the façade for different insulation thicknesses and window types. It is seen that the investment cost increases with the increase in insulation thickness. In addition, a single glazed window façade has the minimum, and the triple glazed façade has the maximum investment cost.

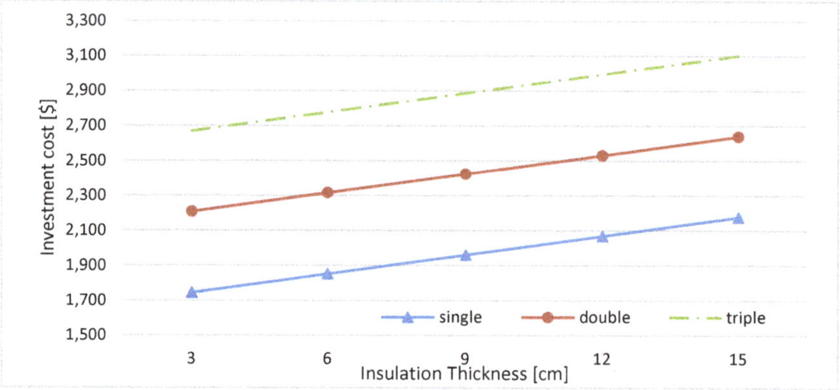

Figure 12. Effect of insulation thickness and window type on the insulation cost of the façade.

Figure 13 shows the annual natural gas consumption of the zone versus different insulation thicknesses and window types. Results show that for the façade with a single glazed window, changing the insulation thickness from 3 to 15 cm decreases the amount of fuel consumption by around 10–12%. For the double glazed and triple glazed façades, increasing the insulation thickness from 3 to 15 cm decreases the fuel consumption by 22–26% and 25–30%, respectively.

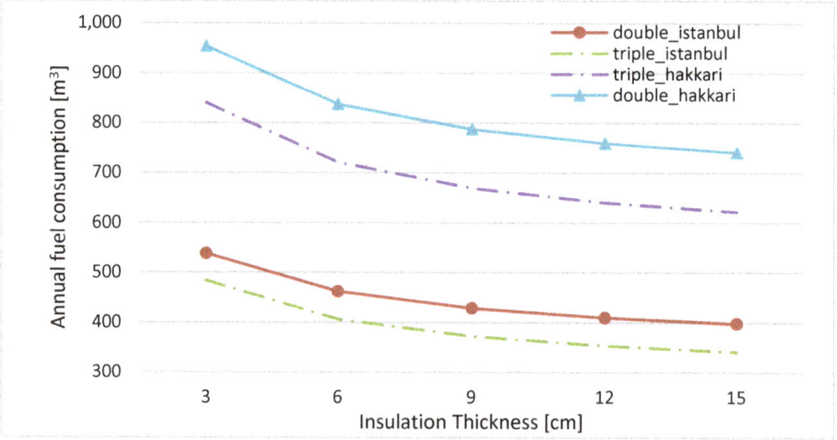

Figure 13. Effect of insulation thickness and window type on fuel consumption.

In Figures 14 and 15, the energy savings ratio is presented for varying insulation thicknesses and glazing types. Results show that, especially for single glazed façades, increasing the insulation thickness increases the energy savings ratio at a higher rate in Hakkari compared to Istanbul. Increasing the thermal insulation thickness of the external walls with a triple glazed façade changes the energy savings ratio between 0.25 and 0.48 in both locations. For a double glazed façade, the energy savings ratio changes between 0.20 and 0.41.

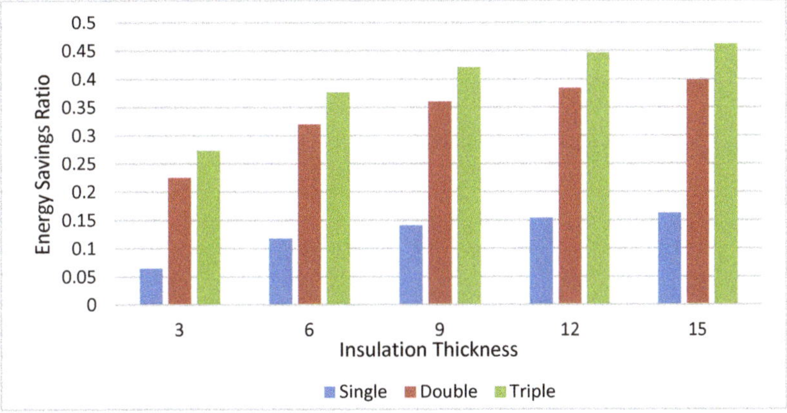

Figure 14. Energy savings ratio versus varying insulation thicknesses and glazing types for the designed zone in Hakkari (40% WWR).

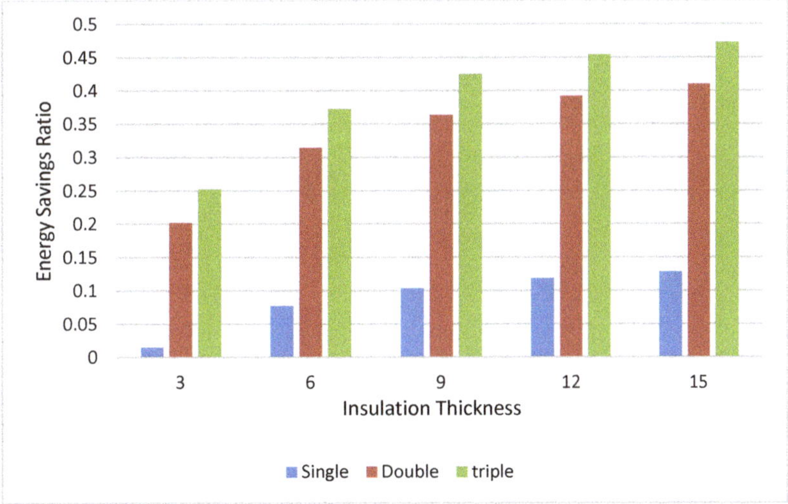

Figure 15. Energy savings ratio versus varying insulation thicknesses and glazing types for the designed zone in Istanbul (40% WWR).

4.4. Infiltration Rate

Infiltration is the uncontrolled movement of air through the building envelope. Heating load is strongly influenced by the air infiltration rate [17,18]. Infiltration has a negative impact on the heating load as it will leak the heat contained in the building to the outside environment [17]. For the base scenario, a relatively low infiltration rate was applied (0.2 ACH). In this part of the study, the effect of the infiltration rate on the net present cost was investigated. The values of the infiltration rate were incremented between 0.2 and 1 ACH. The results of the simulation process are displayed in Figures 16 and 17 for Istanbul and Hakkari, respectively. It can be seen that varying the infiltration rate from 0.2 to 1 ACH increases the heating energy requirement intensively; therefore, the net present cost rises. Between 0.2 and 1 ACH, the total net present cost increases from 36% to 48% in Istanbul (Figure 16) and 40 to 55% in Hakkari (Figure 17). In both locations, the designed zone favors an air-tight construction; however, due to colder climatic conditions, this is more apparent in Hakkari. Appropriate solutions to reduce heat loss through infiltration should be carefully considered to keep infiltration at a lower rate.

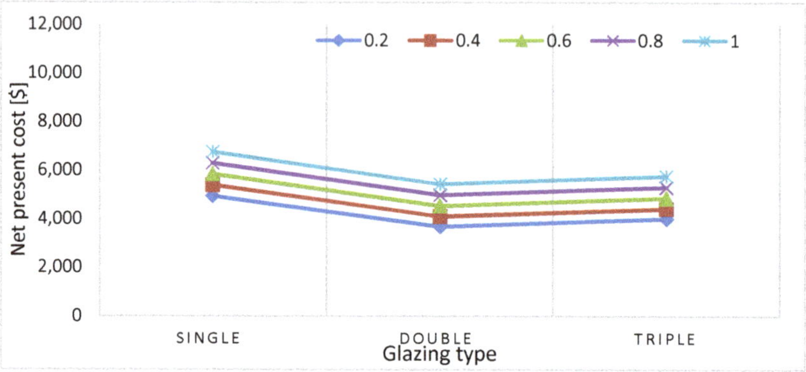

Figure 16. The influence of the infiltration rate on the net present cost for Istanbul.

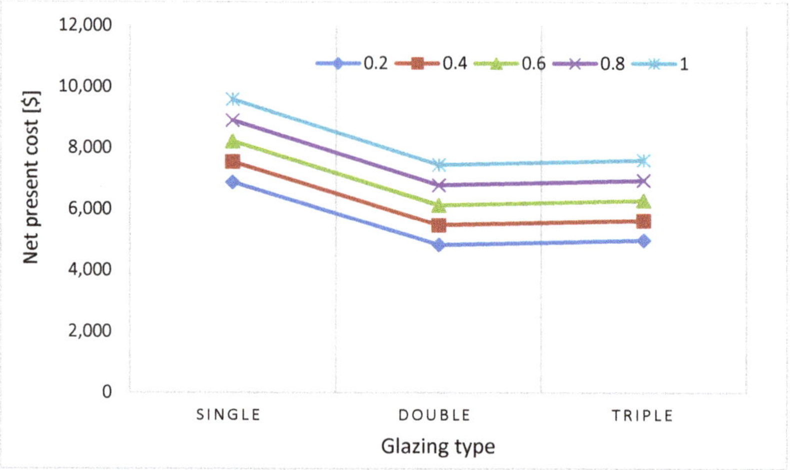

Figure 17. The influence of the infiltration rate on the net present cost for Hakkari.

4.5. Room Set-Point Temperature

The heating equipment works to bring the zone temperature to the identified set-point temperature. The room set-point temperature impacts the heating load. To see the effect of each 2 °C adjustment, the room set-point temperature was varied from 18 to 26 °C. As depicted in Figures 18 and 19, decreasing the room set-point temperature decreases the annual heating energy consumption; therefore, it reduces the net present cost. This tendency is more visible for the zone with single glazed windows. For a 2 °C change in the temperature, the total net present cost increases from 7% to 15%. Therefore, increasing the occupants' awareness regarding temperature control and obtaining energy control mechanisms based on building occupancy is very crucial for energy economics.

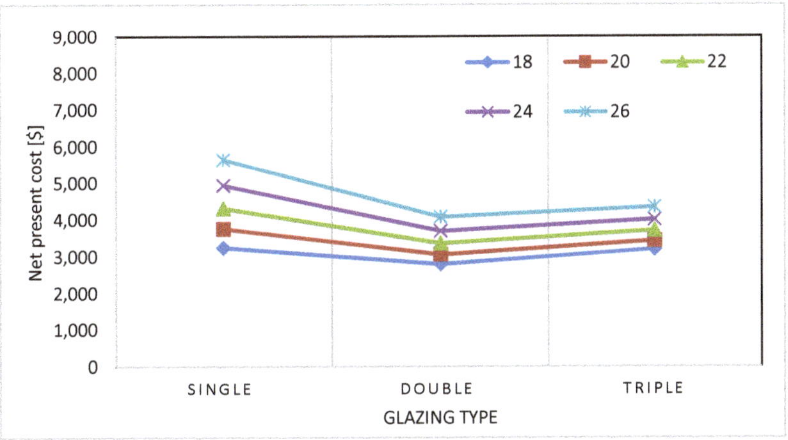

Figure 18. The influence of the heating set-point temperature on the net present cost for Istanbul.

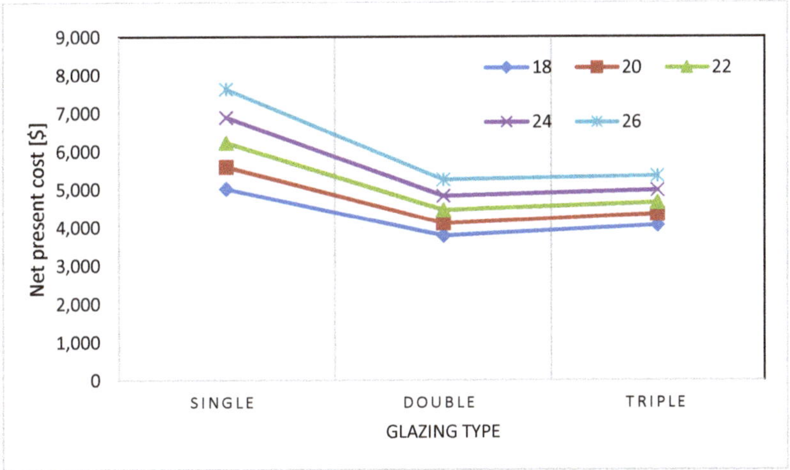

Figure 19. The influence of the heating set-point temperature on the net present cost for Hakkari.

4.6. Fuel Type

Fuel type has a major impact on the annual heating cost and the net present cost. In Figures 20 and 21, the net present cost of the building envelope is presented for various energy types. Natural gas, electricity, coal, LPG, and fuel-oil are the most widely used energy sources for heating; therefore, they were selected for the parametrical study. The lowest net present cost is for natural gas, followed by coal and fuel oil, respectively. Electricity has the highest net present cost. Since electricity generation in Turkey mostly depends on natural gas and coal, the unit electricity cost is high; therefore, electricity for heating is more expensive than other energy sources. The most suitable energy source for heating was found to be natural gas compared to the other selected energy sources. Preferring natural gas instead of electricity decreases the net present cost by 70% to 85% in both locations, depending on the glazing type.

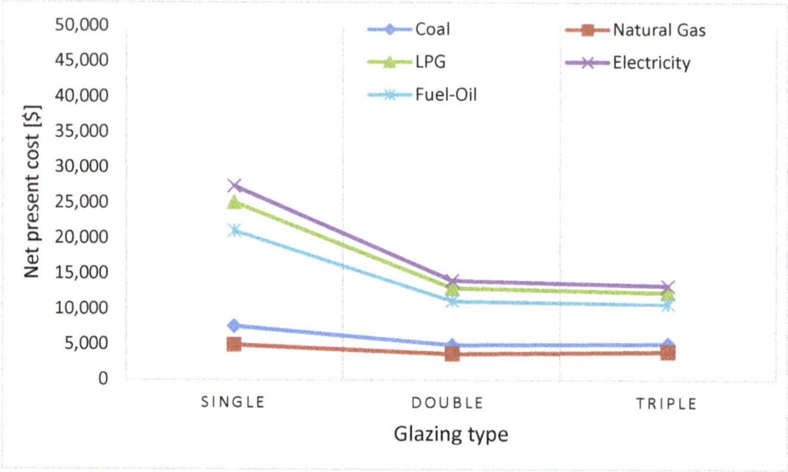

Figure 20. The influence of the fuel type on the net present cost for Istanbul.

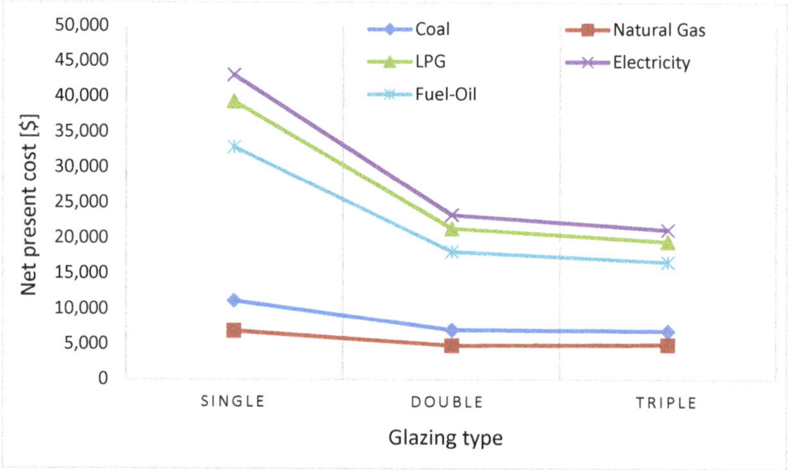

Figure 21. The influence of the fuel type on the net present cost for Hakkari.

5. Conclusions

The energy efficiency of the buildings is a crucial matter, and mostly depends on early design decisions. Therefore, many have researchers investigated optimum insulation thickness to maximize economic profit and minimize energy consumption. Previous studies have investigated the optimum insulation thickness of the buildings without considering the façade as external walls and windows together. Moreover, parameters such as actual meteorological data (solar radiation, environmental temperature, etc.), infiltration rate, and occupant density influence the heating load of a building; therefore, they should not be neglected. This study evaluated the thermo-economic performance of a zone to determine the optimum façade parameters, such as insulation thickness and window type, using dynamic modelling. The optimum insulation thickness of a zone was chosen over a 20-year lifetime for two different locations. The main results of the study are as follows.

Despite having the lowest initial investment cost, the façade with single glazed windows had the highest net present worth in both locations. The main reason behind that trend is the high cost of fuel due to the more significant heating load for single glazed windows.

It was found that the effect of the façade orientation on the net present cost of the system is significant. Results show that the net present cost of the system can be decreased between 11% and 17% if the orientation of the façade is changed from north-east to south-west.

The surface area of the windows has a significant influence on the net present cost, fuel cost, and the investment cost of the façade. For the façade with single glazed windows, the heating cost increases for greater window areas, and this trend is more dramatic in Hakkari due to colder climatic conditions. For the double glazed and triple glazed windows, the heating energy cost decreases with increasing window areas. However, the net present cost of the façade increases with the growing window area for all cases. For the same window size and window type, in south-oriented façades, the net present cost is always lower than the north-oriented façades because of the passive heating and higher solar radiation levels.

Results show that increasing the insulation thickness of the external walls increases the investment cost and decreases the natural gas consumption. The façade with a single glazed window and 3 cm insulation has the minimum investment cost. The optimum insulation thickness of the building façade was determined based on the minimum net present cost. In Hakkari and Istanbul, the net present cost of the building façade is minimum for 9 cm and 6 cm insulation thicknesses, respectively. The minimum net present cost was found for the façade with double glazed windows in both locations. The optimum insulation thickness of the selected areas can decrease annual fuel consumption by between 14% and 18%.

The infiltration rate is a highly influential parameter for the net present cost of the buildings, especially for the colder climatic regions. To keep infiltration heat losses at the minimum, appropriate solutions should be carefully considered.

Room set-point temperature is another essential parameter for energy-efficient building design. Results of the sensitivity analysis revealed that only a 2 °C change in the temperature increases the net present cost between 7% and 15%. This tendency is more apparent for the single glazed façades due to the greater risk of heat loss.

Finally, sensitivity analysis was also conducted to determine the influence of the fuel type on the net present cost of buildings. The five most common energy resources for heating were evaluated and compared in terms of heating cost and the net present cost. Natural gas was found to be the most suitable energy source for heating, followed by coal and fuel oil. Since national electricity is mainly generated from natural gas and coal, the unit price of the electricity cannot compete with that of coal and natural gas. However, with the broader application of renewable-based power generation systems, it is expected to decrease in the future.

The presented results emphasize the importance of the design parameters, such as orientation, glazing type, insulation thickness, and window area, on the energy efficiency and economy; therefore, designers and engineers should consider all these aspects.

Funding: This research received no external funding.

Institutional Review Board Statement: Not applicable.

Informed Consent Statement: Not applicable.

Data Availability Statement: Not applicable.

Conflicts of Interest: The author declares no conflict of interest.

References

1. Akan, A.E. Determination and Modeling of Optimum Insulation Thickness for Thermal Insulation of Buildings in All City Centers of Turkey. *Int. J. Thermophys.* **2021**, *42*, 49. [CrossRef]
2. Mirabella, N.; Röck, M.; Saade, M.R.M.; Spirinckx, C.; Bosmans, M.; Allacker, K.; Passer, A. Strategies to improve the energy performance of buildings: A review of their life cycle impact. *Buildings* **2018**, *8*, 105. [CrossRef]
3. Shehadi, M. Energy consumption optimization measures for buildings in the midwest regions of USA. *Buildings* **2018**, *8*, 170. [CrossRef]
4. Kaynakli, O. Optimum thermal insulation thicknesses and payback periods for building walls in Turkey. *J. Therm. Sci. Technol.* **2013**, *33*, 45–55.

5. Kaynakli, O. Parametric Investigation of Optimum Thermal Insulation Thickness for External Walls. *Energies* **2011**, *4*, 913–927. [CrossRef]
6. Kurekci, N.A. Determination of optimum insulation thickness for building walls by using heating and cooling degree-day values of all Turkey ' s provincial centers. *Energy Build.* **2016**, *118*, 197–213. [CrossRef]
7. Bektas Ekici, B.; Aytac Gulten, A.; Aksoy, U.T. A study on the optimum insulation thicknesses of various types of external walls with respect to different materials, fuels and climate zones in Turkey. *Appl. Energy* **2012**, *92*, 211–217. [CrossRef]
8. Yuan, J.; Farnham, C.; Emura, K. Optimum Insulation Thickness for Building Exterior Walls in 32 Regions of China to Save Energy and Reduce CO_2 Emissions. *Sustainability* **2017**, *9*, 1711. [CrossRef]
9. Canbolat, A.S.; Bademlioglu, A.H.; Saka, K.; Kaynakli, O. Investigation of parameters affecting the optimum thermal insulation thickness for buildings in hot and cold climates. *Therm. Sci.* **2020**, *24*, 2891–2903. [CrossRef]
10. Alsayed, M.F.; Tayeh, R.A. Life cycle cost analysis for determining optimal insulation thickness in Palestinian buildings. *J. Build. Eng.* **2019**, *22*, 101–112. [CrossRef]
11. Özel, G.; Açikkalp, E.; Görgün, B.; Yamik, H.; Caner, N. Optimum insulation thickness determination using the environmental and life cycle cost analyses based entransy approach. *Sustain. Energy Technol. Assessments* **2015**, *11*, 87–91. [CrossRef]
12. Açıkkalp, E.; Yerel, S. A method for determining optimum insulation thickness: Combined economic and environmental method. *Therm. Sci. Eng. Prog.* **2019**, *11*, 249–253. [CrossRef]
13. Barrau, J.; Ibañez, M.; Badia, F. Impact of the optimization criteria on the determination of the insulation thickness. *Energy Build.* **2014**, *76*, 459–469. [CrossRef]
14. Ashouri, M.; Astaraei, F.R.; Ghasempour, R.; Ahmadi, M.H.; Feidt, M. Optimum insulation thickness determination of a building wall using exergetic life cycle assessment. *Appl. Therm. Eng.* **2016**, *106*, 307–315. [CrossRef]
15. Dombayci, O.A.; Ulu, E.Y.; Guven, S.; Atalay, O.; Ozturk, H.K. Determination Of Optimum Insulation Thickness For Building External Walls With Different Insulation Materials Using Environmental Impact Assessment. *Therm. Sci.* **2020**, *24*, 303–311. [CrossRef]
16. Keçebaş, A. Determination of optimum insulation thickness in pipe for exergetic life cycle assessment. *Energy Convers. Manag.* **2015**, *105*, 826–835. [CrossRef]
17. Albatayneh, A. Optimising the parameters of a building envelope in the east mediterranean Saharan, cool climate Zone. *Buildings* **2021**, *11*, 43. [CrossRef]
18. Hachem-Vermette, C.; MacGregor, A. Energy optimized envelope for cold climate indoor agricultural growing center. *Buildings* **2017**, *7*, 59. [CrossRef]
19. Kalua, A. Envelope thermal design optimization for urban residential buildings in Malawi. *Buildings* **2016**, *6*, 13. [CrossRef]
20. Atmaca, A.; Gedik, G.; Wagner, A. Determination of Optimum Envelope of Religious Buildings in Terms of Thermal Comfort and Energy Consumption. *Energies* **2021**, *14*, 6597. [CrossRef]
21. Xu, X.; Feng, G.; Chi, D.; Liu, M.; Dou, B. Optimization of Performance Parameter Design. *Energies* **2020**, *11*, 3252. [CrossRef]
22. Chiesa, G.; Acquaviva, A.; Grosso, M.; Bottaccioli, L.; Floridia, M.; Pristeri, E.; Sanna, E.M. Parametric Optimization of Window-to-Wall Ratio for Passive Buildings Adopting A Scripting Methodology to Dynamic-Energy Simulation. *Sustainability* **2019**, *11*, 3078. [CrossRef]
23. Wang, Z.; Zhao, J. Optimization of Passive Envelop Energy Efficient Measures for Office Buildings in Different Climate Regions of China Based on Modified Sensitivity Analysis. *Sustainability* **2018**, *10*, 907. [CrossRef]
24. Axaopoulos, I.; Axaopoulos, P.; Gelegenis, J.; Fylladitakis, E.D. Optimum external wall insulation thickness considering the annual CO_2 emissions. *J. Build. Phys.* **2019**, *42*, 527–544. [CrossRef]
25. Usman, M.; Frey, G. Multi-Objective Techno-Economic Optimization of Design Parameters for Residential Buildings in Different Climate Zones. *Sustainability* **2022**, *14*, 65. [CrossRef]
26. Altun, A.F.; Kiliç, M. Influence of window parameters on the thermal performance of office rooms in different climate zones of Turkey. *Int. J. Renew. Energy Res.* **2019**, *9*, 226–243.
27. Djokovic, J.M.; Bujnak, J.; Hadzima, B.; Pastorek, F.; Dwornicka, R.; Ulewicz, R. Selection of the Optimal Window Type and Orientation for the Two Cities in Serbia and One in Slovakia. *Energies* **2022**, *15*, 323. [CrossRef]
28. Gasparella, A.; Pernigotto, G.; Cappelletti, F.; Romagnoni, P.; Baggio, P. Analysis and modelling of window and glazing systems energy performance for a well insulated residential building. *Energy Build.* **2011**, *43*, 1030–1037. [CrossRef]
29. Tsikaloudaki, K.; Laskos, K.; Theodosiou, T.; Bikas, D. Assessing cooling energy performance of windows for office buildings in the Mediterranean zone. *Energy Build.* **2012**, *49*, 192–199. [CrossRef]
30. Tsikaloudaki, K.; Theodosiou, T.; Laskos, K.; Bikas, D. Assessing cooling energy performance of windows for residential buildings in the Mediterranean zone. *Energy Convers. Manag.* **2012**, *64*, 335–343. [CrossRef]
31. Kon, O. Calculation of fuel consumption and emissions in buildings based on external walls and windows using economic optimization. *J. Fac. Eng. Archit. Gazi Univ.* **2018**, *33*, 101–113. [CrossRef]
32. Ozel, M. Influence of glazing area on optimum thickness of insulation for different wall orientations. *Appl. Therm. Eng.* **2019**, *147*, 770–780. [CrossRef]
33. Karabay, H.; Arici, M. Multiple pane window applications in various climatic regions of Turkey. *Energy Build.* **2012**, *45*, 67–71. [CrossRef]

34. Derradji, L.; Imessad, K.; Amara, M.; Boudali Errebai, F. A study on residential energy requirement and the effect of the glazing on the optimum insulation thickness. *Appl. Therm. Eng.* **2017**, *112*, 975–985. [CrossRef]
35. Gelegenis, J.; Axaopoulos, P. A multi-parametric mathematical approach on the selection of optimum insulation thicknesses in buildings. *Buildings* **2017**, *7*, 15. [CrossRef]
36. Akan, A.P.; Akan, A.E. Modeling of CO_2 emissions via optimum insulation thickness of residential buildings. *Clean Technol. Environ. Policy* **2021**. [CrossRef]
37. Dylewski, R.; Adamczyk, J. Optimum thickness of thermal insulation with both economic and ecological costs of heating and cooling. *Energies* **2021**, *14*, 3835. [CrossRef]
38. Şencan Şahin, A.; Kovacı, T.; Dikmen, E. Determination and economic analysis of the optimum insulation thickness of building walls, considering annual CO_2 emission. *Pamukkale Univ. J. Eng. Sci.* **2021**, *27*, 60–69. [CrossRef]
39. Hamelin, M.C.; Zmeureanu, R. Optimum envelope of a single-family house based on life cycle analysis. *Buildings* **2014**, *4*, 95–112. [CrossRef]
40. Jaber, S.; Ajib, S. Optimum, technical and energy efficiency design of residential building in Mediterranean region. *Energy Build.* **2011**, *43*, 1829–1834. [CrossRef]
41. Central Bank of Turkey Central Bank of Turkey. Available online: https://www.tcmb.gov.tr/ (accessed on 14 March 2022).
42. Turkish Statistical Institute (TURKSTAT). Available online: https://www.tuik.gov.tr/ (accessed on 14 March 2022).
43. Comparison of the unit energy prices in Turkey. Available online: https://www.enerji-dunyasi.com/belge-indir/4/730/yakit-fiyatlari-karsilastirma-tablosu-02-03-2022.xlsx/ (accessed on 14 March 2022).
44. Ertürk, M. Optimum insulation thicknesses of pipes with respect to different insulation materials, fuels and climate zones in Turkey. *Energy* **2016**, *113*, 991–1003. [CrossRef]
45. Ozel, M. Determination of optimum insulation thickness based on cooling transmission load for building walls in a hot climate. *Energy Convers. Manag.* **2013**, *66*, 106–114. [CrossRef]
46. Çağlayan, S.; Özorhon, B.; Özcan-deniz, G.; Yiğit, S. A life cycle costing approach to determine the optimum insulation thickness of existing buildings. *J. Therm. Sci. Technol.* **2020**, *40*, 1–14.
47. Altun, A.F.; Kilic, M. Design and performance evaluation based on economics and environmental impact of a PV-wind-diesel and battery standalone power system for various climates in Turkey. *Renew. Energy* **2020**, *157*, 424–443. [CrossRef]
48. Ozel, M.; Ozel, C. Effect of window-to-wall-area ratio on thermal performance of building wall materials in Elazığ, Turkey. *PLoS ONE* **2020**, *15*, e0237797. [CrossRef]

buildings

MDPI

Article

Thermo-Energy Performance of Lightweight Steel Framed Constructions: A Case Study

Ligia Moga [1], Ioan Petran [2], Paulo Santos [3] and Viorel Ungureanu [4,*]

1 Department of Civil Engineering and Management, Technical University of Cluj-Napoca, 400027 Cluj-Napoca, Romania; ligia.moga@ccm.utcluj.ro
2 Department of Structures, Technical University of Cluj-Napoca, 400027 Cluj-Napoca, Romania; ioan.petran@dst.utcluj.ro
3 Institute for Sustainability and Innovation in Structural Engineering (ISISE), Department of Civil Engineering, University of Coimbra, 3030-788 Coimbra, Portugal; pfsantos@dec.uc.pt
4 Department of Steel Structures and Structural Mechanics, Politehnica University of Timisoara, 300224 Timisoara, Romania
* Correspondence: viorel.ungureanu@upt.ro

Abstract: The building sector continues to play an essential role in reducing worldwide energy consumption. The reduced consumption is accompanied by stricter regulation for the thermotechnical design of the building envelope. The redefined nearly Zero Energy Building levels that will come into force for each member state will pressure designers to rethink the constructive details so that mandatory levels can be reached, without increasing the construction costs over an optimum level but at the same time reducing greenhouse gas emissions. The paper aims to illustrate the main conclusions obtained in assessing the thermo-energy performance of a steel-framed building representing a holistically designed modular laboratory located in a moderate continental temperate climate, characteristic of the south-eastern part of the Pannonian Depression with some sub-Mediterranean influences. An extensive numerical simulation of the main junctions was performed. The thermal performance was established in terms of the main parameters, the adjusted thermal resistances and global thermal insulation coefficient. Further on, the energy consumption for heating was established, and the associated energy rating was in compliance with the Romanian regulations. A parametric study was done to illustrate the energy performance of the investigated case in the five representative climatic zones from Romania. An important conclusion of the research indicates that an emphasis must be placed on the thermotechnical design of Light Steel Framed solutions against increased thermal bridge areas caused by the steel's high thermal conductivity for all building components to reach nZEB levels. Nevertheless, the results indicate an exemplary behaviour compared to classical solutions, but at the same time, the need for an iterative redesign so that all thermo-energy performance indicators are achieved.

Keywords: LSF constructions; thermal bridging; thermal resistance; thermal transmittance; numerical simulation; thermal performance; energy performance; parametric study; nZEB; energy rating

Citation: Moga, L.; Petran, I.; Santos, P.; Ungureanu, V. Thermo-Energy Performance of Lightweight Steel Framed Constructions: A Case Study. *Buildings* **2022**, *12*, 321. https://doi.org/10.3390/buildings12030321

Academic Editor: Bo Yang

Received: 17 February 2022
Accepted: 6 March 2022
Published: 8 March 2022

1. Introduction

The building sector is still one of the largest energy consumers worldwide, responsible for around 40% of the European Union's (EU) energy consumption and 36% of greenhouse gas emissions coming from construction, usage, renovation and demolition phases [1]. Through the help of The Green Deal program, a very ambitious target was set of going carbon-neutral by 2050 [2]. Besides the already mandatory actions imposed by the 2018/844 Energy Performance of Buildings Directive (EPBD) [3] (e.g., reaching nearly Zero Energy Buildings (nZEB) for both new and existing buildings, thus providing healthier buildings, more robust implementation of the Energy Performance Certificate (EPC)), each member state has to present a strategy through the National Energy and Climate Plans (NECP) for tackling energy consumptions in buildings in the period 2021–2030. The aim consists in reaching the goal of reducing energy consumption by 32.5% by 2030 [4]. Considering that the NECP objectives need to be consolidated for reaching the 2030 targets, a review and revision of the Energy Efficiency Directive (EED) [5] took place hand in hand with several targeted provisions of the EPBD. A proposal for the recast on the EU Directive on energy efficiency was published in 2021 [6], which includes reducing the net greenhouse gas emission by at least 55% by 2030 [7] to become climate-neutral by 2050. At the same time, it proposes higher reductions for primary energy consumption -39%, and final energy consumption −36% by 2030. Therefore, the Green Deal initiatives, the Renovation Wave and the Strategy for Energy Sector Integration represent essential programs in promoting energy efficiency [8].

As it was mentioned by the International Energy Agency (IEA) [9], during 2019, the building sector has deviated from the path towards the Paris agreement objectives. A slowing rate of energy efficiency improvement has been observed since 2015 by the IEA [9]. The final energy consumption grew by 2.2% in 2018, and by 2019 the global energy consumption in the building sector remained at the same level compared to previous years [10], as shown in Figure 1. However, the CO_2 emissions from building operations increased around 38% of the global energy-related CO_2 emissions (see Figure 2), including indirect and direct emissions from non-residential buildings and residential buildings as well as the construction industry [10]. Similar conclusions were highlighted back in 2015 by Ürge-Vorsatz in [11], mentioning that commercial heating and cooling was expected to grow until 2050 with an 84% percentile compared to 2010, while residential heating and cooling with a 79%. It was also concluded that the role of electricity is continuously growing, considering that by 2015 it represented around over half for commercial building energy use. At the same time, in the global electricity consumption, the quantity associated with buildings' operation represents around 55%. Therefore, the building sector should decrease emissions by 6% each year until 2030 to remain on track towards the 2050 objectives [10].

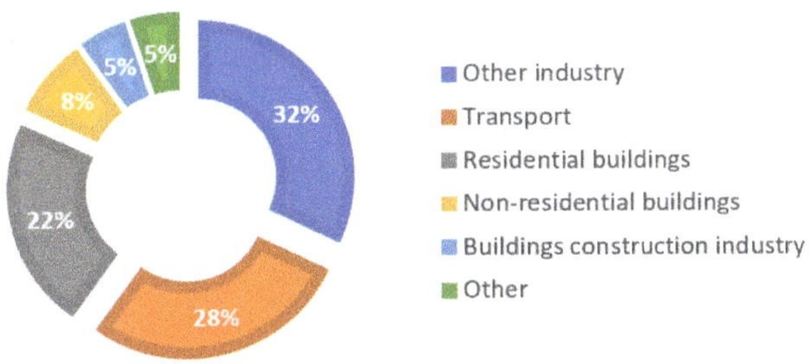

Figure 1. Global share of buildings and construction final energy, 2019. Adapted from [10].

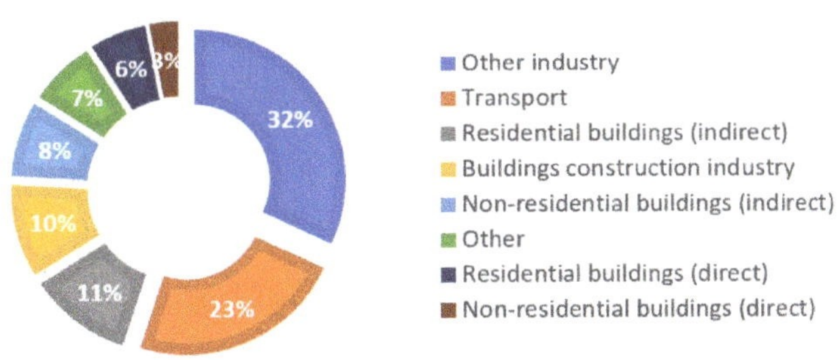

Figure 2. Global share of buildings and construction final CO_2 emissions 2019. Adapted from [10].

Nevertheless, the global COVID-19 pandemic crisis, along with the housing crisis and the economic crisis, produced disturbances in the building sector. At the same time, the longer the health and economic crisis will last, the greater the impact on buildings' energy intensity. As mentioned by IEA [9], an estimation was made regarding a decrease in the global energy demand by 5% to 7% due to the global COVID-19 pandemic.

Although the energy efficiency indicators have slowed down on their path, buildings are still the leading sector where investments will be available to keep the direction towards 2050 objectives. Research is done to identify more innovative and feasible approaches to provide sustainable and energy-efficient technical solutions and construction technology.

Considering all the above mentioned, the project CIA_CLIM "Smart buildings adaptable to the climate change effects" [12], aimed to define a prototype solution described by a lightweight steel-framed (LSF) construction that can tackle in the same time energy efficiency and sustainability issues. The achievement of energy-efficient buildings requires an integrated design concerning various factors such as climate, occupant behaviour, technology, operation, and maintenance.

In the energy design of buildings, an important role is defined by how the building envelope can provide the internal comfort conditions resulting from a compliant thermotechnical design. In the case of LSF constructions placed in cold-dominated climates, the need for an accurate design becomes even more critical. Although the number of LSF buildings is increasing around the globe [13,14] due to their advantages compared to heavyweight constructions, the high thermal conductivity of steel elements may lead to significant thermal bridges, which must be well tackled at the design stage to decrease their negative impact on the energy demand for space heating and cooling [15]. Compared to conventional constructive details, the poor thermal performance of the steel elements can be offset by employing adequate thermal insulation solutions. Nevertheless, a significantly increased thickness is needed when conventional thermal insulating materials are used. Thus, an alternative can be the use of nano-insulation materials, also called super insulation materials (SIM), that are defined by a remarkable reduced thermal conductivity (e.g., a thermal conductivity around 15 mW/(m·K) or even lower). As it was demonstrated by Rajanayagam et al. in [15], when using SIMs the same thermal performance will be reached at lower thicknesses compared to conventional thermal insulation materials thicknesses. The paper also demonstrated that the implemented SIM solution was able to reach the imposed building requirements and help address solutions that can be defined by constructive constraints.

Kempton et al. [16] present various solutions for mitigating thermal bridges, starting from slotting the steel frame members to placing sheets of insulation materials or thermal break strips between the steel frame. In this regard, it was proved by Santos et al. in Reference [17] that for steel frames that do not have exterior thermal insulation, the thermal performance can be increased by 16% for one strip of recycled rubber/cork and by 42% in case of two strips placed at opposite sides of the frame. It is worth mentioning that two strips of aerogel were able to mitigate the thermal bridge effect of the steel frame fully. At the same time, in Reference [18], Santos et al. demonstrated the positive impact of the placement of exterior continuous thermal insulation on an LSF wall's overall performance.

Thereby, the objective of our research focused on four directions that aim to provide answers for both designers and builders. First, the paper highlights the thermo-energy performance of the LSF construction, identified as an experimental module. An extensive bidimensional numerical study is presented to evaluate the proposed solutions' thermal bridges and implicitly thermal behaviour. Second, it provides an overview of to what extent this design approach can reach the nZEB levels defined by the Romanian legislation and simultaneously fulfil the European 2030 targets [4]. Thus, the overall energy performance is assessed against the Romanian national regulations [19]. Third, the parametric study results are highlighted to identify the building performance in all representative climatic zones from Romania and their associated energy class for heating. The study includes parametrisation in terms of climatic zone placement, curtain wall orientation, ventilation rate, type of heating system. Fourth, a parametrisation is done to identify to which level the thermal performance of a building envelope component can impact the energy consumption level. Several preliminary conclusions are drawn as a starting point for continuing the research.

2. The Case Study

2.1. Site and Climate

Five climatic zones define the exterior climate of Romania (see Figure 3), starting from the 1st zone with an exterior temperature $\theta_e = -12$ °C, up to the 5th zone characterised by a $\theta_e = -24$ °C [20]. The city of Timișoara is located in the second climatic zone defined by an exterior temperature in the winter period $\theta_e = -15$ °C. The annual average temperature in Timișoara is 11.4 °C, with an average exterior relative humidity of 72.1% [21]. The coldest winter day is defined by a daily average minimum exterior temperature of −12.6 °C. For the summer period, the average maximum daily exterior temperature is 29 °C.

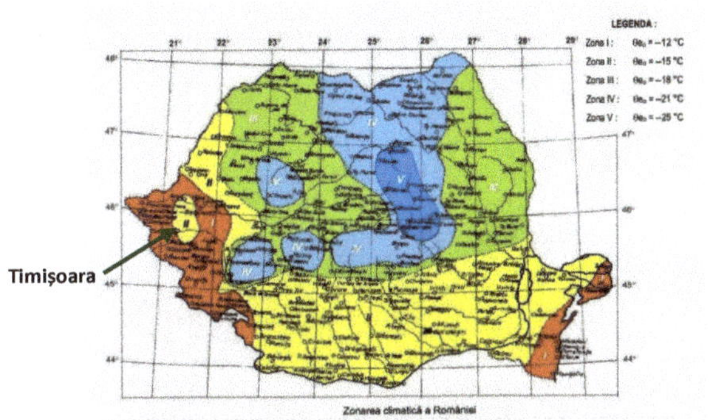

Figure 3. Location of the case study considering the Romanian map for the winter climatic zones. Reprinted from [20].

Buzatu et al. [21] have conducted an analysis with the aim of identifying the impact of Timisoara's climate in terms of internal comfort conditions and energy consumption. The author concluded that according to IWEC weather data and ASHRAE 55 prescriptions in the case of a residential building located in Timişoara at which no design strategies (i.e., heating cooling, natural ventilation or fan-forced ventilation cooling, humidification, dehumidification, shading device, and others) are considered, only 14% (i.e., 1226 h) of the yearly hours are indoor comfortable. Therefore, in order to ensure indoor comfort for a larger period (i.e., over 90% of the annual 8760 h), one must consider several design strategies to provide 7047 h of heating and humidification, along with 387 h cooling and dehumidification (if needed). This leads to a significant increase in the energy demand for the entire year and the building's lifespan. Therefore, integrating several passives and active design strategies for the examined case leads to an annual heating and humidification demand of 4424 h, along with an annual of 31 h for cooling and dehumidification (if needed). That is translated into a reduction of 38% for the annual heating hours and 92% for the annual cooling hours.

One of the passive approaches on which the paper focuses consists of how well the building envelope design can provide proper thermal insulation levels with respect to the thermal performance levels stipulated by design norms [19].

2.2. The Experimental Module

The experimental module, presented in Figures 4 and 5, was designed following sustainability design criteria: material and resources procurement and efficiency, health and well-being, energy and cost-efficiency. Some of the sustainability aspects of LSF constructions include the speed of construction, possibility of prefabrication, architectural flexibility in building retrofit, small weight with increased mechanical strength, significant potential for recycling and reuse, transportation and handling cost savings, and others [13,22–24].

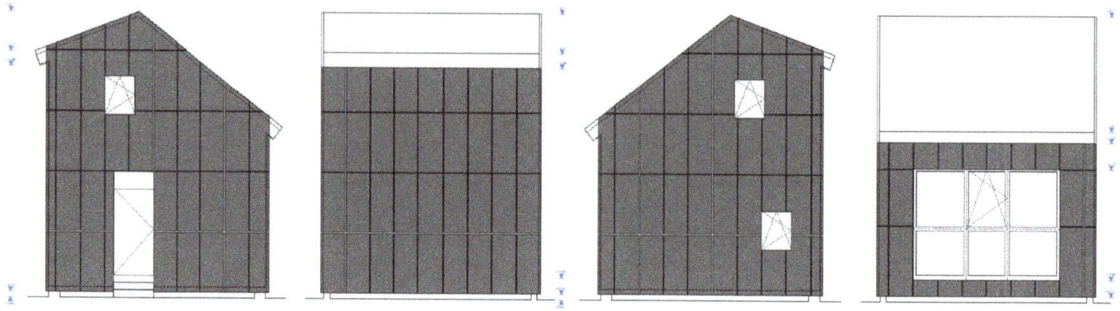

(**a**) (**b**) (**c**) (**d**)

Figure 4. Exterior views of the designed experimental module: (**a**) E orientation; (**b**) N orientation; (**c**) W orientation; (**d**) S orientation.

Figure 5. South view of the built LSF experimental module.

The LSF structure is a two-story modular construction, with a 5 m long span, a 5 m long bay, 3.80 m eave height (on the southern side), and 6.10 m eave height (on the northern side) 6.95 m ridge height (see Figure 4). The eastern façade has two 0.76 m × 0.96 m window openings, the southern façade integrates a 3.56 m × 2.73 m glass curtain opening, while the western façade has a 0.76 m × 0.96 m window opening and a 0.97 m × 2.73 m door opening. There are no openings on the northern side of the building. The access to the second floor is ensured by a 1 m × 1 m attic scuttle door. External photo-voltaic shading lamellae will protect the curtain wall from the sun. The southern side of the roof was designed with a roof pitch of 42° to gain an optimal performance of a roof-mounted solar energy system.

A precast wedge foundation system was adopted, designed as a quick foundation system, easy to handle and install, fully recoverable at the end-of-life of the building and suitable for reuse [25].

Several studies were made to identify the proper materials for the building envelope elements from a holistic design perspective and ease for deconstruction and future reuse of components [26,27]. The chosen thermal insulation is the recycled-PET thermal wadding, fabricated using polyester fibres recycled from post-consumer polyethylene terephthalate

(PET) bottles. This material has a low environmental impact [22,26], high mechanical resistance, and good physical properties [28]. Another reason for this choice was to stimulate the local economy and the recycling and reuse of materials.

Table 1 presents the thermophysical characteristics of the materials used in the LSF experimental module.

Table 1. Thermal properties of the building envelope materials.

Material	Thermal Conductivity [W/(m·K)]	Specific Heat [J/kg·°C]	Density [kg/m^3]
Steel profiles (C150/3, C200/3)	50.00	420	7800
OSB [1]	0.130	1700	620
Recycled-PET [2] thermal wadding	0.054	1350	20
Wood fibreboard	0.050	2100	270
Vapor barrier	0.220	1700	130
Aluminium sheet	160.00	880	2800
XPS [3]	0.035	1450	35
PIR [4] sandwich panel	0.023	1400	30

[1] OSB: oriented strand board; [2] PET: polyethylene terephthalate; [3] XPS: extruded polystyrene; [4] PIR: polyisocyanurate.

The structure is proper for various building envelope configurations. The unidirectional thermal resistances R_{tot} and unidirectional thermal transmittances U_{tot} were calculated based on the constructive details for the envelope components presented in Figure 6. The solutions were chosen with respect to local sourcing and production of building materials, thus reducing transport emissions and associated costs.

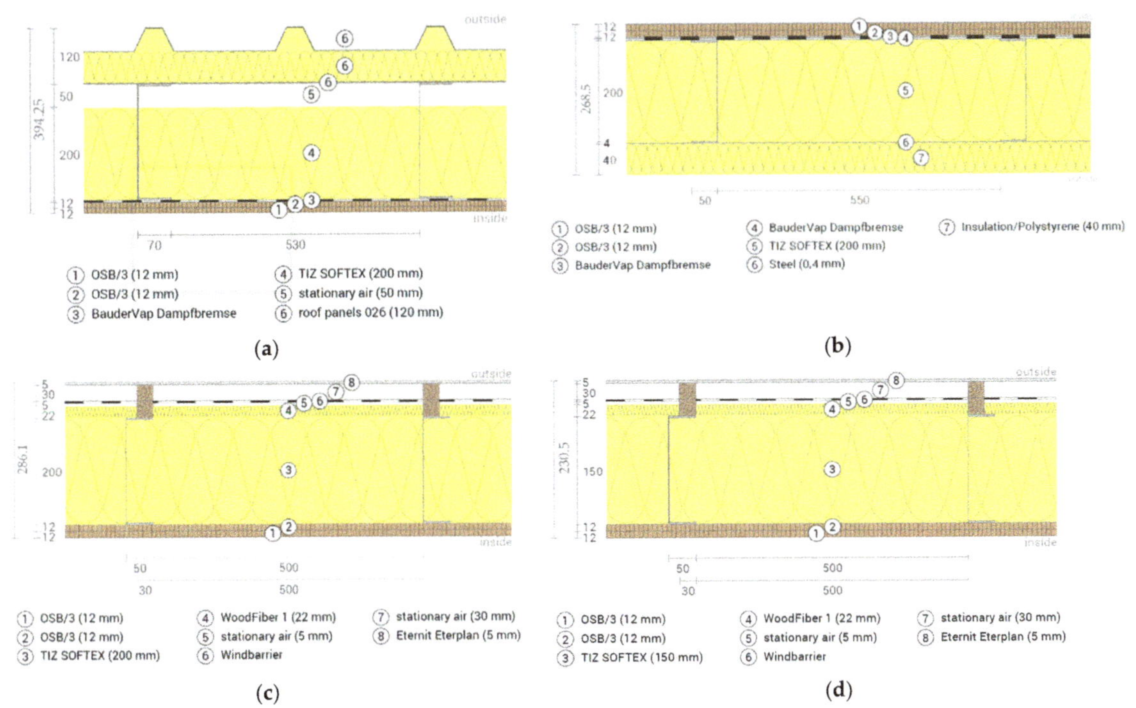

Figure 6. LSF construction elements cross section: (**a**) roof; (**b**) ground floor above crawl space; (**c**) exterior wall north; (**d**) exterior wall—east and west. Reprinted from [22].

An extended presentation of the details from Figure 6 is provided in the next paragraph. Oriented strand board (OSB) panels (24 mm thick) were used as an inner sheeting layer for walls, ceilings and floors. As a thermal insulation material, the recycled-PET thermal wadding was placed between the steel frame, with a thickness of 150 mm, except for the north exterior wall, with 200 mm thick. The exterior wall constructive detail includes a layer of wood fibreboards of 22 mm thick next to the recycled PET thermal wadding, finished by a layer of fiber cement plate of 5 mm thick. The ground floor was elevated 400 mm from the terrain, avoiding moisture retention from the ground. The constructive detail is designed with 200 mm thick recycled PET thermal wadding. A trapezoidal steel sheet of 4 mm thick is placed underneath, and another continuous exterior layer of 40 mm of extruded polystyrene (XPS) closes the element at the exterior bottom side. Polyvinyl chloride (PVC) membranes waterproofed both the floor and roof. The thermal insulation system was completed on the roof in the exterior with PIR sandwich panels of 120 mm thick.

The experimental module was designed following a holistic approach to significantly reduce the energy demand of the building in its operation phase. Therefore, the building envelope design is crucial to meet this objective. At the same time, the design considered the natural light intake and the additional artificial lighting covered by the available LED light sources. The employed renewable energy solutions are based on harvesting solar and wind energy. Therefore, there were installed twelve 250 W polycrystalline cell panels that intake solar energy, with an estimated amount of solar energy produced on-site of 1269 kWh/year (the potential production of the installed polycrystalline cell panels under ideal conditions is 3427.29 kWh/year [26]), and a 1 kW vertical wind turbine.

The LSF experimental module includes a monitoring energy management system that offers a solid overview of the building's performance during the operational phase. The module's functioning is based on the energy provided by the on-site generation technologies, the construction being a non-grid connected building. The data acquisition infrastructure consists of 3 CO_2 sensors, 14 humidity sensors and 53 temperature sensors distributed, as previously presented in reference [22]. Sensors were placed on the inner and outer face of the exterior wall and between the layers.

3. Materials and Methods

3.1. Numerical Approach

It is well known that heat transmission increases significantly through the steel components areas; therefore, even in a thermal insulation layer, the steel element acts as a strong thermal bridge. This phenomenon leads to significant reductions in the global thermal resistance of the building element. Ignoring the negative effect that steel has on the thermal performance of the building envelope can lead to an overestimation of the thermal resistance by up to 50%, as mentioned by Gorgolewski [23]. Simultaneously, the improper temperature profile in the mass of the element can lead to adverse effects, such as condensations and wall staining that occurs on cold spots.

The EN ISO 6946 standard [29] offers the combined method for calculating such constructive details, also known as the simplified method. An upper and lower limit for the thermal resistance is established, also defined as the parallel path and isothermal path method [30]. Based on these two values, their average gives the final thermal resistance. This method is considered to be a simplified one. Other calculation methods were developed based on the presented approach. Nevertheless, an essential prescription of ISO 6946, does not allow the calculation of the thermal performance of a wall with an insulation layer crossed by metal studs due to the increased difference between the upper and lower resistance values which define the combined method.

However, before obtaining the adjusted thermal resistance value R' as defined by the Romanian norms [31], the unidirectional total thermal resistance R_{tot} must be calculated using Equation (1), as prescribed by EN ISO 6946 [26]:

$$R_{tot} = R_{si} + \sum_{j=1}^{n} \frac{d_j}{\lambda_j} + R_{se} \quad (1)$$

where R_{si} and R_{se} are the interior and exterior surface resistances [(m^2·K)/W], d is the thickness of a homogenous layer [m] and λ is the thermal conductivity of the material [W/(m·K)].

The R_{tot} provides an image of the constructive details' thermal performance that describes the building envelope element without considering the weak thermal areas, i.e., assuming homogeneous layers.

As previously mentioned, each building envelope component is also defined by areas where the heat flow increases due to the significant differences between materials' thermal conductivities. Therefore, the calculations for establishing the adjusted thermal resistance value need to be done at least following a 2D modelling and simulation approach [32], also known as the detailed calculation method. A 3D approach is sometimes mandatory for complex steel junctions to accurately identify the thermal performance [33,34]. Therefore, in order to get an accurate understanding of the thermal performance of the assessed constructive details, extensive numerical modelling and simulation approach was employed, following the prescriptions of the EN ISO 10211 standard [35].

The numerical computation tool used for the analysis is the 2D software called PSI-PLAN [36], which is based on solving the plane heat transfer differential equation in steady-state thermal regime:

$$\frac{\partial}{\partial x} \cdot \left[\lambda(x,y) \cdot \frac{\partial \theta(x,y)}{\partial x} \right] + \frac{\partial}{\partial y} \cdot \left[\lambda(x,y) \cdot \frac{\partial \theta(x,y)}{\partial y} \right] = 0 \quad (2)$$

where θ is the temperature in the node (x,y), and $\lambda(x,y)$ has constant values for the materials describing the detail.

Based on a 2D assessment approach, one can establish the value of the linear heat transfer coefficient defined by ψ [W/(m·K)] in order to identify how well from the thermal point of view, the constructive details and implicitly the junction was designed. The coefficient is calculated as described next [35]:

$$\psi = L_{2D} - \sum_{j=1}^{N} U_j \cdot l_j \quad (3)$$

where L_{2D} is the two-dimensional thermal coupling coefficient [W/m·K], U_j is the unidirectional thermal transmittance [W/(m^2·K)] (i.e., the opposite of R_{tot}) of the component j separating the two environments defined by the internal and external temperature, l_j is the length [m] of the two-dimensional geometrical model over which the U_j value is applied. The ψ value was calculated considering the overall internal dimensions [31].

For the intended research, the ψ value was calculated for each junction as a total value for the assessed junction and also divided in two values, i.e., ψ_1 and ψ_2, each being allocated to one of the modelled wings of the junctions. The need to calculate the two ψ values is following the next calculation steps when the thermal performance of each building envelope element is calculated. This approach is defined in Annex G and Annex J of the Romanian design norm C107/3 [31], which describes the proper approach in reaching the two values. An example of this approach is given in Figure 7.

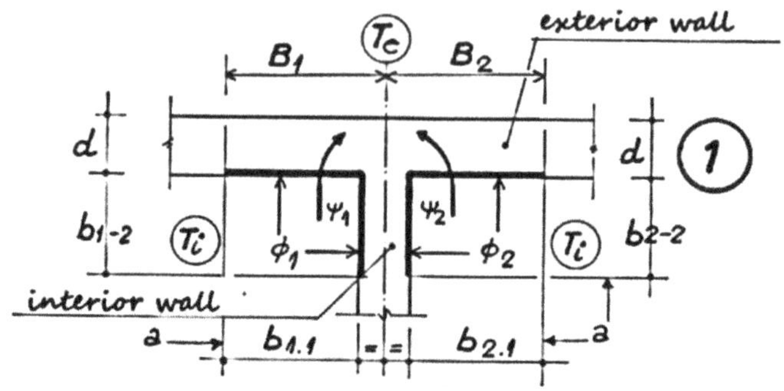

Figure 7. ψ value calculation following the Annex J approach from C107/3. Adapted from [31].

The thermal coupling coefficient L_{2D} is a significant parameter because it shows the heat losses through a building component caused by the temperature difference between the two environments in direct contact with the element. The L_{2D} was obtained based on the heat flow rate ϕ_l [W/m] resulted from the bi-dimensional calculation divided to the temperature difference. The next formula was applied:

$$L_{2D} = \frac{\phi_l}{(\theta_i - \theta_e)} \tag{4}$$

where θ_i is the internal temperature and θ_e the external temperature.

Knowing the ψ value, the adjusted thermal transmittance U' [W/(m^2·K)] of the component is calculated [31]. The adjusted value considers all the weak thermal areas of the defined constructive detail by introducing in the calculation formula the ψ values. The Romanian design norms define a minimum imposed value for the adjusted thermal transmittance and the adjusted thermal resistance for each constructive element that defines the building envelope [19]. The minimum imposed values represent one of the design indicators that must be met in designing new buildings and the energy retrofit of the existing ones. The formula for the U' calculations is given next:

$$U' = \frac{1}{R'} = \frac{1}{R_{tot}} + \frac{\Sigma(\psi \cdot l)}{A} + \frac{\Sigma\chi}{A} \tag{5}$$

where A is the area of the assessed element of the building envelope [m^2], l is the length [m] over which the linear heat transfer coefficient ψ is applied, χ is the point thermal transmittance [W/K] obtained through a 3D simulation, and R_{tot} and ψ were previously defined. The included ψ and χ values are only for the linear and point thermal bridges identified over the A surface of the element. Considering that the approach for the assessed case is 2D, the point thermal transmittance was not considered in calculations. Based on the U' obtained value, the adjusted thermal resistance R' is calculated as an opposite of U'.

The boundary conditions were set for the internal and external environment according to EN ISO 6946 [29]. The surface thermal resistances considered in calculations were R_{si} = 0.13 (m^2·K)/W, and R_{se} = 0.04 (m^2·K)/W. The interior temperature considered in calculations was θ_i = +20 °C, while for the exterior temperature θ_e the studied junctions were modelled in four of the representative climatic zones in Romania starting with a temperature of θ_e = −12 °C for the 1st climatic zone up to θ_e = −21 °C for the 4th climatic zone [20,21]. The aim was to identify the temperature distribution variation of the assessed junctions in each of the four climatic zones. Nevertheless, for the calculation of the ψ, R' and U' values in the bi-dimensional numerical simulations, the temperature difference had a unitary value according to ISO 10211 prescription [35].

The data input in the PSIPLAN software is done graphically by using a graphical module. The spatial geometrical and thermotechnical characteristics, the boundary conditions defined by the superficial thermal resistances, the interior and exterior temperature and relative humidity represent the input data in the program. The meshing of the thermal bridge junction is performed, and with the help of the finite difference method, the temperature values in each node of the discretization network are obtained. The modelling stipulations mentioned in EN ISO 10211 [35] are used.

The program performs the meshing automatically, with respect to the numerical validation as mentioned in Annex C of the Reference [35], point C.2. l that describes the approach for the number of subdivisions and C.2. m that discusses the mandatory convergence value.

As it was previously mentioned in our papers [37–39], the modelling software follows code prescription regarding the calculation of the linear thermal transmittance ψ [W/(m·K)] as well as the design temperature factor at the internal surface f_{Rsi}, also known as the condensation resistance factor [35]:

$$f_{Rsi} = \frac{\theta_{si,min} - \theta_e}{\theta_i - \theta_e} \quad (6)$$

where: $\theta_{si,min}$ is the minimum superficial interior temperature, θ_i is the interior air temperature, and θ_e is the exterior air temperature.

The need to establish the f_{Rsi} value is connected to the aim of identifying the thermal bridge performance. A smaller ψ value indicates a decrease in heat losses and a reduced risk for mold growth, resulting in a higher f_{Rsi} value. Nevertheless, compliance with the mould growth criteria does not necessarily mean a minimized heat flow. Situations are often met in practice when a thermal bridge indicated increased transmission losses, although the values were compliant in terms of mould control.

Based on the obtained results, the global thermal insulation coefficient denoted by G [W/(m^3·K)] can be calculated, according to design norm C107 [31]. The G parameter is the first index that provides an overall image of the thermal performance of the building envelope. Therefore, for residential buildings, the following formula is applied:

$$G = \frac{\Sigma(L_j \cdot \tau_j)}{V} + 0.34 \cdot n \quad (7)$$

where L is the thermal coupling coefficient [W/K], τ_j is the temperature correction coefficient [-], V is the volume of the building envelope [m^3], 0.34 is the ratio between the air density and the specific heat of the air at θ_i = 20 °C [Wh/(m^3·K)], n is the number of air changes per hour due to natural ventilation [h^{-1}] [31]. Its value is compared to the normed value denoted by GN given in reference [19], a value accepted as a maximum for a given case. The reference value for residential buildings is considered in calculations n = 0.5 [h^{-1}].

The L thermal coupling coefficient value is calculated for each element of the building envelope component by using the following formula:

$$L = \frac{A}{R'} \quad (8)$$

where A is the area of the element of the building envelope [m^2], R' is the adjusted thermal resistance of the building envelope element [(m^2·K)/W] calculated as a reversed U' value based on Expression (4). The L value is similar to the transmission heat transfer coefficient H_{tr} coefficient [W/K] that is calculated with the following formula [40],

$$H_{tr} = H_D + H_g + H_u + H_a \quad (9)$$

where H_{tr} is the transmission heat transfer coefficient [W/K], H_D is the direct heat transfer coefficient between the heated or cooled space and the exterior through the building envelope [W/K], H_g transfer coefficient through the ground [W/K], H_u is the transmission heat transfer coefficient through unconditioned spaces [W/K], H_a is the transmission heat transfer coefficient to adjacent buildings, [W/K].

In the assessed case $L = H_{tr} = H_D$, where H_D is defined by:

$$H_D = \sum_i A_i \cdot U_i + \sum_k l_k \cdot \psi_k + \sum_j \chi_j \tag{10}$$

where all the parameters are the ones previously defined. The only difference between H_D. and L is that the summations are done over all the building components separating the internal and the external environments for H_D. In the case of L calculation, the value is calculated per building envelope component, and at the end, all L values are added up for the entire building envelope

The temperature correction coefficient τ_j is calculated by [31]:

$$\tau_j = \frac{\theta_i - \theta_j}{\theta_i - \theta_e} \tag{11}$$

where θ_i is the conventional interior temperature for calculation for each space of the assessed building, while θ_j can either be the exterior temperature or the temperature of the unconditioned interior space. θ_e is the exterior temperature according to the climatic zone, for Timişoara being equal to -15 °C. The aim of τ is to bring a correction for each temperature difference identified for the assessed building by dividing it to the predominant temperature difference described by $(\theta_i - \theta_e)$. Thus, the total coupling coefficient H_{tr} can be multiplied with the predominant temperature difference to obtain the total heat transmission as described in Equation (12). The τ coefficient is similar to the b coefficient as described by ISO 52010-1 [41].

Further on, one can calculate $Q_{H;tr}$ [kWh] as given in ISO 52010-1 [41], describing the total heat transfer by transmission through the building envelope, as described below:

$$Q_{H;tr} = H_{H;tr} \cdot (\theta_{int,calc,H} - \theta_e) \cdot t \tag{12}$$

where $\theta_{int;calc;H}$ is the calculation temperature of the zone for heating [°C], t is the duration for the entire heating period, in [h].

The total heat transfer by ventilation $Q_{H;ve}$ [kWh] and the total heat gains $Q_{H,gn}$ described by the solar and internal gains are calculated following reference [42]. The energy need for heating is calculated using Equation (13), as it follows:

$$Q_{H;nd} = Q_{H;ht} - \eta_{H;gn} \cdot Q_{H;gn} \tag{13}$$

where $Q_{H;nd}$ is the energy need for heating defined by the sum of $Q_{H;tr}$ and $Q_{H;ve}$ [kWh], $\eta_{H;gn}$ is the dimensionless gain utilization factor [-].

The energy consumption for heating from heat delivery to heat production as well the energy rating is established based on Equation (14) and the energy scale for heating (i.e., as defined by the Romanian norm) [43]:

$$q_{heat} = \frac{Q_{H;nd}}{\eta \cdot A_u} \tag{14}$$

where η is the efficiency of the heating systems [-], A_u is the useful heated area of the building [m^2]. The heating system's efficiency takes into account losses from heat delivery, regulation, temperature layers, and heat distribution.

3.2. Modelled Cases

To evaluate the thermal performance of the building envelope elements, the representative junctions in the layouts and vertical sections of the building needed to be identified.

Figure 8 provides a detailed image of the building envelope components in layout and cross-sectional view.

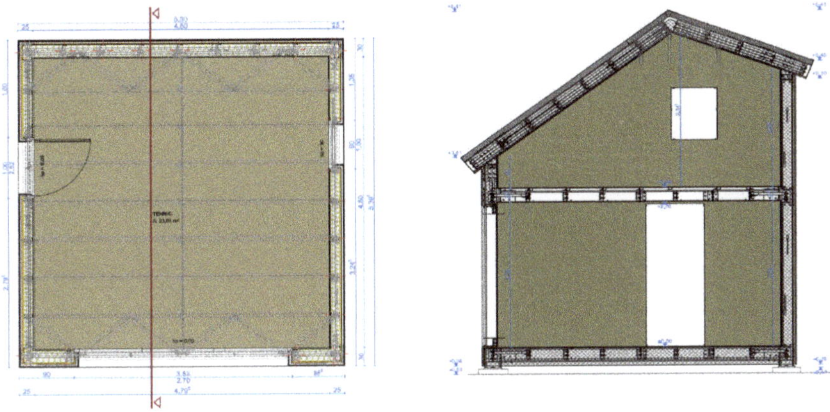

(a) (b)

Figure 8. Detailed view of the building envelope: (**a**) ground floor layout; (**b**) vertical cross-section.

Based on them, the more thermally permeable junctions were identified, as illustrated in Figure 9. The main constructive details were defined, and the geometrical models were created following each layer and component, their dimensions and thermal conductivities. At the same time, the modelling lengths, as well as the boundary conditions, were set.

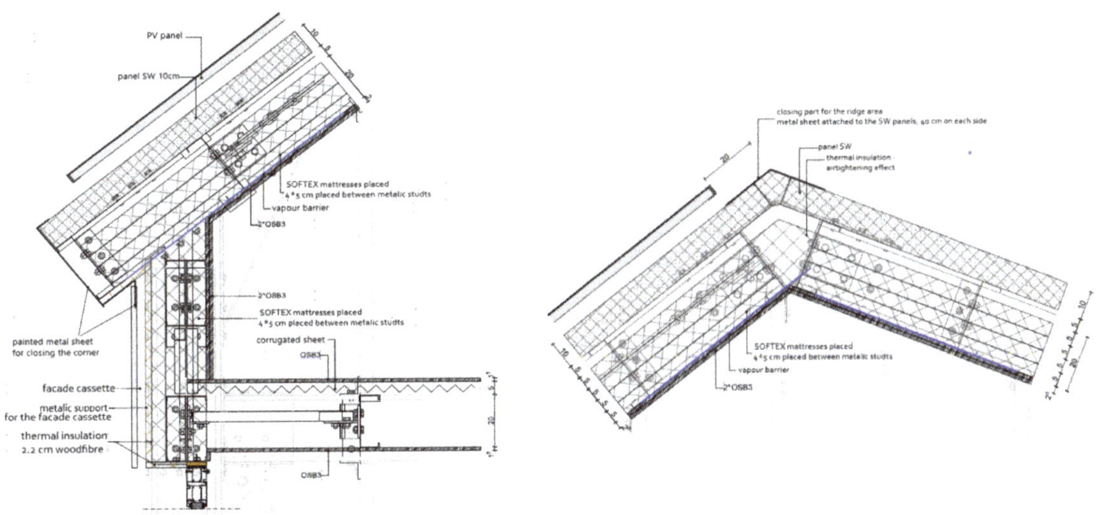

(a) (b)

Figure 9. *Cont.*

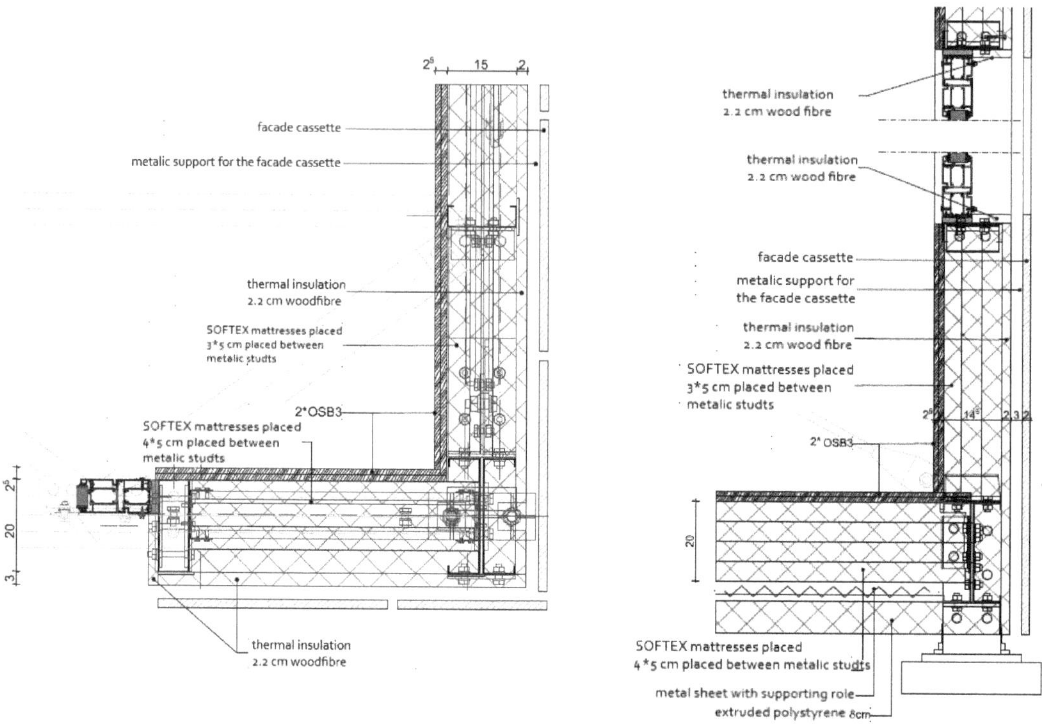

(c) (d)

Figure 9. Constructive details at junctions: (**a**) Roof—exterior wall-intermediate floor; (**b**) Ridge zone; (**c**) Exterior corner—window connection, (**d**) Exterior wall—floor in contact with the exterior.

4. Results and Discussion

4.1. Thermal Performance per Element

Table 2 displays the LSF building envelope elements, such as materials, thicknesses, number of layers. The thermal conductivities for each material layer are given in Table 1. The adjusted thermal resistance and adjusted thermal transmittance were calculated at an intermediate stage of development of the CIA_CLIM project [12] using simplified preliminary calculations as provided by the Ubakus tool [42]. The considered constructive details are presented in Figure 6.

Table 2. Materials, thicknesses (d), adjusted thermal resistance (R') and adjusted thermal transmittances (U').

Element	Material Layers (from Inside to Outside)	d [mm]	R'-Value [(m^2·K)/W]	U'-Value [W/(m^2·K)]
Ground floor above the crawl space	OSB	24	3.677	0.272
	Recycled-PET thermal wadding TIZ SOFTEX	200		
	Steel sheet	4		
	XPS	40		
	Total thickness	268.5		

Table 2. *Cont.*

Element	Material Layers (from Inside to Outside)	d [mm]	R'-Value [(m^2·K)/W]	U'-Value [W/(m^2·K)]
Exterior walls (north)	OSB	24	3.185	0.314
	Recycled-PET thermal wadding-TIZ SOFTEX	200		
	Wood fibreboard	22		
	Stationary air	5		
	Wind barrier			
	Rear ventilated level (outside air)	30		
	Fiber cement plate	5		
	Total thickness	286.1		
Exterior walls (east and west)	OSB	24	2.817	0.355
	Recycled-PET thermal wadding TIZ SOFTEX	150		
	Wood fibreboard	22		
	Stationary air	5		
	Wind barrier			
	Rear ventilated level (outside air)	30		
	Fiber cement plate	5		
	Total thickness	230.5		
Roof	OSB	24	5.208	0.192
	Recycled-PET thermal wadding TIZ SOFTEX	200		
	Stationary air	50		
	PIR sandwich panel	120		
	Total thickness	394.25		
Door and windows	Glass with argon filling	24	1.136	0.880
	PVC casement	92		
Glass Curtain	Glass with argon filling	44	1.351	0.740
	PVC casement	92		

A second analysis included the 2D comprehensive numerical approach. The results are the unidirectional thermal resistance R_{tot}, adjusted thermal resistance R', adjusted thermal transmittance U', the linear thermal transmittance ψ, and the temperature factor f_{Rsi}. The results for the 2D approach using PSIPLAN software are listed in Table 3 for all the analysed junctions. The modelled and simulated results are presented in Figures A1–A8 in the Appendix A.

As mentioned in Section 3.1, the adjusted thermal resistance R', adjusted thermal transmittance U', the linear thermal transmittance ψ and the temperature factor f_{Rsi} were also calculated for each wing, especially for the junction defined by two different building envelope components. The results per modelled wing are presented in Table 4.

As one can see in Tables 3 and 4, the adjusted values per element vary according to the length of the modelled case and modelled wing, constructive detail and position in the building envelope. Starting from 2022, a design requirement that will come into force is that the average ψ value for the building envelope must meet the design criteria $\psi_m \leq 0.15$ W/(m·K) as described in Reference [44]. In this regard, all assessed junctions meet the design criteria with one exception for the Roof eaves—exterior wall intersection. Nevertheless, the average result for ψ_m is met for the building envelope. However, although the reference value was met, it does not implicitly ensure the recommended adjusted thermal resistance.

Table 3. Modelled and simulated junctions-results for the entire length of the modelled cases.

Modelled Cases	R_{tot}-Value [(m^2·K)/W]	R'-Value [(m^2·K)/W]	U'-Value [W/(m^2·K)]	ψ-Value [W/(m·K)]	f_{Rsi}-Value [-]
Exterior wall current field (east and west)	3.660	2.637	0.379	0.066	0.802
Exterior walls current field (north)	4.586	2.981	0.335	0.073	0.823
Exterior corner	4.071	2.059	0.486	0.120	0.690
South Exterior wall-curtain wall left margin	4.586	1.748	0.572	0.136	0.827
South Exterior wall-curtain wall right margin	4.586	1.6542	0.218	0.148	0.826
Exterior wall-window	3.660	1.841	0.543	0.104	0.796
Curtain glass—Ground floor above the crawl space	2.053	1.714	0.583	0.072	0.748
Exterior wall (E,W)—Ground floor above the crawl space	4.584	2.817	0.355	0.082	0.807
Exterior wall (N)—Ground floor above the crawl space	5.247	3.140	0.319	0.077	0.822
Ground floor above the crawl space—current field	6.453	3.951	0.253	0.034	0.918
Exterior wall (E,W)—intermediate floor	3.660	2.476	0.404	0.101	0.922
Exterior wall (N)—intermediate floor	4.586	3.454	0.290	0.055	0.945
Roof—current field	7.987	5.924	0.169	0.0216	0.953
Roof ridge	7.987	3.981	0.251	0.086	0.943
Roof eaves—exterior wall	5.621	2.845	0.351	0.170	0.938

Table 4. Modelled and simulated junctions-results obtained separately per wing of the modelled junctions.

Modelled Cases	Subdivisions	R_{tot}-Value [(m^2·K)/W]	R'-Value [(m^2·K)/W]	U'-Value [W/(m^2·K)]	ψ-Value [W/(m·K)]	f_{Rsi}-Value [-]
Curtain glass—Ground floor above the crawl space	Curtain glass wall	1.351	1.384	0.723	−0.007	0.748
	Ground floor above the crawl space	6.132	2.467	0.405	0.079	0.636
Exterior wall (E,W)—Ground floor above the crawl space	Exterior wall (E,W)	3.660	2.510	0.398	0.038	0.807
	Ground floor above the crawl space	6.132	3.211	0.311	0.045	0.777
Exterior wall (N)—Ground floor above the crawl space	Exterior wall (N)	4.586	2.897	0.345	0.038	0.794
	Ground floor above the crawl space	6.132	3.426	0.292	0.039	0.822
Roof eaves—exterior wall	Roof eaves	7.987	2.510	0.398	0.116	0.895
	Exterior wall	4.586	3.167	0.316	0.054	0.938

Therefore, using the ψ values for each element, Equation (5) and the unidirectional thermal resistances, the final overall adjusted thermal resistances R' for each building envelope element was calculated.

Figure 10 offers an overview of the obtained results, the final R' per building envelope element and the range of R' for the modelled cases, plotted against the normed values for residential buildings according to reference [19]; i.e., exterior walls R'_{min} = 1.80 (m^2·K)/W, roof R'_{min} = 5.00 (m^2·K)/W, the ground floor above the crawl space R'_{min} = 4.50 (m^2·K)/W. As it can be observed, although the range of $R'_{modelled}$ obtained values has a greater domain for each element, the final R' results are achieved just for the exterior wall. In the case of glazing surfaces, the R'_{min} = 0.77 (m^2·K)/W is satisfied by the curtain wall, the windows and the door. Regardless of each element's significantly high thermal resistance in its current field, due to the complexity of the details that include steel joining components and the thermal interaction between different elements, a reduction effect of the thermal performance is obtained.

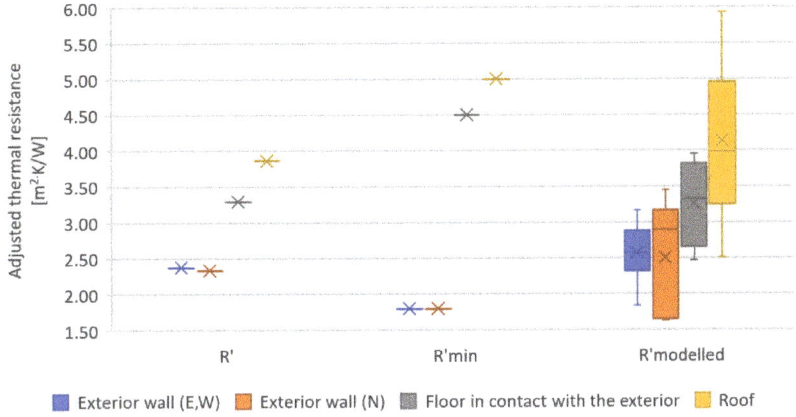

Figure 10. Adjusted thermal resistance comparison: opaque elements.

In the light of reaching the new nZEB criteria that will come into force in Romania starting with 2022 [44], and the increase in the proposed R' values, the design approach must be focused on reaching a unidirectional thermal resistance of at least a R = 10 ($m^2 \cdot K$)/W for horizontal or inclined elements and a R = 7 ($m^2 \cdot K$)/W for the exterior walls. In this way, after applying the linear and point heat transfer coefficients, the imposed R' values, respectively U' values, are met. A significant increase in the unidirectional thermal resistance value can be reached by thicker thermal insulation or by using super insulating materials, i.e., vacuum insulation panels and aerogel insulation [45–47], preponderantly recommended for LSF buildings.

Regarding the f_{Rsi} factor, the value is considered respected when the one obtained from calculations is greater or equal to a set value [48].

$$f_{Rsi} \geq f_{Rsi,limit} \quad (15)$$

Design criteria may vary from country to country but, in general, a lower limiting value of 0.7 is accepted for the temperature factor to reduce the risk of mould and condensation growth in buildings. Thus, for Romania, a $f_{Rsi,limit}$ = 0.7 is considered as the limit value. For the analysed building, values smaller than 0.7 are identified for the floor in contact with the exterior—curtain wall joining and for the case of the exterior corner. Therefore, a thermal redesign of the two junctions could lead to a value greater than 0.7.

4.2. Thermal Performance of the Building Envelope

The thermal coupling coefficient value per element, computed as previously described in Equation (8), is presented in Table 5. The final results per building envelope are indicated in Table 6. Considering the primary function of the building, the interior temperature value at θ_i = 20 °C and the fact that all elements are in contact with the exterior environment, the calculation hypothesis was τ = 1. The geometric characteristic for each building envelope component was calculated considering the overall internal dimensions.

The R'_m and U'_m, respectively, indicate the average adjusted thermal resistances and average thermal transmittance per building envelope.

Table 5. The thermal coupling coefficient value (L) per element j.

Type	Building Envelope Element	A_{Total} [m^2]	R' [m^2 K/W]	$L_j \cdot \tau_j$ [W/K]
1	Exterior Wall	76.57	2.36	32.49
2	Roof	27.61	3.86	7.15
3	Ground floor above the crawl space	23.04	3.30	6.99
4	Curtain wall	9.53	1.35	7.06
5	Glazing surfaces	4.92	1.14	4.33

Table 6. The thermal coupling coefficient value per building envelope.

Building Envelope	A_{Total} [m^2]	R'_m [m^2 K/W]	$L_j \cdot \tau_j$ [W/K]	U'_m [W/m^2 K]
Total B. Env.	141.67	2.69	58.02	0.37

Figure 11 provides a comparison between the heat losses and the surface areas relative to the entire building envelope results. As one can see, the exterior wall is predominant from both Surface (S) and Heat Losses (HL). In the case of the curtain wall, although its area is smaller than the one of the roof and the floor in contact with the exterior environment, the quantity of heat losses is in the same range as the two mentioned elements. The glazing surfaces exhibit the smallest quantity of HL and a smaller surface.

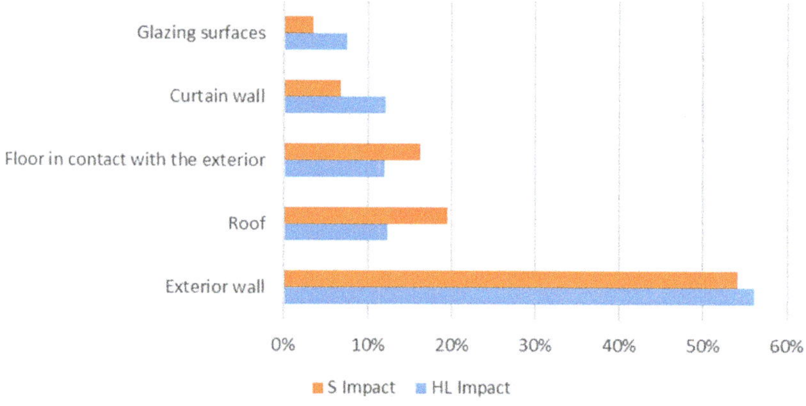

Figure 11. Overview of the Heat Losses (HL) and Surfaces (S) of building envelope elements.

In terms of the global thermal insulation coefficient, the resulting value is $G = 0.661$ W/(m^3·K), where $n = 0.5$ h^{-1} and the volume of building envelope $V = 118.11$ m^3. The resulted value is higher compared to the maximum admissible value $GN = 0.540$ W/(m^3·K) as extracted from [19]. The result follows previous results (i.e., thermal resistance), indicating the necessity of increasing the adjusted thermal resistance for the roof and floor in contact with the exterior at least equal to the normed value. Simultaneously, the results are consistent with the findings from the literature [49,50], mentioning that residential buildings' performance can be highly variable, and even similar houses could have dramatically different performance levels. Nevertheless, as it was mentioned in Reference [22], the building's envelope plays a pivotal role in reducing the energy demand for heating or cooling.

4.3. Energy Performance for Heating of the Building

In terms of energy performance, the energy consumption for heating denoted by q_{heat} is calculated, and an energy rating for heating is established following Equation (14) [43]. After evaluating the heat losses through transmission and the heat losses through ventilation, the internal heat gains and the solar gains were established. The heating system's efficiency was considered η = 85% thus resulting in a B energy class for heating (Figure 12) characterized by a q_{heat} = 110.33 kWh/(m^2·yr). It should be mentioned that the calculations do not consider the overall embodied energy of the building. Therefore, it is highlighted the need to provide energy-efficient solutions [22] described by alternative energy sources to reduce the overall energy consumption.

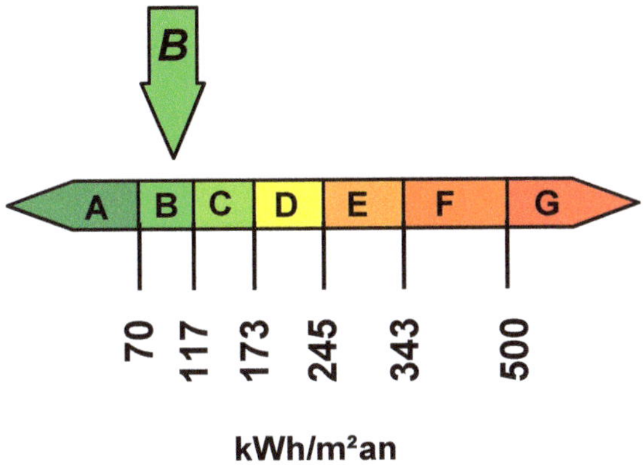

Figure 12. Energy class based on the energy consumption for heating.

4.4. Parametric Study for the Energy Performance for Heating of the Building

In order to offer an image of the energy consumption for heating of the experimental module placed in all five climatic zones from Romania, a parametric study was performed, resulting in 360 case scenarios. The data considered in the study is briefly presented in Table 7.

Table 7. The parametric data for the considered case scenarios.

Parameters	Examined Range
Climatic zone	1st, 2nd, 3rd, 4th and 5th zone
Curtain wall orientation	S, SW, W
Ventilation rate, n [h^{-1}]	0.5, 0.4, 0.3
Heating system efficiency, η [%]	75—Hot-water floor heating system 40°/30 °C 80—Water radiator 70°/40 °C with manifold 82—Electric floor heating 83—Water radiator 45°/35 °C with manifold 86—Roof heating (i.e., electric) 87—Water radiator 70°/40 °C 89—Water radiator 45°/35 °C 95—Electric heater

Due to its larger surface, the curtain wall orientation was considered the main façade from the solar gains perspective. Thus, the other two positions for the main orientation were considered (i.e., SW and W) in order to have a significant solar gain still.

For the ventilation rate n, the chosen reference value was 0.5 as prescribed by the thermotechnical design norm [31] The two other values provided in Table 7 indicate other cases with better control of the ventilation rate and implicitly of the heat losses

For the heating system efficiency, several case scenarios were considered, starting from a η = 75% specific for hot-water floor heating system 40°/30 °C up to a η = 95% that defines the case of an electric heater, as provided by the methodology Mc001 [43].

In terms of G value, the smallest result obtained for the studied cases was $G = 0.593$ W/(m^3·K), which is still higher compared to the reference value of $GN = 0.540$ W/(m^3·K).

In terms of energy performance for heating, one can observe in Figure 13 that q_{heat} for all examined cases ranges from a value greater than $q_{heat} = 70$ kWh/(m^2·yr), to one smaller than $q_{heat} = 190$ kWh/(m^2·yr). The extreme values (i.e., smallest and highest) are identified for the 1st climatic zone $q_{heat} = 71.09$ kWh/(m^2·yr), $n = 0.3$ h^{-1}, S orientation, $\eta = 95\%$ and for the 5th climatic zone $q_{heat} = 211.99$ kWh/(m^2·yr), $n = 0.5$ h^{-1}, W orientation, $\eta = 75\%$. Therefore, the energy class for heating ranges from B up to a D class for the experimental module, as illustrated in Table 8. An overview of the assessed case scenarios is also provided in Figure 14.

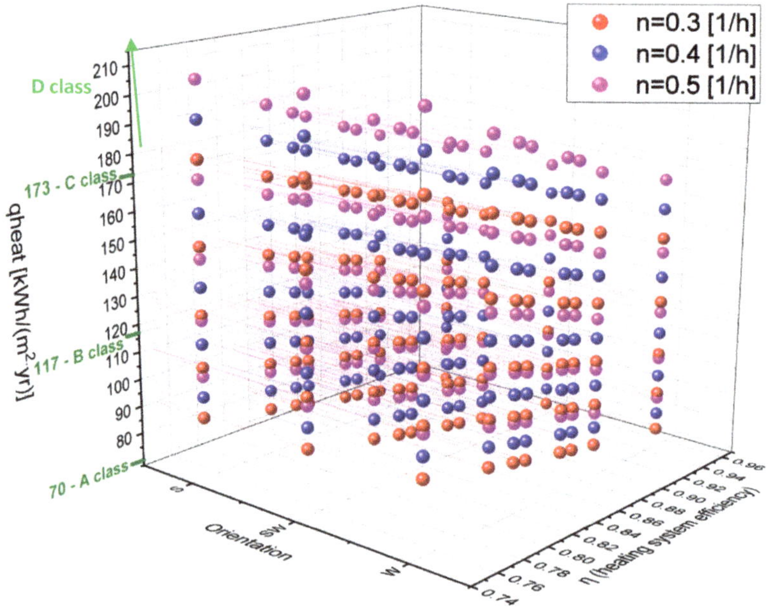

Figure 13. The energy consumption for heating for the 360 case scenarios.

Table 8. Obtained minimum and maximum values for the heating energy consumption, q_{heat}, and corresponding energy class for heating per climatic zone.

Values	**q_{heat} [kWh/(m^2·yr)]—Energy Class per Climatic Zone**				
	1st	**2nd**	**3rd**	**4th**	**5th**
Minimum	71.09—B	85.46—B	100.17—B	119.21—C	142.77—C
Maximum	109.38—B	128.62—B	150.37—C	178.40—C	211.99—D

As illustrated in Table 8, the 1st and 2nd climatic zone are the most favourable locations for the investigated building in terms of reduced energy consumption for heating. However, the A energy class is still not met.

In terms of the ventilation rate (n), the value equal $n = 0.3$ h^{-1} and a heating system efficiency $\eta = 95\%$ which define a best-case scenario for the building, results fall within B and C energy class, respectively $q_{heat} = 71.09$ kWh/(m^2·yr), 1st climatic zone, and $q_{heat} = 142.77$ kWh/(m^2·yr), 5th climatic zone.

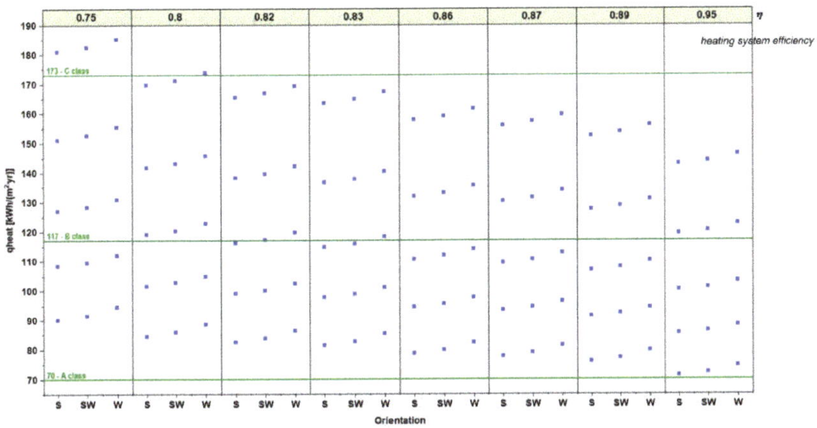

(**a**) Ventilation rate $n = 0.3\ \text{h}^{-1}$

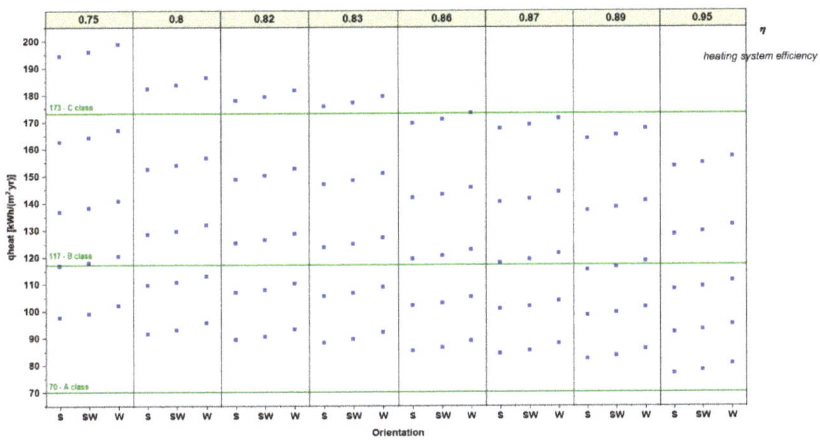

(**b**) Ventilation rate $n = 0.4\ \text{h}^{-1}$

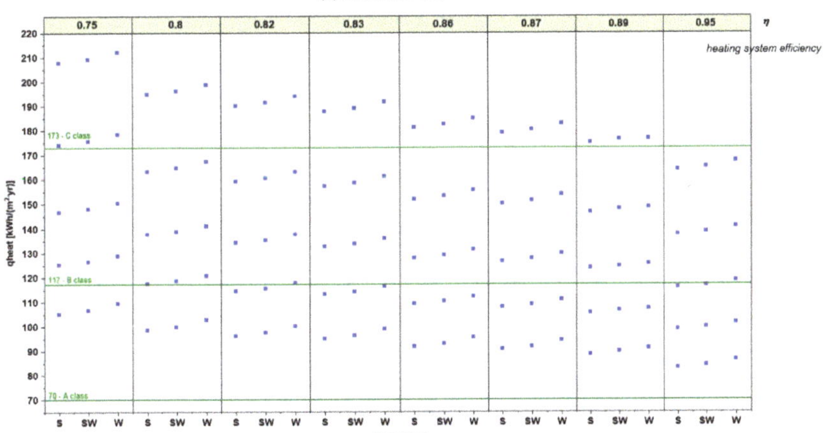

(**c**) Ventilation rate $n = 0.5\ \text{h}^{-1}$

Figure 14. The energy consumption for heating for the case scenario (**a**) $n = 0.3\ \text{h}^{-1}$, (**b**) $n = 0.4\ \text{h}^{-1}$, (**c**) $n = 0.5\ \text{h}^{-1}$ for the five climatic zones and for the heating system efficiency raging from $\eta = 75\%$ up to $\eta = 95\%$.

The South orientation is the most favourable for the considered location and façade design when looking at solar gains. Nevertheless, even for a maximum solar gain for the 5th climatic zone, in the case of a n = 0.5 h^{-1} and η = 95%, the q_{heat} = 163.99 kWh/(m^2·yr) corresponds to a C energy class for heating. In contrast, for a poorer heating system efficiency of η = 75%, the q_{heat} = 207.72 kWh/(m^2·yr) describes a D energy class for heating.

In terms of energy consumptions values for the same orientation for all assessed cases per climatic zone, the difference between the worst and the best results is around 21%. When comparing energy consumption in one climatic zone, the results vary between the considered orientations from 1% to 5%. Instead, when considering the results for the 1st climatic zone as the reference values, the increase in the energy consumptions for the same orientation starts from 16% for the 2nd climatic zone to 50% for the 5th climatic zone. Between two consecutive climatic zones for the same orientation, the values vary between 14–16%, while the values vary around 28–29% between every second climatic zone.

Thus, from the obtained results, it is clear that strategy of heat recovery for the assessed building as well as the use of alternative energy sources for energy production will help the building in meeting the existing A class defined range, as well as the upcoming values for A and A+ energy classes for heating as mentioned in Reference [44]. At the same time, the heating energy demand can be significantly reduced by an optimum orientation, in terms of optimising the passive solar gain, of the façade defined by a large glazing area (i.e., windows), a reduced ventilation rate according to the code requirements, and by selecting a heating system having a higher energy efficiency.

4.5. Parametric Study for the Building Envelope Thermal Performance Impact on the Energy Consumption for Heating

As it is previously shown in Table 5, the roof, as well as the ground floor above the crawl space, are underperforming with a direct negative impact on the energy performance consumption and energy rating. A parametrisation was done to assess the impact of each building envelope component on the energy performance in terms of heating. The parametrisation was done for the case described by a South orientation of the curtain-wall, a ventilation rate n = 0.5 h^{-1} and a heating system defined by a water radiator 70°/40 °C with a η = 87%. The considered case scenarios and the associated results are presented in Table 9.

Table 9. Results in the parametrisation of R' value for each building envelope component–correlation with the energy consumption for heating q_{heat}.

Type	Building Envelope Element	R' [m^2 K/W]				
		Ref.	(1)	(2)	(3)	(4)
1	Exterior Wall	2.36	2.36	2.36	2.36	***4.00***
2	Roof	3.86	***5.00***	3.86	***5.00***	***6.67***
3	Ground floor above the crawl space	3.30	3.30	***4.50***	***4.50***	***5.00***
4	Curtain wall	1.35	1.35	1.35	1.35	1.35
5	Glazing surfaces	1.14	1.14	1.14	1.14	1.14
	q_{heat} [kWh/(m^2·yr)]	107.79	104.70	104.25	101.14	72.12
	Reduction		3%	3%	6%	33%

The reference case scenario provides the resulting values presented in Table 5 for the as-designed and as-built building. The parametrisation does not include variations for the curtain wall and glazing surface due to already met code values. Case scenarios (1), (2) and (3) highlight the change in respect to reaching the code adjusted thermal resistances values. As demonstrated before, the roof and the ground floor above the crawl space illustrate the same thermal performance. Consequently, they display the same reduction when the code value is met, i.e., case 1 and case 2. For case scenario (3), both construction elements meet the code requirements. However, the decrease is still small compared to the reference case and the 4th case scenario, where it was considered that the proposed reference values

R' [44] are met for the opaque construction elements. All four case scenarios provide a B energy class for heating in terms of energy rating. Nevertheless, for some of the assessed case scenarios defined at 4.4, an A energy class for heating could be met for climatic zones 1 and 2 if the proposed code requirements are met.

5. Conclusions

The path toward the decarbonization of the building stock started with the recast of the Energy Performance of Buildings Directive back in 2010 and when the policies for climate and energy for 2020 were defined. This path was defined by short, medium and long-term strategies that should lead us to have an entire building stock up to nearly Zero Energy Buildings levels and even lower. Reaching nZEB levels will generate a reduction in energy consumption that will contribute to the reduction of GHG emissions resulting from the same sector. Positive results are expected from residential and non-residential buildings (i.e., governmental buildings, public buildings, office buildings).

However, achieving buildings with a notable reduced impact on the environment during the operational phase depends on the proper use of the holistic approach from the initial design stages. For the investigated case, in terms of mandatory adjusted thermal resistances (i.e., adjusted thermal transmittances), it can be concluded that the values can be met for typical vertical opaque elements (i.e., exterior walls). In contrast, a thermomechanical redesign is necessary for horizontal or inclined elements.

At the same time, the choice in thermal insulation material must be redirected towards nano-insulation materials. Thus, instead of using 20–30 or even more centimetres of typical thermal insulation, one can use a thinner layer to obtain the same thermal effect. Another aspect that must be considered is preserving the thermal properties of these materials, which might be negatively affected when penetrated by fixing components. The position of the thermal insulation in the constructive detail and the negative impact of the point thermal bridges that shape such structures must also be addressed. Although those LSF buildings indicate a proper thermal performance compared to other existing solutions, more research can be done on the passive measures by which the energy consumption can be reduced in direct correlation with the building envelope design criteria, i.e., shape, orientation, compactness, window to wall ratio, constructive details, ψ and χ magnitude, reduced negative impact caused by the area of the thermal bridge, and others.

Regarding the energy performance for heating, the results are somehow favourable only for 2 of the 5 climatic zones from Romania. However, the experimental module should be redesigned for an A and A+ energy class. Simultaneously, considering the new nZEB criteria that will come into force starting with 2022, iterative calculations will be emphasized so that both the mandatory thermal performance per building envelope element and global energy performance per type of building are achieved.

The novelty of the study lies in the fact that an extensive bidimensional numerical study was carried out to assess the building envelope's junctions. In current design practise, steel structures are assessed by employing unidirectional calculations. The resulting R_{tot} values for each building component are later adjusted by using a predefined reduction coefficient "r" that often does not comply with the assessed performance. In comparison, the bidimensional modelling and simulation of the junction can also provide the thermal interaction between various building components. Also, a comparison is made for the experimental module placed in various climatic zones from Romania. Thus, an image of the energy consumption profile can be illustrated, based on which design conclusions can be highlighted regarding the thermal design of building envelope components, orientation, ventilation rate and heating system efficiency. However, the study's limitations are the lack of experimental validation for the building envelope thermal performance at this stage and the impossibility of assessing all considered case scenarios for the heating system in real operation of the building.

Furthermore, for the examined experimental module, a monitoring phase will provide answers regarding the extent to which the employed alternative systems will impact the improvement of the total energy performance of the building. The results interpretation will indicate if the LSF prototype solution can comply with the new nZEB energy consumptions levels and the new incoming definition that will set out the nearly Zero Emissions Building.

Author Contributions: Conceptualization, L.M., P.S., I.P. and V.U.; methodology, L.M.; software, L.M.; modelling and simulation, L.M.; resources, L.M., V.U. and I.P.; writing—original draft preparation, L.M.; writing—review and editing, L.M., P.S., I.P. and V.U.; supervision, P.S., I.P. and V.U.; project administration, V.U.; funding acquisition, V.U. All authors have read and agreed to the published version of the manuscript.

Funding: This research was funded by the Romanian Ministry of Research and Innovation, CCCDI—UEFISCDI, project number PN-III-P1-1.2-PCCDI-2017-0391/*CIA_CLIM—Smart buildings adaptable to the climate change effects*, within PNCDI III and by a grant of the Romanian Ministry of Research, Innovation and Digitalization, project number PFE 26/30.12.2021, *PERFORM-CDI@UPT100- The increasing of the performance of the Polytechnic University of Timișoara by strengthening the research, development and technological transfer capacity in the field of "Energy, Environment and Climate Change" at the beginning of the second century of its existence*, within Program 1—Development of the national system of Research and Development, Subprogram 1.2—Institutional Performance—Institutional Development Projects—Excellence Funding Projects in RDI, PNCDI III.

Institutional Review Board Statement: Not applicable.

Informed Consent Statement: Not applicable.

Data Availability Statement: The data presented in this study are available on request from the corresponding author.

Acknowledgments: This work was supported by a grant of the Romanian Ministry of Research and Innovation, CCCDI—UEFISCDI, project number PN-III-P1-1.2-PCCDI-2017-0391/*CIA_CLIM –Smart buildings adaptable to the climate change effects*, within PNCDI III and by a grant of the Romanian Ministry of Research, Innovation and Digitalization, project number PFE 26/30.12.2021, *PERFORM-CDI@UPT100- The increasing of the performance of the Polytechnic University of Timișoara by strengthening the research, development and technological transfer capacity in the field of "Energy, Environment and Climate Change" at the beginning of the second century of its existence*, within Program 1—Development of the national system of Research and Development, Subprogram 1.2—Institutional Performance—Institutional Development Projects—Excellence Funding Projects in RDI, PNCDI III.

Conflicts of Interest: The authors declare no conflict of interest. The funders had no role in the design of the study; in the collection, analyses, or interpretation of data; in the writing of the manuscript, or in the decision to publish the results.

Appendix A

Figure A1. Exterior wall current field-Geometrical models (**a**) east and west, (**c**) north; Isothermal surfaces (**b**) east and west, (**d**) north.

(a) (b)

Figure A2. Exterior corner: (**a**) Geometrical models; (**b**) Isothermal surfaces.

(a) (b)

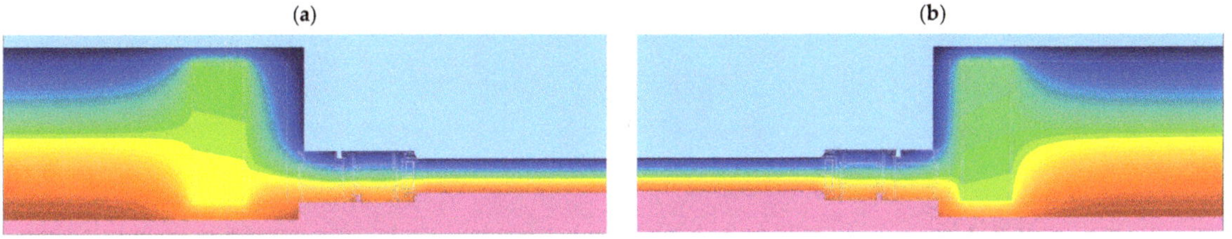

(c) (d)

Figure A3. South exterior wall-curtain wall horizontal connection. Geometrical models: (**a**) left margin; (**b**) right margin; Isothermal surfaces: (**c**) left margin; (**d**) right margin.

(a) (b)

Figure A4. Exterior wall-window horizontal connection: (**a**) Geometrical models; (**b**) Isothermal surfaces.

(**a**) (**b**)

Figure A5. Curtain glass-floor in contact with the exterior: (**a**) Geometrical models; (**b**) Isothermal surfaces.

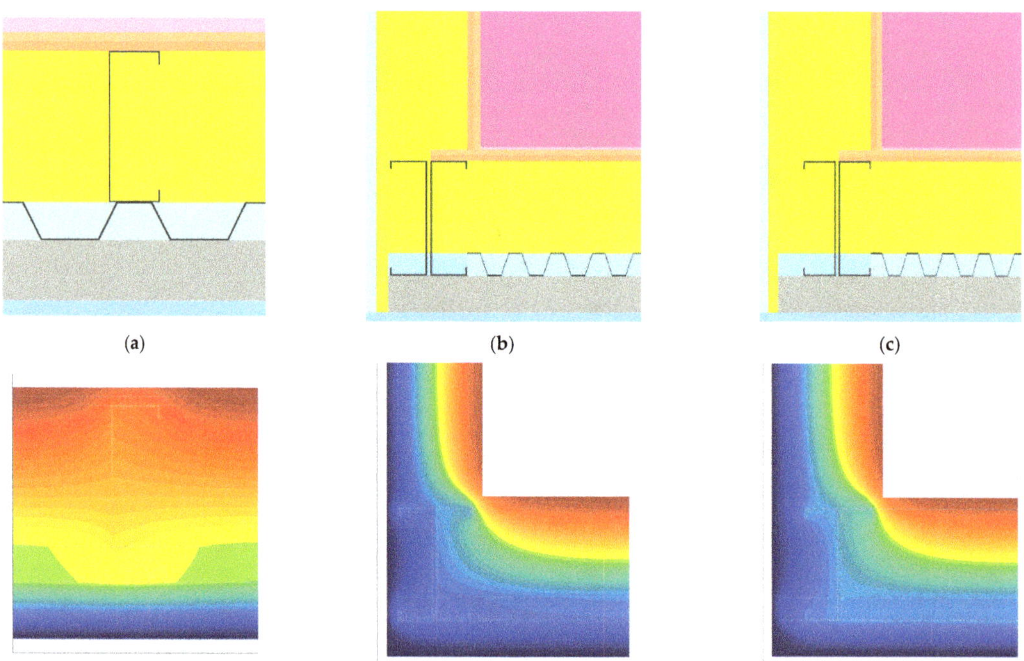

(**a**) (**b**) (**c**)

(**d**) (**e**) (**f**)

Figure A6. The floor in contact with the exterior-Geometrical models: (**a**) current field, (**b**) exterior wall (E, W)—floor in contact with the exterior, (**c**) exterior wall (N)—floor in contact with the exterior; Isothermal surfaces (**d**) current field, (**e**) exterior-wall-floor in contact with the exterior, (**f**) exterior wall (N)—floor in contact with the exterior.

(**a**)

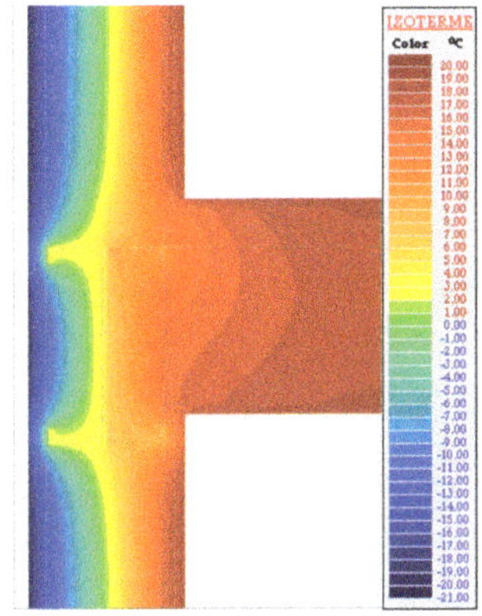

(**b**)

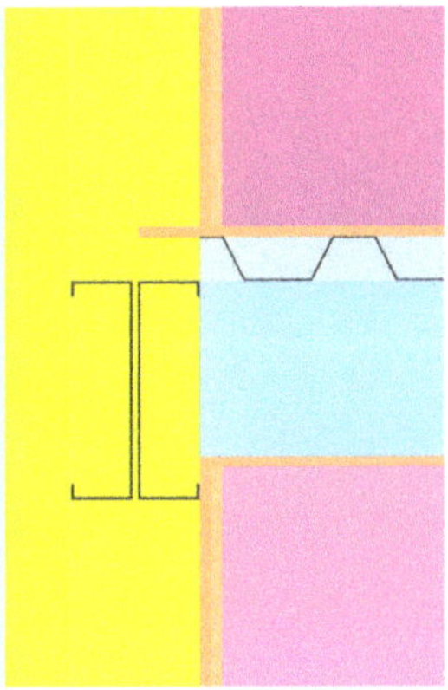

(**c**)

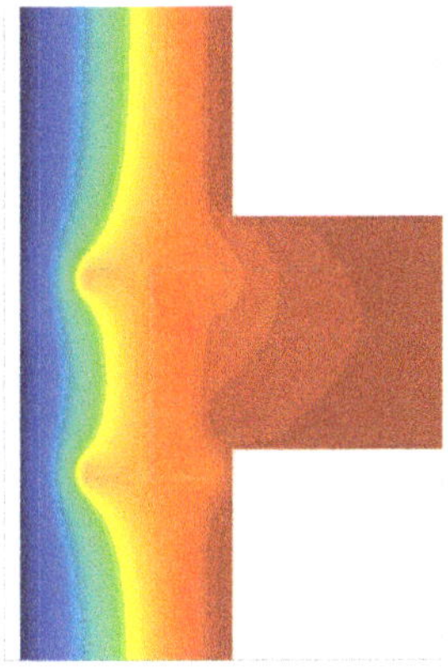

(**d**)

Figure A7. Exterior wall-intermediate floor-Geometrical models (**a**) E-W wall (**c**) N wall; Isothermal surfaces (**b**) E-W wall (**d**) N wall.

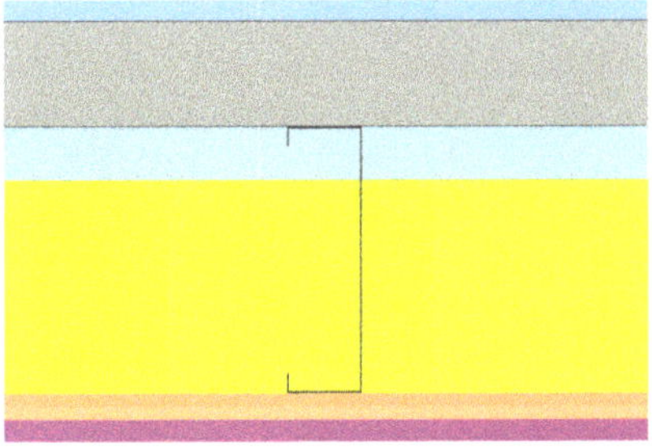

(**a**)

(**b**)

Figure A8. *Cont.*

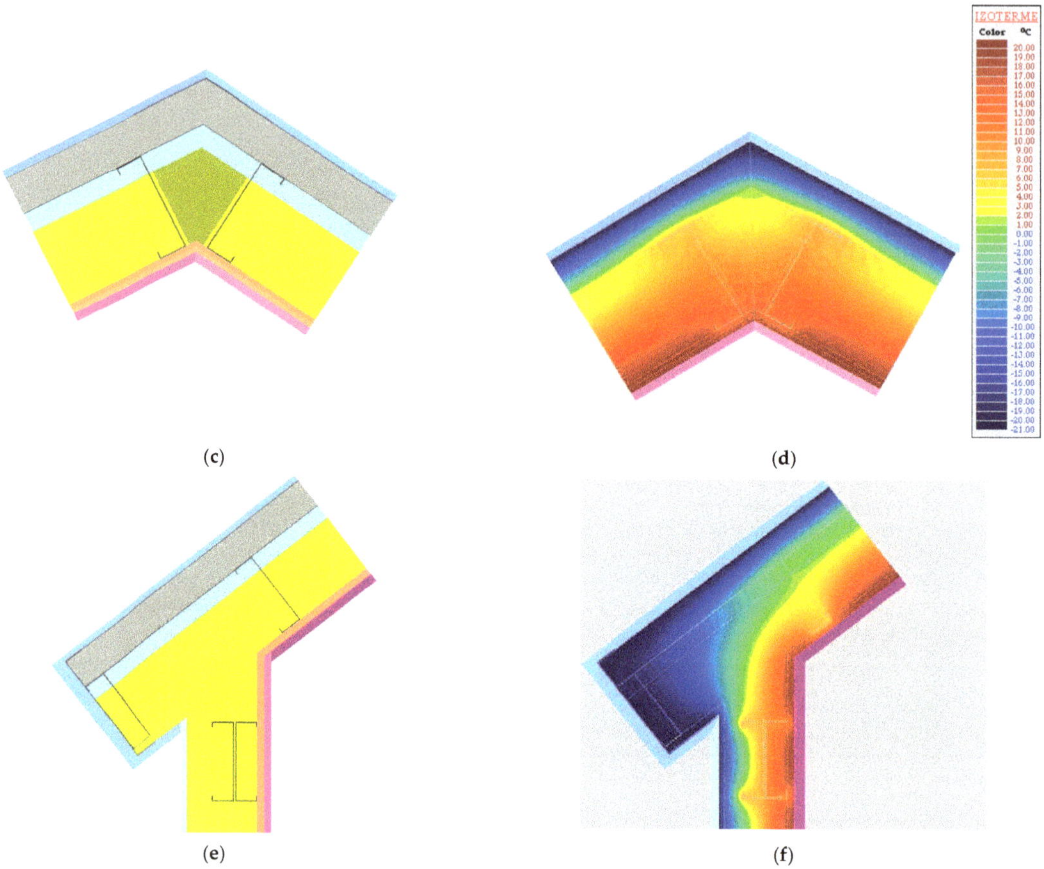

Figure A8. Roof-Geometrical models (**a**) current field, (**c**) ridge, (**e**) eaves; Isothermal (**b**) current field, (**d**) ridge, (**f**) eaves.

References

1. EU Commission. *In Focus: Energy Efficiency in Buildings*; EU Commission: Brussels, Belgium, 2020.
2. EU Commission. *A European Green Deal. Striving to be the First Climate-Neutral Continent*; EU Commission: Brussels, Belgium, 2020.
3. EU Commission. *Directive (EU) 2018/844 of the European Parliament and of the Council of 30th May 2018 Amending Directive 2010/31/EU on the Energy Performance of Buildings and Directive 2012/27/EU on Energy Efficiency*; EU Commission: Brussels, Belgium, 2018.
4. EU Commission. *National Energy and Climate Plans EU Countries' 10-year National Energy and Climate Plans for 2021–2030*; EU Commission: Brussels, Belgium, 2018.
5. EU Commission. *Directive 2012/27/EU of the European Parliament and of the Council of 25th October 2012 on Energy Efficiency, Amending Directives 2009/125/EC and 2010/30/EU and Repealing Directives 2004/8/EC and 2006/32/ECText with EEA Relevance*; EU Commission: Brussels, Belgium, 2012.
6. EU Commission. *Proposal for a Directive of The European Parliament and of the Council on Energy Efficiency (Recast)*; EU Commission: Brussels, Belgium, 2021.
7. EU Commission. *'Fit for 55': Delivering the EU's 2030 Climate Target on the Way to Climate Neutrality*; EU Commission: Brussels, Belgium, 2021.
8. EU Commission. *An EU-Wide Assessment of National Energy and Climate Plans Driving Forward the Green Transition and Promoting Economic Recovery through Integrated Energy and Climate Planning*; EU Commission: Brussels, Belgium, 2020.
9. International Energy Agency. Energy Efficiency 2019. The Authoritative Tracker of Global Energy Efficiency Trends. 2019. Available online: https://www.iea.org/reports/energy-efficiency-2019 (accessed on 14 October 2021).
10. United Nations Environment Programme. *2020 Global Status Report for Buildings and Construction. Towards a Zero-Emissions, Efficient and Resilient Buildings and Construction Sector*; United Nations Environment Programme: Nairobi, Kenya, 2020.

11. Ürge-Vorsatz, D.; Cabeza, L.F.; Serrano, S.; Barreneche, C.; Petrichenko, K. Heating and cooling energy trends and drivers in buildings. *Renew. Sustain. Energy Rev.* **2015**, *41*, 85–98. [CrossRef]
12. Romanian Ministry of Research and Innovation, CCCDI—UEFISCDI, project number PN-III-P1-1.2-PCCDI-2017-0391/CIA_CLIM–Smart Buildings Adaptable to the Climate Change Effects, within PNCDI III, 2018–2021. Available online: https://www.icer.ro/cercetare/proiecte-de-cercetare/cia-clim (accessed on 14 October 2021).
13. Soares, N.; Santos, P.; Gervásio, H.; Costa, J.J.; Simões da Silva, L. Energy efficiency and thermal performance of lightweight steel-framed (LSF) construction: A review. *Renew. Sustain. Energy Rev.* **2017**, *78*, 194–209. [CrossRef]
14. Santos, P.; Ribeiro, T. Thermal Performance Improvement of Double-Pane Lightweight Steel Framed Walls Using Thermal Break Strips and Reflective Foils. *Energies* **2021**, *14*, 6927. [CrossRef]
15. Rajanayagam, H.; Upasiri, I.; Poologanathan, K.; Gatheeshgar, P.; Sherlock, P.; Konthesingha, C.; Nagaratnam, B.; Perera, D. Thermal Performance of LSF Wall Systems with Vacuum Insulation Panels. *Buildings* **2021**, *11*, 621. [CrossRef]
16. Kempton, L.; Kokogiannakis, G.; Green, A.; Cooper, P. Evaluation of thermal bridging mitigation techniques and impact of calculation methods for lightweight steel frame external wall systems. *J. Build. Eng.* **2021**, *43*, 102893. [CrossRef]
17. Santos, P.; Mateus, D. Experimental assessment of thermal break strips performance in load-bearing and non-load-bearing LSF walls. *J. Build. Eng.* **2020**, *32*, 101693. [CrossRef]
18. Santos, P.; Lemes, G.; Mateus, D. Thermal Transmittance of Internal Partition and External Facade LSF Walls: A Parametric Study. *Energies* **2019**, *12*, 2671. [CrossRef]
19. Ministry of Regional Development. Public Administration and European Funds, Order 2641 regarding the modification and completion of the technical regulation. In *Methodology for Calculating the Energy Performance of Buildings*; Approved by the Order of the Minister of Transport, Construction and Tourism no. 157/2007; Ministry of Regional Development: Bucharest, Romania, 2017.
20. Ministry of Regional Development and Public Administration. Order 386 for the modification and completion of the Technical Regulation. In *Normative Regarding the Thermotechnical Calculation of the Construction Elements of the Buildings*; Indicative C 107-2005; Ministry of Regional Development and Public Administration: Bucharest, Romania, 2016.
21. Ministry of Regional Development and Public Administration. *Indicative Mc001 Part 6 Methodology for Calculating the Energy Performance of Buildings*; Ministry of Regional Development and Public Administration: Bucharest, Romania, 2013.
22. Buzatu, R.; Ungureanu, V.; Ciutina, A.; Gireadă, M.; Vitan, D.; Petran, I. Experimental Evaluation of Energy-Efficiency in a Holistically Designed Building. *Energies* **2021**, *14*, 5061. [CrossRef]
23. Gorgolewski, M. Developing a simplified method of calculating U-values in light steel framing. *Build. Environ.* **2007**, *42*, 230–236. [CrossRef]
24. Santos, P.; Gonçalves, M.; Martins, C.; Soares, N.; Costa, J.J. Thermal transmittance of lightweight steel framed walls: Experimental versus numerical and analytical approaches. *J. Build. Eng.* **2019**, *25*, 100776. [CrossRef]
25. Ciutina, A.; Mirea, M.; Ciopec, A.; Ungureanu, V.; Buzatu, R.; Morovan, R. Behaviour of wedge foundations under axial compression. *IOP Conf. Ser. Earth Environ. Sci.* **2021**, *664*, 012036. [CrossRef]
26. Buzatu, R.; Muntean, D.; Ciutina, A.; Ungureanu, V. Thermal Performance and Energy Efficiency of Lightweight Steel Buildings: A Case-Study. *IOP Conf. Ser. Mater. Sci. Eng.* **2020**, *960*, 032099. [CrossRef]
27. Ciutina, A.; Buzatu, R.; Muntean, D.M.; Ungureanu, V. Heat transfer vs environmental impact of modern façade systems. *E3S Web Conf.* **2019**, *111*, 03078. [CrossRef]
28. Intini, F.; Kühtz, S. Recycling in buildings: An LCA case study of a thermal insulation panel made of polyester fiber, recycled from post-consumer PET bottles. *Int. J. Life Cycle Assess.* **2011**, *16*, 306–315. [CrossRef]
29. *EN ISO 6946*; European Committee for Standardization. Building Components and Building Elements—Thermal Resistance and thermal transmittance—Calculation Methods. CEN: Brussels, Belgium, 2017.
30. de Angelis, E.; Serra, E. Light Steel-frame Walls: Thermal Insulation Performances and Thermal Bridges. *Energy Procedia* **2014**, *45*, 362–371. [CrossRef]
31. C107 Norm for Thermotechnical Calculation of the Construction Elements of Buildings, Bucharest, Romania. 2005. Available online: https://ec.europa.eu/growth/tools-databases/tris/en/index.cfm/search/?trisaction=search.detail&year=2010&num=465&mLang=SV (accessed on 14 October 2021).
32. Santos, P.; Poologanathan, K. The Importance of Stud Flanges Size and Shape on the Thermal Performance of Lightweight Steel Framed Walls. *Sustainability* **2021**, *13*, 3970. [CrossRef]
33. Moga, L.; Moga, I. Considerations on the Thermal Modelling of Insulated Metal Panel Systems. In Proceedings of the International Building Physics Conference, Syracuse, NY, USA, 26 September 2018. [CrossRef]
34. Lupan, L.; Manea, D.; Moga, L. Improving Thermal Performance of the Wall Panels Using Slotted Steel Stud Framing. *Proc. Technol.* **2016**, *22*, 351–357. [CrossRef]
35. *EN ISO 10211*; European Committee for Standardization. Thermal Bridges in Building Construction—Heat Flows and Surface Temperatures—Detailed Calculations. CEN: Brussels, Belgium, 2017.
36. PSIPLAN Software. *Two-Dimensional Heat Transfer for Building Modelling Software for Calculating the Linear Thermal Transmittance Value*; Technical University of Cluj-Napoca: Cluj-Napoca, Romania, 2021.
37. Moga, L.; Moga, I. Evaluation of Thermal Bridges Using Online Simulation Software. *E3S Web Conf.* **2020**, *172*, 08010. [CrossRef]
38. Moga, L. Analytic Study of Thermal Bridges Met at High Performance Energy Efficient Buildings. *Inter. Multidis. Scien. GeoConf. SGEM* **2018**, *18*, 621–626. [CrossRef]

39. Moga, L.; Moga, I. *Specific Thermal Bridges at Load Bearing Masonry Buildings*; UTPRESS: Cluj-Napoca, Romania, 2013.
40. *ISO 13789*; Thermal Performance of Buildings–Transmission and Ventilation Heat Transfer Coefficients—Calculation Method. ISO: Brussels, Belgium, 2015.
41. *ISO 52016-1*; Energy Performance of Buildings—Energy Needs for Heating and Cooling, Internal Temperatures and Sensible and Latent Heat Loads—Part 1: Calculation Procedures. ISO: Brussels, Belgium, 2017.
42. Ubakus—Online U-Wert Calculator. Available online: https://www.ubakus.de/u-wert-rechner/ (accessed on 5 October 2021).
43. Ministry of Transportation. *Constructions and Tourism, Indicative Mc001 Parts 1,2,3 Methodology for Calculating the Energy Performance of Buildings*; Ministry of Transportation: Bucharest, Romania, 2006.
44. Ministry of Development, Public Works and Administration Indicative Mc001 2021 Methodology for Calculating the Energy Performance of Buildings-Updated Version Preprint, Bucharest, Romania. Available online: https://epbd-ca.eu/ca-outcomes/outcomes-2015-2018/book-2018/countries/romania (accessed on 14 October 2021).
45. Vajó, B.; Lakatos, Á. Super Insulation Materials—An Application to Historical Buildings. *Buildings* **2021**, *11*, 525. [CrossRef]
46. Lakatos, Á. Thermophysical investigations of nanotechnological insulation materials. *AIP Conf. Proc.* **2017**, *1866*, 030003. [CrossRef]
47. Lakatos, Á.; Csarnovics, I.; Csík, A. Systematic Analysis of Micro-Fiber Thermal Insulations from a Thermal Properties Point of View. *Appl. Sci.* **2021**, *11*, 4943. [CrossRef]
48. *ISO 13788*; Hygrothermal Performance of Building Components and Building Elements—Internal Surface Temperature to Avoid Critical Surface Humidity and Interstitial Condensation—Calculation Methods. ISO: Brussels, Belgium, 2012.
49. Șoimoșan, T.M.; Moga, L.M.; Anastasiu, L.; Manea, D.L.; Căzilă, A.; Zeljković, Č. Overall Efficiency of On-Site Production and Storage of Solar Thermal Energy. *Sustainability* **2021**, *13*, 1360. [CrossRef]
50. Jack, R. *Building Diagnostics: Practical Measurement of the Fabric Thermal Performance of House*; Figure share; Loughborough University: Loughborough, UK, 2015.

buildings MDPI

Article

Circular Building Process: Reuse of Insulators from Construction and Demolition Waste to Produce Lime Mortars

Daniel Ferrández [1], Engerst Yedra [1], Carlos Morón [1,*], Alicia Zaragoza [1] and Marta Kosior-Kazberuk [2]

[1] Departamento de Tecnología de la Edificación, Escuela Técnica Superior de Edificación, Universidad Politécnica de Madrid, 28040 Madrid, Spain; daniel.fvega@upm.es (D.F.); e.yedra@alumnos.upm.es (E.Y.); alicia.zaragoza@alumnos.upm.es (A.Z.)
[2] Department of Building Structures, Bialystok University of Technology, 15-351 Bialystok, Poland; m.kosior@pb.edu.pl
* Correspondence: carlos.moron@upm.es; Tel.: +34-91-336-7583; Fax: +34-91-336-7637

Abstract: This research aims to revalue the possibilities presented by lime mortars for use in renovation and as cladding material on facades. The study focuses on analyzing the technical feasibility of lime mortars with the incorporation of residues from three types of thermal insulation materials: expanded polystyrene with graphite especially suitable for use on facades; expanded polystyrene for use indoors; and insulating mineral wool. The incorporation of these construction and demolition residues makes it possible to improve several technical performance aspects of lime mortars, and to incorporate circular economy criteria in the manufacturing process of these materials. The results showed that the incorporation of mineral wool improves the mechanical resistance to the bending of mortars, increases their durability against freeze–thaw cycles and salt crystallization, and reduces the final shrinkage of mortars. For their part, mortars with the addition of polystyrene-insulating residues reduce mechanical resistance, but also reduce thermal conductivity, and are lighter, which is why they are shown as a possible alternative for use in precast.

Keywords: lime mortars; thermal insulation materials; rehabilitation; construction and demolition waste; sustainable building

Citation: Ferrández, D.; Yedra, E.; Morón, C.; Zaragoza, A.; Kosior-Kazberuk, M. Circular Building Process: Reuse of Insulators from Construction and Demolition Waste to Produce Lime Mortars. *Buildings* **2022**, *12*, 220. https://doi.org/10.3390/buildings12020220

Academic Editors: Paulo Santos and Mark Bomberg

Received: 18 January 2022
Accepted: 10 February 2022
Published: 16 February 2022

1. Introduction

Currently, in the European Union, approximately 40% of total energy consumption is caused by residential buildings, tertiary buildings, shops, offices, and other auxiliary buildings [1]. In Spain, this percentage is reduced to 27.7% due to the better climatic conditions offered by being located in southern Europe [2]. However, this is still a high, and not inconsiderable percentage, meaning that every day more researchers seek to take action to evolve towards a more sustainable building sector [3]. If we continue to limit the problem to Spain, the gradual aging of the building stock increasingly forces autonomous governments to implement energy rehabilitation policies [4]. In Spain alone, more than half of the 25.7 million existing homes and buildings—according to the Ministry of Transport, Mobility and Urban Agenda (MITMA) [5]—were built before the first energy regulation was approved [6], NBE-CT 79 [7]. In this sense, any action aimed at reducing energy consumption, the emission of greenhouse gases, the environmental impact of buildings, and the excessive consumption of natural resources in construction [8–10], is in line with the Goals for Sustainable Development (SDG) set by the United Nations [11]. More specifically, this work is directly linked to SDGs 11 and 12: "Sustainable Cities and Communities" and "Responsible Production and Consumption", although it is a cross-sectional investigation shared with other goals set for 2030.

Together with the problem of high energy consumption in buildings, the European continent faces the challenge of carrying out adequate management of the construction and demolition waste (CDW) that is generated annually [12], and that represents between

25–30% of the solid waste produced (approximately 800 million tons per year) [13]. For this purpose, the European Commission has developed a series of guidelines to improve the identification of CDW and enhance their separation and collection at source (European Commission, 2018 [14]), so that it is possible to recycle, reuse and recover at least 70% by weight of the non-hazardous CDW generated [15,16]. For this reason, within the framework of the European Green Deal [17], the CDW is placed as a priority axis to achieve the elaboration of a circular economy action plan in the European Union [18], due to the great potential that these residues present for their reincorporation into the production process and the generation of new raw materials [19].

Traditionally, hydraulic lime mortars made with river sand have been used in façade rehabilitation works using continuous granulometry with aggregate fractions between 0–4 mm [20,21]. The importance of maintaining a careful granulometry and a clean aggregate becomes even greater if it is intended to produce precast lime [22], since an adequate proportion of fines provides a compact structure and better mechanical performance to the constructive element [23]. It should be noted that lime mortars fell into disuse from the 19th century with the industrial revolution and the cement boom [24], although there are certain properties such as their high plasticity and good workability that make them essential for some forms of restoration and building rehabilitation works [25,26]. Hydraulic lime mortars, in turn, have good permeability and great water retention; these properties are especially relevant when you want to avoid humidity by condensation [27]. Furthermore, as some researchers have verified, hydraulic lime mortars have higher mechanical strength at early ages and are set faster than area lime mortars [28]. From an environmental point of view, some researchers have observed how lime can absorb CO_2 from the air gradually, increasing its rigidity and improving the building's carbon footprint during its useful life [29,30].

In recent decades, studies have been carried out aimed at incorporating recycled materials from CDW in the manufacture of mortars, to reduce the volume that this type of waste occupies in the landfill and give them a new useful life [31]. In Spain, more than 21.5 million tons/year of these wastes are produced [32]. The most common way to incorporate these CDWs in the manufacture of mortars is by grinding and cleaning to use them as recycled aggregates in total or partial replacement of natural aggregate [33,34]. However, some wastes, such as those from thermal insulation materials for façades, can be used to lighten the weight of the binder materials and improve their thermal conductivity [35]. Following this line of research, Milling et al. investigated the possibility of replacing part of the cement content of masonry mortars with a mixture of expanded polystyrene (EPS) waste dissolved in acetone, where the results show that it is possible to obtain technically feasible mortars that significantly reduce the carbon footprint of these materials [36]. This type of EPS waste has also been studied by other researchers to develop low-density precast blocks that meet the minimum requirements for durability and mechanical resistance for use in buildings [37]. In any case, it is a waste that can be incorporated into building conglomerate materials to improve its thermal conductivity and promote the application of this type of mortar as an insulation system [38]. On the other hand, other authors have opted for the use of crushed mineral wool in cement mortars, mainly due to the good cohesion between the wool fibers and the mortar matrix, obtaining good flexural tensile results and improved thermal insulation properties [39]. In addition, this type of insulating mineral wool used as reinforcing fibers serves to significantly improve other properties of mortars, such as resistance to cracking, decreases in shrinkage, or improvements to the modulus of elasticity of mortars [40].

The main objective of this work was to study the technical feasibility of lime mortars with the incorporation of CDW from out-of-use thermal insulation materials. More specifically, an experimental campaign was developed with mechanical, physical, and durability tests, to find out the most relevant properties of lime mortars with the incorporation of three types of recycled insulators (expanded polystyrene recommended for use on façades, expanded polystyrene for interior use, and mineral wool) in two different proportions each.

2. Materials and Methods

This section describes the materials used for the preparation of lime mortars, as well as the dosages used, and the experimental campaign carried out.

2.1. Materials Employed

For this work, the following raw materials were used in the preparation of mortars: hydraulic lime, natural aggregate (NA), expanded polystyrene insulation with graphite for use in facades (EPS-F), expanded polystyrene insulation for indoor use (EPS-I), mineral wool insulation (MW), and water.

2.1.1. Hydraulic Lime

The conglomerate material chosen in this investigation was hydraulic lime. This material has the property of being able to set and harden with water, thanks to its composition generally made up of $Ca(OH)_2$, calcium silicates, and calcium aluminates [41]. More specifically, a hydraulic lime of the NHL-5 type was used, following the classification included in the UNE-EN 459-1: 2002 standard [42].

The technical characteristics of this raw material according to the manufacturer Chaux de Saint-Astier are shown in Table 1. Furthermore, according to the procedure included in the UNE 80103: 2013 [43] standard, it was determined that the real density of the lime used in this study would be 2750 kg/m^3.

Table 1. Technical characteristics of NHL 5 hydraulic lime [44].

Start of Setting (min)	Fineness Blaine (cm^2/g)	Expansion (mm)	Free Water (%)	Compression Resistance (MPa)
≥60	7882	0.80	0.55	≥5.00

2.1.2. Natural Aggregate

The aggregate used to make the mortars was river natural sand. Figure 1 shows the granulometry of the sands used in this investigation. The granulometric curve was obtained using the dried aggregate and a series of standardized sieves according to UNE-EN 933-2 with a mesh opening between 4.000 and 0.063 mm [45]. It was sought to obtain a continuous granulometry in order to improve the compactness of hardened lime mortars [46].

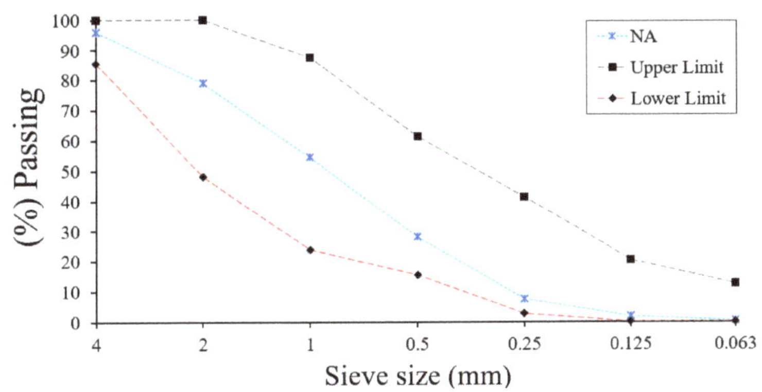

Figure 1. Recycled aggregate size distribution curve compared to the limits of NBE FL-90 [47] adapted to sieve size established by UNE-EN-933-2 [45].

On the other hand, Table 2 shows the results of the physical characterization of the sands used, following the recommendations of the UNE-EN 13139: 2002 standard [48].

Table 2. Physical characterization of the natural river sand.

Test	Fine Content (%)	Fineness Modulus (%)	Friability (%)	Bulk Density (kg/m^3)	Dry Density (kg/m^3)	Water Absorption (%)
Standard	UNE EN 933-1 [49]	UNE-EN 13139 [48]	UNE-EN 146404 [50]	UNE-EN 1097-3 [51]	UNE-EN 1097-6 [52]	UNE-EN 1097-6 [52]
NA	1.92	4.21	20.76	1569	2517	0.87

2.1.3. Waste from Thermal Insulation Materials

In this work, three different types of insulation materials from construction and demolition waste were used: expanded polystyrene insulation with graphite for use on facades (EPS-F); expanded polystyrene insulation for indoor use (EPS-I); and mineral wool insulation (MW), as shown in Figure 2.

(a) (b) (c)

Figure 2. Waste insulation used. (**a**) Expanded polystyrene with graphite for use in facades; (**b**) expanded polystyrene for indoor use; (**c**) mineral wool.

These wastes were obtained from a recycling plant located in El Molar (Madrid, Spain) and had to be collected and separated manually. In the case of polystyrene insulators, they were prepared manually according to the procedure used in other investigations to break them down into small spheres that could be added during the manufacture of mortars [53,54]. For its part, the mineral wool insulation was manually shredded and separated into fibers of 12 mm length, following the recommendations found in the literature [55,56]. Its most relevant characteristics are collected in Table 3.

Table 3. Physical characteristics of the insulation waste used (URSA Ibérica Aislantes, S.A., Madrid, Spain).

Type	Density (kg/m^3)	Thermal Conductivity (W/mK)	Water Diffusion Resistance Factor (Dimensionless)	Geometric Characteristics
EPS-F	30	0.032	20–100	Ømed = 5 mm
EPS-I	15	0.038	20–100	Ømed = 3 mm
MW	40	0.040	1	Lmed = 12 mm

2.1.4. Water

Safe drinking water from the Canal de Isabel II of the Community of Madrid (Spain) was used to mix the different mortar samples. This water has been used previously in research work [57], due to its good characteristics for making mortars. Its main properties include its soft hardness (25 mg $CaCO_3$/L) and neutral pH between 7 and 8 recommended to avoid setting alterations [58].

2.2. Dosages Used

In this work, we intended to study the effect that the incorporation of thermal insulation materials in different proportions has on lime mortars, analyzing their physical-mechanical properties for their use as construction material. For this, two different proportions of insulating waste of each type and a single lime/aggregate ratio were incorporated. The mixing process was carried out following the recommendations of the UNE-EN 196-1 [59] standard, with the help of an IBERTEST planetary mixer model IB32-040V01, and always following the same techniques and methods. The following nomenclature was defined for the different batches:

LM—Type—Quantity

where LM refers to the type of material used, lime mortars with a cement/aggregate ratio 1:3; Type refers to the insulation material incorporated in the manufacturing, which can be expanded polystyrene recommended for use on facades (EPS-F), expanded polystyrene for interior use (EPS-I), mineral wool (MW) or reference without insulation (REF); and finally, Quantity indicates the incorporated mass in grams of each type of waste, which can be 2.5 or 4.5 g.

Table 4 shows the different proportions of each material used to make the lime mortars used in this research.

Table 4. Mixtures proportions by weight and workability of the test mortars.

Mortar Type	Lime (g)	Sand (g)	Water (g)	Insulating Waste (g)	Flow Table Workability (mm) UNE-EN 1015-3:2000/A2:2007 [60]
LM-REF	450	1350	333	–	152
LM-EPS-I-2.5	450	1350	369	2.5	156
LM-EPS-I-4.5	450	1350	396	4.5	147
LM-EPS-F-2.5	450	1350	378	2.5	153
LM-EPS-F-4.5	450	1350	405	4.5	151
LM-MW-2.5	450	1350	360	2.5	152
LM-MW-4.5	450	1350	387	4.5	146

As shown in Table 4, the water content was set experimentally following the indications of the UNE-EN 1015-3:2000/A2:2007 [60] standard, until achieving a workable consistency corresponding to a diameter of the 150 ± 10 mm mortar cake in the flow table workability. It can also be seen that mortars with higher incorporation of insulating residue (4.5 g) required a higher content of mixing water, which had a negative influence on the mechanical resistance of these materials [61]. Finally, it should be noted that the different samples produced were cured in a humid chamber for 28 days, at a temperature of 22 ± 2 °C and relative humidity of 95% [62].

2.3. Experimental Study

The experimental program included in this research work includes a mechanical, physical, and durability characterization of lime mortars with the incorporation of three types of thermal insulation materials.

Regarding the mechanical characterization, flexural and compression failure tests were carried out on standardized $4 \times 4 \times 16$ cm RILEM specimens, following the recommendations of the UNE-EN 1015-11:2000/A1:2007 standard [63]. The results were obtained with the help of a model AUTETEST 200-10SW press, and the fractography derived from the breakage of the specimens in the bending test was analyzed.

For its part, the physical characterization included the performance of five tests. Shore D surface hardness was used to determine the wear resistance of mortars and the possibility of being scratched on their surface: for this we used a Shore D durometer and 4 × 4 × 16 cm test tubes. We also tested the adhesion of lime mortars made on ceramic surfaces, for which the recommendations of the UNE-EN 1015-2:1999/A1:2007 [64] standard were followed, applying a mortar mass of 10 ± 1 mm thick on a previously moistened ceramic scraper and using direct traction equipment with cylindrical specimens with a diameter of 50 mm. Next, we assessed water absorption by capillarity in mortars, following the recommendations of the UNE-EN 1015-18 [65] standard, and the apparent density of mortars in the hardened state, using a precision balance with three significant figures and 4 × 4 × 16 cm test tubes. Finally, we assessed the coefficient of thermal conductivity of mortars, determined by the heat flow meter method and using Fourier's Law.

Finally, the durability tests that were carried out were: resistance to freeze–thaw cycles; salt crystallization; and shrinkage measurement. In the first two, freezing and crystallization of salts, the standards UNE-EN 12371 [66] and UNE-EN 12370 [67] were adapted for natural stone, respectively, using 4 × 4 × 16 cm series of test tubes and taking for each sample subjected to cycles, another equivalent without accelerated aging. The freeze–thaw cycles were completed 25 times (6 h freezing at −12 °C and 18 h saturation in water at 20 °C) and the salt crystallization cycles were completed 15 times (2 h in a solution of sodium sulfate decahydrate saturated with water), with 14% at 22 °C (drying in an oven for 16 h at 105 °C and cooling at 22 °C for 6 h). On the other hand, the shrinkage test was carried out on 25 × 25 × 287 mm specimens for 180 days following the UNE 80-112-89 standard [68].

The statistical analysis of the results was carried out with the statistical package IBM SPSS Statistics v25 ©. To know if the results obtained for each property analyzed between the different types of lime mortar produced differed statistically significantly, an analysis was used employing comparison tests. To carry out the complimentary discussion of the results, six samples of each type of mortar were used for each test.

In the discussion of the physical and mechanical properties, mean comparison tests were carried out for independent samples, previously analyzing the fulfillment of the necessary assumptions for the application of parametric tests: (a) normality, (b) homoscedasticity, and (c) independence [69,70]. In the cases in which these assumptions are verified, the Student's *t*-test (for two independent samples) or ANOVA (for more than two independent samples) were used. In addition, to understand the comparisons within each group when there were more than two independent samples, the Dunn–Bonferroni test was used [70]. On the other hand, if the necessary assumptions for the parametric tests (Student's *t*-test or ANOVA) were not fulfilled, the analysis was carried out using non-parametric Mann–Whitney U tests (for two independent samples), and Kruskal–Wallis (for more than two independent samples).

Finally, in the statistical discussion corresponding to the durability tests (resistance to frost and salt crystallization), comparison tests of related samples were used, as they are measurements carried out on the same type of material before and after being subjected to accelerated aging cycles. For this reason, the Student's *t*-test was used for related samples in cases where the necessary assumptions were met for the performance of parametric tests, and, otherwise, the non-parametric test for Wilcoxon related samples was used [69].

3. Results

Next, the results obtained for the different tests of mechanical, physical, and durability characterization of the studied mortars are presented.

3.1. Mechanical Characterization Tests

Hardened mortars are subjected to a great variety of mechanical stresses depending on their location on site. They experience compression forces when used as joining elements in masonry walls, and they also experience stresses and deformations inherent to the construction system of which they are a part, such as the case of lime mortars used as coatings that may experience deformations as a consequence of mechanical bending stresses [71].

3.1.1. Flexural Strength

The mechanical resistance to bending was determined in standard 4 × 4 × 16 cm specimens that were filled with the mortar in two approximately equal layers, homogenizing each layer of the mixture with 25 strokes in the compactor following the UNE-EN 1015-11:2000/A1:2007 standard [63]. This test is of great relevance for lime mortars that are going to be used in restoration works as coatings, since the ability to withstand deformations without cracks appearing on their surface is greatly important [72]. To carry out the test, a press model AUTETEST 200-10SW was used with a uniform application of the load at a speed between 10–50 N/s. Figure 3 shows the test equipment and the results obtained for flexural strength in each of the mortars studied in this investigation.

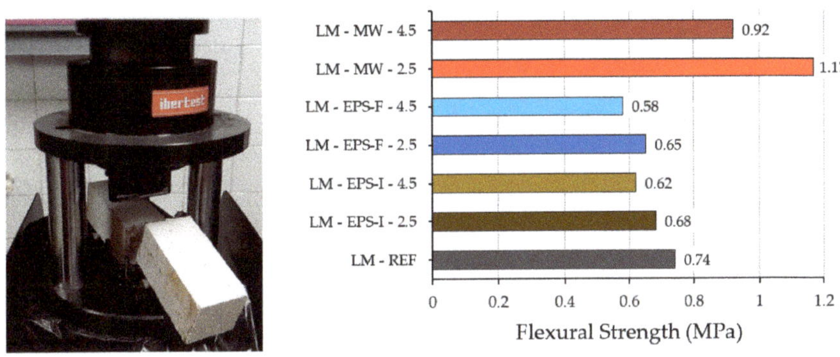

(a) (b)

Figure 3. Flexural strength test. (**a**) IBERTEST press model AUTOTEST 200-10SW; (**b**) results obtained for the different batches.

Table 5 shows the results of the statistical discussion of flexural strength by groups of mortars. First, the means and standard deviations (SD) are reflected, together with the confidence interval for the mean with a significance level of 95%. The second part of Table 5 shows the comparison of means between materials and their statistical significance.

Table 5. Statistical analysis for the flexural strength test (MPa) in each group of mortar and its comparison.

Type	Average		SD		Lower/Upper Limits	
REF	0.7422		0.04055		[0.7111; 0.7734]	
EPS-I	0.6511		0.04484		[0.6288; 0.6734]	
EPS-F	0.6139		0.04804		[0.5900; 0.6378]	
MW	1.0456		0.14456		[0.9737; 1.1174]	
Comparison	REF vs. EPS-I	REF vs. EPS-F	REF vs. MW	EPS-I vs. EPS-F	EPS-I vs. MW	EPS-F vs. MW
K	17.056	24.750	−14.778	7.694	−31.833	−39.528
p-value	0.135	0.006	0.289	1.000	0.000	0.000

According to the results obtained and the statistical analysis carried out, it can be seen that the lime mortars made with mineral wool fibers were the ones with the highest flexural strengths (1.0456 MPa on average). This result is because the incorporation of this insulating residue is similar to the effect caused by the incorporation of reinforcing fibers in masonry mortars [73]. On the other hand, mortars reinforced with polystyrene fibers were the ones that obtained the worst resistance, the flexural resistance of mortars that incorporated EPS-F being worse, mainly due to the greater particle diameter of this recycled material. In addition, in the case of mortars that incorporated polystyrene residues, it can be seen in Figure 4 that after the test tubes broke there was no good cohesion between the mortar matrix and the recycled EPS insulation spheres. In the comparison of means for flexural strength using the Kruskal–Wallis test, statistical significance was observed with a level of confidence higher than 95% between the mortars made with MW and the mortars with EPS-F and EPS-I.

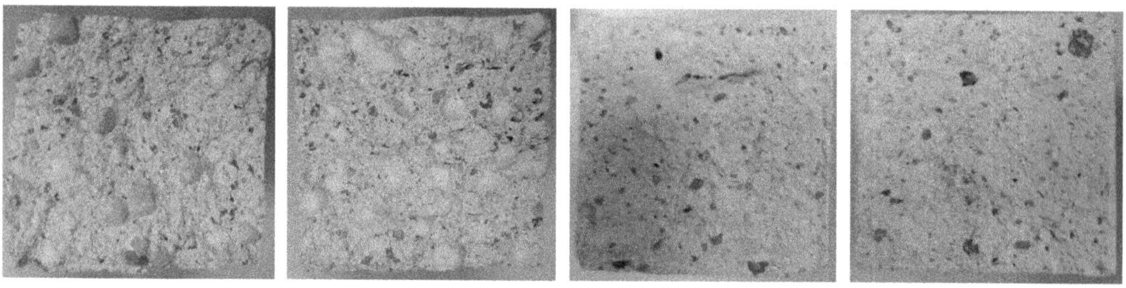

(a) (b) (c) (d)

Figure 4. Breakage of the specimens after the flexural test. (**a**) LM-EPS-F-2.5; (**b**) LM-EPS-I-2.5; (**c**) LM-MW-2.5; (**d**) LM-REF.

To complete the discussion of these results, a comparison has been made in Table 6 between the flexural strength means of the test tubes that incorporated 2.5 g of insulating residue, and those that incorporated 4.5 g.

Table 6. Comparison for flexural strength (MPa) between mortars with incorporation of insulating residue.

Type	Residue (g)	Average	SD	Interval	K	*p*-Value
EPS-I	2.5	0.6833	0.03354	(0.6576; 0.7091)	16.167	1.000
	4.5	0.6189	0.02848	(0.5970; 0.6480)		
EPS-F	2.5	0.6500	0.03317	(0.6245; 0.6755)	15.111	1.000
	4.5	0.5778	0.02949	(0.5551; 0.6004)		
MW	2.5	1.1711	0.07322	(1.1148; 1.2274)	9.000	1.000
	4.5	0.9200	0.05979	(0.8740; 0.9660)		

It can be seen in the results shown in Table 6 that in all the test tubes with the incorporation of 2.5 g of thermal insulation, higher bending strengths were obtained, with the increase in the content of residue incorporated in the matrix impairing this property. However, in no case were these results statistically significant.

3.1.2. Compressive Strength

Compressive strength is one of the most decisive parameters when selecting a type of mortar for application on-site. This property depends on the internal cohesion of the mortar and reflects its ability to support loads without disintegrating [74]. In the case of hydraulic lime mortars, it is known that the mechanical resistance to compression increases with time [75], with the binder content, and with the use of continuous granulometry for aggregates [76]. This property is determined by applying a constant load of 1 mm/min without acceleration on each of the half-samples obtained after the flexural strength test. In Figure 5, the test method used and the results obtained are shown.

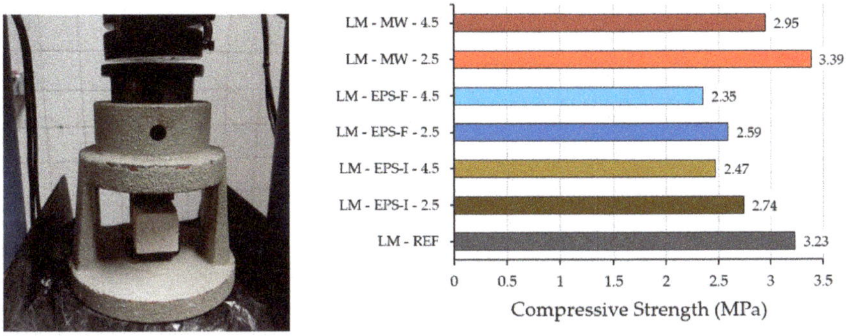

(a) (b)

Figure 5. Compressive strength test. (**a**) IBERTEST press model AUTETEST 200-10SW; (**b**) results obtained for the different batches.

Table 7 shows the statistical discussion for the compressive strength in each group of lime mortar produced. To carry out this discussion, the ANOVA and Dunn–Bonferroni tests have been used, as they meet the necessary assumptions for parametric samples.

Table 7. Statistical analysis for the compressive strength test (MPa) for each group of mortar and its comparison.

Type	**Average**		**SD**		**Lower/Upper Limits**	
REF	3.2367		0.17313		(3.1036; 3.3697)	
EPS-I	2.6061		0.18706		(2.5131; 2.6991)	
EPS-F	2.4650		0.20624		(2.3642, 2.5658)	
MW	3.1744		0.24575		(3.0522; 3.2967)	
Comparison	REF vs. EPS-I	REF vs. EPS-F	REF vs. MW	EPS-I vs. EPS-F	EPS-I vs. MW	EPS-F vs. MW
Difference of Means	0.63056	0.77167	0.06222	0.14111	−0.56833	−0.70944
p-value	0.000	0.000	1.000	0.280	0.000	0.000

As can be seen in Table 7 and Figure 5, the reference mortars without additions were the ones with the highest mechanical resistance to compression (3.2367 MPa), because they presented a more compact matrix and there were no preferred breakage points, as they did not incorporate insulation materials. For this property, it is observed that the addition of mineral wool fibers did not imply an improvement in compressive strength, as has been observed in other previous studies of fiber-reinforced mortars [77]. Even so, mortars with the addition of mineral wool insulation present higher resistance than mortars with the incorporation of expanded polystyrene granules. Thus, the differences in means between mortars with the incorporation of recycled EPS insulation compared to the reference mortars or with the addition of mineral wool were statistically significant for a 95% confidence level.

Table 8 shows the mean difference in compressive strength for the same type of lime mortar as a function of the insulating residue content.

From the analysis of Table 8, it can be deduced that in the three types of mortar with the incorporation of thermal insulation materials analyzed, those with lower residue content (2.5 g) presented higher mechanical resistance to compression, being, in turn, the differences of means obtained for each of the statistically significant groups for a confidence level of 95%. Both the reference mortar, as well as all mortars incorporating 2.5 g of thermal insulation, had a resistant class M2.5 at 28 days according to the UNE-EN 998-2: 2012 standard [78].

Table 8. Comparison for compressive strength (MPa) between lime mortars with incorporation of thermal insulation residues.

Type	Residue (g)	Average	SD	Interval	Difference of Means	*p*-Value
EPS-I	2.5 4.5	2.7411 2.4711	0.13062 0.12762	(2.6407; 2.8415) (2.3730; 2.5692)	0.27000	0.003
EPS-F	2.5 4.5	2.5844 2.3456	0.20653 0.11182	(2.4257; 2.7432) (2.2596; 2.4315)	0.23889	0.013
MW	2.5 4.5	3.3956 2.9533	0.08918 0.10186	(3.3270; 3.4641) (2.8750; 3.0316)	0.44222	0.000

3.2. *Physical Characterization Tests*

For the physical characterization of the prepared mortars, the following properties were studied: bonding strength, apparent density, Shore D surface hardness, water absorption by capillarity, and thermal conductivity.

3.2.1. Bonding Strength

The bonding strength of a mortar can be defined as the property that allows it to remain attached to the ceramic surface that makes up the factory [79]. To determine this property, the UNE-EN 1015-12:2016 standard clarifies that it is necessary to measure the maximum pull-out traction by direct load perpendicular to the surface of the mortar for rendering or plastering that has been applied on certain support [80]. When the adherence is greater, the mechanical energy of the interface between the mortar and the application surface can absorb; this property depends on various factors such as the type of binder, the dosage used, the degree of wetting of the support, and its type or the presence of salts in the masonry factory [81]. The results derived from this test and its method of performance are shown in Figure 6.

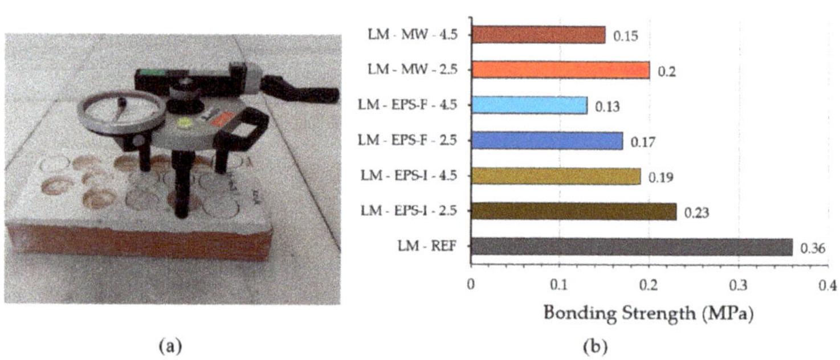

Figure 6. Bonding strength test. (**a**) Adhesion measuring equipment and tested sample; (**b**) results obtained for the different batches.

For the statistical discussion of this property, non-parametric tests were used, as this variable does not comply with the normality assumptions. Table 9 shows the results for each group of mortars and the comparison of means between the different groups using the Kruskal–Wallis test for independent samples.

As shown in Table 9, and in accordance with the results obtained in Figure 6, the highest average adherence was obtained for the reference mortars without additions (0.3567 MPa), the group of mortars with incorporation of EPS-F insulation, and those that had a lower resistance to adhesion (0.1522 MPa). Thus, it can be stated that the incorporation of residues from thermal insulation materials into the matrix of lime mortars decreases its adhesion capacity in ceramic walls. These results follow other previous studies where the resistance to adhesion of masonry mortars incorporates CDW [82]. The results of the

Kruskal–Wallis mean comparison test can also be seen in the lower part of Table 9, where it can be seen that the difference in means between the reference mortar and each of the mortars with incorporation of thermal insulation (EPS-I, EPS-F, and MW) are statistically significant. The same occurs with the difference in adherence means between the samples that incorporate EPS-I and EPS-F, which turned out to be statistically significant for a confidence level greater than 95%.

Table 9. Statistical analysis for the bonding strength test (MPa) for each group of mortar and its comparison.

Type	Average		SD		Lower/Upper Limits	
REF	0.3567		0.03428		(0.3303; 0.3830)	
EPS-I	0.2100		0.04366		(0.1883; 0.2317)	
EPS-F	0.1522		0.03859		(0.1330; 0.1714)	
MW	0.1750		0.04120		(0.1545; 0.1955)	
Comparison	REF vs. EPS-I	REF vs. EPS-F	REF vs. MW	EPS-I vs. EPS-F	EPS-I vs. MW	EPS-F vs. MW
K	20.778	40.861	32.861	20.083	12.083	−8.000
p-value	0.033	0.000	0.000	0.006	0.286	1.000

For its part, Table 10 shows the difference in adhesion means for the same mortar with different proportions of insulating residue (2.5 g versus 4.5 g).

Table 10. Comparison for the adherence (MPa) in the different mortars with the incorporation of CDW.

Type	Residue (g)	Average	SD	Interval	K	*p*-Value
EPS-I	2.5	0.2322	0.02489	(0.2131; 0.2514)	14.667	1.000
	4.5	0.1878	0.04816	(0.1508; 0.2248)		
EPS-F	2.5	0.1744	0.03468	(0.1478; 0.2011)	13.722	1.000
	4.5	0.1300	0.02915	(0.1076; 0.1524)		
MW	2.5	0.2011	0.03180	(0.1767; 0.2256)	17.833	0.813
	4.5	0.1489	0.03257	(0.1238; 0.1739)		

According to the results collected in Table 10, in all cases, a higher adhesion was obtained in mortars that incorporated a lower amount of insulation materials (2.5 g). However, in this case, the differences between the adhesion mean obtained between the same mortars with a different proportion of thermal insulation in their matrix were not statistically significant in any of the cases.

3.2.2. Apparent Density

The density of hardened mortars depends on the individual densities of the different components that make them up. In this sense, one of the greatest advantages of incorporating expanded polystyrene insulation waste is the reduction in final density compared with traditional lime mortars [54]. To determine this property, the recommendations of the UNE-EN 1015-10/A1 [83] standard were followed. Figure 7 shows the hydrostatic balance weighing method and the results derived from this test.

As in the previous case, for the discussion of the results obtained, non-parametric tests were used as this property does not meet the necessary assumptions to perform an ANOVA analysis. Table 11 shows the results obtained for each group of mortars and the comparison of measurements made between the different groups using the Kruskal–Wallis test for independent samples.

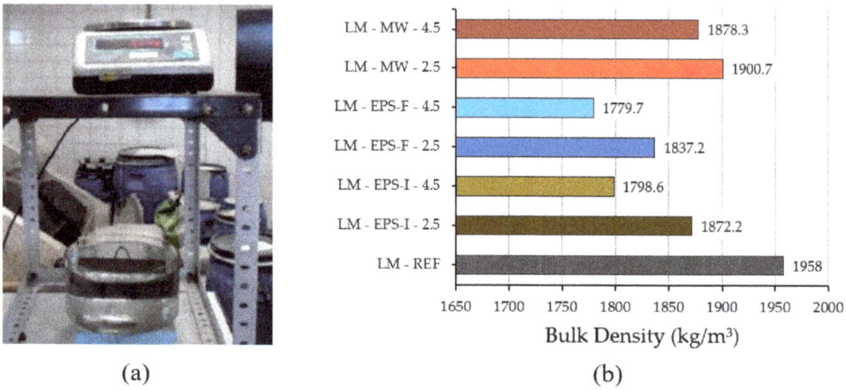

Figure 7. Apparent density. (**a**) Hydrostatic balance and taking measurements; (**b**) results obtained for the different mortars.

Table 11. Statistical analysis for the density test (kg/m^3) for each group of mortar and its comparison.

Type	Average	SD	Lower/Upper Limits
REF	1958.000	21.7313	(1941.296; 1974.704)
EPS-I	1835.389	44.7005	(1813.160; 1857.618)
EPS-F	1808.444	36.7891	(1790.150; 1826.739)
MW	1889.500	23.8605	(1877.634; 1901.366)

Comparison	REF vs. EPS-I	REF vs. EPS-F	REF vs. MW	EPS-I vs. EPS-F	EPS-I vs. MW	EPS-F vs. MW
K	34.583	43.194	15.944	8.611	−18.639	−27.250
p-value	0.000	0.000	0.199	0.952	0.014	0.000

According to the results obtained in Figure 7 and analyzed in Table 11, the incorporation of CDW from thermal insulation materials reduces the final density of hardened lime mortars. More specifically, mortars with the incorporation of expanded polystyrene are those with the greatest reduction in density, with the group of mortars incorporating EPS-F being the lightest (1808.444 kg/m^3). The mean differences found between the reference mortar with EPS-I and EPS-F both turned out to be statistically significant. The same happened in the case of the MW mortar, whose mean differences with both EPS-I and EPS-F were statistically significant for a confidence level greater than 95%. These results are in agreement with those obtained by other authors who have used expanded polystyrene to reduce the density of plaster precast [53].

On the other hand, Table 12 shows the difference in adhesion means for the same mortar with the two proportions of insulating residue used (2.5 g vs. 4.5 g).

Table 12. Comparison for densities (kg/m^3) between mortars with incorporation of insulation.

Type	Residue (g)	Average	SD	Interval	K	*p*-Value
EPS-I	2.5	1872.222	17.9846	(1858.398; 1886.046)	23.167	0.154
	4.5	1798.556	29.4962	(1775.883; 1821.228)		
EPS-F	2.5	1837.222	22.8242	(1819.678; 1854.766)	15.389	1.000
	4.5	1779.667	22.1754	(1762.621; 1796.712)		
MW	2.5	1900.667	22.1303	(1883.656; 1917.678)	9.111	1.000
	4.5	1878.333	20.9643	(1862.219; 1894.448)		

As can be seen in Table 12, mortars with a higher dosage of insulating residue (4.5 g) obtained a lower density than their counterparts with a lower dosage (2.5 g). However, the differences in density mean found between the different dosages used were not statistically significant in any of the types of mortars analyzed with the incorporation of these CDWs. This lower density of mortars with the incorporation of 4.5 g of insulating residues follows the lower mechanical resistance to compression obtained by this type of lime mortar.

3.2.3. Surface Hardness

The surface hardness of mortars represents the ability of these materials to resist being scratched by another material on their surface [84]. These types of tests are of great importance in lime mortars that are going to be used as coatings in rehabilitation works since they give us an idea of the resistance to surface wear that they offer over time [85]. To determine the values of this property, a Shore D durometer was used and the indications of the UNE 102042: 2014 [86] standard were adapted, taking, as a result, the average of five random measurements on two flat-parallel lateral faces of the specimens, with 4 × 4 × 16 cm hardened mortar. Figure 8 shows the test method and the results obtained for each of the dosages used.

(a)

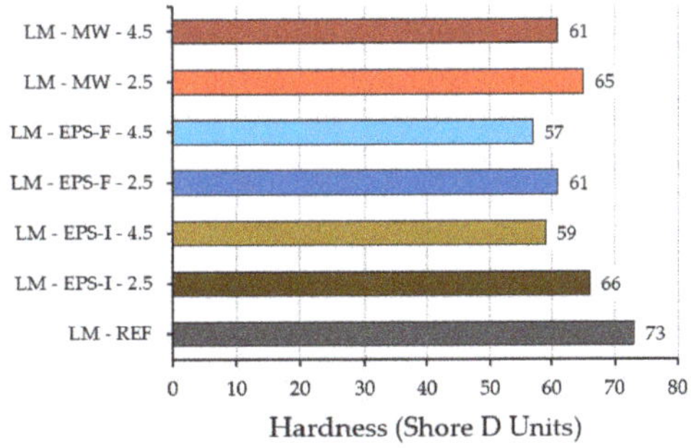

(b)

Figure 8. Surface hardness. (**a**) Taking measurements with a Shore D Durometer; (**b**) results obtained for the different mortars.

The analysis of the results of the Shore D hardness tests for each group of mortar is presented in Table 13. Due to the fulfillment of the necessary assumptions for the application of parametric tests, the comparison of means between groups of mortars was carried out employing the ANOVA and Dunn–Bonferroni tests.

Table 13. Statistical analysis for the surface hardness tests (Ud. Shore D) in each group of mortar and its comparison.

Type	Average		SD		Lower/Upper Limits	
REF	72.89		3.756		(68; 79)	
EPS-I	62.72		5.233		(51; 69)	
EPS-F	58.94		6.102		(50; 68)	
MW	63.44		4.997		(50; 71)	
Comparison	REF vs. EPS-I	REF vs. EPS-F	REF vs. MW	EPS-I vs. EPS-F	EPS-I vs. MW	EPS-F vs. MW
Difference of Means	10.167	13.944	9.444	3.778	−0.722	−4.500
p-value	0.000	0.000	0.000	0.213	1.000	0.078

As can be seen in the results shown in Table 13, the reference mortar without additions was the one with the highest Shore D surface hardness. In addition, the differences in means between the reference mortar and the other three types of mortar with the incorporation of thermal insulation materials were statistically significant. It should be noted that among the mortars with the incorporation of shredded insulation in their composition, the lime mortars with insulating mineral wool fiber were the ones with the highest surface resistance, and the mortars with the incorporation of expanded polystyrene were the ones with the worst performance against superficial deterioration.

Table 14 shows the comparative analysis of each type of mortar with CDW incorporation according to the amount of residue added in its manufacture.

Table 14. Comparison for the Shore D surface hardness tests between mortars with the incorporation of thermal insulation materials.

Type	Residue (g)	Average	SD	Interval	Difference of Means	p-Value
EPS-I	2.5	66.22	2.489	(64.31; 68.14)	7.000	0.065
	4.5	59.22	4.944	(55.42; 63.02)		
EPS-F	2.5	60.67	6.325	(55.81; 65.53)	3.444	1.000
	4.5	57.22	5.696	(52.84; 61.60)		
MW	2.5	65.44	3.283	(62.92; 67.97)	4.000	1.000
	4.5	61.44	5.769	(57.01; 65.88)		

From the analysis of Table 14, it can be seen that all mortars with the incorporation of 2.5 g of recycled thermal insulation were those with the highest average surface hardness. However, the difference in means for the counterparts that incorporated 4.5 g of residue were not statistically significant in any of the cases analyzed.

3.2.4. Water Absorption by Capillarity

When working with lime mortars for rehabilitation and their use as coatings, it is essential to know the mechanisms that cause the suction of water by porous materials [87]. This phenomenon is mainly produced by the surface attraction between liquid and solid, which causes a decrease in the free surface energy of the system if a drop of water spreads inside the porous body [88]. The absorption of water by capillarity depends on the diameter of the capillary pore and on how the network of pores inside the mortar are interconnected [89], although the type of aggregate used in its composition, the binder/aggregate ratio, and the difference also influence pressure between the penetrating water and the inner surface [90]. The test method that allows understanding of the coefficient of water absorption by capillarity in mortars is the one described in the UNE-EN 1015-18:2003 [65] standard, using half-rods of mortar submerged in water, one centimeter from the lower face, and fractured for 90 min. Figure 9 shows the test method and the results obtained for the different mortars used.

(a)

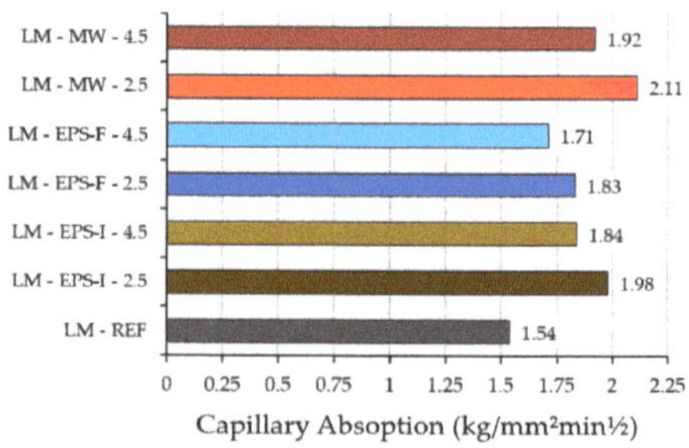

(b)

Figure 9. Water absorption by capillarity. (**a**) Standardized test according to UNE-EN 1015-18: 2003; (**b**) results obtained for the different absorption coefficients.

For the statistical discussion of the results obtained in the water absorption test by capillarity, Table 15 is used. This physical property met the criteria of normality and homoscedasticity required for the application of parametric tests, for this reason, a comparison was used for the mean difference using ANOVA and Dunn–Bonferroni analysis.

Table 15. Statistical analysis for absorption by capillarity (kg/mm^{2}min$^{0.5}$) in each group of mortar and its comparison.

Type	Average		SD		Lower/Upper Limits
REF	1.5378		0.16939		(1.4076; 16680)
EPS-I	1.9128		0.18905		(1.8188; 2.0068)
EPS-F	1.7739		0.17790		(1.6854; 1.8625)
MW	2.0150		0.17644		(1.9273; 2.1027)
Comparison	REF vs. EPS-I	REF vs. EPS-F	REF vs. MW	EPS-I vs. EPS-F	EPS-I vs. MW	EPS-F vs. MW
Difference of Means	−0.37500	−0.23611	−0.47722	0.13889	−0.10222	−0.24111
p-value	0.000	0.013	0.000	0.143	0.001	0.000

As can be seen in Table 15, the reference dosage is the one that shows the best behavior to prevent pathologies derived from the absorption of water by capillarity, since it has the lowest absorption coefficient (1.5378 kg/mm^{2}min$^{0.5}$). For their part, the dosages with the incorporation of mineral wool fiber thermal insulation materials were the ones that presented the worst results for this property, having an average water absorption by capillarity higher than that of the other mortars included in this study. All the analyzed mean differences turned out to be statistically significant for a confidence level greater than 95%, except for the comparison between EPS-I and EPS-F, which returned a non-statistically significant difference.

Table 16 shows the results obtained for the comparison between mortars with the incorporation of thermal insulation based on residue content.

Table 16. Comparison for the absorption by capillarity (kg/mm^{2}min$^{0.5}$) between mortars with incorporation of thermal insulation materials.

Type	Residue (g)	Average	SD	Interval	Difference of Means	*p*-Value
EPS-I	2.5	1.8944	0.17650	(1.8488; 2.1201)	0.14333	1.000
	4.5	1.8411	0.18231	(1.7010; 1.9812)		
EPS-F	2.5	1.8333	0.14925	(1.7186; 1.9481)	0.11889	1.000
	4.5	1.7144	0.19243	(1.5665; 1.8624)		
MW	2.5	2.1122	0.15959	(1.9895; 2.2349)	0.19444	0.360
	4.5	1.9178	0.13935	(1.8107; 2.0249)		

In all the cases analyzed in Table 16, the mortars with the highest proportion of thermal insulation material in their composition (4.5 g) presented less water absorption by capillarity, which may be due to the lower compactness of these mortars [91]. However, as reflected in the *p*-value collected in Table 16, the mean difference with the mortars that incorporated 2.5 g of insulating residue could not be considered statistically significant in any of the tires analyzed.

3.2.5. Thermal Conductivity

The building sector is moving towards an efficient use of natural resources and a reduction in the environmental impact generated by industrial activity; thereby, the sector is committed to the design of new eco-efficient materials that minimize energy consumption and incorporate CDW in composition [92]. In this sense, the study of the thermal performance of lime mortars for use in rehabilitation and restoration works is oriented towards the general idea of using durable materials that improve the starting conditions of the initial construction system [93]. For this reason, one of the greatest possibilities offered by the use of thermal insulation materials in the manufacture of this type of mortar is the reduction of the thermal conductivity of the materials made with these raw materials [94]. For this test, the standard UNE-EN 12667: 2002 [95] was used as a reference.

Figure 10 shows the equipment used to determine the thermal conductivity through the heat flow that passed through the mortar specimen, and the results obtained for this physical property. In addition, Table 17 presents the analysis of the results for each mortar group using the ANOVA and Dunn–Bonferroni tests.

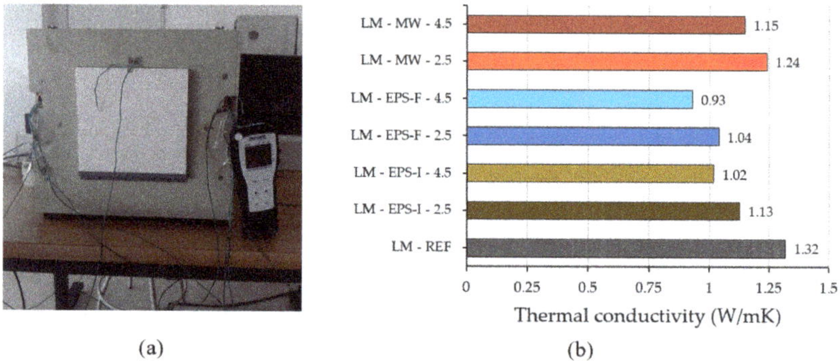

Figure 10. Thermal conductivity. (**a**) Test equipment; (**b**) results obtained for the different mortars with incorporation of thermal insulation material.

Table 17. Statistical analysis for thermal Conductivity (W/mK) in each group of mortar and its comparison.

Type	Average		SD		Lower/Upper Limits	
REF	1.3200		0.07382		(1.2633; 1.3767)	
EPS-I	1.0750		0.07876		(1.0358; 1.1142)	
EPS-F	0.9833		0.10210		(0.9326; 1.0341)	
MW	1.1956		0.08740		(1.1521; 1.2390)	
Comparison	REF vs. EPS-I	REF vs. EPS-F	REF vs. MW	EPS-I vs. EPS-F	EPS-I vs. MW	EPS-F vs. MW
Difference of Means	0.24500	0.33667	0.12444	0.09167	−0.12056	−0.21222
p-value	0.000	0.000	0.006	0.016	0.001	0.000

According to the results shown in Table 17, mortars with the incorporation of thermal insulation material reduce thermal conductivity compared to traditional lime mortars. The incorporation of EPS-F residue is the one that presents the best results on average (0.9833 W/mK), followed by the other residue of expanded polystyrene (EPS-I, 1.075 W/mK). Furthermore, when carrying out the mean comparison study, it was obtained that all the mean differences found in each of the comparisons were statistically significant for confidence levels above 95%.

Table 18 shows an exhaustive analysis of the results obtained for the comparison between mortars with the incorporation of thermal insulation as a function of residue content.

From the analysis of Table 19, it can be deduced that in the three types of mortars with the incorporation of CDW studied in this work, the thermal conductivity was reduced in the samples that contained a greater amount of thermal insulation material. Additionally, the difference in means obtained between mortars with the incorporation of foam was statistically significant. This does not happen in the same way with mortars made with mineral wool fiber, where the differences in the mean thermal conductivity between the mix containing 2.5 g and the one containing 4.5 g were not statistically significant.

Table 18. Comparison for thermal conductivity (W/mK) between mortars with incorporation of the different thermal insulation materials.

Type	Residue (g)	Average	SD	Interval	Difference of Means	*p*-Value
EPS-I	2.5 4.5	1.1311 1.0189	0.05231 0.05798	(1.0909; 1.1713) (0.9743; 1.0635)	0.11222	0.046
EPS-F	2.5 4.5	1.0400 0.9267	0.09042 0.08216	(0.9705; 1.1095) (0.8635; 0.9898)	0.11333	0.042
MW	2.5 4.5	1.2389 1.1522	0.07474 0.08012	(1.1814; 1.2963) (1.0906; 1.2138)	0.08667	0.340

Table 19. Analysis of flexural strength in the freezing test by Student's *t*-test.

Mortar	Test	Average	SD	Confidence Interval *	t	*p*-Value
LM-REF	Not cycles Cycles	1.7467 1.5578	0.10886 0.09365	(1.6630; 1.8303) (1.4858; 1.6298)	4.083	0.004
LM-EPS-I-2.5	Not cycles Cycles	1.4622 1.2689	0.09897 0.08838	(1.3861; 1.5383) (1.2010; 1.3368)	5.082	0.001
LM-EPS-I-4.5	Not cycles Cycles	1.2489 1.0633	0.08507 0.08031	(1.1835; 1.3143) (1.0016; 1.1251)	5.580	0.001
LM-EPS-F-2.5	Not cycles Cycles	1.3956 1.1811	0.16607 0.13923	(1.2679; 1.5232) (1.0741; 1.2881)	2.221	0.057
LM-EPS-F-4.5	Not cycles Cycles	1.1267 0.9722	0.09695 0.09576	(1.0521; 1.2012) (0.8986; 1.0458)	9.655	0.000
LM-MW-2.5	Not cycles Cycles	2.0467 1.7911	0.14874 0.09185	(1.9323; 2.1610) (1.7205; 1.8617)	4.659	0.002
LM-MW-4.5	Not cycles Cycles	1.8578 1.6211	0.13160 0.10325	(1.7566; 1.9589) (1.5417; 1.7005)	3.461	0.009

* Confidence intervals for the mean have been constructed for $\alpha = 0.05$.

3.3. Durability Tests

We understand the durability of mortars as their ability to withstand the physicochemical conditions to which they are exposed throughout their useful life, and that in some way may cause their deterioration. In this sense, durability tests are of vital importance when it comes to knowing the behavior of mortars over time. For this reason, this section contains three tests that evaluate the durability of the processed samples: freezing; salt crystallization; and shrinkage.

3.3.1. Freeze–Thaw Test

The freeze–thaw test evaluates the suitability of the mortars that are going to be used outdoors in climates that reach temperatures below 0 °C, determining the loss of mass and mechanical resistance that occurs in the samples when subjected to constant dimensional changes as a consequence of intermittent icing and thawing inside [96]. To carry out this test, the adopted recommendations of the UNE-EN 12371 standard for natural stone [66] were followed. The results shown in Figure 11 show the variations produced in flexural strength and loss of mass after the test.

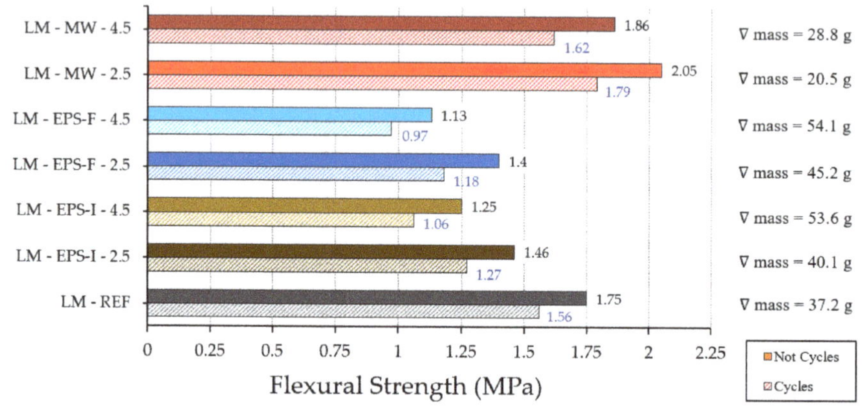

Figure 11. Results of the freeze–thaw test. Flexural strength of samples subjected to freezing cycles (striped stripes) and reference samples (solid stripes), including loss of mass.

As can be seen in Figure 11, lime mortars that incorporated 4.5 g of recycled insulating material had lower flexural strengths and greater loss of mass after the test than mortars that contained 2.5 g of insulation in their composition. Furthermore, it can be seen that the mortars that incorporated polystyrene insulation had a worse performance than the reference mortars without additions, which resulted in a high loss of mass after the test, together with a strong decrease in their mechanical resistance to bending. However, it can be seen that mortars that incorporated mineral wool insulation had greater stability and experienced less degradation when subjected to freeze–thaw cycles. These results are in agreement with those obtained by other authors who have studied the beneficial effect of the incorporation of fibers in the mortar matrix to increase the durability of these construction materials [97].

Table 19 shows the results derived from the statistical comparison between the flexural strengths obtained for the samples subjected to durability cycles versus the reference samples tested at the same age. As they were related samples and complied with the necessary assumptions, the Student's *t*-test was applied for related samples.

In Table 19, it can be seen that in all the cases studied, except in the sample LM-EPS-F-2.5, the difference in means between the flexural strength in the reference specimens and the specimens subjected to freeze–thaw cycles were statistically significant. In addition, it is observed that, in all cases, the mean flexural strength was greater in the specimens with no cycles, as expected, with the sample LM-MW-2.5 being the one with the highest flexural strength, and therefore the one with the best performance presented before this durability test.

3.3.2. Salt Crystallization

The crystallization of salts, of different nature and formation depending on the environmental parameters, is one of the most frequent and aggressive deterioration mechanisms that construction materials can suffer, especially if they are porous [98]. Although there is no specific standard to carry out this test on mortars, the UNE-EN 12370 standard for application to the natural stone was adapted [67]. This is a test that generally causes intense deterioration, since once the water from the solution evaporates and the salt crystallizes, it remains lodged inside the mortar and occupies a volume 17 times greater than in solution [99]. The results of the flexural test on the mortar samples with and without crystallization cycles are shown in Figure 12, together with the loss of mass of the cycled specimens.

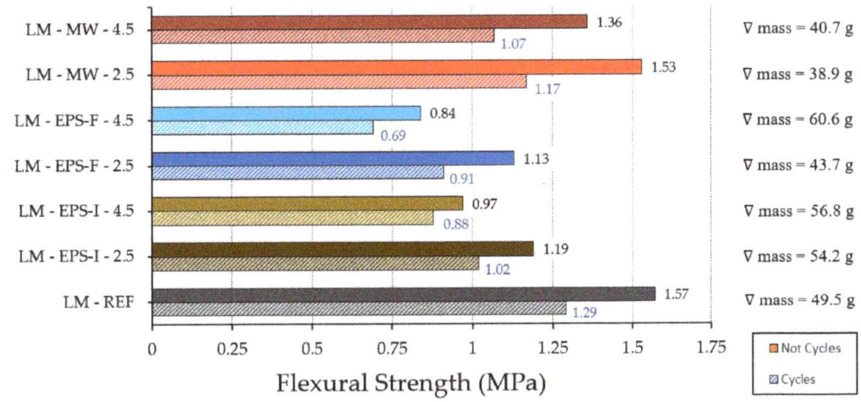

Figure 12. Results of the salt crystallization test. Flexural strength of cycled samples (scratched stripes) and reference samples (solid stripes), including loss of mass.

In Figure 12, it can be seen that the deterioration suffered by the mortar specimens after having been subjected to the salt crystallization test was greater than that obtained in the freezing test. It can be seen that there was a strong loss of mass in all the samples subjected to crystallization cycles, the loss of mass being greater in mortars that incorporated polystyrene insulation materials in their composition. It is also observed that the samples that contained a higher content of insulating residue (4.5 g) showed worse performance in this test, and lower bending strengths than the samples with less incorporation of residue (2.5 g). In addition, as was the case with the freezing test, mortars with the incorporation of mineral wool residue presented less mass loss than the rest when subjected to crystallization cycles, even reaching mechanical bending strengths close to that of lime mortars without additions.

Next, Table 20 presents the results derived from the statistical comparison between the flexural strengths obtained for samples subjected to salt crystallization cycles versus the reference samples tested at the same age. As was the case for the freezing test, these are related samples, and met the necessary assumptions to apply the Student's *t*-test.

Table 20. Analysis for flexural strength in the salt crystallization test by Student's *t* test.

Mortar	Test	Average	SD	Confidence Interval *	t	*p*-Value
LM-REF	Not cycles Cycles	1.5744 1.2856	0.11126 0.10760	(1.4889; 1.6600) (1.2028; 1.3683)	8.131	0.000
LM-EPS-I-2.5	Not cycles Cycles	1.1900 1.0244	0.07550 0.11780	(1.1320; 1.2480) (0.9339; 1.1150)	2.786	0.024
LM-EPS-I-4.5	Not cycles Cycles	0.9733 0.8811	0.11011 0.08950	(0.8887; 1.0580) (0.8123; 0.9499)	1.900	0.094
LM-EPS-F-2.5	Not cycles Cycles	1.1278 0.9133	0.06140 0.11045	(1.0806; 1.1750) (0.8284; 0.9982)	4.217	0.003
LM-EPS-F-4.5	Not cycles Cycles	0.8411 0.6822	0.11656 0.26570	(0.7515; 0.9307) (0.4880; 0.8965)	1.637	0.140
LM-MW-2.5	Not cycles Cycles	1.5311 1.1667	0.09880 0.08818	(1.4552; 1.6071) (1.0989; 1.2344)	7.197	0.000
LM-MW-4.5	Not cycles Cycles	1.3600 1.0700	0.12550 0.16432	(1.2635; 1.4565) (0.9437; 1.1963)	3.886	0.005

* Confidence intervals for the mean have been constructed for $\alpha = 0.05$.

As can be seen in Table 20, except for the mortars with the incorporation of 4.5 g of polystyrene, all the tested samples presented a significant mean difference between the flexural strength of the reference specimens and those subjected to crystallization cycles. It can be seen that, in all cases, the specimens without cycles that did not suffered deterioration or loss of mass showed greater resistance to bending and, as was the case with the freezing test, the LM-MW-2.5 dosage was the sample with the incorporation of insulation material that had the best performance when subjected to this durability test. It should be noted that these types of tests are of special relevance for the preparation of restoration mortars that are to be applied in urban environments, where it has been verified that run-off water can have a high salt content that is detrimental to the lime mortars applied outdoors [100].

3.3.3. Shrinkage

Shrinkage is the process of volumetric contraction that mortars undergo during their setting process [101]. This phenomenon has two distinct origins, which can lead to two types of shrinkage: thermal or hydraulic. In any case, it is a phenomenon that directly affects the durability of mortars, generating internal stresses that, depending on the modulus of elasticity and the deformation capacity of the mortar, can range from volume reduction to cracking of the samples. [102]. For this reason, it was decided that we would study the shrinkage produced in the lime mortars made for 180 days, using 25 × 25 × 287 mm samples. The results derived from this trial are shown in Figure 13.

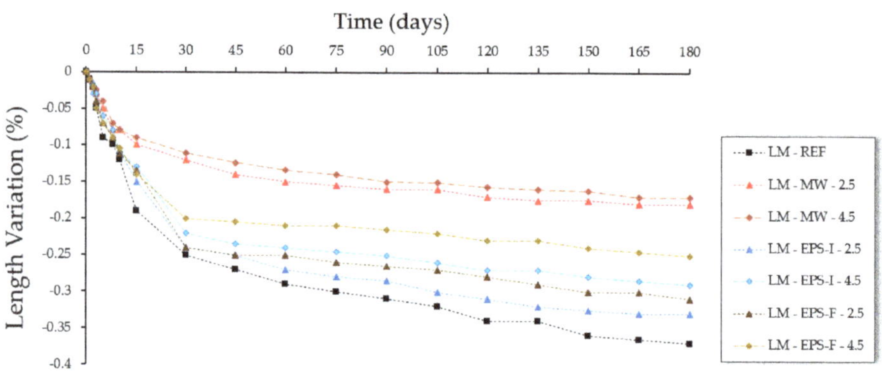

Figure 13. Shrinkage evolution. Length variation (%) during setting time (days).

The results obtained in Figure 13 show that the incorporation of the different isolates from CDW used in this investigation reduced the shrinkage of the lime mortars. In the case of the samples with the incorporation of EPS, it can be seen that the incorporation of an amount of 4.5 g decreased the shrinkage more than in the case of mortars that contained 2.5 g of this material, with the expanded polystyrene specially indicated for use, presenting better results on facades. However, and in agreement with other previous studies [103], the incorporation of mineral wool fibers was the one that reported greater dimensional stability for mortars.

4. Conclusions

This research delves into the possibility of incorporating waste from three different types of thermal insulation materials (EPS-F, EPS-I, and MW) as raw material in the manufacture of lime mortars. In this way, this research is intended to assist in moving towards a more sustainable management of construction and demolition waste, exploring the many possibilities offered by this type of waste to improve the technical performance of the lime mortars traditionally used in restoration and rehabilitation works.

In the mechanical characterization of mortars, it was possible to observe the beneficial effects of the incorporation of mineral wool insulation fibers in the production of these materials. In this way, an increase in the flexural strength of lime mortars was observed with the incorporation of this residue, compared with the reference mortars, not observing large differences between the two in compressive strengths. However, the incorporation of the expanded polystyrene granular residue was detrimental to the mechanical strength of the mortars, regardless of the residue content and the type of polystyrene. Finally, it was possible to observe that the best results were obtained in mortars with the incorporation of 2.5 g of residue, compared to mortars with 4.5 g of residue.

Regarding the physical characterization of the prepared mortars, both the Shore D surface hardness and the adhesion on ceramic surfaces were negatively affected as a result of the incorporation of these insulating residues. On the other hand, the incorporation of expanded polystyrene materials made it possible to reduce the final density of the mortars and their thermal conductivity, with mortars that incorporated 4.5 g of this type of residue in their composition having a better performance, and the prepared mortars having better characteristics with EPS-F. Finally, the absorption of water by capillarity was negatively affected, since in all the analyzed mortars higher absorption coefficients were obtained than in the reference lime mortars.

Finally, durability tests showed that mortars made with this type of thermal insulation waste can be designed for use in exterior cladding. Specifically, after freeze–thaw cycles and salt crystallization, it was possible to verify that lime mortars with the incorporation of mineral wool residue had good stability and better resistance after accelerated aging tests than traditional mortars. In all cases, it was observed that the content of 4.5 g of residue reduced the final mechanical strength of the material and led to a greater loss of mass after durability tests, compared with samples containing 2.5 g of thermal insulation materials. Furthermore, both in the freezing test and in the salt crystallization test, the mortars with the incorporation of EPS residue were the ones that showed the worst results. In the end, the shrinkage test showed the beneficial effect of incorporating these residues in the lime mortar matrix to provide them with greater dimensional stability, obtaining the best results after incorporating mineral wool fiber residues.

Author Contributions: Conceptualization, D.F. and E.Y.; methodology, D.F. and A.Z.; software, E.Y.; validation, D.F., E.Y. and C.M.; formal analysis, D.F.; investigation, A.Z., D.F. and E.Y.; resources, M.K.-K.; data curation, D.F. and E.Y.; writing—original draft preparation, D.F.; writing—review and editing, E.Y.; visualization, C.M.; supervision, C.M. and M.K.-K.; project administration, C.M.; funding acquisition, D.F. and M.K.-K. All authors have read and agreed to the published version of the manuscript.

Funding: This research received no external funding.

Institutional Review Board Statement: Not applicable.

Informed Consent Statement: Not applicable.

Data Availability Statement: Not applicable.

Acknowledgments: The authors would like to acknowledge the collaboration of the company URSA Ibérica Aislantes, SA, through the project P2054090068 "Thermo-acoustic solutions in housing renovation, simulation, and monitoring", which has served as a support and initiative framework for the realization of this research. They also want to thank the collaboration of Irantzu Recalde-Esnoz, who has helped enormously in the statistical treatment of the data.

Conflicts of Interest: The authors declare no conflict of interest.

References

1. D'Agostino, D.; Tsemekidi Tzeiranaki, S.; Zangheri, P.; Bertoldi, P. Assessing Nearly Zero Energy Buildings (NZEBs) development in Europe. *Energy Strategy Rev.* **2021**, *36*, 100680. [CrossRef]
2. Porras-Amores, C.; Martin Garcia, P.; Villoria Sáez, P.; del Rio Merino, M.; Vitielo, V. Assessing the Energy Efficiency Potential of Recycled Materials with Construction and Demolition Waste: A Spanish Case Study. *Appl. Sci.* **2021**, *11*, 7809. [CrossRef]

3. Cerezo-Narváez, A.; Piñero-Vilela, J.M.; Rodríguez-Jara, E.A.; Otero-Matero, M.; Pastor-Fernández, A.; Ballesteros-Pérez, P. Energy, emissions and economic impact of the new nZEB regulatory framework on residential buildings renovation: Case study in southern Spain. *J. Build. Eng.* **2021**, *42*, 103054. [CrossRef]
4. Bustos García, A.; Cobo Escamilla, A.; González Yunta, F. Mejora de las propiedades mecánicas de los morteros de cal hidráulica por adición de fibras de basalto. In Proceedings of the 2° Congreso Internacional de Innovación Tecnológica en Edificación (CITE 2017), Madrid, Spain, 8–10 March 2017; pp. 260–261, ISBN 978-84-16397-56-3.
5. Ministerio de Transporte, Movilidad y Agenda Urbana (MITMA). Available online: https://www.mitma.gob.es/ (accessed on 29 August 2021).
6. Pérez Fargallo, A.; Calama Rodríguez, J.M.; Flores Alés, V. Comparativa de resultados de rehabilitación energética para viviendas en función del grado de mejora. *Inf. Construcción* **2016**, *68*, 541. [CrossRef]
7. De Gobierno, P. *Real Decreto 2429/1979, de 6 de Julio, por el Que se Aprueba la Norma Básica de Edificación NBE-CT-79, Sobre Condiciones Térmicas en los Edificios*; No. 253; BOE: Madrid, Spain, 1979; pp. 24524–24550.
8. Lei, L.; Chen, W.; Wu, B.; Chen, C.; Liu, W. A building energy consumption prediction model based on rough set theory and deep learning algorithms. *Energy Build.* **2021**, *240*, 110886. [CrossRef]
9. Huang, B.; Gao, X.; Xu, X.; Song, J.; Geng, Y.; Sarkis, J.; Fishman, T.; Kua, H.; Nakatani, J. A Life Cycle Thinking Framework to Mitigate the Environmental Impact of Building Materials. *One Earth* **2020**, *3*, 564–573. [CrossRef]
10. Thormark, C. Conservation of energy and natural resources by recycling building waste. *Resour. Conserv. Recycl.* **2001**, *33*, 113–130. [CrossRef]
11. Objetivos Para el Desarrollo Sostenible (Naciones Unidas). Available online: https://www.un.org/sustainabledevelopment/es/objetivos-de-desarrollo-sostenible/ (accessed on 31 August 2021).
12. Villoria Sáez, P.; Omani, M. A diagnosis of construction and demolition waste generation and recovery practice in the European Union. *J. Clean. Prod.* **2019**, *241*, 118400. [CrossRef]
13. González, M.D.; Plaza Caballero, P.; Fernández, D.B.; Jordán Vidal, M.M.; del Bosque, I.F.S.; Medina Martínez, C. The Design and Development of Recycled Concretes in a Circular Economy Using Mixed Construction and Demolition Waste. *Materials* **2021**, *14*, 4762. [CrossRef]
14. European Commission (2018): Guidelines for the Waste Audits before Demolition and Renovation Works of Buildings. Available online: https://ec.europa.eu/growth/content/eu-construction-and-demolition-waste-protocol-0_en (accessed on 5 September 2021).
15. Villoria Sáez, P.; del Rio, M.; San-Antonio González, A.; Porras-Amores, C. Best practice measures assessment for construction and demolition waste management in building constructions. *Resour. Conserv. Recycl.* **2013**, *75*, 52–62. [CrossRef]
16. Gharfalkar, M.; Court, R.; Campbell, C.; Ali, Z.; Hillier, G. Analysis of waste hierarchy in the European waste directive 2008/98/EC. *Waste Manag.* **2015**, *39*, 305–313. [CrossRef] [PubMed]
17. European Commission (2019b), COM(2019) 640. The European Green Deal. Available online: https://eur-lex.europa.eu/resource.html?uri=cellar:b828d165-1c22-11ea-8c1f-01aa75ed71a1.0002.02/DOC_1&format=PDF (accessed on 31 August 2021).
18. Superti, V.; Houmani, C.; Binder, C.R. A systemic framework to categorize Circular Economy interventions: An application to the construction and demolition sector. *Resour. Conserv. Recycl.* **2021**, *173*, 105711. [CrossRef]
19. López-Ruiz, L.A.; Roca Ramón, X.; Gassó Domingo, S. The circular economy in the construction and demolition waste sector—A review and an integrative model approach. *J. Clean. Prod.* **2020**, *248*, 119238. [CrossRef]
20. Stefanidou, M.; Anastasiou, E.; Georgiadis Filikas, K. Recycled sand in lime-based mortars. *Waste Manag.* **2014**, *34*, 2595–2602. [CrossRef] [PubMed]
21. Faria, P.; Henriques, F.; Rato, V. Comparative evaluation of lime mortars for architectural conservation. *J. Cult. Herit.* **2008**, *9*, 338–346. [CrossRef]
22. Saeli, M.; Senff, L.; Tobaldi, D.M.; Seabra, M.P.; Labrincha, J.A. Novel biomass fly ash-based geopolymeric mortars using lime slaker grits as aggregate for applications in construction: Influence of granulometry and binder/aggregate ratio. *Constr. Build. Mater.* **2019**, *227*, 116643. [CrossRef]
23. Torres, I.; Matias, G.; Faria, P. Natural hydraulic lime mortars—The effect of ceramic residues on physical and mechanical behaviour. *J. Build. Eng.* **2020**, *32*, 101747. [CrossRef]
24. Pozo-Antonio, J.S. Evolution of mechanical properties and drying shrinkage in lime-based and lime cement-based mortars with pure limestone aggregate. *Constr. Build. Mater.* **2015**, *77*, 472–478. [CrossRef]
25. Apostolopoulou, M.; Bakolas, A.; Kotsainas, M. Mechanical and physical performance of natural hydraulic lime mortars. *Constr. Build. Mater.* **2021**, *290*, 123272. [CrossRef]
26. González-Cortina, M.; Villanueva-Domínguez, L. Aired ime and chamotte hydraulic mortars. *Mater. Constr.* **2002**, *52*, 65–76.
27. Bustos-García, A.; Moreno-Fernández, E.; Yunta-González, F.; Cobo-Escamilla, A.A. Influencia de la adición de fibras en las propiedades de los morteros de cal hidráulica. *DYNA* **2018**, *93*, 228–232. [CrossRef]
28. Maravelaki-Kalaitzaki, P. Hydraulic lime mortars with siloxane for waterproofing historic masonry. *Cem. Concr. Res.* **2017**, *37*, 283–290. [CrossRef]
29. Forster, A.M.; Válek, J.; Hughes, J.J.; Pilcher, N. Lime binders for the repair of historic buildings: Considerations for CO_2 abatement. *J. Clean. Prod.* **2020**, *252*, 119802. [CrossRef]

30. Ergenç, D.; Fort, R. Accelerating carbonation in lime-based mortar in high CO_2 environments. *Constr. Build. Mater.* **2018**, *188*, 314–325. [CrossRef]
31. Chen, K.; Wang, J.; Yu, B.; Wu, H.; Zhang, J. Critical evaluation of construction and demolition waste and associated environmental impacts: A scientometric analysis. *J. Clean. Prod.* **2021**, *287*, 125071. [CrossRef]
32. Instituto Nacional de EstadísticaEstadística Sobre Generación de Residuos en el Sector Servicios Y Construcción. Año 2017. Cantidad de Residuos Generados por Actividad Económica CNAE-2009, Clase de Residuo y Tipo de Peligrosidad. 2017. Available online: https://www.ine.es/jaxi/Tabla.htm?path=/t26/e068/p03/a2017/l0/&file=01001.px&L=0 (accessed on 16 July 2021).
33. Peixoto-Rosado, L.; Vitale, P.; Penteado, C.S.G.; Arena, U. Life cycle assessment of natural and mixed recycled aggregate production in Brazil. *J. Clean. Prod.* **2017**, *151*, 634–642. [CrossRef]
34. Etxeberría, M.; Vázquez, E.; Marí, A.; Barra, M. Influence of amount of recycled coarse aggregates and production process on properties of recycled aggregate concrete. *Cem. Concr. Res.* **2007**, *37*, 735–742. [CrossRef]
35. Piña Ramírez, C. Comportamiento físico-mecánico y térmico de los morteros de cemento aditivados con fibras minerales procedentes de residuos de construcción y demolición. Ph.D. Thesis, E.T.S. de Edificación (UPM), Madrid, Spain, 2018. [CrossRef]
36. Milling, A.; Mwasha, A.; Martin, H. Exploring the full replacement of cement with expanded polystyrene (EPS) waste in mortars used for masonry construction. *Constr. Build. Mater.* **2020**, *253*, 119158. [CrossRef]
37. Ali, Y.A.Y.; Fahmy, E.H.A.; Abouzeid, M.N.; Shaheen, Y.B.I.; Abdel Mooty, M.N. Use of expanded polystyrene wastes in developing hollow block masonry units. *Constr. Build. Mater.* **2020**, *241*, 118149. [CrossRef]
38. Koksal, F.; Mutluay, E.; Gencel, O. Characteristics of isolation mortars produced with expanded vermiculite and waste expanded polystyrene. *Constr. Build. Mater.* **2019**, *236*, 117789. [CrossRef]
39. Piña, C.; Vidales, A.; Serrano-Somolinos, R.; del Rio, M.; Atanes-Sánchez, E. Analysis of fire resistance of cement mortars with mineral wool from recycling. *Constr. Build. Mater.* **2020**, *265*, 120349. [CrossRef]
40. Fantilli, A.P.; Sicardi, S.; Dotti, F. The use of wool as fiber-reinforcement in cement-based mortar. *Constr. Build. Mater.* **2017**, *139*, 562–569. [CrossRef]
41. Boynton, R.S. *Chemistry and Technology of Lime and Limestone*; John Wiley & Sons Inc.: New York, NY, USA, 1980.
42. UNE-EN 459-1:2002. Building Lime—Part 1: Definitions, Specifications, and Conformity Criteria. Available online: https://www.en.une.org/encuentra-tu-norma/busca-tu-norma/norma?c=N0027076 (accessed on 18 January 2022).
43. UNE 80103:2013. Test Methods of Cements. Physical Analysis. Actual Density Determination. Available online: https://www.une.org/encuentra-tu-norma/busca-tu-norma/norma?c=N0052011 (accessed on 18 January 2022).
44. Saint-Astier. Data Sheet for NHL 5. Available online: http://www.calhidraulica.es/gama-completa/ (accessed on 18 September 2021).
45. UNE-EN 933-2/1M:1999. Test for Geometrical Properties of Aggregates. Part 2: Determination of Particle Size Distribution. Test Sieves, Nominal Size of Apertures. Available online: https://www.une.org/Buscador (accessed on 18 January 2022).
46. Ontiveros-Ortega, E.; Rodríguez-García, R.; González-Serrano, A.; Molina, L. Evolution of mechanical properties in aerial lime mortars of traditional manufacturing, the relationship between putty and powder lime. *Constr. Build. Mater.* **2018**, *191*, 575–589. [CrossRef]
47. NBE FL-90. Norma Básica de Edificación. Muros Resistentes de Fábrica de Ladrillo. Available online: http://www.madrid.org/bdccm/normativa/PDF/Ladrillos%20y%20bloques/Normas%20Tratadas/ESRd172390.pdf (accessed on 18 January 2022).
48. UNE-EN 13139: 2003. Aggregates for Mortar. Available online: https://standards.iteh.ai/catalog/standards/cen/d1169c31-b23e-41fa-8f1d-cf3a9b0ce8dd/en-13139-2002 (accessed on 18 January 2022).
49. UNE-EN 933-1: 2012. Tests for Geometrical Properties of Aggregates—Part 1: Determination of Particle Size Distribution—Sieving Method. Available online: https://www.en-standard.eu/une-en-933-1-2012-tests-for-geometrical-properties-of-aggregates-part-1-determination-of-particle-size-distribution-sieving-method/ (accessed on 18 January 2022).
50. UNE-EN 146404: 2018. Aggregates for Concrete. Determination of the Coefficient of Friability of the Sands. Available online: https://www.mystandards.biz/standard/une-146404-2018-26.12.2018.html (accessed on 18 January 2022).
51. UNE-EN 1097-3: 1999. Tests for Mechanical and Physical Properties of Aggregates—Part 3: Determination of Loose Bulk Density and Voids. Available online: https://www.en-standard.eu/une-en-1097-3-1999-tests-for-mechanical-and-physical-properties-of-aggregates-part-3-determination-of-loose-bulk-density-and-voids/ (accessed on 18 January 2022).
52. UNE-EN 1097-6: 2014. Tests for Mechanical and Physical Properties of Aggregates—Part 6: Determination of Particle Density and Water Absorption. Available online: https://www.en-standard.eu/une-en-1097-6-2014-tests-for-mechanical-and-physical-properties-of-aggregates-part-6-determination-of-particle-density-and-water-absorption/ (accessed on 18 January 2022).
53. González Madariaga, F.J.; Lloveras Macia, J. EPS (expanded poliestyrene) recycled bends mixed with plaster or stucco, some applications in building industry. *Inf. Construcción* **2008**, *60*, 35–43. [CrossRef]
54. Ferrándiz-Mas, V.; García-Alcocel, E. Physical and mechanical characterization of Portland cement mortars made with expanded polystyrene particles addition (EPS). *Mater. Construcción* **2012**, *62*, 547–566. [CrossRef]
55. del Río Merino, M.; Santa Cruz Astorqui, J.; González Cortina, M. Morteros aligerados con arcilla expandida: Influencia de la granulometría y la adicción de fibras de vidrio AR en el comportamiento mecánico. *Inf. Construcción* **2005**, *57*, 39–46. [CrossRef]
56. Piña-Ramírez, C.; del Rio Merino, M.; Viñas Arrebola, C.; Vidales Barriguete, A.; Kosior-Kazberuk, M. Analysis of the mechanical behaviour of the cement mortars with additives of mineral wool fibres from recycling of CDW. *Constr. Build. Mater.* **2019**, *210*, 56–62. [CrossRef]

57. Morón, A.; Ferrández, D.; Saiz, P.; Morón, C. Experimental Study with Cement Mortars Made with Recycled Concrete Aggregate and Reinforced with Aramid Fibers. *Appl. Sci.* **2021**, *11*, 7791. [CrossRef]
58. Canal de Isabel II. *Informe Anual Sobre la Calidad del Agua en Madrid*; Canal de Isabel II: Madrid, Spain, 2012.
59. UNE-EN 196-1:2018. Methods of Testing Cement—Part 1: Determination of Strength. Available online: https://www.en-standard.eu/une-en-196-1-2018-methods-of-testing-cement-part-1-determination-of-strength/ (accessed on 18 January 2022).
60. UNE-EN 1015-3:2000/A2:2007. Methods of Test for Mortar for Masonry—Part 3: Determination of Consistence of Fresh Mortar (by Flow Table). Available online: https://www.en-standard.eu/une-en-1015-3-2000-a2-2007-methods-of-test-for-mortar-for-masonry-part-3-determination-of-consistence-of-fresh-mortar-by-flow-table/ (accessed on 18 January 2022).
61. Vaishnavi, M.; Aswathi, A.; Sri Saarani, S.; Varghese, A.; Sathyan, D.; Mini, K.M. Strength and workability characteristics of coir and nylon fiber reinforced self-compacting mortar. *Mater. Today Proc.* **2021**, *46*, 4696–4701. [CrossRef]
62. Serrano, R.; Cobo, A.; Prieto, M.I.; Nieves-González. M. Analysis of fire resistance of concrete with polypropylene or steel fibers. *Constr. Build. Mater.* **2016**, *122*, 302–309. [CrossRef]
63. UNE-EN 1015-11:2000/A1:2007. Methods of Test for Mortar for Masonry—Part 11: Determination of Flexural and Compressive Strength of Hardened Mortar. Available online: https://www.en.une.org/encuentra-tu-norma/busca-tu-norma/norma/?c=N0039892 (accessed on 18 January 2022).
64. UNE-EN 1015-2:1999/A1:2007. Methods of Test for Mortar for Masonry—Part 2: Bulk Sampling of Mortars and Preparation of Test Mortars. Available online: https://www.en-standard.eu/une-en-1015-2-1999-a1-2007-methods-of-test-for-mortar-for-masonry-part-2-bulk-sampling-of-mortars-and-preparation-of-test-mortars/ (accessed on 18 January 2022).
65. UNE-EN 1015-18:2003. Methods of Test for Mortar for Masonry—Part 18: Determination of Water Absorption Coefficient due to Capillary Action of Hardened Mortar. Available online: https://www.en-standard.eu/une-en-1015-18-2003-methods-of-test-for-mortar-for-masonry-part-18-determination-of-water-absorption-coefficient-due-to-capillary-action-of-hardened-mortar/ (accessed on 18 January 2022).
66. UNE-EN 12371:2011. Natural Stone Test Methods—Determination of Frost Resistance. Available online: https://www.en-standard.eu/une-en-12371-2011-natural-stone-test-methods-determination-of-frost-resistance/ (accessed on 18 January 2022).
67. UNE-EN 12370:20201. Natural Stone Test Methods—Determination of Resistance to Salt Crystallisation. Available online: https://www.en-standard.eu/une-en-12370-2020-natural-stone-test-methods-determination-of-resistance-to-salt-crystallisation/ (accessed on 18 January 2022).
68. UNE 80-112-89 Experimental: Métodos de Ensayo de Cementos: Ensayos Físicos: Determinación de la Retracción de Secado y del Hinchamiento en Agua. Available online: https://books.google.com/books/about/UNE_80_112_89_experimental.html?id=Os39cQAACAAJ (accessed on 18 January 2022).
69. Pardo, A.; Ruiz, M.A.; San Martín, R. *Análisis de Datos en Ciencias Sociales y de la Salud I*; Editorial Síntesis: Madrid, Spain, 2015.
70. Pardo, A.; San Martín, R. *Análisis de Datos en Ciencias Sociales y de la Salud II*; Editorial Síntesis: Madrid, Spain, 2015.
71. Martínez, W.; Alonso, E.M.; Rubio, J.C.; Bedolla, J.A.; Velasco, F.A.; Torres, A.A. Comportamiento mecánico de morteros de cal apagada artesanalmente, adicionados con mucílago de cactácea y ceniza volcánica, para uso en restauración y conservación de monumentos coloniales. *Rev. Construcción* **2008**, *7*, 93–101.
72. Alejandre Sánchez, F.J.; Flores Alés, V. *Blasco López, F.J. Martín del Río, J.J. La Cal: Investigación, Patrimonio y Restauración*; Universidad de Sevilla: Sevilla, Spain, 2014.
73. Piña, C.; del Rio, M.; Viñas, C.; Vidales, A.; Aguilera, P. Durability of cement mortars reinforced with insulation waste from the construction industry. *J. Build. Eng.* **2021**, *40*, 102719. [CrossRef]
74. Apostolopoulou, M.; Armaghani, D.J.; Bakolas, A.; Douvika, M.G.; Moropoulu, A.; Asteris, P.G. Compressive strength of natural hydraulic lime mortars using soft computing techniques. *Procedia Struct. Integr.* **2019**, *17*, 914–923. [CrossRef]
75. Rigopoulus, I.; Kyriakou, L.; Vasiliades, M.A.; Kyratsi, T.; Efstathiou, A.M.; Ioannou, I. Improving the carbonation of air lime mortars at ambient conditions via the incorporation of ball-milled quarry waste. *Constr. Build. Mater.* **2021**, *301*, 124073. [CrossRef]
76. Rita-Santos, A.; Rosario-Veiga, M.; Santos-Silva, A.; de Brito, J.; Ignacio Álvarez, J. Evolution of the microstructure of lime-based mortars and influence on the mechanical behaviour: The role of the aggregates. *Constr. Build. Mater.* **2018**, *187*, 907–922. [CrossRef]
77. Morón, A.; Ferrández, D.; Saiz, P.; Vega, G.; Morón, C. Influence of Recycled Aggregates on the Mechanical Properties of Synthetic Fibers-Reinforced Masonry Mortars. *Infrastructures* **2021**, *6*, 84. [CrossRef]
78. UNE-EN 998-2:2012. Specification for Mortar for Masonry—Part 2: Masonry Mortar. Available online: https://www.en-standard.eu/une-en-998-2-2018-specification-for-mortar-for-masonry-part-2-masonry-mortar/ (accessed on 18 January 2022).
79. García, G.; González, M.; del Rio, M.; Magdalena, F. Morteros adicionados con metacaolín: Efecto de la proporción del agregado. *DYNA* **2017**, *92*, 155–157. [CrossRef]
80. UNE-EN 1015-12:2016. Methods of Test for Mortar for Masonry—Part 12: Determination of Adhesive Strength of Hardened Rendering and Plastering Mortars on Substrates. Available online: https://www.en-standard.eu/une-en-1015-12-2016-methods-of-test-for-mortar-for-masonry-part-12-determination-of-adhesive-strength-of-hardened-rendering-and-plastering-mortars-on-substrates/ (accessed on 18 January 2022).
81. García López de la Osa, G. Análisis de la Adherencia Entre Morteros y Piezas Cerámicas. Ph.D. Thesis, Universidad Politécnica de Madrid, Madrid, Spain, 2020.

82. Sáiz Martínez, P.; González Cortina, M.; Fernández Martínez, F.; Rodríguez Sánchez, A. Comparative study of three types of fine recycled aggregates from construction and demolition waste (CDW), and their use in masonry mortar fabrication. *J. Clean. Prod.* **2016**, *118*, 162–169. [CrossRef]
83. UNE-EN 1015-10:2000/A1:2007. Methods of Test for Mortar for Masonry—Part 10: Determination of Dry Bulk Density of Hardened Mortar. Available online: https://www.en-standard.eu/une-en-1015-10-2000-a1-2007-methods-of-test-for-mortar-for-masonry-part-10-determination-of-dry-bulk-density-of-hardened-mortar/ (accessed on 18 January 2022).
84. Sáiz Martínez, P. Utilización de arenas procedentes de Residuos de Construcción y Demolición, RCD, en la fabricación de morteros de albañilería. Doctoral Thesis, Universidad Politécnica de Madrid, Madrid, Spain, 2015.
85. Vidales Barriguete, A. Caracterización Fisicoquímica y Aplicaciones de yeso con Adición de Residuo Plástico de Cables Mediante Criterios de Economía Circular. Ph.D. Thesis, Universidad Politécnica de Madrid, Madrid, Spain, 2019. [CrossRef]
86. UNE 102042:2014. Gypsum Plasters. Other Test Methods. Available online: https://infostore.saiglobal.com/en-us/Standards/UNE-102042-2014-26752_SAIG_AENOR_UNE_AENOR_UNE_58700/en-us/contact_us/ (accessed on 18 January 2022).
87. Silva, B.A.; Ferreira-Pinto, A.P.; Gomes, A.; Candeias, A. Short- and long-term properties of lime mortars with water-reducers and a viscosity-modifier. *J. Build. Eng.* **2021**, *43*. [CrossRef]
88. García Morales, S. Metodología de Diagnóstico de Humedades de Capilaridad Ascendente y Condensación Higroscópica, en Edificios Históricos. Ph.D. Thesis, Universidad Politécnica de Madrid, Madrid, Spain, 1995.
89. Escudero Lafont, M.E.; García Morales, S. Historical flat roofs of earth in the Mediterranean: Research for the characterization of the water behavior of the materials of the roofs of Ibiza. *Inf. Construcción* **2019**, *71*, 301. [CrossRef]
90. Morón, C.; Ferrández, D.; Saiz, P.; Yedra, E. Measuring system of capillary rising damp in cement mortars. *Measurement* **2019**, *135*, 252–259. [CrossRef]
91. Zhao, H.; Ding, J.; Juang, Y.; Xu, G.; Li, W.; Zhang, S.; Wang, P. Investigation on sorptivity and capilalarity coefficient of mortar and their relationship based on microstructure. *Constr. Build. Mater.* **2020**, *265*, 120332. [CrossRef]
92. San Antonio González, A.D. Caracterización de Compuestos Eco-Eficientes de Yeso Aligerado con Residuo de Poliestireno extruido (XPS). Ph.D. Thesis, Universidad Politécnica de Madrid, Madrid, Spain, 2017. [CrossRef]
93. Varas-Muriel, M.J.; Fort, R.; Gómez-Heras, M. Assessment of an underfloor heating system in a restored chapel: Balancing thermal comfort and historic heritage conservation. *Energy Build.* **2021**, *251*, 111361. [CrossRef]
94. Bicer, A.; Kar, F. Thermal and mechanical properties of gypsum plaster mixed with expanded polystyrene and tragacanth. *Therm. Sci. Eng. Progress* **2017**, *1*, 59–65. [CrossRef]
95. 2. UNE-EN 12667:200. Thermal Performance of Building Materials and Products. Determination of Thermal Resistance by Means of Guarded Hot Plate and Heat Flow Meter Methods. Products of High and Medium Thermal Resistance. Available online: https://www.en-standard.eu/une-en-12667-2002-thermal-performance-of-building-materials-and-products-determination-of-thermal-resistance-by-means-of-guarded-hot-plate-and-heat-flow-meter-methods-products-of-high-and-medium-thermal-resistance/ (accessed on 18 January 2022).
96. Defus, A.; Sansonetti, A.; Possetini, E.; Tedeschi, C.; Vettori, S.; Realini, M. The effectiveness of di-ammonium hydrogen phosphate (DAP) consolidation treatment on lime-based mortars weathered by freeze-thaw cycles. *J. Cult. Herit.* **2021**, *50*, 1–12. [CrossRef]
97. Dong, F.; Wang, H.; Yu, J.; Liu, K.; Guo, Z.; Duan, X.; Quiong, X. Effect of freeze–thaw cycling on mechanical properties of polyethylene fiber and steel fiber reinforced concrete. *Constr. Build. Mater.* **2021**, *295*, 123427. [CrossRef]
98. Franzoni, E.; Santandrea, M.; Gentilini, C.; Fregni, A.; Carloni, C. The role of mortar matrix in the bond behavior and salt crystallization resistance of FRCM applied to masonry. *Constr. Build. Mater.* **2019**, *209*, 592–605. [CrossRef]
99. Syed Ahmed Kabeer, K.I.; Bisht, K.; Jothi Saravanan, T.J.; Kumar Vyas, A. Effect of marble slurry on the microstructure of cement mortars subjected to salt crystallization and alternate wetting and drying cycles. *J. Build. Eng.* **2021**, *44*, 103342. [CrossRef]
100. García-Talegón, J.; Vicente, M.A.; Molina, A. Decay of granite monuments due to salt crystallization in a non-polluted urban environment. *Mater. Construcción* **1999**, *49*. [CrossRef]
101. Lu, T.; Li, Z.; Huang, H. Restraining effect of aggregates on autogenous shrinkage in cement mortar and concrete. *Constr. Build. Mater.* **2021**, *289*, 123166. [CrossRef]
102. Tang, F.; Li, Z.; Tang, Y.; Chen, Y.; Li, H.N. Simultaneous measurement of shrinkage and coefficient of thermal expansion of mortar based on EFPI sensors with nanometer resolution. *Measurement* **2020**, *152*, 107376. [CrossRef]
103. Morón, C.; Saiz, P.; Ferrández, D.; García-Fuentevilla, L. New System of Shrinkage Measurement through Cement Mortars Drying. *Sensors* **2017**, *17*, 522. [CrossRef] [PubMed]

buildings

MDPI

Article

Renovation of Modernist Architecture Study Based on Selected Cases

Joanna Jablonska * and Lukasz Wojciechowski

Faculty of Architecture, Wroclaw University of Science and Technology, 50-370 Wroclaw, Poland; lukasz.wojciechowski@pwr.edu.pl
* Correspondence: joanna.jablonska@pwr.edu.pl

Abstract: Modernist architecture is one of the most significant movements which serve as grounds for contemporary creations. At the beginning of the 20th century, the impact of new structural and building technologies allowed designers to find new ways of architectural expression. It is rooted in thought-out composition and simplicity which serve as a background for the exposition of structure. The geometrical interplay of a building's elements highlights the impressively thin structures and raw surfaces of novel building materials. Nowadays, in selected regions, the architecture of the Modernist Period is neglected or loosely refurbished. As an effect of this phenomenon, buildings are demolished or deprived of aesthetical values. Thus, this article aims at showing an overview of sectional research on good practice. We base the study of on-site design implementations of several modernist architecture refurbishments and focus on technology and design assumptions with optimised thermal modernisation. The paper reveals a set of examples for the refurbishment of modernist architecture, with calculations of heat energy coefficients of the initial and design phase. The presented thermal modernisations aim at adjusting buildings to new requirements concerning energy without any loss of initial architectural expression.

Keywords: modernist architecture; thermal refurbishment; monument preservation; energy reduction; architectural up-cycling

Citation: Jablonska, J.; Wojciechowski, L. Renovation of Modernist Architecture Study Based on Selected Cases. *Buildings* **2022**, *12*, 195. https://doi.org/10.3390/buildings12020195

Academic Editors: Paulo Santos and Mark Bomberg

Received: 10 December 2021
Accepted: 5 February 2022
Published: 8 February 2022

1. Introduction

The need for post-war modernist architecture preservation is a relatively new concept (especially in Poland, where socialist architecture is still neglected), which arose together with an increase of awareness on sustainability and appreciation for iconic solutions [1,2]. The Modern Period, based on a separation between past and present or enlightenment and individualism [3], made a significant impact on European society. The architecture responding to logical engineering [1] offered more than living and workspace [4]. Nowadays, we admire the interplay of forms and proportions, intelligent space management, and the beauty rooted in modesty [5,6]. Though occasionally problematic, modernist buildings require skillful preservation highlighting their qualities.

As Giulian and Bucchignani [7] indicate, the post-war modernist architecture initially was not planned for durability in many cases–especially in the socialist countries where the building techniques, economy and low quality materials and bad craftsmanship lead to many faults of ready buildings (i.e., problems with insulation, overheating, soundproofing, structural joints and reinforcement protection). Thus, nowadays these buildings are perceived as badly ageing and easy to demolish [4]. Post-war modern architecture is a fragile heritage. Most of the buildings in Poland were made of low-quality materials without precision or proper attention to detail. In the minds of people, they represent the creation of the faulty political and economic system. The public does not value this heritage, and some historians and architects neglect it. Architectural values of the Modernist Period in Poland require a sensitive eye, imagination and knowledge to see the initial ideas due to the damages in these buildings–done in the process of careless maintenance and renovation (most of which took place in the 1990s and the 2000s).

Also, a misconception of initial ideas and the apartment crisis in Europe after WWII led to building a lot and poorly [8]. There are a lot of issues involved in the preservation of modernist architecture, starting from novel changes in the lifestyle of users [7], through imperfect technology of flat roofs [1], lack of shading [2], to excessive energy consumption [9]. Especially the latter is crucial in terms of European Union policies, like the New European Bauhaus [10] or Renovation wave [11]. Moreover, the European Green Deal initiative aims at the reduction of greenhouse gas emissions by 55% by the year 2030, and totally by 2050 [12]. All recalled documents highlight that refurbishment investment must be sustainable [9]. According to Sonnleithner [13] and European Union recommendations [10] of high quality, such an approach is advisable, for it favours architectural recycling, instead of raising new, and at occasions redundant, buildings [7,14]. Also crucial may be the economic aspect, bringing substantial savings [14].

The refurbishment issue of modern architecture is present in scientific discourse. Seen from many angles, it always shows a local reference. A good example is an elaboration of Mulfarth et al. [2] on tall buildings from Sao Paulo; Urbanik and Tomaszewicz's [1] article on flat roof renovation in Wroclaw; and Peters' [14] publication on housing in Denmark. Yet, Pikas et al. [9] indicate that lack of knowledge is one of the barriers to the renovation of modernist (standardised) apartment buildings. The literature insufficiency is understandable, for each region (climatic zone) requires different handling the refurbishment design. Therefore, each published case study provides valuable information for understanding the complex issues of renovation (whether successful or not) and allows for the exchange of knowledge. Sonnleithner [13], p. 5 states: "In the search for goal-oriented approaches to solutions and specific action steps, the development of new concepts and research into previously unknown facts are essential. The experience and familiarity with previous activities, experiments, initiatives and projects from history are valuable. Particularly in the field of building renovation [. . .], a look into the past might help to ensure a clear view of possible solutions for the future." Thus, in the article, we present a practical (implemented) approach from Wroclaw (Poland) regarding the renovation of a few modernist buildings. In this way, we add another portion of practical knowledge into the discourse. The crucial aspect of the presentation is to propose solutions not interfering with the architectural and aesthetic values of modernist architecture.

The article aims at presenting designed and implemented solutions for the thermal renovation of modernist architecture. This aspect is especially valid because 75% of buildings in Europe are not energy efficient [11]. Moreover, the literature describes successful modernist implementations where energy consumption decreased significantly, i.e., Brazilian cases by Mulfarth et al. [2]. In general, our goal was following Boza-Kiss et al. [15], p. 7. "decarbonizing the European building stock"—at least a part of it.

2. Materials and Methods

The case studies for the article are the works of the architectural office VROA Architekci co-owned by one of the authors of this publication. Lukasz Wojciechowski is the co-author of all the presented projects (as listed in each case). There were several criteria for case studies selection. First of all, these are the examples aiming at keeping the material and formal integrity of modernist architecture—the main goal of the renovations is not to interfere with the original modernist expression of the buildings. We call this aspect architectural criteria. In this respect, the cultural value of the renovated building was crucial [13]. High aesthetics is as important as sustainability and functionality [10,15,16]. Other selections criteria were affordability [10], availability of materials, and building solutions that maintain the high quality [13]. The cost issue is crucial for as Nowogońska and Mielczarek [17], p. 1 state: "Unfortunately, due to the higher costs of renovation and the need to supervise work in historic buildings, many valuable buildings are damaged." To summarise, the criteria are as follows:

- architectural:
    - preservation of the modernist character of a building,
    - importance of a building to culture,
    - functionality,
    - aesthetics,
    - high quality;
- sustainable:
    - reducing energy use,
    - affordable,
    - and material availability (local products).

Based on the above, the selected case studies are as listed:

- A housing block section by Nankiera Square [18], original design (1969): Włodzimierz Czerechowski, Ryszard Natusiewicz, Anna and Jerzy Tarnawski; renovation design (2015): Agnieszka Hałas, Grzegorz Kaczmarowski, Marta Mnich, and Łukasz Wojciechowski.
- The complex of the residential towers with the commercial pavilion at Grunwaldzki Square [19,20], original design: Jadwiga Grabowska-Hawrylak, Krzysztof Sasiadek (1968–1978), renovation design (2012): Mnich, Marek Lamber, Natalia Rowińska, Łukasz Wojciechowski, Agnieszka Hałas, Hubert Rozewicz, consultants: Jadwiga Grabowska-Hawrylak, Andreas Wolf.
- The restaurant pavilion by the Centennial Hall (UNESCO heritage object) [21], original design (1913): Max Berg; renovation and extension design: Agnieszka Chrzanowska, Marta Mnich, Łukasz Wojciechowski, Wojtek Chrzanowski, in collaboration with Andrzej Chrzanowski, Juliusz Erdman, Grzegorz Kaczmarowski, Danuta Katarasińska, Agata Kurto, Natalia Rowińska, Sebastian Stanisławski.

All buildings are located in Wroclaw (Poland). Selected projects took place in the range of the last ten years.

Thermal refurbishments respond to local law regulations based on the local climate. Citing Climate-data.org [22], conditions in Wroclaw, Poland, are mild and warm, with average temperatures yearly of 10.0 °C, and 700 mm of rainfall. The average temperature ranges between 20.1 °C in July and −0.4 °C in January.

First we made a literature and documents review to institute requirements and methods for modern architecture preservation (see the introduction section). Second, we reviewed existing design documentation [19–21]. To make the article universal, we summarised the local climate based on recent data from Climate-data.org [22]. Next, based on current documents like European Union policies [10–12], commissioned energy audits by Bilka [23], and Żurawski [24]–external auditors, also our calculations, we have established the required thermal coefficients for all partitions. Audits were performed by local, licensed professionals, to provide a high quality of elaboration. In calculations, they included the initial structure of compartments and afterwards designed solutions. According to Sonnleithner [13], p. 6, high quality of investment is as important as, following present law demands. The thermal modernization followed Polish thermal regulations–so-called Ordinance Minister of Infrastructure from 2002 [25], and standard [26] these regulations refer to general building solutions. However the renovated heritage-listed buildings are excluded from energy efficiency requirements.

The law novelty is crucial as Boza-Kiss et al. [15], p. 11 indicate: "the theoretical consumption of a new building today is about 40% less than for dwellings built before 1990".

We compared data before and after planned remodeling based on the figures included in energy audits for each building, done before and after the refurbishment process (first two examples). For other cases, we made before and after calculations. All values before and after presented in this article are based on pre-design and post-design calculations. We plan to evaluate buildings further after compilation of all construction works. Presentation of each implementation follows the scheme:

- location,
- authors of original project and refurbishment,
- history,
- renovation assumptions,
- remodeling solutions,
- calculations of energy savings.

There are several factors recalled in the article. Shape factor A/V means, after Lylykangas [27], p. 4: "the ratio between the outside surface area of the thermal insulation in the building envelope (A) and the heated volume (V)."

In this article, European Standard PN-EN ISO 6946:2017 [26] defines the thermal transmittance U. Its value should be as low as possible and is the reciprocal of the thermal resistance of the entire partition [26]:

$$U = \frac{1}{R_{tot}} \left[\frac{\mathrm{W}}{\mathrm{m}^2}\mathrm{K}\right]$$

where:

W—unit: Watt;

m—unit: meter;

K—unit: Kelvin;

U—thermal transmittance $\left[\frac{\mathrm{W}}{\mathrm{m}^2}\mathrm{K}\right]$;

R_{tot}—total thermal resistance $\left[\frac{\mathrm{m}^2 * \mathrm{K}}{\mathrm{W}}\right]$;

Thermal total resistance R_{tot} determines the following formula [26]:

$$R_{tot} = R_{si} + R_1 + R_2 + \ldots + R_n + R_{se} \left[\frac{\mathrm{m}^2 * \mathrm{K}}{\mathrm{W}}\right]$$

where:

R_{si}—is the internal surface resistance $\left[\frac{\mathrm{m}^2 * \mathrm{K}}{\mathrm{W}}\right]$;

$R_1 + R_2 + \ldots + R_n$—are the design thermal resistance of each layer $\left[\frac{\mathrm{m}^2 * \mathrm{K}}{\mathrm{W}}\right]$;

n—is a total number of the designed layers;

R_{se}—is the external surface resistance $\left[\frac{\mathrm{m}^2 * \mathrm{K}}{\mathrm{W}}\right]$;

Surface resistance values depend on air convection through the compartment and are as follows: for Rsi–upwards: 0.1; downwards: 0.17; horisontal: 0.17; for Rse–upwards: 0.04; downwards: 0.04; horizontal: 0.04, and R the design thermal resistance of layer determines formula:

$$R = \frac{d}{\lambda} \left[\frac{\mathrm{m}^2 * \mathrm{K}}{\mathrm{W}}\right]$$

where:

d—thickness of the layer [m];

λ—the design thermal conductivity of material [W/mK];

The internal temperature adopted is equal to or above 16 °C. The design did not contain air layers over 0.3 m. Therefore, we used a simplified method of calculations according to European Standard PN-EN ISO 6946:2017.

3. Results

3.1. The Block-of-Flats by Nankiera Square-Completed

Włodzimierz Czerechowski, Ryszard Natusiewicz, Anna and Jerzy Tarnawscy were the authors of the initial design implemented in 1969, which was refurbished by Agnieszka Hałas, Grzegorz Kaczmarowski, Marta Mnich, and Łukasz Wojciechowski in 2015 [18]. The complex by Nankiera and Nowy Targ (the New Market) squares occupies lines of the previous historic buildings (demolished during WWII), typical for this part of Wroclaw Old Town. Yet, the past urban tissue replaces now late-modern layouts of free-standing elongated blocks and core buildings (Figure 1). The design was an architectural and political manifestation of freedom from social realism doctrine. However, it still was an element of socialist propaganda and social engineering. Hence, the Old Town area was then occupied by residents of the elite, while blocks were supposed to serve the working class. The thermal modernization of residential substances is valuable, because over 120 million buildings of this type exist in the European Union [15], p. 12. Initially, the buildings had an open plan with movable partition walls. Reinforced structure supplemented brick so-called 'zreanska' and aerated concrete blocks, while the ceilings were constructed from hollow core slabs and slab-on-grade cinder blocks. There were no thermal insulations, and plaster covered the façades. Glass tiles surfaced the inter-widow stripes, which were reminiscent of op-art graphics. The plinths at the service points on the first floor finished with pebbles fixed in mortar, and glass blocks covered the windows in the staircases.

Figure 1. The blocks by Nankiera square—original and refurbished with recreated details (photography by the curtesy of Patryk Kusz/VROA Architekci).

New development founded the municipal company called 'Revitalizations of Wroclaw' (in the Polish language 'Wrocławskie Rewitalizacje'; no longer existing). The renovation included one section of the building and serves as an example for other parts governed by local housing associations. The main goals were thermal modernization and reconstruction of the original block's appearance. However, adding layers, i.e., insulation to the façade, would cause a change in the initial depth of the windows' mounting and would affect in disadvantageous 'puffing' of the architecture. Therefore, the original depth of openings was a priority, and the small windows occupy space on the external wall surface (Figure 2). Other glazed elements mount 15 cm deep into the façade (also according to the original solution). Flashings, balustrades and glass tiles in the strips between the windows are as close to the original ones as possible. The cladding from pebbles covers the ground floor area as in the original. The ironwork has a graphite colour, and the gable elevation composition crowns a newly designed neon above the entrance to the service point (Figure 1).

We gave detailed original and implemented sections for compartments, and the heat transfer coefficients through partitions—initial and design phase comparison is shown in the below juxtaposition (Table 1) [18].

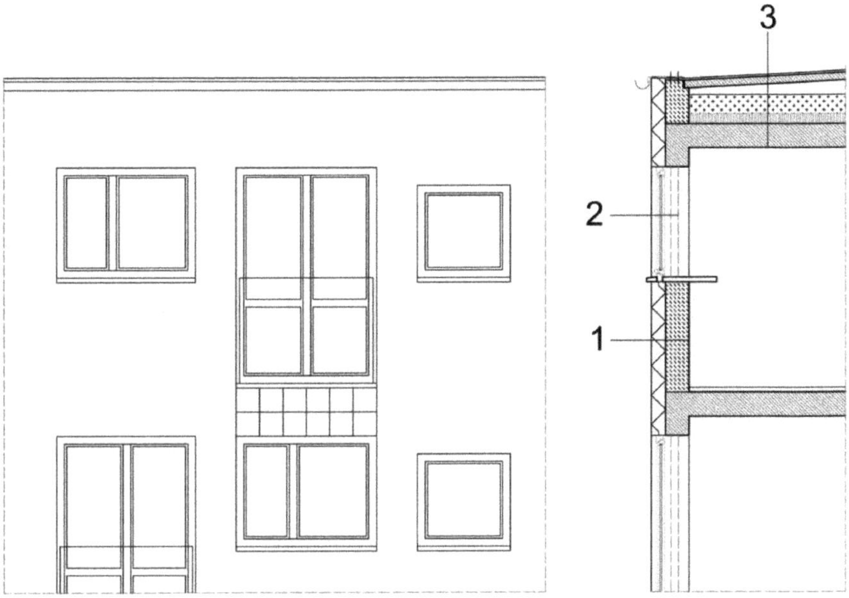

Figure 2. The blocks by Nankiera square—elevation and section. 1. The designed external wall: silicone plaster 1.5 cm, EPS polystyrene 15.0 cm, aerated concrete blocks 24.0 cm, internal plaster 1.5 cm. 2. The depth of the window fixing changed with insulation to preserve the original ratio. 3. The designed roof slab: 2× roofing felt, levelling layer 2.0 cm, slag concrete slabs in decline 8.0 cm, ventilation gap 10.0, the blow of mineral wool granules 20.0 cm, prefabricated channel slabs 24.0 cm, internal plaster 1.5 cm (drawn by the Authors).

3.2. The Façades of Residential Buildings at Grunwaldzki Square–Project, Completed

Jadwiga Grabowska-Hawrylak and Krzysztof Sąsiadek designed the complex of residential buildings with commercial pavilions, plaza and parking spaces (1968–1978). Renovation and thermal modernisation was performed by Marta Mnich, Marek Lamber, Natalia Rowińska, Łukasz Wojciechowski, and Agnieszka Hałas in 2012 with consultations from Jadwiga Grabowska-Hawrylak and Andreas Wolf [19,28]. The document so-called 'The Study of the Conditions and Directions of Spatial Development in Wrocław' protects the complex as a heritage structure of contemporary culture. Also, the List of Monuments of the City of Wrocław includes this development, so design documentation was subject to approval by the Monument Conservator.

The complex consists of six sixteen-story residential towers and service pavilions partially accessible from a raised platform above the garage. A characteristic element of the existing urban layout is a modular 6x6m grid, which supports all build substances. The following distinctive element of the project is the façades of residential high-rise buildings, made of individually planned, oval, reinforced concrete prefabricated elements. They are attached to a reinforced concrete skeleton structure of the so-called 'H-frame' type. The space between the prefabricated elements of the façade and the external wall is filled in with loggias. Pots with greenery were supposed to complete this solution, yet this element was omitted in the initial stage of construction [29], (Figure 3).

Table 1. The thermal modernisation building solutions for the block by Nankiera square (elaborated by the Authors based on [18]).

No	Name of the Layer	The Thickness of the Layer [cm]	The Design Thermal Conductivity of a Material	Heat Transfer Coefficients through Partition [$\frac{W}{m^2}K$]
The existing external wall (before the mal modernisation)				
1	The external plaster	2.0	0.82	0.39
2	The existing brick 'zeranska'/aerated concrete blocks (porous concrete blocks possible)	24.0	0.105	
3	The internal plaster	1.5	0.70	
The designed external wall				
1	The silicone plaster	1.5	0.7	0.15
2	The EPS polystyrene	15.0	0.038	
3	The aerated concrete blocks	24.0	0.105	
4	The internal plaster	1.5	0.7	
5	The silicone plaster	1.5	0.7	
The existing roof slab				
1	2× roofing felt (bituminous felt and tar paper)	-	-	0.34
2	The levelling layer	2.0	1.4	
4	The slag concrete slabs in decline	8.0	1.70	
5	The ventilation gap	5.0–30.0	0.16 [1]	
6	The prefabricated channel slabs	24.0	0.18	
7	The internal plaster	1.5	1.70	
The designed roof slab				
1	2× roofing felt (bituminous felt and tar paper)	-	-	0.04
2	The levelling layer	2.0	1.4	
4	The slag concrete slabs in decline (existing)	8.0	1.70	
5	The ventilation gap (existing)	10.0	0.15	
6	The mineral wool granules injected into the ventilation gap	20.0	0.038	
7	The prefabricated channel slabs (existing)	24.0	0.18	
8	The internal plaster	1.5	1.70	

[1] Averaging 15 cm; value from PN-EN ISO 6946:2017.

Figure 3. The renovation concept of the façades of residential buildings at Grunwaldzki square—a pre-refurbishment condition in 2007 and remodelling visualisation (photograph by the Authors).

The renovation aimed to improve the functioning of selected elements of the initial design. This need resulted from new functional requirements not present or predictable in the 1970s. The complex lacked services on the ground floor, required separation of pedestrians and traffic, supply zones and utility yards adjacent to apartment houses. The design included the original proprietary assumptions and consistently used them when expanding, i.e., all additional cubatures were kept in the original 6x6m grid. They also adapted to the existing heights of service pavilions and the pedestrian platform [19].

The most complicated task was to improve the thermal parameters of the residential buildings and their adaptation to the applicable regulations while maintaining the specific architectural values. The concept assumed the necessity of fully preserving the character of the façade made of prefabricated reinforced concrete elements. Thus, external insulation was applied only on the outer walls in the background and gable walls and ceilings. In the case of other parts of the building, the designers proposed inner thermal insulation (Figure 4). This approach enabled the preservation of the initial concept. The façade carvings highlight the colour scheme with white prefabricated elements against a dark background. This decision comes from in-depth analyses of the initial design and consultations with the architect Jadwiga Grabowska-Hawrylak [19].

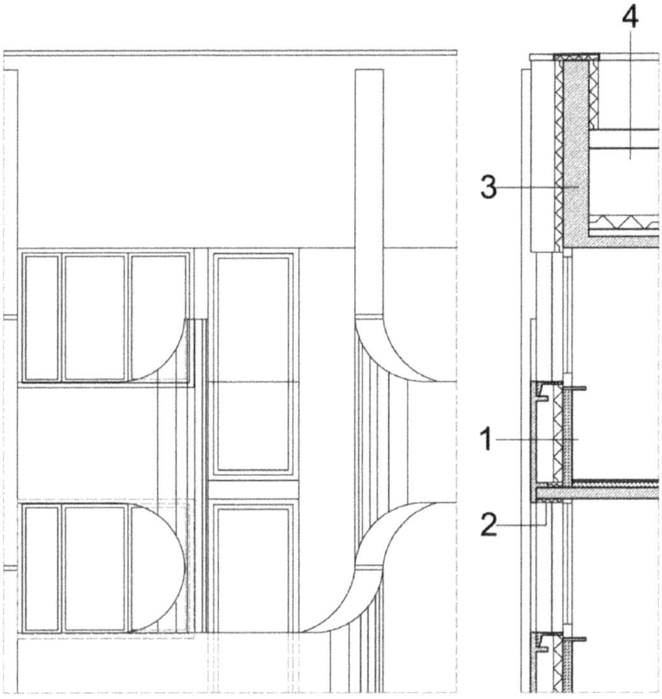

Figure 4. The façades of residential buildings at Grunwaldzki square–elevation and section. 1. The designed external wall (prefabricated): repaired and painted white prefabricates 6.0 cm, external void, mineral wool 8.0 cm, internal walls made of asbestos removed and replaced with light concrete blocks 12.0 cm, internal plaster 1.5 cm. 2. Insulation of loggias–mineral wool. 3. The designed external wall: silicone plaster 1.5 cm, mineral wool 8.0 cm, aerated concrete blocks or reinforced concrete wall 24.0 cm, internal plaster 1.5 cm (between windows). 4. The designed roof slab: 2x roofing felt, levelling layer 2.0 cm, slag concrete slabs in decline 8.0 cm, ventilation gap 5.0–30.0 cm, blown-in mineral wool granules in the air cavity 20.0 cm, prefabricated channel slabs 24.0 cm, internal plaster 1.5 cm (drawn by the Authors).

The leading assumption of the renovation concept of reinforced concrete facade elements was to restore them to their original appearance. The architects aimed at repairing defects and protection against further destruction and corrosion. It was necessary to strengthen the connections of prefabricated elements with the building structure. The conceptual design assumed cleaning of the prefabricated concrete elements, e.g., by sandblasting, repairing and filling the existing damages. The wall surfaces are painted a light grey colour (Figure 2). The existing clinker tiles are no longer visible, screened with insulation and plaster due to the low budget of the renovation. However, Jadwiga Grabowska-Hawrylak confirmed that she did not plan tiles in the original design of the towers. This solution restores the shade of the light-ash architectural concrete used in the original projects and protects the surface from the harmful effects of weather conditions. However, the assumptions are unfulfilled. The oval prefabricated elements were plastered, which deprived them of the concrete texture. The background wall was covered with paint which is too lightly coloured. We present detailed original and implemented sections for compartments (Table 2) and the changes in their heat transfer coefficients (under the table), [19].

Table 2. The thermal modernisation building solutions for the residential buildings at Grunwaldzki square (elaborated by the Authors based on [19]).

No	Name of the Layer	The Thickness of the Layer [cm]	The Design Thermal Conductivity of a Material	Heat Transfer Coefficients through Partition [$\frac{W}{m^2}K$]
	The existing external wall (between windows)			
1	The clinker tile	2.0	0.67	0.39
2	The ventilation gap	2.0	0.00	
3	The aerated concrete blocks	24.0	0.105	
4	The internal plaster	1.5	0.7	
	The designed external wall (between windows)			
1	The silicone plaster	1.5	0.7	0.21
2	The mineral wool	8.0	0.035	
3	The aerated concrete blocks	24.0	0.105	
4	The internal plaster	1.5	0.7	
	The existing roof slab			
1	2× roofing felt (bituminous felt and tar paper)	-	-	0.34
2	The levelling layer	2.0	1.4	
3	The slag concrete slabs in decline	8.0	1.70	
4	The ventilation gap	5.0–30.0	0.16	
5	The prefabricated channel slabs	24.0	0.18	
6	The internal plaster	1.5	1.7	
	The designed roof slab			
1	2× roofing felt (bituminous felt and tar paper)	-	-	0.04
2	The levelling layer	2.0	1.4	
3	The slag concrete slabs in decline	8.0	1.70	
4	The ventilation gap	10	0.15	
5	The mineral wool granules injected into the ventilation gap	20.0	0.038	
6	The prefabricated channel slabs	24.0	0.18	
7	The internal plaster	1.5	1.70	

3.3. The Commercial Pavillon by Grunwaldzki Square–Project, Not Completed

The building is a part of the mentioned residential complex. Jadwiga Grabowska-Hawrylak, Krzysztof Sąsiadek (1968–1978) performed the original design, while Marta Mnich, Łukasz Wojciechowski, Hubert Różycki (2020) drew renovation plans [20]. The pavilion occupies a plot near Grunwaldzki bridge among several commercial spots elevated on a pedestrian platform. The two-story building with a basement has a reinforced concrete structure. Its characteristic spaceship-like façades constitute precast decorative elements with external spiral stairs. The steps wrap around the cylindrical shaft of mechanical ventilation. Glass covers the ground floor façades, divided into three sections with aluminum frames. Above there is the characteristic ornament of an extended curved precast panel with openwork. Circular openings form a repetitive rhythm. Other external walls are opaque and made from aerated concrete, fixed with ceramic tiles and trapezoidal metal sheets (Figure 5), [20].

Figure 5. The commercial pavilion by Grunwaldzki Square—the existing condition (photograph by the Authors).

To preserve the initial character of the building, the design of thermal modernisation assumes the internal isolation of opaque walls with mineral boards, the so-called "Multipor" of 5 cm. Concrete pillars—part of the "H-frame" structure—are isolated externally with a 15 cm layer of Styrofoam boards. The project assumes cleaning of the precast decorative covers and external walls to the original white colour of the concrete. The pavilion's flat roof slab received an additional thermal layer of 20 cm from Styrodur and required sealing. Architects proposed to keep the original and leaky aluminum frames with glazing by creating an internal glass wall. The solution provides proper thermal conditions, enabling the preservation of the modernist character. Old and new façades will divide a gap for cleaning and the accumulation of passive heat. The aluminum frames on the second floor replace new ones resembling the original (Table 3), (Figure 6), [20].

Table 3. The thermal modernisation building solutions for the commercial pavilion at Grunwaldzki square (elaborated by the Authors based on [20]).

No	Name of the Layer	The Thickness of the Layer [cm]	The Design Thermal Conductivity of a Material	Heat Transfer Coefficients through Partition [$\frac{W}{m^2}K$]
	The existing prefabricated walls			
1	The reinforced concrete prefabricates with an internal void of 5.0 cm	15.0 (5 + 5 + 5)	1.7 + 0.11 + 1.7	1.38
	The designed walls (decorative)			
1	The reinforced concrete prefabricates with an internal void of 5.0 cm–cleaning, fulfilling subsidence	15.0	1.7 + 0.11 + 1.7	0.5
2	The internal isolation of Multipor type boards	5.0	0.040	
3	The Heradesign type boards in natural color on a wooden grid	2.5	-	
	The existing roof slab			
1	3× jute felt paper (waterproof insulation)	-	-	0.045
2	The cement screed	1.0	1.4	
3	The sloped roof panels (channelled, reinforced concrete)	25.0	0.18	
4	The ventilation gap	5.0–30.0	0.16	
5	The slag wool	6.0	0.045	
6	The structural ceiling (reinforced concrete beams and trough slabs, reinforced concrete)	30.0	1.7	
	The designed roof slab			
1	The EPDM film	-	-	0.036
2	The Styrodur	20.0	0.035	
3	The vapor-permeable foil	-	-	
4	The cement screed	1.0	1.4	
5	The sloped roof panels (channelled, reinforced concrete)	25.0	0.18	
6	The ventilation gap	5.0–30.0		
7	The slag wool	6.0	0.045	
8	The structural ceiling (reinforced concrete beams and trough slabs, reinforced concrete)	30.0	1.7	

3.4. Renovation and Extension of the Restaurant Pavilion in the Centennial Hall Complex–Completed

Max Berg designed the original complex of the Centennial Hall (1913, now under UNESCO protection), while the remodeling and extension (2007–2010) project had the following authors: Agnieszka Chrzanowska, Marta Mnich, Łukasz Wojciechowski, Wojtek Chrzanowski, in collaboration with Andrzej Chrzanowski, Juliusz Erdman, Grzegorz Kaczmarowski, Danuta Katarasińska, Agata Kurto, Natalia Rowińska, and Sebastian Stanisławski [21].

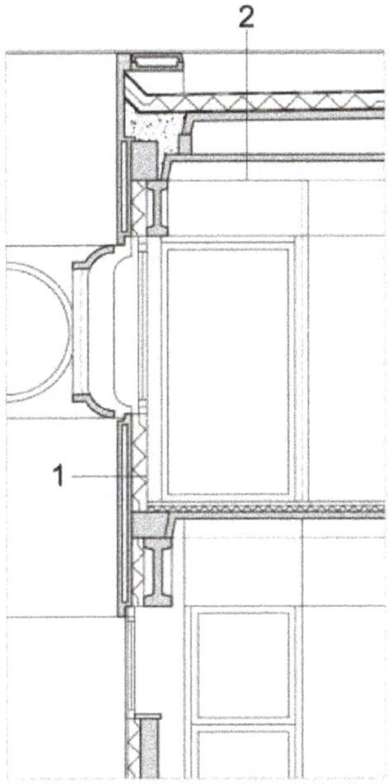

Figure 6. The commercial pavilion by Grunwaldzki Square—section. 1. The designed wall: reinforced concrete prefabricates with an internal void of 5.0 cm—cleaning, fulfilling subsidence 15.0 cm, internal isolation of Multipor-type boards 5.0 cm, and Heradesign-type boards in natural color on a wooden grid 2.5 cm. 2. The designed roof: EPDM film, Styrodur 20.0 cm, vapor-permeable foil, cement screed 1.0 cm, sloped roof panels (channeled, reinforced concrete) 25.0 cm, ventilation gap 5.0–30.0 cm, slag wool 6.0 cm, and structural ceiling (reinforced concrete beams and trough slabs, reinforced concrete) 30.0 cm (drawn by the Authors).

The pavilion dating to 1913 served as a temporary restaurant and was burnt during WWII. The surviving structure served as an office building in the late 1940s. In 2009 there was a competition held to rebuild and create an extension for a new function as a conference center (Figure 7). The winning concept assumed the preservation of existing and original pavilion elements. Preservation included:

- decorative structural pillars,
- a central lobby with oval skylight,
- a horizontally composed facade across the Centennial Hall.

Figure 7. Renovation and refurbishment of the restaurant pavilion of the Centennial Hall complex in Wroclaw—windows and glazed curtain wall (photographs by the Authors).

Two new and fully glazed wings extend the preserved structure. Cubatures accommodate an auditorium and multifunctional room, while the central part contains a foyer and restaurants, and above—on the mezzanine—office rooms. All technical equipment, storage rooms, service areas, and sanitary areas occupy the underground floor. While the original structure is part load-bearing brick masonry and partly reinforced concrete. It required strengthening of both walls and foundations and a partial change of existing slabs. The new structure was also added, made from steel and founded separately, in the two-story part of the building. The first floor has a light steel structure suspended from the beams of the slab over the floor. The skylight occupying the building's central part is a lightweight structure, as are the new extensions. Plate girders of variable height crown these parts. They have cast foundations of monolith reinforced concrete due to the high levels of groundwater. Opaque walls of existing and new extensions received original materials and colouring. A light-grey structural plaster covers them, polished into an even surface. Preservation of the original composition of façades was a key factor—the columns lean against the external wall. Thus, the remodeling assumed internal insulation. The method also allowed the original depth of the windows to be kept [21].

Due to the high (then) costs of internal thermal insulation systems, the architects proposed individual solutions. These consist of internal gypsum–cardboard walls, vapour barrier foil, mineral wool boards on a steal support sub-structure (8 cm width), a ventilation gap, and dimpled foil. Mentioned layers cover an external wall plastered from the outside. The roof slab is insulated transitionally from the exterior (Table 4), (Figure 4), [21].

Table 4. The thermal modernisation building solutions for the restaurant pavilion in the Centennial Hall complex (elaborated by the Authors based on [21]).

No	Name of the Layer	The Thickness of the Layer [cm]	The Design Thermal Conductivity of a Material	Heat Transfer Coefficients through Partition [$\frac{W}{m^2}K$]
		The existing external wall		
1	The external plaster	2.0	1.70	1.38
2	The full brick wall	32.0	0.77	
3	The internal plaster	1.5	1.70	
		The designed external wall		
1	The thin-layer plaster	0.3	1.70	0.36
2	The external cement-lime plaster	2.0	1.70	
3	The masonry brick wall	32.0	0.77	
4	The dimpled foil	-	-	
5	The ventilation gap	2.0	-	
6	The mineral wool	8.0	0.038	
7	The vapour barrier foil	-	-	
		The existing roof slab		
1	The existing wooden structure–for dismantling	-	-	-
		The designed roof slab		
1	The SBS modified tar paper	-	-	0.04
2	The underlay felt paper	-	-	
3	The hardboard of mineral wool–for inclination	5.0–20.0 (avg. 12.5)	0.038	
4	The hardboard of mineral wool	18.0	0.038	
5	The vapor barrier	-	-	
6	The trapezoidal sheet T55x18 0.75 cm	-	-	
7	The steel structure	30.0	-	
8	The fittings gap	112	-	

4. Discussion

The article focuses on the thermal modernisation of modernist architecture while preserving its original aesthesis and composition. This heritage deserves upcycling aiming at the reduction of energy consumption and adjustment to nowadays living standards. We present and analyse case studies of implemented thermal refurbishment designs, which preserve the initial aesthetics of buildings. Selected designs come from the portfolio of one of the article's authors. Another element of the research is the study of energy audits used to establish energy savings. They were as follows:

We calculated exemplary (presented in the article) heat transfer coefficients through construction partitions for the block by Nankiera reduction. We used a simple proportion formula to calculate the thermal transmittance U reduction–between initial and designed values:

$$\begin{array}{lcc} \textit{initial U} & - & 100\% \\ \textit{the designed U} & - & x\% \end{array}$$

And

$$x = \frac{\textit{the designed U} * 100\%}{\textit{initial U}}$$

Afterwards, we obtained the final value by subtracting x from 100%. The outcome was as follows:

- external walls—61.53%
- roof—88.2%.
    - Bilka [23] conducted the external energy audit for the initial and design phases. It concerned modification of possible building compartments, ventilation, change of form of heating, etc. The Author [23] analyzed different variants and selected the best one for the investor and the design studio. Based on this document, the shape factor A/V was (and is) 0.47. Improvements of mentioned parameters influenced calculated thermal power of the heating system, which initially was 46.66 [kW], and after designed refurbishment achieves 17.52 [kW]. The annual heat demand index to heat the building (without taking into account the efficient heating system and heating interruptions) was 224.54 [kWh/(m^2 year)] and is 74.57 [kWh/(m^2 year)], [23]. The Author shows other savings from decreasing heat transfer coefficients for windows—avg. 37%, doors and gates—34.6%, the slab over a passage—91.6%. We expect these parameters once all the construction works are finished according to the recommendations. Data from our calculations and external audits show substantial improvement in building energy performance. At the same time, due to the preservation of the initial structure and adding heat isolation and new coatings, the cost is low. The heat transfer coefficients, through construction partitions, were reduced by a range of 34.6% to 91.6% (dependent on the element).
    - The calculated thermal power of the heating system was reduced by a range of 26.6–62.4% (dependent on the case study).
    - The annual heat demand index to heat the building (without taking into account the efficient heating system and heating interruptions) reduced by a range of 33.4–66.8% (dependent on the case study).

These numbers prove that a thermal refurbishment of modernist heritage buildings holds great potential to reduce overall energy consumption in the built environment. Moreover, we show the building solutions that led to the substantial reductions.

For the commercial pavilion at Grunwaldzki square, the values of the heat transfer coefficients' decrease are as listed:

- external walls—46.15%;
- roof—88.2%.

As in the previous case for the residential blocks at Grunwaldzki Square, the energy audit was conducted by Żurawski [24]. It considered both the pre-design and post-design states, including heating and ventilation cases. Initial shape factor A/V was (and is) 0.26, and the heat transfer coefficients through partitions construction (if construction finishes according to plan), should be reduced by: wooden windows—48.4%; steel doors (communication)—57.4%; the slab over a passage (boards)—79%.

It will influence the calculated thermal power of the heating system, which initially was 425.83 [kW] and after refurbishment achieved 313.20 [kW]. The annual heat demand index to heat the building (without taking into account the efficient heating system and heating interruptions) was 158.92 [kWh/(m^2 year)] and will be 105.90 [kWh/(m^2 year)], [24]. The data show improvement in building energy performance.

For the residential buildings at Grunwaldzki Square, values of the heat transfer coefficients' decrease are listed:

- external walls—63.78%;
- roof—20%.

For the restaurant pavilion in the Centennial Hall complex the changes are the following:

- external walls—73.91%;
- Roof was re-designed.

5. Conclusions

As stated in the paper, thermal insulation is a crucial issue in the renovation process of modern façades. For instance, the thickness of the insulation layer can change a depth ratio of a wall and a window. The incorrect proportion has a damaging effect on the formal expression of architecture. Some available insulating systems may be more effective than traditional insulations, but new solutions are usually still too expensive to apply and therefore are rejected by clients—as was the case in all the analysed examples. The architects need to consider not only the low budgets offered for the insulation of post-war buildings, but mainly the reluctance of the public and clients to treat them as a proper heritage buildings.

We believe that a substantial number of buildings from the Modern Period deserve renewal. They need inclusion in the architectural and cultural heritage of European cities. What is more, minding that many of that substance is residential, it can fulfil its purpose for years to come. We hope that the article can become part of the discussion and an example of practical solutions in line with sustainable striving and current European Union policies.

Author Contributions: J.J. and L.W. conceptualization, methodology, software, validation, formal analysis, investigation, resources, data curation, writing—original draft preparation, writing—review and editing, visualization, supervision, project administration, funding acquisition. All authors have read and agreed to the published version of the manuscript.

Funding: Subsidy under the program "Initiative of Excellence – Research University (IDUB), budget No: 821 120 46 01 N_RRB_PODST_BAD_EXP_BAM.

Institutional Review Board Statement: Not applicable.

Informed Consent Statement: Not applicable.

Data Availability Statement: Not applicable.

Conflicts of Interest: The authors declare no conflict of interest.

References

1. Urbanik, J.; Tomaszewicz, A. Flat Roof—Advantage or Disadvantage of Modern Movement Buildings. In Proceedings of the SAHC2014—9th International Conference on Structural Analysis of Historical Constructions, Mexico City, Mexico, 14–17 October 2014; Peña, F., Chávez, M., Eds.; 2014. Available online: https://www.researchgate.net/publication/343294962_Flat_Roof_-_Advantage_or_Disadvantage_of_Modern_Movement_Buildings (accessed on 8 September 2021).
2. Mulfarth, R.C.K.; Gonçalves, J.C.S.; Michalski, R.L.X.N.; Shimomura, A.R.P.; e Souza, B.N.; Cunha, G.R.M.; Marcondes-Cavaleri, M.P.; Monroy, M.A.G. The environmental response of office buildings from the Brazilian bioclimatic modernism in São Paulo: The case study of Complexo Conjunto Nacional—CCN (1962). *Sol. Energy* **2021**, *220*, 617–634. [CrossRef]

3. Kahraman, A.D. Relationship of modernism, postmodernism and reflections of it on education. *Procedia Soc. Behav. Sci.* **2015**, *174*, 3991–3996. [CrossRef]
4. Springer, F. *Źle Urodzone (in English Language: Badly Borne)*; Karakter: Cracov, Poland, 2017.
5. Gabiś, A. *Całe Morze Budowania. Wrocławska architektura 1956–1970 (in English Language: The Whole Sea of Buiding. The Architecture of Wroclaw 1956–1970)*; Museum of Architecture in Wroclaw: Wroclaw, Poland, 2018.
6. Elser, O.; Kurz, P.; Schmal, P. *SOS Brutalism. A Global Survey*; Park Books: Zürich, Switzerland, 2018.
7. Giuliani, M.V.; Bucchignani, V. Preservation through change: Renovating modern architecture. *J. Archit. Plan. Res.* **2000**, *17*, 34–46.
8. Mika, P. The nature of the technology in which the basic executive component is the precasted element. In Proceedings of the 4th International Multidisciplinary Scientific Conference on Social Sciences and Arts SGEM 2017, Albena, Bulgaria, 24–30 August 2017; Volume 17, pp. 823–830. [CrossRef]
9. Pikas, E.; Seppänen, O.; Koskela, L.; Peltokorpi, A. Challenges in Industrialized Renovation of Apartment Buildings. In Proceedings of the 29th Annual Conference of the International Group for Lean Construction (IGLC29), Lima, Peru, 14–17 July 2021; Alarcon, L.F., González, V.A., Eds.; pp. 985–994. [CrossRef]
10. European Commission. A European Green Deal. 2021. Available online: https://ec.europa.eu/info/strategy/priorities-2019-2024/european-green-deal_en (accessed on 8 September 2021).
11. European Comission. Renovation Wave. 2021. Available online: https://ec.europa.eu/energy/topics/energy-efficiency/energy-efficient-buildings/renovation-wave_en (accessed on 8 September 2021).
12. European Commission. New European Bauhaus: Commission Launches Design Phase. 2021. Available online: https://ec.europa.eu/commission/presscorner/detail/en/IP_21_111 (accessed on 8 September 2021).
13. Sonnleithner, M. New Opportunities for Increasing the Renovation Rate of Buildings. *Archit. Pap. Fac. Archit. Des. STU* **2021**, *26*, 2–9. [CrossRef]
14. Peters, T. Architectural Interventions For Social Sustainability: The Renovation Of Modern Housing. In Proceedings of the Regenerative and Resilient Urban Environment, Sustainable Built Environment, Toronto, ON, Canada, 19–20 September 2016; Available online: https://www.researchgate.net/publication/308702873_Architectural_Interventions_For_Social_Sustainability_The_Renovation_Of_Modern_Housing (accessed on 21 September 2021).
15. Boza-Kiss, B.; Bertoldi, P.; Nives, D.V.; Economidou, M. One-stop shops for residential building energy renovation in the EU. In *Analysis & Policy Recommendations*; JRC Science for Policy Report [JRC125380]; Publications Office of the European Union: Luxembourg, 2021. [CrossRef]
16. Sugár, V.; Talamon, A.; Horkai, A.; Kita, M. Energy saving retrofit in a heritage district: The case of the Budapest. *J. Build. Eng.* **2020**, *27*, 100982. [CrossRef]
17. Nowogońska, B.; Mielczarek, M. Renovation Management Method in Neglected Buildings. *Sustainability* **2021**, *13*, 929. [CrossRef]
18. Mnich, M.; Hałas, A.; Kaczmarowski, G.; Wojciechowski, Ł.; (Wrocław, Poland). Renovation and Reconstruction of a Residential and Service Building Nankiera 14, Wrocław. Personal communication, 2015.
19. Mnich, M.; Wojciechowski, Ł.; Hałas, A.; Rowińska, N.; Wolf, A.; (Wrocław, Poland). The Concept of Revalorization of a Residential and Service Complex Plac Grunwaldzki, Wrocław. Personal communication, 2012.
20. Mnich, M.; Wojciechowski, Ł.; Różycki, H.; (Wrocław, Poland). Renovation and Reconstruction of the Office and Commercial Building with the Conversion of the Basement into an Office and Training Function and Land Development Plac Grunwaldzki, Wrocław. Personal communication, 2020.
21. Mnich, M.; Wojciechowski, Ł.; Chrzanowska, A.; Chrzanowski, W.; Chrzanowski, A.; (Wrocław, Poland). Establishment of the Regional Center for Business Tourism, ul. Wystawowa 1, Wrocław. Personal communication, 2008.
22. Climate-data.org. Climate Wrocław (Poland). 2021. Available online: https://en.climate-data.org/europe/poland/lower-silesian-voivodeship/wroc%C5%82aw-4531/ (accessed on 21 September 2021).
23. Bilka, P.; (Wrocław, Poland). Pracownia Projektowa (transl. from Polish Language: The Design Studio) BILAN Paweł Bilka: The Energy Audit for Nankiera 14 Building in Wroclaw. Personal communication, 2015.
24. Żurawski, J.; (Wrocław, Poland). The Energy Audit for Thermal Modernization Enterprise for Implementation at the Basis of Act at 11 November 2008. Personal communication, 2012.
25. Ordinance Minister of Infrastructure of April 12, 2002 on the Technical Conditions to Be Met by Buildings and Their Location, Journal of Laws 2019.1065, i.e of 2019.06.07, Based on Article. 7 sec. 2 point 1 of the Act of July 7, 1994—Construction Law (Journal of Laws of 2018, item 1202, as Amended); Valid at the Date of the Project. Available online: https://www.gov.pl/web/infrastruktura/warunki-techniczne2 (accessed on 14 October 2021).
26. PN-EN ISO 6946:2017 Building Components and Building Elements—Thermal Resistance and Thermal Transmittance—Calculation Methods, European Standard, Available by: Polish Normalisation Committee (Polski Komitet Normalizacyjny). Available online: https://sklep.pkn.pl/pn-en-iso-6946-2017-10e.html (accessed on 15 January 2022).
27. Lylykangas, K. Shape Factor as an Indicator of Heating Energy Demand, 15. Internationales Holzbau-Forum 09. 2009. Available online: https://www.forum-holzbau.com/pdf/ihf09_Lylykangas.pdf (accessed on 10 November 2021).
28. Wolf, A. *Post-war Modernism 2.0 von VROA-Architekten Wrocław. I. Sonntag (Hr. sgb)*; Sächsische Akademie der Künste: Dresden, Germany, 2014; pp. 202–207.
29. Duda, M. *Patchwork. Architektura Jadwigi Grabowskiej-Hawrylak (Patchwork, The Architecture of Jadwiga Grabowska-Hawrylak)*; Museum of Architecture in Wroclaw: Wrocław, Poland, 2016.

buildings

MDPI

Article

Evaluation of Non-Autoclaved Aerated Concrete for Energy Behaviors of a Residential House in Nur-Sultan, Kazakhstan

Chang-Seon Shon, Inzhu Mukangali, Dichuan Zhang *, Anuar Ulykbanov and Jong Kim

Department of Civil and Environmental Engineering, School of Engineering and Digital Sciences, Nazarbayev University, Nur-Sultan 020000, Kazakhstan; chang.shon@nu.edu.kz (C.-S.S.); inzhu.mukangali@alumni.nu.edu.kz (I.M.); aulykbanov@nu.edu.kz (A.U.); jong.kim@nu.edu.kz (J.K.)
* Correspondence: dichuan.zhang@nu.edu.kz

Abstract: Autoclaved aerated concrete (AAC) is commonly used as a modern, energy-efficient construction material in Nur-Sultan, Kazakhstan—the second-coldest national capital in the world after Ulaanbaatar, Mongolia. The autoclave curing method used to manufacture the AAC has potential risks and is environmentally costly because of its high-pressure and -temperature operation. Therefore, for phase I and II studies, non-autoclaved aerated concrete (NAAC) was cast, and its properties were evaluated in terms of compressive strength, density, porosity, and thermal conductivity. Moreover, the thermal conductivity prediction model of NAAC was successfully developed. In this Phase III study, the energy behavior of the NAAC was evaluated by energy simulation for a typical two-story residential house model in Kazakhstan. Different wall materials, such as fired brick and normal concrete, were adapted to compare the energy performance of NAAC. Finally, the annual heat loss and amount of heat transferred through the wall of the house were calculated to cross-check the energy-saving effect of NAAC. It was found that the NAAC conserved energy, because the heating and cooling loads, annual heat loss, and amount of heat transfer of NACC were lower than those of fired brick and normal concrete.

Keywords: non-autoclaved aerated concrete; thermal conductivity; heating and cooling loads; annual heat loss; heat transfer

Citation: Shon, C.-S.; Mukangali, I.; Zhang, D.; Ulykbanov, A.; Kim, J. Evaluation of Non-Autoclaved Aerated Concrete for Energy Behaviors of a Residential House in Nur-Sultan, Kazakhstan. *Buildings* **2021**, *11*, 610. https://doi.org/10.3390/buildings11120610

Academic Editors: Paulo Santos and Mark Bomberg

Received: 11 November 2021
Accepted: 25 November 2021
Published: 4 December 2021

1. Introduction

In the past decade, sustainable or green building technology in the construction industry has been adapted and has grown continuously. Green building is defined as the practice of creating structures and using environmentally responsible and resource-efficient processes throughout a building's life cycle, from the planning stage to the demolition of the building. This practice includes not only the classical building design criteria such as economy, utility, durability, and comfort, but also the efficient use of land, water, resources, and energy in and around the building, with a low environmental impact [1–3].

The demand for constructing energy-efficient residential houses and buildings is gradually rising, especially in Nur-Sultan, the capital city of Kazakhstan. Nur-Sultan has a significant temperature difference between seasons, with long, harsh winters and short, hot summers. While summer temperatures occasionally reach +35 °C, the temperature between mid-December and early March usually ranges from −20 to −35 °C, along with an average wind velocity of 5.2 m/s, reaching as high as 31 m/s [1,4]. Because of such severe weather conditions, heating and cooling costs of residential houses and buildings form the bulk of operating expenses in the residential houses and buildings in Nur-Sultan. For example, more than 30% of total energy is consumed by residential buildings, and the heat energy consumption in Nur-Sultan increased from 4963 MW to 6401 MW between 2010 and 2014 [5,6]. Moreover, up to 35% of heat loss is induced through the walls in the existing conventional houses built in the 1990s. Therefore, designing an energy-efficient building with the proper construction materials can significantly save homeowners in terms of energy-related operations and maintenance costs.

Aerated concrete (AC) is a modern energy-efficient construction material classified as lightweight concrete due to its low density and strength [7]. According to the production method, the AC can be divided into cellular concrete (CC) and autoclaved aerated concrete (AAC). CC is produced using an organic or synthetic foaming agent and a normal curing method. In contrast, AAC is manufactured using an expansion agent such as aluminum (Al) powder and an autoclaved curing process [8,9]. As a building material, AAC is typically used in concrete masonry units, such as blocks. The typical mixture composition of AAC includes binders (cement and lime), silica-rich supplementary cementitious material, fine aggregates (silica and quartz mineral aggregates), an expansion agent (Al), and water [10,11]. The unique property of AAC is its low thermal conductivity coefficient. The λ of AAC is attributed to the millions of evenly distributed, uniformly sized, and entrapped air voids caused by the chemical reaction between Al powder and alkalis in the cementitious mixtures, producing hydrogen gas [12,13]. Despite depending on the mixture proportions, the typical porosity of AC ranges from 75 to 90% [14]. This unique property gives AAC a thermal conductivity (λ) as low as 0.085–0.30 W/(m·K), depending on density, curing method, moisture content, mixture proportions, and ingredients [13,15]. For example, Walczak et al. [13] reported that the λ value of sand-based AAC is approximately 0.15 W/(m·K), while the λ value of fly-ash-based AAC is 0.085 W/(m·K) at the same density (400 kg/m^3).

For energy conservation, Walczak et al. [13] reported that the λ value of materials used for the construction of buildings should be lower than 0.23 W/(m·K) in order to reduce energy consumption and utility bills; AAC can fulfill this condition. Several researchers have studied the energy efficiency quality of AC materials. Radhi [16] reported that the use of AC materials in the wall layer of buildings reduced energy use by 7%. Narayanan and Ramamurthy [7] described how AAC provides better thermal insulation than conventional concrete blocks, and is considered to be an energy-efficient material, conserving temperature and reducing energy consumption. From the viewpoint of sustainability, approximately 350 kg of CO_2 emissions could be saved by each 1 m^2 of AAC wall throughout the life cycle of a building [17]. According to the research survey conducted by the Portland Cement Association (PCA), 77% of design professionals assert that AAC can be considered a sustainable material that corresponds to all sustainability requirements [18].

Several researchers have compared different building materials, including AAC, in terms of energy consumption. For example, Kaşka and Yumrutaş [19] examined various multilayer building walls consisting of materials commonly used in Turkey, which include briquettes, bricks, blokbims, and AAC. They found that AAC is a more suitable wall material than the other materials because it has a lower temperature at the inner surface, and heat flows through the wall when the external air temperature is high. Heathcote [20] determined the internal temperature of a building without an air conditioner during the summer days, which was constructed with brick veneer, mud bricks, and AAC wall panels; his results indicated that the internal temperature of the building constructed with AAC was 25.0 °C, while the those constructed with brick veneer and mud bricks were 25.4 °C and 26.6 °C, respectively; he concluded that using walls with AAC leads to a more comfortable room than the other materials. Aybek [18] also conducted building energy simulations with AAC, wooden frames, and metal frames; he found that the building model made with AAC consumed 14% and 11.6% less energy than the wood-framed and metal-framed models, respectively.

As previously stated, AC produced using an expansion agent, such as Al powder, typically uses an autoclaved curing method. The autoclave curing used for the AC has potential risks and is environmentally costly because of its high-pressure and -temperature operation. Therefore, a non-autoclaved aerated concrete (NAAC) cured in the air, or in a moist room at 100% RH, was developed. The authors' previous work successfully assessed the properties of NAAC in terms of compressive strength, porosity, and λ, and found that

NAAC could have enough strength and a similar λ to AAC [10]. However, when NAAC is used as a building wall material in a residential house in Nur-Sultan, Kazakhstan, it is interesting how much the NAAC contributes to energy conservation. Therefore, in this research, the energy-saving potential of NAAC to improve the energy performance of a residential house was assessed throughout the simulation with DesignBuilder software tools. Finally, both simple annual heat loss and heat transfer through the walls of the building were calculated.

2. Research Scope and Summary of Phases I and II Work

2.1. Research Scope

This research aims to evaluate how much NAAC contributes to the energy conservation of a residential house in Nur-Sultan, Kazakhstan. As presented in Figure 1, this goal was accomplished by three principal works: (1) the manufacture of NAAC, (2) the development of a performance-based model to predict the thermal conductivity of NAAC, and (3) evaluation of the energy conservation of a residential house containing NAAC walls.

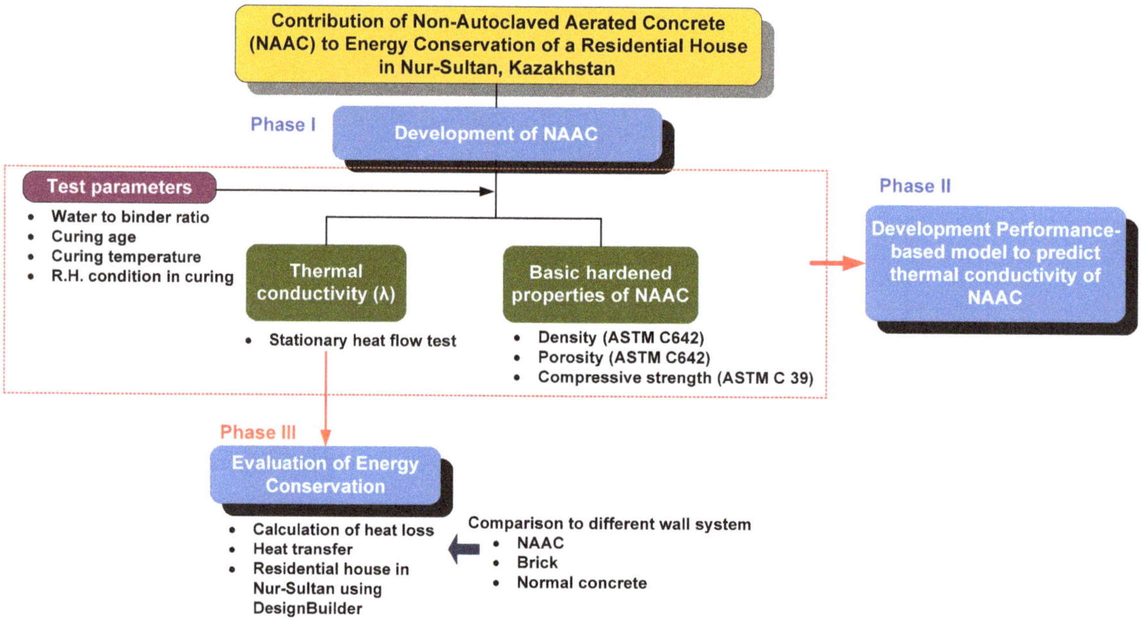

Figure 1. Diagram of research scope and experimental program.

In phases I and II studies, manufacturing of NAAC and development of a prediction model of NAAC's thermal conductivity were successfully conducted and reported [9,20]. The mix design formulation(s) was accomplished by determining density, porosity, compressive strength, and λ under various curing conditions and different water-to-binder ratios (w/b). Furthermore, based on the linear relationship between λ and test parameters and the normalized curves for each parameter, the performance-based mathematical model to predict the λ of NAAC was developed. The advantage of this modeling is that it considers all mixture design parameters and curing conditions. However, there is room to improve the performance-based model to predict the λ of NACC. For example, taking the moisture content of the NAAC mixture as an input parameter for the prediction model may increase the model's reliability.

As a phase III work, this paper mainly focuses on evaluating the contribution of NAAC to energy conservation. The obtained λ values of NAAC mixtures from phase I and II studies were used to calculate the heat loss and transfer for a typical residential house with NAAC walls. Finally, the energy-saving potential of building walls consisting of NAAC was compared to that of the brick and regular concrete commonly used in Kazakhstan throughout the simulation, using DesignBuilder software (V7, DesignBuilder Software Ltd., Stroud, UK).

2.2. *Summary of Phase I and II Work*

2.2.1. Mixture Proportions and Levels of Test Parameters

Nine NAAC mixtures were cast to evaluate key parameters influencing the λ of NAAC in the phase I study. Test parameters included the w/b, curing age, curing temperature, and moisture condition of the cured specimen. Compressive strength, porosity, density, and λ value of NAAC were obtained from the tests. The NAAC was produced with 42.0% sand, 12.9% cement, 9.0% lime, 0.06% aluminum powder, and 36.04% water. Mixture proportions and levels of test parameters are presented in Tables 1 and 2, respectively.

Table 1. Mixture proportions of non-autoclaved aerated concrete.

Mixture	Unit Weight (kg/m^3)						
	Sand 1	Sand 2	Sand 3	Cement	Lime	Water	Aluminum Powder
M1_w/b = 0.6_T20_D	336	379	339	405	198	360	1.62
M2_w/b = 0.6_T40_D	336	379	339	405	198	360	1.62
M3_w/b = 0.6_T60_D	336	379	339	405	198	360	1.62
M4_w/b = 0.6_T20_S	336	379	339	405	198	360	1.62
M5_w/b = 0.6_T40_S	336	379	339	405	198	360	1.62
M6_w/b = 0.6_T60_S	336	379	339	405	198	360	1.62
M7_w/b = 0.5_T20_D	336	379	339	440	229	335	1.62
M8_w/b = 0.7_T20_D	336	379	339	318	220	378	1.62
M9_w/b = 0.8_T20_D	336	379	339	320	176	397	1.62

Note: M#: mixture number; w/b: water-to-binder ratio; T: curing temperature; D: dried condition; S: saturated condition.

Table 2. Test parameter variables.

Test Parameter	Levels
Curing age	3, 7, 14, or 28 days
Curing temperature	20 °C, 40 °C, or 60 °C
Humidity conditions	Dry or moist (saturated)
Water-to-binder ratio	0.5, 0.6, 0.7, or 0.8

2.2.2. Materials and Test Methods

Ordinary Portland cement (OPC-ASTM Type I), lime, three types of sand with different Al_2O_3 and SiO_2 contents, and aluminum (Al) powder were used to cast NAAC. The Al_2O_3 and SiO_2 contents were equal to 11.27% and 56.54%, respectively, in Sand 1; 0.64% and 78.83%, respectively, in Sand 2; and 16.57% and 72.92%, respectively, in Sand 3. Al powder with a medium size of 44–53.5 μm was used as an expansion agent, producing the volume-increased portion of the concrete as a result of many hydrogen–air holes caused by a chemical reaction between the water, Al powder, and alkalis released from cementitious materials (binders). The OPC with specific gravity of 3.14 had the following chemical composition: 18.20% SiO_2, 4.01% Al_2O_3, 11.79% Fe_2O_3, 53.8% CaO, 0.79% K_2O, 0.16% Na_2O, 0.58% MgO, 0.20% TiO_2, 0.46% MnO, and 4.32% SO_3.

Cubic specimens with dimensions of 50 mm × 50 mm × 50 mm were cast for the compressive strength tests, while prismatic-shaped samples with dimensions of 150 mm × 150 mm × 30 mm were cast for the thermal conductivity measurements. During the casting process for the thermal conductivity samples, the concrete mixture was placed into a mold with two layers, and 25 strokes of the tamping rod per layer were applied by a metal rod. Surplus concrete from the mold was removed, and then the mold containing concrete was put into an oven maintained at 40 °C and 40 ± 5% relative humidity (RH), and then covered with plastic sheets for 2 h. After 24 h oven curing, the NAAC specimens were demolded and placed under various curing conditions, as described in Table 2. Three samples per mixture were made in order to increase the reliability of the test results.

The λ of NAAC was measured using a thermal conductivity measuring equipment based on stationary heat flow. This device can measure three thermal properties—thermal conductivity (W/mK), thermal insulance (m^2·K/W), and heat flux (W/m^2)—using the specimen, which had dimensions of 150 mm × 150 mm and thickness ranging from 3 mm to 40 mm.

Based on the ASTM C642-13 Standard Test Method for Density, Absorption, and Voids in Hardened Concrete, the density and porosity of the NAAC sample were determined.

2.2.3. Test Results

Figure 2 presents the development of compressive strength over time for each mixture. Regardless of mixture type, the compressive strength increased as curing age increased. Moreover, the comparison of strength development in mixtures 7–9 clearly shows that the increase in w/b led to lower compressive strength.

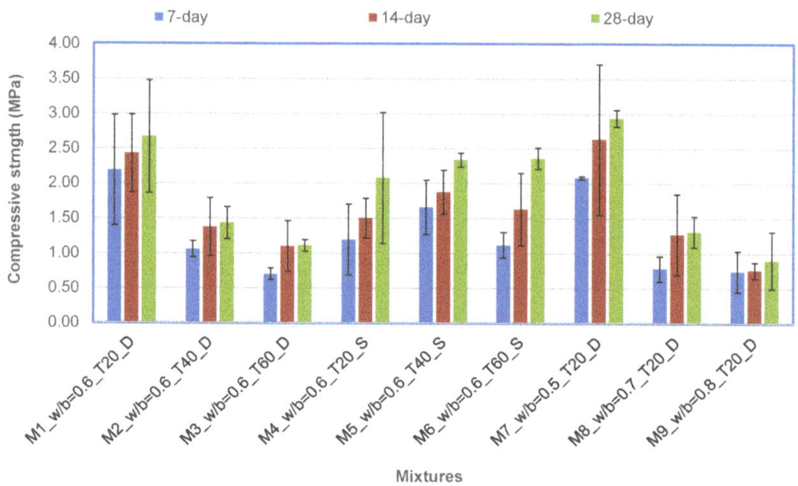

Figure 2. Compressive strength development of each mixture.

In moist curing, the compressive strength generally increased as the curing temperature increased. Interestingly, the compressive strength decreased as the curing temperature increased for the mixtures cured in dry conditions. This result was due to the loss of water from the drying surface of the concrete, which did not contain enough water for further continuation of the hydration process.

Figure 3 shows the relationship between the λ and porosity-to-density ratio (P/D) of NAAC mixtures at 14 and 28 days. In general, the concrete mixtures with higher porosity and lower density have a low λ value, regardless of curing age; in other words, the higher the P/D, the lower the λ [1].

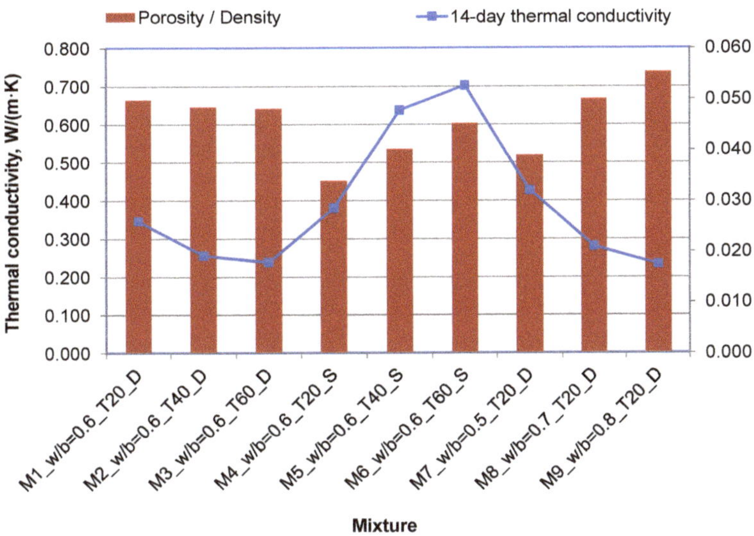

(**a**) 14-day test results

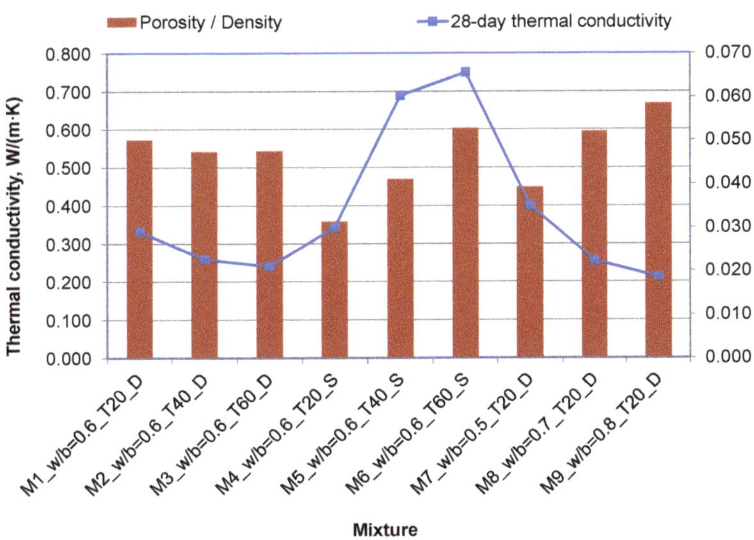

(**b**) 28-day test results

Figure 3. Relationship between thermal conductivity and porosity-to-density ratio.

As expected, the λ values of the NAAC mixture increased as the P/D increased, except for the moist-cured (saturated) specimens. For example, the λ value of mixture 9, with a P/D of 0.055, was 0.233, while the λ value of mixture 7, with a P/D of 0.039, was 0.426; 28-day-cured specimens followed the same trend. Interestingly, the moist-cured sample followed the opposite trend: the higher the P/D, the higher the λ. This result may be attributed to the curing temperature, which affects the formation of large and open-connected pores in the NAAC. Machrafi and Lebon [21] reported that the λ was considerably increased when pore size increased from nano- to macro-pores. Bhattacharjee and Krishnamoorthy [22] reported that concrete with open-pore cells had higher λ than that with enclosed pores, because open-cell concrete has more chance of being saturated. It should be noted that a higher curing temperature at an early age produces coarse and open-cell pores that are filled with free water, consequently resulting in larger λ values.

Phase I research has also investigated how different parameters—such as the w/b, curing age, curing temperature, and moisture condition of the cured specimen—influence the λ of the NAAC mixture. In summary, the λ of the NAAC mixture decreased with the increase in w/b, curing age, and air-dried curing temperature, whereas the λ increased with the rise in the curing temperature under saturated conditions. For example, Figure 4 presents the effect of w/b on the λ of NAAC specimens cured for 14 and 28 days. As previously stated, the λ of the NAAC mixture decreased as the w/b increased. This result can be attributed to the fact that higher water content increases the intensity of pore formation, which creates more pores in the NAAC and, eventually, leads to a lower value of λ.

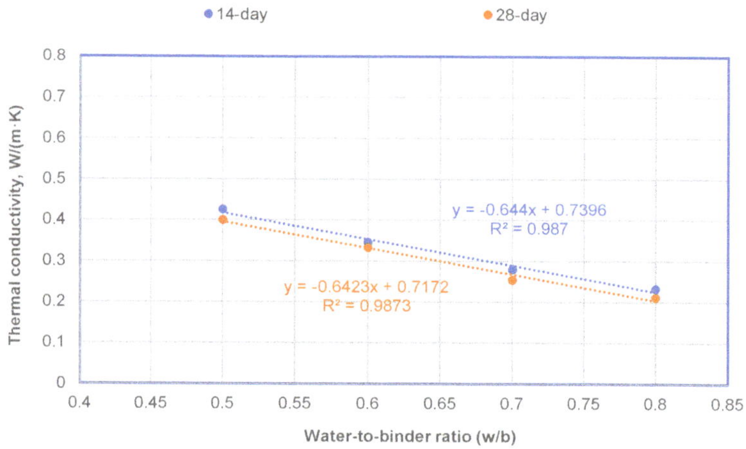

Figure 4. Effect of water-to-binder ratio (w/b) on thermal conductivity at 14 days and 28 days.

Based on the phase I study results, a performance-based model to predict the λ of NAAC mixture was successfully developed (phase II work). The model showed good reliability to predict 28-day λ of NAAC mixtures, along with an R^2 value of 0.875 [10]. This model was based on the relationship between λ and each test parameter, as described above and in Table 2.

3. Evaluation of Energy Behavior in a Residential House (Phase III Study)

3.1. Building Characteristics, Material Properties, and Simulation Software

To determine the contribution of NAAC to the energy conservation of a residential house in Nur-Sultan, Kazakhstan, the geometry of the house was first established. As presented in Figure 5a,c,d, a typical two-story conventional house with a total area of 191.5 m^2 was used as the simulation model. The first floor of the house consisted of a hall, kitchen/dining room, bathroom, guest room, and family room, with areas of 19 m^2, 24 m^2, 8.75 m^2, 20 m^2, and 24 m^2, respectively. The second floor comprised a hall, master bedroom (including a bathroom), 2 bedrooms, and 2 bathrooms, with areas of 16 m^2, 20 m^2, 36 m^2, and 16.1 m^2, respectively. House geometry components for the simulation are given in Table 3.

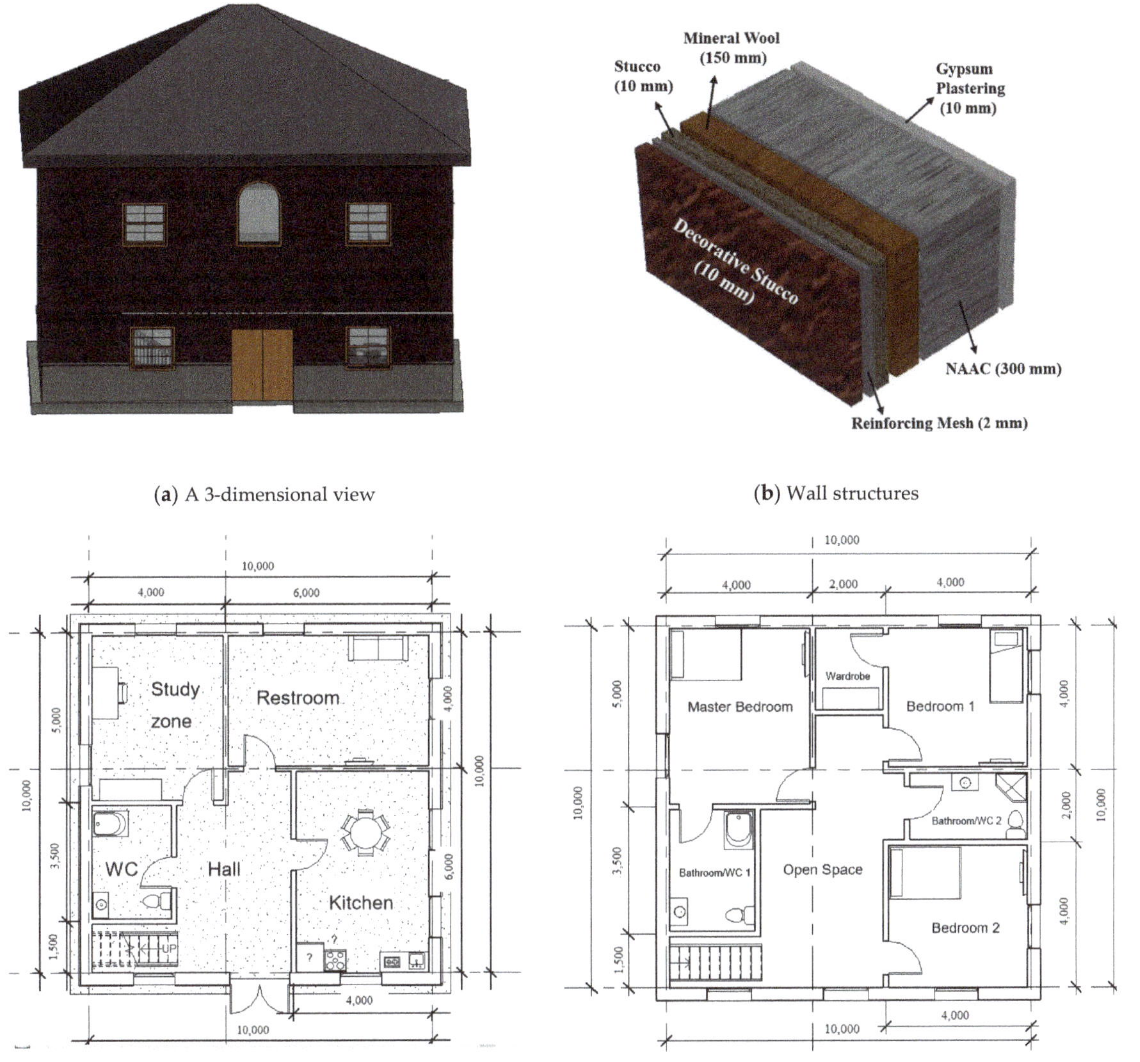

(**a**) A 3-dimensional view (**b**) Wall structures

(**c**) First floor plan (**d**) Second floor plan

Figure 5. A typical 2-story residential house model in Nur-Sultan for energy simulation.

Table 3. House geometry components for the simulation.

Component	Value
Floor height	7 m (22.97 ft)
Perimeter of house	40 m (131.23 ft)
Total area of house	280 m^2 (3013.89 ft^2)
Thickness of wall layer	300 mm (11.81 in.)
Thermal conductivity of mixture 3	0.2315 W/m·K (1.605 Btu·in/h·ft^2·F)

Materials for six wall layers consisted of 10 mm decorative Munich stucco, 2 mm reinforcing mesh, 10 mm stucco, 150 mm mineral wool, 300 mm NAAC, and 10 mm gypsum plastering, as shown in Figure 5b. As previously stated, the other two sets of wall structures, with brick and normal concrete—the most common materials for constructing conventional houses in Kazakhstan—instead of NAAC, were also selected in order to evaluate the differences in energy-saving performance between different materials. The properties of each material in the wall structure—such as density, specific heat, and thermal conductivity—are given in Table 4.

Table 4. Properties of wall structure materials.

Materials	Thickness (mm)	Thermal Conductivity (W/m·K)	Specific Heat (J/kg·K)	Density (kg/m^3)	Cost (GBP/kg)
Decorative stucco	10	0.6918	840	1858	5.26
Reinforcing mesh	2	0.3	1000	1000	0.68
Stucco	10	0.6918	840	1858	2.02
Mineral wool	150	0.0550	840	140	37.90
NAAC	300	0.2315	600	1113	32.16
Brick	300	0.72	840	1920	31.01
Normal concrete	300	1.8	1000	2400	28.31
Gypsum plastering	10	0.16	830	785	0.081

The software simulation used these geometric models and material properties to evaluate how much energy could be saved for the residential house with different materials. While the DesignBuilder software with a cross-platform use of EnergyPlus (V9, U.S. Department of Energy, Washington DC, USA) as the main simulation engine was used for whole-house energy simulation, the Autodesk Revit software (Revit 2021, Autodesk, San Rafael, CA, USA) was used to quickly create the geometry and floor plans needed for simulating the house.

3.2. Calculation of Annual Heat Loss

To cross-check the energy-saving effect of NAAC, the annual heat loss (AHL) through the wall of the house was calculated using the American Society of Heating, Refrigerating, and Air-Conditioning Engineers (ASHRAE) method [23], considering the same house geometry mentioned above. In the ASHRAE method, it is necessary to know the terminology of heating degree days (HDD), cooling degree days (CDD), and degree days. Whereas HDD is the amount of energy that is required to heat the building during the cold season, CDD is the amount of air conditioning required during the hot season. The degree day is the index of fuel consumption demonstrating how many °F the mean temperature fell below 65 °F for the day. Because Nur-Sultan is located in the seventh climate zone, with 9000 < HDD65 °F $\leq$ 12,600, the annual HDD65 value of 10,291 was used. In order to calculate the AHL, this value was multiplied by the heat loss per degree day (HLPDD), which is the heat loss per day with one degree between the internal and external temperatures. The HLPDD can be obtained from Equation (1):

$$HLPDD = \frac{A \times (T_{inside} - T_{outside})}{R} \times 24\frac{\text{h}}{\text{day}} \quad (1)$$

where HPLDD = heat loss per degree day; A = total wall area (ft^2); $(T_{inside} - T_{outside}) = 1$ °F; and R = thermal resistance (h·ft^2·F/Btu).

Thermal resistance was calculated using Equation (2):

$$R = \frac{\Delta x}{\lambda} \quad (2)$$

where Δx = the thickness of the wall layer (ft); and λ = thermal conductivity (Btu·in/h·ft^2·F).

3.3. Estimation of Heat Transfer

To further validate the effect of NAAC on energy saving, heat transfer through the wall of the house was estimated based on a one-dimensional heat transfer method developed by Thomas [24]. Figure 6 shows a steady-state one-dimensional heat transfer model; it is assumed that the wall is a homogeneous material with constant thickness and constant thermal conductivity, and that each face of the wall is held at a continuous uniform temperature. The wall heat transfer system consists of two convection elements on the outside and inside of the wall, and one heat conduction element inside the wall.

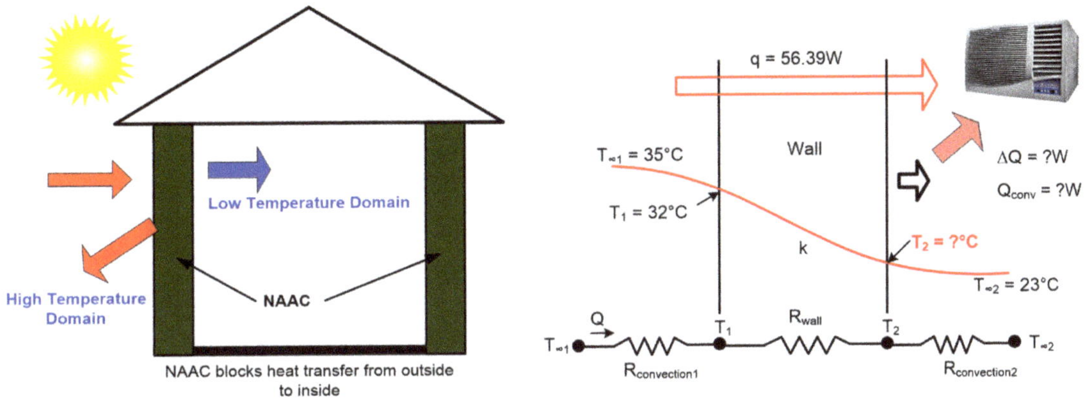

(**a**) Concept of the NAAC wall insulator (**b**) Heat transfer of NAAC wall loads structures

Figure 6. One-dimensional steady-state model.

In general, the one-dimensional heat transfer model uses the following concept: rate of heat convection into the wall = rate of heat conduction through the wall = rate of heat convection from the wall. Table 5 presents the elements used to calculate the heat transfer through the wall, including the air conditioner and inner surface temperature. Therefore, the heat transfer of the wall was calculated using Equation (3).

$$Q_w = h_1 \times A \times (T_{\infty 1} - T_1) = \frac{\lambda \times A \times (T_1 - T_2)}{\Delta x} = h_2 \times A \times (T_2 - T_{\infty 2}) \quad (3)$$

where Q_w = heat transfer of wall; A = area of the wall (m^2); $T_{\infty 1}$ = outdoor temperature; T_1 = outer wall surface temperature; T_2 = inner wall surface temperature; $T_{\infty 2}$ = indoor temperature; λ = thermal conductivity (W/(m·K)); Δx = wall thickness (m); h = heat transfer coefficient (W/m^2·K).

Table 5. Conditions of one-dimensional heat transfer model.

Element		Condition
Outdoor temperature, $T_{\infty 1}$ (°C)		35
Indoor temperature, $T_{\infty 2}$ (°C)		23
Outer wall surface temperature, T_1 (°C)		32
Length of the wall (1 room), L (m)		4
Height of the wall, h (m)		3.5
Thickness of the wall, Δx (m)		0.3
Thermal conductivity, λ (W/(m·K))	NAAC	0.2315
	Brick	0.72
	Normal concrete	1.8

4. Results and Discussion

4.1. Evaluation of Energy Conservation from DesignBuilder Simulation

4.1.1. Energy Consumption of 9 NAAC Mixtures

Figure 7 shows the simulation results of monthly heating and cooling loads for nine different NAAC mixtures. It was observed that NAAC mixture 3, with the lowest λ value, had a significant reduction in heating and cooling loads for each month. For example, the heating and cooling loads in January and July for mixture 3 were 5346 kWh and 2253 kWh, respectively, when these months recorded the lowest and highest outdoor temperatures of the year, respectively. These values were 1.6% and 6.3% lower than the heating and cooling loads in the same months, respectively, for mixture 6, which had the highest λ value. Therefore, it can be concluded that NAAC with a lower λ value leads to lower heating and cooling energy needs, depending on the mixture parameters used in the NAAC mixture.

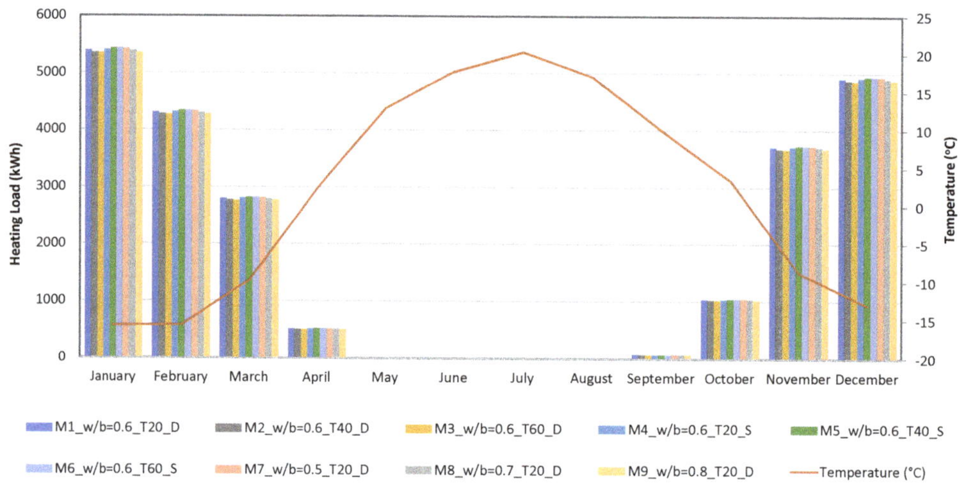

(**a**) Heating loads

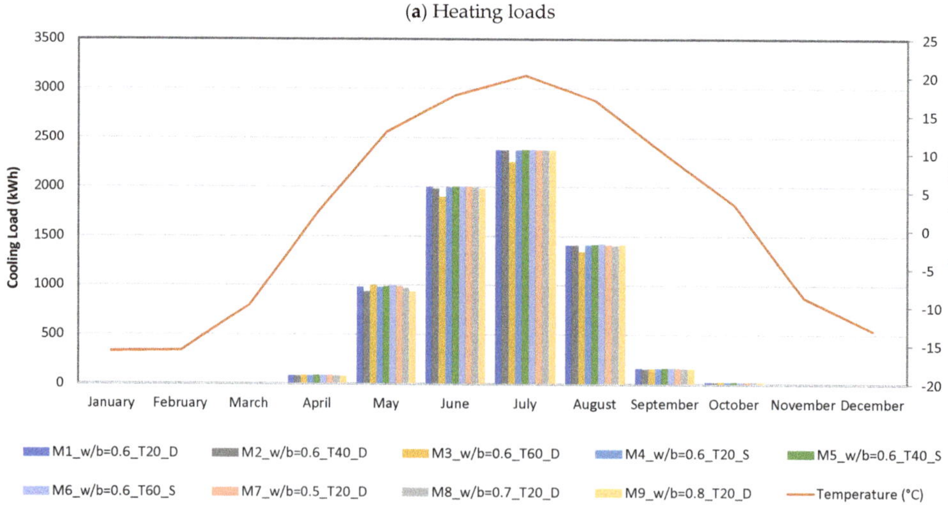

(**b**) Cooling loads

Figure 7. Monthly heating and cooling loads of 9 NAAC mixtures.

In the meantime, Table 6 indicates the energy consumption for heating and cooling of the individual rooms when mixture 3, with the lowest λ value, was used. The annual total energy consumption of heating and cooling for both the first and second floors was 22,515.2 kWh and 6760.8 kWh, respectively. The kitchen, dining room, and hall space recorded the maximum heating (1204.5 kWh) and cooling (445.7 kWh) consumption because of its having the largest area (43 m^3). This area also had the most significant 12-month average heating and cooling energy consumption, with 419.0 kWh and 109.5 kWh, respectively.

Table 6. Heating and cooling energy consumption for each room using mixture 3.

Floor	Room	Heating Energy (kWh)			Cooling Energy (kWh)		
		Annual	Maximum	Average	Annual	Maximum	Average
First floor	Guest room	3142	691.1	261.8	403.0	157.1	33.6
	Bathroom/WC	831.7	177.2	69.3	77.0	33.8	6.4
	Kitchen, dining room, and hall	6026.8	1495.1	502.2	826.5	387.3	68.9
	Family room	3595.3	788.8	299.6	405.5	169	33.8
Second floor	Master bedroom (MB)	3081.3	696.4	256.8	380.4	156.2	31.7
	Bathroom/WC 1	1060.0	225.5	88.3	122.7	51.8	10.2
	Bedroom 1	2711.1	594.9	225.9	345.5	145.5	28.8
	Wardrobe	686.7	143.2	57.2	79.5	35.8	6.6
	Bedroom 2	2374.9	550.9	197.9	511.7	194.6	42.6
	Bathroom/WC 2	773.9	167.2	64.5	104.3	45.7	8.7
	Open space	3353.6	800.6	279.5	601.2	254.7	50.1
Total energy consumption		27,637.7	6330.7	2303.1	3857.4	1631.5	321.5

4.1.2. Energy Consumption of Three Different Wall Materials

The monthly heating and cooling loads for NAAC (mixture 3), brick, and normal concrete are shown in Figure 8. The heating and cooling loads of NAAC were lower than those of brick and normal concrete for the entire year. For example, the heating load of NAAC in January was 5346 kWh, while the heating loads of brick and normal concrete were 5421 kWh and 5445 kWh, respectively. This result indicates that the use of NAAC in wall structure layers leads to a 1.4% and 1.8% reduction in heating load in January compared to brick and normal concrete, respectively. However, the heating load from May to August presents almost zero reading because of higher seasonal temperature, regardless of wall materials.

On the other hand, the maximum cooling load—irrespective of the type of wall material—was recorded in July, which has the highest outdoor temperature. In July, the energy consumption for cooling with NAAC was 2253 kWh, whereas brick and normal concrete had energy demands for cooling of 2379 kWh and 2381 kWh, respectively. Again, this result indicates that the NAAC reduces the cooling load by 5.3% and 5.4% compared to brick and normal concrete, respectively. Therefore, these results clearly indicate that the NAAC has higher energy conservation than brick and normal concrete.

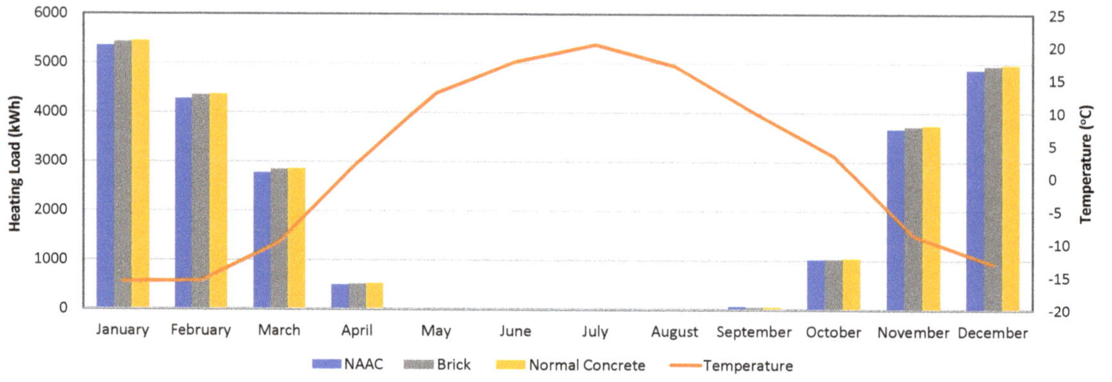

(**a**) Heating loads

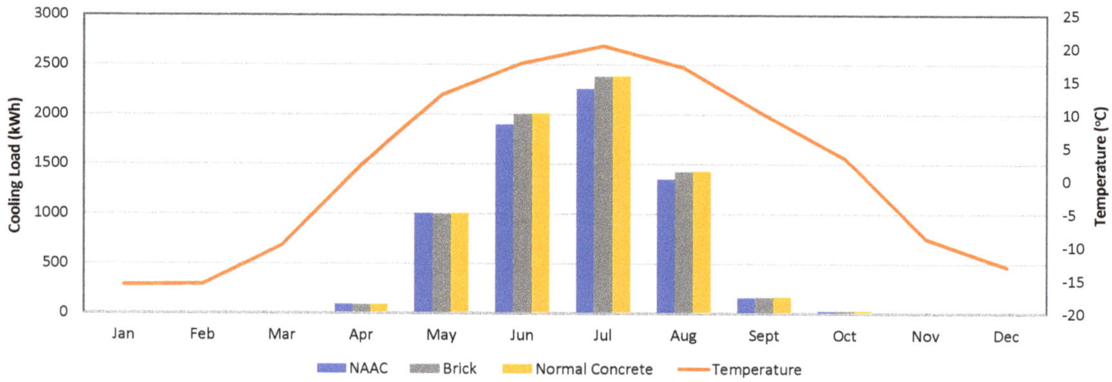

(**b**) Cooling loads

Figure 8. Comparison of monthly heating and cooling loads of NAAC, brick, and normal concrete.

4.1.3. Analyses of Cost, Site Energy, and Source Energy

Table 7 compares the cost, site energy, and source energy between three different wall materials. The site energy is the energy consumed at the final destination of the power generation cycle to maintain the desired conditions; in other words, it is all of the energy consumed by the house (building), and is typically expressed in utility bills. The source energy contains site energy as well as all energies that are produced; this includes the raw materials (e.g., coal) that are consumed to create the power that is distributed to consumers, as well as the power (e.g., electricity) that enters the distribution system [25].

Table 7. Cost, total site energy, and total source energy.

Materials	Cost (GBP)	Site Energy (kWh)	Site Energy Saving (%)	Source Energy (kWh)	Source Energy Saving (%)
NAAC	27813	36210.21		118553.77	
Brick	41440	37402.73	3.29	123273	3.98
Normal concrete	41160	38103.0	5.23	125338	5.72

Both site energy and source energy of two-story houses using NAAC are lower than those of brick and normal concrete materials. In fact, a building made with NAAC, which has lower a material cost than brick or normal concrete, leads to 1.08% and 1.38% lower source energy savings than similar houses constructed with brick and normal concrete, respectively. In addition, the use of NAAC walls can lead to site energy savings of of 0.67% and 0.91% compared to brick and normal concrete, respectively. Overall, the simulation results of DesignBuilder demonstrate that NAAC is a more effective material than conventional brick and normal concrete in terms of heating load, cooling load, site energy, and source energy. This result is consistent with previous research finding that autoclaved aerated concrete can be the most efficient material in terms of operational energy for any building, and can reduce heating and cooling losses if the material's thermal properties are used appropriately [26].

4.2. Evaluation of Annual Heat Loss

To check the energy-saving effect of NAAC, the annual heat loss (AHL) through a wall of the house was calculated as described in Section 3.2. The following is an example of AHL calculation using mixture 3 and the house geometry listed in Table 3.

Using Equation (2), thermal resistance was first calculated:

$$R = \frac{11.811\ \text{in}}{1.605\ \text{Btu} \cdot \frac{\text{in}}{\text{h}} \cdot \text{ft}^2 \cdot \text{F}} = 7.358\ \text{h} \cdot \text{ft}^2 \cdot \frac{\text{F}}{\text{Btu}} \tag{4}$$

Next, heat loss per degree day was calculated by substituting all of the values into Equation (1):

$$Q = \frac{3013.89\ \text{ft}^2 \times 1^\circ\text{F}}{7.358\ \text{h} \cdot \text{ft}^2 \cdot \frac{\text{F}}{\text{Btu}}} \times 24\ \frac{\text{h}}{\text{day}} = 9830\ \frac{\text{Btu}}{\text{degree day}} \tag{5}$$

Finally, the AHL was calculated by multiplying HLPDD and annual degree days in Nur-Sultan city:

$$AHL = Q = 9830\ \frac{\text{Btu}}{\text{degree day}} \times 10{,}291\ \text{degree day} = 101.16\ \frac{\text{MBtu}}{\text{year}} \tag{6}$$

The same procedure was used to identify the AHL for the other NAAC mixtures, brick, and normal concrete. As shown in Figure 9, the AHL was proportionally reduced as the λ value in each mixture was reduced. Moreover, the NAAC mixture produced a significant energy saving in heat loss for a conventional house in Nur-Sultan. For instance, the AHL of mixture 3 has the lowest λ value of 101.2 MBtu, while brick and normal concrete are 314.6 MBtu and 559.3 MBtu, respectively. This means that the NAAC mixture has approximately 3 and 5 times less heat loss than brick and normal concrete, respectively. In other words, this result indicates that the NAAC reduces the AHL by 67.8% and 81.9% compared to brick and normal concrete, respectively. This result follows the same trend as the heating load obtained from the DesignBuilder simulation, but presents more energy conservation. It should be noted that the different energy-saving results between the DesignBuilder simulation and the simple AHL calculation are due to the fact that the building simulated on DesignBuilder comprises six layers of material, while the hand calculation considers only a single material layer.

4.3. Evaluation of Heat Transfer

The DesignBuilder simulation results and annual heat loss calculations clearly indicate that NAAC has a better energy conservation effect than brick or normal concrete, due to thermal conductivity reduction. To support these results, the heat transfer through a wall of the house was obtained from Equation (3). Figure 10 shows the amounts of heat transferred by convection inside the wall, and by the air conditioner and the temperature of the inner surface of the wall, for each mixture. NAAC mixture 3, with the lowest λ value, had the lowest amount of heat transferred inside the wall by convection, as well as the lowest inner surface temperature. For instance, at a given room temperature of 23 °C, 7.51 W of heat was transferred by convection, and the calculated inner surface temperature was 23.4 °C, when mixture 3 was used. In this case, the air conditioner transferred very little heat to maintain its steady state.

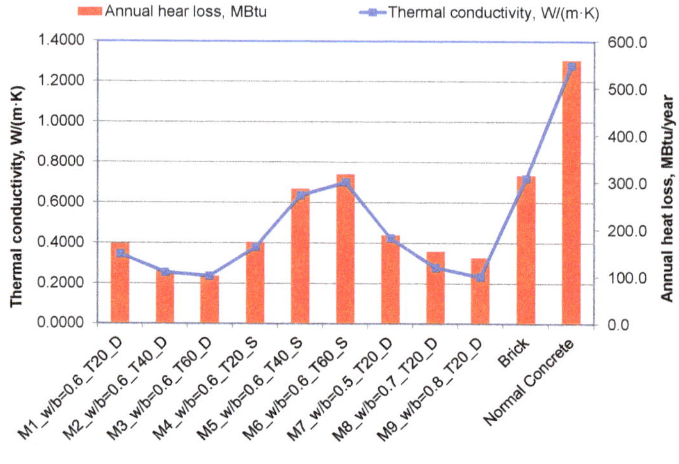

Figure 9. Annual heat loss of NAAC mixtures, brick, and normal concrete.

On the other hand, the results of heat transfer to the wall by convection (Q1), heat transfer inside the wall by convection (Q2), heat transfer by the air conditioner ΔQ (W), and inner surface temperature (T2) are summarized in Table 8. The heat transferred to the wall by convection was constant, because the conditions for external and outer-wall surface temperature were the same. Moreover, NAAC mixture 3, which has the lowest thermal conductivity value, demonstrated the lowest amount of heat transferred inside the wall by convection, along with the lowest inner surface temperature. For example, at a given room temperature of 23 °C, 7.5 W of heat was transferred inside the wall by convection when NAAC mixture 3 was used. Therefore, the air conditioner should transfer 84.9 W of heat in order to maintain the steady-state conditions in the room.

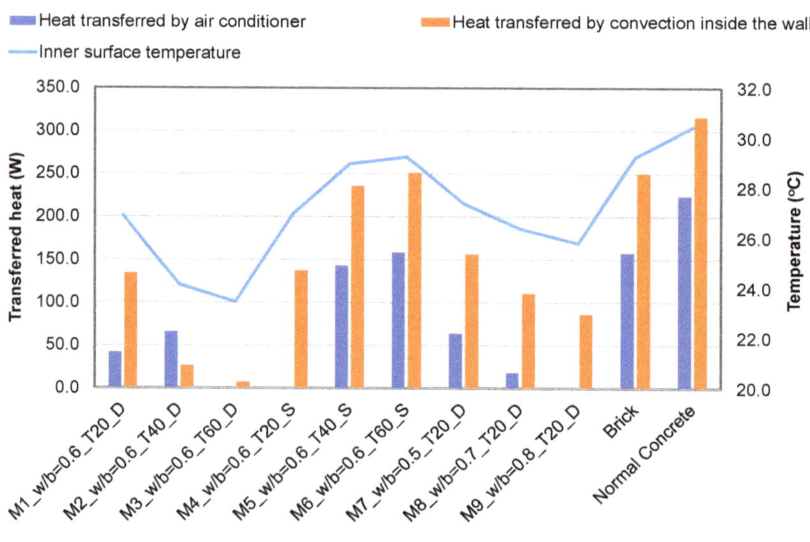

Figure 10. Heat transfer results through a house wall.

Table 8. Heat transfer results.

Mixture	λ (W/(m·K))	Q1 (W)	Q2 (W)	ΔQ (W)	T2 (°C)
M1_w/b = 0.6_T20_D	0.3891	92.4	134.3	41.9	26.9
M2_w/b = 0.6_T40_D	0.2524	92.4	26.5	65.9	24.2
M3_w/b = 0.6_T60_D	0.2315	92.4	7.5	84.9	23.4
M4_w/b = 0.6_T20_S	0.3941	92.4	137.2	44.8	27.0
M5_w/b = 0.6_T40_S	0.6532	92.4	235.4	143.0	29.0
M6_w/b = 0.6_T60_S	0.7234	92.4	250.9	158.5	29.3
M7_w/b = 0.5_T20_D	0.4287	92.4	156.1	63.7	27.4
M8_w/b = 0.7_T20_D	0.3524	92.4	110.6	18.3	26.4
M9_w/b = 0.8_T20_D	0.3197	92.4	86.4	6.0	25.8
Brick	0.7200	92.4	250.2	157.8	29.3
Normal concrete	1.800	92.4	341.83	249.44	30.9

Note: M#: mixture number; w/b: water-to-binder ratio; T: curing temperature; D: dried condition; S: saturated condition.

On the other hand, 250.2 W and 341.8 W of heat were transferred inside the wall by convection when brick and normal concrete were used, respectively. This means that air conditioners should transfer 157.8 W and 249.4 W of heat, respectively, in order to maintain the steady-state conditions in the room. According to the calculation results, it is clear that the use of air-dried NAAC mixture 3 with curing temperature = 60 °C and 0.2315 W/(m*K), w/c = 0.6, results in 97.0% and 97.8% energy cost reduction compared to brick and normal concrete, respectively.

In addition, the inner surface temperature was lower when NAAC mixture 3 was used. For example, the inner surface temperatures of NAAC mixture 3, brick, and normal concrete were 23.4 °C, 29.3 °C, and 30.9 °C, respectively. According to Nicol and Humphreys [27], the most likely comfort thermal temperature can be identified using Equation (7):

$$T_c = 13.5 + 0.54T_0 \quad (7)$$

where T_c = comfort temperature (°C) and T_0 = monthly average outdoor temperature (°C). For example, since the average outdoor temperature in Nur-Sultan in July 2020 was 21 °C, Tc was estimated as 24.8 °C. The inner surface temperature of normal concrete and brick walls is much higher than this value, whereas that of NAAC walls is roughly the same.

5. Conclusions

The paper presents how much NAAC contributes to the energy conservation of a typical residential two-story house in Kazakhstan. The energy conservation potential of NAAC was evaluated throughout the study in comparison to brick and normal concrete, via simulation using DesignBuilder software, the calculation of annual heat loss, and assessment of the amount of heat transferred through the walls of the house. After comparing the results of NAAC, brick, and normal concrete, the following conclusions can be drawn:

(1) The compressive strength of NAAC generally increased with moist curing conditions, the increase in curing temperature, and the decrease in the water-to-binder ratio;
(2) The concrete mixtures with higher porosity and lower density had low λ values, regardless of curing age;
(3) DesignBuilder simulation results indicate that NAAC with lower λ values led to lower heating and cooling energy needs;
(4) The largest area in the house consumed more heating and cooling energy, regardless of mixture type;
(5) Because the heating and cooling loads of NACC were lower than those of brick and normal concrete, the house with NACC walls was more efficient in saving source energy and site energy, as well as in reducing material costs;
(6) Evaluation results of annual heat loss and heat transfer were consistent with DesignBuilder simulation results, showing that the use of NAAC conserves more energy than brick or normal concrete.

The thermal properties of NAAC are remarkably affected by various ingredients of concrete and hardened properties of concrete, such as density and porosity. The NAAC in this study was made from conventional materials such as cement, lime, silica-rich sand, and aluminum powder. Therefore, a more sustainable approach could be adopted if the NAAC was made of industrial byproducts such as fly ash and ground granulated blast-furnace slag (GGBFS), or solid municipal waste materials such as waste glass bottles.

Author Contributions: Conceptualization, C.-S.S. and D.Z.; data curation, A.U.; formal analysis, I.M. and A.U.; funding acquisition, C.-S.S. and J.K.; investigation, C.-S.S., I.M., and A.U.; methodology, C.-S.S., I.M., and D.Z.; project administration, J.K.; resources, J.K.; software, I.M.; supervision, C.-S.S., D.Z., and J.K.; validation, I.M.; visualization, I.M.; writing—original draft, C.-S.S.; writing—review and editing, C.-S.S., D.Z., and J.K. All authors have read and agreed to the published version of the manuscript.

Funding: Nazarbayev University funded this research under Faculty Development Competitive Research Program Grant No. 021220FD1351; the authors are grateful for this support. Any opinions, findings, conclusions, or recommendations expressed in this material are those of the author(s), and do not necessarily reflect the views of Nazarbayev University.

Institutional Review Board Statement: Not Applicable.

Informed Consent Statement: Not Applicable.

Data Availability Statement: The data used to support the findings of this study are included within the article. The original details of the data presented in this study are available on request from the corresponding author.

Acknowledgments: The authors appreciate Ecoton Company and the National Laboratory of Astana (NLA) at Nazarbayev University for their provision of the raw materials and equipment used in this work.

Conflicts of Interest: The authors declare no conflict of interest. The funders had no role in the design of the study, in the collection, analyses, or interpretation of data, in the writing of the manuscript, or in the decision to publish the results.

References

1. Shon, C.-S.; Mukashev, T.; Lee, D.; Zhang, D.; Kim, J.R. Can Common Reed Fiber Become an Effective Construction Material? Physical, Mechanical, and Thermal Properties of Mortar Mixture Containing Common Reed Fiber. *Sustainability* **2019**, *11*, 903. [CrossRef]
2. Real, S.; Gomes, M.D.G.; Rodrigues, A.M.; Bogas, J.A. Contribution of structural lightweight aggregate concrete to the reduction of thermal bridging effect in buildings. *Constr. Build. Mater.* **2016**, *121*, 460–470. [CrossRef]
3. Sobhy, I.; Brakez, A.; Benhamou, B. Analysis for thermal behavior and energy savings of a semi-detached house with different insulation strategies in a hot semi-arid climate. *J. Green Build.* **2017**, *12*, 78–106. [CrossRef]
4. Astana Current Weather Report, World Weather Online. Available online: https://www.worldweatheronline.com/lang/ru/astana-weather-averages/astana/kz.aspx (accessed on 26 September 2021).
5. Committee of Atomic and Energy Supervision and Control. Available online: http://en.kaenk.energo.gov.kz/ (accessed on 26 September 2021).
6. Energy Situation in Astana. Urban Corners. Available online: https://www.urbaneecken.org/en/energy-situation-in-astana/ (accessed on 26 September 2021).
7. Narayanan, N.; Ramamurthy, K. Structure and properties of aerated concrete: A review. *Cem. Concr. Compos.* **2000**, *22*, 321–329. [CrossRef]
8. Qu, M.-L.; Tian, S.-Q.; Fan, L.-W.; Yu, Z.-T.; Ge, J. An experimental investigation and fractal modeling on the effective thermal conductivity of novel autoclaved aerated concrete (AAC)-based composites with silica aerogels (SA). *Appl. Therm. Eng.* **2020**, *179*, 115770. [CrossRef]
9. Soultana, A.; Galetakis, M. Utilization of quarry dust and calcareous fly ash for the production of lightweight cellular micro-concrete—synthesis and characterization. *Buildings* **2020**, *10*, 214. [CrossRef]
10. Ulykbanov, A.; Sharafutdinov, E.; Chung, C.-W.; Zhang, D.; Shon, C.-S. Performance-based model to predict thermal conductivity of non-autoclaved aerated concrete through linearization approach. *Constr. Build. Mater.* **2019**, *196*, 555–563. [CrossRef]
11. ACI Committee 523. *Guide for Design and Construction with Autoclaved Aerated Concrete Panels, ACI 523.4R-09*; American Concrete Institute: Farmington Hills, MI, USA, 2009.
12. Hlaváček, P.; Šmilauer, V.; Škvára, F.; Kopecký, L.; Šulc, R. Inorganic foams made from alkali-activated fly ash: Mechanical, chemical and physical properties. *J. Eur. Ceram. Soc.* **2015**, *35*, 703–709. [CrossRef]

13. Walczak, P.; Szymański, P.; Różycka, A. Autoclaved aerated concrete based on fly ash in density 350kg/m3 as an environmentally friendly material for energy-efficient constructions. *Procedia Eng.* **2015**, *122*, 39–46. [CrossRef]
14. Ropelewski, L.; Neufeld, R. Thermal inertia properties of autoclaved aerated concrete. *J. Energy Eng.* **1999**, *125*, 59–75. [CrossRef]
15. Jones, M.; McCarthy, A. Utilising unprocessed low-lime coal fly ash in foamed concrete. *Fuel* **2005**, *84*, 1398–1409. [CrossRef]
16. Radhi, H. A systematic methodology for optimising the energy performance of buildings in Bahrain. *Energy Build.* **2008**, *40*, 1297–1303. [CrossRef]
17. Drochytka, R.; Zach, J.; Korjenic, A.; Hroudová, J. Improving the energy efficiency in buildings while reducing the waste using autoclaved aerated concrete made from power industry waste. *Energy Build.* **2013**, *58*, 319–323. [CrossRef]
18. Aybek, H. *Energy Performance Evaluation of AAC*; The University of Alabama at Birmingham: Birmingham, AL, USA, 2013.
19. Kaşka, Ö.; Yumrutaş, R. Comparison of experimental and theoretical results for the transient heat flow through multilayer walls and flat roofs. *Energy* **2008**, *33*, 1816–1823. [CrossRef]
20. Heathcote, K. Comparison of the Summer Thermal Performance of Three Test Buildings with that Predicted by the Admittance Procedure. *Arch. Sci. Rev.* **2008**, *51*, 31–38. [CrossRef]
21. Machrafi, H.; Lebon, G. Size and porosity effects on thermal conductivity of nanoporous material with an extension to nanoporous particles embedded in a host matrix. *Phys. Lett. A* **2015**, *379*, 968–973. [CrossRef]
22. Bhattacharjee, B.; Krishnamoorthy, S. Permeable Porosity and Thermal Conductivity of Construction Materials. *J. Mater. Civ. Eng.* **2004**, *16*, 322–330. [CrossRef]
23. ASHRAE. ASHRAE Handbook-Fundamentals. In *American Society of Heating, Refrigerating, and Air-Conditioning Engineers, USA*; ASHRAE: Atlanta, GA, USA, 2017.
24. Thomas, L.C. *Mass Transfer Supplement—Heat Transfer*; Prentice Hall: Hoboken, NJ, USA, 1991.
25. Site vs. Source Energy. ArchToolBox. 2021. Available online: https://www.archtoolbox.com/materials-systems/sustainability/site-vs-source-energy.html (accessed on 26 September 2021).
26. Zhao, Z.; Yang, X.; Qu, X.; Zheng, J.; Mai, F. Thermal insulation performance evaluation of autoclaved aerated concrete panels and sandwich panels based on temperature fields: Experiments and simulations. *Constr. Build. Mater.* **2021**, *303*, 124560. [CrossRef]
27. Nicol, J.F.; Humphreys, M.A. Adaptive thermal comfort and sustainable thermal standards for buildings. *Energy Build.* **2002**, *34*, 563–572. [CrossRef]

buildings MDPI

Article

Laboratory Measurement and Boundary Conditions for the Water Vapour Resistivity Properties of Typical Australian Impermeable and Smart Pliable Membranes

Toba Samuel Olaoye [1,*], Mark Dewsbury [1] and Hartwig Künzel [2]

[1] Architecture and Design, University of Tasmania, Inveresk, Launceston 7250, Australia; mark.dewsbury@utas.edu.au
[2] Fraunhofer Institute for Building Physics IBP, Fraunhoferstr. 10, 83626 Valley, Germany; hartwig.kuenzel@ibp.fraunhofer.de
* Correspondence: toba.olaoye@utas.edu.au; Tel.: +61-4-0627-7304

Abstract: The duo of better insulated and more air-tight envelopes without appropriate consideration of water vapour diffusion and envelope moisture management has often demonstrated an increased potential of moisture accumulation, interstitial condensation, and mould growth within the building envelope. To inform a resilient, energy efficient, and healthy building design, long-term transient hygrothermal modelling are required. Since 2008, concern has been raised to the Australian building regulators regarding the need to establish the vapour diffusion properties of construction materials, in order to develop a hygrothermal regulatory framework. This paper discusses the results from laboratory testing of the vapour diffusion properties of two common reflective pliable membranes, and one smart pliable membrane. The two reflective pliable membranes are often used within the exterior walls of Australian buildings. The smart pliable membrane is a relatively new, internationally available product. The three membranes were tested as per ISO 12,572 at 23 °C and 50% RH. To establish if the vapour resistivity properties were constant, under different relative humidity conditions, the membranes were further tested at 23 °C and relative humidity values of 35%, 65%, and 80%. The results of the three pliable membranes show that the vapour resistivity properties varied in a non-linear (dynamic) manner subject to relative humidity. In conclusion, this research demonstrates that regardless of the class, each of the tested membrane types behaved differently under varying relative humidity and pressure gradients within the testing laboratory.

Keywords: vapour resistivity; hygrothermal modelling; energy efficient; airtightness; condensation; hygrothermal boundary curve; moisture management; impermeable membrane; diffusion; smart membrane; relative humidity

Citation: Olaoye, T.S.; Dewsbury, M.; Künzel, H. Laboratory Measurement and Boundary Conditions for the Water Vapour Resistivity Properties of Typical Australian Impermeable and Smart Pliable Membranes. *Buildings* **2021**, *11*, 509. https://doi.org/10.3390/buildings11110509

Academic Editor: Paulo Santos

Received: 24 September 2021
Accepted: 25 October 2021
Published: 27 October 2021

1. Introduction

The aim of this research was to investigate whether the single point vapour resistivity test method as described in ISO 12572 and ASTM E 96m provides adequate data to inform building envelope hygrothermal simulation. The current standard only requires a single point measurement for materials tested at 23 °C and 50% relative humidity (RH). The incorporation of high-quality material property inputs in the hygrothermal simulation has been identified by many as critical, which may significantly impact the moisture and mould risk calculations. This article reports on the observed water vapour resistivity properties of two impermeable pliable membranes, and one smart pliable membrane, when tested under different relative humidity conditions (humidity-dependent), in order to plot multiple point, rather than single point, hygrothermal boundary curves for each of these materials.

The combination of thermal insulation and airtightness without appropriate consideration of the external envelope's ability to manage water vapour diffusion and moisture have been identified as key contributors to moisture, moisture accumulation, and mould growth within internal and interstitial spaces in energy efficient buildings [1–4]. Since the 1990s, design processes have increasingly included complex hygrothermal simulation methods, which calculate moisture accumulation and mould growth. In 2019, Australia's national building regulations, the National Construction Code (NCC), included the first performance requirement regarding condensation. These regulations were the result of significant market-based concerns about the visible presence of condensation and mould prevalent in many Australian new buildings. The performance requirement states, "*Risks associated with water vapour and condensation must be managed to minimise their impact on the health of occupants*" [5]. Reflecting the universal acceptance that mould spores affect human health, the new clauses were included within the Health and Amenity section [6–8]. In Australia, the development of guidelines, policy, and the regulatory framework regarding material data and bio-hygrothermal simulation methods is of importance due to the increased requirements for highly energy-efficient buildings and the corresponding development of condensation-related regulations for 2019, 2022, and 2025.

In this research, the water vapour resistivity properties of two types of impermeable membrane, and one type of smart pliable membrane were measured under different relative humidity conditions. Following international methods, the data was then modified to obtain variable hygrothermal boundary curves for inclusion in hygrothermal simulation. The effect of the differences between hygrothermal simulations completed using a single point, or a multi-point hygrothermal boundary curve will be reported in a future article. Within this context, the research collaboration between the University of Tasmania and Fraunhofer Institute of Building Physics (IBP) Germany, also addressed the matter of how to establish the hygrothermal boundary curves for materials tested via the gravimetric cup method and under different relative humidity conditions. This boundary curve is of great importance, especially if it is identified that construction materials have variable characteristics subject to different relative humidity conditions [9–11]. This is because the vapour resistivity properties of construction materials is not constant across the boundary curve but varies continuously along the cross-section of such materials. Previous publications about this research project have described the advanced method for construction material water vapour resistivity testing that has been completed to assist in providing high quality data for hygrothermal simulation.

There has been a growing concern that the standard testing method normally adopted in North America and Europe may provide insufficient hygrothermal guidance for the design guidance of building envelopes [2,12–15]. The current methods were developed at a time of limited computer processing capacity and for non-transient simulation methods. This method has been critiqued internationally, due to its focus on a single environmental condition, namely, 23 °C and 50% RH, which may not adequately represent the conditions within the external envelope [2,13,16]. Currently, Australia has adopted the North American Testing Method, ASTM E 96m, to obtain water vapour diffusion properties, as referenced in Australia Standard, AS4200:1, which deals with the application of pliable membranes in buildings [17,18]. As Australia seeks a more detailed climatically- based approach to address surface and interstitial moisture problems in buildings, it has been acknowledged that there is a need to quantify construction material's hygrothermal diffusion characteristics under different relative humidity conditions that may better represent the boundary conditions within external walls [13]. Relative humidity considerations are important, as the normal level of relative humidity varies significantly between climate types, whether they be desert, cool/temperate, maritime, or hot and humid climates. ASHRAE Standard 160 recommends interior relative humidity (RH) conditions be maintained between 30 and 60% [19]. Several researchers have identified internal relative humidity levels within occupied, conditioned, and unconditioned housing well above the 50% RH level prescribed in the single point vapour resistivity test method [16]. To adequately complete hygrothermal simulations requires appropriately detailed interior and exterior climatic data and precise water vapour diffusion properties for each material in the external envelope. A common standpoint for any simulation-based decision making is: garbage in—garbage out. Similarly, as buildings become more complex, apply increased air tightness and insulation levels, increase the use of sustainable/renewable construction materials, and include many materials within an envelope system, the sensitivity of the hygrothermal model to any change may significantly impact on the simulation result [20]. To enhance and give confidence to the accuracy of hygrothermal simulations, which predict the wetting, drying, and mould growth potential, a more comprehensive understanding of the vapour resistivity properties of all materials used in the external envelope of buildings is required under different temperate and relative humidity conditions.

The application of pliable membranes has a long history and is common in many sectors. Increasingly, the advancement of pliable membranes will continue to usher new opportunities within the construction sector [21–23]. However, in the building design and construction industry, pliable membranes used to assist building energy efficiency have gone through many stages of product development, and manufacturers have used many terminologies to drive the marketing values of different products. These terms include vapour retarder, thermal barrier, vapour barrier, water resistive barrier, air barrier, damp proof membrane, weather barrier, and breathable building wrap. In Australia, these terms are often misunderstood and confused when considering the selection for building design and construction. The most acceptable way to define and classify different types of pliable membranes is according to their degree of resistance to gas or vapour transport, which is either permeable, semi permeable, semi-impermeable, and impermeable [13,21]. Furthermore, climate change is driving national demand for more airtight and energy-efficient buildings, progressively requiring manufacturers to create more innovative products. This is expected to allow new buildings to be more resilient, supportive, and responsive to changing interior and exterior climates, and improve the building envelope thermal performance and indoor air quality. Depending on the polymer, or other technology, of these products, manufacturers are now shifting towards the use of the terms such as smart or variable vapour diffusion/resistivity membranes to describe these new generations of product lines. An impermeable membrane is usually referred to as a vapour barrier because they do not allow water vapour diffusion to occur through the membrane. On the other hand, semi-impermeable membranes do allow moderate vapour diffusion through the membrane. However, their vapour resistance factor and air layer equivalent thickness values are moderate, but still high in comparison to a permeable membrane, which is very open to the vapour diffusion process [24].

Previously, this research program has measured and reported the vapour diffusion properties of other types of pliable building membranes. Like the research reported in this article, the water vapour diffusion properties were measured under different relative humidity conditions. This research revealed that vapour permeable pliable membranes which were identified as Class 3 and Class 4, within the Australian Standards (AS4200), behaved differently and in a non-linear manner under different relative humidity conditions [13].

Therefore, this paper reports on the results obtained from laboratory gravimetric cup measurements and the method employed to plot the hygrothermal boundary curves of two types of vapour impermeable, and reflective, pliable membranes, and one type of semi-impermeable (commonly called variable) pliable membrane. The reflective and vapour impermeable pliable membrane products are often used within the exterior timber and steel-framed walls of Australian buildings, and classified as a class 1 vapour control membrane (0.0 to 0.0022 μg/N/s) in AS4200 [17]. Similarly, the smart pliable membrane is a relatively new, internationally available product and is classified as a class 2 vapour control membrane (0.022 to 0.1429 μg/N/s) in AS4200:1.

2. Materials and Methods

The methodology involved undertaking gravimetric testing of the water vapour resistivity characteristics of two types of impermeable membranes and one type of smart membrane in a laboratory environment. For easy identification purposes in this paper, the sample from the smart membrane is tagged as specimen C, while samples from the two impermeable reflective pliable membranes are tagged specimen D and E, respectively. These membranes are commonly applied to the external face of the timber or steel frame (outside the insulation layer), of low-rise buildings, to improve air tightness (thermal performance), as a water barrier (façade durability), and as a vapour control layer (condensation and mould) in walls and roofs of buildings. The two types of impermeable membrane were reflective foil products from different manufacturers. They both had similar polymer characteristics and were coated with aluminium foil on the surface of one side. The smart membrane, which is commonly used in walls and roofs is a polyethylene copolymer product with varying water vapour diffusion properties, subject to the air moisture content. The total measurement period for all pliable membrane testing took fourteen months, which involved four experiments during which the temperature was maintained at 23 °C and the relative humidity conditions of the hygrothermally-controlled room were maintained at 35%, 50%, 65%, and 80%, respectively. Since the three membrane types discussed here are not readily open to the water vapour diffusion process, it took an average of three and a half months to establish a moisture equilibrium state for each specimen for each period of relative humidity conditioning.

2.1. Boundary Conditions

The experiments are based on the principle of diffusion of water vapour from an area with higher partial pressure of water vapour to an area with lower partial pressure. In these experiments, the vapour drive was established as a result of the pressure differences between the relative humidity within the internal air space of the test dish and the relative humidity within the internal condition of the hygrothermal testing room. This creates partial vapour pressure differences which cause water vapour diffusion to occur, which is similar to what the external building envelope may be experiencing in real life. The production and control of the testing room for these experiments has been discussed in previous publications. To avoid both air temperature and humidity stratification within the test room, a fan was in operation throughout the entire measurement period. This fan generated an air velocity of approximately 0.2 m/s which stabilised the pressure within the room. The average air temperature of the room was maintained at 23 °C throughout the four periods of measurement with ±1 °C variation. The relative humidity and their vapour pressure differences, which represent the targeted boundary condition for the four periods of the experiment, are shown in Table 1. This table was adapted from ISO 12,572 with additional testing points that were identified based on common climatic condition adopted for the four testing periods in this research. This table is important as it provides input data for the calculation and tabulation of the vapour resistivity properties of materials involving multiple moisture-dependent variables, which are needed to plot the hygrothermal curves that reflect different boundary conditions. In each of the experiments, the relative humidity in the hygrothermally-controlled room was carefully controlled. The relative humidity was maintained between 35.0% to 36.9%, with an average of 36% in the first test period. In the second test period, the relative humidity was maintained between 49.8% to 50.8%, with the average humidity of 50.4%. The relative humidity was maintained between 64.5% to 65.2%, with an average relative humidity of 65.12% in the third test period. During the fourth test period, the relative humidity was maintained between 77.84% to 83.2% with an average relative humidity of 80.29%. The details about the temperature and relative humidity performance of the chamber with respective variation has been reported in the previous publication [13].

Table 1. ΔP values for each of the test conditions.

Experimental Period	Temperature (°C)	Relative Humidity (Dry Side) (%)	Relative Humidity (Wet Side) (%)	Water Vapour Pressure Difference ΔP (Pa)
1 dry test	23	3	35	898
1 wet test	23	35	93	1628
2 dry test	23	3	50	1319.8
2 wet test	23	50	93	1207.25
3 dry test	23	3	65	1740.74
3 wet test	23	65	93	786.12
4 dry test	23	3	80	2161.83
4 wet test	23	80	93	364.99

2.2. Experimenal Procedure

The gravimetric procedure employed in this research followed the guidelines of the international standard ISO 12,572 [25]. The procedure involved completing both wet cup and dry cup gravimetric water vapour transmission measurement for samples from the three membranes. In each testing period, ten specimens were cut from the smart membrane (C), and another ten specimens from each of the two types of impermeable membrane (D). For each of the pliable membranes, five of these specimens were for wet cup testing and five specimens were for dry cup testing. Due to the amount of sample material provided, only six specimens were prepared for the second type of impermeable membrane (E), representing three specimens for wet cup and three specimens for the dry cup test. The test dishes comprised a round glass dish with 200 mm diameter and a 60 mm depth. The thickness of each specimen was measured by a digital micrometre screw gauge over ten different points on the surface of each specimen and the mean value was determined and recorded for later use during the water vapour diffusion calculation process. The specimens were precisely cut to the dish size (200 mm) such that they fit to the edge of the mouth of the dishes. The desired relative humidity within the cups were achieved by the use of dry and wet substrates. This research adopted silica gel as dry cup substrate because it was very stable at 3% relative humidity. The wet cup test achieved the desired relative humidity of 93% through the use of anhydrous ammonium dihydrogen phosphate solution at 23 °C. The substrates were gently poured into the dishes until an airspace of 20 mm was established between the top of the substrate and the top of the test dish. The specimens were then sealed to the edge of the test dishes with glue, followed by tightly wrapping the edge of the dishes with paper tape. To achieve adequate vapour seal, paraffin wax of 6:4 ratio and melting at 58–60 °C was gently applied around the paper tape with an artist's brush until the paper tape was no longer visible and the molten wax was allowed to harden. The gravimetric measurement began once a set of five specimens were completed for a particular test. Given that the temperatures inside the cup and outside the cup are the same, partial vapour pressure differential was achieved by the difference between the conditioned room's relative humidity and the wet or dry substrate within each dish, causing water vapour diffusion through the test specimen. The amount of water vapour diffusion was established by periodically measuring the weight of the cup assembly. The measurement of the dish weight continued until a steady state was reached. However, due to these membranes being vapour impermeable, after an initial period of regular measurements, the time between measurements was expanded to once a week, for a period of three months, and the diffusion properties of the specimens were determined through mathematical calculations.

3. Results

The result for the water vapour resistivity properties for each specimen tested was calculated from the mathematical equations provided by the ISO 12,572 [25]. The mathematical calculation and procedure for obtaining various iterative resistivity properties using specimen C 1 as an example is presented in the Appendix A. This calculation procedure was repeated for all the specimens, for all the boundary conditions (35%, 50%, 65%, and 80%). The detail results for the properties of each membrane tested and the statistical analysis for all the four testing periods are tabulated in Appendix C Tables A2–A13. Table 2, below, shows the average barometric pressure and the calculated vapour permeability of air for the four testing periods. Table 3 shows the summary of results for the water vapour permeability and permeance, in dry cup and wet cup scenarios, from the 26 tested materials. Similarly, the summary of the average water vapour resistance factor μ and diffusion air layer thickness S_D for all the tested specimens is tabulated in Appendix B. The air gap resistance used for calculating the water vapour resistance factor for each of these membranes was calculated by multiplying the initial resistance factor by the mean thickness of each membrane to get the initial equivalent air layer thickness S_D. This is then followed by deducting 20 mm air gap from the initial equivalent air layer thickness to obtain the final S_D values. This final S_D value is in turn divided by the mean thickness resulting into the final vapour resistance factor (also see Appendix A for details). Tables 4 and 5 shows the summary of hygrothermal water vapour resistance factor and equivalent air layer thickness across the boundary conditions over the average relative humidity for the three tested materials respectively, which were used to plot the hygrothermal boundary curves for specimen C, D, and E after harmonic adjustment.

Table 2. Average barometric pressure (Pa) and water vapour permeability of air for each experiment.

Method	Test Period 1 (35%)	Test Period 2 (50%)	Test Period 3 (65%)	Test Period 4 (80%)
		Material 1 (C)		
Dry test	1017.92	1018.04	1009.79	1010.85
Wet test	1017.92	1018.04	1009.79	1026.31
Calculated water vapour permeability of air	1.95×10^{-10} kg/(m. s. Pa)	1.944×10^{-10} kg/(m. s. Pa)	1.96×10^{-10} kg/(m. s. Pa)	1.9579×10^{-10} kg/(m. s. Pa)
		Material 2 (D)		
Dry test	1013.38	1023.31	1015.45	1017.85
Wet test	1013.38	1023.31	1015.45	1017.85
Calculated water vapour permeability of air	1.9533×10^{-10} kg/(m. s. Pa)	1.933×10^{-10} kg/(m. s. Pa)	1.9492×10^{-10} kg/(m. s. Pa)	1.9492×10^{-10} kg/(m. s. Pa)
		Material 3 (E)		
Dry test	1017.92	1023.07	1013.06	1014.52
Wet test	1017.92	1023.07	1013.06	1014.52
Calculated water vapour permeability of air	1.950×10^{-10} kg/(m. s. Pa)	1.9348×10^{-10} kg/(m. s. Pa)	1.9539×10^{-10} kg/(m. s. Pa)	1.9512×10^{-10} kg/(m. s. Pa)

Table 3. Summary of average water vapour permeance and water vapour permeability of samples.

RH%	Dry Test Water Vapour Permeance kg/(s·m^2·Pa)	Wet Test Water Vapour Permeance kg/(s·m^2·Pa)	Dry Test Water Vapour Permeability kg/(s·m·Pa)	Wet Test Water Vapour Permeability kg/(s·m·Pa)
		Sample C		
35	4.2×10^{-9}	5.8×10^{-11}	1.0×10^{-15}	1.5×10^{-14}
50	8.4×10^{-12}	1.0×10^{-10}	2.0×10^{-15}	2.7×10^{-14}
65	1.2×10^{-11}	8.2×10^{-11}	3.0×10^{-15}	2.2×10^{-14}
80	7.1×10^{-11}	1.0×10^{-10}	1.9×10^{-15}	2.8×10^{-14}
		Sample D		
35	3.95×10^{-12}	7.32×10^{-12}	8.79×10^{-16}	1.62×10^{-15}
50	6.15×10^{-12}	4.17×10^{-12}	1.39×10^{-15}	9.23×10^{-16}
65	3.25×10^{-12}	3.94×10^{-12}	7.35×10^{-16}	8.73×10^{-16}
80	2.22×10^{-12}	9.63×10^{-12}	5.02×10^{-16}	2.14×10^{-15}
		Sample E		
35	1.71×10^{-12}	2.00×10^{-12}	5.20×10^{-16}	6.41×10^{-16}
50	1.20×10^{-12}	8.32×10^{-12}	3.64×10^{-16}	2.67×10^{-15}
65	1.35×10^{-12}	6.76×10^{-12}	4.13×10^{-16}	2.15×10^{-15}
80	1.70×10^{-12}	1.28×10^{-11}	5.15×10^{-16}	4.06×10^{-15}

Table 4. Water vapour resistance factor with different boundary conditions.

Temperature	Dry Side RH (%)	Wet Side RH (%)	Average RH (%)	Specimen C Resistance Factor (μ)	Specimen D Resistance Factor (μ)	Specimen E Resistance Factor (μ)
23	3	35	19	189,398	222,099	383,221
23	3	50	26.5	94,895	139,160	530,781
23	3	65	34	64,499	265,273	472,951
23	3	80	41.5	10,005	388,586	378,743
23	35	93	64	13,043	120,265	304,191
23	50	93	71.5	7180	210,033	71,909
23	65	93	79	7640	223,063	90,899
23	80	93	81.5	6918	91,312	47,612

Table 5. Equivalent air layer thickness with different boundary conditions.

Temperature	Dry Side RH (%)	Wet Side RH (%)	Average RH (%)	Specimen C Sd (m)	Specimen D Sd (m)	Specimen E Sd (m)
23	3	35	19	46.21	49.5	116.4
23	3	50	26.5	23.25	31.4	160.8
23	3	65	34	16.04	59.9	144.6
23	3	80	41.5	2.72	87.7	114.9
23	35	93	64	3.4	26.6	97.5
23	50	93	71.5	2.5	46.5	23.07
23	65	93	79	2.37	40.19	30.9
23	80	93	81.5	1.98	20.3	15.2

4. Discussion

The aim of this research was to investigate the diffusion behaviour of water vapour impermeable, semi-impermeable, and permeable pliable membranes under different relative humidity boundary (moisture-dependent). This article focusses on the measured and calculated water vapour permeability values from two types of water vapour impermeable and one type of semi-impermeable membrane. A previous article reported the results from the measured and calculated water vapour diffusion properties of tested water vapour permeable membranes [13]. Firstly, it is important to state that this research successfully kept the temperature in the hygrothermal chamber stable at 23 °C with ±1 °C variation throughout all the four testing periods. Therefore, the shape of all the samples from the three tested pliable membranes observed was flat throughout the four testing periods, suggesting that there were no significant impact caused by temperature variation. Variation in temperature may alter the shape of membrane, which would affect the surface area calculations, which is very important when considering temperature-dependent measurement.

In Figure 1, the blue colour in the plot show the values for the water vapour resistance factor against the hygrothermal boundary conditions established from the average relative humidity conditions from the four laboratory measuring periods for Specimen C, which is classified as a class 2 (semi-impermeable membrane). In these measurements, the results indicate that the specimen was exposed to different relative humidities on both dry and wet sides. This difference in vapour pressure normally generates water vapour flow through the specimens, and as the humidity varies along the cross-section of the specimens, say from wet side to the dry, the vapour resistance also varies. Figure 1 shows that specimen C has different μ-values. From approximately 20% RH to 40% RH, the rate of water vapour diffusion varies significantly. Whilst from 40% RH to 80% RH, the water vapour diffusion rate is relatively constant. The vapour flow thus contains information about all the μ-values corresponding to the applied humidity interval as shown in Figure 2. This indicates that the water vapour diffusion resistivity properties of specimen C are non-linear as the amount of water vapour diffusion which passed through the specimen encounters different resistances, subject to the level of humidity. The graph also shows that the μ-values indicate an inverse relationship as the strength of the resistance factor decreased with increases in relative humidity.

Similar behaviour is observed from Figure 2, the blue colour is the plot of values of the equivalent air layer thickness against the established boundary conditions of the four experiment testing periods for specimen C. This also indicates that the water vapour flow across the cross-section of the specimen causes the equivalent air layer thickness to decrease in higher relative humidity, as the S_D value is high at lower relative humidity and lower in higher relative humidity in an inverse and dynamic pattern. Even though this dynamic behaviour is consistent with previous findings, it further supports the suggestion that the hygrothermal properties of construction materials should be considered under different relative humidity conditions. In Australia, hygrothermal diffusion function of pliable membrane is not yet defined under the current AS4200:1, and determining the appropriateness of pliable membranes according to their diffusion properties in different relative humidity conditions has become contestable among design professionals and researchers. Moreover, there are no appropriate and recent hygrothermal data for construction materials, typically used in Australia. Therefore, the ability to determine the hygrothermal boundary conditions of specimen C, with the resistance factor and equivalent layer thickness, is essential for accurately fulfilling the hygrothermal modelling pathway for moisture management introduced to NCC in 2019.

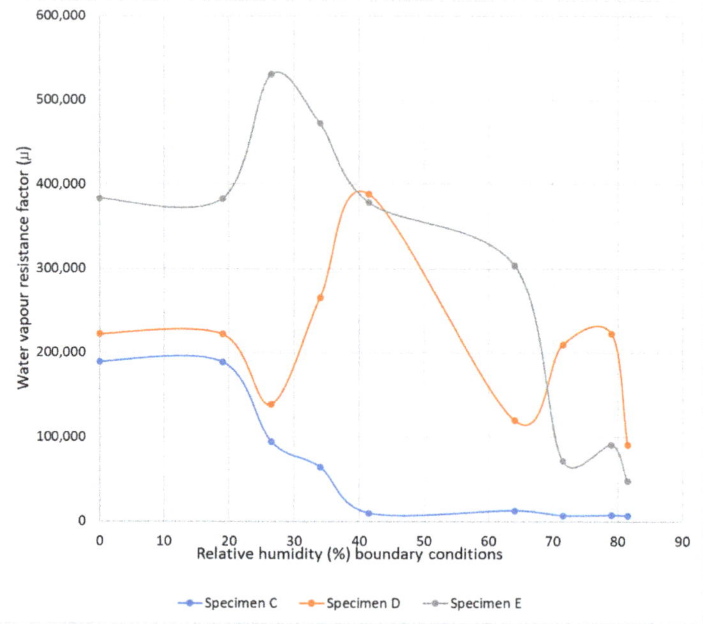

Figure 1. Plot of resistance factor against relative humidity prior harmonic curve adjustment for the three specimens.

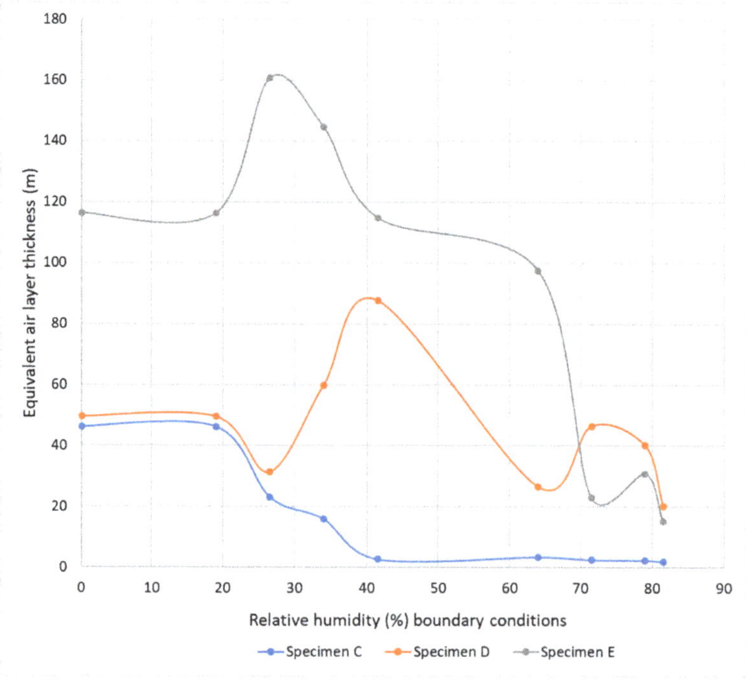

Figure 2. Plot of equivalent air layer thickness against relative humidity prior to harmonic curve adjustment for the three specimens.

Similarly, the plot of the vapour resistance values against the relative humidity boundary condition for specimen D and E were indicated by the red colour and green, respectively, in Figure 1. The red and the green colour in Figure 2 also show the plot of equivalent air layer thickness against the humidity boundary conditions of specimen D and E respectively. Generally, impermeable pliable membranes usually have a very high vapour resistance factor and air layer equivalent thickness values as observed in Figure 2. However, it is essential to point out that even these graphs further document dynamic behaviour, similar to what this research has observed with other pliable membranes under varying relative humidity boundary conditions. The expectation was that these reflective impermeable pliable membranes would have a constant high resistance to water vapour diffusion. On the contrary, this did not occur as these reflective foil products have a rather unusual water vapour diffusion behaviour as the relative humidity conditions increased. The reason for the dynamic patterns observed in these graphs has resulted from the changes in air permeability and vapour pressure that has occurred under the four different relative humidity conditions. This level of water vapour diffusion resistance will cause these impermeable pliable membranes to have narrow usability potential for moisture-permeable construction systems.

As previously mentioned, measurements were usually taken weekly because of high water vapour diffusion resistance of these specimens. In this scenario, this research observed that when the specimens from the two types of reflective membranes were tested at 80% RH, the specimen were swelling within a week. Additionally, by the second week of weighing, there was a visible formation of mould growth on the reflective surface of all the specimens (see Figures 3 and 4). The program for this research involved testing specimens from all the classes of pliable membranes at 80% RH during the same period. The yellow-brownish colouration of mould formation on these reflective materials was not observed on any of the other materials tested.

Comparing the result of these three membranes with each other, specimen C, which is semi-impermeable and belongs to Class 2, appears to be more open to vapour diffusion process with higher vapour resistance values when the humidity was less than 40%. Specimen D and E, which are classified as Class 1, impermeable pliable membranes, have very high vapour resistance factors and equivalent air layer thickness despite the increase in relative humidity conditions.

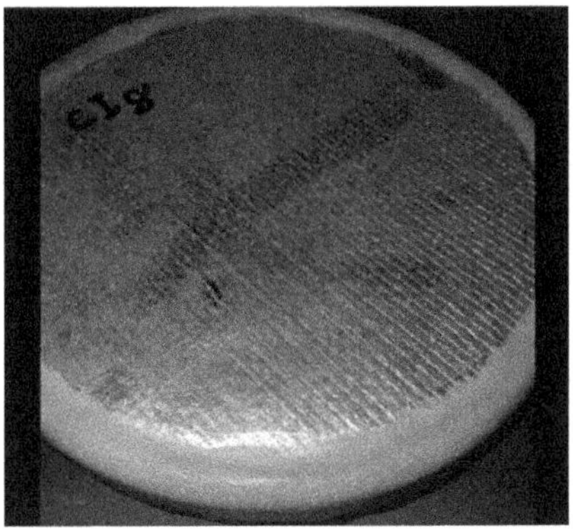

Figure 3. Mould formation on specimen D.

Figure 4. Mould growth on specimen E.

Harmonic Adjusment of Hygrothermal Boundary Curve

The cup measurements are used to determine the boundary condition of the water vapour diffusion characteristics of a material by plotting a curve after harmonic adjustment of the data from dry and wet cup gravimetric measurement. This is because there is the need to establish the different values of vapour diffusion at any given specified relative humidity ϕ, as the cup measurements do not directly provide these values. During the gravimetric measurement, the specimen is exposed to different relative humidities on both sides, for example, 50% on one side and 80% on the other side. The humidity varies continuously along the cross-section of the specimen, from 50% RH on one side to 80% RH on the other side, with all intermediate values also occurring somewhere in the specimen. Furthermore, the material in different parts of the specimen also has different μ-values, and all μ-values from μ (50%) to μ (80%) occur simultaneously somewhere in the specimen. However, the cup measurement can only tell us the strength of the vapour flow which diffused through the specimen. The vapour flow thus contains information about all the μ-values corresponding to the applied humidity interval, which are encoded in a single number. The task now is to determine μ-values for individual humidities. This can be done by combining the results of cup measurements using different humidity ranges. Hence, analysing gravimetric cup measurements as if the material had a constant μ-value, provides the effective μ-value of the specimen in such states of moisture. Therefore, the effective μ-value for a given humidity range is the harmonic mean of the $\mu(\phi)$-curve, taken over the humidity range. This is why it may be misleading to draw a $\mu(\phi)$-curve by simply plotting the effective μ-values against the mean applied humidity ranges from purely gravimetric measurement.

Figure 5 shows a $\mu(\phi)$-curve (grey) and a series of mean values (orange bars) of this curve, taken over humidity ranges with the same midpoint but with increasing widths. First, due to the curvature of the μ-curve, all those mean values are lower than the curve itself at the midpoint. The curve point is the desired μ-value corresponding to the midpoint humidity. The mean values correspond to the results of cup measurements performed by applying the respective humidity ranges. The results of all cup measurements are lower than the μ-curve point corresponding to the midpoint humidity, so it would be inappropriate to simply substitute a cup measurement result for this μ-value. Secondly, the mean values for humidity ranges with the same midpoint are lower for wider ranges. This means that when cup measurements with different wide humidity ranges are combined into one diagram (plotting them against the midpoint humidity), a misleading zigzag-shape of the diagram may result, even though the underlying true $\mu(\phi)$-curve may be continuously falling for increasing humidity.

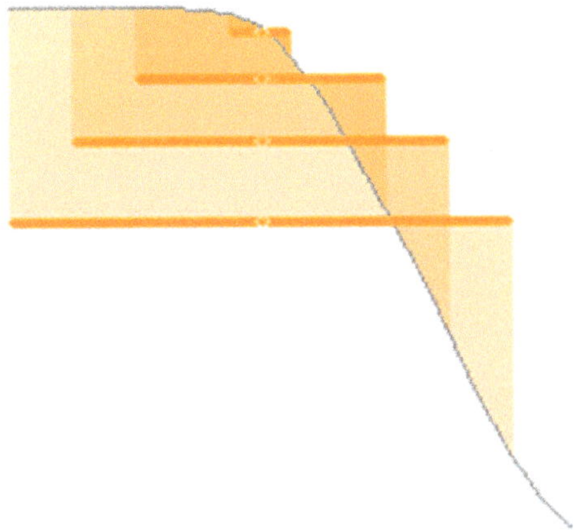

Figure 5. Plotting mean of values with the same midpoint from cup measurements.

A better approach than plotting the cup results against the midpoints of the applied humidity ranges is to determine a curve which has the properties that the harmonic mean values provide, taken over the humidity ranges which were used for the cup measurements, and are identical with the cup results. Figure 6 illustrates the procedure using data from WUFI for an adaptive vapour retarder, with the green bars representing a set of cup measurements, the grey curve being the adjusted test curve, and the thin orange bars being the mean values of the test curve, taken over the indicated ranges. The test curve is adjusted until the orange bars coincide as well as possible with the green bars.

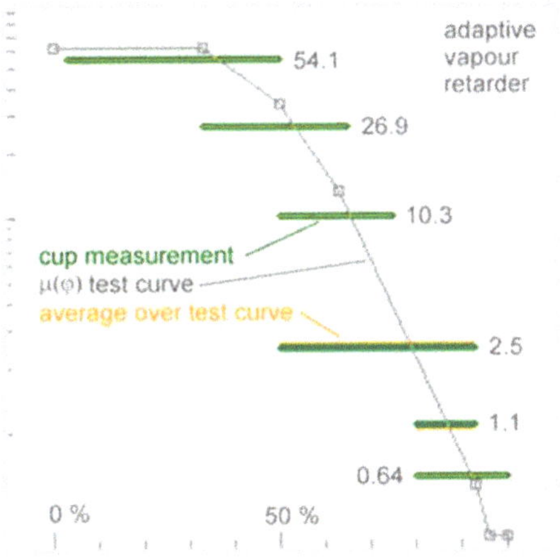

Figure 6. Example of plotted hygrothermal curve boundary of an adaptive vapour retarder from WUFI.

The above principle was applied to the three materials measured in this research, to plot the final hygrothermal boundary curve for the equivalent air layer thickness for specimen C, D, and E (see Figure 7) after the harmonic adjustment. These harmonically balanced values could be applied to the material properties within the transient hygrothermal simulation software to provide a more accurate simulation of the water vapour diffusion through an external envelope.

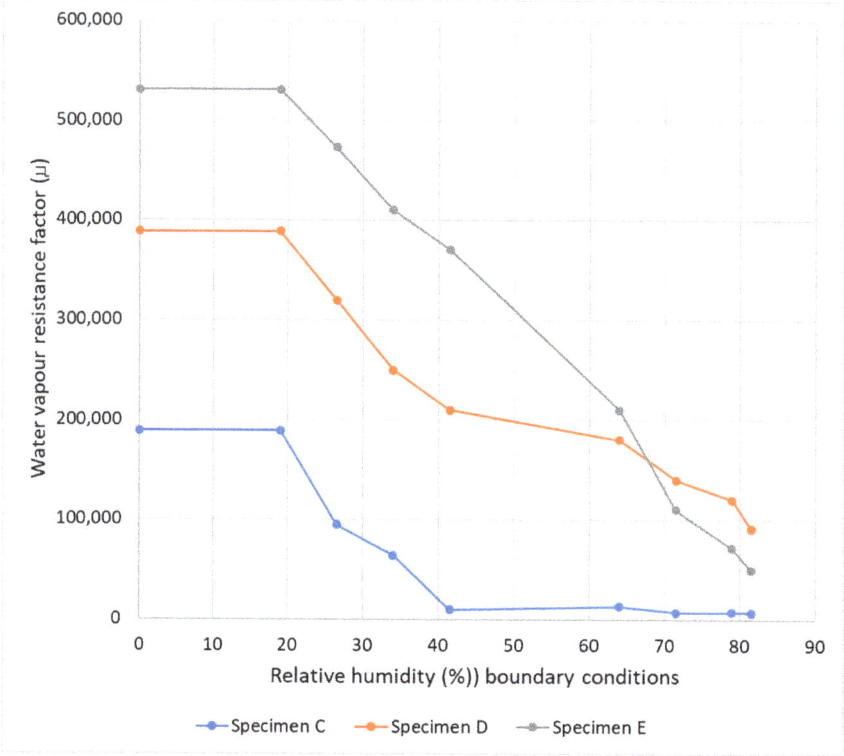

Figure 7. Final plot of the resistance factor μ against boundary conditions after harmonic curve adjustment for specimen C.

5. Conclusions

This research has investigated through laboratory measurement the water vapour diffusion characteristics of two types of water vapour impermeable reflective pliable membranes and one type of smart membrane. The research included measuring their water vapour diffusion behaviour under varying boundary conditions, with specific attention to different relative humidity conditions.

The results from the cup measurements show that the two vapour impermeable pliable membrane products are not open to vapour diffusion even at higher relative humidity conditions, as their resistance factor and air layer thickness are too high to allow for vapour diffusion. On the other hand, the result from the smart pliable membrane product indicates that vapour diffusion is possible at higher relative humidity, as the resistance factor and air layer thickness decreases with increases in relative humidity. Contrary to the single point test method applied typically to construction materials, the graphs for all the three types of pliable membrane material show that they behave in a dynamic non-linear manner subject to the relative humidity conditions. This finding is similar in nature to the previously published findings which reported on the non-linear vapour diffusion properties for tested vapour permeable pliable membranes [13]. Alarmingly, the composite materials that make up the vapour impermeable pliable foil-faced membranes supported mould growth during the test period, whilst the non-foil polyethylene copolymer products did not support mould growth. Further analysis based on the harmonic adjustment approach was then employed to plot the hygrothermal boundary curves for each of these pliable membranes. This is because the results from the gravimetric cup measurement may not be enough to determine the effectiveness of a materials' moisture behaviour along the cross-section of these materials, which is needed for hygrothermal modelling.

In conclusion, this research demonstrates that regardless of the water vapour diffusion class, each of the tested pliable membrane types behaved differently under different relative humidity conditions and vapour pressure gradients within the testing laboratory. This may indicate that the current single point value for construction material vapour diffusion properties used in hygrothermal simulation may be inadequate and may provide inaccurate guidance regarding moisture, moisture accumulation, and mould growth.

Author Contributions: T.S.O.—Main author carried out the all the experiments at UTAS, involved in the conceptualization; collects data; analyze data; graphs and visualization; provide the original draft manuscript, and revised manuscript. M.D.—Second author provides guidance to experiment, source for funding to procure equipment; project administration; contribute to data analysis; data curation; contribute to graphs and visualization, edit, and provided revision to the manuscript. H.K.—Provides guidance for the laboratory operation, supervision, data curation; contribute to graphs and visualization and revision of manuscript. All authors have read and agreed to the published version of the manuscript.

Funding: This research was co-funded by Commonwealth Scientific Industrial Research Organization (CSIRO), grant number 00004612.

Institutional Review Board Statement: Not applicable.

Informed Consent Statement: Not applicable.

Data Availability Statement: Not apllicable.

Acknowledgments: This research acknowledges the support received from the team at Fraunhofer Institute for Building Physics, Department Hygrothermics, Holzkirchen, Germany, and mathematical support provided by Des Fitz-Gerald UTAS.

Conflicts of Interest: The authors declare no conflict of interest.

Appendix A

Calculation procedure for various water vapour resistivity properties for Specimen C1.

The mass change G for specimen C1 is calculated as:

$$G = \frac{change\ in\ mass}{total\ time_{(S)}}\ \text{kg}$$

Therefore, for specimen C1:

$$\frac{1743.45-1743.24}{1{,}735{,}200_{(S)}} = \frac{2.1\times10^{-4}}{1{,}735{,}200_{(S)}}\ \text{kg}$$
$$= 1.21\times10^{-10}\ \text{kg/s}$$

The vapour flux g is calculated $= \frac{G}{A}$ kg/s.m^2.

Where A is the arithmetic mean of the exposed area of the test specimen in m^2, for specimen C1, the diameter of the specimen after sealing is 190 mm:

$$\therefore A = \pi r^2 \text{ where } r = \tfrac{d}{2} = \tfrac{190}{2} = 95\ \text{mm}$$
$$r = 0.095\ \text{m}$$
$$\therefore A = 3.14\times0.095^2 = 0.0283$$
$$\therefore g = \frac{G}{A} = \frac{1.21\times10^{-10}\ \text{kg/s}}{0.0283\ \text{m}^2} = 4.28\times10^{-9}\ \text{kg/s.m}^2$$

The water vapour permeance is then calculated as:

$$W = \frac{g}{\Delta P_V}\ \text{in kg/(s·m}^2\text{·Pa)},$$

To calculate $\Delta P_V = P_{satwetside} - P_{satdryside}$

Considering the relative humidity at 35% (0.35) for wet side and 3% (0.03) for dry side:

$$\varnothing_{wetside} = 0.35 \times 610\, e^{1.5258} = 982.65 \text{ Pa}$$

$$\varnothing_{dryside} = 0.03 \times 610.5\, e^{1.5258}$$
$$= 0.03 \times 2807.58 = 84.19 \text{ Pa}$$

$$\Delta P_V = 982.65 - 84 = 898.46 \text{ Pa}$$

$$W = \frac{g}{\Delta P_V} = \frac{4.28 \times 10^{-9} \text{ kg/s.m}^2}{898.46} = 4.74 \times 10^{-12} \text{ kg/s·m}^2\text{·Pa}$$

The water vapour resistance is calculated as follows:

$$Z = \frac{1}{W} = \frac{1}{4.74 \times 10^{-12} \text{ kg/s.m}^2\text{.Pa}}$$

$$Z = 2.11 \times 10^{11} \text{ s.m}^2\text{.Pa/kg}$$

If the thickness of the specimen is $d = 2.13 \times 10^{-4}$ m, permeability of the specimen is calculated as $\delta = W.d$:

$$\delta = W.d = 2.13 \times 10^{-4} \text{ m} \times 4.72 \times 10^{-12} \text{ kg/s.m}^2\text{.Pa}$$

$$\delta = 1.00536 \times 10^{-15} \text{ kg/s·m·Pa}$$

Therefore, the resistance factor $\mu = \frac{\delta_a}{\delta}$ where δ_a is the vapour permeability of air around the laboratory site at 23 °C. This can either be calculated from Schirmer equation or extrapolated from Figure 2 in ISO 12,572.

Therefore, $\delta_a = 1.95 \times 10^{-10}$ kg/s·m·Pa:

$$\therefore \mu = \frac{\delta_a}{\delta} = \frac{1.95 \times 10^{-10} \text{ kg/s·m·Pa}}{1.00536 \times 10^{-15} \text{ kg/s·m·Pa}} = 193{,}960.37$$

S_D (equivalent air layer thickness) = $\mu \times$ d:

$$S_D = 193{,}960.37 \times 2.13 \times 10^{-4} = 41.31 \text{ m}$$

Recall that air gap is 20 mm = 0.02 m. Final $S_D = 41.31 - 0.02 = 41.29$ m:

$$\therefore Final\ \mu = \frac{41.29}{2.13 \times 10^{-4}} = 193{,}666$$

Appendix B

Table A1. Summary of average resistance factor μ and diffusion air layer thickness Sd of samples.

RH%	Dry Test Resistance Factor μ	Wet Test Resistance Factor μ	Dry Test S_D (m)	Wet Test S_D (m)
		Sample C		
35	189,398	13,043	46.21	3.4
50	94,895	7180	23.25	2.5
65	64,499	7640	16.04	2.37
80	10,005	6918	2.72	1.98

Table A1. *Cont.*

RH%	Dry Test Resistance Factor μ	Wet Test Resistance Factor μ	Dry Test S_D (m)	Wet Test S_D (m)
		Sample D		
35	222,099	120,265	49.5	26.82
50	139,160	210,033	31.4	48.99
65	265,273	223,063	59.9	50.44
80	388,586	91,312	87.7	20.27
		Sample E		
35	383,221	304,191	116.4	97.5
50	530,781	71,909	160.8	23.07
65	472,951	90,899	144.6	30.9
80	378,743	47,612	114.9	15.2

Appendix C

Table A2. Water vapour resistivity properties for sample C at 35% RH 23 °C.

Sample	Mean Thickness	Mass Change Rate	Area	Water Vapour Flux	Water Vapour Permeance	Water Vapour Resistance	Water Vapour Permeability	Water Vapour Resistance	Diffusion-Equivalent
at 23 °C 35% RH	d (m)	/time (G in kg/s)	m^2	g = G/A in kg/(s·m^2)	W = g/dp in kg/(s·m^2·Pa)	Z = 1/W in (s·m^2·Pa)/kg	δ = W × d in kg/(s·m·Pa)	factor μ	air layer thickness Sd
Dry cup test									
C-1	0.000213	1.2×10^{-10}	0.02830	4.3×10^{-9}	4.7×10^{-12}	2.1×10^{11}	1.0×10^{-15}	193,866.0000	41.2900
C-2	0.000248	9.8×10^{-11}	0.02630	3.7×10^{-9}	4.2×10^{-12}	2.4×10^{11}	1.0×10^{-15}	189,233.8700	46.9300
C-3	0.000256	9.2×10^{-11}	0.02540	3.6×10^{-9}	4.0×10^{-12}	2.5×10^{11}	1.0×10^{-15}	188,509.8800	48.2600
C-4	0.000256	9.8×10^{-11}	0.02750	3.6×10^{-9}	4.0×10^{-12}	2.5×10^{11}	1.0×10^{-15}	191,794.3500	49.1000
C-5	0.000252	9.8×10^{-11}	0.02600	3.8×10^{-9}	4.2×10^{-12}	2.3×10^{11}	1.1×10^{-15}	183,888.8900	46.3600
Mean	0.000245	1.0×10^{-10}	0.02670	3.8×10^{-9}	4.2×10^{-12}	2.4×10^{11}	1.0×10^{-15}	189,458.5980	46.3880
Wet cup test									
C-6	0.000246	2.8×10^{-9}	0.02630	1.1×10^{-7}	6.6×10^{-11}	1.5×10^{10}	1.6×10^{-14}	11,910.5800	2.9300
C-7	0.000287	2.3×10^{-9}	0.02780	8.2×10^{-8}	5.0×10^{-11}	2.0×10^{10}	1.4×10^{-14}	13,491.7600	3.8700
C-8	0.000255	2.2×10^{-9}	0.02690	8.2×10^{-8}	5.1×10^{-11}	2.0×10^{10}	1.3×10^{-14}	15,034.3300	3.8340
C-9	0.000251	2.6×10^{-9}	0.02690	9.6×10^{-8}	5.9×10^{-11}	1.7×10^{10}	1.5×10^{-14}	13,149.6300	3.3000
C-10	0.000253	2.9×10^{-9}	0.02780	1.0×10^{-7}	6.3×10^{-11}	1.6×10^{10}	1.6×10^{-14}	12,090.9500	3.0600
Mean	0.0002584	2.6×10^{-9}	0.02714	9.4×10^{-8}	5.8×10^{-11}	1.8×10^{10}	1.5×10^{-14}	13,135.4500	3.3988

Table A3. Water vapour resistivity properties of sample C at 23 °C 50% RH.

Sample Tested	Mean Thickness	Mass Change Rate	Area	Water Vapour Flux	Water Vapour Permeance	Water Vapour Resistance	Water Vapour Permeability	Water Vapour Resistance	Diffusion-Equivalent
at 23 °C 50% RH	d (m)	/time (G in kg/s)	m^2	g = G/A in kg/(s·m^2)	W = g/dp in kg/(s·m^2·Pa)	Z = 1/W in (s·m^2·Pa)/kg	δ = W × d in kg/(s·m·Pa)	factor μ	air layer thickness Sd
Dry cup test									
C-1	0.000213	3.2×10^{-10}	0.02830	1.1×10^{-8}	8.6×10^{-12}	1.2×10^{11}	1.8×10^{-15}	106,291.0800	22.6600
C-2	0.000248	2.4×10^{-10}	0.02630	9.3×10^{-9}	7.0×10^{-12}	1.4×10^{11}	1.7×10^{-15}	111,688.4600	27.7000
C-3	0.000256	2.9×10^{-10}	0.02550	1.2×10^{-8}	8.8×10^{-12}	1.1×10^{11}	2.2×10^{-15}	86,641.7800	22.1800
C-4	0.000256	3.2×10^{-10}	0.02750	1.2×10^{-8}	8.8×10^{-12}	1.1×10^{11}	2.3×10^{-15}	85,962.4200	22.0100
C-5	0.000252	3.0×10^{-10}	0.02600	1.1×10^{-8}	8.6×10^{-12}	1.2×10^{11}	2.2×10^{-15}	89,654.7600	22.5900
Mean	0.000245	2.9×10^{-10}	0.02672	1.1×10^{-8}	8.4×10^{-12}	1.2×10^{11}	2.0×10^{-15}	96,047.7000	23.4280
Wet cup test									
C-6	0.000246	2.9×10^{-9}	0.02630	1.1×10^{-7}	9.1×10^{-11}	1.1×10^{10}	2.2×10^{-14}	8570.3800	2.1100
C-7	0.000287	2.6×10^{-9}	0.02780	9.2×10^{-8}	7.6×10^{-11}	1.3×10^{10}	2.2×10^{-14}	8810.0270	2.5300
C-8	0.000251	3.8×10^{-9}	0.02690	1.4×10^{-7}	1.2×10^{-10}	8.5×10^{9}	3.0×10^{-14}	6494.0200	1.6300
C-9	0.000251	3.9×10^{-9}	0.02550	1.5×10^{-7}	1.3×10^{-10}	8.0×10^{9}	3.2×10^{-14}	6091.5500	1.5300
C-10	0.000253	3.8×10^{-9}	0.02780	1.4×10^{-7}	1.1×10^{-10}	8.9×10^{9}	2.8×10^{-14}	6758.0000	1.7100
Mean	0.0002576	3.4×10^{-9}	0.02686	1.3×10^{-7}	1.0×10^{-10}	9.9×10^{9}	2.7×10^{-14}	7344.7954	1.9020

Table A4. Water vapour resistivity properties of sample C at 23 °C 65% RH.

Sample Tested	Mean Thickness	Mass Change Rate	Area	Water Vapour Flux	Water Vapour Permeance	Water Vapour Resistance	Water Vapour Permeability	Water Vapour Resistance	Diffusion-Equivalent
at 23 °C 65% RH	d (m)	/time (G in kg/s)	m^2	g = G/A in kg/(s·m^2)	W = g/dp in kg/(s·m^2·Pa)	Z = 1/W in (s·m^2·Pa)/kg	δ = W × d in kg/(s·m·Pa)	factor μ	air layer thickness Sd
Dry cup test									
C-1	0.000229	5.6×10^{-10}	0.02720	2.1×10^{-8}	1.2×10^{-11}	8.5×10^{10}	2.7×10^{-15}	72,237.3900	16.5400
C-2	0.000248	5.9×10^{-10}	0.02750	2.1×10^{-8}	1.2×10^{-11}	8.2×10^{10}	3.0×10^{-15}	64,399.4000	15.9700
C-3	0.000256	6.1×10^{-10}	0.02690	2.3×10^{-8}	1.3×10^{-11}	7.7×10^{10}	3.3×10^{-15}	59,677.9700	15.2800
C-4	0.000258	5.8×10^{-10}	0.02750	2.1×10^{-8}	1.2×10^{-11}	8.3×10^{10}	3.1×10^{-15}	63,056.0000	16.2700
C-5	0.000252	5.6×10^{-10}	0.02689	2.1×10^{-8}	1.2×10^{-11}	8.3×10^{10}	3.0×10^{-15}	64,375.2400	16.2200
Mean	0.0002486	5.8×10^{-10}	0.02720	2.1×10^{-8}	1.2×10^{-11}	8.2×10^{10}	3.0×10^{-15}	64,749.2000	16.0560
Wet cup test									
C-6	0.000277	1.9×10^{-9}	0.02750	6.8×10^{-8}	8.6×10^{-11}	1.2×10^{10}	2.4×10^{-14}	8158.8500	2.2600
C-7	0.000284	1.7×10^{-9}	0.02750	6.1×10^{-8}	7.8×10^{-11}	1.3×10^{10}	2.2×10^{-14}	8802.8200	2.5000
C-8	0.000278	2.0×10^{-9}	0.02720	7.2×10^{-8}	9.1×10^{-11}	1.1×10^{10}	2.5×10^{-14}	7645.3800	2.1300
C-9	0.000262	1.7×10^{-9}	0.02660	6.2×10^{-8}	7.9×10^{-11}	1.3×10^{10}	2.1×10^{-14}	9395.5200	2.4600
C-10	0.000263	1.6×10^{-9}	0.02660	5.9×10^{-8}	7.5×10^{-11}	1.3×10^{10}	2.0×10^{-14}	9809.4800	2.5800
Mean	0.0002728	1.7×10^{-9}	0.02708	6.4×10^{-8}	8.2×10^{-11}	1.2×10^{10}	2.2×10^{-14}	8762.4100	2.3860

Table A5. Water vapour resistivity properties of sample C at 23 °C 80% RH.

Sample Tested at 23 °C 80% RH	Mean Thickness d (m)	Mass Change Rate/Time (G in kg/s)	Area m^2	Water Vapour Flux g = G/A in kg/(s·m^2)	Water Vapour Permeance W = g/dp in kg/(s·m^2·Pa)	Water Vapour Resistance Z = 1/W in (s·m^2·Pa)/kg	Water Vapour Permeability δ = W × d in kg/(s·m·Pa)	Water Vapour Resistance Factor μ	Diffusion Equivalent Air Layer Thickness Sd
Dry cup test									
C-1	0.000284	4.2×10^{-9}	0.02720	1.5×10^{-7}	7.1×10^{-11}	1.4×10^{10}	2.0×10^{-14}	9628.2300	2.7300
C-2	0.000277	4.3×10^{-9}	0.02750	1.5×10^{-7}	7.2×10^{-11}	1.4×10^{10}	2.0×10^{-14}	9797.8500	2.7100
C-3	0.000278	4.1×10^{-9}	0.02690	1.5×10^{-7}	7.0×10^{-11}	1.4×10^{10}	1.9×10^{-14}	10,035.9700	2.7900
C-4	0.000262	4.2×10^{-9}	0.02750	1.5×10^{-7}	7.0×10^{-11}	1.4×10^{10}	1.8×10^{-14}	10,584.5300	2.7700
C-5	0.000263	4.3×10^{-9}	0.02690	1.6×10^{-7}	7.4×10^{-11}	1.4×10^{10}	1.9×10^{-14}	10,030.1700	2.6400
Mean	0.0002728	4.2×10^{-9}	0.02720	1.5×10^{-7}	7.1×10^{-11}	1.4×10^{10}	1.9×10^{-14}	10,015.3500	2.7280
Wet cup test									
C-6	0.000277	9.3×10^{-10}	0.02750	3.4×10^{-8}	9.2×10^{-11}	1.1×10^{10}	2.6×10^{-14}	7476.8900	2.0700
C-7	0.000284	1.1×10^{-9}	0.02750	4.1×10^{-8}	1.1×10^{-10}	8.9×10^{9}	3.2×10^{-14}	5954.2400	1.6900
C-8	0.000278	1.2×10^{-9}	0.02720	4.3×10^{-8}	1.2×10^{-10}	8.6×10^{9}	3.2×10^{-14}	5882.1300	1.6400
C-9	0.000262	9.3×10^{-10}	0.02660	3.5×10^{-8}	9.5×10^{-11}	1.0×10^{10}	2.5×10^{-14}	7643.6600	2.0000
C-10	0.000263	8.5×10^{-10}	0.02660	3.2×10^{-8}	8.7×10^{-11}	1.1×10^{10}	2.3×10^{-14}	8314.1600	2.9000
Mean	0.0002728	1.0×10^{-9}	0.02708	3.7×10^{-8}	1.0×10^{-10}	1.0×10^{10}	2.8×10^{-14}	7054.2160	2.0600

Table A6. Water vapour resistivity properties for sample D at 23 °C 35% RH.

Sample Tested at 23 °C 35% RH	Mean Thickness d (m)	Mass Change Rate/Time (G in kg/s)	Area m^2	Water Vapour Flux g = G/A in kg/(s·m^2)	Water Vapour Permeance W = g/dp in kg/(s·m^2·Pa)	Water Vapour Resistance Z = 1/W in (s·m^2·Pa)/kg	Water Vapour Permeability δ = W × d in kg/(s·m·Pa)	Water Vapour Resistance Factor μ	Diffusion-Air Layer Thickness Sd
Dry cup test									
D-1	0.000224	7.23×10^{-11}	2.60×10^{-2}	2.78×10^{-9}	3.10×10^{-12}	3.23×10^{11}	6.94×10^{-16}	281,432.34	63.04
D-2	0.00022	9.65×10^{-11}	2.57×10^{-2}	3.75×10^{-9}	4.18×10^{-12}	2.39×10^{11}	9.19×10^{-16}	212,427.32	46.73
D-3	0.000221	8.20×10^{-11}	2.49×10^{-2}	3.29×10^{-9}	3.66×10^{-12}	2.73×10^{11}	8.10×10^{-16}	241,020.61	53.27
D-4	0.000227	1.01×10^{-10}	2.32×10^{-2}	4.37×10^{-9}	4.86×10^{-12}	2.06×10^{11}	1.10×10^{-15}	176,990.51	40.18
D-5	0.000221	8.20×10^{-11}	2.32×10^{-2}	3.53×10^{-9}	3.93×10^{-12}	2.54×10^{11}	8.69×10^{-16}	224,580.78	49.63
Mean	0.000223	8.68×10^{-11}	0.0246	3.54×10^{-9}	3.95×10^{-12}	2.59×10^{11}	8.79×10^{-16}	2.27×10^{5}	50.57
Wet cup test									
D-6	0.00022	3.71×10^{-10}	0.0275	1.35×10^{-8}	8.24×10^{-12}	1.21×10^{11}	1.81×10^{-15}	1.08×10^{5}	23.69
D-7	0.00022	3.08×10^{-10}	0.0269	1.14×10^{-8}	6.98×10^{-12}	1.43×10^{11}	1.54×10^{-15}	1.27×10^{5}	27.96
D-8	0.00023	2.86×10^{-10}	0.0263	1.09×10^{-8}	6.64×10^{-12}	1.51×10^{11}	1.50×10^{-15}	1.31×10^{5}	29.38
D-9	0.00022	2.97×10^{-10}	0.0255	1.16×10^{-8}	7.11×10^{-12}	1.41×10^{11}	1.57×10^{-15}	1.24×10^{5}	27.46
D-10	0.00022	3.29×10^{-10}	0.0263	1.25×10^{-8}	7.63×10^{-12}	1.31×10^{11}	1.69×10^{-15}	1.15×10^{5}	25.59
Mean	0.00022	3.18×10^{-10}	0.0265	1.20×10^{-8}	7.32×10^{-12}	1.37×10^{11}	1.62×10^{-15}	1.21×10^{5}	26.816

Table A7. Water vapour resistivity properties for sample D at 23 °C 50% RH.

Sample Tested at 23 °C 50% RH	Mean Thickness d (m)	Mass Change Rate/Time (G in kg/s)	Area m^2	Water Vapour Flux g = G/A in kg/(s·m^2)	Water Vapour Permeance W = g/dp in kg/(s·m^2·Pa)	Water Vapour Resistance Z = 1/W in (s·m^2·Pa)/kg	Water Vapour Permeability δ = W × d in kg/(s·m·Pa)	Water Vapour Resistance Factor μ	Diffusion-Air Layer Thickness Sd
Dry cup test									
D-1	0.000230	2.02×10^{-10}	0.02750	7.33×10^{-9}	5.56×10^{-12}	1.80×10^{11}	1.28×10^{-15}	151,165	34.77
D-2	0.000223	3.17×10^{-10}	0.02776	1.14×10^{-8}	8.65×10^{-12}	1.16×10^{11}	1.93×10^{-15}	100,117.68	22.33
D-3	0.000228	2.59×10^{-10}	0.02750	9.43×10^{-9}	7.14×10^{-12}	1.40×10^{11}	1.63×10^{-15}	118,581.32	27.04
D-4	0.000227	2.02×10^{-10}	0.02750	7.33×10^{-9}	5.56×10^{-12}	1.80×10^{11}	1.26×10^{-15}	153,171.81	34.77
D-5	0.000221	1.38×10^{-10}	0.02750	5.04×10^{-9}	3.82×10^{-12}	2.62×10^{11}	8.43×10^{-16}	229,117.18	50.64
Mean	0.000226	2.236×10^{-10}	0.02755	8.108×10^{-9}	6.145×10^{-12}	1.755×10^{11}	1.388×10^{-15}	1.504×10^{5}	33.91
Wet cup test									
D-6	0.00022	2.04×10^{-10}	0.02750	7.42×10^{-9}	6.15×10^{-12}	1.63×10^{11}	1.35×10^{-15}	143,277.73	31.52
D-7	0.000221	1.15×10^{-10}	0.02690	4.29×10^{-9}	3.55×10^{-12}	2.82×10^{11}	7.85×10^{-16}	246,887.09	54.56
D-8	0.000225	1.06×10^{-10}	0.02630	4.05×10^{-9}	3.35×10^{-12}	2.98×10^{11}	7.55×10^{-16}	256,844.44	57.79
D-9	0.000221	1.15×10^{-10}	0.02780	4.15×10^{-9}	3.44×10^{-12}	2.91×10^{11}	7.60×10^{-16}	255,149.38	56.39
D-10	0.000222	1.42×10^{-10}	0.02690	5.28×10^{-9}	4.37×10^{-12}	2.29×10^{11}	9.62×10^{-16}	201,486.17	44.73
Mean	0.000222	1.37×10^{-10}	0.02708	5.04×10^{-9}	4.17×10^{-12}	2.52×10^{11}	9.23×10^{-16}	220,728.96	48.998

Table A8. Water vapour resistivity properties for sample D at 23 °C 65% RH.

Sample Tested at 23 °C 65% RH	Mean Thickness d (m)	Mass Change Rate/Time (G in kg/s)	Area m^2	Water Vapour Flux g = G/A in $kg/(s \cdot m^2)$	Water Vapour Permeance W = g/dp in $kg/(s \cdot m^2 \cdot Pa)$	Water Vapour Resistance Z = 1/W in $(s \cdot m^2 \cdot Pa)/kg$	Water Vapour Permeability $\delta = W \times d$ in $kg/(s \cdot m \cdot Pa)$	Water Vapour Resistance Factor μ	Diffusion-Air Layer Thickness Sd
				Dry cup test					
D-1	0.000230	1.434×10^{-10}	0.0275	5.216×10^{-9}	2.996×10^{-12}	3.338×10^{11}	6.892×10^{-16}	282,754.23	65.03
D-2	0.000223	1.149×10^{-10}	0.2780	5.358×10^{-9}	3.078×10^{-12}	3.249×10^{11}	6.864×10^{-16}	283,905.36	63.31
D-3	0.000228	1.545×10^{-10}	0.0275	5.617×10^{-9}	3.227×10^{-12}	3.099×10^{11}	7.357×10^{-16}	264,860.83	60.39
D-4	0.000225	1.545×10^{-10}	0.0241	6.409×10^{-9}	3.682×10^{-12}	2.716×10^{11}	8.284×10^{-16}	235,199.55	52.92
D-5	0.000223	1.379×10^{-10}	0.0241	5.722×10^{-9}	3.287×10^{-12}	3.042×10^{11}	7.331×10^{-16}	265,798.54	59.27
Mean	0.000226	1.410×10^{-10}	0.0762	5.664×10^{-9}	3.254×10^{-12}	3.089×10^{11}	7.345×10^{-16}	266,503.70	60.184
				Wet cup test					
D-6	0.000220	1.048×10^{-10}	0.02750	3.811×10^{-9}	4.848×10^{-12}	2.063×10^{11}	1.067×10^{-15}	182,658.01	40.19
D-7	0.000221	7.723×10^{-11}	0.02690	2.871×10^{-9}	3.652×10^{-12}	2.714×10^{11}	8.071×10^{-16}	241,413.14	53.35
D-8	0.000225	6.620×10^{-11}	0.02630	2.517×10^{-9}	3.202×10^{-12}	3.123×10^{11}	7.204×10^{-16}	270,479.00	60.86
D-9	0.000221	7.723×10^{-11}	0.02630	2.937×10^{-9}	3.735×10^{-12}	2.677×10^{11}	8.255×10^{-16}	236,027.34	52.16
D-10	0.000222	8.826×10^{-11}	0.02630	3.356×10^{-9}	4.269×10^{-12}	2.343×10^{11}	9.477×10^{-16}	205,582.47	45.64
Mean	0.000222	8.275×10^{-11}	0.02666	3.098×10^{-9}	3.941×10^{-12}	2.584×10^{11}	8.735×10^{-16}	227,231.99	50.44

Table A9. Water vapour resistivity properties for sample D at 23 °C 80% RH.

Sample Tested at 23 °C 80% RH	Mean Thickness d (m)	Mass Change Rate/Time (G in kg/s)	Area m^2	Water Vapour Flux g = G/A in $kg/(s \cdot m^2)$	Water Vapour Permeance W = g/dp in $kg/(s \cdot m^2 \cdot Pa)$	Water Vapour Resistance Z = 1/W in $(s \cdot m^2 \cdot Pa)/kg$	Water Vapour Permeability $\delta = W \times d$ in $kg/(s \cdot m \cdot Pa)$	Water Vapour Resistance Factor μ	Diffusion-Air Layer Thickness Sd
Dry cup test									
D-1	0.000230	1.412×10^{-10}	0.0275	5.135×10^{-9}	2.375×10^{-12}	4.210×10^{11}	5.463×10^{-16}	356,892.45	82.09
D-2	0.000223	1.540×10^{-10}	0.0278	5.541×10^{-9}	2.563×10^{-12}	3.902×10^{11}	5.716×10^{-16}	341,081.84	76.06
D-3	0.000228	1.284×10^{-10}	0.0275	4.668×10^{-9}	2.159×10^{-12}	4.632×10^{11}	4.923×10^{-16}	396,028.31	90.3
D-4	0.000225	1.091×10^{-10}	0.0269	4.056×10^{-9}	1.876×10^{-12}	5.330×10^{11}	4.222×10^{-16}	461,832.23	103.91
D-5	0.000223	1.284×10^{-10}	0.0278	4.617×10^{-9}	2.136×10^{-12}	4.682×10^{11}	4.763×10^{-16}	409,333.34	91.3
Mean	0.000226	1.322×10^{-10}	0.0275	4.803×10^{-9}	2.222×10^{-12}	4.551×10^{11}	5.017×10^{-16}	393,033.63	88.732
Wet cup test									
D-6	0.000220	1.012×10^{-10}	0.0275	3.368×10^{-9}	1.009×10^{-11}	9.909×10^{-10}	2.220×10^{-15}	87,878.64	19.33
D-7	0.000221	9.662×10^{-11}	0.0269	3.592×10^{-9}	9.847×10^{-12}	1.016×10^{11}	2.176×10^{-15}	89,638.00	19.81
D-8	0.000225	9.202×10^{-11}	0.0263	3.499×10^{-9}	9.592×10^{-12}	1.043×10^{11}	2.158×10^{-15}	90,400.00	20.34
D-9	0.000221	9.662×10^{-11}	0.0278	3.476×10^{-9}	9.528×10^{-12}	1.050×10^{11}	2.106×10^{-15}	92,658.10	20.48
D-10	0.000222	8.742×10^{-11}	0.0263	3.324×10^{-9}	9.113×10^{-12}	1.097×10^{11}	2.023×10^{-15}	96,454.65	21.41
Mean	0.000222	9.478×10^{-11}	0.0270	3.452×10^{-9}	9.635×10^{-12}	8.409×10^{10}	2.137×10^{-15}	91,405.88	20.274

Table A10. Water vapour resistivity properties for sample E at 23 °C 35% RH.

Sample Tested at 23 °C 35% RH	Mean Thickness d (m)	Mass Change Rate/Time (G in kg/s)	Area m^2	Water Vapour Flux g = G/A in $kg/(s{\cdot}m^2)$	Water Vapour Permeance W = g/dp in $kg/(s{\cdot}m^2{\cdot}Pa)$	Water Vapour Resistance Z = 1/W in $(s{\cdot}m^2{\cdot}Pa)/kg$	Water Vapour Permeability $\delta = W \times d$ in $kg/(s{\cdot}m{\cdot}Pa)$	Water Vapour Resistance Factor μ	Diffusion-Air Layer Thickness Sd
Dry cup test									
E-1	0.00031	4.306×10^{-11}	0.02780	1.549×10^{-9}	1.724×10^{-12}	5.800×10^{11}	5.345×10^{-16}	389,498.37	120.75
E-2	0.00029	4.037×10^{-11}	0.02460	1.640×10^{-9}	1.827×10^{-12}	5.475×10^{11}	5.297×10^{-16}	368,056.99	106.74
E-3	0.00031	4.038×10^{-11}	0.02840	1.424×10^{-9}	1.584×10^{-12}	6.312×10^{-11}	4.960×10^{-16}	393,081.26	123.03
Mean	0.00030	4.127×10^{-11}	0.02693	1.538×10^{-9}	1.712×10^{-12}	3.758×10^{11}	5.201×10^{-16}	383,545.54	116.84
Wet cup test									
E-4	0.000335	9.802×10^{-11}	0.0278	3.553×10^{-9}	2.165×10^{-12}	4.618×10^{11}	7.26×10^{-16}	268,715.42	90.04
E-5	0.000324	8.524×10^{-11}	0.0269	3.169×10^{-9}	1.946×10^{-12}	5.139×10^{11}	6.30×10^{-16}	309,231.34	100.19
E-6	0.000301	8.524×10^{-11}	0.0278	3.066×10^{-9}	1.883×10^{-12}	5.311×10^{11}	5.67×10^{-16}	344,000.60	103.54
Mean	0.000320	8.950×10^{-11}	0.0275	3.262×10^{-9}	1.998×10^{-12}	5.023×10^{11}	6.41×10^{-16}	307,315.79	97.92

Table A11. Water vapour resistivity properties for sample E at 23 °C 50% RH.

Sample Tested at 23 °C 50% RH	Mean Thickness d (m)	Mass Change Rate/Time (G in kg/s)	Area m^2	Water Vapour Flux g = G/A in $kg/(s \cdot m^2)$	Water Vapour Permeance W = g/dp in $kg/(s \cdot m^2 \cdot Pa)$	Water Vapour Resistance Z = 1/W in $(s \cdot m^2 \cdot Pa)/kg$	Water Vapour Permeability δ = W × d in $kg/(s \cdot m \cdot Pa)$	Water Vapour Resistance Factor μ	Diffusion-Air Layer Thickness Sd
Dry cup test									
E-1	0.00031	3.851×10^{-11}	0.0278	1.385×10^{-9}	1.050×10^{-12}	9.526×10^{11}	3.254×10^{-16}	594,508.48	184.3
E-2	0.00029	4.630×10^{-11}	0.0246	1.882×10^{-9}	1.426×10^{-12}	7.012×10^{11}	4.136×10^{-16}	467,726.01	135.64
E-3	0.00031	4.243×10^{-11}	0.0284	1.494×10^{-9}	1.132×10^{-12}	8.832×10^{11}	3.544×10^{-16}	545,857.49	170.85
Mean	0.00030	4.241×10^{-11}	0.0269	1.587×10^{-9}	1.203×10^{-12}	8.457×10^{11}	3.645×10^{-16}	536,030.66	163.5966667
Wet cup test									
E-4	0.00034	3.110×10^{-10}	0.0278	1.119×10^{-8}	9.267×10^{-12}	1.079×10^{11}	3.105×10^{-15}	62,260.73	20.86
E-5	0.00032	2.534×10^{-10}	0.0269	9.421×10^{-9}	7.804×10^{-12}	1.281×10^{11}	2.528×10^{-15}	74,814.43	24.24
E-6	0.00030	2.650×10^{-10}	0.0278	9.531×10^{-9}	7.894×10^{-12}	1.267×10^{11}	2.376×10^{-15}	81,357.68	24.49
Mean	0.00032	2.765×10^{-10}	0.0275	1.005×10^{-8}	8.322×10^{-12}	1.209×10^{11}	2.670×10^{-15}	72,810.95	23.19666667

Table A12. Water vapour resistivity properties for sample E at 23 °C 65% RH.

Sample Tested at 23 °C 65% RH	Mean Thickness d (m)	Mass Change Rate/Time (G in kg/s)	Area m^2	Water Vapour Flux g = G/A in $kg/(s \cdot m^2)$	Water Vapour Permeance W = g/dp in $kg/(s \cdot m^2 \cdot Pa)$	Water Vapour Resistance Z = 1/W in $(s \cdot m^2 \cdot Pa)/kg$	Water Vapour Permeability $\delta = W \times d$ in $kg/(s \cdot m \cdot Pa)$	Water Vapour Resistance Factor μ	Diffusion-Air Layer Thickness Sd
Dry cup test									
E-1	0.00031	9.181×10^{-11}	0.0278	3.303×10^{-9}	1.897×10^{-12}	5.271×10^{11}	5.881×10^{-16}	332,157.95	102.97
E-2	0.00029	4.391×10^{-11}	0.0246	1.784×10^{-9}	1.025×10^{-12}	9.757×10^{11}	2.973×10^{-16}	657,256.52	190.60
E-3	0.00031	5.588×10^{-11}	0.0284	1.968×10^{-9}	1.130×10^{-12}	8.846×10^{11}	3.538×10^{-16}	552,166.05	172.83
Mean	0.00030	6.387×10^{-11}	0.0269	2.351×10^{-9}	1.351×10^{-12}	7.958×10^{11}	4.131×10^{-16}	513,860.17	155.47
Wet cup test									
E-4	0.00034	1.357×10^{-10}	0.0278	4.882×10^{-9}	6.210×10^{-12}	1.610×10^{11}	2.080×10^{-15}	93,864.25	40.19
E-5	0.00032	1.198×10^{-10}	0.0269	4.452×10^{-9}	5.663×10^{-12}	1.766×10^{11}	1.835×10^{-15}	106,429.44	34.48
E-6	0.00030	1.836×10^{-10}	0.0278	6.605×10^{-9}	8.402×10^{-12}	1.190×10^{11}	2.529×10^{-15}	77,196.40	23.24
Mean	0.00032	1.464×10^{-10}	0.0275	5.313×10^{-9}	6.758×10^{-12}	1.522×10^{11}	2.148×10^{-15}	92,496.70	32.64

Table A13. Water vapour resistivity properties for sample E at 23 °C 80% RH.

Sample Tested at 23 °C 80% RH	Mean Thickness d (m)	Mass Change Rate/Time (G in kg/s)	Area m^2	Water Vapour Flux g = G/A in kg/(s·m^2)	Water Vapour Permeance W = g/dp in kg/(s·m^2·Pa)	Water Vapour Resistance Z = 1/W in (s·m^2·Pa)/kg	Water Vapour Permeability δ = W × d in kg/(s·m·Pa)	Water Vapour Resistance Factor μ	Diffusion-Air Layer Thickness Sd
Dry cup test									
E-1	0.00031	1.158×10^{-10}	0.0278	4.164×10^{-9}	1.926×10^{-12}	5.191×10^{11}	5.972×10^{-16}	326,677.42	101.27
E-2	0.00029	9.710×10^{-11}	0.0246	3.945×10^{-9}	1.825×10^{-12}	5.479×10^{11}	5.293×10^{-16}	368,603.69	106.90
E-3	0.00031	8.216×10^{-11}	0.0284	2.893×10^{-9}	1.338×10^{-12}	7.473×10^{11}	4.188×10^{-16}	465,794.19	145.79
Mean	0.00030	9.834×10^{-11}	0.0269	3.667×10^{-9}	1.697×10^{-12}	6.048×10^{11}	5.151×10^{-16}	387,025.10	117.99
Wet cup test									
E-4	0.00034	1.232×10^{-10}	0.0278	4.433×10^{-9}	1.215×10^{-11}	8.233×10^{10}	4.069×10^{-15}	47,894.29	16.05
E-5	0.00032	1.083×10^{-10}	0.0269	4.026×10^{-9}	1.103×10^{-11}	9.066×10^{10}	3.574×10^{-15}	54,534.09	17.67
E-6	0.00030	1.531×10^{-10}	0.0278	5.508×10^{-9}	1.509×10^{-11}	6.627×10^{10}	4.542×10^{-15}	42,029.55	12.65
Mean	0.00032	1.282×10^{-10}	0.0275	4.656×10^{-9}	1.276×10^{-11}	7.975×10^{10}	4.062×10^{-15}	48,152.64	15.46

References

1. Olaoye, T.S.; Mark, D. Establishing an Environmentally Controlled Room to Quantify Water Vapour Resistivity Properties of Construction Materials. In *Revisiting the Role of Architecture for 'Surviving' Development*; Avlokita, A., Rajat, G., Eds.; Architectural Science Association (ANZAScA): Roorkee, India, 2019; pp. 675–684.
2. Olaoye, T.S.; Dewsbury, M.; Kunzel, H.; Nolan, G. An Empirical Measurement of the Water Vapour Resistivity Properties of Typical Australian Pliable Membrane. In Proceedings of the 54th International Conference of the Architectural Science Association (ANZAScA), Auckland, New Zealand, 25–28 November 2020.
3. Künzel, H.M. Adapted Vapour Control for Durable Building Enclosures. In Proceedings of the 10th International Conference on Durability of Building Materials & Components, Lyon, France, 17–20 April 2005.
4. Künzel, H.M.; Zirkelbach, D.; Sedlbauer, K. Predicting Indoor Temperature and Humidity Conditions Including Hygrothermal Interactions with the Building Envelope. In Proceedings of the 1st International Conference on Sustainable Energy and Green Architecture, Building Scientific Research Center (BSRC), King Mongkut's University, Thonburi, Bangkok, 8–10 October 2003.
5. Australia Building Code Board (ABCB). *National Construction Code Ncc 2019 Building Code of Australia—Volume One*; ABCB: Canberra, Australia, 2019.
6. WHO. *Who Guidelines for Indoor Air Quality: Dampness and Mould*; WHO Regional Office for Europe: Copenhagen, Denmark, 2009.
7. US EPA, IAQ. *Moisture Control Guidance for Building Design, Construction and Maintenance*; U.S. Environmental Protection Agency: Washington, DC, USA, 2013.
8. Institute of Medicine IOM. *Damp Indoor Spaces and Health*; Institute of Medicine, National Academies Press: Washington, DC, USA, 2004.
9. Maurice, D.; Lacasse, M.; Laouadi, A. A Comparison of Hygrothermal Simulation Results Derived from Four Simulation Tools. *J. Build. Phys.* **2021**. [CrossRef]
10. Maref, W.; Tariku, F.; Di Lenardo, B.; Gatland, S.D. Hygrothermal Performance of Exterior Wall Systems Using an Innovative Vapor Retarder in Canadian Climate. In Proceedings of the 4th International Building Physics Conference, Istanbul, Turkey, 15–18 June 2009.
11. John, S.; Burnett, E. Review of Modeling Methods for Building Enclosure Design. Ph.D. Thesis, University of Waterloo, Waterloo, ON, Canada, 1999.
12. John, C.; Roger, G.; Mavinkal, K.K. A Logical Extension of the Astm Standard E96 to Determine the Dependence of Water Vapour Transmission on Relative Humidity. In *Insulation Materials: Testing and Applications*, 3rd ed.; ASTM International: West Conshohocken, PA, USA, 1997.
13. Samuel, T.; Dewsbury, M.; Künzel, H. Empirical Investigation of the Hygrothermal Diffusion Properties of Permeable Building Membranes Subjected to Variable Relative Humidity Condition. *Energies* **2021**, *14*, 4053.
14. Valovirta, I.; Juha, V. Water Vapor Permeability and Thermal Conductivity as a Function of Temperature and Relative Humidity. In Proceedings of the IX International Conference, Oak Ridge, FL, USA, 5–10 December 2004.
15. Bomberg, M.; Marcin, P. Methods to Check Reliability of Material Characteristics for Use of Models in Real Time Hygrothermal Analysis. In Proceedings of the First Central European Symposium on Building Physics, Lodz, Poland, 13–15 September 2010.
16. Olaoye, T.S.; Mark, D. Australian Building Materials and Vapour Resistivity. In *Building Physics Forum*; AIRAH: Wollogong, Australia, 2018.
17. AS/NZS. Pliable Building Membranes and Underlays. In *AS/NZS 4200:1*; Council of Standards Australia: Sydney, Australia, 2017.
18. ASTM. Standard Test Methods for Water Vapor Transmission of Materials. In *E96/E96M*; ASTM International: West Conshohocken, PA, USA, 2010.
19. ANSI/ASHRAE. Criteria for Moisture-Control Design Analysis in Buildings. In *ASHRAE Standard 160*; American Society of Heating, Refrigerating and Air-Conditioning Engineers: Atlanta, GA, USA, 2009.
20. Künzel, H.M.; Daniel, Z. Advances in Hygrothermal Building Component Simulation: Modelling Moisture Sources Likely to Occur Due to Rainwater Leakage. *J. Build. Perform. Simul.* **2013**, *6*, 346–353. [CrossRef]
21. Feldman, D. Polymer Barrier Films. *J. Polym. Environ.* **2001**, *9*, 49–55. [CrossRef]
22. May-Britt, H. Membranes in Chemical Processing a Review of Applications and Novel Developments. *Sep. Purif. Methods* **1998**, *27*, 51–168. [CrossRef]
23. Tabor, J.; Tushar, G. Building and Construction Textiles. *High Perform. Tech. Text.* **2019**. [CrossRef]
24. Lstiburek, J. Moisture Control for Buildings. *ASHRAE J.* **2002**, *44*, 36–41.
25. ISO, International Standard Organization. Hygrothermal Performance of Building Materials and Products—Determination of Water Vapour Transmission Properties Cup Method Iso 12572. In *EVS-EN ISO 12572:2016*; Estonian Centre or Standardization: Brussels, Belgium, 2016.

buildings

MDPI

Article

Thermal Performance of Double-Pane Lightweight Steel Framed Walls with and without a Reflective Foil

Paulo Santos * and Telmo Ribeiro

ISISE, Department of Civil Engineering, University of Coimbra, 3030-788 Coimbra, Portugal; telmo.ribeiro@dec.uc.pt
* Correspondence: pfsantos@dec.uc.pt

Abstract: One strategy to increase energy efficiency of buildings could be the reduction of undesirable heat losses by mitigating the heat transfer mechanisms across the building envelope. The use of thermal insulation is the simplest and most straightforward way to promote thermal resistance of building elements by reducing the heat transfer by conduction. However, whenever there is an air cavity, radiation heat transfer could be also very relevant. The use of thermal reflective insulation materials inside the air gaps of building elements is likewise an effective way to increase thermal resistance without increasing weight and wall thickness. Some additional advantages are its low-cost and easy installation. In this work, the performance of a thermal reflective insulation system, constituted by an aluminium foil placed inside an air cavity between a double pane lightweight steel framed (LSF) partition, is experimentally evaluated for different air gap thicknesses, ranging from 0 mm up to 50 mm, with a step increment of 10 mm. We found a maximum thermal resistance improvement of the double pane LSF walls due to the reflective foil of around +0.529 $m^2 \cdot °C/W$ (+21%). The measurements of the *R*-values were compared with predictions provided by simplified models (CEN and NFRC 100). Both models were able to predict with reasonable accuracy (around ±5%) the thermal behaviour of the air cavities within the evaluated double pane LSF walls.

Keywords: thermal performance; experimental assessment; simplified models; double-pane; lightweight steel frame (LSF); partition walls; aluminium reflective foil

Citation: Santos, P.; Ribeiro, T. Thermal Performance of Double-Pane Lightweight Steel Framed Walls with and without a Reflective Foil. *Buildings* **2021**, *11*, 301. https://doi.org/10.3390/buildings11070301

Academic Editor: Cinzia Buratti

Received: 25 May 2021
Accepted: 5 July 2021
Published: 8 July 2021

1. Introduction

Buildings are a major key sector regarding energy consumption. In the European Union (EU), almost 50% of final energy consumption is used for heating and cooling, of which 80% is used in buildings [1]. Moreover, the building stock is responsible for approximately 36% of all CO_2 emissions in the EU. Consequently, in order to achieve highly energy efficient and decarbonised buildings, European countries have established ambitious climate and energy targets, guaranteeing the conversion of existing building stock into nearly zero-energy buildings (NZEB's) [1], and developing the use of renewable energy sources (e.g., solar) [2], thus implementing long-term refurbishment strategies.

One possible strategy to improve energy efficiency of buildings is the reduction of undesirable heat losses, by mitigating each heat transfer mechanism across the building envelope [3]: conduction, convection and radiation. The simplest and most straightforward way to promote thermal resistance of building elements is the use of thermal insulation, which significantly reduces heat transfer by conduction, although its effectiveness depends also on their position within the building element [4]. Moreover, this thermal insulation material also promotes sound insulation, mainly when fibrous insulation materials are used inside the air gaps [5].

Nowadays, highly efficient insulation materials are emerging with very low thermal conductivities, which are designated as super insulating materials (SIMs) [6]. Two common examples are vacuum insulating panels (VIPs) [7] and aerogels [8]. Nevertheless, the inadequate building design and use of these SIMs may contribute to increasing thermal bridges' importance. In fact, thermal bridges are very relevant for thermal behaviour and energy efficiency of buildings, and could be responsible for up to almost one third of the heating energy needs [9]. As such, when using structural solutions containing materials with high thermal conductivity, such as steel, the significance of thermal bridges could be even greater [10].

Over recent years, the lightweight steel frame (LSF) construction system has attracted attention from buildings' stockholders, mainly for low-rise residential houses, due to their inherent advantages [11]. Among these benefits are low weight, high mechanical strength, fast construction, reduced on-site disruption, high potential for recycling and reuse, high architectural flexibility, suitability for retrofitting, easy prefabrication, economical transportation and handling, superior quality, precise tolerances, high quality standards, humidity stability shape, and insect damage resistance.

Currently, there are several strategies to mitigate thermal bridges in LSF buildings' components, such as thermal break strips [12–14], slotted steel studs [12,15,16] and continuous external thermal insulation composite system (ETICS) [17,18]. Notice that even slight modifications in the steel frame, such as in the stud flanges size and shape, may have a significative influence on LSF elements [19].

As mentioned before, another important heat transfer mechanism is convection. Therefore, another strategy to retain heat inside buildings, particularly in cold climates, is ensuring good airtightness of the building envelope [20], as well as inside the building elements, by using a continuous air barrier. Notice that besides thermal bridging, thermal bypass of cavities (convective heat flows through air leakage) could be also a weakness of LSF walls.

Moreover, regarding radiation thermal transfer, the use of thermal reflective insulation materials, such as a reflective low-emissivity foil or paint inside the air gaps of building elements (e.g., roofs and facade walls) is also an effective way to increase thermal resistance without increasing weight and wall thickness of these components, with the additional advantages of low cost and easy installation [21]. Jelle et al. [22] performed a very interesting state-of-the-art review about low emissivity materials for building applications for both opaque and transparent envelopes, such as windows, walls, roofs and floors.

Regarding the opaque building envelope, there are several strategies to take advantage of the radiation heat transfer mitigation, with the use of reflective foils [23] and paints [21] being the most established. In fact, the use of double pane elements (e.g., walls) with an air cavity, besides the extra thermal resistance provided by the air enclosure and a better humidity infiltration control, also allows to reduce the heat transfer by radiation whenever a low-emissivity surface is provided inside.

Bruno et al. [23] investigated the use of reflective thermal insulation in non-ventilated air gaps for refurbishment purposes, making use of exterior insulation panels. They measured an increase of the air gap thermal resistance up to seven times when commercial reflective panels were placed inside a non-ventilated air gap for the same wall sample thickness. Additionally, their numerical simulation results showed, for real-scale air cavities, that thermo-reflective panels can produce the same effect of at least 6 cm of traditional insulating materials (thermal conductivity of 0.030 W/(m·K)), with the advantage of avoiding the need of additional space. Moreover, they found an optimal air gap thickness of 4–5 cm when using low-emissivity reflective foils.

Fantucci and Serra [21] presented several solutions regarding the use of low-emissivity paints in opaque building envelopes, with the purpose of improving their thermal performance by reducing heat losses, given the smaller radiation heat exchange; this research work was mainly experimental. In this work, several case studies were presented, such as: (i) hollow bricks coated with low-emissivity paint; (ii) low-emissivity paint coating below roof tiles, and; (iii) low-emissivity paint on the walls behind radiators. Regarding these case studies, their experimental results allowed the following conclusions: (i) the equivalent thermal conductivity of the hollow bricks was reduced by 18%; (ii) the summer heat loads across a roof component were reduced by 19%, and; (iii) the heat loss from the wall behind radiators was reduced by 25%.

As recently stated by Bruno et al. [23], there is a lack of investigations about heat transfer inside enclosures equipped with low-emissivity materials. This lack is even more noticeable in LSF double pane elements containing an air cavity with a reflective foil; no research work addressing this issue was found in the literature.

In this work, the performance of a thermal reflective insulation system, constituted by an aluminium foil placed inside an air cavity between a double pane LSF partition, is experimentally evaluated for different air gap thicknesses, ranging from 0 mm up to 50 mm, with a step increment of 10 mm. First, the description of materials and methods are performed, including the LSF walls description and materials characterization. In this section, the experimental lab tests are also presented, together with the experimental setup, set points, test procedures and verifications. Moreover, the description of the numerical simulations used to verify and compare with the measurement results are carried out, namely the geometry, domain discretization, boundary conditions and unventilated airspaces models. Next, the obtained results are presented and discussed, and related to both lab measurements (thermal resistances and infrared thermography) and numerical simulations. Finally, the key conclusions of the present work are listed.

2. Materials and Methods

The aim of this section is to describe the tested LSF double pane walls and characterize the used materials. Moreover, the experimental setup is explained, and the test procedures described. Finally, a numerical simulation verification is performed to ensure the accuracy and reliability of the achieved experimental results.

2.1. Walls Description

Figure 1 displays a horizontal cross-section of the double pane LSF wall internal partitions evaluated in the lab experiments. Each LSF wall pane has steel studs (C48 × 37 × 4 × 0.6 mm) with 400 mm spacing, filled between with mineral wool (MW) (48 mm thick). On the outer surface there are two gypsum plasterboard (GPB) panels with a total thickness of 25 mm. The air cavity thickness is variable and changes from 50 mm down to 0 mm, with 10 mm step. Moreover, another variable parameter was the use of a reflective aluminium foil (emissivity of 0.05) on the outer surface of the air cavity.

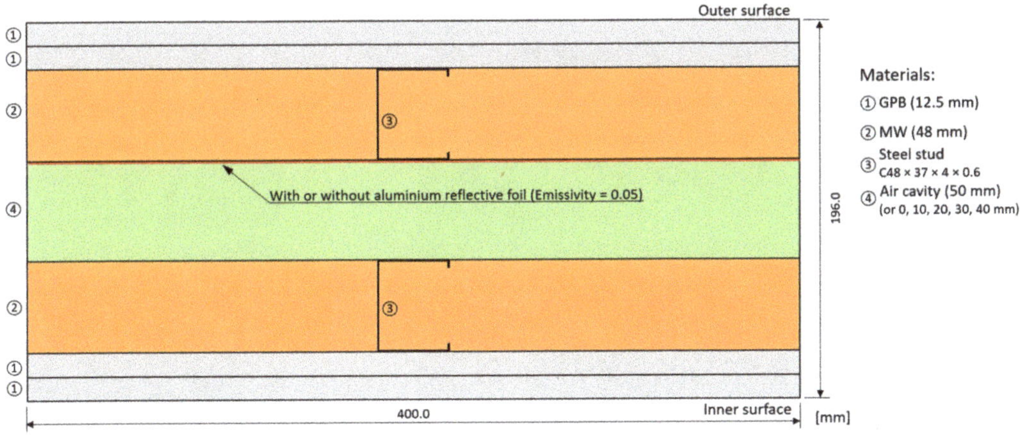

Figure 1. Ideal geometry and materials for a double pane LSF wall with different air cavity thicknesses, with and without a reflective aluminium foil.

2.2. Materials Characterization

Table 1 shows the thickness and the materials' thermal conductivities considered in the double pane LSF wall internal partitions tested in this research work. Notice that aluminium reflective foil thickness is very reduced (around 0.1 mm) and, therefore, its thermal conductive resistance was neglected. The emissivity of this aluminium reflective foil is 0.05 [23], the cold-formed galvanized steel is 0.23 [19], while for the remaining materials it is 0.90 [21].

Table 1. Double pane LSF wall materials, thickness (d) and thermal conductivities (λ).

Material (Outer to Inner Layer)	d [mm]	λ [W/(m·K)]	Reference
GPB [1] (2 × 12.5 mm)	25.0	0.175	[24]
MW [2]	48.0	0.035	[25]
Steel stud (C48 × 37 × 4 × 0.6 mm)	-	50.000	[26]
Air cavity	0, 10, 20, 30, 40, 50	-	-
MW [2]	48.0	0.035	[25]
Steel stud (C48 × 37 × 4 × 0.6 mm)	-	50.000	[26]
GPB [1] (2 × 12.5 mm)	25.0	0.175	[24]
Total Thickness	**146.0–196.0**	-	-

[1] GPB—Gypsum Plaster Board; [2] MW—Mineral Wool.

2.3. Experimental Lab Tests

2.3.1. Experimental Setup

To perform the lab measurements, a mini hot box apparatus was used, as illustrated in Figure 2a. The heating of the hot box and the cooling of the cold box were carried out using an electrical resistance and a refrigerator, respectively. The double pane LSF wall test sample (Figure 2b) was placed between these two chambers. The heat loss that occurred through the lateral surfaces of the test sample was minimized by covering the perimeter with polyurethane foam insulation (80 mm thick), as showed in Figure 2a. The two LSF wall panes were separated by an EPS frame as illustrated in Figures 2a and 3b. Notice that the perimeter thickness of this EPS frame (Figure 3b) is equal to the thermal insulation thickness of the cold/hot box walls and slab envelopes.

(a) (b)

Figure 2. Mini hot box apparatus used in the lab measurements. (**a**) Lateral view with sample perimeter insulation; (**b**) LSF wall test sample.

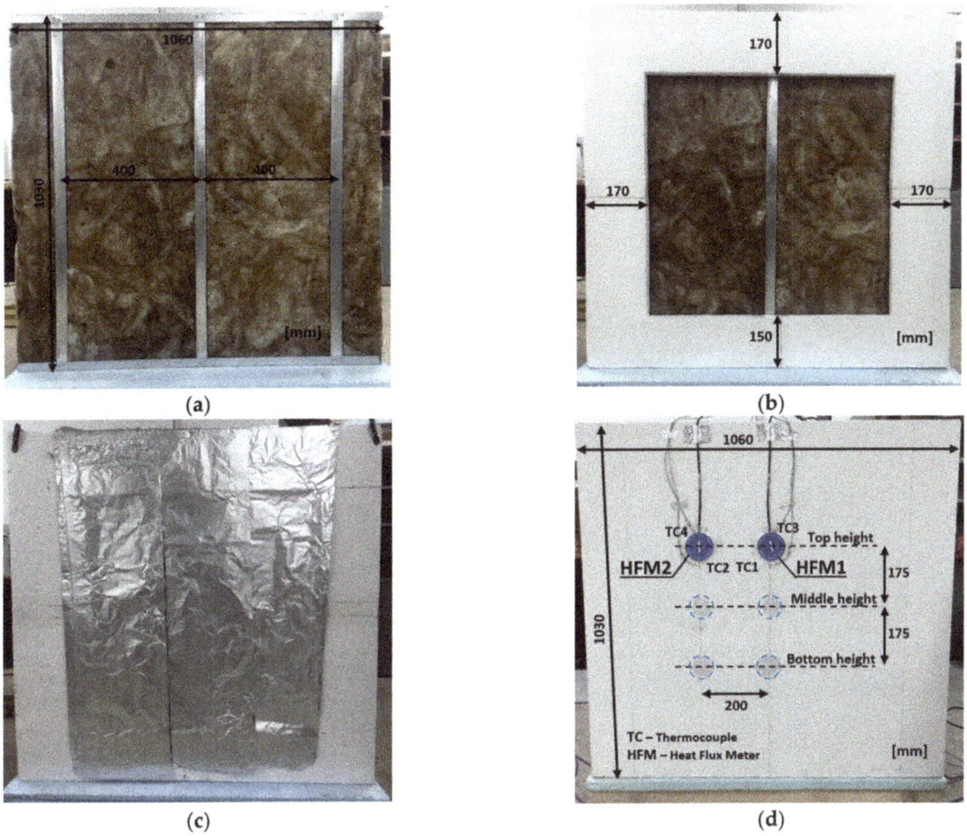

Figure 3. Double pane LSF wall test sample. (**a**) Hot LSF wall pane; (**b**) EPS perimeter frame; (**c**) Aluminium reflective foil; (**d**) Sensors (cold surface).

Figure 3a displays a frontal view of the LSF wall pane (hot side), where the vertical steel studs spacing is visible (400 mm), which is filled with mineral wool. Notice that among the three vertical steel studs, only the central one is visible (Figure 3b) and exposed to the temperature gradient established by the hot and cold boxes. The aluminium reflective foil was placed on the exterior (cold) side of the air cavity, as illustrated in Figure 3c. To promote internal air circulation and minimize the probability of air temperature stratification inside the cold and hot boxes, small interior fans were used. Moreover, two black radiation shields were used, one each side of the LSF wall test sample (10 cm apart).

Regarding the monitoring system, the heat flux through the test sample was measured using four heat flux meters (Hukseflux model HFP01, precision: ±3%), of which two were placed on the hot wall surface and the another two were placed on the cold wall surface (Figure 3d). On the hot and cold wall surface, to measure the two distinct thermal behaviour zones that exist on the LSF wall sample, two locations for the application of the HFMs were considered: (1) in the central vertical steel stud zone (HFM1); and (2) midway through the insulation cavity (HFM2). To measure temperatures, twelve type K (1/0.315) PFA insulated thermocouples (TCs) were used, presenting class one precision certification. Furthermore, the calibration of these TCs was set in the temperature range [5 °C; 45 °C], with a 5 °C increment, by immerging the TCs in a thermostatic stirring water bath (Heto CB 208).

For each side of the experiment, six TCs were used to perform the measurements, and the following configuration has been defined: two measured the environment air temperature inside each box (TC5 and TC6), another two measured the air temperature between the radiation shield and the wall surface (TC3 and TC4), and the remaining two measured the wall surface temperatures (TC1 and TC2), as illustrated in Figure 3d. For each side of the wall test sample (hot and cold), one PICO TC-08 data logger (precision: ±0.5 °C) was used to record the temperature and heat flux data measured during the tests. The two data loggers were connected to a laptop and the data were managed using the PicoLog version 6.1.10 software. The main features of the measurement equipment used in the lab experiments are listed in Table 2.

Table 2. Features of the measurement equipment used in the lab experiments.

Equipment	Brand	Model	Measurement Range	Precision
Heat Flux Meter	Hukseflux	HFP01	−2 to +2 kW/m^2	±3%
Thermocouples	LabFacility	Type K * (1/0.315)	−75 to +260 °C	±1.5 °C
Data-logger	PICO	TC-08	−270 to +1820 °C	±0.5 °C

* Tolerance Class 1 certified.

Given the expansible behaviour of the MW batt insulation and since there was no confinement on the air cavity side, an increased thickness of this insulation layer was observed. As illustrated in Figure 4a, this increment in the MW insulation thickness was around 7.5 mm on each wall pane. Moreover, it was also observed that when there is no air cavity, the steel studs were not perfectly joined because the space is occupied by the heads of the connecting screws, as illustrated in Figure 4b. Therefore, it was assumed that the two contiguous steel studs were separated by 4 mm, when theoretically there is no air cavity between the LSF wall panes.

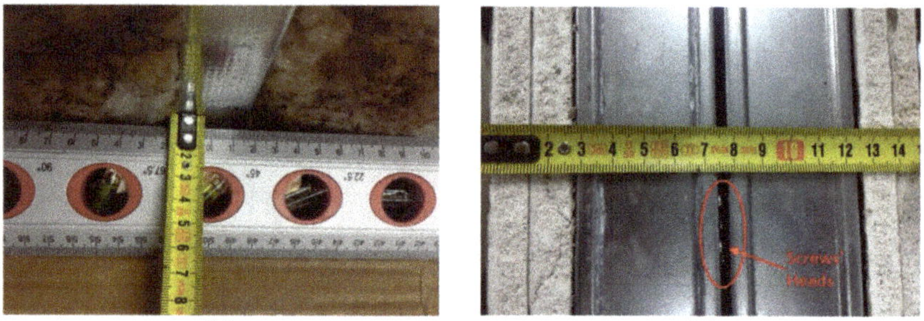

(a) (b)

Figure 4. Geometry details of the measured double pane LSF wall test sample. (**a**) Expansible MW around steel stud; (**b**) Steel track distance.

2.3.2. Set-Points and Test Procedures

The measurement of the thermal performance of the LSF test samples was performed using the heat flow meter (HFM) method [27]. However, in order to increase the precision and reduce the test duration, instead of measuring on only one side (as prescribed by ISO 9869-1 [27]), the measurements were performed, simultaneously, at both wall surfaces (cold and hot), implementing the improvement suggested by Rasooli and Itard [28]. The hot and cold boxes were programmed to maintain set point temperatures of 40 °C and 5 °C, respectively, being the measurements performed in a quasi-steady-state heat transfer condition.

The convergence criteria prescribed in ASTM C1155–95 [29] were adopted for the "summation technique", i.e., assuming a maximum admissible convergence factor equal to 10%. Thus, only the estimated hourly *R*-values with an absolute difference, in relation to the previous time obtained R-value, lower than 10% were considered in the measurements. Each measurement test was carried out for a minimum of 24 h.

To ensure the repeatability of the experimental measurements, for each wall three tests were performed corresponding to three high locations, as illustrated in Figure 3d, that is: (1) top, (2) middle, and (3) bottom. Furthermore, the average of these three tests were considered the measured overall conductive R-value of the LSF wall. From the heat fluxes and temperatures recorded for each test and applying the HFM method [27], two distinct conductive local R-values were obtained, corresponding to the two locations considered for the HFMs (Figure 3d): (1) a lower value in the central vertical steel stud zone (R_{stud}), and (2) a higher value in a midway through the insulation cavity (R_{cav}). To obtain the overall surface-to-surface -value of the wall (R_{global}), an area-weighted average of both measured conductive R-values were considered. Following the ASHRAE zone method [30], the steel stud influence area (A_{stud}), was defined considering a zone factor (zf) value of 2.0 [31]. More details about these measurements can be found in reference [14].

2.3.3. Test Procedures Verification

With the aim of ensuring the accuracy of the experimental apparatus (e.g., data-loggers and sensors) used in this work, a homogeneous XPS panel (Topox® Cuber SL), with a thickness of 60 mm and a thermal conductivity of 0.034 W/(m·K), were tested under the same conditions. The thermal conductive resistance measured (1.784 m^2·K/W) indicates that the thermal conductivity value calculated experimentally is equal to the provided by the XPS manufacturer (0.034 W/(m·K)), confirming the good working conditions of the sensors and data acquisition system.

2.4. Numerical Simulation Verification

The finite element method (FEM) software THERM® (version 7.6.1) [32] was used to perform another verification of the measured R-value results, as explained next. This software is a well-known state-of-the-art freeware computer program for a two-dimensional heat-transfer analysis. It was developed in the United States of America (USA) by the Department of Energy (DoE), through the Lawrence Berkeley National Laboratory (LBNL).

2.4.1. Geometry and Domain Discretization

Since it is a bi-dimensional FEM numerical simulation due to the existence of only vertical steel studs (Figure 3a), only a 2D representative part of the double-pane LSF wall cross-section (400 mm width) was modelled, as previously illustrated in Figure 1. Nevertheless, in a previous research work [19] a similar 2D THERM model were compared with a 3D ANSYS model and, as expected, the thermal resistance difference between both models was very small, i.e., only 0.002 m^2 K/W. Therefore, in this work it was decided not to present a similar redundant comparison.

As previously mentioned in Section 2.3.1 and illustrated in Figure 4a, the MW insulation naturally expands. Since there was no confinement on the air cavity side, the ideal geometry model (Figure 1) was discarded. A new and more realistic model was adopted, as illustrated in Figure 5a, for a nominal 50 mm air cavity thickness, assuming a MW expansion of 7.5 mm for each insulation layer. Similarly, as previously illustrated in Figure 4b, the real steel studs are not perfectly united even for a nominal 0 mm air cavity thickness. Therefore, the correspondent THERM model assumes a 4 mm separation between the vertical steel studs and the consequent MW expansion, as showed in Figure 5b. Notice that all other LSF wall models for nominal air cavities of 10, 20, 30 and 40 mm are derived from the previous two, making use of similar geometric rules (not illustrated here for sake of brevity).

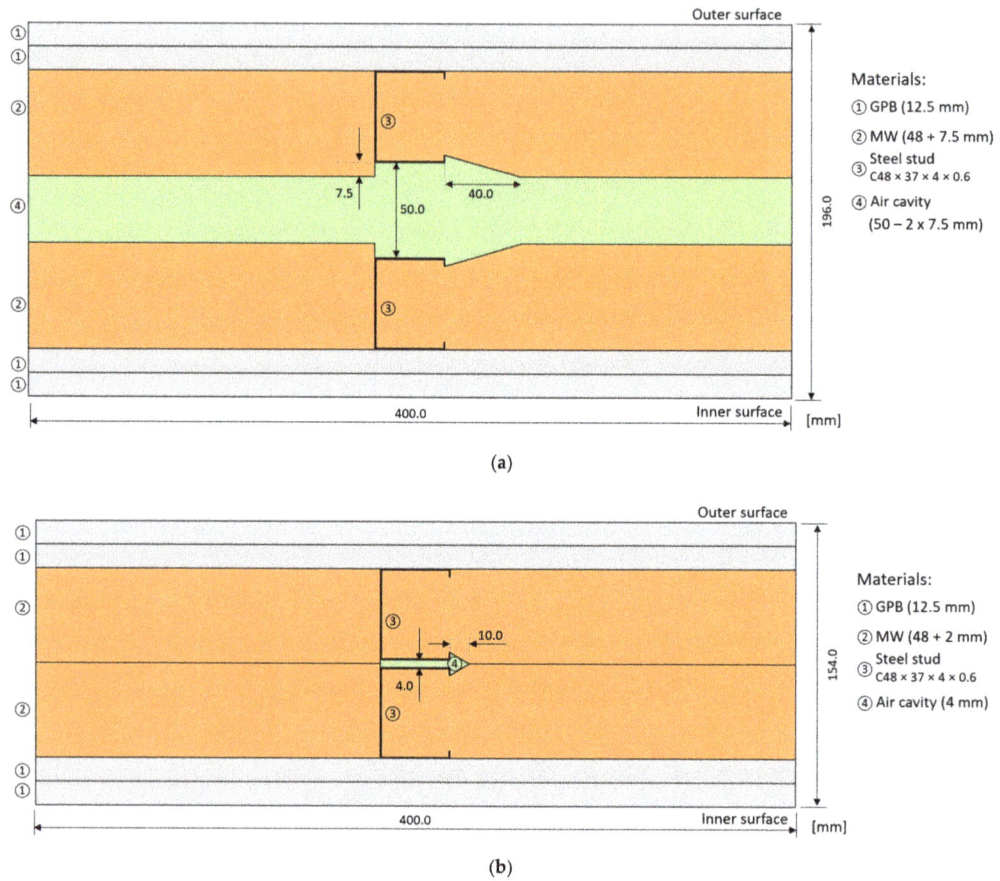

Figure 5. Modelled geometry and used materials for the double pane LSF walls. (**a**) Nominal 50 mm air cavity thicknesses; (**b**) Nominal 0 mm air cavity thicknesses.

Notice that the self-drilling screws used to assemble the steel frame (Figure 4b) and to fix the sheathing GPB to the studs were not modelled here. However, as demonstrated in previous research works [4], its relevance in the obtained thermal resistance value is neglectable.

The thermal properties of the materials used in these simulations were previously presented in Section 2.2. Additionally, the FEM mesh was refined to achieve a maximum error of 3% in these computations. The used quad tree mesh parameter was set to 6 and the maximum number of iterations was 100, the convergence tolerance was equal to 1×10^{-6}, and the mesh void tolerance was 1 mm^2. Using this mesh configuration, the maximum number of finite elements was 15,429.

2.4.2. Boundary Conditions

Regarding the boundary conditions, the air temperature was set to 40 °C and 5 °C, for the inner and outer environments, respectively. Notice that these values are equal to the set points defined for hot and cold boxes, as previously described in Section 2.3.2. Moreover, the surface thermal resistances were modelled using the average values measured for each LSF wall surface and for each test, considering the air and surface temperature differences and the surface heat fluxes. The measured surface thermal resistances vary within the interval [0.06; 0.13] $m^2 \cdot K/W$, thereby respecting the range defined by EN ISO 6946 [33] for horizontal heat flow, i.e., between 0.04 m^2·K/W for external surface resistance (R_{se}) and 0.13 m^2·K/W for internal surface resistance (R_{si}).

In this work, since only the surface-to-surface (or conductive) R-values are considered to evaluate the thermal performance of the test samples, the surface thermal resistances are not included. However, they should be defined in the numerical simulations performed by THERM software, considering a film coefficient ($1/R_s$) being later deducted.

2.4.3. Unventilated Airspaces Models

The airspaces between the two LSF wall panes were modelled as unventilated making use of two different approaches available in the THERM software: CEN simplified and NFRC 100. Both methods make use of a solid-equivalent effective thermal conductivity of the airspace and incorporates the convective and radiative heat transfer effects by taking into account the geometry, heat flow direction, surface emissivity and temperature of the surrounding surfaces of the cavity area. The CEN simplified method is based on the calculation procedures defined in the standard EN ISO 6946 [33] for unventilated airspaces, while the NFRC 100 method is based on the simplified radiation model defined in the standard ISO 15099 [34].

3. Results and Discussion

3.1. Lab Measurements

In this subsection, the lab-measured values are presented and discussed, including thermal resistances and the infrared thermography images.

3.1.1. Thermal Resistance Values

Figure 6 displays the measured conductive R-values of double pane LSF walls with different air cavity thicknesses (0–50 mm with an increment of 10 mm), with and without an aluminium reflective foil. Looking first to the -values without a reflective foil (black line), it is visible that the increment in the air cavity thickness is useful up to 30 mm, with the measured thermal resistance being nearly constant from there, i.e., for 40 and 50 mm. Adding a reflective aluminium foil (red line), there is a significant increase in the achieved R-value, particularly for higher thicknesses of the air cavity.

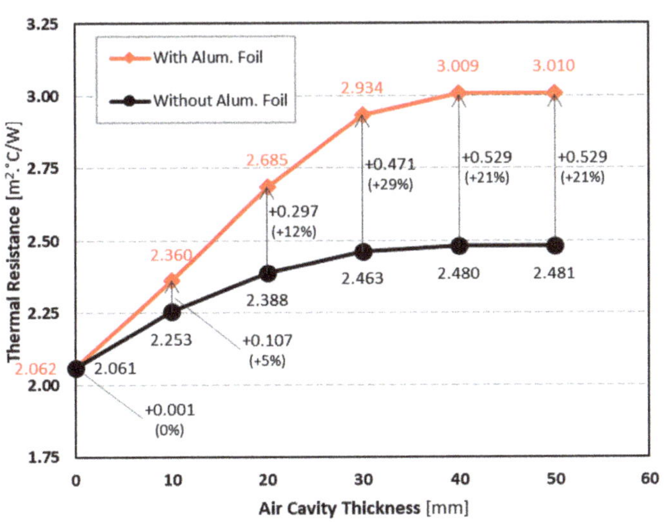

Figure 6. Measured conductive thermal resistances of a double pane LSF wall, with different air cavity thicknesses, with and without a reflective aluminium foil.

As also displayed in Figure 6, this thermal resistance increase ranges from only +0.001 $m^2 \cdot °C/W$ for a null air cavity thickness, up to + 0.529 $m^2 \cdot °C/W$ (+21%) for 50 mm air cavity thickness. When there is a reflective aluminium foil (red plot), the increment of the measured *R*-values with the increase in the air cavity thickness is almost linear until 30 mm, where it is nearly constant after 40 mm thick. Thus, it can be concluded that 30 mm is the recommended air cavity thickness when there is no reflective foil, while when using an aluminium reflective foil, the suggested thickness is increased to 40 mm. These results are in line with the conclusions achieved by Bruno et al. [23], which reported a maximum air gap thermal resistance, when using a reflective foil, for an air cavity thickness of 4 cm.

Notice that the maximum increased *R*-value measured (+0.529 $m^2 \cdot °C/W$), when using an aluminium reflective foil, is equivalent to an 18.5 mm mineral wool layer (thermal conductivity equal to 0.035 $W/(m \cdot K)$), but without the need of an increased wall thickness, whenever an air gap exists before the application of the foil. Additionally, the reflective aluminium foil thermal resistance increment (+0.529 $m^2 \cdot °C/W$) is higher than the -value increase due to a 50 mm air gap when there is no reflective foil (+0.420 $m^2 \cdot °C/W$), highlighting the importance of the radiative heat transfer. Furthermore, filling the cavity with mineral wool appears to have a comparable effect to the aluminium foil for small cavities. However, the existence of an air gap has several advantages, mainly related with humidity control in case of water infiltration.

3.1.2. InfraRed Thermography

Figure 7 shows the infrared (IR) images captured on the cold surface of the double pane LSF walls with and without a reflective aluminium foil, when using two extreme air cavity thicknesses: 0 and 50 mm. In these IR images, the central vertical steel stud is quite well visible, and it is more evident in the two LSF walls with null air cavity thicknesses. This is due to an increased localised heat transfer near the vertical steel stud, originating an augmented temperature in the cold surface of the wall.

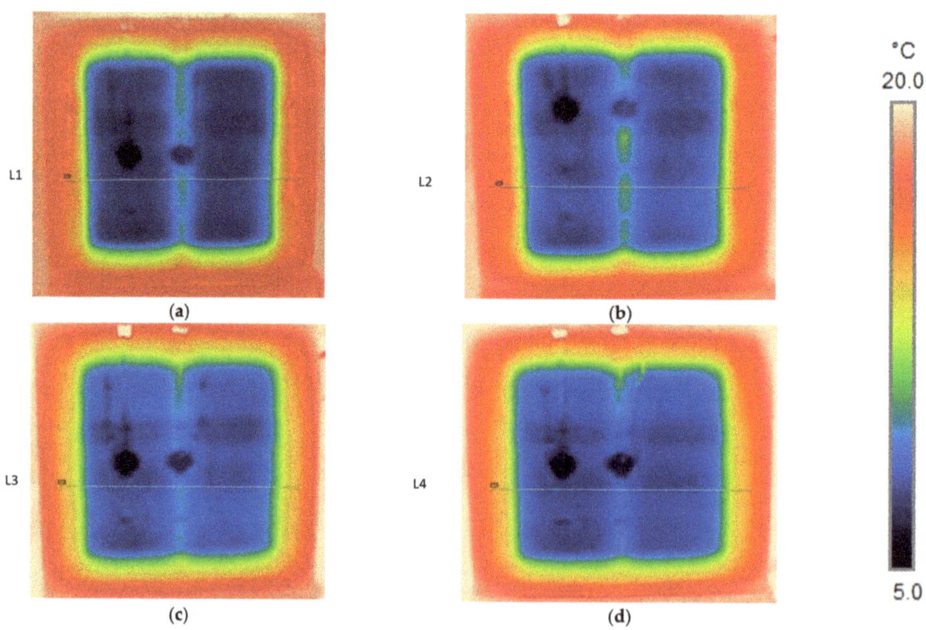

Figure 7. Infrared images of the double pane LSF walls, with and without reflective aluminium foil, for two air cavity's thicknesses (0 and 50 mm), on the cold surface. (**a**) 0 mm without reflective foil; (**b**) 0 mm with reflective foil; (**c**) 50 mm without reflective foil; (**d**) 50 mm with reflective foil.

To better visualize this steel stud thermal bridge effect, Figure 8 illustrates the surface temperatures recorded along horizontal lines in the IR images plotted in Figure 7. The highest peak temperatures were recorded near the steel stud for a null air cavity thickness, 10.6 °C and 10.0 °C, with and without an aluminium reflective foil, respectively. Another interesting feature visible in Figures 7 and 8 is that impact or influence of the thermal bridge originated by the vertical steel studs is well beyond the measurement circular area of the HFM1, which is only 8×10^{-4} m^2 (diameter equal to 32 mm).

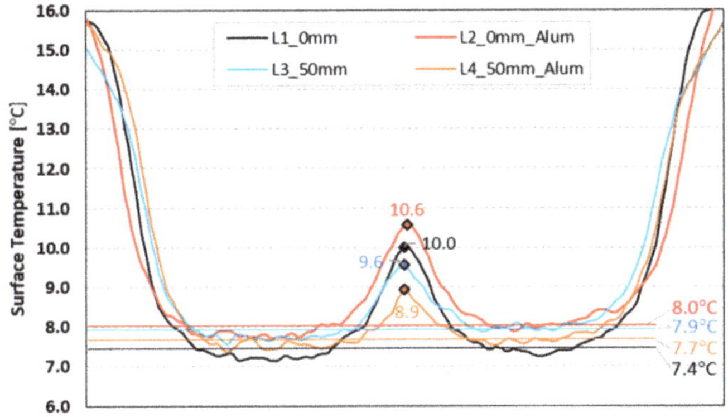

Figure 8. Horizontal temperature lines from infrared images of the double pane LSF walls, with and without reflective aluminium foil, for two air cavity's thicknesses (0 and 50 mm), on the cold surface.

Figure 9 shows the relative steel stud surface temperature increase along the horizontal lines on the cold surface of these four double pane LSF walls. Now, the higher surface temperature increase for the null air cavity thickness LSF walls is even better visualized due to the thermal bridge effect. In this case, the surface temperature increase near the vertical steel studs is +2.6 °C and +2.5 °C, with and without an aluminium reflective foil, respectively. As expected, for a 50 mm air cavity thickness the steel stud thermal bridge effect is quite smaller, exhibiting a surface temperature increase of only +1.6 °C and 1.3 °C, with and without aluminium reflective foil, respectively.

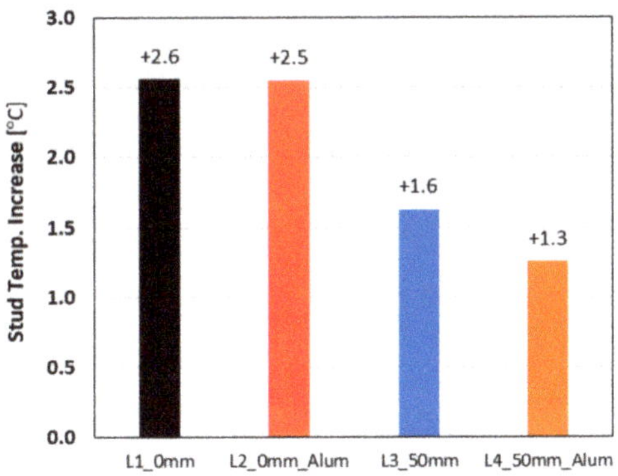

Figure 9. Steel stud surface temperature increase along horizontal lines on the cold surface of the double pane LSF walls, with and without reflective aluminium foil, for two air cavity's thicknesses (0 and 50 mm).

3.2. Numerical Simulations

As previously mentioned in Section 2.4, the experimental lab measurements were also compared with finite element numerical simulations predicted values making use of two different approaches to model the air cavities: CEN simplified and NFRC 100. Table 3 displays the predicted *R*-values for both methodologies, with and without an aluminium reflective foil, for the evaluated air cavity thicknesses (0–50 mm), as well as the differences to the measured values. To better visualise this info, Figure 10 graphically illustrates the measured and the predicted CEN and NFRC *R*-values. The first remark is that both models can reproduce with quite good accuracy a similar trend to the measured *R*-values, by predicting for some air cavities higher and for others smaller *R*-values.

Table 3. Predicted conductive thermal resistances (*R*-values) of a double pane LSF wall, with different air cavity thicknesses, with and without a reflective aluminium foil, and the thermal resistance difference to the measured values.

	Air Cavity	Without Aluminium Foil			With Aluminium Foil		
	Thickness	*R*-Value	Difference		*R*-Value	Difference	
	[mm]	[$m^2 \cdot °C/W$]	[$m^2 \cdot °C/W$]	[%]	[$m^2 \cdot °C/W$]	[$m^2 \cdot °C/W$]	[%]
CEN Simplified	**0**	2.109	+0.048	+2%	2.132	+0.070	+3%
	10	2.242	−0.011	0%	2.424	+0.064	+3%
	20	2.391	+0.003	0%	2.840	+0.155	+5%
	30	2.391	−0.072	−3%	2.993	+0.059	+2%
	40	2.367	−0.113	−5%	2.973	−0.036	−1%
	50	2.354	**−0.127**	**−5%**	2.960	−0.050	−2%
NFRC 100	**0**	2.105	+0.044	+2%	2.125	+0.063	−3%
	10	2.229	−0.024	−1%	2.375	+0.015	−1%
	20	2.388	0.000	0%	2.826	+0.141	−5%
	30	2.376	−0.087	−4%	2.904	−0.030	−1%
	40	2.350	−0.130	−6%	2.881	−0.128	−4%
	50	2.337	**−0.144**	**−6%**	2.863	**−0.147**	**−5%**

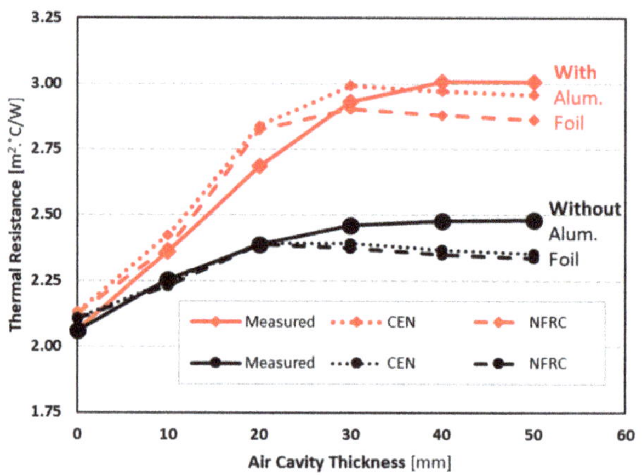

Figure 10. Measured and predicted conductive thermal resistances of a double pane LSF wall, with different air cavity thicknesses, with and without a reflective aluminium foil.

Without a reflective foil, both models predict very similar values, which are smaller than the measured ones for air cavities higher or equal to 30 mm. In this case, the major difference occurs for a 50 mm air gap, which is $-0.127\ m^2 \cdot °C/W$ (−5%) and $0.144\ m^2 \cdot °C/W$ (−6%) for the CEN simplified and NFRC 100 models, respectively.

When there is an aluminium reflective foil, the differences between both models enlarge, particularly for higher thicknesses of the air cavity (30–50 mm). In this thickness range, the CEN simplified methodology provides results closer to the measured *R*-values. For instance, for 50 mm the predictions are $-0.050\ m^2 \cdot °C/W$ (−2%) and $-\ 0.147\ m^2 \cdot °C/W$

(−5%) for CEN and NFRC, respectively. Moreover, for smaller air gaps (0–20 mm), these models exhibit a trend to slightly overestimate the *R*-values, ranging between +1% up to +5%.

4. Conclusions

In this work, the thermal performance of double-pane LSF walls with and without an aluminium reflective foil was assessed. This assessment was based in lab measurements and the recorded conductive *R*-values were compared with two different approaches to model the air cavities: "CEN simplified" [33] and "NFRC 100" [34], which were implemented in a 2D Finite Element software [32]. The thicknesses of these air gaps varied from 0 mm up to 50 mm, with an increment of 10 mm.

The major conclusions of this study could be summarized as follows:

- The use of a reflective foil is a very effective way to increase the thermal resistance of double pane LSF walls, without increasing the wall thickness and weight; the maximum improvement is around +0.529 $m^2 \cdot °C/W$ (+21%) in the achieved conductive *R*-value.
- It is not worthy to increase the air gap to values higher than 30 mm or 40 mm with and without a reflective foil, respectively.
- The *R*-value increase due to the reflective aluminium foil (+0.529 $m^2 \cdot °C/W$), is higher than the *R*-value raise due to a 50 mm air gap when there is no reflective foil (+0.420 $m^2 \cdot °C/W$).
- Besides the thermal resistance increase, the use of an aluminium reflective foil is also favourable to reduce the thermal bridge effect due to the steel studs' high thermal conductivity, whenever there are both a continuous air cavity and an aluminium foil.
- Both "CEN simplified" and "NFRC 100" models were able to predict with reasonable accuracy the thermal behaviour of the air cavities within the evaluated double pane LSF walls, ranging the obtained differences around ±5%.
- The major differences between both air cavity models arises for bigger air gaps (30–50 mm) when an aluminium reflective foil (0.05 emissivity) is used.

Notice that in this study the air cavity evaluated was continuous. Therefore, any decrease in thermal resistance due to wall ties, connectors or other bridging elements inside the cavity was not considered.

Author Contributions: Conceptualization, P.S.; methodology, P.S.; validation, P.S. and T.R.; formal analysis, P.S.; investigation, P.S. and T.R.; writing—original draft preparation P.S.; writing—review and editing, P.S. and T.R.; visualization, P.S.; supervision, P.S.; project administration, P.S.; funding acquisition, P.S. All authors have read and agreed to the published version of the manuscript.

Funding: This research was funded by FEDER funds through the Competitivity Factors Operational Programme—COMPETE and by national funds through FCT—Foundation for Science and Technology within the scope of the project POCI-01-0145-FEDER-032061.

Cofinanciado por: POCI-01-0145-FEDER-032061

Institutional Review Board Statement: Not applicable.

Informed Consent Statement: Not applicable.

Acknowledgments: The authors also want to thank the support provided by the following companies: Pertecno, Gyptec Ibéria, Volcalis, Sotinco, Kronospan, Hulkseflux, Hilti and Metabo.

Conflicts of Interest: The authors declare no conflict of interest.

References

1. European Union. Directive (EU) 2018/844 of the European Parliament and of the Council of 30 May 2018 amending Directive 2010/31/EU on the energy performance of buildings and Directive 2012/27/EU on energy efficiency. *Off. J. Eur. Union* **2018**, *156*, 75–91.
2. European Union. Directive (EU) 2018/2001 of the European Parliament and of the Council on the promotion of the use of energy from renewable sources. *Off. J. Eur. Union* **2018**, *328*, 82–209.
3. Santos, P.; Simões da Silva, L.; Ungureanu, V. *Energy Efficiency of Light-weight Steel-framed Buildings*, 1st ed.; European Convention for Constructional Steelwork (ECCS), Technical Committee 14—Sustainability & Eco-Efficiency of Steel Construction: Brussels, Belgium, 2012; ISBN 978-92-9147-105-8.
4. Roque, E.; Santos, P. The effectiveness of thermal insulation in lightweight steel-framed walls with respect to its position. *Buildings* **2017**, *7*, 13. [CrossRef]
5. Roque, E.; Santos, P.; Pereira, A.C. Thermal and sound insulation of lightweight steel-framed façade walls. *Sci. Technol. Built Environ.* **2019**, *25*, 156–176. [CrossRef]
6. Berardi, U.; Sprengard, C. An overview of and introduction to current researches on super insulating materials for high-performance buildings. *Energy Build.* **2020**, *214*, 109890. [CrossRef]
7. Zach, J.; Peterková, J.; Dufek, Z.; Sekavčnik, T. Development of vacuum insulating panels (VIP) with non-traditional core materials. *Energy Build.* **2019**, *199*, 12–19. [CrossRef]
8. Lamy-Mendes, A.; Dora, A.; Pontinha, R.; Alves, P.; Santos, P.; Durães, L. Progress in silica aerogel-containing materials for buildings' thermal insulation. *Constr. Build. Mater.* **2021**, *286*, 122815. [CrossRef]
9. Erhorn-Klutting, H.; Erhorn, H. *ASIEPI P148—Impact of Thermal Bridges on the Energy Performance of Buildings*; Buildings Platform, European Communities: Brussels, Belgium, 2009. Available online: https://www.semanticscholar.org/paper/Impact-of-thermal-bridges-on-the-energy-performance-Erhorn-kluttig-Erhorn/d208711a5785eb5dedd8b4ae7fe4013d3640f81d (accessed on 8 July 2021).
10. Santos, P.; Martins, C.; Simões da Silva, L. Thermal performance of lightweight steel-framed construction systems. *Metall. Res. Technol.* **2014**, *111*, 329–338. [CrossRef]
11. Soares, N.; Santos, P.; Gervásio, H.; Costa, J.J.; Simões da Silva, L. Energy efficiency and thermal performance of lightweight steel-framed (LSF) construction: A review. *Renew. Sustain. Energy Rev.* **2017**, *78*, 194–209. [CrossRef]
12. Martins, C.; Santos, P.; Simoes da Silva, L. Lightweight steel-framed thermal bridges mitigation strategies: A parametric study. *J. Build. Phys.* **2016**, *39*, 342–372. [CrossRef]
13. Santos, P.; Lemes, G.; Mateus, D. Thermal transmittance of internal partition and external facade LSF walls: A parametric study. *Energies* **2019**, *12*, 2671. [CrossRef]
14. Santos, P.; Mateus, D. Experimental assessment of thermal break strips performance in load-bearing and non-load-bearing LSF walls. *J. Build. Eng.* **2020**, *32*, 101693. [CrossRef]
15. Varadi, J.; Toth, E. Thermal improvement of lightweight façades containing slotted steel girders. In Proceedings of the Twelfth International Conference on Civil, Structural and Environmental Engineering Computing, Funchal, Madeira, Portugal, 1–4 September 2009; p. 107.
16. Lupan, L.M.; Manea, D.L.; Moga, L.M. Improving thermal performance of the wall panels using slotted steel stud framing. *Procedia Technol.* **2016**, *22*, 351–357. [CrossRef]
17. Santos, P.; Gonçalves, M.; Martins, C.; Soares, N.; Costa, J.J. Thermal Transmittance of Lightweight Steel Framed Walls: Experimental Versus Numerical and Analytical Approaches. *J. Build. Eng.* **2019**, *25*, 100776. [CrossRef]
18. Kapoor, D.R.; Peterman, K.D. Quantification and prediction of the thermal performance of cold-formed steel wall assemblies. *Structures* **2021**, *30*, 305–315. [CrossRef]
19. Santos, P.; Poologanathan, K. The importance of stud flanges size and shape on the thermal performance of lightweight steel framed walls. *Sustainability* **2021**, *13*, 3970. [CrossRef]
20. Mélois, A.B.; Moujalled, B.; Guyot, G.; Leprince, V. Improving building envelope knowledge from analysis of 219,000 certified on-site air leakage measurements in France. *Build. Environ.* **2019**, *159*, 106145. [CrossRef]
21. Fantucci, S.; Serra, V. Low-E paints enhanced building components: Performance, limits and research perspectives. *Energy Procedia* **2017**, *126*, 274–281. [CrossRef]
22. Jelle, B.P.; Kalnæs, S.E.; Gao, T. Low-emissivity materials for building applications: A state-of-the-art review and future research perspectives. *Energy Build.* **2015**, *96*, 329–356. [CrossRef]
23. Bruno, R.; Bevilacqua, P.; Ferraro, V.; Arcuri, N. Reflective thermal insulation in non-ventilated air-gaps: Experimental and theoretical evaluations on the global heat transfer coefficient. *Energy Build.* **2021**, *236*, 110769. [CrossRef]
24. Gyptec Ibérica. Technical Sheet: Standard Gypsum Plasterboard. 2021. Available online: https://www.gyptec.eu/documentos/Ficha_Tecnica_Gyptec_A.pdf (accessed on 19 April 2021). (In Portuguese)
25. Volcalis Mineral Wool. Technical Sheet: Alpha Panel Mineral Wool. 2021. Available online: https://www.volcalis.pt/categoria_file/fichatecnica_volcalis_alphapainel-777.pdf (accessed on 19 April 2021). (In Portuguese)
26. Santos, C.; Matias, L. *ITE50—Coeficientes de Transmissão Térmica de Elementos da Envolvente dos Edifícios*; LNEC—Laboratório Nacional de Engenharia Civil: Lisboa, Portugal, 2006. (In Portuguese)

27. *ISO 9869-1, Thermal Insulation—Building Elements—In-Situ Measurement of Thermal Resistance and Thermal Transmittance. Part 1: Heat Flow Meter Method*; ISO—International Organization for Standardization: Geneva, Switzerland, 2014.
28. Rasooli, A.; Itard, L. In-situ characterization of walls' thermal resistance: An extension to the ISO 9869 standard method. *Energy Build.* **2018**, *179*, 374–383. [CrossRef]
29. *ASTM-C1155–95(Reapproved-2013), Standard Practice for Determining Thermal Resistance of Building Envelope Components from the In-Situ Data*; ASTM—American Society for Testing and Materials: Philadelphia, PA, USA, 2013.
30. ASHRAE. *Handbook of Fundamentals (SI Edition)*; ASHRAE—American Society of Heating, Refrigerating and Air-Conditioning Engineers: Atlanta, GA, USA, 2017.
31. Santos, P.; Lemes, G.; Mateus, D. Analytical methods to estimate the thermal transmittance of LSF walls: Calculation procedures review and accuracy comparison. *Energies* **2020**, *13*, 840. [CrossRef]
32. *THERM*; Software Version 7.6.1; Lawrence Berkeley National Laboratory: Berkeley, CA, USA; United States Department of Energy: Washington, DC, USA, 2017. Available online: https://windows.lbl.gov/software/therm (accessed on 14 February 2019).
33. *EN ISO 6946, Building Components and Building Elements—Thermal Resistance and Thermal Transmittance—Calculation Methods*; CEN—European Committee for Standardization: Brussels, Belgium, 2017.
34. *ISO_15099, Thermal Performance of Windows, Doors and Shading Devices—Detailed Calculations*; ISO—International Organization for Standardization: Geneva, Switzerland, 2003.

buildings MDPI

Article

BIM-Based Energy Analysis and Sustainability Assessment—Application to Portuguese Buildings

José Pedro Carvalho [1,*], Manuela Almeida [1,2], Luís Bragança [1,2] and Ricardo Mateus [1,2]

1 Institute for Sustainability and Innovation in Structural Engineering (ISISE), University of Minho, 4800-058 Guimarães, Portugal; malmeida@civil.uminho.pt (M.A.); braganca@civil.uminho.pt (L.B.); ricardomateus@civil.uminho.pt (R.M.)
2 School of Engineering, Civil Engineering Department, University of Minho, 4800-058 Guimarães, Portugal
* Correspondence: jpcarvalho@civil.uminho.pt; Tel.: +351-253-510-200

Abstract: Buildings are responsible for several negative impacts on the environment, most of them related to nonrenewable energy consumption, increasing the concern regarding buildings energy efficiency. In this context, computer software has been used to estimate the energy needs of the built environment, and the Building Information Modelling (BIM) methodology can be used to simplify this process. This study aims to validate a BIM-based framework to streamline the energy analysis of Portuguese buildings, based on the method of the national regulation for the thermal performance of residential buildings. Currently, designers need to spend considerable time assessing all the building characteristics and performing the mandatory calculations for energy performance analysis. It is also intended to link the results of the energy simulation with a Building Sustainability Assessment method—SBTool^{PT}-H. The purpose is to demonstrate how it is possible to benefit from this approach to simultaneously improve building sustainability during the design stage. To do so, different case studies were modelled in Autodesk Revit and exported to a BIM energy tool to perform energy simulation analysis. The results were validated against the official assessment method of the Portuguese thermal regulation and were successfully used to assess the SBTool^{PT}-H energy efficiency category. The research outcomes provide design teams with a reliable BIM-based framework to improve building energy performance and to develop thermal projects while enhancing building sustainability. It also increases the knowledge about the integration of sustainability assessment in the BIM environment, providing new insights for complete integration.

Keywords: Building Information Modelling (BIM); Building Energy Modelling (BEM); energy efficiency; Building Sustainability Assessment (BSA); sustainability

Citation: Carvalho, J.P.; Almeida, M.; Bragança, L.; Mateus, R. BIM-Based Energy Analysis and Sustainability Assessment—Application to Portuguese Buildings. *Buildings* **2021**, *11*, 246. https://doi.org/10.3390/buildings11060246

Academic Editors: Paulo Santos and Mark Bomberg

Received: 11 May 2021
Accepted: 6 June 2021
Published: 9 June 2021

1. Introduction

As society is growing, there is an increasing concern about building occupants' comfort demand and energy consumption. The main reasons are directly related to weak buildings energy performance and irrational energy use [1]. Energy efficiency is an essential factor to achieve sustainable development. It is necessary to optimise energy use without compromising the indoor environmental quality, using efficient technologies and passive and active construction solutions [1,2]. Energy efficiency is related to the building's performance in the three sustainability dimensions:

- Environment—due to resources use and carbon emissions;
- Society—due to indoor environmental comfort;
- Economy—due to energy cost and its impact on household income.

Hence, Building Sustainability Assessment (BSA) schemes usually evaluate a set of building energy-related characteristics and performance. Such schemes can provide a decision support framework for designers to improve the sustainability of buildings, as well as to evaluate them according to local standards and regulations. Effectively acting on energy use and building sustainability is an essential path to achieve better, ecological, comfortable and cost-effective buildings.

Facing the increasing capabilities of Building Information Modelling (BIM) for the construction industry, designers and researchers are extensively applying it to manage building data and improve efficiency and global quality [3]. During the last five years, the use of BIM for sustainable construction purposes has also witnessed exponential growth [4]. BIM allows storing multi-disciplinarily information into a single model, promoting a real-time collaboration environment among stakeholders through the building life cycle [5]. Between BIM most known applications, its connection with Building Energy Modelling (BEM) has been used to improve buildings energy performance. BEM is a powerful tool to analyse and enhance building energy performance and thermal comfort, providing project teams with concise data to evaluate the performance and environmental impacts of different design solutions [6,7]. Despite the recognised advantages during the design phase, there is still a great scalability potential, as several designers still do not use BIM for energy analysis due to the required knowledge and time to prepare the energy model and interpret the results [8].

Regardless of the benefits and different BIM approaches to perform energy analysis, BIM is not being used in the Portuguese context to develop mandatory thermal projects. Major BIM energy tools are usually region-oriented, and calculation engines are not, according to the Portuguese energy regulations. Furthermore, Portuguese building technologies and indoor environmental quality standards are quite different from most European countries [9]. Portuguese designers can only benefit from those tools in the optimisation of the building design. Additionally, Portuguese legislation requires a thermal project for every building to issue construction permits. Nowadays, designers are required to manually fill a set of calculation spreadsheets with the building's characteristics, requiring in-depth knowledge about the building and the energy calculation method, as well as substantial time to carry out the analysis.

Facing the knowledge gap, this study primarily aims to define and apply a BIM process that can support and optimise the thermal project of Portuguese buildings. Thus, the first Research Question (RQ) of this research arises—"Can BIM support and optimise Portuguese buildings thermal project?". To provide an answer, a suitable BIM-based method will be identified and applied to Portuguese residential buildings case studies. The BIM software results will face a conventional approach to prove the reliability of the method and answer the second RQ—"Is the BIM-based method reliable and according to the Portuguese standards?". Since energy efficiency is a standard sustainability indicator in BSA methods [10,11], the development of a new BIM-based approach creates the opportunity to link energy simulation results with a local BSA method, leading to the last RQ—"Can the results be used for a sustainability assessment scheme?". Therefore, it will be possible to effectively demonstrate how it is feasible to benefit from this approach in the process of improving the sustainability of a building during the design stage. Additionally, it will also enhance the integration of BSA in the BIM collaborative process and promote the use of BSA methods by the Portuguese designers.

2. Literature Review

2.1. Buildings Energy Performance

The European Union (EU) authorities and the society have raised awareness about the negative impacts of buildings on the environment. The relation between the construction industry and environmental problems has already been proven by the scientific community [12–14]. The European building stock is still responsible for 40% of the total energy use and 36% of CO_2 emissions [15]. To improve buildings efficiency and reduce the energy demand, the EU has approved a set of standards. In this context, the most important is the Energy Performance of Buildings Directive (EPBD), updated in 2010 [16]. The principal goal was to create a main legal instrument to improve EU buildings energy performance.

In Portugal, the EPBD and other Directives were transposed to the Portuguese legal framework by the Decree Law 118/2013 in 2013, composed by the Buildings Energy Certification System (SCE), by the Residential Buildings Energy Performance Regulation (REH) and by the Services and Commercial Buildings Energy Performance Regulation (RECS) [17]. In 2018, Portugal was the seventh European country with more dependence on energy imports, with 75.9% of the consumed energy being imported [18]. The Portuguese building sector (residential and service buildings) was responsible for 31.9% of the country energy demand, exceeded by the industry and transportation sectors. According to the SCE, in 2019, almost 61% of the 1.78 million Portuguese energy classified buildings were rated less than B- (minimum requirement for new buildings). This was justified, since the majority of the Portuguese buildings were built before the first national thermal regulation in 1990. A 2020 energy report [18] highlighted the potential savings of 60% on the building energy demand by adopting energy efficiency measures. More than half of them are related to interventions on the building envelope, i.e., in external walls (24.7%), roofs (13.3%), ground floors (4.4%) and windows (10.4%).

2.2. BIM and BEM

Construction projects are becoming more difficult and complicated. New approaches, like BIM, have been introduced in construction companies to support designers in managing all the project information [4,19,20]. BIM can be defined as a working methodology that makes it possible to manage the 3D drawing and the project data in a digital format during the entire building life cycle [21]. BIM can improve design and management processes productivity, with stakeholders working in real-time collaboration [4,22]. BIM implies a building information model, which is an object-oriented parametric model, with all the project data. The model Level of Development (LOD), which ranges from 100 to 500, specifies and articulates the content and reliability of the BIM model [23].

BIM creates an excellent opportunity to incorporate sustainable measures throughout the design process [24]. Some authors have already identified the preconstruction and project phases as the critical ones, where the main decisions regarding building sustainability must be taken [20,25,26]. Since that is also the phase when projects can most benefit from BIM, the influence that it can have in enhancing buildings sustainability becomes clear [27]. BIM can provide information about the estimated building performance even in the very early design stages [8].

In 2008, Krygiel and Nies [28] recognised seven aspects where BIM can be used to support sustainable design. Five of those—Building Orientation, Building massing, Daylight Analysis, Renewable Energy and Energy Modelling—were directly energy-related criteria. Therefore, an energy analysis can significantly benefit from BIM [10], leading to several BIM-based energy-related research—BEM [6,7]. BIM and BEM allow designers to evaluate different design options during the project early stages, creating the opportunity to develop optimised buildings with higher energy efficiency and comfort. However, such application is frequently performed in an isolated manner, empowered by the use of energy simulations in the early phases of the project. Furthermore, BEM integration in BIM collaborative workflow is also not sufficiently developed and synchronised, and energy-efficient design strategies are often not well-implemented [7].

Within the context of BIM, model interoperability between software is usually made with Industry Foundation Classes (IFC), Green Building XML (gbXML) or direct plug-ins. These are the most common open standard data schemas, which are commonly used for information exchange between BIM and BEM tools [6,29,30]. Kamel and Memari [31] highlighted the differences between both when used for energy simulation purposes. Despite the use of the Green Building XML schema (gbXML) mainly for the energy simulation domain, only rectangular geometry is readable, and it does not allow the targeting of specific areas of a project. The IFC is commonly used for different domains of application (a more comprehensive type of data), and it is capable of reading any geometry. Nevertheless, some data is still not transferred appropriately, leading to the development of corrective add-ins.

Depending on the energy analysis type and intended accuracy, the model must have a certain LOD [32]. Farzaneh et al. [33] suggested that the LOD should be defined according to the information requirements. However, the design process must also be "user-friendly", with a LOD that allows quick modelling and provides reliable results [5]. An accurate energy assessment requires, at least, a LOD of 300, while a simpler analysis, such as, i.e., solar exposure, only requires a LOD 200, without materials and spaces characteristics [34].

On the application field, several reviews have been made comparing the analysis capabilities and end-users of energy analysis tools, highlighting the capabilities of Autodesk Green Building Studio (GBS) and Integrated Environmental Solutions Virtual Environment (IESVE) [35,36]. Regarding the modulation software, Autodesk Revit is one of the most used, capable, and embracing BIM platforms [37–39]. Existing reviews pointed out Autodesk Revit as the most used BIM platform by researchers when concerning producing sustainable and efficient buildings.

Using Autodesk Revit and GBS, Abanda and Byres [40] concluded that a building's orientation can have a considerable impact on the building energy consumption. Gourlis and Kovacic [39] applied BIM to analyse the energy efficiency of industrial buildings using EnergyPlus. They highlighted that the BIM and BEM approach is still not mature enough, requiring a significant amount of time, assumptions and remodelling. Ryu et al. [11] presented a simulation process based on Autodesk Revit to assess energy-related BSA credits. They concluded that BIM could produce significant time savings, but considerable time was wasted in double-checking and geometry correction. Montiel-Santiago et al. [19] submitted a hospital BIM model to a set of analyses on Insight 360, suggesting an energy renovation package able to save 47% of the actual building energy demand. Carvalho et al. [41] conducted an energy renovation of a Portuguese dwelling using GBS and DesignBuilder. The main benefit of BIM in the Portuguese context was the decision support provided to designers in the early project stages, as none of the BIM tools considered the Portuguese thermal regulation (REH).

Despite the advantages, most researchers have reached a common conclusion about BIM for energy purposes—there is still an interoperability gap between BIM modulation platforms and BIM energy tools [26,42,43]. BEM is not integrated correctly into the BIM environment, and often, a continuous information flow in the digital modelling is not possible. The lack of interoperability between BIM and BEM makes it challenging to create projects that are seeking sustainable and efficient energy performances [30]. Gao et al. [7] argued that BIM models usually contain high-level data that is too complicated for the BIM energy tools to understand. Designers are also required to assume a set of parameters for the simulations, and human behaviour is usually treated as robots [44]. Furthermore, the existence of several BIM energy tools, parameters and approaches [45] makes it difficult to establish a common procedure, usually making unfeasible comparisons between buildings. There is a need to establish common procedures and standards to perform an energy analysis and to characterise the information exchanges between BIM platforms and energy tools [10,31,45].

2.3. Building Sustainability Assessment

2.3.1. Overview

In the last two decades, several worldwide entities have developed BSA methods. These methods aim at implementing and spreading sustainable principles and evaluating and monitoring buildings performance and gathering information to support decision-making [46–48]. Some of their benefits include environmental conservation; improved building performance and occupants' comfort, health and safety [49].

BSA methods are usually characterised by assessing some partial building features and aggregating the results into a sustainability score. They provide an opportunity for projects to demonstrate their ecological, economic and social benefits to the local community [50]. The most known BSA methods, which provide a basis to all the other approaches, are the Building Research Establishment Environmental Assessment Methodology (BREEAM) from the Building Research Establishment (BRE), Leadership in Energy and Environmental Design (LEED) from the United States Green Building Council (USGBC) and Sustainable Building Tool (SBTool) from the international initiative for a Sustainable Built Environment (iiSBE) [51,52].

Nowadays, BSA methods are usually applied after the design is completed (or even after the building construction), turning possible modifications to improve the building sustainability unbearable or too expensive [53]. This can be justified due to the amount and complexity of data and documentation required for the evaluation. The application of BSA methods is also a voluntary approach worldwide, with the absence of mandatory legislation. Additionally, the assessment procedure is a time-consuming process—particularly in performance-based methods—which is usually incompatible with project companies' short deadlines [54,55], making it necessary to search for more efficient and expeditious approaches. For instance, Zhang et al. [53] suggested a real-time green building rating method that can identify potential ways to optimise the final rating during the design process. Some other constraints were also identified by researchers, such as errors from manual and traditional measuring tools or through calculations or even during data collection [56,57]. These types of errors can have great impact on the environment, since these methods are the main source of shaping sustainable decisions in projects [8]. User-friendly restrictions, the complex credit structure and the required user knowledge are also common issues that hinder the use of BSA methods [49].

The Portuguese scenario for BSA methods still has a long run. Portuguese designers often neglect sustainability assessment schemes, since there is not any mandatory sustainable evaluation for buildings. Additionally, performing a BSA is a time-consuming and complex process. To date, Portugal has not had an official BSA method, and only a couple of building sustainability rating schemes have been explicitly developed for the country conditions, such as the SBTool^{PT}-H, LiderA and Domus Natura [58,59]. According to Pires and Fidélis [60], the development of sustainable indicators in Portugal lacks political commitment and vision, as well as poor stakeholder involvement. Nevertheless, existing methods have already been adapted to embrace different Portuguese building types, such as residential houses, hospitals, schools, offices, or even urban areas [48,61,62].

2.3.2. BIM Integration

Despite the recent trend on the use of BIM for sustainability purposes, there is still a lack of research considering all the sustainability dimensions [4]. Regardless of the potential benefits, so far, BIM has not been used comprehensively to obtain the required information for a sustainability assessment [27]. The existing BIM software stills lacks sustainability issues, and exchange format files are still in need of further developments [52,63]. Thus, the opportunity for the BSA methods to benefit from BIM capabilities has emerged, as well as the possibility of integrating the different BSA methods in the BIM collaborative process. Beyond the direct benefits for project teams and buildings occupants, significant advantages are expected to the construction industry, such as more sustainable and ecological buildings and a reduction in the potential environmental impacts. With the increasing maturity of BIM, a higher integration of building sustainability is also expected [4].

Currently, BIM is most commonly used to support the assessment of LEED, mainly in the categories of energy and atmosphere and materials and resources [29,37]. Nevertheless, several authors [63–67] focused their research on assessing different credits from all LEED categories. BREEAM has also attracted researchers' attentions. BIM has been used to support the assessment of BREEAM credits on the Materials, Energy, Water, Land use and Ecology, Health and Wellbeing and Waste and Pollution categories [8,66,68,69]. Edwards et al. [8] identified which criteria from BREEAM and LEED can be assessed with some recognised BIM tools. Most of them can often provide data to assess energy-related and indoor comfort-related categories of both schemes.

Other attempts for different BSA schemes have also been made, with Wong and Kuan [24] assessing 26 out of 56 criteria of BEAM Plus method with BIM in a faster and more accurate way. On the Australian Green Star Building Certification, Gandhi and Jupp [70] used BIM to evaluate 66% of the office building scheme credits. Hoseini et al. [71] suggested that BIM can also support the assessment of 75% of New Zeeland Green Star Certification with the development of proper guidelines. Concerning SBTool, Carvalho et al. [27,72] developed a conceptual framework for the integration of BIM with the Portuguese BSA method for residential buildings—SBTool^{PT}-H. The assessment procedure of almost all the SBTool^{PT}-H criteria can significantly benefit from the use of BIM. They also compared the feasibility of using BIM in SBTool^{PT}-H with its use in other BSA methods, such as LEED and BREEAM [37].

Despite the benefits, a common conclusion is that BIM is not properly oriented and has not achieved its full potential to sustainable building design [8,27,73]. There are also frequent interoperability issues between BIM platforms and tools, with some information lost, requiring additional time for model checking and corrections. The common standards for data exchange must also be further developed to include more sustainability issues [4,74,75]. Chong et al. [74] argued that future BIM standards should consider a requirement for sustainability assessment, while Gandhi and Jupp [70] asked for specific BIM coordination and execution plans.

2.3.3. SBTool^{PT}-H

The BSA method SBTool is considered the most comprehensive of all the methods due to its flexibility to be adjusted to the region local context [48,52]. SBTool has already influenced Austria's, Spain's, Japan's and Korea's national rating systems, and custom versions are in use in Italy, the Czech Republic and Portugal [76,77]. The transposition of the SBTool to the Portuguese residential scenario (SBTool^{PT}-H) was done to create a generic method to assess the sustainability of existing, new and renovated Portuguese buildings. This method aims at supporting design teams since the early design stages and raising awareness to adopt more sustainable construction solutions. In SBTool^{PT}-H, there are a total of 25 parameters, sorted by nine categories, which are divided into three sustainability dimensions: Environment, Society and Economy. The scheme structure is presented in Appendix A. Each parameter has a different weight according to the national standards and practices, and it is classified with a quantitative "score" that results from the comparison between the performance of the analysed building and two benchmarks: best and conventional sustainable practice. Each "score" is normalised to establish a dimensionless value that expresses the building performance in comparison to the benchmarks [78]. In this dimensionless scale, 0 corresponds to the conventional practice and 1 to the best practice. The normalised value is then converted into a qualitative scale that ranges from E to A+, where D corresponds to the conventional practice and A to the best practice.

Since the aim of this research is to use BIM-based energy simulation results to support the assessment of BSA methods, it is required to further investigate SBTool^{PT}-H energy efficiency-related criteria.

The energy efficiency category (C3) from SBTool^{PT}-H gathers two sustainability parameters related to building energy efficiency: P7—Primary Energy and P8—On-site energy production from renewables. To obtain high scores in these parameters, it is essential to optimise the building energy efficiency and on-site energy production from renewable sources by improving the building envelope and energy systems. Both parameters are based on calculation methods defined in the Portuguese regulation REH, forcing designers to perform different and time-consuming calculations to achieve the required data for the assessment.

The assessment of the energy efficiency category follows the general pattern of the remaining categories, where the building performance is compared with the national benchmarks. Thus, for the assessment of P7, the primary energy needs of the case study ($P_{ENR} = N_{tc}$) are compared with two benchmarks: the Portuguese conventional practice ($P_{ENR*} = N_t$) and best practice ($P_{ENR}^* = 0.25 \times N_t$); according to the current thermal regulation, the national best practice corresponds to a building that consumes 25% or less energy than a conventional building. The comparison is carried out using a quantitative normalised value, which results from the application of Equation (1). Then, the final result ($\overline{P_{ENR}}$) is reached by converting the normalised value into a qualitative scale, according to Table 1.

$$\overline{P_{ENR}} = \frac{P_{ENR} - P_{ENR*}}{P_{ENR}* - P_{ENR*}} \tag{1}$$

where:

$\overline{P_{ENR}}$—case study normalised result for P7;
P_{ENR}—case study result for P7;
P_{ENR}^*—national best practice for P7;
P_{ENR*}—national conventional practice for P7.

Table 1. Conversion from the quantitative to the qualitative performance scales in SBTool^{PT}-H.

Qualitative Level	Quantitative Value
A+	$p > 1.00$
A	$0.70 < p \leq 1.00$
B	$0.40 < p \leq 0.70$
C	$0.10 < p \leq 0.40$
D	$0.00 \leq p \leq 0.10$
E	$p < 0.00$

For the assessment of P8, the renewable energy production (P_{ER}) of the case study is compared again with two benchmarks: the Portuguese conventional practice (P_{ER*} = production from renewables of 50% of the energy needs for Domestic Hot Water—DHW) and best practice (P_{ER}^* = production from renewables of 90% of the total primary energy needs). Once again, the comparison is made using the normalised value (Equation (2)), and the final result ($\overline{P_{ER}}$) is according to the qualitative scale of Table 1.

$$\overline{P_{ER}} = \frac{P_{ER} - P_{ER*}}{P_{ER}* - P_{ER*}} \tag{2}$$

where:

$\overline{P_{ER}}$—case study normalised result for P8;
P_{ER}—case study result for P8;
P_{ER}^*—national best practice for P8;
P_{ER*}—national conventional practice for P8.

3. Materials and Methods

The primary purpose of this study is to define and apply a BIM process that can support and optimise the mandatory energy performance analysis of Portuguese buildings. To date, despite the several studies about using BIM to assess the energy performance of buildings, none of those has defined and identified a suitable method for Portuguese dwellings that is according to the Portuguese legislation. Furthermore, the process should also provide data for a sustainability assessment since energy performance simulation results are usually valuable insights for the energy performance category of BSA methods. Nowadays, the application of BSA methods is not a standard procedure between Portuguese construction companies, due to the required time, knowledge and resources for the assessment. It is then necessary to take essential steps for the integration of BIM and BSA methods to effectively improve building sustainability. To guide the research, the following RQ were defined:

- RQ1: Can BIM enhance and optimise Portuguese buildings energy efficiency and buildings thermal project?
- RQ2: Is the BIM-based method reliable and according to the Portuguese standards?
- RQ3: Can the results be used for sustainability assessments?

As a first step, it necessary to identify a suitable process and software to perform an energy analysis according to the Portuguese regulations. Additionally, the process should consider Autodesk Revit as a BIM platform, since it was identified as the most commonly used software.

A study from Carvalho et al. [27] developed a conceptual framework for the integration of BIM in SBTool and suggested BIM approaches for each criteria assessment. For the energy-related criteria (P7—Primary Energy and P8—On-site energy production from renewables), they identified Cypetherm REH as the most suitable tool to estimate the energy performance of Portuguese residential buildings. In this tool, the calculation method is based on the Portuguese thermal regulation (REH), and it automatically produces the necessary information for the mandatory analysis of the building energy performance and national energy label. Cypetherm REH is one of the several software from CYPE Ingenieros, a Spanish company that develops computer software to support the AEC industry stakeholders. This software is adequately adapted to the national standards, and it is the most used among the Portuguese building design offices. Furthermore, Cype software has an add-in for BIM platform Autodesk Revit that allows exporting BIM models using IFC format. This process uses the BIMServer.center (https://bimserver.center) cloud, which acts as an intermediary platform between the selected software.

Following the previous goals, Cypetherm REH (version 2020d, Cype, Portugal) was selected to carry out the energy analysis for this research. However, the method of the Portuguese energy regulation requires as the input the amount of on-site renewable energy production, and the Cypetherm REH does not allow to perform this simulation. Therefore, the user must perform the simulation in an external software tool and input it into the software. To avoid the use of different climate databases, an official spreadsheet from the Portuguese Directorate-General for Energy and Geology (DGEG) was used to assess the case study's minimum requirements for renewable energy production. The simulation results from Cypetherm REH will be validated against the official Portuguese assessment method, i.e., using the official REH Excel spreadsheet (version V3.15 of 23 July 2020) for the evaluation of the building's thermal performance, which was developed by IteCons and University of Coimbra [79]. Currently, the method commonly used by designers requires considerable time to assess the building envelope characteristics, to select the calculation parameters and to perform a set of calculations. From the results of this stage, it will be possible to answer both RQ1 and RQ2.

After the simulation procedure, the results will be used to evaluate the two parameters of the SBTool^{PT}-H energy efficiency category: P7 and P8. Cypetherm REH provides enough data for a comprehensive evaluation of both energy-related parameters. The assessment is made by linking the results from Cypetherm REH with SBTool^{PT}-H evaluation spreadsheet, without performing any other calculations. Here, it will be possible to provide an answer for RQ3. The research procedure is summarised in Figure 1.

To apply the suggested procedure a case study is required. It must be framed under the scope of both REH and SBTool^{PT}-H, i.e., must be a residential building. Thus, two different case studies were selected for the analysis: an existing building and a new building project (Figure 2).

The existing building is a 3-bedroom single-family house located in Porto, Portugal. It is representative of most of the common characteristics from Portuguese residential buildings built during the 1970s [80,81] in terms of: thermal resistance of the envelope elements (no insulation); construction materials (brick wall, sloped roof with ceramic tile, prestressed slab and wooden frame windows); typology (3-bedroom) and floor area (less than 100 m^2). Since the first Portuguese thermal regulation was only introduced in 1990 [82], the dwelling has no insulation materials, creating a need for improving the thermal behaviour of the building envelope. The dwelling total net floor area is 74.92 m^2, the interior floor to ceiling height is 3.05 m and the glazed area is 6.3 m^2. The building is at an altitude of 155 m and located 5 km from the coastline. This case study will be submitted to an energy analysis as it is—the reference Model—and then, the building envelope will be optimised—the optimised Model. After, another simulation is going to be carried out to demonstrate how the simulation tool can support the designer's decision-making in comparing the performance of different design scenarios. The optimisation is made by only improving the building envelope thermal resistance, and the aim is to reduce at least 70% of the building energy demand and meet the current national standards. The optimised model concerns an energy renovation scenario, and it was defined according to the actual thermal requirements.

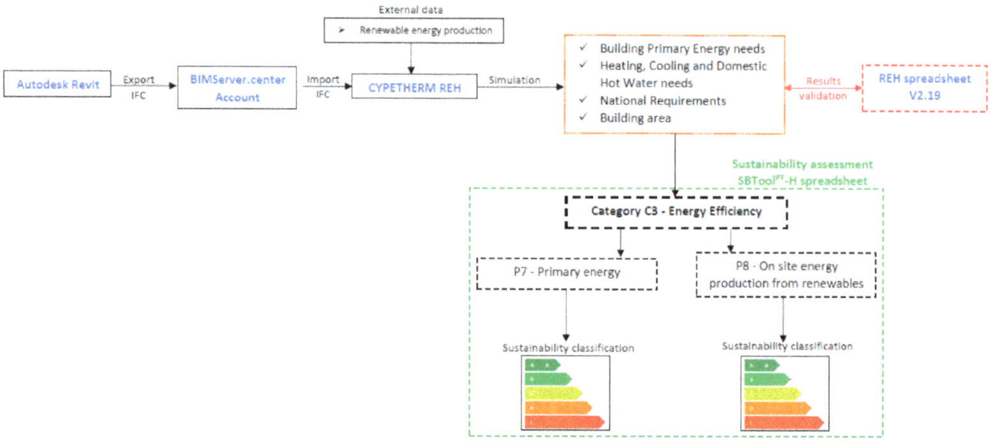

Figure 1. Simulation procedure.

The new building project case study is also a 3-bedroom single-family house located in Porto, Portugal (altitude of 155 m and located 5 km from the coastline). The building has not yet been built and is representative of almost all characteristics of the Portuguese dwellings nowadays [81]: construction materials (double-brick wall, flat roof, lightweight block and beam slab and aluminium window frame with thermal break); typology (3-bedroom); windows area (window-to-floor ratio of 15–20%) and floor area (average of 150 m^2). The building total net area is 143.53 m^2, the interior floor-to-ceiling height is 2.60 m and the glazed area is 39.66 m^2. Since, in this case study, it is necessary to fulfil the REH minimum energy requirements, only one virtual model was made and analysed. This design scenario was already defined to match the regulation best practices.

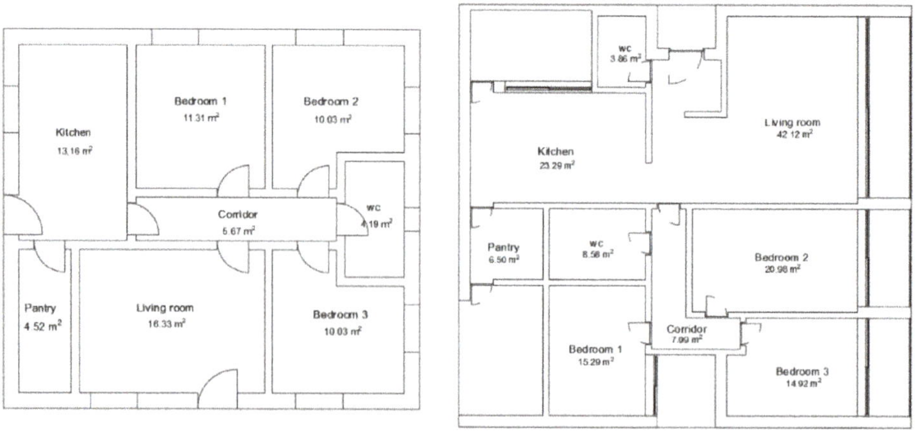

Figure 2. Floor plan of the case studies: existing model and new building model.

The thermal characteristics of both case studies scenarios are presented in Table 2. The surface albedo for walls, doors and roofs was defined as a bright colour (D = 0.4), which influenced the building summer gains (more gains with darker colours and fewer with brighter colours). The efficiencies of all the systems (heating, cooling, DHW and solar) were kept constant for all the simulations. The simulations also considered gains and losses by natural ventilation, and the adopted ventilation renovation rates (air changes per hour—ach) were according to the recommendations of the Portuguese regulation REH (winter = 0.4 ach, summer = 0.6 ach).

Table 2. Construction solutions and systems.

Element	Existing Building		New Building
	Reference Model	Optimised Model	
Exterior walls	Simple brick wall (15 cm) U = 1.69 W/(m^{2}K)	Double brick wall (15 + 11) with XPS insulation (6 cm) in the middle U = 0.40 W/(m^{2}K)	Double brick wall (15 + 11) with XPS insulation (5 cm) in the air cavity (2 cm) U = 0.38 W/(m^{2}K)
Interior walls	Simple brick wall (11 cm) U = 1.78 W/(m^{2}K)		
Floor slab	Lightweight block and beam slab (20 cm) and wooden floor finishing (3 cm) U = 1.60 W/(m^{2}K)	Lightweight block and beam slab (20 cm), with interior XPS insulation (7 cm) and wooden floor finishing (3 cm) U = 0.34 W/(m^{2}K)	Lightweight block and beam slab (25 cm), with interior XPS insulation (4 cm) and wooden floor finishing (3 cm) U = 0.30 W/(m^{2}K)
Roof slab	Lightweight block and beam slab (25 cm) U = 2.02 W/(m^{2}K)	Lightweight block and beam slab (25 cm), with exterior XPS insulation (8.5 cm) and waterproof membrane U = 0.35 W/(m^{2}K)	Lightweight block and beam slab (25 cm), with exterior XPS insulation (8.5 cm), waterproof membrane and gravel (5 cm) U = 0.35 W/(m^{2}K)
Sloped roof	Lightweight block and beam slab (25 cm) with ceramic tile U = 2.02 W/(m^{2}K)	Lightweight block and beam slab (25 cm) with ceramic tile U = 2.02 W/(m^{2}K)	-
Windows–glass	Single glass 6 mm (Solar factor 0.85) U = 2.50 W/(m^{2}K)	Double-glass 6 mm + 4 mm (Solar factor 0.78) U = 1.50 W/(m^{2}K)	
Windows–frame	Wooden frame (w/ exterior shutter) U = 2.50 W/(m^{2}K)	Aluminium frame with thermal break (w/ exterior shutter) U = 1.50 W/(m^{2}K)	
Exterior doors	French wooden door U = 1.50 W/(m^{2}K)		
Interior doors	Wooden light door U = 2.50 W/(m^{2}K)		

4. BIM Model for Energy Simulation

4.1. Building Modelling

The first research step consisted of creating the BIM models in the BIM platform Autodesk Revit (version 2019, Autodesk, United States of America). By allowing a parametric modulation, building elements automatically recognise each other, and the related parameters are established between them. After setting the building location and orientation, it was necessary to define the thermal characteristics of the building. For this purpose, the materials Heat Transfer Coefficients (U) and density were set, according to ITE 50 from the Portuguese national civil engineering laboratory (LNEC) [83], so they could be representative of the Portuguese context. Then, the simulation tool recognises every building compartment; the function of each element was set (interior or exterior) for walls, floors, doors and roofs and the "room" function applied to every compartment. At this step, the compartment-specific properties such as name, dimensions and boundaries were organised. Finally, the model energy settings, interferences and errors were also rapidly checked by using the integrated functions of Autodesk Revit. Overall, the models can be classified as a LOD 300—they contain data about the building form, openings/windows, interior spaces and partitions, floors, walls, dimensions and material properties [34]. Note that the model creation may be performed in another BIM platform, such as ArchiCAD or Bentley, since the only requirement is the capacity to export the model in the IFC format.

4.2. IFC Upload Using Open BIM Collaboration Add-In

Then, BIM models were uploaded into the BIMServer.center account (through IFC). For this specific case, this step was made using the Open BIM collaboration add-in for Autodesk Revit, avoiding the need to save the model in IFC and uploading it to the web page of BIMServer.center. Note that the add-in requires the use of Autodesk Revit. However, it is still possible to directly upload IFC models to the platform. After that, a new project was created in Cypetherm REH and the IFC file imported from the BIMServer.center account. The software allows to import materials libraries from the user's directory, and therefore, it was linked to the materials library from Autodesk Revit.

4.3. Building Envelope, Systems and Thermal Bridges

When the BIM models were uploaded into Cypetherm REH, a couple of setbacks were faced. First, it was noticed that the materials library was not successfully linked. According to Cype technical support, it only allows linking materials libraries from other Cype software. The materials were defined again according to ITE 50, which are available on the Cype database. The building's location was also not accurately transmitted, and it was necessary to define it again. Since the software considers REH parameters, the climatic zones are automatically defined (which are used for the determination of the building envelope thermal quality requirements). Despite the general interoperability quality of the building's geometries and compartment identities, in these specific case studies, it was possible to notice slight differences in the compartment's floor-to-ceiling heights. This may be caused in Autodesk Revit upon the creation of the energy analytical model. The compartment volumes were adjusted according to the case studies models. A synthesis of the interoperability is presented in Table 3. To avoid some of these issues, it is recommended to double-check the BIM models first in IFC Builder software, which is used to create and adapt BIM models for Cype software use. Thus, geometry and identity errors can be easily corrected before importing the model into the energy analysis tool.

Table 3. Interoperability synthesis.

Successfully Transmitted	Not Transmitted
Building geometry	Building location
Building orientation	Material characteristics
Walls and floors thickness	Compartment's height
Opening's identity, size and location	Building energy system
Compartment's area	
Compartment's identity	

Next, it was necessary to complete the information regarding the building envelope. For every building element, i.e., external and internal walls, ground floor and roof, it was essential to confirm if all the imported data was in accordance with the BIM model and specify the absorption coefficient and the case study's typology (number of rooms). Cypetherm REH will then automatically calculate the thermal capacity and the U-values of the building elements, according to the selected materials. Regarding the interior and exterior doors, only the U-value (if not correctly transmitted) and the absorption coefficient must be defined.

Concerning the windows, some more adjustments were necessary. Since Autodesk Revit does not allow designers to define all the parameters that influence the simulation, some actions are required in Cypetherm REH. The glazing type must be correctly defined according to ITE 50 from LNEC, such as the thickness of the air cavity, the glass solar factor and the glazed fraction (0.7 for all the simulations). The window frame characteristics must also be completed in terms of typology and U-value. Finally, the accessories, such as shading devices, must be defined in terms of colour, position and solar reduction factor (0.07 for all the simulations).

At last, it was necessary to define the systems for the dwellings—namely, the heating, cooling, DHW and the on-site systems—to produce energy from the renewables (e.g., solar thermal collectors). Since Autodesk Revit does not allow designers to set all this information, it must be directly defined in Cypetherm REH. For the heating, cooling and DHW systems, it is necessary to select the energy source (e.g., electricity or fuel) and the equipment type (e.g., split, multi-split, chiller or boiler). The natural ventilation renovation rates are also defined here. Additionally, the system wattage and efficiency must be stated, as well as the heated floor area. Concerning the renewable energy source, which is mandatory for new buildings and major renovations, Cypetherm REH allows choosing between solar and photovoltaic (PV) panels. For these case studies, solar panels were selected and the panel area, production, efficiency and losses defined.

Regarding the thermal bridges, since Autodesk Revit performs parametric modelling, Cypetherm REH recognises the connections between different elements and automatically defines the existing linear thermal bridges. The user must only define how to calculate the correspondent transmittances according to the selected standard. For these case studies, REH-simplified values were used, according to Table 7, from Despacho n° 15793-K/2013 [84].

5. Results and Discussion

5.1. BIM-Based Energy Simulation

Before performing any simulation in Cypetherm REH, it was necessary to estimate the amount of energy produced on-site from renewables. The use of on-site renewable energy is mandatory according to the Portuguese thermal regulation, and, therefore, Cypetherm REH does not allow performing any simulation without defining a renewable energy system. According to the Portuguese thermal regulation, the minimum amount of renewable energy to be produced on-site must be equivalent to half of the building DHW needs. In this study, it was assumed the introduction of a solar thermal collector with an area of 2 m^2 in each of the building's roofs, with a 35° slope and oriented to the south quadrant. The amount of renewable energy that can be produced was defined according to the national minimum requirements. By selecting the building location, altitude and main obstructions in the official spreadsheet from the DGEG, the minimum requirements calculated for the location were 1366 kWh per year. This value will be used for all the case studies simulations, both in Cypetherm REH and in the official REH Excel spreadsheet (version V3.15). Within the BIM environment, it would be easy to export the model to other energy simulation tool to assess renewable energy production. However, energy simulation software usually has different climatic databases, and using both would result in inaccurate outcomes.

With all the required data, it is possible to conduct energy performance simulations. First, the existing reference model was simulated in Cypetherm REH. Then, the existing optimised model was created in Autodesk Revit by introducing insulation layers into the building envelope and by replacing the windows with more efficient ones. The model was quickly updated into BIMServer.center using the Autodesk Revit Open BIM add-in, which automatically updates it in Cypetherm REH. The new building project model was also created in Autodesk Revit and simulated once in Cypetherm REH. Only one scenario was considered, as the goal was to demonstrate the method applicability for new buildings.

The simulation results—heating needs (N_{ic}), cooling needs (N_{vc}), primary energy needs (N_{tc}) and DHW needs (Q_a and N_{ac}), as well as the regulation (REH) limit values—are presented in Table 4. The existing reference model (without insulation) achieved heating, cooling and total needs far above the national limits. The results were as expected, due to the inexistence of any insulation material in the building. The scenario of the energy renovation—optimised model—accomplished the REH requirements and reached a reduction of 78.5% on the building total primary energy needs. This was mainly achieved by reducing the heating needs in over 80%. For the new building project, the current construction techniques and materials were used, reaching primary energy needs below the existing model (despite its larger area). This difference was mainly due to the building heating needs, which were almost 30% lower than the existing optimised model. The primary energy consumption (N_{tc}) results from the application of Equation (3) are

$$N_{tc} = \frac{1 * N_{ic}}{\eta i} * fpui + \frac{1 * N_{vc}}{\eta v} * fpuv + \frac{1 * N_{ac}}{\eta a} * fpua - \frac{E_{ren} + E_{solar}}{A_p} \quad (3)$$

where:

N_{tc}—building primary energy (PE) needs ($kWh_{PE}/(m^2.year)$);
N_{ic}—building heating needs ($kWh/(m^2.year)$);
N_{vc}—building cooling needs ($kWh/(m^2.year)$);
N_{ac}—building DHW needs ($kWh/(m^2.year)$);
fpui—conversion factor to convert the final heating energy into PE;
fpuv—conversion factor to convert the final cooling energy into PE;
fpua—conversion factor to convert the final DHW energy into PE;
ηi—heating system effieicny;
ηv—cooling sysrem efficiency;
ηa—DHW system efficiency;
E_{ren}—renewable energy produced for electric use (kWh/year);
E_{solar}—renewable energy produced for DHW use (kWh/year);
A_p—total net floor area (m^2).

Table 4. Cypetherm REH results.

	Energy Needs	Existing Building		New Building
		Reference Model	Optimised Model	
	N_{ic} $kWh/(m^2 \cdot year)$	201.21	38.61	27.12
Limit value	N_i $kWh/(m^2 \cdot year)$	49.00		68.07
	N_{vc} $kWh/(m^2 \cdot year)$	4.98	4.35	7.91
Limit value	N_v $kWh/(m^2 \cdot year)$	9.15		9.15
	N_{tc} $kWh_{PE}/(m^2 \cdot year)$	518.29	111.26	80.17
Limit value	N_t $kWh_{PE}/(m^2 \cdot year)$	165.79		195.61
	Q_a (kWh/year)		2139.85	
	N_{ac} $kWh/(m^2 \cdot year)$		28.56	

The conversion factors fpui, fpuv and fpua take the value of 2.5 if the energy source is electricity and the value of 1 for the fuel sources. The efficiency of the systems is represented by ηi for the heating system (with a value of 1), ηv for the cooling system (with a value of 3) and ηa for the DHW system (with a value of 0.93). E_{ren} and E_{solar} concern the amount of renewable energy proceeded by the building (1366 kWh/year). All these values were used for all the simulations with both models.

According to the REH calculation procedure, the existing reference model achieved an F mark on the energy label resulting from the relation between N_{tc} and N_t, while the existing optimised model achieved a B mark. The new building project model reached an A level for the national energy label.

With this BIM-based method, the process to perform and compare energy performance simulations is enhanced. The requested amount of time is considerably reduced, and designers can quickly understand the impact of the rehabilitation scenario or other construction solutions on the building energy demand. Furthermore, Cypetherm REH automatically creates all the documentation and drawings for the building thermal project. Users can select documents such as "Energy performance label", "Compliance with REH requirements", "Thermal inertia", "Thermal bridges description", "Obstruction factors" and "Elements and material description", as well as personalise their own drawings. This is mandatory data to deliver within the building thermal project to obtain a construction permit. Regarding the comparison of scenarios, Cypetherm REH also allows introducing improvement measures directly. However, the BIM model in Autodesk Revit must be manually updated later.

5.2. BIM-Based Energy Simulation Validation

At this stage, with the performance of such analysis, it was feasible to assume that RQ1 can be positively answered. Nevertheless, it is essential to validate this tool and process against a conventional and official method—RQ2. Thus, the Cypetherm REH simulation results were compared with the ones obtained from an official REH spreadsheet (conventional calculation process).

The simulation parameters for the REH spreadsheet were set equal to the ones of Cypetherm REH, namely: Building location, building typology, interior height, altitude, orientation, compartments and elements area, elements characteristics, systems efficiency and type, as well as thermal bridges (which were obtained from Cype's "thermal bridges description" and later defined in the REH spreadsheet). Overall, all the input data in both methods were kept constant. During the validation simulation with the REH spreadsheet, the conventional procedure was adopted, but the BIM model was used to quickly assess the building envelope characteristics, such as the dimensions and U-values. The achieved results and comparison of both the energy performance simulation tools are presented in Tables 5 and 6 for the existing building and new building, respectively.

Table 5. Comparison between the Cypetherm REH results and REH spreadsheet results—existing building.

	Reference Model		Difference (%)	Optimised Model		Difference (%)
	Cypetherm REH	REH Spreadsheet		Cypetherm REH	REH Spreadsheet	
N_{ic}	201.21	200.68	0.26	38.61	38.54	0.18
N_i	49.00	45.97	6.18	49.00	45.97	6.18
N_{vc}	4.98	4.61	7.43	4.35	4.35	0.00
N_v	9.15	9.13	0.22	9.15	9.13	0.22
N_{ac}	28.56	28.56	0.00	28.56	28.56	0.00
N_{tc}	518.29	516.66	0.32	111.26	111.08	0.16
N_t	165.79	158.18	4.59	165.79	158.18	4.59

Table 6. Comparison between the Cypetherm REH results and REH spreadsheet results—new building.

	New Building		Difference (%)
	Cypetherm REH	REH Spreadsheet	
N_{ic}	27.12	26.78	1.27
N_i	68.07	65.36	4.14
N_{vc}	7.91	7.91	0.00
N_v	9.15	9.13	0.22
N_{ac}	28.56	28.56	0.00
N_{tc}	80.17	81.11	1.17
N_t	195.61	189.61	3.16

Concerning the calculated energy needs, both the Cypetherm REH and REH spreadsheet showed similar values, with deviations below 7.43%. However, the existing optimised model reached fewer differences with the validation engine than the existing reference model. The building performance maximum difference on the optimised model was only 0.18% (N_{ic}), while, in the reference model, was 7.43% (N_{vc}). A similar scenario was found for the new building project model, with a 1.27% difference between both methods' heating needs (N_{ic}).

One of the most notorious difference was registered on the reference/limit value for the heating needs (N_i) for both models—6.18% for the existing building and 4.14% for the new building. These values did not affect the simulation results, since they only represented the Portuguese reference value for this type of building. Although, by analysing the calculation methods, it was noticed that Cypetherm REH considers a reference U-value of 2.40 for exterior doors, while the REH spreadsheet considers 0.40. Thus, the calculation engine of Cypetherm REH considers that the heat transference by elements transmission is higher, and, therefore, the reference heating limit (N_i) for the building is also higher. The reference U-value is given by REH; more specifically, from Portaria 379-A/2015 [85]. According to it, the reference U-value for the windows and doors for Porto's climatic zone (I2) should take a value of 2.40. Additionally, it also defines the reference U-value for opaque vertical and horizontal elements (0.40 and 0.35, respectively). Therefore, it can be concluded that the REH spreadsheet considers exterior doors as "opaque vertical elements". This leads to a more conservative value on the heating needs for the national reference/limit. With such a difference in the Ni-value, a difference of the same magnitude in the Nt-value was also expected, since it depends on it. Nevertheless, the difference was not so significant, reaching only 4.59% for the existing building and 3.16% for the new building.

Another significant difference was achieved in the cooling needs (N_{vc}) only for the existing reference model (7.43%). To understand the reasons, once again, the calculation methods were carefully analysed. The problem was found in the windows' effective heating collection area. Since Cypetherm REH contains precise geometric data, the software can easily and accurately recognise which net glazed area is effectively facing south. Moreover, it also considers any existing obstructions or shading elements with higher accuracy. In the REH spreadsheet, the user must introduce all this data manually, and the orientations are fixed according to the cardinal axis. Overall, a slight difference of 0.1 m^2 on the solar collection was found between both methods. Since the existing reference model did not have any kind of insulation, this factor provided such a significant difference.

An additional setback was related to the number of decimal digits used in the calculations. Once again, the dimensions and U-values in Cypetherm REH are defined with several decimal digits (but only two or three decimal digits are displayed). At the same time, in the REH spreadsheet, it is the user's decision (according to availability and preference). This issue was the reason for the other small differences achieved between Cypetherm REH and the REH spreadsheet. As an example, Table 7 presents the same calculations but using Cypetherm REH and the REH spreadsheet. As it is possible to understand, the same calculation provided different results (0.1% difference).

Table 7. Rounding differences between Cypetherm REH and the REH spreadsheet.

	Cypetherm REH	**REH Spreadsheet**
Effective glazed area facing south (m^2)	6.75	6.75
	×	×
Average south radiation (kWh/m^2·month)	130.00	130.00
	×	×
Heating season duration (months)	6.23	6.23
	=	=
Gross solar gains (kWh/year)	5466.83	5472.29

Nevertheless, some notes must be made regarding the input parameters. There are still some different inputs for the selected simulation tools, such as in the building exterior windows and building systems. For windows, Cypetherm REH allows defining some more simulation parameters, such as the frame type and performance, as well as different shading devices and their influence on the thermal performance of the window. For the building systems, Cypetherm REH also allows defining more types of efficiency, namely the seasonal Coefficient of Performance (COP) efficiency and the renewable energy system losses.

With these results, it is possible to provide an answer for RQ2. The identified BIM-based method can globally be accepted to carry out energy simulations and thermal projects in the Portuguese context. Moreover, the BIM-based process provides more reliable and precise results than the existing calculation spreadsheets. Still, there is some space for improvement, namely on the predefined reference U-values. Two calculation engines based on the same method must consider the same reference U-values for the exterior doors. Given that, and the small differences achieved, the BIM-based process can be successfully validated to be used in the assessment of the energy performance of residential buildings, according to the Portuguese thermal regulation.

5.3. *Sustainability Assessment*

Finally, it is possible to approach RQ3 and look for a sustainability assessment. Following the conceptual framework from Carvalho et al. [27], the energy simulation results from Cypetherm REH were used to assess the SBTool^{PT}-H parameters P7—Primary Energy and P8—On-site energy production from renewables. With the energy simulation results, all the required information to assess the energy efficiency category is collected. The following sections present the assessment procedure for the SBTool^{PT}-H parameters P7 and P8. For timesaving, the results were exported from Cypetherm REH to an XML document and then linked to the SBTool^{PT}-H Excel spreadsheet.

5.3.1. Parameter 7

The Cypetherm REH results were linked to the SBTool^{PT}-H spreadsheet, and it was possible to reach an automatic assessment. For parameter P7—Primary energy demand, the required information for the evaluation is the total net floor area (74.92 m^2 for the existing model and 143.53 m^2 for the new building model), primary energy needs (and respective limit) and dwelling typology (three-bedroom dwelling). According to the SBTool^{PT}-H evaluation guide, the building primary energy needs ($P_{ENR} = N_t$) must be compared with two benchmarks: the national best ($P_{ENR}{}^* = 0.25 \times N_t$) and conventional ($P_{ENR*} = N_t$) practice.

Therefore, using the Cypetherm REH results, these data assume the following values for the existing reference model:

$$P_{ENR} = N_{tc} = 518.29\ kWh_{PE}/(m^2{\cdot}year)$$

$$P_{ENR*} = N_t = 165.79\ kWh_{PE}/(m^2{\cdot}year)$$

$$P_{ENR}{}^{*} = 0.25 \times N_t = 41.45\ kWh_{PE}/(m^2{\cdot}year)$$

The comparison against the benchmarks for the existing reference model is made using a normalised value, provided by Equation (1).

$$\overline{P_{ENR}} = \frac{518.29 - 165.79}{41.45 - 165.79} = -2.84$$

To reach the final score for P7, the normalised value must be converted into a qualitative scale, as presented in Table 8. The existing reference model achieved an E level for P7. This was an expected result for a non-insulated building.

The same assessment procedure was adopted for the existing optimised model, where the variables took the following values:

$$P_{ENR} = N_{tc} = 111.26\ kWh_{PE}/(m^2{\cdot}year)$$

$$P_{ENR*} = N_t = 165.79\ kWh_{PE}/(m^2{\cdot}year)$$

$$P_{ENR}{}^{*} = 0.25 \times N_t = 41.45\ kWh_{PE}/(m^2{\cdot}year)$$

The normalised value for the existing optimised model is also provided by Equation (1).

$$\overline{P_{ENR}} = \frac{111.26 - 165.79}{41.45 - 165.79} = 0.4$$

Converting the normalised value into a qualitative score, the existing optimised model achieved a higher mark than the reference model—a B level for P7. Thus, the existing optimised model performance is between the national best and conventional practices for new buildings.

The same procedure was adopted for the new building to reach the analysis benchmarks.

$$P_{ENR} = N_{tc} = 80.17\ kWh_{PE}/(m^2{\cdot}year)$$

$$P_{ENR*} = N_t = 195.61\ kWh_{PE}/(m^2{\cdot}year)$$

$$P_{ENR}{}^{*} = 0.25 \times N_t = 48.90\ kWh_{PE}/(m^2{\cdot}year)$$

By normalising the results using Equation (1) and converting them into a qualitative score, the new building model reached an A level for P7. The results from all the simulated models are presented in Table 8.

$$\overline{P_{ENR}} = \frac{80.17 - 195.61}{48.90 - 195.61} = 0.79$$

The results showed that, for parameter P7, all the required data for the assessment can be quickly obtained using BIM methodology. By exporting the BIM model from Autodesk Revit to Cypetherm REH, the building area elements (as walls, slabs, windows and doors) and material characteristics are automatically recognised. In the conventional assessment procedure, identifying all these characteristics is one of the most time-consuming tasks, which was almost instantaneous assessed using BIM. Then, just by adjusting/defining some simulation parameters, both the energy primary needs and limits are automatically calculated according to REH in the required units for SBTool^{PT}-H use.

In what concerns the optimisation or adjustments of the building design, this task is also simple, since modifications in the BIM model (in Autodesk Revit) can be automatically updated in Cypetherm REH, allowing designers to compare the performances of different designs scenarios.

5.3.2. Parameter 8

Concerning SBTool^{PT}-H parameter P8—On-site energy production from renewables, Cypetherm REH is not able to estimate the required renewable energy production (for electrical use—E_{ren}—or for DHW production—E_{solar}). An official spreadsheet from DGEG was used to assess the minimum national requirements for renewable energy production, resulting in a value of 1366 kWh/year. All this energy was specified for the production of DHW—E_{solar}.

The required data for SBTool^{PT}-H to perform this evaluation are: dwelling typology (three-bedroom), total net floor area (74.92 m^2 for the existing model and 143.53 m^2 for the new building model), heating, cooling and DHW needs and energy production (E_{ren} and/or E_{solar}). To assess parameter P8, the building energy production (P_{ER})—calculated through Equation (4)—must be compared with the conventional national practice (P_{ER*})—Equation (5)—and best practice (P_{ER}*)—Equation (6).

The following calculations were performed for the existing reference model:

$$P_{ER} = \frac{E_{solar} + E_{ren}}{A_p} \quad (4)$$

$$P_{ER} = \frac{0 + 1366}{74.92} = 18.23\ kWh_{PE}/\left(m^2 \cdot year\right)$$

$$P_{ER*} = 0.5 \times \frac{Q_a}{0.95 \times A_p} \quad (5)$$

where:

P_{ER*}—national conventional practice for P8;
Q_a—building DHW needs (kWh/year);
A_p—building total net area (m^2).

$$P_{ER*} = 0.5 \times \frac{2139.85}{0.95 * 74.92} = 14.35\ kWh_{PE}/\left(m^2 \cdot year\right)$$

$$P_{ER}* = N_{tc}' = 1.2 \times \left(\frac{N_{ic}}{1} \times 2.5 + \frac{N_{vc}}{2.8} \times 2.5 + \frac{Q_a}{(0.95 \times A_p)} \times 1.0\right) \quad (6)$$

$$P_{ER}* = 1.2 \times \left(\frac{201.21}{1} \times 2.5 + \frac{4.98}{3} \times 2.5 + \frac{2139.85}{(0.95 \times 74.92)} \times 1.0\right) = 644.69/2.5 = 257.88\ kWh_{PE}/\left(m^2 \cdot year\right)$$

With all the variables, it is possible to compare the building's renewable energy production with both benchmarks using Equation (2). As a normalised value of 0 corresponds to the conventional national practice, the analysed building is slightly better than that.

$$\overline{P_{ER}} = \frac{18.23 - 14.35}{257.88 - 14.35} = 0.016$$

Converting the normalised score into a qualitative scale, the existing reference model achieved a D level for parameter P8, as presented in Table 8.

The same procedure was adopted to assess parameter P8 for the existing optimised model. As the renewable energy production was kept the same, as well as the DHW needs, both calculations for the building energy production (P_{ER}) and the conventional national practice (P_{ER*}) were equivalent. It was only necessary to calculate the national best practice (P_{ER}*)—Equation (6)—by introducing the existing optimised model heating and cooling needs.

$$P_{ER} = 18.23\ kWh_{PE}/\left(m^2 \cdot year\right)$$
$$P_{ER*} = 14.35\ kWh_{PE}/\left(m^2 \cdot year\right)$$

$$P_{ER}* = 1.2 \times \left(\frac{38.61}{1} \times 2.5 + \frac{4.35}{3} \times 2.5 + \frac{2139.85}{(0.95 \times 74.92)} \times 1.0\right) = 156.26/2.5 = 62.50\ kWh_{PE}/\left(m^2 \cdot year\right)$$

The comparison between the building renewable energy production and both benchmarks was performed by using the normalisation Equation (2).

$$\overline{P_{ER}} = \frac{18.23 - 14.35}{62.50 - 14.35} = 0.081$$

The qualitative score for the existing optimised model is presented in Table 8. Only by improving the thermal insulation of the building envelope was it possible to improve the classification of this parameter slightly. Even though the existing optimised model achieved a D level, the classification was closer to the bottom border of the upper qualitative level. The results were as expected, since the renewable energy production was set according to the regulation reference (minimum requirements). The same amount of renewable energy production was considered for both models.

Finally, the energy simulation results from the new building model were also linked to SBTool^{PT}-H for the assessment of P8. As the building area is different, the benchmarks must be defined again.

$$P_{ER} = \frac{E_{solar} + E_{ren}}{A_p}$$

$$P_{ER} = \frac{0 + 1366}{143.53} = 9.52\ kWh_{PE}/\left(m^2 \cdot year\right)$$

$$P_{ER*} = 0.5 \times \frac{Q_a}{0.95 \times A_p}$$

$$P_{ER*} = 0.5 \times \frac{2139.85}{0.95 * 143.53} = 7.85\ kWh_{PE}/\left(m^2 \cdot year\right)$$

$$P_{ER}* = N_{tc}' = 1.2 \times \left(\frac{N_{ic}}{1} \times 2.5 + \frac{N_{vc}}{2.8} \times 2.5 + \frac{Q_a}{(0.95 \times A_p)} \times 1.0\right)$$

$$P_{ER}* = 1.2 \times \left(\frac{27.12}{1} \times 2.5 + \frac{7.91}{3} \times 2.5 + \frac{2139.85}{(0.95 \times 143.53)} \times 1.0\right) = 108.1/2.5 = 43.24\ kWh_{PE}/\left(m^2 \cdot year\right)$$

The normalised value for P8 is given by the application of Equation (2). The conversion into a qualitative score resulted in a D level for the new building model. Since the renewable energy is the same as for the existing building, a similar result was also expected. All the sustainability scores are presented in Table 8.

$$\overline{P_{ER}} = \frac{9.52 - 7.85}{43.24 - 7.85} = 0.05 \quad (7)$$

Table 8. Final score for parameters P7 and P8.

Qualitative Level	Quantitative Value	P7			P8		
		Existing Building		New Building	Existing Building		New Building
		Reference	Optimised		Reference	Optimised	
A+	$p > 1.00$						
A	$0.70 < p \leq 1.00$			X			
B	$0.40 < p \leq 0.70$		X				
C	$0.10 < p \leq 0.40$						
D	$0.00 \leq p \leq 0.10$				X	X	X
E	$p < 0.00$	X					

The use of Autodesk Revit and Cypetherm REH can provide almost all the necessary data for the assessment of parameter P8. The exception goes for renewable energy production that needs to be previously calculated, which is the main setback. Nevertheless, Cypetherm REH allows designers to assess quickly (and simultaneous with the primary energy needs for P7) the building heating, cooling and DHW needs according to REH, which are necessary for the P8 assessment.

However, the fact that Cypetherm REH cannot estimate the amount of renewable energy production only allows optimising parameter P8 by reducing the building energy needs. To further improve the P8 evaluation, it was necessary to select, for example, a higher area of solar thermal collectors to estimate the impact of a possible increase on renewable energy production. However, it was not the goal of this research, and therefore, the renewable energy production was kept constant.

Overall, a positive reply can be given to RQ3, as the simulation results from Cypetherm REH can be directly used in the assessment of the energy efficiency category of SBTool^{PT}-H.

6. Conclusions

The use of BIM in the construction industry may be an essential path to optimise buildings' energy performances and the occupants' comfort requirements. BIM can significantly minimise the resources of Building Energy Modelling to analyse different design alternatives and improve building performances. This research validated a BIM-based process to carry out energy analyses and develop building thermal projects in the Portuguese context. The simulation results were also linked to the energy efficiency category of the Portuguese BSA method SBTool^{PT}-H to analyse how the results can be used to assess the sustainability of buildings. This study analysed the possibility of using a BIM-based framework to improve buildings' energy performances and to develop mandatory thermal projects while improving buildings' sustainability. Designers can compare the impact of different energy solutions on the sustainability level of their buildings during the early design stage without spending too much time, money and other resources. As a case study, the presented framework was applied, discussed and validated in the Portuguese context but can, however, be extrapolated to other countries. This research also contributed to improving the knowledge about the integration of BSA in the BIM project workflow, providing new insights for complete implementation. BIM can significantly reduce the efforts for BSA application in project early design stages, bringing the opportunity to create more sustainable buildings.

When faced with the conventional assessment procedure, the applied BIM process improved the assessment of the building energy needs in terms of reliability and time. Less human errors are expected in assessing the building characteristics, in selecting simulation parameters according to REH and in defining thermal bridges. Less time was also required, as most of the building features were automatically recognised, the primary energy needs and limits were automatically calculated and information extraction happened faster. Moreover, it automatically provided mandatory documentation for the building thermal project. The simulation results were also revealed to be reliable. The differences from the REH spreadsheet were only noticeable due to the number of decimal digits considered for the several parameters, data accuracy and a predefined reference U-value for the exterior doors. Regarding the sustainability assessment, a single energy simulation in Cypetherm REH provided results for both of the SBTool^{PT}-H energy efficiency criteria. It reduced the required efforts and time to carry out the BSA, encouraging designers to apply it in their projects.

Still, some limitations were also found during the BIM-based process. Renewable energy production must be assessed before the energy simulation using external tools. Some interoperability constraints were also noticed in the transmission of the building features (as presented in Table 3), which required a double-check revision with the energy simulation tool. Nevertheless, by using IFC builder, designers can check and correct the model's geometry and data before exporting to the energy simulation tool. Other issues were related to the model materials, which were not possible to transmit; the simulation results must always be connected to the SBTool^{PT}-H spreadsheet, and the improvement measures must be manually introduced in Autodesk Revit. Overall, the

BIM-based process to carry out the energy analysis and thermal projects for Portuguese buildings was successfully validated, but it still requires further maturity.

It must be noted that this study was conducted with region-specific factors, which were established considering the Portuguese context. A specific oriented simulation tool was used, as well as a suitable BSA method. Nevertheless, SBTool^{PT}-H is the adaptation to the Portuguese context of the international SBTool method, which has already been adapted to other countries' specificities. Additionally, Cype software (or other equivalent software) is also available in several countries, which make it possible to retain and export valuable insights about the applied process in other regions, especially about the interoperability between Autodesk Revit and Cype software. Nevertheless, other BIM platforms may be used, as long as they allow exporting the model IFC for BIMServer.center. This same procedure can also be applied for service buildings (using Cypetherm RECS instead of Cypetherm REH) and different parameters of SBTool^{PT}-H, e.g., in the assessment of the acoustic performance (using Cypesound RRAE).

Author Contributions: Conceptualisation, J.P.C.; methodology, J.P.C.; software, J.P.C.; validation, J.P.C., M.A., L.B. and R.M.; formal analysis, J.P.C.; investigation, J.P.C.; writing—original draft preparation, J.P.C.; writing—review and editing, M.A., L.B. and R.M.; visualisation, M.A., L.B. and R.M.; supervision, M.A., L.B. and R.M. and funding acquisition, J.P.C. All authors have read and agreed to the published version of the manuscript.

Funding: This research was funded by the Portuguese Foundation for Science and Technology through the Regional Operation Programme of North (grant number SFRH/BD/145735/2019).

Institutional Review Board Statement: Not applicable.

Informed Consent Statement: Not applicable.

Conflicts of Interest: The authors declare no conflict of interest.

Appendix A

Table A1. SBTool^{PT}-H list of categories and parameters.

Dimension	Category	Parameters	Category Weight (%)	Dimension Weight (%)
Environment	C1—Climate change and outdoor air quality	P1—Construction materials embodied environmental impact	12	40
	C2—Land use and biodiversity	P2—Urban density	19	
		P3—Soil sealing index of the development		
		P4—Use of precontaminated land		
		P5—Use of native plants		
		P6—Heat-island effect		
	C3—Energy Efficiency	P7—Nonrenewable primary energy consumption	39	
		P8—In situ energy production from renewables		
	C4—Materials and waste management	P9—Materials and products reused	22	
		P10—Use of materials with recycled content		
		P11—Use of certified organic materials		
		P12—Use of cement substitutes in concrete		
		P13—Waste management during operation		
	C5—Water efficiency	P14—Water consumption	8	
		P15—Reuse of grey and rainwater		

Table A1. *Cont.*

Dimension	Category	Parameters	Category Weight (%)	Dimension Weight (%)
Social	C6—Occupant's health and comfort	P16—Natural ventilation efficiency	60	30
		P17—Indoor air quality		
		P18—Thermal comfort		
		P19—Natural lighting performance		
		P20—Acoustic comfort		
	C7—Accessibilities	P21—Accessibility to public transport	30	
		P22—Accessibility to urban amenities		
	C8—Education and awareness of sustainability	P23—Occupant's awareness and education regarding sustainability issues	10	
Economic	C9—Life cycle costs	P24—Capital costs	100	30
		P25—Operation costs		

References

1. Horta, R.M.C. Construção Sustentável de Edifícios de Balanço Energético Quase Zero (Sustainable Construction of Nearly Zero-Energy Buildings). Master's Thesis, Faculdade de Ciências e Tecnologia, Universidade Nova de Lisboa, Lisboa, Portugal, 2012.
2. Ruparathna, R.; Hewage, K.; Sadiq, R. Improving the energy efficiency of the existing building stock: A critical review of commercial and institutional buildings. *Renew. Sustain. Energy Rev.* **2016**, *53*, 1032–1045. [CrossRef]
3. Santos, R.; Costa, A.A.; Silvestre, J.D.; Pyl, L. Integration of LCA and LCC analysis within a BIM-based environment. *Autom. Constr.* **2019**, *103*, 127–149. [CrossRef]
4. Santos, R.; Costa, A.A.; Silvestre, J.D.; Pyl, L. Informetric analysis and review of literature on the role of BIM in sustainable construction. *Autom. Constr.* **2019**, *103*, 221–234. [CrossRef]
5. Soust-Verdaguer, B.; Llatas, C.; García-Martínez, A. Critical review of bim-based LCA method to buildings. *Energy Build.* **2017**, *136*, 110–120. [CrossRef]
6. Chong, A.; Xu, W.; Chao, S.; Ngo, N.T. Continuous-time Bayesian calibration of energy models using BIM and energy data. *Energy Build.* **2019**, *194*, 177–190. [CrossRef]
7. Gao, H.; Koch, C.; Wu, Y. Building information modelling based building energy modelling: A review. *Appl. Energy.* **2019**, *238*, 320–343. [CrossRef]
8. Edwards, R.E.; Lou, E.; Bataw, A.; Kamaruzzaman, S.N.; Johnson, C. Sustainability-led design: Feasibility of incorporating whole-life cycle energy assessment into BIM for refurbishment projects. *J. Build. Eng.* **2019**, *24*, 100697. [CrossRef]
9. Bragança, L.; Mateus, R.; Koukkari, H. Building Sustainability Assessment. *Sustainability* **2010**, *2*, 2010–2023. [CrossRef]
10. Andriamamonjy, A.; Saelens, D.; Klein, R. A combined scientometric and conventional literature review to grasp the entire BIM knowledge and its integration with energy simulation. *J. Build. Eng.* **2019**, *22*, 513–527. [CrossRef]
11. Ryu, H.-S.; Park, K.-S.; Ryu, H.-S.; Park, K.-S. A Study on the LEED Energy Simulation Process Using BIM. *Sustainability* **2016**, *8*, 138. [CrossRef]
12. Dong, Y.H.; Ng, S.T. A life cycle assessment model for evaluating the environmental impacts of building construction in Hong Kong. *Build. Environ.* **2015**, *89*, 183–191. [CrossRef]
13. Wen, T.J.; Siong, H.C.; Noor, Z.Z. Assessment of embodied energy and global warming potential of building construction using life cycle analysis approach: Case studies of residential buildings in Iskandar Malaysia. *Energy Build.* **2015**, *93*, 295–302. [CrossRef]
14. Li, Y.; Chen, X.; Wang, X.; Xu, Y.; Chen, P.-H. A review of studies on green building assessment methods by comparative analysis. *Energy Build.* **2017**, *146*, 152–159. [CrossRef]
15. European Comission. Climate Strategies and Amp Targets. Available online: https://ec.europa.eu/clima/policies/strategies_en (accessed on 27 December 2018).
16. C of the EU European Parliament. Directive 2010/31/EU of the European Parliament and of the Council of 19 May 2010 on the energy performance of buildings (recast). *Off. J. Eur. Union* **2010**, *153*, 13–35.
17. MEE. *Regulamento dos Edifícios de Habitação-REH (Residential Buildings Regulation). Decreto-Lei nº 159/2013*; Série I; Diário da República: Lisboa, Portugal, 2013.
18. Observatório da Energia, DGEG, and ADENE, Energia em Números 2020. Lisboa. 2020. Available online: https://www.dgeg.gov.pt/media/43zf5nvd/energia-em-n%C3%BAmeros-edi%C3%A7%C3%A3o-2020.pdf (accessed on 21 October 2020).
19. Montiel-Santiago, F.J.; Hermoso-Orzáez, M.J.; Terrados-Cepeda, J. Sustainability and Energy Efficiency: BIM 6D. Study of the BIM Methodology Applied to Hospital Buildings. Value of Interior Lighting and Daylight in Energy Simulation. *Sustainability* **2020**, *12*, 5731. [CrossRef]
20. Reizgevičius, M.; Ustinovičius, L.; Cibulskienė, D.; Kutut, V.; Nazarko, L. Promoting Sustainability through Investment in Building Information Modeling (BIM) Technologies: A Design Company Perspective. *Sustainability* **2018**, *10*, 600. [CrossRef]

21. Succar, B. Building information modelling framework: A research and delivery foundation for industry stakeholders. *Autom. Constr.* **2009**, *18*, 357–375. [CrossRef]
22. Merschbrock, C.; Munkvold, B.E. Effective digital collaboration in the construction industry-A case study of BIM deployment in a hospital construction project. *Comput. Ind.* **2015**, *73*, 1–7. [CrossRef]
23. AIA. *AIA Document G202-2013, Digital Practice Documents, E Series*; Project Building Information Modeling Protocol Form; AIA Trust: Washington, DC, USA, 2013.
24. JWong, K.-W.; Kuan, K.-L. Implementing 'BEAM Plus' for BIM-based sustainability analysis. *Autom. Constr.* **2014**, *44*, 163–175. [CrossRef]
25. Najjar, M.; Figueiredo, K.; Palumbo, M.; Haddad, A. Integration of BIM and LCA: Evaluating the environmental impacts of building materials at an early stage of designing a typical office building. *J. Build. Eng.* **2017**, *14*, 115–126. [CrossRef]
26. Zhang, L.; Chu, Z.; He, Q.; Zhai, P. Investigating the Constraints to Buidling Information Modeling (BIM) Applications for Sustainable Building Projects: A Case of China. *Sustainability* **2019**, *11*, 1896. [CrossRef]
27. Carvalho, J.P.; Bragança, L.; Mateus, R. Optimising building sustainability assessment using BIM. *Autom. Constr.* **2019**, *102*, 170–182. [CrossRef]
28. Krygiel, E.; Nies, B. *Green BIM: Successful Sustainable Design with Building Information Modelling*; John Wiley & Sons: Indianaplois, IN, USA, 2008.
29. Ansah, M.K.; Chen, X.; Yang, H.; Lu, L.; Lam, P.T.I. A review and outlook for integrated BIM application in green building assessment. *Sustain. Cities Soc.* **2019**, *48*, 101576. [CrossRef]
30. Pezeshki, Z.; Soleimani, A.; Darabi, A. Application of BEM and using BIM database for BEM: A review. *J. Build. Eng.* **2019**, *23*, 1–17. [CrossRef]
31. Kamel, E.; Memari, A.M. Review of BIM's application in energy simulation: Tools, issues, and solutions. *Autom. Constr.* **2019**, *97*, 164–180. [CrossRef]
32. Shalabi, F.; Turkan, Y. Bim-energy simulation approach for detecting building spaces with faults and problematic behaviour. *J. Inf. Technol. Constr.* **2020**, *25*, 342–360. [CrossRef]
33. Farzaneh, A.; Monfet, D.; Forgues, D. Review of using Building Information Modeling for building energy modeling during the design process. *J. Build. Eng.* **2019**, *23*, 127–135. [CrossRef]
34. Aksamija, A. BIM-Based Building Performance Analysis: Evaluation and Simulation of Design Decisions. In Proceedings of the 17th ACEEE Summer Study on Energy Efficiency in Buildings, Online Event, 12–17 August 2012.
35. Azhar, S.; Brown, J. BIM for Sustainability Analyses. *Int. J. Constr. Educ. Res.* **2009**, *5*, 276–292. [CrossRef]
36. Lu, Y.; Wu, Z.; Chang, R.; Li, Y. Building Information Modeling (BIM) for green buildings: A critical review and future directions. *Autom. Constr.* **2017**, *83*, 134–148. [CrossRef]
37. Carvalho, J.P.; Bragança, L.; Mateus, R. A Systematic Review of the Role of BIM in Building Sustainability Assessment Methods. *Appl. Sci.* **2020**, *10*, 4444. [CrossRef]
38. Eleftheriadis, S.; Mumovic, D.; Greening, P. Life cycle energy efficiency in building structures: A review of current developments and future outlooks based on BIM capabilities. *Renew. Sustain. Energy Rev.* **2017**, *67*, 811–825. [CrossRef]
39. Gourlis, G.; Kovacic, I. Building Information Modelling for analysis of energy efficient industrial buildings—A case study. *Renew. Sustain. Energy Rev.* **2017**, *68*, 953–963. [CrossRef]
40. Abanda, F.H.; Byers, L. An investigation of the impact of building orientation on energy consumption in a domestic building using emerging BIM (Building Information Modelling). *Energy* **2016**, *97*, 517–527. [CrossRef]
41. Carvalho, J.P.; Silva, S.M.; Mateus, R. Using BIM to streamline the energy renovation processes of residential buildings during the early design stages. In Proceedings of the International Conference on Sustainable Housing Planning, Management and Usability, Porto, Portugal, 16–18 November 2016; Available online: http://repositorium.sdum.uminho.pt/handle/1822/43388 (accessed on 13 March 2019).
42. Olawumi, T.O.; Chan, D.W.M. Identifying and prioritizing the benefits of integrating BIM and sustainability practices in construction projects: A Delphi survey of international experts. *Sustain. Cities Soc.* **2018**, *40*, 16–27. [CrossRef]
43. Kim, H.; Shen, Z.; Kim, I.; Kim, K.; Stumpf, A.; Yu, J. BIM IFC information mapping to building energy analysis (BEA) model with manually extended material information. *Autom. Constr.* **2016**, *68*, 183–193. [CrossRef]
44. Reinhart, C.F.; Davila, C.C. Urban building energy modelling—A review of a nascent field. *Build. Environ.* **2016**, *97*, 196–202. [CrossRef]
45. Reeves, T.; Olbina, S.; Issa, R.R.A. Guidelines for using building information modeling for energy analysis of buildings. *Buildings* **2015**, *5*, 1361–1388. [CrossRef]
46. Ardda, N.; Mateus, R.; Bragança, L. Methodology to Identify and Prioritise the Social Aspects to Be Considered in the Design of More Sustainable Residential Buildings—Application to a Developing Country. *Buildings* **2018**, *8*, 130. [CrossRef]
47. Awadh, O. Sustainability and green building rating systems: LEED, BREEAM, GSAS and Estidama critical analysis. *J. Build. Eng.* **2017**, *11*, 25–29. [CrossRef]
48. Mateus, R.; Bragança, L. Sustainability assessment and rating of buildings: Developing the methodology SBToolPT–H. *Build. Environ.* **2011**, *46*, 1962–1971. [CrossRef]
49. Varma, C.R.S.; Palaniappan, S. Comparision of green building rating schemes used in North America, Europe and Asia. *Habitat Int.* **2019**, *89*, 101989. [CrossRef]

50. Wong, J.K.W.; Zhou, J. Enhancing environmental sustainability over building life cycles through green BIM: A review. *Autom. Constr.* **2015**, *57*, 156–165. [CrossRef]
51. Alyami, S.H.; Rezgui, Y. Sustainable building assessment tool development approach. *Sustain. Cities Soc.* **2012**, *5*, 52–62. [CrossRef]
52. Kamaruzzaman, S.N.; Salleh, H.; Lou, E.C.W.; Edwards, R.; Wong, P.F. Assessment Schemes for Sustainability Design through BIM: Lessons Learnt. *MATEC Web Conf.* **2016**, *66*, 00080. [CrossRef]
53. Zhang, D.; Zhang, J.; Guo, J.; Xiong, H. A Semantic and Social Approach for Real-Time Green Building Rating in BIM-Based Design. *Sustainability* **2019**, *11*, 3973. [CrossRef]
54. Ade, R.; Rehm, M. The unwritten history of green building rating tools: A personal view from some of the 'founding fathers'. *Build. Res. Inf.* **2020**, *48*, 1–17. [CrossRef]
55. Raouf, A.M.I.; Al-Ghamdi, S.G. Building information modelling and green buildings: Challenges and opportunities. *Archit. Eng. Des. Manag.* **2019**, *15*, 1–28. [CrossRef]
56. Haapio, A.; Viitaniemi, P. A critical review of building environmental assessment tools. *Environ. Impact Assess. Rev.* **2008**, *28*, 469–482. [CrossRef]
57. Santos, R.; Costa, A.A.; Silvestre, J.D.; Vandenbergh, T.; Pyl, L. BIM-based life cycle assessment and life cycle costing of an office building in Western Europe. *Build. Environ.* **2020**, *169*, 106568. [CrossRef]
58. Caetano, N.S.; Carvalho, R.R.; Franco, F.R.; Afonso, C.A.R.; Felgueiras, C. Sustainable engineering labs-A Portuguese perspective. *Energy Procedia.* **2018**, *153*, 455–460. [CrossRef]
59. Ferreira, J.; Pinheiro, M.D.; de Brito, J. Portuguese sustainable construction assessment tools benchmarked with BREEAM and LEED: An energy analysis. *Energy Build.* **2014**, *69*, 451–463. [CrossRef]
60. Pires, S.M.; Fidélis, T. Local sustainability indicators in Portugal: Assessing implementation and use in governance contexts. *J. Clean. Prod.* **2015**, *86*, 289–300. [CrossRef]
61. Bragança, L. SBTool Urban: Instrumento Para a Promoção da Sustentabilidade Urbana. 2017. Available online: http://civil.uminho.pt/urbenere/wp-content/uploads/2018/05/E28-SINGEURB-2017.pdf (accessed on 28 December 2018).
62. Castro, M.d.; Mateus, R.; Bragança, L. Development of a healthcare building sustainability assessment method–Proposed structure and system of weights for the Portuguese context. *J. Clean. Prod.* **2017**, *148*, 555–570. [CrossRef]
63. Azhar, S.; Carlton, W.A.; Olsen, D.; Ahmad, I. Building information modeling for sustainable design and LEED® rating analysis. *Autom. Constr.* **2011**, *20*, 217–224. [CrossRef]
64. Jalaei, F.; Jrade, A. Integrating building information modeling (BIM) and LEED system at the conceptual design stage of sustainable buildings. *Sustain. Cities Soc.* **2015**, *18*, 95–107. [CrossRef]
65. Marzouk, M.; Azab, S.; Metawie, M. BIM-based approach for optimizing life cycle costs of sustainable buildings. *J. Clean. Prod.* **2018**, *188*, 217–226. [CrossRef]
66. Salgueiro, I.B.; Ferries, B. An 'Environmental BIM' Approach for the Architectural Schematic Design Stage. *Int. J. Archit. Comput.* **2015**, *13*, 299–312. [CrossRef]
67. Li, J.; Li, N.; Afsari, K.; Peng, J.; Wu, Z.; Cui, H. Integration of Building Information Modeling and Web Service Application Programming Interface for assessing building surroundings in early design stages. *Build. Environ.* **2019**, *153*, 91–100. [CrossRef]
68. Ilhan, B.; Yaman, H. Green building assessment tool (GBAT) for integrated BIM-based design decisions. *Autom. Constr.* **2016**, *70*, 26–37. [CrossRef]
69. Oti, A.H.; Tizani, W.; Abanda, F.H.; Jaly-Zada, A.; Tah, J.H.M. Structural sustainability appraisal in BIM. *Autom. Constr.* **2016**, *69*, 44–58. [CrossRef]
70. Gandhi, S.; Jupp, J. BIM and Australian Green Star Building Certification. In Proceedings of the 2014 International Conference on Computing in Civil and Building Engineering, Orlando, FL, USA, 23–25 June 2014; pp. 275–282. [CrossRef]
71. GhaffarianHoseini, A.; Doan, D.T.; Naismith, N.; Tookey, J.; GhaffarianHoseini, A. Amplifying the practicality of contemporary building information modelling (BIM) implementations for New Zealand green building certification (Green Star). *Eng. Constr. Archit. Manag.* **2017**, *24*, 696–714. [CrossRef]
72. Carvalho, J.P.; Bragança, L.; Mateus, R. Sustainable building design: Analysing the feasibility of BIM platforms to support practical building sustainability assessment. *Comput. Ind.* **2021**, *127*, 103400. [CrossRef]
73. El-Diraby, T.; Krijnen, T.; Papagelis, M. BIM-based collaborative design and socio-technical analytics of green buildings. *Autom. Constr.* **2017**, *82*, 59–74. [CrossRef]
74. Chong, H.-Y.; Lee, C.-Y.; Wang, X. A mixed review of the adoption of Building Information Modelling (BIM) for sustainability. *J. Clean. Prod.* **2017**, *142*, 4114–4126. [CrossRef]
75. Liu, Z.; Osmani, M.; Demian, P.; Baldwin, A. A BIM-aided construction waste minimisation framework. *Autom. Constr.* **2015**, *59*, 1–23. [CrossRef]
76. Larsson, N. SB Method and SBTool for 2011-Overview. Available online: http://www.iisbe.org/system/files/private/NilsLarsson_NewSBToolGenericFramework.pdf (accessed on 11 December 2018).
77. Vonka, M.; Hajek, P.; Lupisek, A. SBToolCZ: Sustainability rating system in the Czech Republic. *Int. J. Sustain. Build. Technol. Urban. Dev.* **2013**, *4*, 46–52. [CrossRef]
78. Castro, M.d.; Mateus, R.; Serôdio, F.; Bragança, L. Development of Benchmarks for Operating Costs and Resources Consumption to be Used in Healthcare Building Sustainability Assessment Methods. *Sustainability* **2015**, *7*, 13222–13248. [CrossRef]

79. IteCons and University of Coimbra, Plataforma para a Eficiência Energética de Edifícios. Available online: http://www.itecons.uc.pt/p3e/index.php (accessed on 19 November 2020).
80. Mateus, R.; Silva, S.M.; de Almeida, M.G. Environmental and cost life cycle analysis of the impact of using solar systems in energy renovation of Southern European single-family buildings. *Renew. Energy* **2019**, *137*, 82–92. [CrossRef]
81. INE and LNEC, O Parque Habitacional e a sua Reabilitação-Análise e Evolução: 2001–2011 (Housing Stock and Rehabilitation–Analysis and Evolution: 2001–2011). Lisboa. 2013. Available online: https://www.ine.pt/xportal/xmain?xpid=INE&xpgid=ine_publicacoes&PUBLICACOESpub_boui=165231362&PUBLICACOEStema=00&PUBLICACOESmodo=2 (accessed on 11 May 2021).
82. Sousa, J.; Bragança, L.; Almeida, M.; Silva, P. Research on the Portuguese Building Stock and Its Impacts on Energy Consumption–An Average U-Value Approach. *Arch. Civ. Eng.* **2013**, *59*, 523–546. [CrossRef]
83. Luís Matias e Pina dos Santos. *Coeficientes de Transmissão Térmica de Elementos da Envolvente dos Edifícios (Heat Transfer Coefficients from Building Envolope Elements)*; Laboratório Nacional de Engenharia Civil (LENC): Lisboa, Portugal, 2009.
84. Ministério do Ambiente, Ordenamento do Território e Energia-Direção-Geral de Energia e Geologia. *Despacho no 15793-K/2013 Diário da República no 234/2013, 3o Suplemento, Série II*; Ministério do Ambiente: Lisboa, Portugal, 2013.
85. Ministério do Ambiente, Ordenamento do Território e Energia. Portaria 379-A/2015, 2015-10-22-DRE. Available online: https://dre.pt/web/guest/pesquisa/-/search/70789581/details/normal?l=1 (accessed on 19 November 2020).